PAUL A. SAVORY

PAUL A. SAVORY

Introduction to
Simulation and SLAM II

Introduction to Simulation and SLAM II

Third Edition

A. Alan B. Pritsker

President, *Pritsker & Associates, Inc.*
Adjunct Professor, *Purdue University*

First Edition

A. Alan B. Pritsker
Claude Dennis Pegden

A Halsted Press Book, John Wiley & Sons
New York • Chichester • Brisbane

Systems Publishing Corporation
West Lafayette, Indiana

Distributed by:

Halsted Press, a Division of
John Wiley & Sons, Inc., New York

and

Systems Publishing Corporation
P.O. Box 2161
West Lafayette, Indiana 47906

Library of Congress Cataloging in Publication Data
Pritsker, A   Alan B   1933-
    Introduction to Simulation and SLAM II

    1. Digital computer simulation.   2. SLAM II (Computer
program language)
I. Title.
QA 76.9.C65P75   1986   001.4′34   86-15860
ISBN 0-470-20292-0

Printed in the United States of America
10 9 8 7 6 5 4 3 2

To Loring G. Mitten, a professor's professor;
C. B. Gambrell, Jr., a leader of men; and
my family.

AABP

Preface

This textbook presents a simulation language, SLAM II®, and information on the use of simulation for problem solving. The book can be used in courses in business, industrial engineering, management, operations research, or computer science at the senior or first year graduate level depending on the goal of the course. Emphasis is placed on developing modeling skills. Many examples are provided that illustrate procedures for modeling systems using SLAM II. Chapters on applications, probability and statistics, and support systems provide extensive material on problem solving.

SLAM II is an advanced FORTRAN based simulation language that allows models to be built based on three different world views. It provides network symbols for building graphical models that are easily translated into input statements for direct computer processing. It contains subprograms that support both discrete event and continuous model developments. By combining network, discrete event, and continuous modeling capabilities, SLAM II is a *S*imulation *L*anguage for *A*lternative *M*odeling. The interfaces between the alternative modeling approaches are explicitly defined to allow new conceptual views of systems to be explored.

SLAM II has been well received and has contributed to the increased use of modeling and simulation throughout the world. It has been installed in more than 800 industrial, academic, and government installations. The SLAM II program is portable and runs on a wide variety of computing systems including microcomputers and workstations. A key ingredient in SLAM II's success is the maintenance, development, and support provided by Pritsker & Associates (P&A). Currently, P&A is distributing SLAM II Version 3.1. Upward compatibility has been maintained so that models developed with earlier versions can continue to be executed.

SLAM and SLAM II are registered trademarks of Pritsker & Associates, Inc.

This edition of the book updates the language to include the features implemented in SLAM II Version 3.1.

The text introduces general information on the use of simulation in industry and government. Chapters on the simulation approach to problem resolution (Chapter 1), simulation model building (Chapter 3), and applications of simulation (Chapter 4) are included. Chapter 2 provides basic information on probability and statistics useful to a simulation analyst. Chapter 18 describes standard statistical distributions and random sampling procedures. Chapter 19 presents statistical methods for estimating performance from simulation outputs. A review of simulation languages is given in Chapter 15.

Chapters 5 through 14 provide a detailed description of SLAM II. Twenty one complete examples and twenty three illustrations are provided. Input procedures and output reports are described. The organizational structure provided by SLAM II for helping the analyst to build models is explained, illustrated, and applied.

The Materials Handling(MH) Extension to SLAM II facilitates the modeling of manufacturing systems. The capabilities of the MH Extension are described in Chapter 16. Problem-solving support for simulation using databases, interactive processing, and graphics capabilities is discussed in Chapter 17. The capabilities of TESS™ for supporting simulation and SLAM II are also presented in Chapter 17.

Appendices provide ready reference to the SLAM II language elements, subprograms, inputs, and diagnostics. Exercises are given at the end of each chapter which require the application of the material presented.

There are many people who have contributed to the writing of this book. I particularly acknowledge contributions of Nicholas Hurst, Philip J. Kiviat, C. Dennis Pegden, C. Elliott Sigal and Robert F. Young. Other important contributors are: Steven D. Duket, Jean J. O'Reilly, Charles R. Standridge and James Reed Wilson. C. Dennis Pegden led in the development of the original version of SLAM and did the initial conception, design and implementation. Portions of SLAM were based on Pritsker & Associates' proprietary software called GASP and Q-GERT. Since its original implementation, SLAM has been continually refined and enhanced by Pritsker & Associates. In particular, Jean O'Reilly has been instrumental in implementing the new features incorporated into SLAM II. Jean also leads P&A's maintenance, training, and user-support activity. Charlie Standridge is the project leader for the TESS support program. Steve Duket led the development of the original version of the MH Extension. Jim Wilson of Purdue University provided significant contributions to Chapter 19 on the statistical aspects of simulation.

TESS is a trademark of Pritsker & Associates, Inc.

Throughout the years I have been associated with excellent students and colleagues who have contributed directly or indirectly to this textbook. In addition, to those cited above, I want to thank the following: Neal Bengtson, Christopher Clapp, Hank Grant, Mary Grant, Gordon Hazen, James Henriksen, Lawrence Moore, Don T. Phillips, Jerry Sabuda, Bruce Schmeiser, Anne Spinoza, Cathy Stein, William Thompson, Dave Vaughan, Ware Washam, Gary Whitehouse, and David B. Wortman.

This book directly uses material contained in the books by Thomas Schriber and George S. Fishman. I thank these friends for granting me permission to use the material from their books. My thanks go to Neal Bengtson, Steve Duket, Gordon Hazen, Elliott Sigal, and Jim Wilson for reviewing early versions of the manuscript and to Steve Duket, Charlie Standridge, Bill Lilegdon, Christopher Clapp, Cathy Stein, Jean O'Reilly, Mike Sale, Jim Whitford, Ken Musselman, and Connie Busch for reviewing portions of the third edition manuscript. They have made suggestions that helped to clarify the material. I would also like to thank Carole Vasek, Darrell Starks, Carolyn Tobin, and Jean O'Reilly for helping in the presentation of the examples and illustrations. My thanks and appreciation go to Miriam Walters for her efforts in the typing and preparation of numerous drafts of the manuscript; Jennifer Matthews of Photo Comp Corporation for coordinating the typesetting effort, and Anne Pritsker for reviewing and editing the manuscript.

A Solutions Manual for all the exercises at the end of each chapter is available from Systems Publishing Corporation, P.O. Box 2161, West Lafayette, IN 47906. In addition, an educational guide for using the text is available. This guide provides information on course organization, homework assignments, and test questions. I wish you success in mastering the modeling concepts and simulation procedures described in this book. When you do, you will have a modeling framework and a simulation tool to solve meaningful problems.

A. Alan B. Pritsker

West Lafayette, Indiana
May 1986

Table of Contents

APPENDICES

INDEX

Examples

Illustrations

Introduction to
Simulation and SLAM II

CHAPTER 1

Introduction to Modeling
and Simulation

1.1   PROBLEM SOLVING

The problems facing industry, commerce, government, and society in general continue to grow in size and complexity. The need for procedures and techniques for resolving such problems is apparent. This book advocates the use of modeling and, in particular, simulation modeling for the resolution of problems. Simulation models can be employed at four levels:

- As explanatory devices to define a system or problem;
- As analysis vehicles to determine critical elements, components and issues;
- As design assessors to synthesize and evaluate proposed solutions;
- As predictors to forecast and aid in planning future developments.

1

In order to resolve problems using simulation models, it is necessary to understand the systems and to define problems relating to those systems. In our judgment, models should be developed to resolve specific problems. The form of the model, although dependent on the problem solver's background, requires an organized structure for viewing systems. A simulation language provides such a vehicle. It also translates a model description into a form acceptable by a computing system. The computer is used to exercise the model to provide outputs that can be analyzed in order that decisions relating to problem resolution can be made.

The goal of this book is to provide useful information for problem solving. The book is both an introduction to simulation methodology and an introduction to the *s*imulation *l*anguage for *a*lternative *m*odeling, SLAM II. SLAM II supports the modeling of systems from diverse points of view. In this book, we model systems using these points of view and thus the book contains information on different methods of structuring models of systems.

SLAM II has been designed to support engineers, managers, and researchers. To do this it provides, in addition to modeling views, extensive input and output capabilities. Since many of today's problems are statistical in nature, the input and output capabilities require a background in probability and statistics. Thus, parts of this book are devoted to presenting probabilistic and statistical concepts related to problem solving using simulation models.

In this chapter, we present general discussions and definitions of simulation-related topics and our suggested procedure for conducting projects that resolve problems by employing simulation models.

1.2   SYSTEMS

A system is a collection of items from a circumscribed sector of reality that is the object of study or interest. Therefore, a system is a relative thing. In one situation, a particular collection of objects may be only a small part of a larger system—a subsystem; in another situation that same collection of objects may be the primary focus of interest and would be considered as the system. The scope of every system, and of every model of a system, is determined solely by its reason for being identified and isolated. The scope of every simulation model is determined by the particular problems the model is designed to solve.

To consider the scope of a system, one must contemplate its boundaries and contents. The boundary of a system may be physical; however, it is better to think of a boundary in terms of cause and effect. Given a tentative system definition, some external factors may affect the system. If they completely govern its behavior, there is no merit in experimenting with the defined system. If they partially influence the system, there are several possibilities:

The system definition may be enlarged to include them.
They may be ignored.
They may be treated as inputs to the system.

If treated as inputs, it is assumed that the factors are functionally specified by prescribed values, tables, or equations. For example, when defining the model of a company's manufacturing system, if the sales of the company's product are considered as inputs to the manufacturing system, the model will not contain a cause and effect sales relation; it only includes a statistical description of historical or predicted sales, which is used as an input. In such a model of the manufacturing system, the sales organization is outside the boundaries of the "defined" system. In systems terminology, objects that are outside the boundaries of the system, but can influence it, constitute the environment of the system. Thus, systems are collections of mutually interacting objects that are affected by outside forces. Figure 1-1 shows such a system.

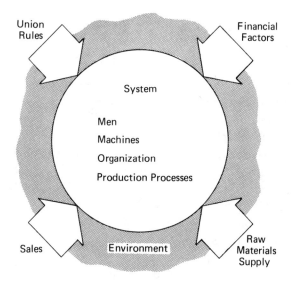

Figure 1-1 Manufacturing system model showing external influences.

1.3   MODELS

Models are *descriptions* of systems. In the physical sciences, models are usually developed based on theoretical laws and principles. The models may be scaled physical objects (iconic models), mathematical equations and relations (abstract models), or graphical representations (visual models). The usefulness of models has been demonstrated in describing, designing, and analyzing systems. Many students are educated in their discipline by learning how to build and use models. Model building is a complex process and in most fields is an art. The modeling of a system is made easier if: 1) physical laws are available that pertain to the system; 2) a pictorial or graphical representation can be made of the system; and 3) the variability of system inputs, elements, and outputs is manageable (10).

The modeling of complex, large-scale systems is often more difficult than the modeling of physical systems (12) for the following reasons: 1) few fundamental laws are available; 2) many procedural elements are involved which are difficult to describe and represent; 3) policy inputs are required which are hard to quantify; 4) random components are significant elements; and 5) human decision making is an integral part of such systems. Through the use of a simulation approach, we will illustrate methods for alleviating these difficulties.

1.4   MODEL BUILDING

Since a model is a description of a system, it is also an *abstraction* of a system. To develop an abstraction, a model builder must decide on the elements of the system to include in the model. To make such decisions, a purpose for model building should be established. Reference to this purpose should be made when deciding if an element of a system is significant and, hence, should be modeled. The success of a modeler depends on how well he or she can define significant elements and the relationships between elements.

A pictorial view of our proposed model building approach is shown in Figure 1-2. A system as discussed in Section 1.2 is considered as a set of interdependent objects united to perform a specified function. The concept of a system is not well-defined. A particular definition of a system's objects, and their function is subjective and depends on the individual who is defining the system. Because of this, the first step of our approach is the development of a purpose for modeling that is based on a stated problem or project goal. Based on this purpose, the boundaries of the system and a level of modeling detail are established. This abstraction results in a model that smooths out many of the rough ill-defined edges of the actual system. We also include in the model the desired performance measures and design alternatives to be evaluated. These can be considered as part of the model or as inputs to the model. Assessments of design alternatives in terms of the specified performance measures are considered as model outputs. Typically, the assessment process requires redefinitions and redesigns. In fact, the entire model building approach is performed iteratively. When recommendations can be made based on the assessment of alternatives, an implemenation phase is initiated. Implementation should be carried out in a well-defined environment with an explicit set of recommendations. Major decisions should have been made before implementation is attempted.

Simulation models are ideally suited for carrying out the problem-solving approach illustrated in Figure 1-2. Simulation provides the flexibility to build either aggregate or detailed models. It also supports the concepts of iterative model building by allowing models to be embellished through simple and direct additions. These aspects of simulation models are described in the next section.

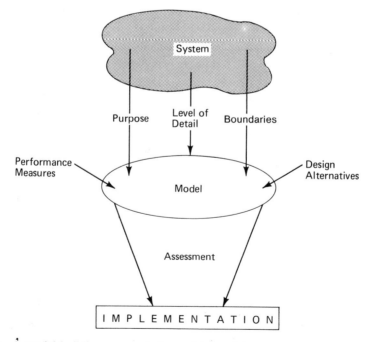

Figure 1-2 A model building approach for problem solving.

1.5   DEFINITION OF SIMULATION

In its broadest sense, computer simulation is the process of designing a mathematical-logical model of a real system and experimenting with this model on a computer (9, 11). Thus simulation encompasses a model building process as well as the design and implementation of an appropriate experiment involving that model. These experiments, or simulations, permit inferences to be drawn about systems

- Without building them, if they are only proposed systems;
- Without disturbing them, if they are operating systems that are costly or unsafe to experiment with;
- Without destroying them, if the object of an experiment is to determine their limits of stress.

In this way, simulation models can be used for design, procedural analysis, and performance assessment (9).

Simulation modeling assumes that we can describe a system in terms acceptable to a computing system. In this regard, a key concept is that of a *system state description*. If a system can be characterized by a set of variables, with each combination of variable values representing a unique state or condition of the system, then manipulation of the variable values simulates movement of the system from state to state. A simulation experiment involves observing the dynamic behavior of a model by moving from state to state in accordance with well-defined operating rules designed into the model.

Changes in the state of a system can occur continuously over time or at discrete instants in time. The discrete instants can be established deterministically or stochastically depending on the nature of model inputs. Although the procedures for describing the dynamic behavior of discrete and continuous change models differ, the basic concept of simulating a system by portraying the changes in the state of the system over time remains the same. In the next section, we will illustrate the type of dynamic behavior obtained from a simulation experiment with a discrete change model.

1.6   SIMULATION OF A BANK TELLER

As an example of the concept of simulation, we will examine the processing of customers by a teller at a bank. Customers arrive to the bank, wait for service by the teller if the teller is busy, are served, and then depart the system. Customers arriving to the system when the teller is busy wait in a single queue in front of the teller. For simplicity, we assume that the time of arrival of a customer and the service time by the teller for each customer are known. These values are given in Table 1-1. Our objective is to manually simulate the above system to determine the percent of time the teller is idle and the average time a customer spends at the bank.

Since a simulation is the dynamic portrayal of the changes in the state of a system over time, the states of the system must be defined. For this example, they can be defined by the status of the teller (busy or idle) and by the number of customers at the bank. The state of the system is changed by: 1) a customer arriving to the bank; and 2) the completion of service by the teller and subsequent departure of the customer. To illustrate a simulation, we will determine the state of the system over time by processing the events corresponding to the arrival and departure of customers in a time-ordered sequence.

The manual simulation of this example corresponding to the values in Table 1-1 is summarized in Table 1-2 by customer number. It is assumed that initially there are no customers in the system, the teller is idle, and the first customer is to arrive at time 3.2.

In Table 1-2, columns (1) and (2) are taken from Table 1-1. The start of service time given in column (3) depends on whether the preceding customer has departed the bank. It is taken as the larger value of the arrival time of the customer and the departure time of the preceding customer. Column (4), the departure time, is the sum of the column (3) value and the service time for the customer given in Table 1-1. Values for time in queue and time in bank for each customer are computed as shown in Table 1-2. Average values per customer for these variables are 2.61 minutes and 5.81 minutes, respectively.

Table 1-1 Customer arrival and service times.

Customer Number	Time of Arrival (Minutes)	Service Time (Minutes)
1	3.2	3.8
2	10.9	3.5
3	13.2	4.2
4	14.8	3.1
5	17.7	2.4
6	19.8	4.3
7	21.5	2.7
8	26.3	2.1
9	32.1	2.5
10	36.6	3.4

Table 1-2 presents a good summary of information concerning the customer but does not provide information about the teller and the queue size for the teller. To portray such information, it is convenient to examine the events associated with the situation.

Table 1-2   Manual simulation of bank teller.

Customer Number (1)	Arrival Time (2)	Start Service Time (3)	Departure Time (4)	Time in Queue (5) = (3)−(2)	Time in Bank (6) = (4)−(2)
1	3.2	3.2	7.0	0.0	3.8
2	10.9	10.9	14.4	0.0	3.5
3	13.2	14.4	18.6	1.2	5.4
4	14.8	18.6	21.7	3.8	6.9
5	17.7	21.7	24.1	4.0	6.4
6	19.8	24.1	28.4	4.3	8.6
7	21.5	28.4	31.1	6.9	9.6
8	26.3	31.1	33.2	4.8	6.9
9	32.1	33.2	35.7	1.1	3.6
10	36.6	36.6	40.0	0.0	3.4

The logic associated with processing the arrival and departure events depends on the state of the system at the time of the event. In the case of the arrival event, the disposition of the arriving customer is based on the status of the teller. If the teller is idle, the status of the teller is changed to busy and the departure event is scheduled for the customer by adding his service time to the current time. However, if the teller is busy at the time of an arrival, the customer cannot begin service at the current time and, therefore, he enters the queue (the queue length is increased by 1). For the departure event, the logic associated with processing the event is based on queue length. If a customer is waiting in the queue, the teller status remains busy, the queue length is reduced by 1, and the departure event for the first waiting customer is scheduled. However, if the queue is empty, the status of the teller is set to idle.

An event-oriented description of the bank teller status and the number of customers at the bank is given in Table 1-3. In Table 1-3, the events are listed in chronological order. A graphic portrayal of the status variables over time is shown in Figure 1-3. These results indicate that the average number of customers at the bank in the first 40 minutes is 1.4525 and that the teller is idle 20 percent of the time.

In order to place the arrival and departure events in their proper chronological order, it is necessary to maintain a record or calendar of future events to be processed. This is done by maintaining the times of the next arrival event and next departure event. The next event to be processed is then selected by comparing these event times. For situations with many events, an ordered list of events would be maintained which is referred to as an event file or event calendar.

Table 1-3 Event-oriented description of bank teller simulation.

Event Time	Customer Number	Event Type	Number in Queue	Number in Bank	Teller Status	Teller Idle Time
0.0	—	Start	0	0	Idle	—
3.2	1	Arrival	0	1	Busy	3.2
7.0	1	Departure	0	0	Idle	
10.9	2	Arrival	0	1	Busy	3.9
13.2	3	Arrival	1	2	Busy	
14.4	2	Departure	0	1	Busy	
14.8	4	Arrival	1	2	Busy	
17.7	5	Arrival	2	3	Busy	
18.6	3	Departure	1	2	Busy	
19.8	6	Arrival	2	3	Busy	
21.5	7	Arrival	3	4	Busy	
21.7	4	Departure	2	3	Busy	
24.1	5	Departure	1	2	Busy	
26.3	8	Arrival	2	3	Busy	
28.4	6	Departure	1	2	Busy	
31.1	7	Departure	0	1	Busy	
32.1	9	Arrival	1	2	Busy	
33.2	8	Departure	0	1	Busy	
35.7	9	Departure	0	0	Idle	
36.6	10	Arrival	0	1	Busy	0.9
40.0	10	Departure	0	0	Idle	

There are several important concepts illustrated by this example. We observe that at any instant in simulated time, the model is in a particular *state*. As *events* occur, the state of the model may change as prescribed by the logical-mathematical relationships associated with the events. Thus, the events define the dynamic structure of the model. Given the starting state, the logic for processing each event, and a method for specifying sample values, our problem is largely one of bookkeeping. An essential element in our bookkeeping scheme is an event calendar which provides a mechanism for recording and sequencing future events. Another point to observe is that we can view the state changes from two perspectives:

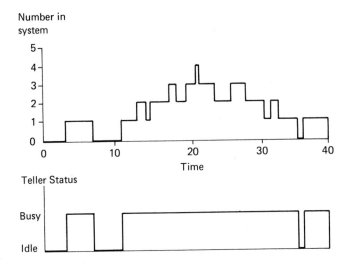

Figure 1-3   Graphic portrayal of bank teller simulation.

the *process* that the customer encounters as he seeks service (the customer's view); or the *events* that cause the state of the teller to change (the teller's or bank's view). These so called world views are described in detail in Chapter 3.

1.7   THE SIMULATION PROCESS

As we alluded to earlier, the process for the successful development of a simulation model consists of beginning with a simple model which is embellished in an evolutionary fashion to meet problem-solving requirements. Within this process, the following stages of development can be identified.†

1. Problem Formulation   The definition of the problem to be studied including a statement of the problem-solving objective.

2. Model Building   The abstraction of the system into mathematical-logical relationships in accordance with the problem formulation.

3. Data Acquisition   The identification, specification, and collection of data.

---

† The stages listed are a slightly modified version of those presented by Shannon (13).

4. Model Translation	The preparation of the model for computer processing.
5. Verification	The process of establishing that the computer program executes as intended.
6. Validation	The process of establishing that a desired accuracy or correspondence exists between the simulation model and the real system.
7. Strategic and Tactical Planning	The process of establishing the experimental conditions for using the model.
8. Experimentation	The execution of the simulation model to obtain output values.
9. Analysis of Results	The process of analyzing the simulation outputs to draw inferences and make recommendations for problem resolution.
10. Implementation and Documentation	The process of implementing decisions resulting from the simulation and documenting the model and its use.

Although some of these steps were discussed in conjunction with model building, we prefer to restate them here due to the importance of the concepts (1, 3, 4, 7, 8).

The first task in a simulation project is the construction of a clear definition of the problem and an explicit statement of the objectives of the analysis. Because of the evolutionary nature of simulation, problem definition is a continuing process which typically occurs throughout the duration of the study. As additional insights into the problem are gained and additional questions become of interest, the problem definition is revised accordingly.

Once an initial problem statement is formulated, the task of formulating a model of the system begins. The model of a system consists of both a static and dynamic description. The static description defines the elements of the system and the characteristics of the elements. The dynamic description defines the way in which the elements of the system interact to cause changes to the state of the system over time.

The actual process of formulating the model is one which is largely an art. The modeler must understand the structure and operating rules of the system and be able to extract the essence of the system without including unnecessary detail. The model should be easily understood, yet sufficiently complex to realistically reflect the important characteristics of the real system. The crucial decisions concern what simplifying assumptions are valid, what elements should be included in the model, and what interactions occur between the elements. The amount of detail included in the model should be based on the purpose for which the model is being built. Only those elements that could cause significant differences in decision-making need be considered.

Both the problem formulation and modeling phases require close interaction among project personnel. "First cut" models should be built, analyzed, and discussed. In many cases, this will require heroic assumptions and a willingness on the part of the modeler to expose his incomplete knowledge of the system under study. However, an evolutionary modeling process allows inaccuracies to be discovered more quickly and corrected more efficiently than would otherwise be possible. Furthermore, the close interaction in the problem definition and model formulation phases creates confidence in the model on the part of the model user and therefore helps to ensure a successful implementation of simulation results.

The model formulation phase will generate data input requirements for the model. Some of the data required may be readily available while other data requirements may involve considerable time and cost for collection. Typically, such data input values are initially hypothesized or based on a preliminary analysis. In some cases, the exact values for one or more of the input parameters may have little effect on the simulation results. The sensitivity of the simulation results to changes in the input data to the model can be evaluated by making a series of simulation runs while varying the input parameter values. In this way, the simulation model can be used to determine how best to allocate money and time in refining the input data to the model.

Once a model has been developed and initial estimates have been established for the input data, the next task is to translate the model into a computer acceptable form. Although a simulation model can be programmed using a general purpose language, there are distinct advantages to using a simulation language. In addition to the savings in programming time, a simulation language also assists in model formulation by providing a set of concepts for articulating the system description. In this text, we describe in detail the use of the SLAM II simulation language which provides a graphical vehicle that combines the model formulation and translation task into a single activity. SLAM II also includes a programming capability to allow models to be embellished in an evolutionary fashion to any level of detail required to reflect the complexities of the system being studied.

The verification and validation stages are concerned with evaluating the performance of the simulation model. The verification task consists of determining that the translated model executes on the computer as the modeler intended. This is typically done by manual checking of calculations. Fishman and Kiviat (5) describe statistical methods which can aid in the verification process. The validation task consists of determining that the simulation model is a reasonable representation of the system (14). Validation is normally performed in levels. We recommend that a validation be performed on data inputs, model elements, subsystems, and interface points. Validation of simulation models, although difficult, is a significantly easier task than validating other types of models, for example, validating a linear programming formulation. In simulation models, there is a correspondence between the model elements and system elements. Hence, testing for reasonableness involves a comparison of model and system structure and comparisons of the number of times elemental decisions or subsystem tasks are performed.

Specific types of validation involve evaluating reasonableness using all constant values in the simulation model or assessing the sensitivity of outputs to parametric variation of data inputs. In making validation studies, the comparison yardstick should encompass both past system outputs and experiential knowledge of system performance behavior. A point to remember is that past system outputs are but one sample record of what could have happened.

Strategic and tactical planning refer to the task of establishing the experimental conditions for the simulation runs (2). The strategic planning task consists of developing an efficient experimental design either to explain the relationship between the simulation response and the controllable variables, or to find the combination of values for the controllable variables which either minimize or maximize the simulation response. In contrast, tactical planning is concerned with how each simulation within the experimental design is to be made to glean the most information from the data. Two specific issues in tactical planning are the starting conditions for simulation runs and methods for reducing the variance of the mean response.

The next stages in the simulation development process are experimentation and analysis of results. These phases of simulation development involve the exercising of the simulation model and the interpretation of the outputs. When simulation results are used to draw inferences or to test hypotheses, statistical methods should be employed.

The final stages in the simulation development process are the implementation of results and the documentation of the simulation model and its use. No simulation project should be considered complete until its results are used in the decision-making process. The success of the implementation task is largely dependent upon the degree to which the modeler has successfully performed the other activities in the simulation development process. If the model builder and model user have worked closely together and both understand the model and its outputs, then it is likely that the results of the project will be implemented with vigor. On the other hand, if the model formulation and underlying assumptions are not effectively communicated, then it is more difficult to have recommendations implemented, regardless of the elegance and validity of the simulation model.

The stages of simulation development outlined above are rarely performed in a structured sequence beginning with problem definition and ending with documentation. A simulation project may involve false starts, erroneous assumptions which must later be abandoned, reformulation of the problem objectives, and repeated evaluation and redesign of the model. If properly done, however, this iterative process should result in a simulation model which properly assesses alternatives and enhances the decision-making process.

1.8 CHAPTER SUMMARY

Simulation is a technique that has been employed extensively to solve problems. Simulation models are abstractions of systems. They should be built quickly, explained to all project personnel, and changed when necessary. The implementation of recommendations to improve system performance is an integral part of the simulation methodology.

1.9 EXERCISES

1-1. Define the elements shown in Figure 1-2 for a specific problem related to your organization. Include proposed design alternatives, assessment procedures, and one possible implementation outcome. Repeat this exercise from your supervisor's (instructor's) perspective.

1-2. Build a diagram similar to Figure 1-1 for the University System. Build an explanatory model of the University System from a student's point of view.

1-3. Explain how a simulation language supports modeling from a problem organization standpoint.

1-4. In the simulation of the bank teller, hypothesize a relation between the average time in bank per customer and average number of customers in the bank. Determine if your hypothesis holds for average time in queue and average number in queue.

1-5. Discuss how a simulation language impacts on each of the stages of the simulation process.

1-6. Describe the operation of a machine tool.

1-7. Describe the events associated with maintaining accounting records.

1-8. Reduce the number of stages in the simulation process from ten to seven by combining the most similar activities. Provide a rationale for your decisions.

1-9. Discuss how a data base system could support your simulation modeling and analysis activities.

1.10   REFERENCES

1. Banks, J. and J. S. Carson, II, *Discrete-Event System Simulation,* Prentice-Hall, 1984.
2. Conway, R. W., B. M. Johnson, and W. L. Maxwell, "Some Problems of Digital Systems Simulation," *Management Science,* Vol. 6, 1959, pp. 92-110.
3. Bratley, P., B. L. Fox, and L. E. Schrage, *A Guide to Simulation,* Springer-Verlag, 1983.
4. Emshoff, J. R. and R. L. Sisson, *Design and Use of Computer Simulation Models,* Macmillan, 1970.
5. Fishman, G. S. and P. J. Kiviat, "Analysis of Simulated Generated Time Series," *Management Science,* Vol. 13, 1967, pp. 525-557.
6. Law, A. M. and W. D. Kelton, *Simulation Modeling and Analysis,* McGraw-Hill, 1982.
7. Mihram, G. A., "The Modeling Process," *IEEE Transactions on Systems, Man and Cybernetics,* Vol. SMC-2, 1972, pp. 621-629.
8. Mihram, G. A., *Simulation: Statistical Foundations and Methodology,* Academic Press, 1972.
9. Pritsker, A. A. B., *The GASP IV Simulation Language,* John Wiley, 1974.
10. Pritsker, A. A. B., *Modeling and Analysis Using Q-GERT Networks,* Halsted Press and Pritsker & Associates, Inc., 1977.
11. Pritsker, A. A. B., "Compilation of Definitions of Simulation," *SIMULATION,* Vol. 33, 1979, pp. 61-63.
12. Pritsker, A. A. B., "Models Yield Keys to Productivity Problems, Solutions," *Industrial Engineering,* October, 1983, pp. 83-87.
13. Shannon, R. E., *Systems Simulation: The Art and Science,* Prentice-Hall, 1975.
14. Van Horn, R. L., "Validation of Simulation Results," *Management Science,* Vol. 17, 1971, pp. 247-258.

CHAPTER 2

Probability and Statistics

2.1   INTRODUCTION

Systems to be simulated are generally composed of one or more elements that have uncertainty associated with them. Such systems evolve through time in a manner that is not completely predictable and are referred to as stochastic systems. The modeling of stochastic systems requires that the variability of the elements in the system be characterized using probability concepts. The outputs from a simulation model are also probabilistic, and therefore statistical interpretations about them are usually required. Although the reader will likely have some familiarity with probability and statistics, this chapter is included to provide a review of important probability and statistics concepts related to simulation modeling and analysis. We presume the reader has had previous exposure to probability and statistics at the level of one of the introductory books listed at the end of the chapter (6, 10, 12, 15).

2.2   EXPERIMENT, SAMPLE SPACE, AND OUTCOMES

An experiment is a well defined procedure or process whose outcome is observable but is not known with certainty in advance. The set of all possible outcomes is called the sample space. If the sample space is finite or countably infinite, it is said to be discrete; otherwise it is continuous.

Outcomes can be combined to form new outcomes† by the set theory operations of union (∪) and intersection (∩). If the outcome C is defined as the union of a set of outcomes A and a set of outcomes B, denoted C = A∪B, then C consists of the set of all outcomes within A or B. If the outcome D consists of the intersection of A and B, denoted D = A∩B, then D consists of the set of outcomes that are in both A and B.

As an example of the concepts discussed above, consider the operation of a single teller bank system. Customers arrive to the bank, possibly wait, and are processed by the teller. Both the time between arrivals of customers to the bank and the service time by the teller will be assumed to exhibit variability. Let us define

---

† Typically, the term "event" is used to describe combinations of outcomes. In simulation terminology, an event is defined differently so we avoid its use here.

our first experiment as observing the time between customer arrivals to the bank. The sample space for the experiment consists of all possible observations for the time between arrivals. Since the time between customer arrivals can be any non-negative real value, the sample space is continuous. An outcome is defined as any subset of the sample space, and therefore one possible outcome could be defined as the occurrence of an interarrival time between 8 and 9 minutes.

As a second example, consider the experiment of observing the number of customers processed during the first hour of operation. The number of customers processed during the first hour can be any of the values 0, 1, 2, 3, . . . , that is, the set of nonnegative integers. In this case, the sample space is discrete. One possible outcome could be defined as the processing of five customers during the one hour period.

2.3   PROBABILITY

The probability of an outcome is a measure of the degree of likelihood that the outcome will occur. More formally, a probability measure is a function $P(\ )$ which maps outcomes into real numbers and satisfies the following axioms of probability:

1. $0 \leqslant P(E) \leqslant 1$ for any outcome E
2. $P(S) = 1$ where S is the sample space or "certain outcome"
3. If $E_1, E_2, E_3, \ldots$ are mutually exclusive outcomes, then

$$P(E_1 \cup E_2 \cup E_3 \ldots) = P(E_1) + P(E_2) + P(E_3) + \ldots$$

From these three axioms and the rules of set theory, the basic laws of probability can be derived. However, these axioms are not sufficient to compute the probability of an outcome. Numerical values for probabilities are usually difficult to obtain; nevertheless it is useful to postulate their existence.

In some simple cases, the exact probability of an outcome can be calculated using combinatorial analysis. Examples of this include determining the probability of h heads in n tosses of a fair coin and computing the probability of three aces in a five card hand. However, in most cases the exact probability of an outcome cannot be calculated. In such cases, an approximate value for the probability of an outcome can sometimes be obtained using the frequency interpretation of probability. If we repeat an experiment n times and outcome E occurs k times, then $\frac{k}{n}$ is the proportion of times that E occurs. The probability of E can be interpreted as

$$\mathbf{P}(E) = \lim_{n\to\infty} \frac{k}{n},$$

assuming the limit exists. By selecting a sufficiently large value of n, the proportion of occurrences, $\frac{k}{n}$, approximates the probability of E. The approximate values for probabilities obtained in this way can be shown to satisfy the axioms of probability stated earlier. The practical limitation of this approach is that it is sometimes not possible or economical to perform the required experimentation.

2.4   RANDOM VARIABLES AND PROBABILITY DISTRIBUTIONS

A function which assigns a real number to each outcome in the sample space is called a random variable. Discrete random variables are those that take on a finite or a countably infinite set of values. Continuous random variables can take on a continuum of values. In our example of the bank teller system, the interarrival time is a continuous random variable and the number of customers processed during the first hour is a discrete random variable.

A probability distribution is any rule which assigns a probability to each possible value of a random variable. The rule for assigning probabilities takes on two distinct forms depending upon whether the random variable is discrete or continuous.

For discrete random variables, the probability associated with each value of the random variable is commonly specified using a probability mass function, p(x), defined as†

$$p(x_i) = \mathbf{P}(X = x_i)$$

---

† When presenting probability and statistics concepts, we will attempt to use a capital letter to indicate a random variable and a lower case letter to indicate an observed potential numerical value. Sometimes this is not feasible; however, the context of the discussion should clarify the situation.

For each possible value x_i, the function assigns a specific probability that the random variable X assumes the value x_i. The axioms of probability impose the following restrictions on $p(x_i)$.

$$0 \leqslant p(x_i) \leqslant 1 \qquad \text{for all i}$$
$$\sum_{\text{all i}} p(x_i) = 1$$

An alternate representation for the probability distribution is the cumulative distribution function, F(x), defined as follows:

$$F(x) = P(X \leqslant x)$$

In this case, the function F(x) specifies the probability that the random variable X assumes a value less than or equal to x. From the axioms of probability, F(x) must have the following properties:

$$0 \leqslant F(x) \leqslant 1 \qquad \text{for all x,}$$
$$F(-\infty) = 0,$$
and
$$F(\infty) = 1.$$

The distribution function is related to the probability mass function by

$$F(x) = \sum_{x_i \leqslant x} p(x_i)$$

As an example of a discrete probability distribution, consider an experiment consisting of three tosses of a fair coin. Let the random variable X denote the number of heads obtained from the three losses. The random variable X can assume the discrete value of 0, 1, 2, or 3. There are eight possible outcomes of which 1 has 0 heads, 3 have 1 head, 3 have 2 heads and 1 has 3 heads. The probability mass function for the random variable X is depicted in Figure 2-1 and the cumulative distribution function is depicted in Figure 2-2.

For continuous random variables, a different form for the probability distribution is required. Since the random variable can assume any of an uncountably infinite number of values, the probability of a specific value is zero. This does not say that the value is impossible, but that the value is extremely unlikely given the infinite number of alternative values. However, the probability that the variable assumes a value in the interval between two distinct points a and b will generally not

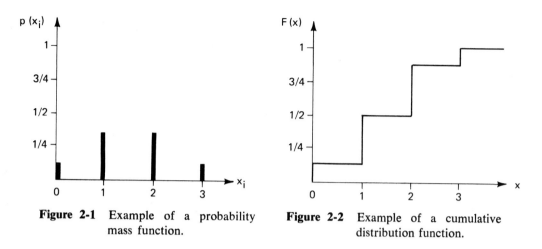

Figure 2-1 Example of a probability mass function.

Figure 2-2 Example of a cumulative distribution function.

be zero. Therefore, the probability mass function as defined for the discrete case is replaced in the continuous case by the probability density function, f(x), defined according to the following relationship

$$P(a \leqslant X \leqslant b) = \int_a^b f(x)dx$$

Thus the probability density function, when integrated between a and b, gives the probability that the random variable will assume a value in the interval between a and b. To be consistent with the axioms of probability, the probability density function must satisfy the following conditions

$$f(x) \geqslant 0$$
$$\int_{-\infty}^{\infty} f(x)dx = 1$$

The cumulative distribution function, F(x), defined for the continuous case is

$$F(x) = \int_{-\infty}^{x} f(y)dy = P(X \leqslant x)$$

The function F(x) defines the probability that the continuous random variable X assumes a value less than or equal to x.

As an example of a continuous probability distribution, consider a random variable X which can assume any value in the range between 0 and 1. Assuming that each of the uncountably infinite number of possible values are equally likely, the corresponding probability density function and cumulative distribution function are shown in Figures 2-3 and 2-4, respectively. The probability that the random variable X assumes a value in the interval between .50 and .75 is the area under the probability density function between .50 and .75. For the random variable depicted in Figure 2-3, this probability is equal to .25.

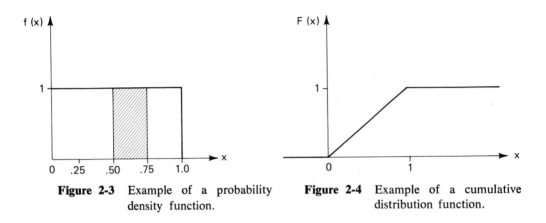

Figure 2-3 Example of a probability density function.

Figure 2-4 Example of a cumulative distribution function.

A random variable can also be both continuous and discrete. Random variables of this type are referred to as having "mixed" distributions. A random variable having a mixed distribution can assume either discrete values with finite probabilities or a continuum of values as prescribed by a probability density function. Figure 2-5 depicts such a distribution where the discrete values 1 and 2 each occur with a probability of 1/3 as denoted by the spikes on the graph. The values between 1 and 2 are governed by the density function f(x) = 1/3.

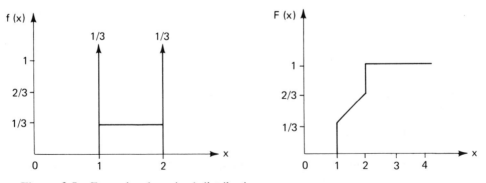

Figure 2-5 Example of a mixed distribution.

Such a distribution would result if samples were drawn from a continuous distribution with equally likely values in the range from 0 to 3, with values greater than 2 set to 2, and values less than 1 set to 1. The equation for the cumulative distribution function for this random variable is

$$F(x) = \begin{cases} 0 & x < 1 \\ \frac{1}{3} + \frac{x-1}{3} & 1 \leqslant x < 2 \\ 1 & x \geqslant 2 \end{cases}$$

From this equation and Figure 2-5, we see that $F(x)$ has discontinuities at $x = 1$ and $x = 2$. At points of discontinuity, $P(X=x)$ is equal to the jump which $F(x)$ makes at the point x. For example, $P(X=1) = 1/3$. However, for $1 < x < 2$, $F(x)$ is continuous at x and $P(X=x) = 0$.

Distribution functions capture the probabilistic characteristic of a variable. Some of the common distribution functions are: normal or Gaussian, uniform, triangular, exponential, lognormal, Erlang, beta, gamma and Poisson. These distributions are presented in graphic form in Chapter 18. Also, a discussion is given of when it is appropriate to use a particular distribution.

Procedures for generating observations for computer simulation from these distributions require the use of a random number which, when generated on a digital computer, is referred to as a pseudorandom number. Pseudorandom numbers are uniformly distributed random samples between the values of 0 and 1 (see Figure 2-3) with successive samples having the property of perceived independence. A random observation generated from a distribution using a pseudorandom number is referred to as a random variate, a random deviate, or a random sample. What is meant by these terms is that a collection of the random observations approximates the underlying distribution from which the random samples were generated, that is, the observations when all taken together characterize the distribution from which the random samples were obtained. The generation of pseudorandom numbers and random samples is described in Chapter 18.

2.5 EXPECTATION AND MOMENTS

It is sometimes desirable to characterize a random variable by one or more values which summarize information contained in its probability distribution function. The expectation or expected value of a random variable X, denoted E[X], is such a value and is defined as follows:

$$E[X] = \sum_{\text{all i}} x_i\, p(x_i) \qquad \text{when X is discrete}$$

$$E[X] = \int_{\text{all x}} x\, f(x)\, dx \qquad \text{when X is continuous}$$

The expectation is a probability weighted average of all possible values of X and therefore a measure of centrality for the distribution. For this reason, it is called the mean value.

Expectations can be taken of functions of random variables. In particular, the expectation of X^n is defined as the nth moment of a random variable and can be expressed as follows

$$E[X^n] = \sum_{\text{all i}} x_i^n p(x_i) \qquad \text{when X is discrete}$$

$$E[X^n] = \int_{\text{all x}} x^n f(x)\, dx \qquad \text{when X is continuous}$$

The expected value is a special case of the above when $n=1$, and, hence, is called the first moment.

A variant of the nth moment is the nth moment about the mean which is defined as

$$E[(X - E[X])^n]$$

In this case, the expected value of X is subtracted from X before computing the nth moment.

A moment of particular importance in probability theory is the second moment about the mean, commonly referred to as the variance of X, and is denoted as σ^2 or Var[X]. The variance of a random variable is a measure of the spread of the probability distribution. If a random variable has a small variance, then samples tend to occur near the expected value. The square root of the variance is referred to as the standard deviation of the random variable.

If X and Y are random variables, then the covariance of X and Y, denoted Cov[X,Y], is defined as

$$Cov[X,Y] = E[(X - E[X])(Y - E[Y])]$$

The covariance is important because it measures the linear association, if any, between X and Y. If the outcome of X has no influence on the outcome of Y, then X and Y are said to be independent and the Cov[X,Y] will be zero. More formally, X and Y are independent if and only if:

$p(y|x) = p(y)$ in the discrete case where $p(y|x)$ is the probability that $Y=y$ given $X=x$

$f(y|x) = f(y)$ in the continuous case where $f(y|x)$ is the conditional density function of Y for $X=x$.

These statements specify that the probability distribution of Y given knowledge of X is the same as the probability distribution of Y without knowledge of X.

A measure of dependence which is related to the covariance is the correlation coefficient, ρ, defined as

$$\rho = \frac{Cov[X,Y]}{\sqrt{Var[X] \cdot Var[Y]}}$$

The correlation coefficient has a range from −1 to 1 with a value of zero indicating no correlation between X and Y. A positive sign indicates that Y tends to be high when X is high, and a negative sign indicates that Y tends to be low when X is high. The magnitude of ρ indicates the degree of linearity of Y plotted against X. If a plot of Y versus X is a straight line, then $\rho = \pm 1$. If X and Y are independent, then a plot of Y versus X will produce random points and ρ will equal zero. Typical scatter diagrams of Y versus X for different values of ρ are depicted in Figure 2-6.

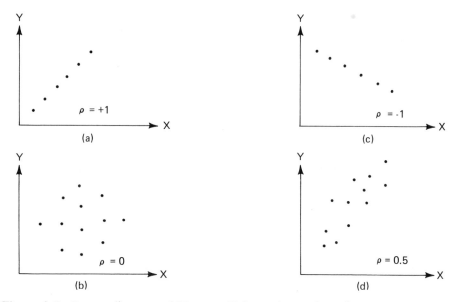

Figure 2-6   Scatter diagrams of Y versus X for various values of ρ.

2.5.1   Theorem on Total Probability (16)

The probability of the outcome B is equal to the sum of the conditional probabilities associated with B given the occurrence of mutually exclusive and exhaustive outcomes A_i weighted by the probability of A_i, that is,

$$P(B) = \sum_i P(B \mid A_i)P(A_i)$$

An analogous result for the expectation of a random variable, Y, is

$$E[Y] = \sum_i E[Y \mid X = x_i]P(X = x_i)$$

2.5.2   Joint Probabilities (16)

The probability of the joint outcome associated with a set of random variables can be expressed as a product of conditional probabilities

$$P(Y_1, Y_2, \ldots, Y_n) = P(Y_1)P(Y_2 \mid Y_1)P(Y_3 \mid Y_1, Y_2) \ldots P(Y_n \mid Y_1 \ldots Y_{n-1})$$

If we assume the random variables have the Markovian property that $P(Y_j \mid Y_1, Y_2, \ldots, Y_{j-1}) = P(Y_j \mid Y_{j-1})$, we have

$$P(Y_1, Y_2, \ldots, Y_n) = P(Y_1)P(Y_2 \mid Y_1)P(Y_3 \mid Y_2) \ldots P(Y_n \mid Y_{n-1})$$

If independence is assumed then

$$P(Y_1, Y_2, \ldots Y_n) = P(Y_1)P(Y_2)P(Y_3) \ldots P(Y_n)$$

2.6   FUNCTIONS OF RANDOM VARIABLES

A function of a random variable is itself a random variable. In this section, we summarize several important properties of functions of random variables.

If X and Y are random variables and k is any arbitrary constant, then the following properties for expectation can be derived:

$$E[X + Y] = E[X] + E[Y]$$
$$E[kX] = kE[X]$$
$$E[X + k] = E[X] + k$$

The similar properties for variances are less obvious than those for expectation and are

$$Var[X + Y] = Var[X] + Var[Y] + 2\,Cov[X,Y]$$
$$Var[kX] = k^2\,Var[X]$$
$$Var[X + k] = Var[X]$$
$$Var[kX + nY] = k^2\,Var[X] + n^2\,Var[Y] + 2kn\,Cov[X,Y]$$

Note that if X and Y are independent random variables, then the $Cov[X,Y]$ is zero, and

$$Var[X + Y] = Var[X] + Var[Y] \qquad \text{for X and Y independent}$$

A random variable of considerable importance in statistics is the sample mean, $\bar{X}_I$, of I samples from a probability distribution. The sample mean is defined as the sum of the samples divided by the number of samples and can be notationally expressed as

$$\bar{X}_I = \frac{1}{I} \sum_{i=1}^{I} X_i$$

Assuming that the X_i are independent and identically distributed (iid) random variables and using the properties of expectation and variance, we can derive the following results:

$$E[\bar{X}_I] = E[X]$$

and
$$Var[\bar{X}_I] = \frac{Var[X]}{I}$$

The variance of the sample mean of I independent samples is a factor $1/I$ smaller than the variance of the random variable from which the samples are drawn. Hence, by selecting a sufficiently large I, the variance of the sample mean can be reduced to an arbitrarily small value.

Note that the relationship given above for the variance of $\bar{X}_I$ applies only if the samples are independent. If the samples are not independent, the calculation of $Var[\bar{X}_I]$ requires the consideration of the covariances between samples. For example, in the bank teller problem, the waiting times for successive customers will be correlated because there is a greater likelihood that the $(i+1)$st customer will be delayed if the ith customer waits than if the ith customer begins service immediately. Hence the variance of the average waiting time cannot be estimated by simply dividing the variance of the waiting time by the number of samples. Such a sequence of correlated samples is referred to as an autocorrelated series. In Section 2.13 and in Chapter 19, we will address the problem of estimating the variance of a sample mean for an autocorrelated series.

2.6.1   Random Sum of Independent Random Variables (18)

If $X_1, X_2, \ldots, X_K$ are independent and identically distributed random variables and K is a discrete random variable independent of X_i then for the sum

$$Y = \sum_{i=1}^{K} X_i \; ,$$

we have

$$E[Y] = E[X]E[K]$$

and

$$Var[Y] = E[K]Var[X] + Var[K]E^2[X]$$

2.6.2   Change of Variables Formula (11,18)

Given:

$$Y_j = g_j(W_1, W_2, \ldots, W_n) \; , \; j = 1, 2, \ldots, n$$

then the joint density function of the $Y_j, f_Y(\cdot)$, is

$$f_Y(y_1, y_2 \ldots, y_n) = f_W(w_1, w_2 \ldots, w_n) \frac{1}{|J|}$$

where $f_W(\cdot)$ is the joint density function for $W_1, W_2, \ldots, W_n$ and J is the Jacobian defined as the determinant of the matrix

$$\begin{pmatrix} \dfrac{\partial g_1}{\partial w_1} & \dfrac{\partial g_1}{\partial w_2} & \cdots & \dfrac{\partial g_1}{\partial w_n} \\ \cdot & \cdot & & \cdot \\ \cdot & \cdot & & \cdot \\ \cdot & \cdot & & \cdot \\ \dfrac{\partial g_n}{\partial w_1} & \dfrac{\partial g_n}{\partial w_2} & \cdots & \dfrac{\partial g_n}{\partial w_n} \end{pmatrix}$$

and $|J|$ is the absolute value of J. This formula prescribes a procedure for making a transformation of random variables.

2.6.3   Rao-Blackwell Theorem (12)

Let X and Y denote random variables such that Y has mean μ and variance $\sigma_Y^2 > 0$. Let $E[Y|x] = \phi(x)$. Then $E[\phi(X)] = \mu$ and $\sigma_{\phi(X)}^2 \leq \sigma_Y^2$.

This theorem states that if we are interested in the statistical properties of the random variable Y and can define a related random variable $\phi(X)$ which is the expected value of Y conditioned on X then we can estimate μ from the expected value of $\phi(X)$ and the variance of this estimate will be at least as small as the variance of a direct estimator. This theorem establishes the worth of using prior information in estimating sample means.†

2.7   GENERATING FUNCTIONS

A commonly referred to function of a random variable is the generating function. Several types of generating functions have been defined and we will discuss only the probability generating function and the moment generating function. The probability generating function for a discrete random variable is defined as

$$A(s) = \sum_i p(x_i)s^i$$

If A(s) is known in a closed form then the probabilities, $p(x_i)$, can be obtained by taking the ith derivative with respect to s and setting s equal to zero. (If A(s) is in a polynomial form, $p(x_i)$ can be seen by inspection.) The expectation of X can be obtained from A(s) by taking the first derivative with respect to s and setting $s = 1$. Higher order moments can be obtained in a similar fashion but require combina-

---

† This observation concerning the Rao-Blackwell Theorem as a basis for use of prior information as a variance reduction technique was pointed out to the author by James R. Wilson.

tions of derivatives. A function related to the probability generating function is the Z-transform.

The moment generating function, MGF, of a random variable X, is defined as

$$M(s) = E[e^{sX}]$$

The nth moment about the origin is obtained by taking the nth derivative with respect to s, that is,

$$\frac{d^n M(s)}{ds^n} = E[X^n e^{sX}]$$

By setting $s = 0$, we have $E[X^n]$. A function related to the moment generating function is the characteristic function.

In addition to obtaining moments of a random variable from generating functions, they are useful for obtaining moments of sums of independent random variables. For example, if $W = X + Y$ and X and Y are independent then

$$E[e^{sW}] = E[e^{s(X+Y)}] = E[e^{sX}]E[e^{sY}]$$

Thus, the MGF of W is the product of the MGFs of X and Y. The moments for W can then be obtained from its MGF. Tables of probability generating functions and moment generating functions for the commonly used random variables are available (3,9).

2.8   LAW OF LARGE NUMBERS AND CENTRAL LIMIT THEOREM

There are two important theorems which characterize the behavior of X_I as the number of samples increases to infinity. The first theorem is the strong law of large numbers and states the intuitive result that as the sample size, I, increases, that with probability one, $\overline{X}_I$ approaches $E[X]$. An associate result referred to as the weak law of large numbers is:

$$\lim_{I \to \infty} \mathbf{P}\{|\overline{X}_I - E[X]| > \epsilon\} = 0 \qquad \text{for any positive } \epsilon$$

This simply says that for any positive value of ϵ, however small, the probability that the difference between $\overline{X}_I$ and $E[X]$ exceeds ϵ approaches zero as I approaches infinity.

The second important theorem which characterizes the behavior of $\overline{X}_1$ is the Central Limit Theorem. This theorem states that under certain mild conditions, the distribution of the sum of I independent samples of X approaches the normal distribution as I approaches infinity, regardless of the distribution of X. Hence, sample means are approximately normally distributed for sufficiently large I. It is difficult to say what sample size is sufficient for assuring normality. However, relatively small sample sizes, like 10 to 15, are often sufficient. Feller (7) presents two theorems that deal with the quality of the normal approximation used in conjunction with the Central Limit Theorem. These theorems are discussed in Bratley, Fox, and Schrage (4). Many variations of the central limit theorem exist. In particular, one variation involves the conditions under which the central limit theorem is applicable for sequences of dependent random variables. These conditions are described in the next section.

2.9   ASYMPTOTIC NORMALITY OF RECURRENT EVENTS (6,7)

If a recurrent event is persistent† and the time between events has a finite mean μ and variance σ^2, then T_r and N_t are asymptotically normally distributed where T_r is the time until the rth event occurrence with $E[T_r] = r\mu$ and $Var[T_r] = r\sigma^2$; and N_t is the number of event occurrences in t time units with $E[N_t] = t/\mu$ and $Var[N_t] = t\,\sigma^2/\mu^3$. For example, if in a simulation the time between arrivals has a mean $\mu = 10$ and variance $\sigma^2 = 4$, then the number of the arrivals in t = 1000 time units is approximately normal with a mean of 100 and a variance of 4.

The above statement is a central limit theorem for a sequence of dependent variables and can be used to check the reasonableness of input generators for a simulation model.

In addition to the above central limit theorem for recurrent events, there are central limit theorems that establish normality conditions for the sample mean or stationary stochastic processes. Moran (13) presents a theorem for moving average (MA) type processes and Dianada (5) for processes in which independence occurs after r lags or time periods.

---

† Feller's terms recurrent and persistent, although out of fashion, provide understandable descriptions of the concepts involved.

2.10   DATA COLLECTION AND ANALYSIS

An essential function in simulation modeling is the collection and analysis of data. This function is required in both defining inputs for the model and in obtaining performance measures from experimentation with the model. In this section, we will review some of the important statistical concepts applicable to data collection and analysis.

2.10.1   Data Acquisition

Data acquisition is the process of obtaining data on a phenomenon of interest. There are a variety of methods by which the data can be acquired. In some cases, the data are available in existing documents, and the problem is that of locating and accessing the data. In other cases, data acquisition may involve the use of questionnaires, field surveys, and physical experimentation.

In aggregate models such as those of urban or economic systems, the required data can frequently be obtained from existing documentation. Common sources of data for these models include census reports, the Statistical Abstract of the United States, United Nations publications, and other publications of governmental and international organizations. Sometimes such data are available in both report form and on computer tape.

In models of business systems, a valuable source of data is the accounting and engineering records of the company. These records are rarely sufficient to form the complete basis for estimating product demand, production cost, and other relevant data. However, they represent a starting point. Questionnaires and field surveys are also potential methods for obtaining data for industrial models.

Physical experimentation is commonly the most expensive and time consuming method for obtaining data. This process includes measurement, recording, and editing of the data. Considerable care must be taken in planning the experiment to assure that the experimental conditions are representative and that the data are recorded correctly. For a discussion of experimental design considerations in data collection, the reader is referred to the text by Bartee (2).

In some cases, there may be no existing data and the available budget or nature of the system may preclude experimentation. An example of such a case would be the use of simulation modeling to compare several proposed assembly line layouts.

A possible approach to data acquisition in such cases is the use of synthetic or predetermined data (1,14). In this method, estimates of activity durations are synthesized by using tables of standard data. Thus, this method permits activity times to be estimated before the process is actually in operation.

2.10.2 Descriptive Statistics

In both collecting data for defining inputs to the model and collecting data on system performance from the model, we encounter the problem of how to convert the raw data to a usable form. Hence, we are interested in treatments designed to summarize or describe important features of a set of data. These treatments normally summarize the data at the expense of a loss of certain information contained within the data.

Grouping Data. One method for transforming data into a more manageable form is to group the data into classes or cells. The data is then summarized by tabulating the number of data points which fall within each class. This kind of table is called a frequency distribution table and normally gives a good overall picture of the data. An example of a frequency distribution table for data collected on customer waiting times is depicted below.

Waiting Time (Seconds)	Number of Customers
0 → 20	21
20 → 40	35
40 → 60	42
60 → 80	35
80 → 100	19
100 → 120	10
> 120	10

The numbers in the right-hand column denote the number of customers falling into each class and are called the class frequencies. The numbers in the left-hand column define the range of values in each class and are referred to as the class limits. The difference between the upper class limit and lower class limit in each case is called the class width. Classes with an unbounded upper or lower class limit are referred to as open. If a class has bounded limits, it is denoted as closed. Frequently the first and/or last class in a frequency distribution will be open.

There are several variations of the class frequency tables which are useful for displaying grouped data. One variation is the cumulative frequency which is obtained by successively adding the frequencies in the frequency table. The cumulative frequency table for the customer waiting time data is depicted below.

Waiting Time Less Than	Cumulative Number of Customers
20	21
40	56
60	98
80	133
100	152
120	162
∞	172

The values in the right hand column represent the cumulative or total number of customers whose waiting time was less than the upper class limit specified in the left hand column. Another variation is obtained by converting the class frequency table (or cumulative table) into a corresponding frequency distribution by dividing each class frequency (cumulative frequency) by the total number of data points. Frequency distributions are particularly useful when comparing two or more distributions.

The frequency and cumulative distribution are sometimes presented graphically in order to enhance the interpretability of the data. The most common among graphical presentations is the histogram which displays the class frequencies as rectangles whose lengths are proportional to the class frequency. Figure 2-7 depicts a histogram for the customer waiting time data.

The primary consideration in the construction of frequency distributions is the specification of the number of classes and the upper and lower class limits for each class. These choices depend upon the nature and ultimate use of the data; however, the following guidelines are offered.

1. Whenever possible, the class widths should be of equal length. Exceptions to this are the first and last classes which are frequently open.

2. Class intervals should not overlap and all data points should fall within a class. In other words, each data point should be assignable to one and only one class.

3. Normally at least five but no more than twenty classes are used.

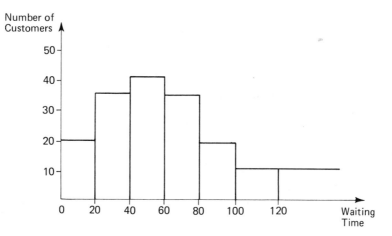

Figure 2-7 Histogram for customer waiting time data.

Parameter Estimation.   If a set of data points consists of all possible observations of a random variable, we refer to it as a *population;* if it contains only part of these observations, we refer to it as a *sample.* Another method for summarizing a set of data is to view the data as a sample which is then used to estimate the parameters of the parent or underlying population. The population parameters of most frequent interest are the mean which provides a measure of centrality, and the variance which provides a measure of dispersion.

To illustrate, consider again the data on customer waiting times. This data can be viewed as a sample from the population which consists of all possible customer waiting times. We can use this sample data to estimate the mean customer waiting time and the variance of the customer waiting time for the population of all possible customers.

Different symbols are used to distinguish between population parameters and estimates of these parameters based upon a sample. The greek letters μ and σ^2 are often used to denote the population mean and variance, respectively. The corresponding estimates of these parameters based upon the sample record $x_1, x_2, \ldots,$ x_I are the average, denoted as $\bar{x}_I$, and the variance estimate, denoted as s_x^2. In order to further distinguish between descriptions of populations and descriptions of samples, the first are referred to as parameters and the second are referred to as statistics.

Before proceeding with this discussion of descriptive statistics, a clarifying point regarding the notation used to describe random samples and experimental values of a random variable or stochastic sequence is necessary. Before an experimental value is observed, it is a random variable denoted by X_i. After a value is observed, it is denoted by x_i. By the sample mean, $\overline{X}_I$, we refer to a random variable that is the sum of I random samples before they are observed divided by I. The average, $\overline{x}_I$, however, is the sum of I observed values x_i divided by I. In an analogous fashion, S_x^2 is the random variable describing an estimate of the sample variance before experimental values are observed and s_x^2 is the estimate of the variance of observed values. This notation conforms to our policy of using capital letters for random variables where possible and lower case letters for numerical quantities.

In constructing estimates of the population parameters from sample data, there are two distinct cases to consider. In the first case, we consider a sample record where we are concerned only with the value of each observation and not the times at which the observations were recorded. The data on customer waiting times is an example of such a record. Statistics derived from a time independent sample record are referred to as *statistics based upon observations*.

The second case to be considered is for variables which have values defined over time. For example, the number of busy tellers in a bank is a random variable that has a value which is defined over time. In this case, we require knowledge of both the values assumed by the random variable and the time periods for which each value persisted. Statistics derived from time dependent records are referred to as *statistics on time-persistent variables*.

The formulas for calculating $\overline{x}_i$ and s_x^2 for both statistics based upon observations and statistics on time-persistent variables are summarized in Table 2-1. For the time-persistent case the sample mean is designated by $\overline{x}_T$ where T is the total time interval observed. Sometimes the formulas for s_x^2 are given in a slightly different form, however the form shown is the most convenient for computational purposes. Note that for statistics based upon observations, the $\sum_{i=1}^{I} x_i$, the $\sum_{i=1}^{I} x_i^2$,

and the number of samples I, are sufficient to compute both $\overline{x}_I$ and s_x^2. Similarly, for statistics on time-persistent variables, $\int_0^T x \, dt$, $\int_0^T x^2 \, dt$, and T are required.

Another statistic which is commonly employed in summarizing a set of data is the coefficient of variation, $s_x/\overline{x}_I$. The coefficient of variation expresses the sample

Table 2-1 Formulas for calculating the average and variance of a sample record.

Statistic	Formula	
	Statistics based upon Observations	Statistics for Time Persistent Variables
Sample Mean	$\bar{x}_I = \dfrac{\sum\limits_{i=1}^{I} x_i}{I}$	$\bar{x}_T = \dfrac{\int\limits_{0}^{T} x(t)\, dt}{T}$
Sample Variance	$s_X^2 = \dfrac{\sum\limits_{i=1}^{I} x_i^2 - I\bar{x}_I^2}{I-1}$	$s_X^2 = \dfrac{\int\limits_{0}^{T} x^2(t)\, dt}{T} - \bar{x}_T^2$

standard deviation relative to the sample mean. The use of the coefficient of variation is advantageous when comparing the variation between two or more sets of
data.

Distribution Estimation. A related but more difficult problem is the use of the
sample record to identify the distribution of the population. This problem frequently arises in modeling because of the need to characterize random elements of
a system by particular distributions. Although an understanding of the properties
of the theoretical distributions described in Chapter 18 will aid the modeler in hy-
pothesizing an appropriate distribution, it is frequently desired to test the hypothesis by applying one or more goodness-of-fit tests to the sample record. The
chi-square and Kolmogorov-Smirnov are probably the best known tests, and descriptions and examples of these can be found in most statistics textbooks. A discussion of identifying a distribution function that fits a sample record and
programs to perform the distribution fitting tasks is given in Chapter 18.

2.11 STATISTICAL INFERENCE

In simulation studies, inferences or predictions concerning the behavior of the
system under study are to be made based on experimental results obtained from

the simulation. Because a simulation model contains random elements, the outputs from the simulation are observed samples of random variables. As a consequence, any assertions which are made concerning the operation of the system based on simulation results should consider the inherent variability of the simulation outputs. This variability is summarized or taken into account by the use of confidence intervals or through hypothesis testing.

2.11.1   Confidence Intervals

In Section 2.10.2, we discussed methods for estimating the mean and variance parameters of a population based on a sample record. The estimates were calculated as a single number from the sample record and are referred to as *point estimates*. In general, an estimate will differ from the true but unknown parameter as the result of chance variations. The use of a point estimate has the disadvantage that it does not provide the decision maker with a measure of the accuracy of the estimate. A probability statement which specifies the likelihood that the parameter being estimated falls within prescribed bounds provides such a measure and is referred to as confidence interval or an interval estimate.

The parameter of primary interest in simulation analysis is the population mean. In the classical development of the confidence interval for the mean, it is assumed that the samples are independent and identically distributed (iid). Hence, by the Central Limit Theorem, the sample mean, $\overline{X}_I$, is approximately normally distributed for sufficiently large I. As stated previously, the assumption of independence is not a necessary condition for the application of the Central Limit Theorem.

If we assume that $\overline{X}_I$ is normally distributed, then the statistic

$$Z = \frac{\overline{X}_I - \mu}{\sigma_{\overline{X}}}$$

is a random variable which is normally distributed with a mean of zero and standard deviation of one. Furthermore,

$$\mathbf{P}[-Z_{a/2} < Z < Z_{a/2}] = 1-\alpha$$

where $Z_{a/2}$ is the value for Z such that the area to its right on the standard normal curve equals $\alpha/2$. Hence, we can assert with probability $1-\alpha$ that

$$\overline{X}_I - Z_{\alpha/2} \cdot \sigma_{\overline{x}} < \mu < \overline{X}_I + Z_{\alpha/2} \cdot \sigma_{\overline{x}} \qquad (2\text{-}1)$$

that is, a proportion, 1–a, of confidence intervals based on I samples of x should contain (cover) the mean m This proportion is called the coverage for the confidence interval.

The above formula assumes knowledge of the standard deviation of the mean, $\sigma_{\overline{x}}$, which is usually unknown. If we use the sample standard deviation of the mean, $S_{\overline{x}}$, to estimate $\sigma_{\overline{x}}$, we can develop a similar relationship by noting that the statistic

$$t = \frac{\overline{X}_I - \mu}{S_{\overline{x}}}$$

is a random variable having a student t-distribution with I−1 degrees of freedom. Hence, a 1−α confidence interval for μ using the estimate $S_{\overline{x}}$ is given by

$$\overline{X}_I - t_{\alpha/2,I-1}S_{\overline{x}} < \mu < \overline{X}_I + t_{\alpha/2,I-1}S_{\overline{x}} \qquad (2\text{-}2)$$

where $t_{\alpha/2,I-1}$ is a critical value of the t-statistic with (I−1) degrees of freedom.

If the samples X_i are iid, the confidence intervals given by 2-1 and 2-2 are modified by the substitutions

$$\sigma_{\overline{x}} = \frac{\sigma_x}{\sqrt{I}} \qquad (2\text{-}3)$$

and

$$S_{\overline{x}} = \frac{S_x}{\sqrt{I}} \qquad (2\text{-}4)$$

respectively. This substitution provides an expression for the confidence interval based on samples. However, this simple relationship between the variance of the samples and the variance of the mean of the samples is valid only if the samples are independent.

Methods for defining $S_{\overline{x}}$ for use in Expression 2-2 in the case of autocorrelated samples are described in Chapter 19. The most direct approach is to organize the experiment to obtain independent observations which can be accomplished through replicating the simulation or organizing the data into batches.

2.11.2 Tolerance Intervals

A tolerance interval provides a range for an observation or for an average of a set of observations. Remember that the observation and its average are random variables. Because we are setting an interval on a random variable, it is necessary to specify the fraction of samples of the random variable that is desired to be within the tolerance interval and then to further specify the confidence with which we desire the fraction of samples to be contained within the interval. Hence, we specify that, with $(1-\delta)$ probability, we desire a range such that $(1-\epsilon)$ fraction of observations of $\overline{X}_I$, each based on a sample size of I, will fall in the tolerance interval. Wilson (19) developed the following formula for such a range:

$$\overline{X}_I \pm Z_{\epsilon/2} \, Q \frac{S_x}{\sqrt{I}}$$

where $Z_{\epsilon/2}$ is a critical value of the normal distribution corresponding to a $(1-\epsilon)$ confidence, and Q is given by

$$Q = \frac{(2I+1)}{2I} \sqrt{\frac{(I-1)}{\chi^2_{\delta,I-1}}}$$

where $\chi^2_{\delta,I-1}$ is a critical value from the chi-square distribution with $(1-\delta)$ confidence and $(I-1)$ degrees of freedom.

In an experiment consisting of I runs, Wilson also derived the following tolerance interval formula that the probability is at least $(1-\delta)(1-\epsilon)$ that a single additional observation X_{I+1} will fall in the interval

$$\overline{X}_I \pm Z_{\epsilon/2} \, Q \, S_x$$

2.12   HYPOTHESIS TESTING

In some applications of simulation, the objective is to decide if a statement concerning a parameter is true or false. For example, we might want to decide whether a change in a dispatching rule for a job shop reduces the average late time for the jobs processed. Due to the experimental nature of simulation, we must account for the chance variation in the estimates of the parameters being compared. This is done using hypothesis testing.

The general procedure of hypothesis testing calls for defining a *null hypothesis* (denoted H_0) and an *alternate hypothesis* (denoted H_1). The null hypothesis is usually set up with the objective of determining whether or not it can be rejected. For example, if we wish to establish that job loading rule A reduces average late time relative to job loading rule B, we would define the null and alternate hypotheses as

H_0: average waiting time for rule A equals average waiting time for rule B
H_1: average waiting time for rule A is less than the average waiting time for rule B

We would then use the experimental results from simulations with rules A and B to attempt to reject H_0 in favor of H_1.

Testing the null hypothesis against the alternate hypothesis involves selecting a decision rule based on the sample data which leads to the acceptance or rejection of the null hypothesis. Acceptance of the null hypothesis does not infer that the null hypothesis is true, but that there is insufficient evidence based on the sample data to reject the hypothesis.

There are two types of errors that can be made in applying the decision criterion. The *Type I* error is to reject the null hypothesis when the hypothesis is true. The *Type II* error is to accept the null hypothesis when it is false. A decision rule can be judged by the probabilities associated with Type I and Type II errors. These probabilities are typically denoted as α and β probabilities, respectively. The probability α of a Type I error is referred to as the *level of significance* of the test.

The decision criterion is established by constructing a *test statistic* which has a known distribution. The test statistic is calculated from the sample data and is compared using a rejection rule. If the test statistic falls within the critical region, then the null hypothesis is rejected.

The test statistic and rejection rule for hypothesis tests concerning means are summarized in Table 2-2. Tests 1 and 2 are for a mean being equal to a given value μ_0. Tests 3 and 4 concern the comparison of two means. The equations for the test statistics are expressed in terms of $\sigma_{\bar{x}}$ and S_x since assumptions of independence cannot be prescribed.

Table 2-2   Hypothesis Tests for Means

Null Hypothesis	Condition	Test Statistic†	Test Statistic Distribution	Degrees of Freedom	Alternative Hypothesis	Null Hypothesis Rejection Rule
1. $\mu = \mu_0$	Known $\sigma_{\bar{X}}$	$Z = \dfrac{\bar{X} - \mu}{\sigma_{\bar{X}}}$	Standard Normal	—	$\mu > \mu_0$ $\mu < \mu_0$ $\mu \neq \mu_0$	$Z > Z_\alpha$ $Z < -Z_\alpha$ $|Z| > Z_{\alpha/2}$
2. $\mu = \mu_0$	Unknown $\sigma_{\bar{X}}$	$t = \dfrac{\bar{X} - \mu}{S_{\bar{X}}}$	Student t	$I - 1$	$\mu > \mu_0$ $\mu < \mu_0$ $\mu \neq \mu_0$	$t > t_\alpha$ $t < -t_\alpha$ $|t| > t_{\alpha/2}$
3. $\mu_X = \mu_Y$	Known $\sigma_{\bar{X}}$ and $\sigma_{\bar{Y}}$	$Z = \dfrac{\bar{X} - \bar{Y}}{\sqrt{\sigma_{\bar{X}}^2 + \sigma_{\bar{Y}}^2}}$	Standard Normal	—	$\mu_X > \mu_Y$ $\mu_X < \mu_Y$ $\mu_X \neq \mu_Y$	$Z > Z_\alpha$ $Z < Z_\alpha$ $|Z| > Z_{\alpha/2}$
4. $\mu_X = \mu_Y$	Unknown $\sigma_{\bar{X}}$ and $\sigma_{\bar{Y}}$	$t = \dfrac{\bar{X} - \bar{Y}}{\sqrt{S_{\bar{X}}^2 + S_{\bar{Y}}^2}}$	Student t	Nearest integer to: $\dfrac{(S_{\bar{X}}^2 + S_{\bar{Y}}^2)^2}{\dfrac{S_{\bar{X}}^4}{I_X+1} + \dfrac{S_{\bar{Y}}^4}{I_Y+1}} - 2$	$\mu_X > \mu_Y$ $\mu_X < \mu_Y$ $\mu_X \neq \mu_Y$	$t > t_\alpha$ $t < -t_\alpha$ $|t| > t_{\alpha/2}$

Legend: I = sample size
I_X = sample size for X
I_Y = sample size for Y
μ_0 = hypothesized mean
α = level of significance

†The subscript I has been dropped from $\bar{X}$ and $\bar{Y}$ for convenience.

2.13 STATISTICAL PROBLEMS RELATED TO SIMULATION

Decision analysis based on the results of a simulation model normally requires an estimate of the average simulation response and an estimate of its variance. Both of these estimators are affected by experimental conditions. The experimental conditions which the modeler must establish include the initial or starting states for the simulation, the time at which statistics collection is to begin, and the run length and number of replications. In this section, we will introduce some of the considerations and problems associated with establishing these conditions. In Chapter 19, we discuss these problems in detail.

2.13.1 Initial Conditions

Implicit in every simulation model is an initial condition or starting state for the simulation. The simplest and probably most commonly used initial state is "empty and idle," in which the simulation begins with no entities in the system and all servers idle. The appropriateness of this starting condition depends on the nature of the system being modeled and whether we are interested in the transient or steady-state behavior† of the system.

When the purpose of our analysis is to study the steady-state behavior of a system, we can frequently improve our estimate of the mean by beginning the simulation in a state other than empty and idle. The starting condition can be established by estimating an initial state which is representative of the long term behavior of the system, perhaps by observing the plotted output from a pilot simulation run. For a transient analysis, the starting condition should reflect the initial status of the system.

---

† Steady-state behavior does not denote a lack of variability in the simulation response, but specifies that the probability mechanism describing this variability is unchanging and is no longer affected by the starting condition.

2.13.2   Data Truncation

A method which is frequently used to reduce any bias in estimating the steady-state mean resulting from the initial conditions is to delay the collection of statistics until after a "warm up" period. This is normally done by specifying a truncation point before which data values are not included in the statistical estimates. The intent is to reduce the initial condition bias in the estimates by eliminating values recorded during the transient period of the simulation. However, by discarding a portion of the data, we are not using observations and, hence, may be increasing the estimated variance of the mean. Thus, by truncation we improve the quality of the estimate of the mean at the possible expense of increased variability in the simulation outputs.

The most common method for determining the truncation point is to examine a plot of the response from a pilot simulation run. The truncation point is selected as the time at which the response "appears" to have reached steady state. There are also methods which attempt to formalize this procedure in the form of a rule which can be incorporated into the simulation program to automatically determine the truncation point during the execution of the simulation. These rules are discussed in Chapter 19. A theorem is presented in the next section that addresses the question of returns to a state or the crossing of an expected value.

2.13.3   The First Arc Sine Law (6)

Consider a binomial random variable Y_n with $P[Y_n = 1] = 1/2$ and $P[Y_n = -1] = 1/2$ for $n = 1, 2, \ldots, N$. Consider the sequence of partial sums

$$Z_n = \sum_{i=1}^{n} Y_i$$

for all times up to time N. For a fixed α ($0 < \alpha < 1$), we focus on the experimental outcome in which $Z_n > 0$ for at most $N\alpha$ time units, that is, we look for experiments in which the sequence $\{Z_n : 1 \le n \le N\}$ up to time N spends at most α percent of the time above the axis. As $N \to \infty$, the probability of observing such a result tends to

$$\frac{2}{\pi} \sin^{-1}(\sqrt{\alpha})$$

For example, the probability that the fraction of time is less than $\alpha=0.976$ is 0.90. Thus, with probability 0.20, the fraction of time spent on one side of zero or the other is 0.976.

Another result is that, in a time period of length 2N, the probability that the number of partial sums $\{Z_n\}$ equal to zero being at most $\alpha\sqrt{2N}$ tends to

$$\sqrt{\frac{2}{\pi}}\int_0^a e^{-q^2/2}dq \quad \text{as } N\to\infty$$

For example, if we drew 10,000 samples of Y_n then there is a probability of 0.50 that there will be fewer than 68 times when $\sum Y_n = 0$. A related result also given by Feller specifies that the number of changes of sign (crossings) in the sequence of partial sums in N time units increases as the $\sqrt{N}$, that is, in 100N time units we should only expect 10 times as many crossings of 0 as in N time units. These theorems illustrate the conceptual difficulties and nonintuitive behavior associated with even simple stochastic processes. These results indicate potential difficulties associated with the use of returns to a state or the crossing of a state in statistical analysis.

2.13.4   Run Length and Number of Replications

An important experimental design decision which the analyst must make is the tradeoff between run length and number of replications of the simulation. The use of a few long runs as opposed to many short runs generally produces a better estimate of the steady state mean because the initial bias is introduced fewer times and less data is truncated. However, the reduced number of samples corresponding to fewer replications may increase our estimate of the variance of the mean. The use of many short runs, on the other hand, may introduce a bias due to the starting conditions. The larger the initial bias, the more important it is to use longer runs to reduce the effects of the starting conditions.

There are several alternate methods for specifying the duration of a simulation. Perhaps the most common method is to specify a time at which the simulation is to end. A disadvantage of this method is that the number of samples collected is a random variable and may differ in each replication. A method which allows us to control the sample size is to specify the number of entities which are to be entered into the model. In this case the simulation executes until the prescribed number of entities which are entered into the model are completely processed through the system. Thus, the simulation stops in the empty and idle state. A similar but different method is to specify the number of entities which are processed through the system. Note that in this case the system is not necessarily empty and idle at the time it is stopped. When using this approach, it is necessary to ensure that the entities remaining to be processed are representative. An example where this stopping method may be inappropriate is when a shortest processing time dispatching rule is employed and, hence, jobs with long processing times may be the ones still remaining in the queues.

Another approach for controlling the duration of a simulation is the use of automatic stopping rules. These methods automatically monitor the simulation results at selected intervals during the execution of the simulation. The simulation is stopped when the estimate of the variance of the mean is within a prescribed tolerance. The use of automatic stopping rules is discussed in more detail in Chapter 19.

If we are estimating the variance of an output variable X by replication and if we assume that X is normally distributed (which if X is a mean value is a good assumption) then the number of independent replications of the simulation required to attain a specified confidence interval for $\overline{X}$ is given by

$$I = \left(\frac{t_{a/2, I-1} S_X}{g}\right)^2$$

where

$t_{a/2, I-1}$ is a value from the table of critical values of the t-statistic with $I-1$ degrees of freedom

g is the half-width of the desired confidence interval

Unfortunately, the use of this formula for I requires knowledge of the t-statistic with I–1 degrees of freedom and S_X. Typically, we must assume a value for I, make the I replications of the simulation, obtain values of t and s_x based on these runs, and then use the above formula with these values inserted to test the sufficiency of our initial assumption or to determine the number of additional replications which are required.

2.14 CHAPTER SUMMARY

This chapter has provided the probability and statistics background required for simulation analysis. Detailed developments have not been presented as the intent was to cover a wide range of simulation related topics. The material introduces sufficient simulation subject matter to permit the understanding of simulation modeling concepts and the experimental nature of simulation analysis. It also provides a basis for understanding the detailed aspects associated with the statistical analysis of simulation results.

2.15 EXERCISES

2-1. Given the following simulation results for customer time-in-system for 20 simulation runs, compute an estimate of the mean, variance, and coefficient of variation. Construct a histogram that has 5 cells, a cell width of 1, and the lower limit of the first cell equal to 0.

1.1, 2.8, 3.7, 1.9, 4.9, 1.6, .4, 3.8, 1.5, 3.4, 1.9,
2.1, 3.8, 1.6, 3.2, 2.9, 3.7, 2.0, 4.2, 3.3

2-2. The following figure depicts the number of customers in a waiting line over a fifteen minute time interval. Calculate the average and standard deviation of the number of customers waiting.

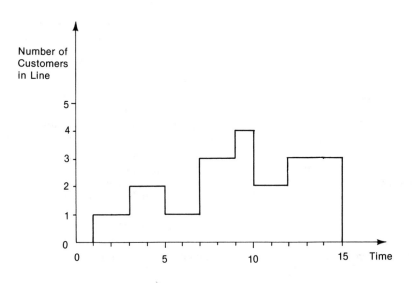

2-3.   A simulation model was employed to study the design of a proposed subway system. The following are average passenger waiting times in minutes based on 20 independent replications of the simulation model. Construct a 99% confidence interval for the average passenger waiting time.

   15, 17, 14, 15, 16, 14, 15, 18, 15, 14,
   15, 20, 17, 14, 16, 16, 15, 18, 14, 15

2-4.   The following are average weekly inventory costs in thousands of dollars for two proposed inventory control policies. The values were obtained from independent replications of a simulation model of the system. Test the hypothesis that the average weekly cost for policy A is less than that for policy B at a 5% significance level.

Policy A

   1.2, 1.3, 1.1, 1.4, 0.9, 1.1, 1.5, 1.2,
   1.3, 1.2, 1.3, 1.4, 1.1, 1.2, 1.0

Policy B

   1.1, 1.5, 1.4, 1.3, 1.6, 1.5, 1.4, 1.5,
   1.7, 1.3, 1.6, 1.5, 1.8, 1.7, 1.6

2-5.   In a simulation study of a proposed assembly line layout, ten independent replications were made to estimate the average hourly production for the line. Based on this data, estimate the number of runs required to determine the average production rate within 2 units at a one percent significance level.

   157, 162, 151, 170, 162, 157, 166, 165, 152, 160

2-6.   Show that the sum of two independent Poisson random variables with mean μ_1 and μ_2 is a random variable which is Poisson distributed with mean $\mu_1 + \mu_2$ (Hint: The moment generating function for the Poisson distribution with mean μ is $e^{\mu(e^t-1)}$.) What is the distribution of the sum of two independent normal random variables?

2-7.   The thief of Baghdad has been placed in a dungeon with three doors. One door leads to freedom, one door leads to a long tunnel, and a third door leads to a short tunnel. The tunnels return the thief to the dungeon. If the thief returns to the dungeon, he attempts to gain his freedom again but his past experiences do not help him in selecting the door that leads to freedom, that is, we assume a Markov thief. The thief's probabilities of selecting the doors are: 0.30 to freedom; 0.20 to the short tunnel; and a 0.50 to the long tunnel. Assuming that the travel times through the long and short tunnels are 6 and 3, respectively, determine the expected time until the thief selects the door which leads to freedom.

2-8.   In a high school track meet, eight heats are to be run with six runners in each heat. The eight heats are to be run consecutively with a 15 minute delay from the end of one heat to the beginning of the next. Each runner's time is approximately uniformly distributed between 8 and 12 minutes. Develop an expression for the time required to complete all eight heats. Specify any assumptions required in your analysis. Suppose the time for each runner was exponentially distributed with a mean of 10, how would this change the expression developed?

2-9.   Discuss the impact of the Rao-Blackwell Theorem on the use of analytic results within a simulation model.

2-10.   Discuss and compare the definitions of the following terms: the mean of a system variable; the mean of a model of a system variable; expected value of a variable in a model; an average of a model variable; and an average from a system variable. Include within your discussion how the concepts of confidence intervals, tolerance intervals, and coverage are used with respect to each concept.

2-11.   If N operations are performed in parallel, the time until the next operation completion Y is the minimum of the operation times, X_i. If the operations times are independent and identically distributed, show that the cumulative distribution function, $F_Y(t)$, for the time until the next event, is equal to $1-(1-F_X(t))^N$. Use this formula to show that the distribution of the time until the next operation completion is exponentially distributed if each operation completion time is exponentially distributed. What is the mean time until the completion of the next operation?

2-12.   If B is exponentially distributed with mean rate λ, the probability of B not occurring in the next Δt time units is $e^{-\lambda \Delta t}$. If B has not occurred up to time t_1, show that the probability that B does not occur in the next Δt time units is also equal to $e^{-\lambda \Delta t}$. This property of the exponential distribution is called the forgetfulness or Markovian property. Discuss how the forgetfulness property of the exponential and the result of Exercise 2.11 that the minimum of a set of exponentials is exponential can be used to reduce the number of events occurring in a simulation.

2.16   REFERENCES

1.   Barnes, R. M., *Motion and Time Study: Design and Measurement of Work,* Sixth Edition, John Wiley, 1968.

2.   Bartee, E. M., *Engineering Experimental Design Fundamentals,* Prentice-Hall, 1968.

3.   Beightler, C. S., L. G. Mitten, and G. L. Nemhauser, "A Short Table of Z-Transforms and Generating Functions," *Operations Research,* Vol. 9, 1961, pp. 576-577.

4.   Bratley, P., B. L. Fox, and L. E. Schrage, *A Guide to Simulation,* Springer-Verlag, 1983.

5.   Diananda, P. H., "Some Probability Limit Theorems with Statistical Applications." *Proceedings, Cambridge Phil. Soc.,* Vol. 49, 1953, pp. 239-246.

6.   Feller, W., *An Introduction to Probability Theory and Its Applications,* John Wiley, (Second Edition), 1957.

7.   Feller, W., *An Introduction to Probability Theory and Its Applications,* John Wiley, Vol. II, 1972.

8.   Fishman, G. S., *Principles of Discrete Event Simulation,* John Wiley, 1978.

9.   Giffin, W., *Transform Techniques for Probability Modeling,* Academic Press, 1975.

10.   Hald, A., *Statistical Theory with Engineering Applications,* John Wiley, 1952.

11.   Hoel, P. G., *Elementary Statistics,* Second Edition, John Wiley, 1966.

12.   Hogg, R. V. and A. T. Craig, *Introduction to Mathematical Statistics,* Macmillan, 1970.

13.   Moran, P. A. P., "Some Theorems on Time Series, I," *Biometrika,* Vol. 34, 1947, pp. 281-291.

14.   Niebel, B. W., *Motion and Time Study,* Fourth Edition, Richard D. Irwin, 1967.

15.   Papoulis, A., *Probability, Random Variables, and Stochastic Processes,* McGraw-Hill, 1965.

16. Parzen, E., *Modern Probability Theory and Its Applications,* John Wiley, 1960.

17. Pritsker, A. A. B., *The GASP IV Simulation Language,* John Wiley, 1974.

18. Ross, S., *A First Course in Probability,* Macmillan, 1976.

19. Wilson, J. R., *Statistical Techniques for Simulation Practitioners,* Course Notes, Pritsker & Associates, 1985.

CHAPTER 3

Simulation Modeling Perspectives

3.1   INTRODUCTION

In developing a simulation model, an analyst needs to select a conceptual framework for describing the system to be modeled. The framework or perspective contains a "world view" within which the system functional relationships are perceived and described. If the modeler is employing a simulation language, then the world view will normally be implicit within the language. However, if the modeler elects to employ a general purpose language such as FORTRAN, PL/I, or C, then the perspective for organizing the system description is the responsibility of the modeler. In either case, the world view employed by the modeler provides a conceptual mechanism for articulating the system description. In this chapter, we summarize the alternative world views for simulation modeling and introduce the unified modeling framework of SLAM II.

3.2   MODELING WORLD VIEWS

Models of systems can be classified as either discrete change or continuous change. Note that these terms describe the model and not the real system. In fact, it may be possible to model the same system with either a discrete change (hereafter referred to simply as discrete) or a continuous change (continuous) model. In most simulations, time is the major independent variable. Other variables included in the simulation are functions of time and are the dependent variables. The adjectives discrete and continuous when modifying simulation refer to the behavior of the dependent variables.

Discrete simulation occurs when the dependent variables change discretely at specified points in simulated time referred to as event times. The time variable may be either continuous or discrete in such a model, depending on whether the discrete changes in the dependent variable can occur at any point in time or only at specified points.

The bank teller problem discussed in Chapter 1 is an example of a discrete simulation. The dependent variables in that example were the teller status and the number of waiting customers. The event times corresponded to the times at which customers arrived to the system and departed from the system following completion of service by the teller. In general, the values of the dependent variables for discrete models do not change between event times. An example response for a dependent variable in a discrete simulation is shown in Figure 3-1.

In *continuous simulation* the dependent variables of the model may change continuously over simulated time. A continuous model may be either continuous or discrete in time, depending on whether the values of the dependent variables are available at any point in simulated time or only at specified points in simulated time. Examples of response measurements for continuous simulations are shown in Figure 3-2 and Figure 3-3.

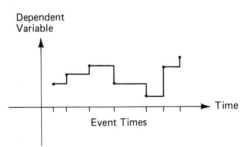

Figure 3-1   Response measurement from a discrete event simulator.

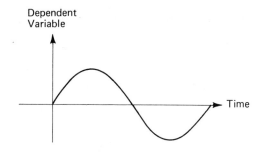

Figure 3-2   Response measurement from a continuous simulator.

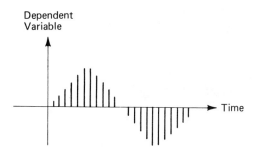

Figure 3-3   Response measurement from a continuous simulator using discrete time steps.

The modeling of the concentration of a reactant in a chemical process or the position and velocity of a spacecraft are illustrations of situations where a continuous representation is appropriate. However, in some cases, it is also useful to model a discrete system with a continuous representation by considering the entities in the system in the aggregate rather than as individual entities. For example, we would probably prefer to model the population of a particular species in a lake using a continuous representation, even though in reality the population changes discretely.

In *combined simulation* the dependent variables of a model may change discretely, continuously, or continuously with discrete jumps superimposed. The time variable may be continuous or discrete. The most important aspect of combined simulation arises from the interaction between discretely and continuously changing variables. For example, when the concentration level of a reactant in a chemical process reaches a prescribed level, the process may be shut down. A combined simulation language must contain provisions for detecting the occurrence of such conditions and for modeling their consequences. An example of a response from a combined simulation model is shown in Figure 3-4.

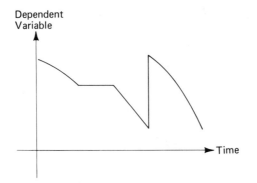

Figure 3-4 Response measurement from a combined simulator.

3.3   DISCRETE SIMULATION MODELING

The objects within the boundaries of a discrete system, such as people, equipment, orders, and raw materials are called entities. There are many types of entities and each has various characteristics or attributes. Although they engage in different types of activities, entities may have a common attribute requiring that they be grouped together. Groupings of entities are called files.† Inserting an entity into a file implies that it has some relation with other entities in the file.

The aim of a discrete simulation model is to reproduce the activities that the entities engage in and thereby learn something about the behavior and performance potential of the system. This is done by defining the states of the system and constructing activities that move it from state to state. The state of a system is defined in terms of the numeric values assigned to the attributes of the entities. A system is said to be in a particular state when all of its entities are in states consonant with the range of attribute values that define that state. Thus, simulation is the dynamic portrayal of the states of a system over time.

In discrete simulation, the state of the system can change only at event times. Since the state of the system remains constant between event times, a complete dynamic portrayal of the state of the system can be obtained by advancing simulated time from one event to the next. This timing mechanism is referred to as the next event approach and is used in most discrete simulation languages.

---

† The word file is used in SLAM II to mean a set of entities or events. The description of systems in terms of entities, attributes and files (sets) was first employed by Markowitz (10).

A discrete simulation model can be formulated by: 1) defining the changes in state that occur at each event time; 2) describing the activities in which the entities in the system engage; or 3) describing the process through which the entities in the system flow. The relationship between the concept of an *event, activity,* and a *process* is depicted in Figure 3-5. An event takes place at an isolated point in time at which decisions are made to start or end activities. A process is a time-ordered sequence of events and may encompass several activities. These concepts lead naturally to three alternative world views for discrete simulation modeling. These world views are commonly referred to as the event, activity scanning, and process orientations, and are described in the following sections. For additional discussions of discrete simulation modeling orientations, the reader is referred to the excellent reviews by Kiviat (7, 8).

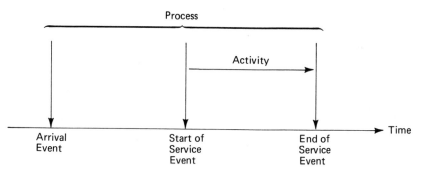

Figure 3-5 Relationship of events, activities, and processes.

3.3.1 Event Orientation

In the event-oriented world view, a system is modeled by defining the changes that occur at event times. The task of the modeler is to determine the events that can change the state of the system and then to develop the logic associated with each event type. A simulation of the system is produced by executing the logic associated with each event in a time-ordered sequence.

To illustrate the event orientation, consider again the bank teller problem discussed in Chapter 1. Customers arrive to the system, possibly wait, undergo service by the teller, and then exit the system. The state of the system is defined by the status of the teller and the number of waiting customers. The state of the system remains constant except when a customer arrives to the system or departs from it. Therefore, the event model for this system consists of describing what happens at a customer arrival time and an end-of-service time. Since a change in the state of the system can occur only at these event times, the customer arrival and end-of-service events can be used to completely describe the dynamic structure of the system. At the event time, the simulation clock stands still and only status changes are made. However, future events are placed on the event calendar which represents times in the future at which status changes may take place.

Consider first the logic associated with the customer arrival event. The first action taken is to schedule the next arrival. This is done to provide a sequence of arrivals. Hence, once the first arrival is scheduled, a continuing stream of arrivals will occur. The disposition of the current customer arrival depends upon the state of the system at the customer arrival time. If the teller is busy, the arriving customer must wait, and therefore the state of the system is changed by increasing the number of waiting customers by one. Otherwise, the arriving customer can be placed immediately into service. In this case, the state of the system is changed by setting the status of the teller to busy. In addition, the end-of-service event for the customer must be scheduled to occur at the current simulated time plus the time it takes the teller to serve the customer.

Next, consider the logic associated with processing the end-of-service event. In this event, both the disposition of the customer and the teller must be specified. We assume the customer departs the bank after being served. For the teller who is completing service on the current customer, we first test to see if additional customers are waiting for service by the teller. If customers are waiting, we reduce the number waiting by 1 and schedule the end-of-service event for the first waiting customer. Otherwise, we set the teller to idle status.

To create a simulation of the bank teller problem using the event orientation, we would maintain a calendar of events and cause their execution to occur at the proper points in simulated time. The event calendar would initially contain an event notice corresponding to the first arrival event. As the simulation proceeds, additional arrival events and end-of-service events would be scheduled onto the calendar as prescribed by the logic associated with the events. Each event would be executed in a time-ordered sequence, with simulated time being advanced from one event to the next. The event orientation of simulation represents a large decomposition of the modeling effort. To obtain the dynamic behavior of a model, the time variable need only be examined at event times. The specification of a limited number of events reduces the modeling effort by allowing the same logic to be used at the event occurrences.

If the modeler employs a general purpose language such as FORTRAN to code a discrete event model, then a considerable amount of programming effort will be directed at developing the event calendar and a timing mechanism for processing the events in their proper chronological order. Since this function is common to all discrete event models, a number of simulation languages including SLAM II have been developed which provide special features for event scheduling, as well as other functions which are commonly encountered in discrete event models.

3.3.2 Activity Scanning Orientation

In the activity scanning orientation, the modeler describes the activities in which the entities in the system engage and prescribes the conditions which cause an activity to start or end. The events which start or end the activity are not scheduled by the modeler, but are initiated from the conditions specified for the activity. As simulated time is advanced, the conditions for either starting or ending an activity are scanned. If the prescribed conditions are satisfied, then the appropriate action for the activity is taken. To insure that each activity is accounted for, it is necessary to scan the entire set of activities at each time advance.

For certain types of problems, the activity scanning approach can provide a concise modeling framework. The approach is particularly well suited for situations where an activity duration is indefinite and is determined by the state of the system satisfying a prescribed condition. However, because of the need to scan each activity at each time advance, the approach is relatively inefficient when compared to the discrete event orientation. As a result, the activity scanning orientation has not been widely adopted as a modeling framework for discrete simulations. However, a number of languages employ specific features which are based on the concept of activity scanning. SLAM II includes two methods for incorporating activities whose start and end times are based on system status.

3.3.3 Process Orientation

Many simulation models include sequences of elements which occur in defined patterns, for example, a queue where entities wait for processing by a server. The logic associated with such a sequence of events can be generalized and defined by a single statement. A simulation language could then translate such statements into the appropriate sequence of events. A process oriented language employs such statements to model the flow of entities through a system. These statements define a sequence of events which are automatically executed by the simulation language as the entities move through the process. For example, the following network could be used to describe the process for the bank teller problem.

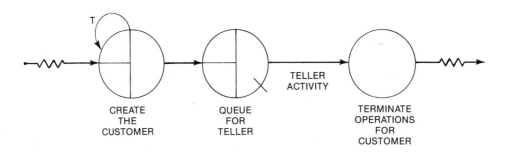

From the above example, we see that the process orientation provides a description of the flow of the entities through a process. Its simplicity is derived from the fact that the event logic associated with the statements is contained within the symbols of the simulation language. However, since we are normally restricted to a set of standardized symbols provided by the simulation language, our modeling flexibility is not as great as with the event orientation.

3.4   CONTINUOUS SIMULATION MODELING

In a continuous simulation model, the state of the system is represented by dependent variables which change continuously over time. To distinguish continuous change variables from discrete change variables, the former are referred to as *state variables.* A continuous simulation model is constructed by defining equations for a set of state variables whose dynamic behavior simulates the real system.

Models of continuous systems are frequently written in terms of the derivatives of the state variables. The reason for this is that it is often easier to construct a relationship for the rate of change of the state variable than to devise a relationship for the state variable directly. Equations of this form involving derivatives of the state variables are referred to as differential equations. For example, our modeling effort might produce the following differential equation for the state variable, s, over time, t, together with an initial condition at time 0.

$$\frac{ds(t)}{dt} = s^2(t) + t^2$$

$$s(0) = k$$

The first equation specifies the rate of change of s as a function of s and t and the second equation specifies the initial condition for the state variable. The simulation analyst's objective is to determine the response of the state variable over simulated time.

In some cases, it is possible to determine an analytical expression for the state variable, s, given an equation for ds/dt. However in many cases of practical importance, an analytical solution for s will not be known. As a result we must obtain the response, s, by integrating ds/dt over time using an equation of the following type:

$$s(t_2) = s(t_1) + \int_{t_1}^{t_2} \left(\frac{ds}{dt}\right) dt$$

How this integration is performed depends upon whether the modeler employs an analog or digital computer.

During the 1950's and 1960's, analog computers were the primary means for performing continuous simulations. An analog computer represents the state variables in the model by electrical charges. The dynamic structure of the system is modeled using circuit components such as resistors, capacitors, and amplifiers. The principal shortcoming of an analog computer is that the quality of these components limits the accuracy of the results. In addition, the analog computer lacks the logical control functions and data storage capability of the digital computer.

A number of continuous simulation languages have been developed for use on digital computers. It is necessary to recognize that a digital computer is technically discrete in its operation. As a practical matter, however, any variable whose possible values are limited only by the word size of the computer is considered continuous.

A digital computer performs the common mathematical operations such as addition, multiplication, and logical testing with great speed and accuracy, and it uses these characteristics to perform the numerical integration required in continuous simulation. Numerical integration methods divide the independent variable (normally time) into small slices referred to as *steps*. The values for the state variables requiring integration are obtained by employing an approximation to the derivative of the state variable over time. The accuracy of these methods depends upon the order of the approximation method and the size of the step, with greater accuracy resulting from higher-order approximations and smaller step sizes. Since higher-order approximations and smaller step sizes result in more computations, a trade-off exists between accuracy of state variable calculations and computer run time. A description of the various numerical integration algorithms can be found in any of the introductory texts for numerical analysis (3,17). The numerical integration scheme employed by SLAM II for simulating continuous models involving differential equations is described in Chapter 13.

Sometimes a continuous system is modeled using difference equations. In these models, the time axis is decomposed into time periods of length Δt. The dynamics of the state variables are described by specifying an equation which calculates the value of the state variable at period k + 1 from the value of the state variable at period k. For example, the following difference equation could be employed to describe the dynamics of the state variable s:

$$s_{k+1} = s_k + r^* \Delta t$$

When using difference equations, the essential structure of a continuous simulation model is often reflected in the relationship between the rate *r* used to project the state variable at period k + 1 from the value s_k of the state variable at period k.

Continuous simulation languages for digital computers normally employ either a block or statement orientation. The block oriented languages employ a set of blocks which functionally emulate the circuit components of an analog computer. Thus the modeler familiar with analog block diagrams would find these languages easy to learn. Most of the recently developed continuous simulation languages employ an equation orientation. In these languages, the differential or difference equations are explicitly coded in equation form. An advantage of the equation orientation is the increased flexibility afforded by the algebraic and logical features of these languages. A committee of the Society for Computer Simulation (16) has developed a set of standards for continuous system-simulation languages (CSSL).

3.5   COMBINED DISCRETE-CONTINUOUS MODELS

In combined discrete-continuous models, the dependent variables may change both discretely and continuously. The world view of a combined model specifies that the system can be described in terms of entities, their associated attributes, and state variables. The behavior of the system model is simulated by computing the values of the state variables at small time steps and by computing the values of attributes of entities at event times.

There are three fundamental interactions which can occur between discretely and continuously changing variables. First, a discrete change in value may be made to a continuous variable. Examples of this type of interaction are: the completion of a new power station which instantanously increases the total energy available within a system; and the chemical spraying of a lake which instantaneously decreases the population of a particular species in the lake. Second, an event involving a continuous state variable achieving a threshold value may cause an event to occur or to be scheduled. As examples consider: a chemical process that is completed when a prescribed concentration level is obtained and the process is shut down for cleaning and maintenance activities; and the shutdown of a refinery when the level of crude oil available for input is below a prescribed value. Third, the functional description of continuous variables may be changed at discrete time instants. Examples of this are: the discharge of a pollutant into an ecosystem that immediately alters the growth relationships governing species populations; and the completion of a docking operation of a space vehicle which requires the use of new equations for simulating the space vehicle's motion.

The interaction between the continuous and discrete change state variables in a combined discrete-continuous change system necessitates a broader interpretation of an event than is normally used in discrete change languages. For combined simulation models (12):

> An event occurs at any point in time beyond which the status of the system cannot be projected with certainty.

Note that this definition allows the system status to change continuously without an event occurring, as long as the change has been prescribed in a well-defined manner.

There are two types of events that can occur in combined simulations. *Time-events* are those events which are scheduled to occur at specified points in time. They are commonly thought of in terms of discrete simulation models. In contrast, *state-events* are not scheduled, but occur when the system reaches a particular state. For example, as illustrated in Figure 3-6, a state-event could be specified to occur whenever state variable SS(1) crosses state variable SS(2) in the positive direction. Note that the idea of a state-event is similar to the concept of activity scanning in that the event is not scheduled but is initiated by the state of the system. The possible occurrence of a state-event must be tested at each time advance in the simulation.

These constructs were implemented in GASP IV (6, 12). The GASP IV language is FORTRAN based and provides a formalized world view which combines the discrete event orientation for modeling discrete systems with the state variable equation orientation for continuous system modeling. The analysis of systems using combined simulation models continues to be a fertile area for research, development and application (11, 18, 19, 20).

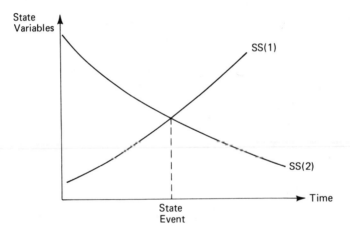

Figure 3-6 Example of a state-event occurrence.

3.6 SLAM II: A UNIFIED MODELING FRAMEWORK

In the preceding sections, we described several alternate world views for simulation modeling. Each of these views presumes a framework within which the system is described. The diversity of world views has persisted because each has certain advantages and disadvantages. For example, the process orientation provides a concise and easy to learn modeling framework, but may lack flexibility. On the other hand, the event orientation is normally more difficult to learn but, once mastered, provides a highly flexible modeling framework.

In SLAM II, the alternate modeling world views are combined to provide a unified modeling framework. A discrete change system can be modeled within an event orientation, process orientation, or *both*. Continuous change systems can be modeled using either differential or difference equations. Combined discrete-continuous change systems can be modeled by combining the event and/or process orientation with the continuous orientation. In addition, SLAM II incorporates a number of features which correspond to the activity scanning orientation.

The process orientation of SLAM II employs a *network* structure as illustrated in Section 3.3.3 which consists of specialized symbols called *nodes* and *branches*. These symbols model elements in a process such as queues, servers, and decision points. The modeling task consists of combining these symbols into a network model which pictorially represents the system of interest. In short, a network is a pictorial representation of a process. The entities in the system (such as people and items) flow through the network model.

In the event orientation of SLAM II, the modeler defines the events and the potential changes to the system when an event occurs. The mathematical-logical relationships prescribing the changes associated with each event type are coded by the modeler as FORTRAN subroutines. A set of standard subprograms is provided by SLAM II for use by the modeler to perform common discrete event functions such as event scheduling, file manipulations, statistics collection, and random sample generation. The executive control program of SLAM II controls the simulation by advancing time and initiating calls to the appropriate event subroutines at the proper points in simulated time. Hence, the modeler is completely relieved of the task of sequencing events to occur chronologically.

A continuous model is coded in SLAM II by specifying the differential or difference equations which describe the dynamic behavior of the state variables. These equations are coded by the modeler in FORTRAN by employing a set of special SLAM II defined storage arrays. The value of the Ith state variable is maintained as variable $SS(I)$ and the derivative of the Ith state variable, when required, is maintained as the variable $DD(I)$. The immediate past values for state variable I and its derivative are maintained as $SSL(I)$ and $DDL(I)$, respectively. When differential equations are included in the continuous model, they are automatically integrated by SLAM II to calculate the values of the state variables within an accuracy prescribed by the modeler.

An important aspect of SLAM II is that alternate world views can be combined within the same simulation model. There are six specific interactions which can take place between the network, discrete event, and continuous world views of SLAM II:

1. Entities in the network model can initiate the occurrence of discrete events.
2. Events can alter the flow of entities in the network model.
3. Entities in the network model can cause instantaneous changes to values of the state variables.
4. State variables reaching prescribed threshold values can initiate entities in the network model.
5. Events can cause instantaneous changes to the values of state variables.
6. State variables reaching prescribed threshold values can initiate events.

3.7   CHAPTER SUMMARY

In this chapter, we described the alternate world views of simulation modeling and introduced the unified modeling framework of SLAM II. In Chapters 5 through 14 we describe the network, event, continuous, and combined modeling features of SLAM II. As the reader masters each of these orientations within the unified framework of SLAM II, the need for a simulation language with the alternative modeling concepts and relationships discussed in this chapter should become clear.

3.8   EXERCISES

3-1. Consider the operation of a physican's office. Specify the boundaries of the system, and describe its operation in terms of entities, attributes, relationships, and activities.

3-2. Give an example of a situation in which the end of an activity cannot be scheduled in advance but must be based on the status of the system.

3-3. A paint shop employs six workers who prepare jobs to be spray painted. The preparation time is lengthy compared to the spraying operation and, hence, only two spraying machines are available. After a worker completes the preparation of a job, he proceeds to a spraying machine where he waits if necessary for a free spraying machine. Jobs to be prepared and painted are always available to the workmen. Describe this system using the event orientation and the process orientation.

3-4. Describe a residential heating and cooling system in terms of state variables, time-events, and state-events.

3-5. Model an elevator system to ascertain energy usage. The elevator serves five floors. Assume the arrival of passengers at each floor is a random variable and that the probability associated with the passenger's floor to floor transition are known. Note that energy use is a function of the dynamic chacteristics of the elevator's motion.

3-6.   Write a simulation model of a single-server queueing system using a general purpose programming language such as FORTRAN, PL/I or C. Assume that the time between customer arrivals is exponentially distributed with mean of 5 time units and the service time is uniformly distributed between 2 and 6 time units. Estimate the average queue length, server utilization, and time-in-system based on 1000 customers.

Embellishment: Change the model to include two parallel servers and a finite queue capacity of 10 customers.

3.9   REFERENCES

1. Birtwhistle, G. M., O. Dahl, B. Myhrhaug, and K. Nygaard, *SIMULA Begin,* Auerbach, 1973.
2. Buxton, J. N. and J. G. Laski "Control and Simulation Language," *Computer Journal,* Vol. 5, 1964, pp. 194-199.
3. Carnahan, B., H. A. Luther, and J. O. Wilkes, *Applied Numerical Methods,* John Wiley, 1969.
4. Franta, W. R., *The Process View of Simulation,* North Holland, 1977.
5. Gordon, G., *The Application of GPSS V to Discrete Systems Simulation,* Prentice-Hall, 1975.
6. Hurst, N. R., *GASP IV: A Combined Continuous/Discrete FORTRAN Based Simulation Language,* Unpublished Ph.D. Thesis, Purdue University, 1973.
7. Kiviat, P. J., *Digital Event Simulation: Modeling Concepts,* The Rand Corporation, RM-5378-PR Santa Monica, Calif., 1967.
8. Kiviat, P. J., *Digital Computer Simulation: Computer Programming Languages,* The Rand Corporation, RM-5883-PR, Santa Monica, Calif., 1969.
9. Kiviat, P. J., R. Villanueva, and H. Markowitz, *The SIMSCRIPT II Programming Language,* Prentice-Hall, 1969.
10. Markowitz, H. M., H. W. Karr, and B. Hausner, *SIMSCRIPT: A Simulation Programming Language,* Prentice-Hall, 1963.
11. Miles, G. E., R. M. Pearl, and A. A. B. Pritsker, "CROPS: A GASP IV Based Crops Simulation Language," *Proceedings, Summer Computer Simulation Conference,* 1976, pp. 921-924.
12. Pritsker, A. A. B., *The GASP IV Simulation Language,* John Wiley, 1974.
13. Pritsker, A. A. B. and R. E. Young, *Simulation with GASP_PL/I,* John Wiley, 1975.
14. Pritsker, A. A. B., *Modeling and Analysis Using Q-GERT Networks,* Halsted Press and Pritsker & Associates, Inc., Second Edition, 1979.
15. Schriber, T., *Simulation Using GPSS,* John Wiley, 1974.
16. SCi Software Committee, "The SCi Continuous-Systems Simulation Language," *Simulation,* Vol. 9, 1967, pp. 281-303.

17. Shampine, L. F. and R. C. Allen, Jr., *Numerical Computing: An Introduction,* W. B. Saunders, 1973.

18. Sigal, C. E. and A. A. B. Pritsker, "SMOOTH: A Combined Continuous-Discrete Network Simulation Language," *Simulation,* Vol. 21, 1974, pp. 65-73.

19. Washam, W., *GASPPI: GASP IV With Process Interaction Capabilities,* Unpublished M.S. Thesis, Purdue University, 1976.

20. Wortman, D. B., S. D. Duket et al, *Simulation Using SAINT: A User-Oriented Instruction Manual,* AMRL-TR-77-61, Aerospace Medical Research Laboratory, Wright-Patterson AFT, Ohio, 1978.

CHAPTER 4

Applications of Simulation

4.1 INTRODUCTION

Simulation has been used to study such wide ranging topics as urban systems, economic systems, business systems, production systems, biological systems, social systems, transportation systems, health care delivery systems, and many more. Table 4-1 presents areas in which simulation methods are currently being used (19). Simulation is the most widely-used management science and operations research technique employed by industry and government.

Table 4-1   Areas in which simulation methods are currently being used†.

Air traffic control queueing	Financial forecasting
Aircraft maintenance scheduling	Insurance
Airport design	Schools
Ambulance location and dispatching	Computer leasing
Assembly line scheduling	Insurance manpower hiring decisions
Bank teller scheduling	Grain terminal operation
Bus (city) scheduling	Harbor design
Circuit design	Industry models
Clerical processing system design	Textile
Communication system design	Petroleum (financial aspects)
Computer time sharing	Information system design
Telephone traffic routing	Intergroup communication (sociological studies)
Message systems	Inventory reorder rule design
Mobile communications	Aerospace
Computer memory-fabrication test-facility design	Manufacturing
Consumer behavior prediction	Military logistics
Brand selection	Hospitals
Promotion decisions	Job shop scheduling
Advertising allocation	Aircraft parts
Court system resource allocation	Metals forming
Distribution system design	Work-in-process control
Warehouse location	Shipyards
Mail (Post office)	Library operations design
Soft drink bottling	Maintenance scheduling
Bank courier	Airlines
Intrahospital material flow	Glass furnaces
Enterprise models	Steel furnaces
Steel production	Computer field service
Hospitals	National manpower adjustment system
Shipping lines	Natural resource (mine) scheduling
Railroad operations	Iron ore
School districts	Strip mining
Equipment scheduling	Parking facility design
Aircraft	Numerically controlled production facility design
Facility layout	Personnel scheduling
Pharmaceutical centers	Inspection departments
	Spacecraft trips

Table 4-1 (continued).

Petrochemical process design	Taxi dispatching
Solvent recovery	Traffic light timing
Police response system design	Truck dispatching and loading
Political voting prediction	University financial and operation forecasting
Rail freight car dispatching	Urban traffic system design
Railroad traffic scheduling	Water resources development
Steel mill scheduling	

---

† Source: Emshoff, J. R. and R. L. Sisson, *Design and Use of Computer Simulation Models,* ©1970, p. 264. Reprinted by permission of The Macmillan Company, New York, N.Y.

The continuing development of simulation languages has been an important factor in this growth. Another major factor is the flexibility of simulation modeling when compared, for example, to the structural restrictions imposed by a mathematical programming formulation of a problem. Even when an analytic model can be applied to a problem, simulation is frequently used to study the practical implications of the assumptions underlying the analytic model. In this chapter, general simulation application areas are described and capsule summaries of specific simulation projects performed at Pritsker & Associates or its clients are given. The summaries show that simulation is a useful technique. They also provide a discussion of problem types for which simulation has been used in actual decision-making situations, and the types of questions that have been answered by using simulation. The procedures employed in building the specific simulation models described are contained in the referenced papers.

This text provides procedures for developing models. The design of models for industrial use involves extensive insight and logical thinking. Such models are larger in size, but conceptually not more complex, than those that are presented in this book. A significant point when developing a simulation model is the understanding that there are alternative approaches to building a model. Thus, as the examples illustrate, it is not necessary to conform to any fixed set of rules when modeling. Using new and novel approaches is encouraged. Remember, it is easy to add to a simulation model or to use a first pass model to set the specifications for a second pass model. Do not hesitate to design when modeling and to make modeling an essential part of design and analysis. In all cases, it is necessary to establish a purpose for modeling and the measures that evaluate performance.

4.2   PERFORMANCE MEASURES

The performance of a system is measured by its effectiveness and efficiency in achieving system objectives. The objectives of different types of systems vary and performance measures across areas of simulation applications are not the same. The applications presented in this chapter provide information on the benefits obtained from modeling and simulation. To provide general information on the inputs to the computation of benefits, we present a discussion of performance measures for manufacturing systems. A taxonomy of performance measures for different classes of systems would be a research contribution to the fields of engineering and management.

4.2.1   Performance Measures for Manufacturing Systems (49)

In many situations, objectives are established in terms of a level of cost effectiveness or system profitability. Inputs to such measures of performance are: price and cost values; and measures of the operations of a system. The calculation of a system performance measure involves the combining of these two types of inputs. We will concentrate on the operations performance measures since simulation is more commonly used to estimate such values. Once obtained, they can be combined with the dollar values and used to estimate system performance.

For manufacturing systems, performance measures of operations can be grouped into four categories:

1. Measures of throughput
2. Measures of ability to meet deadlines
3. Measures of resource utilization
4. Measures of in-process inventory

Throughput is the output produced in a given period of time. Another name for this measure is the *production rate*. *System capacity* is often defined as the maximum throughput that can be obtained.

The ability to meet deadlines is measured by product lateness, tardiness, or flowtime. *Lateness* is the time between when a job is completed and when it was due to be completed. *Tardiness* is the lateness of a job only if it fails to meet its due date; otherwise, it is zero. *Flowtime* is the amount of time a job spends in the system. In some cases, the total time it takes to complete all jobs is of importance. This time is referred to as the *makespan*. These measures are indications of the effectiveness and efficiency of the system in satisfying customer orders.

System resources include personnel, materials, machines, and work space. The utilization of these resources, as measured by the fraction of time they are productive, is another measure of system effectiveness. Measures of resource utilization relate the degree to which a system is operating at capacity.

The final category of manufacturing operations performance measures is concerned with the buildup of raw materials and unfinished parts during production. This buildup, called in-process inventory or work in process (WIP), is usually due to parts waiting for available resources. Since inventory requires storage space, often a critical resource in itself, and also ties up capital, in-process inventory requirements are of great importance in manufacturing operations assessment.

The measures outlined above provide the basis for evaluating diverse objectives. By concentrating on operations performance measures, we bypass the question of identifying the objectives of system planning, which by necessity are situation dependent. However in many situations objectives are satisfied when performance measures reach prescribed levels.

4.3 EVALUATION OF A PROPOSED CAPITAL EXPENDITURE IN THE STEEL INDUSTRY (61)

Bethlehem Steel Corporation was considering a design for new facilities for improving the steel-making process. Included in the design were new operations involving the melting of scrap and the desulfurization of hot metal. The analysis was to determine the need for additional hot metal carriers, called submarines, to support the proposed new operations.

A discrete event simulation model was developed consistent with the objectives of evaluating a proposed capital expenditure and involved modeling the various operations associated with delivering hot metal from a set of blast furnaces (BF) to a set of basic oxygen furnaces (BOF). A schematic diagram of the operations model is shown in Figure 4-1. The submarines serve as materials handling equipment which transport the hot metal through a series of operations before returning to perform the set of tasks again. The demand for hot metal carriers depends on the casting times of the blast furnaces and the new scrap melter. Casting times are

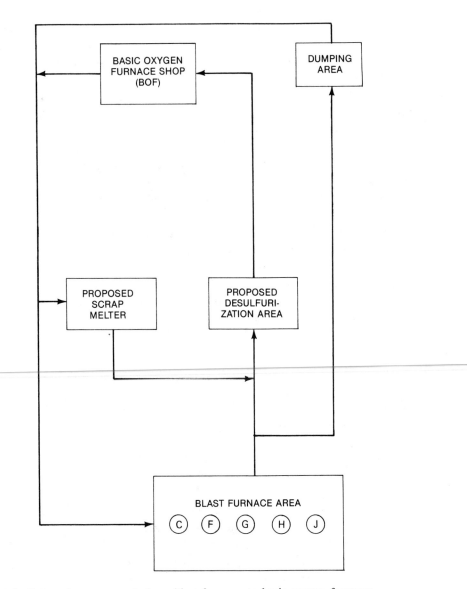

Figure 4-1 Submarine movements from blast furnaces to basic oxygen furnaces.

scheduled but actual performance times depend on the hot metal characteristics. Not having submarines available when a cast is ready is a dangerous and expensive situation and could cause a furnace to be shut down.

Scheduling rules were incorporated for routing the submarines through the de-sulfurization operation, if necessary, and to hot metal ladles which provided in-puts to the basic oxygen furnaces. Submarines were also required to transfer hot metal from the scrap melter. A decision rule was also incorporated into the model for determining if an insufficient number of submarines would be available to ac-commodate the next cast from a blast furnace. When this situation occurs, a sub-marine dumps its hot metal and returns immediately to serve a blast furnace.

The simulation model showed, in contrast to earlier studies recommending the purchase of three submarines, that by altering the scheduling rules, the current number of submarines could support the new operations. Thus, the simulation analysis resulted in a recommendation which led to an avoidance of a capital ex-penditure of over one million dollars. In addition, procedures were suggested by which further improvements in the total steelmaking process could be made.

4.4   INJECTION BODY FMS DESIGN EVALUATION (41, 66)

A manufacturer of castings desired to evaluate alternative milling machine cen-ter configurations to achieve a production goal of 157,000 finished castings per year. A flexible manufacturing system (FMS) was designed to perform machining operations on the castings and is shown in Figure 4-2. Castings are initially loaded onto pallets which can carry 16 parts each and sent to one of two lathes on a con-veyor. Upon completion of the turning operation, the parts are transported, again by conveyor, to a wash/load area before being sent to the machining center on a wire-guided vehicle.

The machining center consists of ten identical horizontal milling machines which can perform any one of three operations. The three operations are referred to as OP10, OP20, and OP30. For a particular part type, the milling machines are set up to be either dedicated to performing one operation or flexible so that any of the three operations can be performed.

Two types of fixtures, A and B, are used in this system. Fixture A is used for OP10 and Fixture B is used for OP20 and OP30.

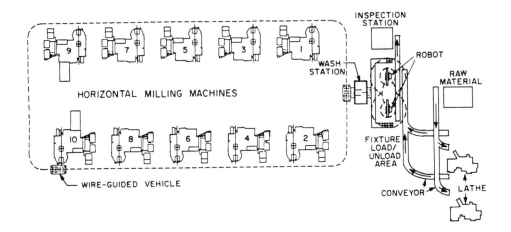

Figure 4-2   A flexible milling machine system.

Before a set of parts is routed for OP10 machining, each part in that set is attached to fixture A and sent through the wash station. When buffer space becomes available at one of the machines capable of performing OP10, this set of parts is transported to that machine's input buffer. After the parts have been machined, they are returned to the wash/load area. The parts are then attached to fixture B to await OP20. The same procedure is used for OP30 except that the parts are rotated 180 degrees on the same fixture. After all three machining operations have been completed, the parts are sent to final inspection before departing the system.

The objectives of the simulation study are to evaluate system balance and productivity, determine the need for additional equipment, determine which resources, if any, could be eliminated and to evaluate the number of dedicated versus flexible machines required. Tooling costs are much higher if a machine is used to perform all three operations.

With a configuration consisting of nine dedicated machines and one flexible machine, simulation results revealed that a wire-guided vehicle, four fixtures and several sets of tools could be eliminated while achieving the desired production goal of 157,000 finished castings per year.

4.5   DESIGN OF A CORN SYRUP REFINERY

In July 1975, A.E. Staley Manufacturing Company broke ground for a new 85 million dollar corn wetmilling plant in Lafayette, Indiana. The plant was engineered with a flexible end-product mix capability which would permit Staley to respond to changes in the demand for various types of corn sweeteners. In 1975, a proposed design was developed that is depicted in Figure 4-3. As can be seen from the figure, the process involves two evaporation steps, one carbon refining step, and one ion exchanger step. Also shown are the required material flows from step to step. Water is used extensively in the process. Since it is recycled and contains sweeteners, it is referred to as sweetwater. In addition, storage tanks are required for maintaining balanced operations through the process. A combined discrete event-continuous model of the system was built to determine the size of the storage tanks, evaporators, and ion exchanger (52).

The model was exercised and the most cost effective size of the various units was determined. Since large, special purpose equipment is involved, a lead time of over eighteen months is typical, and simulation was the only feasible way to evaluate the proposed design. In addition to the sizing study, system parameters were investigated for both a manual and a computerized control system. One control system model included the setting of valves to regulate sweetwater-source flows from different tanks, each having a different concentration level. It was presumed that the concentration level in each sweetwater tank could be assessed hourly. With knowledge of the concentration levels, a linear programming blending model was developed and imbedded in the simulation to set valves in order to maximize the profitability associated with the end-product (the end-product is a function of the component saccharide distributions which can change as a function of the concentration in the sweetwater tanks and the process control parameters). Thus, after using the simulation to finalize the design of the plant, it was used to identify the most economical control strategy compatible with production quality and volume requirements. These results have been incorporated into the refinery's process control procedures which are now in operation (53).

4-6   RISK ANALYSIS OF PIPELINE CONSTRUCTION

The construction of a pipeline basically involves: 1) preparing a site for laying pipe, 2) laying the pipe, and 3) welding sections of the pipe together. Supporting

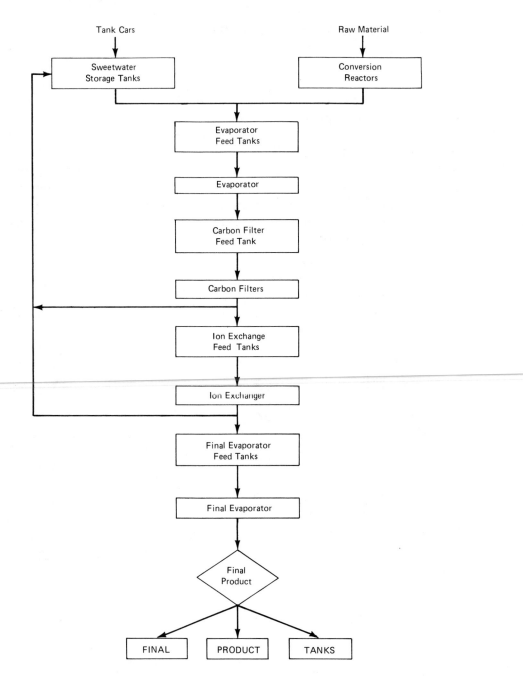

Figure 4-3   Proposed block diagram design of a corn syrup refinery.

operations for pipeline construction involve the building, dismantling, and moving of campsites; the construction of roads and other transportation facilities; and relandscaping the site. When pipeline construction is performed in Alaska, the adverse weather conditions must be considered when planning the construction project. A network model was developed consisting of the pipeline construction activities, and transportation facility development activities (22, 49). The effects of weather conditions on construction activities were also included in the model. A risk analysis was performed to determine the probability of completing pipeline construction by specified due dates. A cost analysis was also performed to determine potential overrun conditions. The analysis indicated that both time and cost overruns could be expected. The effects of changing the activity schedule and construction rates were also evaluated.

4-7 EVALUATING SECURITY FORCES AT A SPORTING EVENT (21)

In conjunction with the Los Angeles Sheriff's Department, a model was developed to determine the number and location of security forces necessary to respond to disturbances that might occur at the Rose Bowl. The model portrayed disturbances such as fights, fires and terrorism as discrete events requiring the services of security forces. Detailed graphics were developed to display the location and routing of the security forces and how they respond to the disturbances. Local law enforcement personnel assisted in developing the algorithms for establishing staff levels and the routing of the security forces. The algorithms were developed on an iterative basis with the modeler and personnel from the Sheriff's Department assessing different strategies. The benefits of the project included a 25 percent reduction in the required security forces at the Rose Bowl and an understanding of the policies and procedures employed in protecting and controlling a large crowd.

4-8 EVALUATION OF THE PRODUCTIVE CAPACITY OF A PROPOSED ROCKET SYSTEM (47)

Vought Corporation of Grand Prairie, Texas, developed a plan to manufacture the Multiple Launch Rocket System (MLRS) for the U.S. Army. The MLRS vehicle is comprised of a tracked vehicle (similar to a personnel carrier) carrying a launch pod container holding six rockets. Vought's manufacturing operations would be located in two buildings, the metal parts building and the load, assembly, and pack building. A schematic diagram of the facility is shown in Figure 4-4. Operations performed in the Metal Parts Building include the fabrication of the necessary detail parts, assembly of the launch pod container, and production of the motor case. Operations performed in the load, assembly and pack building include loading the warhead with grenades, joining the warhead and motor case, and loading each container with six rockets.

The objectives of the analysis were to support Vought's proposal to the Army by accomplishing the following tasks:

- Verify the production rates of both buildings.
- Size storage space requirements for key buffers in the manufacturing process.
- Gather statistics on labor and machine utilization.
- Determine production "dry-up" time for key pieces of equipment.
- Develop a graphic output format to communicate the capabilities of the model to Army and management personnel.
- Provide the capability for follow-up analysis by Vought personnel.

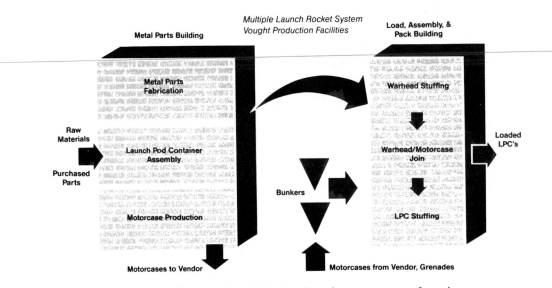

Figure 4-4   Schematic diagram of multiple launch rocket system manufacturing process.

The model of the load, assembly, and pack building identified a potential bottleneck and was used to verify the redesign of the buffer causing the problem. An illustration of the graphic outputs from the SLAM II simulation is shown in Figure 4-5 for two machine operations in series separated by an in-process storage area.

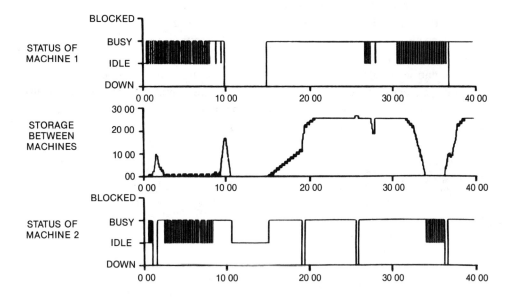

Figure 4-5 Dynamics of machine operations, queues, and buffers.

Another output showed the number of rockets produced as a function of time and the number of rockets shipped. Although these graphs were not used for quantitative analysis, they did portray how the outputs of the model varied over time. They were important in convincing Vought management of the validity of the project. All the objectives of the simulation project were achieved.

The models of the two buildings validated production capabilities for a normal rate of production (two eight-hour shifts a day) and surge production rates (two ten-hour and three eight-hour shifts a day). In April 1980, Vought Corporation was awarded the contract by the Army to produce the rocket system. The computer modeling analysis was cited by the Army as a determinant in awarding the contract to Vought.

4-9   INGOT MOULD INVENTORY AND CASTING ANALYSIS

Bethlehem Steel Corporation has a centralized foundry for casting ingot moulds for all Bethlehem plants. The demand for moulds is based on the usage and condemnation of the moulds inventoried at the various plants. A wide variety of mould types and sizes is required since ingot size is a prime determinant of finished product yield per ingot at a steel mill. To obtain increases in yields, larger ingots are being used. However, the ability of the foundry to produce large ingot moulds is constrained by available equipment.

A simulation model of the casting of ingot moulds was developed in order to determine capacity expansion requirements for the foundry. The operations involved in casting ingot moulds are shown in Figure 4-6. A discrete event model was developed which included the operation of sand slinging in the moulding pits,

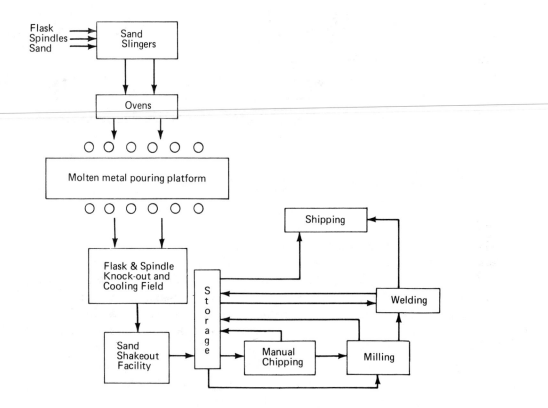

Figure 4-6 Operations involved in casting ingot moulds.

baking in the core ovens, pouring of molten metal, cooling, mould removal, sand shakeout, chipping, milling, and shipping. Data collection, scheduling rules, model verification, and output analyses were performed in conjunction with the foundry supervisor and his chief scheduler along with plant industrial engineers.

As part of the study, the demand for ingot moulds was investigated. Data were collected on mould types for a three year period. Specific data values for inventory levels, steel production quantities by mould type, and condemnation rates for moulds were obtained. Based on this information, a procedure was developed for smoothing the demand for ingot moulds. By smoothing the demand for mould types across all Bethlehem plants, the foundry simulation model showed that no increase in foundry capacity was required and, in fact, excess capacity could be obtained which could be used to satisfy special peak demands. In this application, a capital expenditure for new foundry facilities was avoided and, at the same time, corporate inventory levels were reduced.

4.10   WORK FLOW ANALYSIS OF A PROPERTY AND CASUALTY INSURANCE COMPANY'S REGIONAL SERVICE OFFICE (32, 49)

The work flow system of a regional service office involves the processing of property and casualty insurance claims through a centralized computer information system. The various types of forms are routed through fourteen distinct operating units or departments within the regional service office which employs over 150 personnel. The system was modeled as a complex queueing situation to identify the bottlenecks of the work flow and to assist in investigating the effects of certain managerial decisions. A network model aided in estimating the consequences of specific actions contemplated or anticipated by management. The procedural changes investigated were:

1. Changes in the volume of each type of work handled by the regional service office;
2. Changes in the composition and requirements of the work input;
3. Changes in priority rules for the processing of work in the same operating unit;
4. Changes in the pathways of the work flow system;
5. Reallocation of personnel among the different departments; and
6. Changes in total processing times of customer requests due to training programs in specific areas.

4.11   CHEMICAL MANUFACTURING SYSTEMS ANALYSIS

A discrete event simulation model was developed of a chemical company manufacturing facility (37) consisting of a series of process reactors, pumps, storage tanks, and filtration systems as depicted in Figure 4-7. This manufacturing facility produces batches of different product types to meet customer demands. The model of the process was discrete event oriented with events representing the beginning and ending of each process stage. A discrete event orientation was possible as system status is not altered until a batch is completely processed. Thus, an end of batch processing event can be scheduled from knowledge of when the batch was started and the processing characteristics of a stage. Included in the event routines were complex scheduling procedures to determine the disposition of the batch just completed and the next batch to be started.

The simulation model was used to perform the following types of analyses:

1. Determine the effect of customer demand patterns on the operating requirements of the manufacturing facility.
2. Determine the effect of order lead time requirements on manufacturing costs, inventory levels, and production scheduling.
3. Determine the effect of alternative production scheduling procedures.
4. Determine the effect of alternative system configurations (the number and capacity of storage tanks, filters, reactors, and pipelines).

As an example of the types of results obtained from the model, it was determined that the filtration process was the primary bottleneck. By doubling the capacity of the filtration system, an 80 percent increase in product throughput could be expected. However, additional increases in filtration capacity indicated only marginal gains in throughput. Another analysis indicated that some storage tanks could be eliminated without affecting product throughput. It was also demonstrated how product scheduling procedures impact on product throughput. As is typical of simulation models, the analysis had the added benefit of providing valuable insights into the operation of a complex system.

4.12   SIMULATION OF AN AUTOMATED WAREHOUSE

Philip Morris had built a new manufacturing and warehouse facility containing many computers for controlling product flow. The capacity and potential bottle-

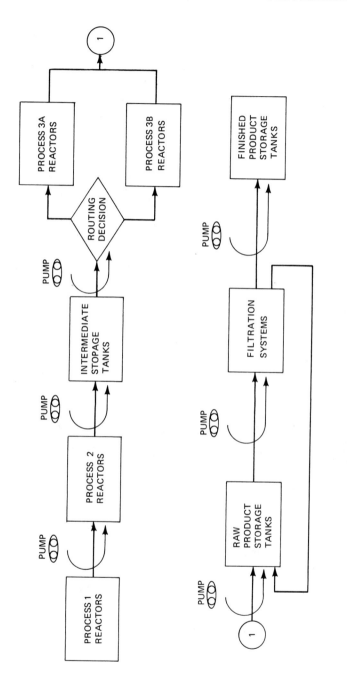

Figure 4-7 General diagram of manufacturing process.

necks of the finished goods warehouse were of concern to management since it involved a hierarchial control system using five minicomputers. A combined discrete event-continuous simulation model of this facility was developed.

Basically, the warehouse receives cases of finished goods from manufacturing and sorts them by brand before they are stored, if necessary, in a high-density stacker storage area for eventual shipment to local customers or distribution warehouses.

Figure 4-8 illustrates the general flow through the "Case Input System." The following discussion is abstracted from Jarvis and Waugh (29). The cases are first received from the manufacturing floor and pass a laser scanner to determine on which tier the product is to be accumulated. Cases then travel on two long conveyors to an input scanner that determines a lane for the case. The case is then tracked by photocells until it arrives at its assigned lane where a computer diverts the case into its assigned lane.

When a sufficient number of cases have been accumulated in a lane (a full pallet load plus five extra cases), the lane is set ready to meter out a load for automatic palletizing. An output belt and a verification scanner are used to verify correct product codes and to build case trains. The case train is routed to a contingency diverter where trains from both tiers can be merged to a single palletizer in the event of a palletizer failure. All cases that were not properly verified at the output scanner are diverted to a manual palletizing area. The verified cases are then fed to the palletizer where they are automatically palletized in a brand-dependent pattern. Considerable logic and a number of conveyor belts between the contingency diverter and the palletizers insure space between cases in several areas where case count is important, and insure the proper merging of case trains when operating in the contingency mode.

An interesting aspect of this simulation involved the modeling of the two generic types of conveyors. The roller conveyors are infrequently turned on and off since case spacing is not too important. Thus, loading and unloading of cases can be modeled as discrete events. Belt conveyors, however, are frequently turned on and off due to downstream conditions and case travel time is not easily projected. These conveyors were modeled using state variables where the belt position is considered as a state variable and pointers are used to locate the cases on the belt.

The results of the study identified that a maximum utilization of 84 percent was obtainable from the palletizers due to conveyor operations. The decrease to be expected in utilization due to down time of specific pieces of equipment was also established. By adding two "zone clear" photocells and changing the location of another photocell, an increase in the maximum utilization was obtained. The results of the study were implemented. The predicted improvement in system performance was observed for the contingency situations involving equipment down time.

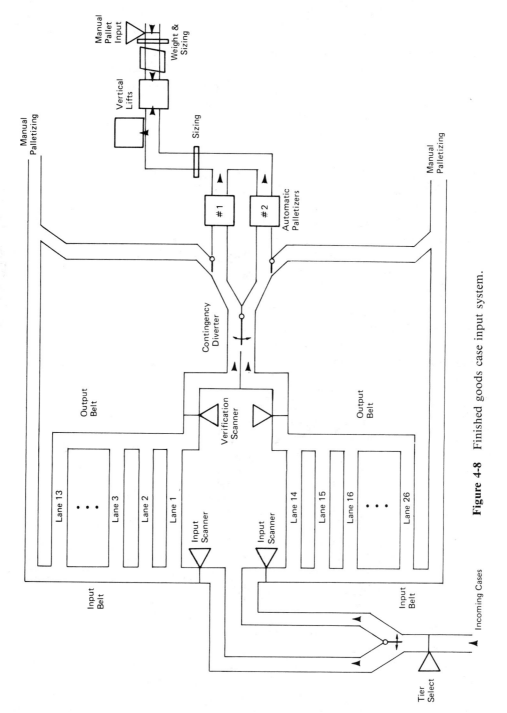

Figure 4-8 Finished goods case input system.

The simulation model has been extended and has been used to evaluate a variety of algorithms for surge controlled interleaving, zoning based on turnover, and load leveling (60). Impressive improvements in system operations have been obtained by employing the simulation model for planning purposes.

4.13   EVALUATION OF AIR TERMINAL CARGO FACILITIES (6, 16, 49)

Managers of the military airlift system need a way of measuring the productive capacity of aerial port cargo processing. Specifically, the managers need to determine the effects of fluctuating demands for airlift cargo on a terminal's ability to meet the demand in a timely manner. Resource utilization is also an important factor.

At a terminal, cargo arrives by truck or by aircraft. The arriving cargo is offloaded and sorted by shipment type, destination, and priority. The sorted cargo is moved to various in-process storage areas where it is held until some form of consolidation is possible. Once consolidated, it is weighed, inspected, and stored. Its

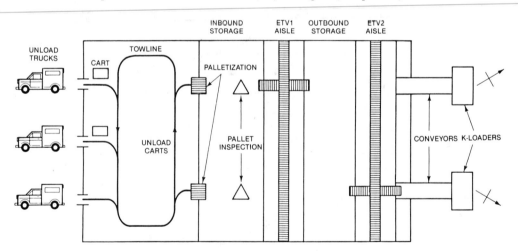

Figure 4-9   Schematic diagram of air terminal.

status can then be classified as "movement-ready." When movement-ready cargo is selected for a mission, it is transferred to a staging area where it is combined with the other cargo assigned to the mission and defined as a load. The load is then processed by cargo loading equipment, K-loaders, and transferred to the aircraft. A schematic diagram of an air terminal is shown in Figure 4-9.

A network model of this situation was constructed to answer the following procedural questions:

1. Is it worthwhile to introduce automation equipment in ports to improve processing capacity?
2. Where should new equipment be located?
3. How many aircraft can a port load simultaneously?
4. During contingencies, what additional resources will be required to support an increase in the level of air traffic?

4.14 OPTIMUM SYSTEMS ALLOCATION

At Western Electric Company, a general simulation model was developed to determine the best man/machine assignment policy consistent with dynamic managerial goals (11). The model has been used to set manpower requirements for several processes. One process involves the twisting of single strand insulated wire to form a twisted conductor. A machine line consisting of sixteen twisting machines, each with two heads, was modeled. Each head on a machine is required to produce the same twist lengths, although in all other aspects heads are independent. Two supply reels feed single strand insulated wire into the machine. The output of the machine is twisted wire which is taken up by reels mounted external to the machine. Ten product twist lengths can be produced on any of the sixteen twisting machines. The machines are shut down for lunch and rest periods, but can be left running unattended for short time intervals.

The simulation model was designed to evaluate system performance given that an operator is assigned to maintain and control a specific number of heads. Events in the simulation model include operator arrival to a head, removal of a full reel, end of installation of an empty take-up reel, and restarting the machine. When a supply reel is empty, the head automatically stops and events associated with the setting up of a full reel are performed. Where operators are assigned to more than two heads, they must travel between machines in order to perform the required functions.

Outputs from the model included the number of twist lengths of each type as a function of the number of heads assigned to each operator. In addition, percent utilization and percent interference as a function of heads assigned to an operator were plotted. From the plots, the optimum allocation of heads to an operator was determined. The simulation model has been used in other situations to optimally assign operators to machines (11).

4.15   DESIGN OF REFINERY OFFSITE SYSTEMS

A simulation program was developed during the initial planning of a significant expansion and upgrading of the Raffinor Refinery at Mongstad in Norway (from 4.5 million tons per year to 6.5 million tons per year). The objective of the project was to design a system of lines, pumps, blenders, component tanks, product tanks and jetties which minimizes the total operating cost for the expected refinery production and shipping pattern. A schematic diagram and an aggregate block diagram of the offsite facility are shown in Figure 4-10. The refinery produces a flow of components which are stored in tanks. With a modern refinery, upgrading processes are employed to generate a large number of blending components that provide a number of different products. For example, the expansion at Raffinor involved over 60 component streams to produce as many as 40 different products. The products were loaded over seven berths to approximately 2000 ships annually.

The project was a large effort and the simulation model was only a part of a total program. The various components of the program are listed below (15, 31). Note that items 5 and 6 below use operations performance measures to calculate management performance measures.

1. Refinery and petro-chemical planning system including a generalized refinery linear programming package.
2. Refinery operating plan preprocessor to develop production schedules.
3. Marine operating plan preprocessor which is a combined distribution linear program with embedded heuristic logic.
4. Refinery offsites operations simulator which is programmed in SLAM II and includes a network model to portray the arrival of ships and the effects of weather. The flow of components in the system is modeled using continuous concepts. The setting and the termination of product blending, shiploading and offloading are modeled as discrete events.

5. Facilities investment cost estimator for the overall determination of the investment cost of the facility.
6. Refinery offsite profitability analysis to perform the profitability analysis including the costs of investment, demurrage, and capital.

The total system has been used for the analysis of a number of different offsite systems with the main objectives to size tankage to decide the number of jetties required, and to evaluate the consequences of bad weather on the performance of different fleets. The proper operation of the model was determined through comparison with existing data. Program results and operating statistics comparisons showed that the program accurately represented the operation of the offsite system.

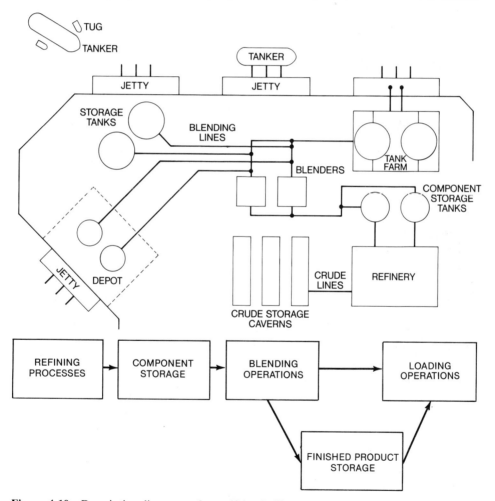

Figure 4-10 Descriptive diagrams of an offsite facility.

4.16   INDUSTRIAL ENGINEERING AND MANAGEMENT SCIENCE APPLICATIONS

In the preceding sections, emphasis was placed on simulation models that were developed by systems analysts and used for specific decision-making. There have been many other models used for analyzing a problem situation, for designing a new system, or for projecting future developments (38, 49, 50, 62). References to papers by areas of application are given below.

Manufacturing Operations
- Plant design (2, 27, 52)
- Productivity improvements (18, 28, 41)
- Manpower assignment (11)
- Computer-aided manufacturing (33, 51)
- Scheduling (1, 34, 37)
- Materials handling (13, 29, 60)

Transportation Systems
- Passenger railroad system performance (50)
- Bus scheduling and routing (50)
- Air traffic control (67)
- Terminal operations (6, 16, 42)

Computer and Communication Systems
- Performance evaluation (5, 64)
- Microprocessor design (30)
- Signal processing (25, 26)

Project Planning and Control (39)
- Product planning (8)
- Marketing (9, 10)
- Research and development (40)
- Construction (22)
- Scheduling (63)

Financial Planning
- Capital investment (61)
- Cash flow analysis (3)
- Corporate models (44)
- Econometric models (43)

Environmental and Ecological Studies
  • Flood control (45)
  • Pollution control (23, 54, 58)
  • Energy flows (14, 57)
  • Agriculture (7, 46)
  • Insect control (36)
  • Reactor maintainability (20)

Health Care Systems
  • Inventory management (17)
  • Hospital planning (4)
  • Manpower planning (55)
  • Materials handling (56)

4.17   CHAPTER SUMMARY

In this chapter the types of applications of simulation have been presented. Thirteen situations in which simulation was useful in problem solving are described. Simulation is a practical tool of widespread use.

4.18   EXERCISES

4-1. Select two of the applications presented in this chapter and describe the modeling approach that you would use in resolving the specified problem. Define the necessary entities, attributes, processes, activities and files that you would employ. List the variables on which you would collect statistical information.

4-2. Categorize the thirteen applications presented in this chapter by class of sponsor, simulation model type, and simulation modeling approach. Develop other categories for simulation models.

4-3. Rank the thirteen applications given in this chapter according to difficulty of the modeling effort; difficulty of obtaining data for the model; difficulty in applying the results; and potential benefit from using the model.

4-4. Develop specifications for a network simulation language which can be used to build simulation models to evaluate materials-handling equipment.

4-5. Categorize the areas listed in Table 4-1 according to the courses you have taken during your academic studies.

4-6. Develop a list of performance measures for the Department of Defense.

4.19   REFERENCES

1. Abbot, R. A. and T. J. Greene, "Determination of Appropriate Dynamic Slack Sequencing Rules for An Industrial Flow Shop Via Discrete Simulation," *Proceedings, 1982 Winter Simulation Conference*, 1982, pp. 223-230.
2. Adam, N. and J. Surkis, "A Comparison of Capacity Planning Techniques in a Job Shop Control System," *Management Science,* Vol. 23, 1977, pp. 1011-1015.
3. Adam, N. R. and A. Dogramaci, Eds., *Current Issues in Computer Simulation,* Academic Press, 1979.
4. Allessandra, A. J., T. E. Grazman, R. Parameswaran, and U. Yavas, "Using Simulation in Hospital Planning," *Simulation,* Vol. 30, 1978, pp. 62-67.
5. Alexander, E. L., Jr., "SCOPE: A Simulation Approach to Computer Performance Evaluation Using GASP IV," *Proceedings, Summer Computer Simulation Conference,* 1976, pp. 6-10.
6. Auterio, V. J., "Q-GERT Simulation of Air Terminal Cargo Facilities," *Proceedings, Pittsburgh Modeling and Simulation Conference,* Vol. 5, 1974, pp. 1181-1186.
7. Barrett, J. R. and R. M. Peart, "Systems Simulation in U.S. Agriculture," published in *Progress in Modelling Simulation,* edited by F. E. Cellier, Academic Press, 1982, pp. 39-59.
8. Bellas, C. J. and A. C. Samli, "Improving New Product Planning with GERT Simulation," *California Management Review,* Vol. XV, 1973, pp. 14-21.
9. Bird, M., E. R. Clayton, and L. J. Moore, "Sales Negotiation Cost Planning for Corporate Level Sales," *Journal of Marketing,* Vol. 37, 1973, pp. 7-13.
10. Bird, M., E. R. Clayton, and L. J. Moore, "Industrial Buying: A Method of Planning for Contract Negotiations," *Journal of Economics and Business,* 1974, pp. 1-9.
11. Bredenbeck, J. E., M. G. Ogdon, III, and H. W. Tyler, "Optimum Systems Allocation: Applications of Simulation in an Industrial Environment," *Proceedings, Midwest AIDS Conference,* 1975, pp. 28-32.
12. Cellier, F. E., Ed., *Progress in Modelling and Simulation,* Academic Press, 1982.
13. DeJohn, F. A., C. W. Sanderson, C. T. Lewis, and J. R. Gross, "The Use of Computer Simulation Programs to Determine Equipment Requirements and Material Flow in the Billet Yard," *Proceedings, 1980 AIIE Spring Annual Conference,* 1980, pp. 402-408.
14. Deporter, E. L., H. A. Kurstedt, Jr., and J. A. Nachlas, "A Combined Simulation Model of the Nuclear Fuel Cycle," *Proceedings, 1977 Winter Simulation Conference,* 1977, pp. 213-216.
15. Duket, S. and C. R. Standridge, "Applications of Simulation: Combined Models," *Modeling,* The Simulation Technical Committee Newsletter (IEEE), No. 19, Dec. 1983.
16. Duket, S. and D. Wortman, "Q-GERT Model of the Dover Air Force Base Port Cargo Facilities," MACRO Task Force, Military Airlift Command, Scott Air Force Base, Illinois, 1976.
17. Dumas, M. B. and M. Rabinowitz, "Policies for Reducing Blood Wastage in Hospital Blood Banks," *Management Science,* Vol. 23, 1977, pp. 1124-1132.
18. Embury, M. C., G. V. Reklaitis, and J. M. Woods, "Simulation of the Operation of a Staged Multi-Product Process with Limited Interstage Storage Buffers," Presented at the Ninth International Conference on Systems, 1976.

19. Emshoff, J. R. and R. L. Sisson, *Design and Use of Computer Simulation Models,* Macmillan, 1970.

20. Engi, Dennis, "Maintainability Analysis Using Q-GERT Simulation," *Simulation,* Vol. 44, 1985, pp. 67-74.

21. Erdbruegger, D. D., W. G. Parmelee and D. W. Starks, "SLAM II Model of the Rose Bowl Staffing Plans," *Proceedings, 1982 Winter Simulation Conference,* 1982, pp. 127-135.

22. Federal Power Commission Exhibit EP-237, "Risk Analysis of the Arctic Gas Pipeline Project Construction Schedule," Vol. 167, Federal Power Commission, 1976.

23. Fehr, R. L., J. R. Nuckols et al., "GASP IV Simulation of Flush Water Recycling Systems," *Proceedings, 1977 Winter Simulation Conference,* 1977, pp. 513-519.

24. Gibson, D. F. and E. L. Mooney, "Analyzing Environmental Effects of Improving System Productivity," *Proceedings, AIIE Systems Engineering Conference,* 1975. pp. 76-82.

25. Gracia, J. A. and R. M. Huhn, "Using Simulation to Analyze Pulse Stuffing Network Jitter," *Proceedings, 1977 Winter Simulation Conference,* 1977, pp. 801-810.

26. Green, R. and M. Fox, "AN-TTC-39 Circuit Switch Simulation," *Proceedings, Winter Simulation Conference,* 1975, pp. 211-216.

27. Gross, J. R., S. M. Hare, and S. Roy, "Simulation Modeling as an Aid to Casting Plant Design for an Aluminum Smelter, *Proceedings, 1982 IMACS Conference,* Vol. 2, 1982, pp. 160-161a.

28. Hancock, W., R. Dissen, and A. Merten, "An Example of Simulation to Improve Plant Productivity," *AIIE Transactions,* Vol. 9, 1977, pp. 2-10.

29. Jarvis, G. L. and R. M. Waugh, "A GASP IV Simulation of an Automated Warehouse," *Proceedings, 1976 Winter Simulation Conference,* 1976, pp. 541-547.

30. Jayakumar, M. S. and T. M. McCalla, Jr., "Simulation of Microprocessor Emulation Using GASP_PL/I," *Computer,* April 1977, pp. 20-26.

31. Kristiansen, T. K. and A. Landsnes, "Design of Refinery Offsite Systems by Simulation," *Proceedings, 1985 IMACS Conference,* Vol. 2, 1985, pp. 293-296.

32. Lawrence, K. D. and C. E. Sigal, "A Work Flow Simulation of a Regional Service Office of a Property and Casualty Insurance Company with Q-GERT," *Proceedings, Pittsburgh Modeling and Simulation Conference,* Vol. 5, 1974, pp. 1187-1192.

33. Lenz, J. E. and J. J. Talavage, *The Optimal Planning of Computerized Manufacturing Systems Simulator (GCMS),* Report No. 7, Purdue University, August 1977.

34. Lilegdon, W. R., C. H. Kimpel and D. H. Turner, "Application of Simulation and Zero-One Programming for Analysis of Numerically Controlled Machining Operations in the Aerospace Industry," *Proceedings, 1982 Winter Simulation Conference,* 1982, pp. 281-289.

35. Lyon, R. B., "Radionuclide Pathway Analysis Calculations Using a Set of Computer Programs Interfaced with GASP IV," *Proceedings, 1976 Winter Simulation Conference,* 1976, pp. 549-557.

36. Miles, G. E. et al., "SIMAWEV II: Simulation of the Alfalfa Weevil with GASP IV," *Proceedings, Pittsburgh Conference on Modeling and Simulation,* 1974, pp. 1157-1161.

37. Miner, R. J., D. B. Wortman, and D. Cascio, "Improving the Throughput of a Chemical Plant," *SIMULATION,* Vol. 35, 1980, pp. 125-132.

38. Moder, J. J. and S. E. Elmaghraby, *Handbook of Operations Research: Models and Applications,* Vol. 2, Van Nostrand Reinhold, 1978.

39. Moore, L. J. and E. R. Clayton, *GERT Modeling and Simulation: Fundamentals and Applications,* Petrocelli/Charter, 1976.

40. Moore, L. J. and B. W. Taylor, III, "MultiTeam, Multiproject Research and Development Using GERT," *Management Science,* Vol. 24, 1977, pp. 401-410.

41. Musselman, K. J., "Computer Simulation: A Design Tool for FMS," *Manufacturing Engineering,* Vol. 93, 1984, pp. 117-120.

42. Nagy, E. A., "Intermodal Transhipment Facility Simulation: A Case Study," *Proceedings, 1975 Winter Simulation Conference,* 1975, pp. 217-223.

43. Naylor, T. H., *Computer Simulation Experiments with Models of Economic Systems,* Wiley, 1971.

44. Naylor, T. H., *Corporate Planning Models,* Addison-Wesley, 1979.

45. Otoba, K., K. Shibatani, and H. Kuwata, "Flood Simulator for the River Kitakami," in J. McLeod, Ed., *Simulation,* McGraw-Hill, 1968.

46. Peart, R. M., and J. R. Barrett, Jr., "Simulation in Crop Ecosystem Management," *Proceedings, 1976 Winter Simulation Conference,* 1976, pp. 389-402.

47. Pritsker, A. A. B., "Applications of SLAM," *IIE Transactions,* Vol. 14, 1982, pp. 70-77.

48. Pritsker, A. A. B., Applications of Simulation, in *OPERATIONAL RESEARCH '84: Proceedings of IFORS,* J. P. Brans, Ed., Elsevier Science Publishers, 1984, pp. 908-920.

49. Pritsker, A. A. B., and C. Elliott Sigal, *Management Decision Making: A Network Simulation Approach,* Prentice-Hall, 1983.

50. Reitman, J., *Computer Simulation Applications,* John Wiley, 1971.

51. Runner, J., *CAMSAM: A Simulation Analysis Model for Computer-Aided Manufacturing Systems,* Unpublished MS Thesis, Purdue University, 1978.

52. Schooley, R. V., "Simulation in the Design of a Corn Syrup Refinery," *Proceedings, 1975 Winter Simulation Conference,* 1975, pp. 197-204.

53. Schuman, R. E., E. L. Janzen, and W. H. Dempsey, "Applications of GASP IV Simulation," Presentation to ORSA/TIMS Combined Chicago Chapters, November 16, 1977.

54. Sigal, C. E., "Designing a Production System with Environmental Considerations," *Proceedings, AIIE Systems Engineering Conference,* 1973, pp. 31-39.

55. Standridge, C., C. Macal, A. Pritsker, H. Delcher, and R. Murray, "A Simulation Model of the Primary Health Care System of Indiana," *Proceedings, 1977 Winter Simulation Conference,* 1977, pp. 349-358.

56. Swain, R. W. and J. J. Marsh, III, "A Simulation Analysis of an Automated Hospital Materials Handling System," *AIIE Transactions,* Vol. 10, 1978, pp. 10-18.

57. Sweet, A. L. and S. D. Duket, "A Simulation Study of Energy Consumption by Elevators in Tall Buildings," *Computing and Industrial Engineering,* Vol. 1, 1976, pp. 3-11.

58. Talavage, J. J. and M. Triplett, "GASP IV Urban Model of Cadmium Flow," *Simulation,* Vol. 23, 1974, pp. 101-108.

59. Triplett, M. B., T. L. Willke, and J. D. Waddell, "NUFACTS: A Tool for the Analysis of Nuclear Development Policies," *Proceedings, 1977 Winter Simulation Conference,* 1977, pp. 793-798.

60. Waugh, R. M. and R. A. Ankener, "Simulation of an Automated Stacker Storage System," *Proceedings, 1977 Winter Simulation Conference,* 1977, pp. 769-776.

61. Weinberger, A., A. Odejimi et al., "The Use of Simulation to Evaluate Capital Investment Alternatives in the Steel Industry: A Case Study," Presented at the 1977 Winter Simulation Conference, 1977.
62. Wilson, J. R. and A. A. B. Pritsker, "Computer Simulation," in *Handbook of Industrial Engineering* ed. by G. Salvendy, John Wiley, 1982.
63. Wilson, J. R., D. K. Vaughan, E. Naylor and R. G. Voss, "Analysis of Space Shuttle Ground Operations," *Simulation,* Vol. 38, 1982, pp. 187-203.
64. Wong, G., "A Computer System Simulation with GASP IV," *Proceedings, Winter Computer Simulation Conference,* 1975, pp. 205-209.
65. Wong, G. A. et al., "A Systematic Approach to Data Reduction Using GASP IV," *Proceedings, 1976 Winter Simulation Conference,* 1976, pp. 403-410.
66. Wortman, D. B., and J. R. Wilson, "Optimizing a Manufacturing Plant by Computer Simulation," *Computer-Aided Engineering,* Vol. 3, 1984, pp. 48-54.
67. Yu, J., W. E. Wilhelm, and S. A. Akhand, "GASP Simulation of Terminal Air Traffic," *Transp. Eng. J.,* Vol. 100, 1974, pp. 593-609.

CHAPTER 5

Basic Network Modeling

5.1   INTRODUCTION

In Chapter 3, we introduced SLAM II as a simulation language which allows the modeler to select the "world view" that is most applicable to the system under study. As such, the SLAM II user can employ any combination of the following perspectives when modeling a system: process, event, continuous, and activity scanning. This chapter begins the description of the process orientation or network modeling procedures available in SLAM II. The task of the analyst is to use network concepts to formulate a model which reflects the important characteristics of the system. In approaching network modeling using SLAM II, the analyst confronts two related problems: 1) deciding what detail to include in the model and 2) deciding how to represent that detail with the SLAM II network framework.

In modeling and simulation, the level of detail to include in a model is relative to the purpose of the model. By knowing the purpose of the model, the relative worth of including specific details can be assessed. Only those elements that could cause significant differences in decision making resulting from the outputs from simulation need be considered. In addition, for larger models it is often advantageous to decompose the models into stages of development. The decomposition could consist of initially developing an aggregate model which crudely approximates the system under study, and then improving the model through embellishments in subsequent stages, or it could consist of segmenting the total system into subsystems each of which are modeled separately and then combined. In any case, the prerequisite step in developing a network model of a system is to construct a problem statement which defines the purpose so that the specific detail level to be included in the model can be decided.

Once the problem statement is complete, the system can be represented as entities which flow through a network of nodes and activities. The first step in this modeling process is to define the elements which are to be represented as entities. Recall that an entity is any object, person, unit of information, or combination thereof which defines or can alter the state of the system. Therefore the entities to be modeled can be identified by defining the variables that represent the system state and determining the changes in state that can occur. For example, in a radio inspection problem, the status of the system could be represented by the number of busy inspectors and the number of radios waiting for inspection. Status changes are due entirely to the movement of a radio through the system; therefore, the entity to be modeled is the radio. For more complicated problems, the entities may be more abstract and there may be more than one entity type within the simulation.

The next and most challenging step in developing a network model is the synthesis of a network of nodes and activities which represents the process through which the entities flow. Although the synthesis can be done using either the graphic or statement form of the network elements, the intitial development most often takes place using the graphic symbols. The advantage of using the graphic symbols is that they provide a pictorial medium for both conceptualization and communication. The graphic symbols of SLAM II play a role for the simulation analyst similar to that of the free body diagram for mechanical engineers or the circuit diagram for electrical engineers. Once the graphic model is complete, the transcribing of the graphic model into the equivalent statement representation is straightfoward. In fact, the TESS program has automated this step (4).

As an introduction to SLAM II network modeling, let us consider a simple queueing system in which items arrive, wait, are processed by a single server, and then depart the system. Such a sequence of events, activities, and decisions is referred to as a *process*. *Entities* flow through a process. Thus, items are considered as entities. An entity can be assigned *attribute* values that enable a modeler to distinguish between individual entities of the same type or between entities of different types. For example, the time an entity enters the system could be an attribute of the entity. Such attributes are attached to the entity as it flows through the network. The resources of the system could be servers, tools, or the like for which entities compete while flowing through the system. A resource is busy when processing an entity, otherwise it is idle.

SLAM II provides a framework for modeling the flow of entities through processes. The framework is a network structure consisting of specialized nodes and branches that are used to model resources, queues for resources, activities, and entity flow decisions. In short, a SLAM II network model is a representation of a process and the flow of entities through the process.

5.2   A SLAM II NETWORK OF A SINGLE SERVER QUEUEING SYSTEM

To illustrate the basic network concepts and symbols of SLAM II, we will construct a model of an inspection process in the manufacturing of transistor radios. In this system, manufactured radios are delivered to an inspector at a central inspection area. The inspector examines each radio. After this inspection, the radio leaves the inspection area. Although we could model the entire manufacturing process, we are only interested in the operations associated with the inspection of radios. Therefore, we concern ourselves with the following three aspects of the system:

1. The arrival of radios to the inspection area;
2. The buildup of radios awaiting inspection; and
3. The activity of inspecting radios by a single inspector.

This is a single resource queueing system, and is similar to the bank teller system described in previous chapters. The radios are the system's entities. The inspector is the resource and will be modeled as a *server*. The *service activity* is the actual inspection, and the buildup of the radios awaiting service is the *queue*.

A pictorial diagram of this inspection system is shown below.

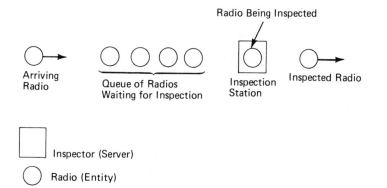

5.2.1   Modeling Queues and Servers

Let us now build a network for this one server system. The passage of time is represented by a *branch*. Branches are the graphical representation of activities. Clearly, the service operation (the inspection of the radios) is an activity and, hence, is modeled by a branch. If the service activity is ongoing, that is, the server (the inspector) is busy, arriving entities (radios) must wait. Waiting occurs at a QUEUE node. Thus, a one-server, single-queue operation is depicted in SLAM II by a QUEUE node and a branch as follows:

In our example, radios wait for service at the queue. When the inspector is free, he removes a radio from the queue and performs the service activity. The procedure for specifying the time to perform the service operation will be discussed later. A wide variety of service time distributions are available for use in SLAM II.

Since there may be many queues and service activities in a network, each can be identified numercially. Entities waiting at queues are maintained in files, and a file number, IFL, is associated with a queue. Service activities are assigned a value to indicate the number, N, of parallel servers described by the branch, that is, the number of possible concurrent processings of entities. Activities can also be given an activity number, A, for identification and statistics collection purposes. The notation shown below is the procedure for labeling these elements of the network.

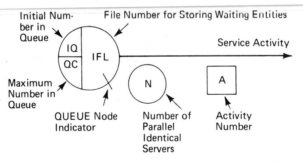

The file number is put on the right-hand side of the node. The procedure for ranking entities in the file is specified separately by a PRIORITY statement and is not shown on the graphic model. Also specified for a QUEUE node are the initial number of entities at the QUEUE node, IQ, and the capacity of the queue, QC. This latter quantity is the largest number of entities that can wait for service at the QUEUE node. Arriving entities to a full queue will either balk or be blocked. A QUEUE node has a "hash" mark in the lower right-hand corner to make the symbol resemble the letter Q. For the service activity, the number of parallel servers is put in a circle below the branch, and the activity number is put in a square below the branch.

5.2.2 Modeling the Arrival of Entities

Turning our attention to the entities (the individual radios), we must model the arrival of radios to the system. In SLAM II, entities are inserted into a network by CREATE nodes. The symbol for the CREATE node is shown below

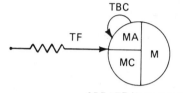

CREATE Node Symbol

where

TF is the time at which the first entity is to be created and sent into the network;

TBC is the time between creations of entities;

MA is the attribute number in which the creation or mark time is to be maintained;

MC is the maximum number of entities that can be created at this node; and

M is the maximum number of branches along which a created entity can be routed from this node (referred to as the M-number or "max take" value).

There are several important features to note about the CREATE node. At a prescribed time, TF, a first entity will be created. If desired, the time at which an entity is created can be assigned to attribute MA of the entity. This time is usually referred to as the "mark" time. The created entity will be routed over the branches emanating from the node in accordance with the M-number. If M is equal to one and there are two branches emanating from the node, the entity will only be routed over one of the two branches. Procedures for selecting which branch over which to route the entity will be described later. If all branches are to be taken, M need not be specified.

The second entity created at the node will occur at time TF + TBC where TBC is the time between creation of entities. For the radio example, TBC is the time between the arrivals of radios which can be a constant, a SLAM II variable, or a random variable. This is described in the section on specifying attribute or duration assignments. The variable MC prescribes a maximum number of entities that can be created at the node. If no limit is specified, entities will continue to be created until the end of the simulation run.

5.2.3   Modeling Departures of Entities

We have now modeled the arrival pattern of entities and the waiting and service operations. All that remains is the modeling of the departure process for the entity. For our simple system, we will let the entities leave the system following the completion of service. The modeling of the departure of an entity is accomplished by a TERMINATE node as shown below.

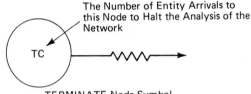

TERMINATE Node Symbol

A squiggly line is used on the output side of a node to indicate that entities are to be terminated or destroyed at the node. TERMINATE nodes are one way to specify the stopping procedure to be used when analyzing a SLAM II network. Each analysis of a network is referred to as a run. The TERMINATE node can be used to specify that TC entity arrivals at the TERMINATE node are required to complete one run. As we shall see, the stopping condition can also be based on a time period. For example, a run could be made for 1000 hours of operation.

5.2.4   Combining Modeling Concepts

We are now ready to combine the arrival, service, and departure operations to obtain a complete network model of the one server, single queue process. This SLAM II model is shown below with data values prescribed for the variables associated with the network symbols. This network depicts the flow of an entity and all the potential processing steps associated with the entity. The first entity arrives at the CREATE node at time 7. The next entity is scheduled to arrive 10 time units later, which would be at time 17. The first entity is routed to the service activity by the branch to the QUEUE node. The branch represents the activity of traveling to the server and is prescribed to be 3 time units in duration. When the entity arrives at the QUEUE node, it will immediately be serviced if server 3 is idle. If this occurs, the entity flows from the QUEUE node to the TERMINATE node in 9 time units. During this time, server 3 is busy.

Other entities will also follow the pattern described above. However, if server 3 is busy when an entity arrives at the QUEUE node, the entity is placed in file 10 which models the queue of entities waiting for server 3. When an entity joins a queue, a rule is used that specifies the order in which the entities are ranked in the queue. (The ranking rule for the queue is characteristic of the file and is not specified on the graphical model.) If no ranking rule is specified, a first-in, first-out (FIFO) procedure is used, that is, entities are taken from the queue in the order in which they arrived to the queue. After entities are served by server 3, they reach the TERMINATE node where the entity is removed from the system since its routing through the process is completed.

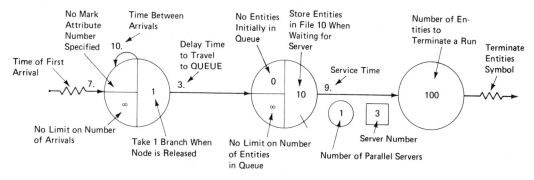

5.3 SLAM II NETWORK MODELING

A SLAM II network model consists of a set of interconnected symbols that depict the operation of the system under study. The symbols can be converted into a form for input to a program that analyzes the model using simulation techniques. The input corresponding to a graphic SLAM II model is in the form of *statements*. To provide an illustration of a statement model, the network model presented previously is given in statement form below. A semicolon is used to indicate the end of the data on a particular record. Comments can be given following a semicolon. The sequence of statements must correspond to the process an entity encounters as it flows through the network.

```
;EXAMPLE OF A SLAM STATEMENT MODEL
NETWORK;                 START OF NETWORK STATEMENTS
      CREATE,10.,7.;      TIME BETWEEN ARRIVALS = 10
      ACTIVITY,3;         TIME TO REACH QUEUE NODE IS 3
      QUEUE(10);          USE FILE 10 FOR QUEUE
      ACTIVITY(1)/3,9;    SERVICE TIME = 9
      TERMINATE,100;      RUN MODEL FOR 100 ENTITIES
      ENDNETWORK;         END OF NETWORK STATEMENTS
```

This illustration is only to indicate the similarity of the graphic model and the statement model that is acceptable as input for computer analysis.† Later in this section, a descripton of the basic SLAM II symbols and statements are presented.

As previously seen, a network consists of an interrelated set of nodes and branches. The nodes and branches can be considered as elements that are combined and integrated into a system description. The task of the modeler is to integrate the elements into a network model for the system of interest.

Before presenting the basic SLAM II symbols and statements, several comments on general sequencing and entity flow are in order. The flow of entities normally follows the directed branches indicated on the network. Node labels are used to identify non-standard flows of entities. In statements, node labels are used as statement labels in a fashion similar to statement numbers in a FORTRAN program. Node labels can be appended to any node. On the graphic model, they are placed below the node symbol. On the statement, they precede the node name. The node or statement name is given starting in column seven or later on the input record, whereas the node label if required is given in columns one through five of the input record.

As described previously, branches are used to depict activities. In some situations, it is desired to have entities flow from one node to another node with no intervening activity. Such transfers are depicted on the network by branches with no specifications or by broken (nonsolid) lines, and are referred to as connectors. No statements are required in the statement model to describe connectors.

5.3.1   Routing Entities from Nodes (Branching)

Entities are routed along the branches emanating from nodes. The maximum number of branches, M, that can be selected is specified on the right-hand side of the node through the value assigned to M. The default value for M is ∞. When M equals 1, at most one branch will be taken. If probabilities are assigned to the branches emanating from a node that has M = 1, then the node is said to have *probabilistic branching*. If no conditions or probabilities are prescribed for the branches, and M equals the number of branches emanating from the node then

---

† The TESS program provides the capability to build SLAM II networks graphically at a terminal. TESS then converts the graphic network to the input required for the SLAM II processor.

deterministic branching is specified. Deterministic branching causes an entity to be duplicated and routed over every branch emanating from the node.

The branching concept prescribed by the value of M is quite general. It allows the routing of entities over a subset of branches for which conditions are prescribed. For example, if M is equal to 2 and there are five branches emanating from the node, then the entity would be routed over the first two branches for which the condition is met.

An even more complex situation involves a combination of probabilistic and conditional branching. Letting p_i be the probability of routing an entity over branch i and letting c_j be the condition for routing over branch j, consider the following situation:

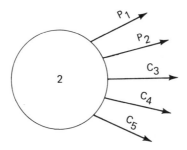

The node type shown above is the GOON node. Every entity that arrives to a GOON node causes it to be "released." The 2 inside the node specifies the M-value previously mentioned. The statement syntax for the GOON node is

GOON,M;

for this example, M is equal to 2 which signifies that at most two of the five branches are to be taken. Assuming the branches are evaluated in the order 1, 2, 3, 4, and 5, a random selection between branches 1 and 2 will be made (p_1 + p_2 must equal 1) and then branch 3 will be taken if condition 3 (c_3) is satisfied. If not, c_4 is tested and then c_5. If c_3 is satisfied, the other branches would not be taken even though c_4 or c_5 were satisfied. In the statement model, when branches (activities) emanate from a node, a statement describing each activity is placed immediately after the statement describing the node. The order of the activity statements defines the order in which the conditions are evaluated. Additional information on the GOON node will be presented in Section 5.11.

5.3.2 SLAM II Variables, Functions, and Random Variables

Table 5-1 presents SLAM II variables and their definitions.

Table 5-1 SLAM II variables.

Variable Name	Definition
ATRIB(I)	Attribute I of a current entity
II	An integer global variable; II is frequently used as an index or argument.
DD(I)	Derivative of SS(I)
SS(I)	State variable I
TNOW	Current time
XX(I)	System or global vector
ARRAY(I,J)	System or global array

The vector ATRIB defines the attributes of an entity as it flows through the network. Each entity has its own attribute vector which is attached to the entity as it moves through nodes and across branches. Reference is sometimes made to the current entity which is the entity that has just been created or has completed an activity and is arriving to a node in the network. The variable II is an integer variable that is frequently used an an index for other SLAM II variables, for example, ATRIB(II). The variable XX(I) is a SLAM II global variable that maintains a value until it is reset by the user within the model. ARRAY(I,J) provides a means for storing tables of values in a row and column format. An ARRAY statement is provided for initializing the ARRAY table. The variables DD(I) and SS(I) have special definitions which are used to model continuous status variables. Specifically, DD(I) is defined as the first derivative of SS(I) when differential equations are used to model system status.

The current simulation time is represented by the variable TNOW. TNOW is maintained automatically in SLAM. II and is advanced in accordance with the time at which variables in the model may change value. TNOW is used to communicate to the modeler the current simulation clock time.

Table 5-2 presents the SLAM II network variables, functions, and random variables which are used to model the logical and stochastic aspects of a system. NNACT, NNCNT, NNGAT, NNQ, NNRSC and NRUSE are SLAM II status variables that are updated automatically by SLAM II as the number of entities and the status of the entities in activities, gates, resources and files changes over time. For example, NNQ(3) describes the number of entities in file 3 at the current time,

TNOW. If the user wants to model a decision which is dependent on the number of entities in file 3, NNQ(3) would be used as part of the logical expression representing the decision.

The function USERF(N) is a function that can be inserted throughout a network model to include user-written FORTRAN code into the model. The argument N is a code that differentiates the various invocations of function USERF by the modeler. Function USERF can be used on a SLAM II network in any location where it is permitted to use a SLAM II variable. This allows extensive modeling flexibility using SLAM II networks.

Function GGTBLN (IRX, IRY, XVALUE) is a SLAM II table lookup function that can be used directly on a network. The first two arguments of function GGTBLN provide pointers to rows of ARRAY which contain values of a dependent variable corresponding to values of an independent variable X. If IRY = 5 and IRX = 3 then the value of Y stored in ARRAY (5, 1) corresponds to the value of X stored in ARRAY (3, 1). XVALUE is the X value for which a corresponding Y value is desired. As an example, consider the following table.

K	ARRAY (3, K)	ARRAY (5, K)
1	100	10
2	200	20
3	300	15
4	400	10
5	1000	0

The specification GGTBLN (3, 5, 400) assigns a value of 10 to the network location in which GGTBLN is placed. Linear interpolation is used for intermediate values so that a value of 5 is obtained when GGTBLN (3, 5, 700) is specified. Values outside of the range of the independent variable are assigned the end values for the dependent variable. For example, an XVALUE less than 100 results in an assignment of 10 and an XVALUE greater than 1000 results in an assignment of 0. Additional discussion on SLAM II table lookup capabilities is presented in Chapter 12.

The remainder of the entries in Table 5-2 represent functions from which samples can be obtained in accordance with a specified distribution function. The equations for the distribution functions associated with the random sampling functions presented in Table 5-2 are included in Chapter 18.

Table 5-2 SLAM II network variables and functions.

Variable/Function	Definition
NNACT(I)	Number of active entities in activity I at current time
NNCNT(I)	The number of entities that have completed activity I
NNGAT(GATE)	Status of gate GATE at current time : 0 → open; 1 → closed
NNQ(I)	Number of entities in file I at current time
NNRSC(RES)	Current number of units of resource type RES available
NRUSE(RES)	Current number of units of resource type RES in use
USERF(N)	A value obtained from the user-written function USERF with user function number N (See Chapter 9)
GGTBLN(IRX,IRY,XVALUE)	A value corresponding to XVALUE obtained from a table lookup with the Kth point in the table defined by (ARRAY(IRX,K) ,ARRAY(IRY,K))
BETA(THETA,PHI,IS)	A sample from a beta distribution with parameters THETA and PHI using random number stream IS
DRAND(IS)	A pseudo-random number obtained from random number stream IS
DPROBN(IC,IV,IS)	A sample from a probability mass function where the cumulative probabilities are in row IC of ARRAY and the corresponding sample values are in row IV of ARRAY using random number stream IS.
ERLNG(EMN,XK,IS)	A sample from an Erlang distribution which is the sum of XK exponential samples each with mean EMN using random number stream IS
EXPON(XMN,IS)	A sample from an exponential distrubution with mean XMN using random number stream IS
GAMA(BETA,ALPHA,IS)	A sample from a gamma distribution with parameters BETA and ALPHA using random number stream IS
NPSSN(XMN,IS)	A sample from a Poisson distribution with mean XMN using random number stream IS
RNORM(XMN,STD,IS)	A sample from a normal distribution with mean XMN and standard deviation STD using random number stream IS
RLOGN(XMN,STD,IS)	A sample from a lognormal distribution with mean XMN and standard deviation STD using random number stream IS
TRIAG(XLO,XMODE,XHI,IS)	A sample from a triangular distribution in the interval XLO to XHI with mode XMODE using random number stream IS
UNFRM(ULO,UHI,IS)	A sample from a uniform distribution in the interval ULO to UHI using random number stream IS
WEIBL(BETA,ALPHA,IS)	A sample from a Weibull distribution with scale parameter BETA and shape parameter ALPHA using random number stream IS

The arguments for the random sampling routines relate to the parameters of the distribution functions associated with the random variables. A description of the random variables and the functional form of the distribution functions describing the random variables is given in Chapter 18. At this time, a brief discussion of the arguments of the functions presented in Table 5-2 is presented. Throughout the discussion, μ denotes the mean of the distribuion and σ denotes the standard deviation of the distribution.

For the Poisson, exponential, normal and lognormal distributions, the parameters are defined in terms of the mean and standard deviation. In the sampling functions, XMN is used as an argument to signify the mean μ and STD is used to signify the standard deviation σ. For the triangular, TRIAG, and the Weibull, WEIBL, distributions, the mean and standard deviation are not normally used to describe the parameters. For the triangular distribution a minimum value, XLO, a modal value, XMODE, and a maximum value, XHI, are typically used. For the Weibull distribution, a scale parameter, BETA, and a shape parameters, ALPHA, are used. These parameters are used directly as the arguments for these two random sampling functions.

An Erlang random variable is the sum of independent and identically distributed exponential random variables. The arguments of function ERLNG are specified in terms of the exponential parameters. EMN is the mean of the underlying exponential random variable and XK is the number of exponential random variables included in the sum. If the mean and standard deviation of the Erlang distribution are known, then $EMN = \sigma^2/\mu$ and $XK = \mu/EMN$. The gamma distribution is a generalization of the Erlang distribution in which the second parameter need not be an integer. The arguments for function GAMA in terms of the mean and standard deviation of the gamma distribution are $BETA = \dfrac{\sigma^2}{\mu}$ and $ALPHA = \dfrac{\mu}{BETA}$.

The uniform distribution is typically specified in terms of a range with a low value, ULO, and a high value, UHI. Samples within this range are equally likely. The arguments for function UNFRM are ULO and UHI. For the uniform distribution, these values can be specified in terms of the mean and standard deviation as follows:

$$ULO = \mu - \sqrt{3}\sigma$$
$$UHI = \mu + \sqrt{3}\sigma$$

The beta distribution is defined over a finite region and can take on many shapes. The SLAM II function BETA generates samples in the range 0 to 1. Thus, it is necessary for the user to make the transformation of a general beta distribution to a beta distribution in the range 0 to 1 through the following equations

$$\mu = \frac{\mu_\beta - \text{BMIN}}{\text{BMAX} - \text{BMIN}}$$

$$\sigma^2 = \frac{\sigma_\beta^2}{\text{BMAX} - \text{BMIN}}$$

where μ_β and σ_β^2 are the untransformed mean and variance of the beta distribution and BMIN and BMAX are the end points for the beta distribution.

With this transformation the arguments for Function BETA are:

$$\text{THETA} = \frac{\mu^2}{\sigma^2}(1-\mu) - \mu$$

$$\text{and PHI} = \text{THETA}\left(\frac{1-\mu}{\mu}\right)$$

In order to obtain a sample in the range BMIN to BMAX, BETAT, the following equation should be used:

BETAT = BETA(THETA,PHI) * (BMAX - BMIN) + BMIN.

SLAM II provides the function DPROBN(IC,IV,IS) to obtain a sample from a probabilty mass function. The cumulative probabilities associated with the probability mass function are stored in row IC of the global variable ARRAY and the corresponding sample values are stored in row IV of ARRAY. IS is a random number stream. The use of function DPROBN is similar to the use of the table lookup function GGTBLN with a random number being used as the independent value for which a random sample is to be obtained.

5.3.3 EQUIVALENCE Statement for SLAM II Variables

SLAM II provides an EQUIVALENCE statement so that textual names can be used for the SLAM II variables on a network model. The format for the EQUIVALENCE statement is

EQUIVALENCE/SLAM II variable,name/repeats;

The variables that can be used on the EQUIVALENCE statement are ATRIB, II, XX, ARRAY, SS, DD, a SLAM II random variable or a constant value. The name can be a maximum of twelve characers beginning with an alphabetic character.

The use of a name for a SLAM II variable in network and statement models provides for increased readability for the non-SLAM II user. For the experienced SLAM II user, knowledge of the type of variable that is being used in a model provides additional information regarding the structure of the model. The degree to which textual names are included in models should be dependent upon the model, the modeler, and the use to which the model is to be made. In this book, the use of the EQUIVALENCE statement will be kept to a minimum so that the basic understanding of the use of entities, attributes, global variables, and the random sampling functions can be seen directly.

The EQUIVALENCE statement

EQUIVALENCE/ATRIB(1),PROC_TIME;

specifies that the name PROC_TIME may be used in place of ATRIB(1).

In the network model, statistics can be collected on PROC_TIME by the following statement.

COLCT,PROC_TIME;

This statement causes PROC_TIME to be collected for each arriving entity, that is, ATRIB(1) of the arriving entity is collected.

Below an EQUIVALENCE statement is used to indicate that INVENTORY can be used in place of the global variable XX(1) and REORDER_PT for XX(2).

EQUIVALENCE/XX(1),INVENTORY/
    XX(2),REORDER_PT/
    UNFRM(4,6),REVIEW TIME:

In addition, the name REVIEW TIME may be used in place of the random sampling function UNFRM(4,6). To illustrate the use of these equivalences, the statement

ACTIVITY,REVIEW TIME,INVENTORY.LE.REORDER_PT;

specifies an activity whose duration is REVIEW TIME, that is, a sample from a uniform distribution between 4 and 6, and whose condition for performing the activity is that the INVENTORY is less than or equal to the REORDER_PT.

5.3.4   ARRAY Statement

The ARRAY statement is used to INITIALIZE a row of the SLAM II global table, ARRAY. The number of elements in a row of ARRAY can vary and, hence, the table is referred to as a ragged table. The format for the ARRAY statement is

ARRAY(IROW,NELEMENTS)/initial values/repeats;

where IROW is an integer constant defining the row for which initial values are being provided; NELEMENTS is the number of elements in this row; and initial values are constants to be inserted in the order of the columns for the row. For example, the statement

ARRAY(2,4)/5,4,2,7.3;

defines ARRAY(2,1) to be 5; ARRAY(2,2) to be 4; ARRAY(2,3) to be 2; and ARRAY(2,4) to be 7.3.

Elements of ARRAY may be referenced on a SLAM II network where a SLAM II variable is allowed. ARRAY subscripts may be constants or the SLAM II variables: II, XX(I), and ATRIB(I) where I is a positive constant. For example, if an entity has ATRIB(1) defined as a job type, ATRIB(2) as its next job step, and ATRIB(3) as the machine number for the next job step and the table ARRAY is organized such that machine numbers are the values included in the table with each row defined by a job type and each column by a job step, then

ATRIB(3) = ARRAY(ATRIB(1),ATRIB(2))

assigns a machine number to attribute 3 based on an entity's job type and current job step value.

ARRAY can be used in conjunction with the EQUIVALENCE statement to make a model more readable. Continuing with the above example, we will equivalence JOBTYPE to ATRIB(1), JOBSTEP to ATRIB(2) and MACHINE to ATRIB(3). These equivalences are shown on the following EQUIVALENCE statement.

EQUIVALENCE/ATRIB(1),JOBTYPE/
          ATRIB(2),JOBSTEP/
          ATRIB(3),MACHINE;

With these equivalences made, the above replacement statement becomes

MACHINE = ARRAY(JOBTYPE,JOBSTEP)

ARRAY elements are accessible from user-written FORTRAN code as well as from network statements. The subprograms GETARY, PUTARY, and SETARY are used to get values of an element of ARRAY: put or insert a new value into an element of ARRAY and to set a complete row of ARRAY. These subprograms are described in Chapter 9.

5.4 INTRODUCTION TO BASIC NETWORK ELEMENTS

xxx

There are seven basic network elements in SLAM II. These network elements are: CREATE node, QUEUE node, TERMINATE node, ASSIGN node, ACTIVITY branches, GOON node and COLCT node. With these basic network elements, many diverse network models can be built.

The CREATE node is a method for creating entities for arrival or insertion into the network. The QUEUE node is used to model the complex decision processes involved when an entity arrives to a service operation where the disposition of the entity is dependent upon the status of the server and the number entities already waiting for the server in a queue. The TERMINATE node is used to delete entities from the network. The ASSIGN node is used to assign new or updated values to SLAM II variables. When an entity arrives to an ASSIGN node, the assignments specified at the ASSIGN node are made. ACTIVITY branches represent explicit time delays for entities traversing the network. Service activities are used to represent machines, operators, and the like which can process a limited number of entities concurrently. Preceding a service activity, a buffer or waiting area must be prescribed which is accomplished through the use of a QUEUE node. Activities which model explicit delays but do not have a limit on the number of concurrent entities are referred to as regular activities. GOON nodes are used to separate activities and can model branching logic to route entities following the completion of an activity. Statistical information on entities and SLAM II variables is obtained through the use of the COLCT node. Each of these basic network elements will now be described in detail. A complete description including parameter options and defaults for every SLAM II network element is given in Appendix B.

5.5 CREATE NODE

The CREATE node generates entities and routes them into the system over activities that emanate from the CREATE Node. A time for the first entity to be created by the CREATE node is specified by the variable TF. The time between creations of entities after the first is specified by the variable TBC. TBC can be specified as a constant, a SLAM II variable, or a SLAM II random variable. Entities will continue to be created until a limit is reached. This limit is specified as MC, the maximum number of creations allowed at the node. When MC entities have been input to the system, the CREATE node stops creating entities.

The time at which the entity is created can be assigned to an attribute of the entity. This time is referred to as the *mark time* of the entity and it is placed in the MAth attribute of the entity. As will be discussed shortly, the variable ATRIB(MA) stores this value. The symbol and statement for the CREATE node are shown below.

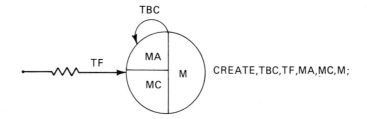

The following are examples of the CREATE node.

1. Create entities starting at time zero and every 10 time units thereafter. Put the mark time into attribute 2 of the entity. Take all branches emanating from the CREATE node.

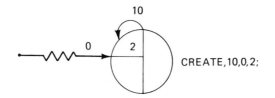

The default values taken are: MC = ∞ and M = ∞.

2. Create fifty entities starting at time 100.0. The time between creations should be 30. Take 2 branches emanating from the node.

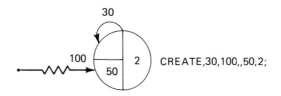

The default value for MA is not to mark the entities.

3. Create 1 entity at time 75 and take all branches emanating from the node.

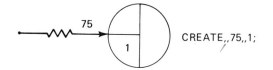

CREATE,,75,,1;

The default values are: no marking; TBC = ∞; and M = ∞.

4. Create entities that model a Poisson arrival process, that is, an exponential time between arrivals, with a mean time of 10.

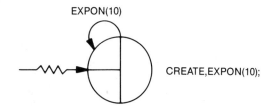

CREATE,EXPON(10);

The default values are taken for all parameters except the time between creations.

5. Create entities based on user written function 1.

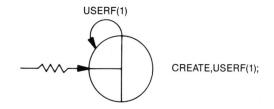

CREATE,USERF(1);

FUNCTION USERF(I) would need to be written to return USERF as the TBC value to be employed. USERF(I) with argument 1 would be called at time 0 since TF=0 and then at times specified by the values given to USERF. In writing FUNCTION USERF(I), all FORTRAN statement types are available. In particular, an external file could be read to obtain arrival times from historical records.

5.6   QUEUE NODE

A QUEUE node is a location in the network where entities wait for service. When an entity arrives at a QUEUE node, its disposition depends on the status of the server that follows the QUEUE node. If the server is idle, the entity passes through the QUEUE node and goes immediately into the service activity. If no server is available, the entity waits at the QUEUE node until a server can process it. When a server does become available, the entity will automatically be taken out of the queue and service will be initiated. SLAM II assumes that no delay is involved from the time a server becomes available and the time service is started on an entity that was waiting in the QUEUE node.

When an entity waits at a QUEUE node, it is stored in a file which maintains the entity's attributes and the relative position of the entity with respect to other entities waiting at the same QUEUE node. The order in which the entities wait in the QUEUE is specified outside the network on a PRIORITY statement which defines the ranking rule for the file associated with the QUEUE. Files can be ranked on: first-in, first-out (FIFO); last-in, first-out (LIFO); low-value first based on an attribute K (LOW(K)); and high-value first based on attribute K (HIGH(K)). FIFO is the default priority for files.

Entities can initially reside at queues, as the initial number of entities at a QUEUE node, IQ, is part of the description of the QUEUE node. These entities all start with attribute values equal to zero. When IQ > 0, all service activities emanating from the QUEUE node are assumed to be busy initially working on entities with all attribute values equal to zero. QUEUE nodes can have a capacity which limits the number of entities that can reside at the queue at a given time. The basic symbol and statement for the QUEUE node are shown below.

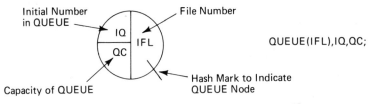

When an entity arrives at a QUEUE node which is at its capacity, its disposition must be determined. This decision is based on a specification at the QUEUE node as to whether the entity should balk or be blocked. In the case of balking, the entity can be routed to another node of the network. This node is specified by providing the label of the node. If no balking node label is specified, the entity is deleted from the system. The symbol for balking is shown below in a network segment involving balking from one QUEUE node to another QUEUE node labeled QUE2. There is no restriction on the type of node to which entities can balk.

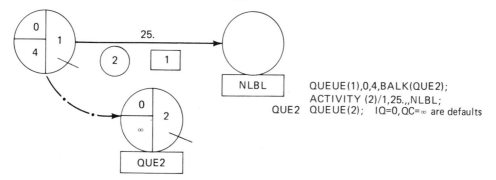

QUEUE(1),0,4,BALK(QUE2);
ACTIVITY (2)/1,25.,,NLBL;
QUE2 QUEUE(2); IQ=0,QC=∞ are defaults

When an entity is blocked by a QUEUE node, it waits until a free space is available in the queue. The activity which just served the entity that is blocked is also considered as blocked. A blocked entity will join the queue when a space is available. At that time, the blocked activity becomes free to process other entities waiting for it. *Queue nodes may only block service activities.* No time delay is associated with these deblocking operations. The symbol and statement for blocking at a QUEUE node are shown below.

QUEUE(3),2,10,BLOCK;

The file number for the QUEUE node can be specified as an attribute of the arriving entity. When this is done, a range of file numbers must be given. This specification for the file number, IFL, is in the form ATRIB(I)=J,K where I is the attribute number and J through K are the allowable file numbers specified by ATRIB(I). As an example, consider the following

QUEUE(ATRIB(2)=3,5),0;

where ATRIB(2) defines the file number which can be 3, 4 or 5. For graphics simplicity, we sometimes abbreviate ATRIB(2)=3,5 to ATR(2).

When more than one service activity follows a queue and the service activities are not identical, a selection of the server to process an entity must be made. This selection is not made at the QUEUE node but at a SELECT node that is associated with the QUEUE node. The label of the SELECT node associated with a QUEUE node is entered on the QUEUE statement. When an entity arrives at a QUEUE node, its associated SELECT node is interrogated. When a SELECT node finds a free server, the entity arriving at the QUEUE node is transferred to the SELECT node and is immediately put into service. Direct transfers of this type are shown in SLAM II through the use of dashed lines. An illustration of a QUEUE node-SELECT node combination is shown below.

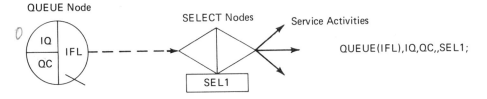

Multiple SELECT nodes may be associated with a QUEUE node. When this occurs, the order in which the SELECT nodes are listed on the QUEUE statement specifies the order in which SELECT nodes are interrogated to find an available server. Multiple servers and a selection rule would be given for each SELECT node. Additional information on SELECT nodes is given in Chapter 7.

The following are examples of QUEUE nodes:

1. Cause arriving entities to wait in file 7 if the number of waiting entities is less than 6 and the following server is busy.

The default values are: IQ = 0 implying no initial entities are at the QUEUE node; no balking node is specified, implying that entities arriving to the queue when it is full are lost to the system; no SELECT nodes are specified, implying a single type of service activity is represented by the branch that follows the QUEUE node.

2. Cause arriving entities to be blocked if two entities are waiting in the QUEUE node called QUE1. Entities at QUE1 wait in file 3. Two servers can process entities arriving to the QUE1. Initially there is one entity waiting at QUE1.

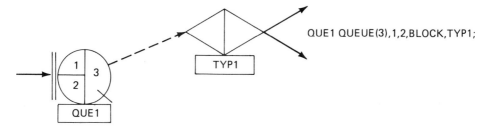

QUE1 QUEUE(3),1,2,BLOCK,TYP1;

In this example, the branch incident to QUE1 represents a service activity that will be blocked if it finishes processing an entity when two entities are already waiting at QUE1 (in file 3). The service activities that process entities waiting in QUE1 are represented by the branches following SELECT node TYP1.

5.7   TERMINATE NODE

The TERMINATE node is used to destroy or delete entities from the network. It can be used to specify the number of entities to be processed on a simulation run. This number of entities is referred to as the termination count or TC value. When multiple TERMINATE nodes are employed, the first termination count reached ends the simulation run. If a TERMINATE node does not have a termination count, the entity is destroyed and no further action is taken. The symbol and statement for the TERMINATE node are shown below.

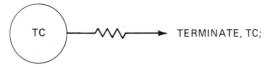

TERMINATE, TC;

As an example, consider the ending of a simulation run after 25 entities arrive to node HELP. This would be modeled as shown below.

HELP TERMINATE, 25;

Every entity routed to a TERMINATE node is destroyed. To facilitate the graphic modeling, the TERMINATE operation can also be displayed as follows:

This symbol allows an entity to be destroyed after it is processed at any node of the network. A separate TERMINATE statement would still be used but would follow the node statement directly.

5.8   ASSIGN NODE

The ASSIGN node is used to prescribe values to the attributes of an entity passing through the ASSIGN node or to prescribe values to system variables that pertain to the network in general. The ASSIGN node can also change values pertaining to discrete and continuous models that can be a part of SLAM II. The variables to which assignments can be made are: ATRIB(I), II, DD(I), SS(I), XX(I), and ARRAY(I,J). In addition, a special assignment can be made to the variable STOPA to cause the completion of an activity. This feature is described further in Section 5.9.1.

A value assigned to a variable can be used as an activity duration, as a routing condition, and in program inserts. In combined simulations, assignments can be used to change the values of variables that are part of the discrete or continuous model. This latter concept will be described in more detail in later chapters where combined discrete and continuous modeling concepts are presented.

The values assigned to variables at an ASSIGN node can take on a wide variety of forms. The value can be a constant, one of the variables described above, a network status variable, the current time, a sample from a probability distribution, or a value set in a user-written function (a program insert). Table 5-2 presented the names of the variables and functions that can be used to obtain values for inclusion in the calculations within the ASSIGN node.

The symbol and statement for the ASSIGN node are shown below.

Basically each line in the ASSIGN node can be considered as a FORTRAN replacement statement with a left-hand side variable being restricted to those presented in Table 5-1 with the exception that assignments to the clock time TNOW are not allowed. The right-hand side expression can involve up to ten arithmetic operations using the constants and variables defined in Tables 5-1 and 5-2. In evaluating the right-hand side expression, multiplication and division are evaluated first, then addition and subtraction. The expression is evaluated from left to right. Parentheses are allowed only to denote subscripts. Note that ASSIGN, XX(3)=5./ 10.*2.; sets XX(3) to 1.0.

The parameters used in the probability functions are those described in the equations presented in Section 5.3.2. Each parameter for a distribution can be specified as either a constant, ATRIB (I) or XX(I). In network assignments, the stream number, IS, can be omitted and a default stream number† is used. An example of the ASSIGN node is shown below.

$$
\begin{array}{|l|}
\hline
\text{ATRIB(2) = 7.0} \\
\hline
\text{ATRIB(3) = ATRIB(3)/XX(2)} \\
\hline
\text{XX(1) = RNORM(4.,2.)} \\
\hline
\end{array}
\quad 1
$$

ASSIGN,ATRIB(2)=7.0,ATRIB(3)=ATRIB(3)/XX(2),
XX(1)=RNORM(4.,2.),1;

In this example, the value of attribute 2 of the entity passing through the node is changed to seven. The value of attribute 3 is replaced by its current value divided by XX(2). The third assignment specifies that XX(1) be given a value that is a sample from a normal distribution whose mean is 4 and whose standard deviation is 2. The value of M is set to 1 to indicate that the entity is to be routed through only one branch. Since the statement is long, it can be divided and placed on separate lines as long as a comma ends the preceding line (the details associated with statement continuation and input procedures are described in Chapter 8). For example, the statement could have been written as

ASSIGN,ATRIB(2) = 7.0,
       ATRIB(3) = ATRIB(3)/XX(2),
       XX(1) = RNORM(4.,2.), 1;

To allow indirect addressing of subscripts or arguments, the variable II is made available. Illustrations of its uses are shown below.

    ASSIGN, II = ATRIB(2), ATRIB(3) = XX(II);
    ASSIGN, II = UNFRM(0.,10.), ATRIB(4) = II;

---

† In SLAM II, there are NNSTR stream numbers available. Typically NNSTR is set to 10. The default stream number is NNSTR−1. Stream NNSTR is used for generating random numbers for probabilistic branching.

In the first illustration, II is set equal to attribute 2 and attribute 3 is set equal to the global variable XX with an index that is obtained from attribute 2.

In the second example, II is set equal to a sample from a uniform distribution in the range (0,10). Since II is integer, the real values will be truncated, and II will take on the values 0 through 9 with an equal probability of 0.10. ATRIB(4) can then be set equal to II as a sample from a discrete uniform distribution.

As another example of the flexibility of the ASSIGN node, consider the setting of XX(1) equal to a sample from an exponential distribution whose mean is taken as ATRIB(1) using stream 2. This is accomplished with the following statement:

ASSIGN,XX(1) = EXPON(ATRIB(1),2);

The variables to which attributes are assigned can be used for many purposes. The primary uses involve the routing of entities and the duration of activities based on assigned values. Specific examples of the use of assignments will be deferred until these concepts are presented.

In summary, attribute I of the entity passing through the ASSIGN node is changed if any of the assignments involve the variable ATRIB(I) on the left-hand side of the replacement statement. Global system variables are not associated with entities but are changed by the passage of an entity through an ASSIGN node. A global variable retains its value until another entity passes an ASSIGN node at which the global variable is recomputed. Following the assignments, the entity arriving to the ASSIGN node is routed in accordance with the M-number prescribed for the node.

5.9   ACTIVITIES

Branches are used to model activities. Only at branches are explicit time delays prescribed for entities flowing through the network. Activities emanating from QUEUE or SELECT nodes are referred to as service activities. Service activities restrict the number of concurrent entities flowing through them to be equal to the number of servers represented by the activity. Activities represented by branches emanating from other node types have no restriction on the number of entities that can simultaneously flow through them. The duration of an activity is the time delay that an entity encounters as it flows through the branch representing the activity.

Each branch has a start node and an end node. When an entity is to be routed from the start node, the branch may be selected as one through which the entity should be routed. The selection can be probabilistic in which case a probability is part of the activity description. The selection can be conditional in which case a condition is specified as part of the activity description. Service activities cannot have prescribed conditions, as their availability is limited and they must be allocated when free. If no probability or condition is specified (a common situation), the activity will be selected unless the M-number associated with its start node has been satisfied.

Activities can be given activity numbers. If the number I is prescribed for an activity then statistics are maintained and reported on the number of entities that are currently being processed through the activity, NNACT(I), and the number of entities that have completed the activity, NNCNT(I). A range of numbers can be prescribed, for example, ATRIB(I), I=3,5.

For service activities, the number of parallel identical servers represented by the activity needs to be specified if different from one. (For non-service activities, the number of parallel processes is assumed as infinite.) For service activities, SLAM II automatically provides utilization statistics.

The symbol for a branch representing an activity is shown below.

DUR, PROB or COND

ACTIVITY (N)/A, DUR, PROB or COND, NLBL; ID

where

    N is the number of parallel servers if the activity represents a set of identical servers;

    A is an activity number (an integer or a range of integers);

    DUR is the duration specified for the activity;

    PROB is a probability specification for selecting the activity;

    COND is a condition for selecting the activity if the activity is a non-server;

    NLBL is the end node label and is only required if the end node is not the next node listed; and

    ID is an activity identification that is part of the comment section of the statement. ID is printed on the SLAM II summary report to provide a textual description of the activity.

5.9.1   Activity Durations

Activity durations (DUR) can be specified by any expression containing the variables described in Tables 5-1 and 5-2. Thus, a duration can be assigned a value in the same way as an attribute or system variable is assigned a value. For example, a duration can be taken as the value of attribute 3 by specifying DUR to be ATRIB(3) or as a sample from an exponential distribution whose mean is ATRIB(3), that is, EXPON(ATRIB(3)). If a sample from a probability distribution is negative, and the sample is used for an activity duration, SLAM II assumes a zero value for the activity's duration.

The duration can be made to depend on the release time of a node of the network by specifying that the activity continue until the next release of the node.† This is accomplished using the REL(NLBL) specification. When the duration is specified in this manner, the activity will continue in operation, holding the entity being processed until the next release of the node with the label NLBL. The REL specification corresponds to an activity scanning orientation as described in Chapter 3.

The duration can also be made to depend on an assignment made at an ASSIGN node. This is accomplished using the STOPA(NTC) specification where NTC is an integer code to distinguish entities in such activities. The value of NTC can be specified as a number, a SLAM II variable or a SLAM II random variable. The value of NTC will be truncated to the nearest integer. The duration of an activity specified by STOPA will continue in operation holding the entity being processed until an assignment is made in which STOPA is set equal to NTC. Thus, STOPA is similar to a *wait until* specification. For example, the activity statement

ACT,STOPA(1);

will cause each entity to wait in the activity until STOPA is set to 1 at an ASSIGN node. Every entity in this activity or other STOPA specified activities with an NTC value of 1 will be allowed to proceed when the assignment is made. By specifying the NTC code as an attribute of an entity or as a random variable, a modeler can assign different entity codes to entities within the same activity or for

---

† By release of a node is meant the act of an entity arriving at the node and the attempt to route it from the node. In Chapter 7, the ACCUMULATE node is described where entities arrive but may not release (pass through) the node.

different activities. One use of the STOPA specification is as a switch for entity types. Let ATRIB(1) be defined as an entity type code. Then

ACT,STOPA(ATRIB(1));

will require an assignment of STOPA = 1 to release type 1 entities from the activity, an assignment of STOPA = 2 to release type 2 entities, and so on. In Chapter 9, additional information on the use of STOPA is given.

5.9.2   Probability Specification for Branches

A probability is specified for a branch as a real value, SLAM II variable, or SLAM II random variable with a value between 0.0 and 1.0. Arithmetic calculations are allowed in the probability specification. The sum of the probabilities of those branches with probabilities emanating from the same node need not be one. Probabilities may be assigned to branches emanating from QUEUE nodes. In this case, the activities emanating from the QUEUE node are assumed to be the same server(s) and the probabilities can be used to obtain different duration specifications or different routings for entities processed by the same server(s).

Examples of probability specifications are: 0.7,XX(3), 1−XX(3), ATRIB(2), and NNQ(2)/100.

5.9.3   Condition Specification for Branches

A condition specification is only allowed for regular activities. Service activities must be initiated if an entity is waiting. Conditions are prescribed in the form:

VALUE. OPERATOR. VALUE.

VALUE can be a constant, a SLAM II variable, or a SLAM II random variable (see Tables 5-1 and 5-2). OPERATOR is one of the standard FORTRAN relational codes defined below.

Relational Code	Definition
LT	Less than
LE	Less than or equal to
EQ	Equal to
NE	Not equal to
GT	Greater than
GE	Greater than or equal to

Examples of condition codes are:

Condition	*Take branch if*
TNOW. GE. 100.0	Current time greater than or equal to 100.
ATRIB(1).LT.DRAND(2)	Attribute 1 of the current entity is less than a random number obtained from stream 2.
NNQ(7).EQ.10	The number of entities in file 7 is equal to 10.

The union and intersection of two or more conditions can be prescribed for an activity using .AND. and .OR. specifications. Thus, a possible conditional expression for a branch could be

TNOW.GE.100.0 .AND. ATRIB(2).LT.5.0

An example of the use of the union of two conditions is

NNQ(7).EQ.10 .OR. II.NE.4

If more than two conditions are combined using the .AND. and .OR. specifications, then the .AND. conditions are tested prior to the .OR. conditions as is done in FORTRAN. Complicated logic testing requiring parentheses is not permitted directly on the network. Such advanced logic is coded in the user-written function USERF, for example, USERF(1).GE.0.

5.10   ILLUSTRATIONS

In this section, we present illustrations that combine the node and activity concepts into networks. Both network and statement models are presented.

5.10.1   Illustration 5-1.   Two Parallel Servers

Consider a situation involving the processing of customers at a bank with two tellers and a single waiting line. A network that models this situation is shown below.

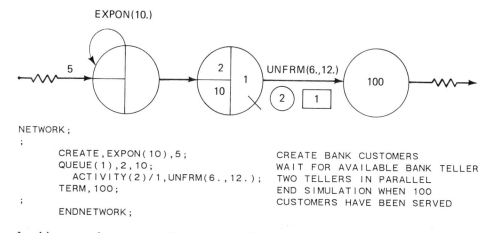

```
NETWORK;
;
        CREATE,EXPON(10),5;              CREATE BANK CUSTOMERS
        QUEUE(1),2,10;                   WAIT FOR AVAILABLE BANK TELLER
          ACTIVITY(2)/1,UNFRM(6.,12.);   TWO TELLERS IN PARALLEL
        TERM,100;                        END SIMULATION WHEN 100
;                                        CUSTOMERS HAVE BEEN SERVED
        ENDNETWORK;
```

In this example, two parallel servers (tellers) are associated with activity 1. The service time of each server is uniformly distributed between 6 and 12 time units. Entities (customers) that arrive at the QUEUE node when both servers are busy wait in file 1. Initially, there are two entities in the queue which causes both servers to be busy initially. Thus, there are four in the system initially; two in service and two waiting at the QUEUE node. A capacity of ten entities has been assigned to the queue.

The TERMINATE node indicates that the model is to be analyzed until 100 entities have completed processing. The time between arrivals is prescribed at the CREATE node as samples from an exponential distribution with a mean of 10. The first entity is scheduled to arrive at time 5.

5.10.2   Illustration 5-2.   Two Types of Entities

Consider a situation involving two types of jobs that require processing by the same server. The job types are assumed to form a single queue before the server. The network and statement model of this situation is shown below.

In this model, one type of entity is scheduled to arrive every 8 time units and only 100 of them are to be created. These entities have a service time estimated to be a sample from an exponential distribution with a mean time of 7. This service time is assigned to attribute 1 at an ASSIGN node. For the other type of entity the time between arrivals is 12 time units and 50 of these entities are to be created. The estimated service time for each of these entities is exponentially distributed with a mean time of 10. Both types of entities are routed to a QUEUE node

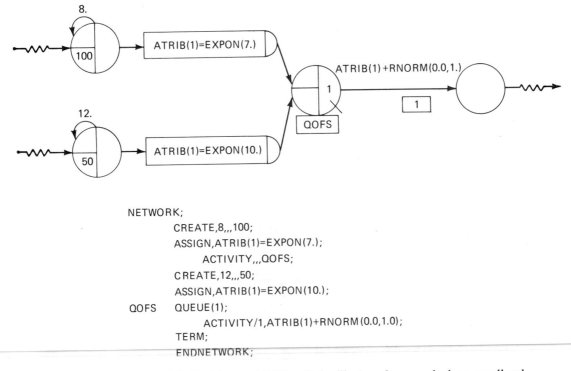

```
NETWORK;
        CREATE,8,,,100;
        ASSIGN,ATRIB(1)=EXPON(7.);
            ACTIVITY,,,QOFS;
        CREATE,12,,,50;
        ASSIGN,ATRIB(1)=EXPON(10.);
QOFS    QUEUE(1);
            ACTIVITY/1,ATRIB(1)+RNORM(0.0,1.0);
        TERM;
        ENDNETWORK;
```

whose label is QOFS. Entities at QOFS wait in file 1 and are ranked on small values of attribute one. This priority specification is made through a PRIORITY statement that will be described in Section 5.13. The server is modeled as activity 1 where the service time is specified as attribute 1 plus a sample from a normal distribution. Thus, the actual processing time is equal to the estimated processing time plus an error term that is assumed to be normally distributed. This model might be used to represent a job shop in which jobs are performed in the order of the smallest estimated service time.

Many default values were assumed in the above statement model. Both entity types have their first arrival at time zero and neither are marked. The M-number for all the CREATE and ASSIGN nodes are defaulted which causes an infinite M-number to be used, that is, take all branches. The default value for the initial number in the QUEUE node is zero and the capacity of the QUEUE node is assumed to be infinite. No specification for the termination count is made at the TERMINATE node and, hence, the run is completed when all entities created have passed through the system, which in this case is 150. (Note: this termination condition cannot be determined absolutely from the model description, as a completion time for the network could have been prescribed. The procedure for specifying a completion time is discussed in Section 5.13.)

To illustrate the use of the EQUIVALENCE statement to make the model more readable, the statement model is redone below.

```
EQUIVALENCE/ATRIB(1),ESERVET/
                    RNORM(0.0,1.0),NOISE;
;
NETWORK;
;
;   CREATE AND ASSIGN ESTIMATES OF PROCESSING TIME
;
        CREATE,8,,,100;
        ASSIGN,ESERVT=EXPON(7);
          ACTIVITY,,,QOFS;              SEND JOB TO PROCESSING QUEUE
        CREATE,12,,,50;
        ASSIGN,ESERVT=EXPON(10.);
QOFS    QUEUE(1);                        PROCESSING QUEUE
;
;   ACTIVITY TIME IS ESTIMATE + ERROR IN ESTIMATE
;
          ACTIVITY/1,ESERVT+NOISE;
        TERM;
        ENDNETWORK;
```

The estimated service time is defined as ESERVET and equivalenced to ATRIB(1). The error of the estimate of the service time is defined as NOISE and equivalenced to RNORM(0.0,1.0). As can be seen, the use of the EQUIVALENCE statement improves the readability of a SLAM II statement model.

5.10.3   Illustration 5-3.   Blocking and Balking from QUEUE Nodes

Consider a company with a maintenance shop that involves two operations in series. When maintenance is required on a machine and four machines are waiting for operation 1, the maintenance operations are subcontracted to an external vendor. This situation is modeled below.

In this model, entities are created every two time units and routed directly to a QUEUE node that has a capacity of four. At this node, entities are stored in file 1 and, if an entity arrives when there are four other entities in file 1, it balks to the TERMINATE node SUBC. When ten entities balk to node SUBC, the run is to be ended. The service time for activity 1 is triangularly distributed with a mode of 0.4 and minimum and maximum values of 0.2 and 0.8, respectively. When service activity 1 is completed, entities are routed directly to a second QUEUE node. File 2 is used to store waiting entities for server 2. If two entities are already waiting

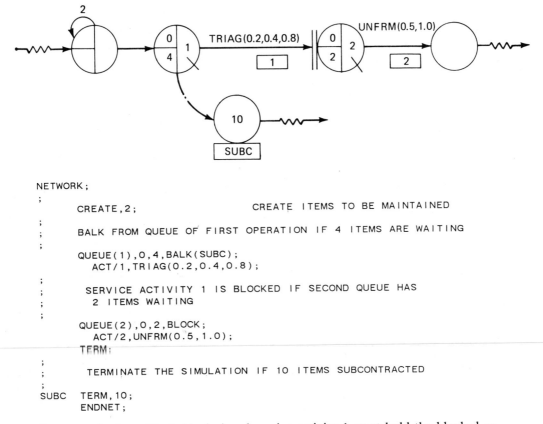

```
NETWORK;
;
        CREATE,2;                    CREATE ITEMS TO BE MAINTAINED
;
;       BALK FROM QUEUE OF FIRST OPERATION IF 4 ITEMS ARE WAITING
;
        QUEUE(1),0,4,BALK(SUBC);
          ACT/1,TRIAG(0.2,0.4,0.8);
;
;        SERVICE ACTIVITY 1 IS BLOCKED IF SECOND QUEUE HAS
;        2 ITEMS WAITING
;
        QUEUE(2),0,2,BLOCK;
          ACT/2,UNFRM(0.5,1.0);
        TERM;
;
;       TERMINATE THE SIMULATION IF 10 ITEMS SUBCONTRACTED
;
SUBC    TERM,10;
        ENDNET;
```

for server 2, the entity is blocked and service activity 1 must hold the blocked entity. No further service activities can be started for server 1 even though entities are waiting in file 1. When the number of entities in file 2 decreases below 2, the blocked entity is routed to file 2 and another service activity for server 1 can be started. The processing time for server 2 is uniformly distributed between 0.5 and 1. The example to be presented in Section 5.14 is similar to this illustration and provides the inputs to and outputs from the SLAM II processor.

5.10.4   Illustration 5-4.   Conditional and Probabilistic Branching

Consider a situation involving an inspector and an adjustor. Presume that seventy percent of the items inspected are routed directly to packing and thirty percent of the items require adjustment. Following adjustment, the items are returned for reinspection. We will let the inspection time be a function of the number of items waiting for inspection (NNQ(1)) and the number waiting for adjustment (NNQ(2)). The model corresponding to this description is shown below.

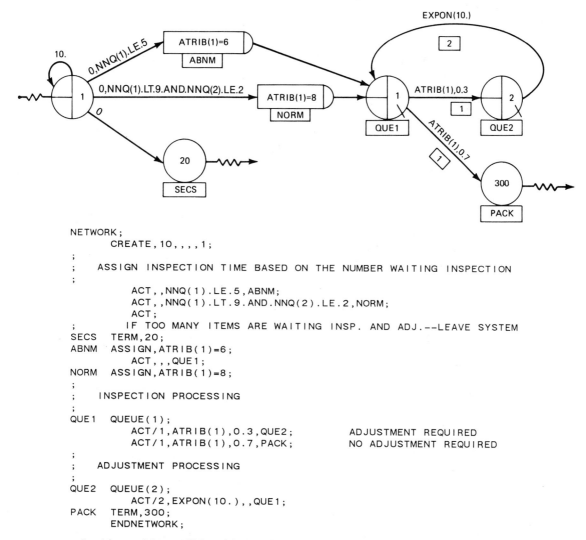

```
NETWORK;
        CREATE,10,,,,1;
;
;     ASSIGN INSPECTION TIME BASED ON THE NUMBER WAITING INSPECTION
;
            ACT,,NNQ(1).LE.5,ABNM;
            ACT,,NNQ(1).LT.9.AND.NNQ(2).LE.2,NORM;
            ACT;
;           IF TOO MANY ITEMS ARE WAITING INSP. AND ADJ.--LEAVE SYSTEM
SECS  TERM,20;
ABNM  ASSIGN,ATRIB(1)=6;
            ACT,,,QUE1;
NORM  ASSIGN,ATRIB(1)=8;
;
;     INSPECTION PROCESSING
;
QUE1  QUEUE(1);
            ACT/1,ATRIB(1),0.3,QUE2;        ADJUSTMENT REQUIRED
            ACT/1,ATRIB(1),0.7,PACK;        NO ADJUSTMENT REQUIRED
;
;     ADJUSTMENT PROCESSING
;
QUE2  QUEUE(2);
            ACT/2,EXPON(10.),,QUE1;
PACK  TERM,300;
            ENDNETWORK;
```

In this model, conditional branching is specified from the CREATE node where the M-number is one, that is, a maximum of one of the three branches emanating from the CREATE node is to be taken. The entity is routed to node ABNM, if the number of entities in file 1 is less than or equal to 5. At ASSIGN node ABNM,

ATRIB(1) is set equal to 6 time units. ATRIB(1) will be used in this illustration to represent the processing time for an entity. The branch from the CREATE node to the ASSIGN node NORM is taken if the number of entities in file 1 is less than 9 and the number of entities in file 2 is less than or equal to 2. When this occurs, a processing time of 8 time units is stored in ATRIB(1) at the ASSIGN node NORM.

No condition is specified on the branch from the CREATE node to the TERM node SECS. Since the conditions are evaluated in the order prescribed by the statement model, this branch will only be taken if the preceding two branches are not taken. This branch represents a non-processing of an entity through the server process. If 20 such occurrences materialize, the run is to be completed.

QUE1 is the QUEUE node for server 1. With probability 0.3, the entity is routed to QUE2. With probability 0.7, it is routed to the TERM node PACK. Both of these activities represent service activity 1. The service time is set equal to attribute 1 previously defined at the ASSIGN nodes. AT QUE2, the entity goes through a second service activity whose service time is exponentially distributed with a mean time of 10. Entities are then routed back to QUE1 for additional processing by server 1. At TERMINATE node PACK, a requirement of 300 entity arrivals is indicated in order to complete one run of the network. Thus, a run can be terminated by an entity arrival to either node PACK or node SECS.

5.10.5   Illustration 5-5.   Service Time Dependent on Node Release

Consider an assembly line that is paced so that units can only be completed at the end of ten-minute intervals. This situation is modeled below.

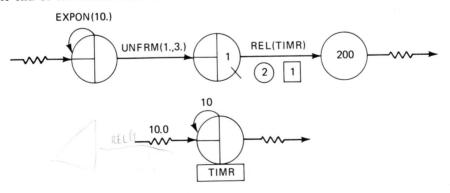

```
NETWORK;
;
;   SUBNETWORK FOR ARRIVAL AND PROCESSING OF PARTS
;
      CREATE,EXPON(10);
          ACT,UNFRM(1.,3.);
      QUEUE(1);
              PROCESSING TIME BASED ON RELEASE OF PACING TIMER
          ACT(2)/1,REL(TIMR);
      TERM,200;
;
;
;   SUBNETWORK FOR PACING SERVER IN ACTIVITY 1
;
TIMR   CREATE,10.,10;
      TERM;
      ENDNET;
```

This model depicts two identical servers with a single queue and is similar to the model presented in Illustration 5-1. However, for this model the service time duration is specified as the next release time of CREATE node TIMR. CREATE node TIMR is released for the first time at time 10 and every 10 time units thereafter. This specifies that the duration for service activity 1 will end at a multiple of 10. If service activity 1 starts at time 13 then the end time will be 20. Thus, the service duration will be 7. If an entity is put into service at time 49, it will complete service at time 50 and its duration will be 1. Other aspects of this illustration were described previously.

5.10.6   Illustration 5-6.   Attribute Specified File and Service Activity Numbers

In this illustration, the use of an attribute to specify a file number and a service activity number will be demonstrated. The situation to be modeled consists of entities that arrive to a system every five time units. Each arriving entity is to be processed sequentially by servers 1,2, and 3. Entities waiting for server I are stored in file I with I equal to 1, 2, or 3. This situation will be modeled by a single QUEUE node-service activity combination where the file number and service activity number are specified by the value of attribute 2 (ATRIB(2)). The value of ATRIB(2) will be set initially to 1 and then indexed each time a service is performed on the entity. When the value of ATRIB(2).EQ.3, the entity will be routed to a TERM node because processing on the entity is completed. The service time for each server is assumed to be exponentially distributed with a different mean service time given by $XX(II), II = 1,2,3$. It will be assumed that these values of XX are set through initial conditions or in a disjoint network not shown in this illustration. The network model and statements are shown below.

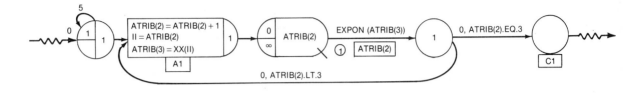

```
NETWORK;
          CREATE,5,0,1,,1;
     ;
     ;ATRIB(2) = NEXT SERVER STATION,ATRIB(3) = MEAN SERVICE TIME
     ;
     ;A1      ASSIGN,ATRIB(2) = ATRIB(2) + 1,II = ATRIB(2),ATRIB(3) = XX(II);
     ;
     ;SET FILE NUMBER = SERVER STATION NO.
     ;
          QUEUE(ATRIB(2) = 1,3);
               ACT(1)/ATRIB(2) = 1,3,EXPON(ATRIB(3));
          GOON,1;
               ACT,0,ATRIB(2).LT.3,A1; GO TO NEXT SERVER STATION
               ACT,0,ATRIB(2).EQ.3,C1; COMPLETED THREE SERVICES
     C1     TERM:
          ENDNETWORK;
```

In the network model, the file number and service activity number are specified as ATRIB(2). For convenience, the range of values for attributes are typically not shown on the network. In the statement model, these values must be included.

Basically the above model consists of a single queue-service activity combination with the use of an ASSIGN node and conditional branching to allow for different definitions of the file and server numbers. For this illustration, it may be just as easy to model the situation with three queue-server combinations in series. However, for a larger number of serial operations, the above procedure for indirectly specifying the file number and server number can condense the size of a network significantly. Although this illustration has entities flowing through servers in a prescribed order, it is easy to modify the network so that the value of ATRIB(2) is changed in accordance with a routing structure defined for each entity type, that is, an attribute could be used to define an entity type and then, based on the entity type, different server numbers can be prescribed for ATRIB(2).

5.11   GOON NODE

The GO ON or GOON node is included as a continue type node. The symbol and statement for the GOON node are:

The GOON node is used in the modeling of sequential activities since two consecutive activity statements are used to model parallel activities. This should be clear from the following two network segments.

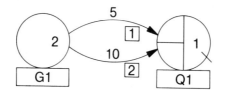

```
; MODEL OF 2 ACTIVITIES IN PARALLEL
  G1 GOON,2;
       ACT/1,5,,Q1;
       ACT/2,10,,Q1;
  Q1 QUEUE(1);
```

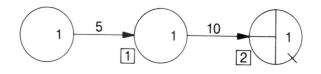

```
; MODEL OF 2 ACTIVITIES IN SERIES
     GOON,1;
        ACT/1,5;
     GOON,1;
        ACT/2,10;
     QUEUE(1);
```

Note that for each entity arriving to GOON node G1, two entities are routed to QUEUE node Q1. The attributes of the two entities are the same as the attributes of the entity arriving to node G1. For the model of two activities in series, each arriving entity is routed over activity 1 first, then activity 2.

5.12   COLCT NODE

Statistics can be collected on five types of variables at a COLCT node. Four of the variables refer to the time or times at which an entity arrives at the COLCT node. The fifth statistic type allows for the collection of SLAM II status variables at specified nodes. The five types of variables are:

1. Time-of-first arrival (FIRST). Only one value is recorded during each run.
2. Time-of-all arrivals (ALL). Every time an entity arrives to the node, its arrival time is added to all previous arrival times. At the end of a run, an average time of arrival is computed.
3. Time between arrivals (BETWEEN). The time of first arrival is used as a reference point. On subsequent arrivals, the time between arrivals is collected as an observation of the statistic of interest.
4. Interval statistics (INT(NATR)). This statistic relates to the arrival time of an entity minus an attribute value of the entity, that is,
   TNOW − ATRIB(NATR).
5. SLAM II variable. The value of a SLAM II variable is recorded as an observation every time an entity arrives to the node. Note that USERF(I) is considered as a SLAM II variable.

For each of the above five types of variables, estimates for the mean and standard deviation of the variable are obtained. In addition, a histogram of the values collected at a COLCT node can be obtained. This is accomplished by specifying on input the number of cells, NCEL; the upper limit of the first cell, HLOW; and a cell width, HWID, for the histogram.

The number of cells specified, NCEL, is the number of interior cells, each of which will have a width of HWID. Two additional cells will be added that contain the interval $(-\infty, \text{HLOW}]$ and $(\text{HLOW} + \text{NCEL}*\text{HWID}, \infty)$. The cells are closed at the high value. Thus, if HLOW = 0, the value 0 will be included in the first cell. For the specification: NCEL = 5; HLOW = 0; and HWID = 10, the number of times the variable on which statistics are being maintained is in the following intervals would be presented as part of the standard SLAM II summary report:

$$(-\infty,0], \ (0,10], \ (10,20], \ (20,30], \ (30,40],, \ (40,50], \ \text{and} \ (50,\infty).$$

If the number of cells is not specified, no histogram will be prepared. A 16 character identifier can be associated with a COLCT node. This identifier, denoted ID, will be printed on the SLAM II summary report to identify the output associated with the COLCT node.

The symbol and statement for the COLCT node are: †

TYPE | ID,H | M   COLCT,TYPE,ID,NCEL/HLOW/HWID,M;

Since histograms are not always requested, a single field identified as 'H' is used on the symbol where it is implied that H represents NCEL/HLOW/HWID. Examples of the COLCT node are given below.

1. Collect statistics on the completion time of a project with no histogram required.

   FIRST | PROJECT COMP   COLCT,FIRST,PROJECT COMP;

2. Collect statistics on the time in the system of an entity whose time of entering the system is maintained in ATRIB(3). No histogram is requested for this COLCT node and the entity is to be routed over one branch.

   INT(3) | SYSTEM TIME | 1   COLCT,INT(3),SYSTEM TIME,,1;

3. Collect statistics on the value of global variable, XX(2), every time an entity passes through the COLCT node. Identify the statistical output with the identifier SAFETY STOCK. Prepare a histogram consisting of 20 cells for which the first cell interval is $(-\infty,10]$ and the width of the next 20 cells is 5. Set the maximum number of branches to be taken to 2.

---

† In order to identify values stored by a particular COLCT node in user written subprograms, an integer index code can be defined in the COLCT statement in the location specified by the letter N below:

COLCT(N),TYPE,ID,NCEL/HLOW/HWID,M;

In this illustration, XX(2) can be considered as an inventory level and the entity arriving to the COLCT node represents a replenishment of the units in inventory.

5.13   INTRODUCTION TO EXAMPLES AND CONTROL STATEMENTS

Throughout this book, a standard format is employed in the presentation of examples. First, the problem statement is presented which describes the system to be modeled including the objective of the analysis. A Concepts Illustrated section is then given which describes the major concepts which the example is intended to illustrate. The SLAM II Model is then presented. Lastly, a Summary of Results is presented including a reproduction of the SLAM II Summary Report where appropriate.

The final step in developing a network simulation is to combine the network description statements with the necessary control statements. Control statements provide information about the simulation experiment to be performed. At a minimum, the control statements GENERAL, LIMITS, and FINISH must be included. Other control statements such as PRIORITY, MONTR, and INITIALIZE are included as required to specify the operation of the simulation runs. The TIMST statement is used to collect statistics on time persistent variables, and the ENTRY statement causes entities to be inserted into files.

In this section, the format of these control statements will be presented in abbreviated form. In Chapter 8 and Appendix C, a complete discussion of the general format conventions for network and control statements is given along with a complete description of each field on a control statement. The abbreviated form for the GENERAL or GEN statement is shown below.

GEN,NAME,PROJECT,MONTH/DAY/YEAR,NNRNS;

The fields on the GEN statement are for the name of the modeler, the project title, and the date of the run. The value specified for NNRNS is the number of runs to be made.

The format of the LIMITS or LIM statement is shown below:

LIM,MFIL,MATR,MNTRY;

where MFIL is the largest file number used, MATR is the largest number of at-tributes prescribed for an entity, and MNTRY is an estimate of the maximum number of concurrent entities in all files. MNTRY is typically estimated from pilot runs.

The format for the INTLC statement is

INTLC,VAR = value,repeats;

INTLC is used to assign initial values to SLAM II variables.

The abbreviated format for the INITIALIZE or INIT statement is shown be-low:

INIT,TTBEG,TTFIN,JJCLR;

where TTBEG is the beginning time for the simulation, TTFIN is the desired end-ing time for the simulation, and JJCLR is used to specify if statistics are to be maintained separately for each run.

The abbreviated format for the TIMST statement is:

TIMST,VAR,ID;

where VAR is a SLAM II variable whose value persists for a time duration, for ex-ample, XX(1). ID is an alphanumeric identifier printed in the SLAM II summary report to identify the summary statistics calculated for VAR.

The format for the ENTRY statement is shown below.

ENTRY/IFILE,ATRIB(1),ATRIB(2), . . . ,ATRIB(MATR)/repeats;

The ENTRY statement is used to place initial entries into files. An entry is speci-fied by entering the file number, IFILE, followed by the attributes of the entry separated by commas. The slash is used to denote the beginning of another speci-fication of a new entry.

The format of the PRIORITY statement is shown below:

PRIORITY/IFILE,ranking/repeats;

where IFILE is the file for which the ranking priority is being specified, and ranking is the priority specification. The options for the ranking are: FIFO, first-in, first-out; LIFO, last-in, first-out; HVF(N), entities with a higher value of attribute N are given priority, that is, a high-value-first priority; and LVF(N), entities with a lower value of attribute N are given priority, that is, a low-value-first priority.

The format for the MONTR statement is shown below:

MONTR,option,TFRST,TSEC,variables;

For the present, only the TRACE and CLEAR options are considered. TRACE specifies that a list of events is to be printed starting at the time TFRST and ending at time TSEC. Variables is a list of SLAM II variables whose value is to be displayed at each event time. The CLEAR option causes statistics to be discarded at time TFRST. CLEAR is used to eliminate statistics collected during a transient period.

The FIN statement consists of a single field as shown below

FIN;

and denotes the end of all SLAM II input statements. The SIMULATE or SIM statement consists of a single field as shown below

SIM;

and denotes the end of the SLAM II input statements for one run. A SIM statement is not required for a run if a FIN statement is the last statement.

5.14   EXAMPLE 5-1.   WORK STATIONS IN SERIES

The maintenance facility of a large manufacturer performs two operations. These operations must be performed in series; operation 2 always follows operation 1. The units that are maintained are bulky, and space is available for only eight units including the units being worked on. A proposed design leaves space for two units between the work stations, and space for four units before work station 1. The proposed design is illustrated in Figure 5-1. Current company policy is to subcontract the maintenance of a unit if it cannot gain access to the in-house facility (1,2).

Historical data indicates that the time interval between requests for maintenance is exponentially distributed with a mean of 0.4 time units. Service times are also exponentially distributed with the first station requiring on the average 0.25 time units and the second station, 0.5 time units. Units are transported automatically from work station 1 to work station 2 in a negligible amount of time. If the queue of work station 2 is full, that is, if there are two units waiting for work station 2, the first station is blocked and a unit cannot leave the station. A blocked work station cannot serve other units.

To evaluate the proposed design, statistics on the following variables are to be obtained over a period of 300 time units:

1. work station utilization;
2. time to process a unit through the two work stations;
3. number of units/time unit that are subcontracted;
4. number of units waiting for each work station; and
5. fraction of time that work station 1 is blocked.

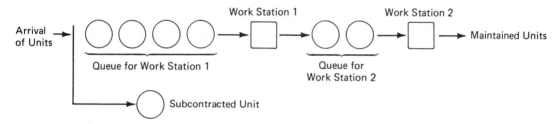

Figure 5-1 Schematic diagram of work stations in series.

Concepts Illustrated. This example will illustrate the general SLAM II network modeling procedure which consists of: (1) identifying the entities to be modeled; (2) constructing a graphical model of the entity flow process through the system; and (3) transcribing the graphical model into the SLAM II statement representation of the system. Specific network modeling concepts illustrated by this example include the creation of entities by a CREATE node, the modeling of a service system with a QUEUE node and ACTIVITY, balking and blocking at QUEUE nodes, and statistics collection at a COLCT node.

SLAM II Model. The maintenance facility described in this example is representative of a large class of queueing-type systems in which units can be represented by entities that flow through the work stations. In this example, the entity flow process can be conveniently modeled by representing the storage area preceding each work station by a QUEUE node. The QUEUE node for work station 2 will be prescribed to have a blocking capability to stop the processing of units by work station 1 when work station 1 has completed processing a unit and the queue before work station 2 is at its capacity. Each work station is represented by a service ACTIVITY with one server associated with each work station.

The SLAM II graphical model for this system is depicted in Figure 5-2. Entities representing the units are created at the CREATE node with the time between entities specified to be exponentially distributed with mean of 0.4 time units. Each entity's first attribute (ATRIB(1)) is marked with its time of creation at the CREATE node. Marking is specified to permit interval statistics to be collected on the time in the system for each entity. The entity is sent to the first QUEUE node which is used to represent the waiting area for work station 1. The parameters for this QUEUE node specify that the queue is initially empty, has a capacity of four, and that entities waiting in the queue are placed in file 1. The subcontracting for the maintenance of units is modeled by the balking option for this QUEUE node. Entities which arrive to the system when the queue is full are denied access to the queue and are routed to the COLCT node labeled SUB. The COLCT node collects values on the time between entity arrivals which corresponds to the time between the subcontracting of units. The first work station is represented by activity 1 emanating from the QUEUE node with the service time specified as exponentially distributed with a mean of 0.25 time units.

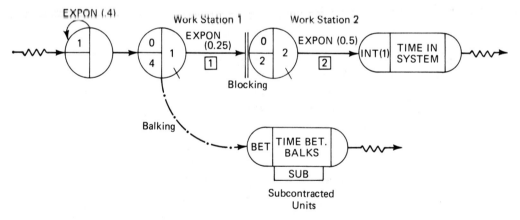

Figure 5-2 SLAM II network model of maintenance facility.

Following completion of service at work station 1, the entity attempts to enter the second QUEUE node which is used to model the storage area for work station 2. The parameters for this node specify that the QUEUE node is initially empty, has a capacity of two, and that entities waiting in the queue are placed in file 2. In addition, the blocking option is specified as indicated by the parallel lines preceding the QUEUE node. Thus, if an entity completes service activity 1 when there are two entities waiting in file 2, work station 1 will be blocked from further processing until work station 2 completes service on an entity. Work station 2 is represented by activity 2 and has a service time specified as exponentially distributed with mean of 0.5 time units.

Following completion of service by work station 2, the entity proceeds to a COLCT node where INT(1) statistics are collected on the interval of time between the time recorded in attribute 1 of the entity and the current simulated time. Since the entity was marked at the CREATE node, attribute 1 is the time of arrival of the entity to the system. Thus, the interval of time represents the total processing time of the entity in the system. Following the COLCT node, the entity is terminated.

Once the graphical representation of the system is completed, the next step in the modeling process is to transcribe the graphical representation of the system into the equivalent SLAM II statement representation according to the format specifications described in Chapter 8. The statement representation for this example is shown in Figure 5-3. Note that the sequence numbers along the left side of the figure are included to simplify statement referencing and are not part of the input data. Note also that the comments appearing on the right of the statement listing occur after the line terminator character (the semicolon) and, except for activity identifiers, are ignored by the SLAM II processor.

```
 1   GEN,PRITSKER,SERIAL WORK STATIONS,7/14/83,1;
 2   LIMITS,2,1,50;
 3   NETWORK;
 4        CREATE,EXPON(.4),,1;                      CREATE ARRIVALS
 5        QUEUE(1),0,4,BALK(SUB);                   STN1 QUEUE
 6        ACT/1,EXPON(.25);                         STN1 SVC TIM
 7        QUEUE(2),0,2,BLOCK;                       STN2 QUEUE
 8        ACT/2,EXPON(.50);                         STN2 SVC TIM
 9        COLCT,INT(1),TIME IN SYSTEM,20/0/.25;     COLLECT STATISTICS
10        TERM;
11   SUB  COLCT,BET,TIME BET. BALKS;                COLLECT STATISTICS
12        TERM;
13        END
14   INIT,0,300;
15   FIN;
```

Figure 5-3 Input statements for maintenance facility model.

The first statement in the simulation program listing has the code GEN in field 1 and is used to provide general project information including the simulation author's name, project title, date, and number of simulation runs. This statement is followed by the LIMITS statement which specifies that the maximum number of files used in the simulation is 2, the maximum number of user attributes is 1, and the maximum number of concurrent entities in the system is 50. (In this example there can be 2 entities in service, 6 in files for QUEUE nodes, 1 in creation, and 2 in the departure process. Thus, a value of 9 could have been used.) This is followed by the NETWORK statement which specifies that the simulation includes a network model with the network statement representation beginning with the next statement.

The network statement representation of the system parallels closely the graphical network representation of the system. The CREATE statement (line 4) specifies that entities are marked in ATRIB(1). The time between entity arrivals is specified to be exponentially distributed with a mean of 0.4 time units. Since the time of first creation is not specified, a zero value is used. Also, there is no limit to the number of entities generated at this node. Each entity proceeds to the QUEUE statement (line 5). The parameters for the QUEUE statement specify that entities residing in the queue waiting for service are stored in file 1, that initially there are no entities in the queue, the queue has a capacity of four, and entities arriving to the queue when the queue is at capacity balk to the node labeled SUB. The ACT statement (line 6) following the QUEUE statement is a service activity representing work station 1 with a service time exponentially distributed with a mean of 0.25 time units.

Following completion of service at work station 1, the entities continue to the second QUEUE node (line 7). The parameters for this QUEUE statement specify that entities waiting in the queue for service are stored in file 2, the queue is initially empty, has a capacity of two, and incoming entities (and service activities) are blocked when the queue is at capacity. The ACT statement (line 8) following the QUEUE statement is a service activity representing work station 2 with a service time that is exponentially distributed with a mean of 0.5 time units.

Following completion of service at work station 2, the entities arrive at the COLCT node (line 9) which causes interval statistics to be collected using the mark time in attribute 1 as a reference time, with the output statistics labeled TIME IN SYSTEM. A histogram is requested with 20 cells, with the upper limit of the first cell set equal to 0, and with a cell width equal to 0.25 units. Entities are then terminated as specified by the TERM statement (line 10).

The subcontracting of units which cannot gain access to the in-house facility is modeled by the balking of entities from the QUEUE statement (line 5) to the COLCT node labeled SUB (line 11). The options listed for the COLCT statement specify that statistics are to be collected on the time BETWEEN entity arrivals to SUB, and that the output statistics are to be labeled TIME BET. BALKS. No histogram is requested. The entities representing the subcontracted units are then destroyed by the TERM statement (line 12). The END statement (line 13) denotes an end to the network description portion of the simulation program.

The simulation is initialized by the INIT statement (line 14) which sets the beginning time of the simulation (TTBEG) to 0 and the ending time of the simulation (TTFIN) to 300 time units. The FIN statement (line 15) denotes an end to all SLAM II simulation input and causes execution of the simulation to begin.

Summary of Results. The results for the simulation are summarized by the SLAM II Summary Report depicted in Figure 5-4. The first category of statistics is for variables based upon observations. For this example, these statistics were collected by the network model at the COLCT nodes and include the interval statistics for TIME IN SYSTEM and the between statistics for TIME BET. BALKS. During the 300 simulated time units, there were a total of 586 units processed by the in-house facility. The average time in the system for these units was 2.761 time units with a standard deviation of 1.278 time units and times ranged from 0.10 to 7.191

```
              S L A M   I I   S U M M A R Y   R E P O R T


        SIMULATION PROJECT SERIAL WORK STATIONS      BY PRITSKER

        DATE  7/14/1983                    RUN NUMBER   1 OF   1


        CURRENT TIME   0.3000E+03
        STATISTICAL ARRAYS CLEARED AT TIME   0.0000E+00
```

STATISTICS FOR VARIABLES BASED ON OBSERVATION

	MEAN VALUE	STANDARD DEVIATION	COEFF. OF VARIATION	MINIMUM VALUE	MAXIMUM VALUE	NUMBER OF OBSERVATIONS
TIME IN SYSTEM	0.2761E+01	0.1278E+01	0.4629E+00	0.1002E+00	0.7191E+01	586
TIME BET. BALKS	0.1545E+01	0.3291E+01	0.2130E+01	0.1288E-01	0.2757E+02	179

FILE STATISTICS

FILE NUMBER	ASSOC NODE LABEL/TYPE	AVERAGE LENGTH	STANDARD DEVIATION	MAXIMUM LENGTH	CURRENT LENGTH	AVERAGE WAITING TIME
1	QUEUE	2.0433	1.5151	4	0	1.0390
2	QUEUE	1.5558	0.7267	2	2	0.7924
3	CALENDAR	2.4126	0.5124	4	3	0.2098

SERVICE ACTIVITY STATISTICS

ACTIVITY INDEX	START NODE OR ACTIVITY LABEL	SERVER CAPACITY	AVERAGE UTILIZATION	STANDARD DEVIATION	CURRENT UTILIZATION	AVERAGE BLOCKAGE	MAXIMUM IDLE TIME/SERVERS	MAXIMUM BUSY TIME/SERVERS	ENTITY COUNT
1	STN1 SVC TIM	1	0.4705	0.4991	1	0.4044	2.2051	4.4780	589
2	STN2 SVC TIM	1	0.9421	0.2336	1	0.0000	2.5795	69.4326	586

Figure 5-4 SLAM II summary report for serial work station model.

time units. The distribution for time in the system is depicted by the histogram generated by SLAM II and included as Figure 5-5. There was a total of 179 observations of time between balks, and therefore, there were 180 units that were subcontracted. Recall that the time of first release of the COLCT node is not included as a value for between statistics because it may not be a representative value.

```
                    **HISTOGRAM NUMBER  1**

                        TIME IN SYSTEM

 OBSV    RELA    CUML      UPPER
 FREQ    FREQ    FREQ    CELL LIMIT      0        20        40        60        80       100
                                        +    +    +    +    +    +    +    +    +    +    +
    0   0.000   0.000   0.0000E+00       +                                               +
    5   0.009   0.009   0.2500E+00       +                                               +
    4   0.007   0.015   0.5000E+00       +C                                              +
   11   0.019   0.034   0.7500E+00       +*C                                             +
   21   0.036   0.070   0.1000E+01       +**C                                            +
   30   0.051   0.121   0.1250E+01       +***   C                                        +
   31   0.053   0.174   0.1500E+01       +***      C                                     +
   38   0.065   0.239   0.1750E+01       +***         C                                  +
   33   0.056   0.295   0.2000E+01       +***           C                               +
   44   0.075   0.370   0.2250E+01       +****             C                            +
   32   0.055   0.425   0.2500E+01       +***                C                          +
   59   0.101   0.526   0.2750E+01       +*****                  C                       +
   55   0.094   0.619   0.3000E+01       +*****                       C                  +
   34   0.058   0.677   0.3250E+01       +***                            C               +
   39   0.067   0.744   0.3500E+01       +***                               C            |
   27   0.046   0.790   0.3750E+01       +**                                  C          +
   27   0.046   0.836   0.4000E+01       +**                                     C       +
   18   0.031   0.867   0.4250E+01       +**                                       C     +
   17   0.029   0.896   0.4500E+01       +*                                          C   +
   12   0.020   0.916   0.4750E+01       +*                                            C +
   16   0.027   0.944   0.5000E+01       +*                                             C+
   33   0.056   1.000     INF            +***                                            C
   ---                                   +    +    +    +    +    +    +    +    +    +    +
   586                                   0        20        40        60        80       100
```

Figure 5-5   SLAM II histogram for time in system, serial work station model.

The second category of statistics for this example is the file statistics. The statistics for file 1 and file 2 correspond to the units waiting for service at work stations 1 and 2, respectively. Thus, the average number of units waiting at work station 1 was 2.0433 units, with a standard deviation of 1.5151 units, a maximum of 4 units waited, and at the end of the simulation there were no units in the queue.

The last category of statistics for this example is statistics on service activities. The first row of service activity statistics corresponds to the server at work station 1 who was busy 47.05 percent of the time and blocked 40.44 percent of the time. Since the capacity of the server is one, the values 2.2051 and 4.4780 refer to the maximum length of the server idle period and busy period, respectively.

A number of additional statistics not provided by the SLAM II Summary Report can be obtained by straightforward analysis using the statistics provided. For example, the average time that units spent waiting for service (2.011) can be obtained by subtracting the sum of the average service time (0.75) from the average time in the system (2.761). The fraction of time idle (.1251) for the server at work station 1 can be determined by simply subtracting the sum of the average utilization (.4705) and average blockage (.4044) from 1.

The main conclusion from the simulation results is that work station 2 does not have the capacity to satisfy the anticipated work load on the facility. The service time for work station 2 must be decreased if less subcontracting of units is desired.

5.15   EXAMPLE 5-2.   INSPECTION AND ADJUSTMENT STATIONS ON A PRODUCTION LINE

The probelm statement for this example is taken from Schriber (3), who presents a GPSS model of the problem. A Q-GERT model of this example is presented by Pritsker (2). Assembled television sets move through a series of testing stations in the final stage of their production. At the last of these stations, the vertical control setting on the TV sets is tested. If the setting is found to be functioning improperly, the offending set is routed to an adjustment station where the setting is adjusted. After adjustment, the television set is sent back to the last inspection station where the setting is again inspected. Television sets passing the final inspection phase, whether for the first time or after one or more routings through the adjustment station, are routed to a packing area.

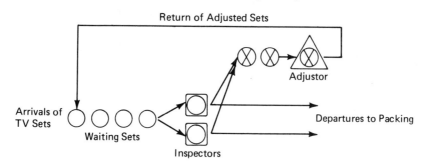

Figure 5-6 Schematic diagram of inspection and adjustment stations.

The situation described is pictured in Figure 5-6 where "circles" represent television sets. "Open circles" are sets waiting for final inspection, whereas "circled x's" are sets whose vertical control settings are improper, and which are either being serviced at the adjustment station or waiting for service there.

The time between arrivals of television sets to the final inspection station is uniformly distributed between 3.5 and 7.5 minutes. Two inspectors work side-by-side at the final inspection station. The time required to inspect a set is uniformly distributed between 6 and 12 minutes. On the average, 85 percent of the sets pass inspection and continue on to the packing department. The other 15 percent are routed to the adjustment station which is manned by a single worker. Adjustment of the vertical control setting requires between 20 and 40 minutes, uniformly distributed.

The inspection station and adjustor are to be simulated for 480 minutes to estimate the time to process television sets through this final production stage and to determine the utilization of the inspectors and the adjustor.

Concepts Illustrated. This example illustrates the uses of a service activity to model parallel identical servers. In addition, this example illustrates the use of regular activities for probabilistic branching. The procedure for obtaining a SLAM II trace through the use of a MONTR statement is illustrated.

SLAM II Model. The entities to be modeled in this system are the television sets. The television sets arrive and are routed to the inspection station. The two inspectors at the inspection station are represented as servers. If both inspectors are busy, a queue of television sets forms. This process can be conveniently modeled in SLAM II with a QUEUE node that precedes a service activity that represents two servers. Following the service activity representing the inspectors, 85 percent of the entities are accepted and depart to packing. The remaining 15 percent of the televisions do not pass inspection and are routed to the adjustor. If the adjustor is busy, a queue would form of televisions waiting for the adjustor. The adjustment process can be modeled as a QUEUE node followed by a service activity with a capacity of one. Following the adjustment operation, the entity is routed back to the queue of the inspectors.

The above describes the complete processing and routing of television sets through the inspection and adjustment stations. The SLAM II graphical model can be built directly from this discussion and is shown in Figure 5-7. Entities representing the television sets are created by the CREATE node with the time between entities uniformly distributed between 3.5 and 7.5 time units. The entity's arrival time is recorded as attribute 1. Each entity proceeds to the QUEUE node labeled INSP, and will proceed directly into service if an inspector represented by the emanating service activity is free. Recall that the 2 in the circle under the service ACTIVITY denotes two parallel identical servers. The service time for each server is specified as uniformly distributed between 6 and 12. Entities which arrive to the QUEUE node when both servers are busy wait in the QUEUE node which prescribes that they be stored in file 1.

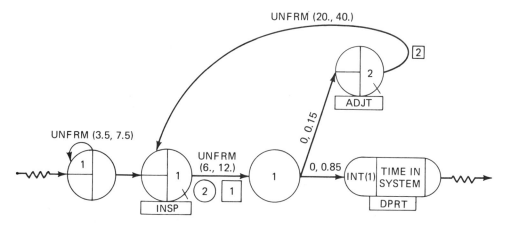

Figure 5-7 SLAM II network of TV inspection and adjustment stations.

Following inspection, the entities arrive at a GOON Node where they are probabilistically routed since the M-number of the GOON node is 1 and probabilities are associated with the two acitivites emanating from the GOON Node. One activity leads to the COLCT node labeled DPRT representing departure to the packing area. The other activity leads to the QUEUE node labeled ADJT representing the waiting line for the adjustment operation. Entities in the ADJT QUEUE wait for the adjustor in file 2. The adjustment operation is represented by the emanating service activity whose duration is uniformly distributed between 20 and 40 time units. Following adjustment, television entities are routed back to the QUEUE node labeled INSP. Entities which are routed to the COLCT node DPRT have interval statistics collected based on the time of creation which was stored in attribute 1 at the CREATE node. This interval of time corresponds to the total time that the television set spends in the inspection and adjustment process. The entities are then terminated.

THE SLAM II statement listing follows directly from the network model and is shown in Figure 5-8. Entities representing televisions are created by the CREATE statement (line 4) with a time between creations that is uniformly distributed in the range (3.5, 7.5) time units. The arrival time is marked as attribute 1 at the

CREATE node. The entity is routed to the INSP QUEUE where they wait in file 1 for a free server. The two parallel inspectors are represented by an ACT statement (line 6). The 2 in parentheses indicates two parallel identical servers and the duration for each is specified as uniformly distributed between 6 and 12. Entities completing inspection arrive at the GOON statement (line 7) which has two emanating activities. The first activity (line 8) has a default duration of zero, and is taken with a probability of 0.85 and routes the entity to the statement labeled DPRT. The second activity (line 9) has a default duration of zero, and is taken with a probability of 0.15 and routes the entity to the statement labeled ADJT. Entities which arrive to this QUEUE node wait in file 2 for the adjustor. The adjustor is represented by the emanating service activity (line 11) which has a duration uniformly distributed from 20 to 40 and routes entities to the statement labeled INSP. Entities which arrive to the DPRT statement (line 12) have interval statistics collected from the mark time recorded in attribute 1, and the statistics are labeled TIME IN SYSTEM. The entities then continue to the TERM statement (line 13) where they are destroyed. The END statement (line 14) denotes an end to the network description. The MONTR statement (line 16) initiates the tracing of events at time 0, stops the tracing of events at time 60, and specifies that the value of attribute 1 be printed along with the number in the inspector's queue, NNQ(1), and the number in the adjustor's queue, NNQ(2). The trace output is shown in Figure 5-9.

```
 1  GEN,OREILLY,TV INSP. AND ADJUST.,6/25/83,1;
 2  LIMITS,2,2,50;
 3  NETWORK;
 4        CREATE,UNFRM(3.5,7.5),,1;               CREATE TELEVISIONS
 5  INSP  QUEUE(1);                               INSPECTION QUEUE
 6        ACT(2)/1,UNFRM(6.,12.);                 INSPECTION
 7        GOON;
 8        ACT,,.85,DPRT;                          85  DEPART
 9        ACT,,.15,ADJT;                          15  ARE RE-ADJUSTED
10  ADJT  QUEUE(2);                               ADJUST QUEUE
11        ACT/2,UNFRM(20.,40.),,INSP;             ADJUSTMENT
12  DPRT  COLCT,INT(1),TIME IN SYSTEM;            COLLECT STATISTICS
13        TERM;
14        END;
15  INIT,0,480;
16  ;
17  ;    WRITE A TEXT TRACE FROM TIME 0 TO TIME 60,
18  ;    PRINT THE VALUE OF ATRIB(1), NNQ(1), AND NNQ(2)
19  MONTR,TRACE,0,60,ATRIB(1),NNQ(1),NNQ(2);
20  FIN;
```

Figure 5-8 Network statement model for television set inspection and adjustment stations.

Summary of Results. The results for this simulation are summarized in the SLAM II Summary Report shown in Figure 5-10. The first category of statistics for this example is statistics for variables based on observations. It consists of the interval

SLAM II TRACE BEGINNING AT TNOW= 0.0000E+00

TNOW	JEVNT	NODE ARRIVAL LABEL TYPE	CURRENT VARIABLE BUFFER			ACTIVITY SUMMARY INDEX	DURATION	END NODE
0.0000E+00		CREATE INSP QUEUE	0.0000E+00 0.0000E+00	0.0000E+00 0.0000E+00	0.0000E+00 0.0000E+00			
0.4711E+01		CREATE INSP QUEUE	0.4711E+01 0.4711E+01	0.0000E+00 0.0000E+00	0.0000E+00 0.0000E+00	1	11.6027	
0.8774E+01		CREATE INSP QUEUE	0.8774E+01 0.8774E+01	0.0000E+00 0.0000E+00	0.0000E+00 0.0000E+00	1	8.8256	
0.1160E+02		GOON	0.0000E+00	0.1000E+01	0.0000E+00			
						1 0 0	8.2106 NOT RELEASED 0.0000	ADJT DPRT
		DPRT COLCT TERM	0.0000E+00 0.0000E+00	0.0000E+00 0.0000E+00	0.0000E+00 0.0000E+00			
0.1354E+02		GOON	0.4711E+01	0.0000E+00	0.0000E+00			
						0 0	NOT RELEASED 0.0000	DPRT ADJT
		ADJT QUEUE	0.4711E+01	0.0000E+00	0.0000E+00	2	26.3948	INSP
0.1572E+02		CREATE INSP QUEUE	0.1572E+02 0.1572E+02	0.0000E+00 0.0000E+00	0.0000E+00 0.0000E+00			
0.1981E+02		GOON	0.8774E+01	0.0000E+00	0.0000E+00	1	7.2015	
						0 0	NOT RELEASED 0.0000	ADJT DPRT
		DPRT COLCT TERM	0.8774E+01 0.8774E+01	0.0000E+00 0.0000E+00	0.0000E+00 0.0000E+00			
0.2268E+02		CREATE INSP QUEUE	0.2268E+02 0.2268E+02	0.0000E+00 0.0000E+00	0.0000E+00 0.0000E+00			
0.2292E+02		GOON	0.1572E+02	0.0000E+00	0.0000E+00	1	7.6142	
						0 0	NOT RELEASED 0.0000	ADJT DPRT
		DPRT COLCT TERM	0.1572E+02 0.1572E+02	0.0000E+00 0.0000E+00	0.0000E+00 0.0000E+00			
0.2880E+02		CREATE INSP QUEUE	0.2880E+02 0.2880E+02	0.0000E+00 0.0000E+00	0.0000E+00 0.0000E+00			
0.3030E+02		GOON	0.2268E+02	0.0000E+00	0.0000E+00	1	10.2201	
						0 0	NOT RELEASED 0.0000	DPRT ADJT
0.3512E+02		ADJT QUEUE CREATE INSP QUEUE	0.2268E+02 0.3512E+02 0.3512E+02	0.0000E+00 0.0000E+00 0.0000E+00	0.0000E+00 0.1000E+01 0.1000E+01			
0.3902E+02		GOON	0.2880E+02	0.0000E+00	0.1000E+01	1	10.9990	
						0 0	NOT RELEASED 0.0000	ADJT DPRT
		DPRT COLCT TERM	0.2880E+02 0.2880E+02	0.0000E+00 0.0000E+00	0.1000E+01 0.1000E+01			
0.3944E+02		CREATE INSP QUEUE	0.3944E+02 0.3944E+02	0.0000E+00 0.0000E+00	0.1000E+01 0.1000E+01			
0.3993E+02		INSP QUEUE	0.4711E+01	0.0000E+00	0.1000E+01	1	9.2284	
0.4299E+02		CREATE INSP QUEUE	0.4299E+02 0.4299E+02	0.1000E+01 0.1000E+01	0.0000E+00 0.0000E+00	2	23.8558	INSP
0.4612E+02		GOON	0.3512E+02	0.2000E+01	0.0000E+00			

Figure 5-9 Trace report for television set inspection and adjustment stations.

statistics on the TIME IN SYSTEM collected by the COLCT statement (line 12). During the 480 minutes of simulated operation, a total of 84 television sets completed processing with the sets spending an average of 26.63 minutes in the system. However, there was a high variability in times in the system between television sets as reflected by the minimum and maximum times and the high standard deviation and coefficient of variation. This is to be expected since a fraction of the television sets have much larger times in the system due to their being adjusted. Also, 480 time units is insufficient to reach steady-state values.

The second category of statistics for this example is the file statistics. The results show that there was an average of .8515 television sets waiting for inspection in file 1 and 1.4651 television sets waiting for adjustment in file 2. Again, the high standard deviations relative to the average indicate a high degree of variation in queue length over time.

```
              S L A M   I I   S U M M A R Y   R E P O R T


        SIMULATION PROJECT TV INSP. AND ADJUST.      BY OREILLY

        DATE  6/25/1983                      RUN NUMBER   1 OF   1


        CURRENT TIME   0.4800E+03
        STATISTICAL ARRAYS CLEARED AT TIME  0.0000E+00
```

STATISTICS FOR VARIABLES BASED ON OBSERVATION

	MEAN VALUE	STANDARD DEVIATION	COEFF. OF VARIATION	MINIMUM VALUE	MAXIMUM VALUE	NUMBER OF OBSERVATIONS
TIME IN SYSTEM	0.2663E+02	0.3591E+02	0.1348E+01	0.6381E+01	0.1622E+03	84

FILE STATISTICS

FILE NUMBER	ASSOC NODE LABEL/TYPE	AVERAGE LENGTH	STANDARD DEVIATION	MAXIMUM LENGTH	CURRENT LENGTH	AVERAGE WAITING TIME
1	INSP QUEUE	0.8515	0.7756	3	0	4.0465
2	ADJT QUEUE	1.4651	1.1945	4	1	46.8822
3	CALENDAR	3.9019	0.4912	6	4	4.7177

SERVICE ACTIVITY STATISTICS

ACTIVITY INDEX	START NODE OR ACTIVITY LABEL	SERVER CAPACITY	AVERAGE UTILIZATION	STANDARD DEVIATION	CURRENT UTILIZATION	AVERAGE BLOCKAGE	MAXIMUM IDLE TIME/SERVERS	MAXIMUM BUSY TIME/SERVERS	ENTITY COUNT
1	INSPECTION	2	1.9059	0.2920	2	0.0000	2.0000	2.0000	99
2	ADJUSTMENT	1	0.8710	0.3352	1	0.0000	48.3651	245.4028	13

Figure 5-10 Summary report for TV inspection and adjustment stations.

The final category of statistics is the service activity statistics. In this example, there are two service activities corresponding to the inspectors and adjustors, respectively. The first service activity represents the inspectors and they had an average utilization of 1.9059. Note that since this activity has a capacity of 2, the maximum idle and busy values refer to the number of servers. The output indicates that both servers were idle at one point during the simulation and that both servers were busy at one point during the simulation. The second service activity represents the adjustor who had an average utilization of .8710. The longest period of time for which the adjustor was idle was 48.3651 minutes and the longest period of time for which the adjustor was busy was 245.4028 minutes. The summary report also indicates that one television is waiting for adjustment and that both inspectors and the adjustor are busy at the end of the simulation run. Adding these four television sets to the 84 that were completely processed indicates that 88 television sets were created. This is comparable to the 480/5.5 or approximately 87 arrivals expected.

The information on the SLAM II summary report can be used to assess system performance relative to questions such as: Are the queue storage areas large enough? Is the allocation of manpower between the inspection station and adjustor station proper? Is the time to process a television too long? In general, "what if" type questions are readily addressed using SLAM II.

5.16   EXAMPLE 5-3.   QUARRY OPERATIONS

In this example, the operations of a quarry are modeled. In the quarry, trucks deliver ore from three shovels to a crusher. A truck always returns to its assigned shovel after dumping a load at the crusher. There are two different truck sizes in use, twenty-ton and fifty-ton. The size of the truck affects its loading time at the shovel, travel time to the crusher, dumping time at the crusher, and return trip time from the crusher back to the appropriate shovel. For the twenty-ton trucks, these loading, travel, dumping, and return trip times are: exponentially distributed with a mean 5; a constant 2.5; exponentially distributed with mean 2; and a constant 1.5. The corresponding times for the fifty-ton trucks are: exponentially distributed with mean 10; a constant 3; exponentially distributed with mean 4; and a constant 2. To each shovel is assigned two twenty-ton trucks and one fifty-ton truck. The shovel queues are all ranked on a first-in, first-out basis. The crusher queue is ranked on truck size, largest trucks first. A schematic diagram of the quarry operations is shown in Figure 5-11. It is desired to analyze this system over 480 time units to determine the utilization and queue lengths associated with the shovels and crusher (2).

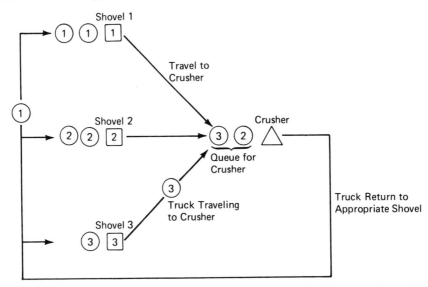

Figure 5-11   Quarry operations schematic diagram.

Concepts Illustrated. In the previous examples, entities have been entered into the network by the CREATE node. In this example, we illustrate an alternate approach of directly inserting entities into the network using the ENTRY statement. The use of an attribute of an entity to define a file number of a QUEUE node and to specify a service activity number is demonstrated. Additional concepts illustrated by this example include the specification of activity durations as a function of the attributes of the entities, the ranking of entities at a QUEUE node based upon an attribute value and EQUIVALENCE statements for relating textual names to SLAM II variables.

SLAM II Model. The network modeling of the quarry operations involves the routing of two distinct entity types representing the twenty-ton and fifty-ton trucks. Each entity is assigned five attribute values consisting of: 1) truck tonnage (either 20 or 50); 2) shovel number to which the truck is assigned; 3) mean shovel loading duration; 4) mean crusher dumping duration; and 5) return trip time. The SLAM II network model for processing the entities through the quarry operations is depicted in Figure 5-12. Note that entities are neither created or terminated within the graphical model. Therefore, to initiate the simulation, we must insert six entities representing the twenty-ton trucks and three entities representing the fifty-ton trucks directly into the network. These entities will continue to cycle through the network until the simulation is terminated. However, before discussing the procedure for inserting the truck entities into the network, we will complete our description of the network model.

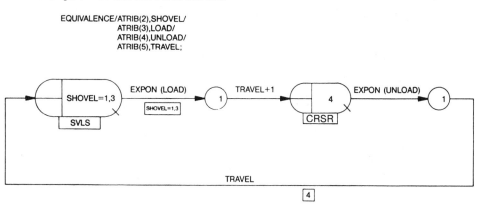

Figure 5-12 Network representation of quarry operations.

We begin our description at the far left QUEUE node labeled SVLS. Entities arriving to the node SVLS represent trucks completing their return trip from the crusher. SHOVEL is an attribute of the truck entity and contains the shovel number to which the truck is assigned. Each shovel queue is represented by the QUEUE node SVLS with the loading operation represented by an emanating ACTIVITY. Through the use of the attribute SHOVEL, the queue file numbers and service activity numbers for all three shovels can be represented in a single QUEUE-ACTIVITY combination. The duration of the service ACTIVITY is specified to be exponentially distributed with a mean given by the attribute LOAD. LOAD is equal to 5 or 10 depending upon whether the entity represents a twenty-ton or fifty-ton truck. Therefore the duration of the ACTIVITY is dependent upon the type of entity being processed. After completing the loading operation, the entity then undertakes the ACTIVITY with duration TRAVEL + 1.0 representing the loaded travel time. The entity continues to the QUEUE node labeled CRSR representing the queue of trucks waiting to dump ore at the crusher. Entities in the QUEUE node are waiting for the CRUSHER in file 4 which is ranked on high values of attribute 1 (HVF(1)). Since the first attribute denotes truck tonnage, priority is given to the fifty-ton trucks. The dumping operation is represented by the service ACTIVITY emanating from the CRSR QUEUE node and is exponentially distributed with a mean given by the UNLOAD attribute of a truck entity. Following completion of the dumping operation, the entity arrives to the GOON node and then continues through the emanating ACTIVITY representing the return trip. This ACTIVITY has a duration given by TRAVEL, an activity index of 4, and routes the entity back to the QUEUE node labeled SVLS.

The statement model for this example is shown in Figure 5-13. The PRIORITY statement (line 3) specifies that file 4 corresponding to the crusher queue is ranked on HVF(1). The network model consists of statements 9 through 21.

```
 1  GEN,PRITSKER,QUARRY OPERATIONS,12/19/85,1;
 2  LIMITS,4,5,75;
 3  PRIORITY/4,HVF(1);
 4  EQUIVALENCE/ATRIB(2),SHOVEL/
 5              ATRIB(3),LOAD/
 6              ATRIB(4),UNLOAD/
 7              ATRIB(5),TRAVEL;
 8  NETWORK;
 9  ;
10  ;      ATRIB(1)=TONNAGE,ATRIB(2)=SVL NO.,ATRIB(3)=SVL TIME
11  ;      ATRIB(4)=CRSR TIME,ATRIB(5)=RETURN TRIP TIME
12  ;
13  SVLS  QUEUE(SHOVEL=1,3);
14        ACT/SHOVEL=1,3,EXPON(LOAD);          LOAD SHOVEL
15  ENDL  GOON;
16        ACT,TRAVEL+1;
17  CRSR  QUEUE(4);
18        ACT,EXPON(UNLOAD);
19        GOON;
20        ACT/4,TRAVEL,,SVLS;                  TRUCK RETURN
21        END;
22  INIT,0,480;
23  ;
24  ;      PLACE TWO 20 TON AND ONE 50 TON TRUCK IN EACH SHOVEL QUEUE
25  ;
26  ;      FILE  TONNAGE  SHOVEL  LOAD  UNLOAD  TRAVEL
27  ;
28  ENTRY/ 1 ,    20 ,     1 ,     5 ,     2 , 1.5 /
29         1 ,    20 ,     1 ,     5 ,     2 , 1.5 /
30         1 ,    50 ,     1 ,    10 ,     4 , 2   /
31         2 ,    20 ,     2 ,     5 ,     2 , 1.5 /
32         2 ,    20 ,     2 ,     5 ,     2 , 1.5 ;
33  ENTRY/ 2 ,    50 ,     2 ,    10 ,     4 , 2   /
34         3 ,    20 ,     3 ,     5 ,     2 , 1.5 /
35         3 ,    20 ,     3 ,     5 ,     2 , 1.5 /
36         3 ,    50 ,     3 ,    10 ,     4 , 2   ;
37  FIN;
```

Figure 5-13 Statement model of quarry operations.

The travel time from a shovel to the crusher is represented by an ACTIVITY (line 14). Following the QUEUE node CRSR is a service ACTIVITY (line 18) representing the dumping operation which ends at the sequential GOON node (line 19). The last ACTIVITY (line 20) is assigned activity index number 4, has a duration given by TRAVEL, and routes the entity back to the QUEUE node labeled SVLS.

The initial insertion of truck entities into the network is accomplished by the ENTRY statements (line 28-36). Recall that each ENTRY statement places an entity into the file specified with attribute values as listed. Multiple entities can be inserted by a single ENTRY statement by separating the entities with a slash. For example, the first ENTRY statement places three entities into file 1 with attribute values of (20, 1, 5, 2, 1.5), (20, 1, 5, 2, 1.5), and (50, 1, 10, 4, 2.0). These entities correspond to two twenty-ton trucks and one fifty-ton truck. Since file 1 is a queue for one of the shovels, the filing of an entity into file 1 is processed as an arrival to node SVLS. Likewise, entities filed in file 2 and file 3 are processed as arrivals to node SVLS. Therefore the ENTRY statements place two twenty-ton and one fifty-ton truck in each shovel queue even though a single QUEUE node is used to model all three shovel queues.

Summary of Results. The results for this example are summarized by the SLAM II Summary Report shown in Figure 5-14. Note that in addition to statistics on queue lengths and server utilizations, statistics are also provided on each numbered ACTIVITY. Activity 4 represents the return trip from the crusher to the shovels. The results show that 152 trucks completed the return trip during the 480 time units of simulation, and that there was an average of 0.5410 trucks in transit between the crusher and the shovels. The maximum number of trucks on the return road at any one time was five.

```
           S L A M   I I   S U M M A R Y   R E P O R T


         SIMULATION PROJECT QUARRY OPERATIONS          BY PRITSKER

         DATE 12/19/1985                       RUN NUMBER   1 OF   1


         CURRENT TIME    0.4800E+03
         STATISTICAL ARRAYS CLEARED AT TIME  0.0000E+00



              **FILE STATISTICS**

     FILE    ASSOC NODE    AVERAGE    STANDARD    MAXIMUM   CURRENT    AVERAGE
    NUMBER   LABEL/TYPE    LENGTH     DEVIATION   LENGTH    LENGTH   WAITING TIME

       1     SVLS QUEUE     0.5159     0.7567        2         1        4.5025
       2     SVLS QUEUE     0.5563     0.7731        2         2        4.9451
       3     SVLS QUEUE     0.5072     0.7457        2         1        4.6821
       4     CRSR QUEUE     3.0464     2.2505        8         0        9.4952
       5        CALENDAR    4.3741     1.3701        9         5        2.4818
```

Figure 5-14 SLAM II Summary report for quarry operations example.

REGULAR ACTIVITY STATISTICS

ACTIVITY INDEX/LABEL	AVERAGE UTILIZATION	STANDARD DEVIATION	MAXIMUM UTIL	CURRENT UTIL	ENTITY COUNT
4 TRUCK RETURN	0.5410	0.7431	5	1	152

SERVICE ACTIVITY STATISTICS

ACTIVITY INDEX	START NODE OR ACTIVITY LABEL	SERVER CAPACITY	AVERAGE UTILIZATION	STANDARD DEVIATION	CURRENT UTILIZATION	AVERAGE BLOCKAGE	MAXIMUM IDLE TIME/SERVERS	MAXIMUM BUSY TIME/SERVERS	ENTITY COUNT
1	LOAD SHOVEL	1	0.7073	0.4550	1	0.0000	20.8085	62.6826	53
2	LOAD SHOVEL	1	0.6423	0.4793	1	0.0000	18.4326	47.8185	51
3	LOAD SHOVEL	1	0.7065	0.4554	1	0.0000	18.5865	57.8931	50
0	CRSR QUEUE	1	0.9113	0.2843	1	0.0000	5.2714	84.1352	

Figure 5-14   (continued).

5.17   SUMMARY OF SYMBOLS AND STATEMENTS

Table 5-3 presents the basic symbols and statements of SLAM II network modeling. The basic node types are: CREATE, QUEUE, TERMINATE, ASSIGN, GOON and COLCT. There is only one activity symbol and it is used to represent both service and non-service activities. An explicit delay in the processing of an entity can only be modeled by using an activity. The capability to model and analyze systems with 7 graphical elements demonstrates the power and flexibility of the SLAM II network approach. The default values associated with SLAM II network statements will be presented in Chapter 8, Table 8-1. A complete description of each network element is provided in Appendix B.

5.18   EXERCISES

5-1.   Develop a SLAM II portion of a network in which the time to traverse an activity is normally distributed with a mean of 10, a standard deviation of 2, a minimum value of 7 and a maximum value of 15. State your assumptions regarding the type of truncation used.

Table 5-3 Basic symbols and statements for SLAM II networks.†

Name	Symbol	Statement
ACTIVITY	DUR, PROB or COND (N) A	ACTIVITY (N)/A, DUR, PROB or COND, NLBL;ID
ASSIGN	VAR=Value ⋮ M	ASSIGN, VAR = Value, VAR = Value, M;
COLCT	TYPE ID,H M	COLCT(N), TYPE or VAR, ID, NCEL/HLOW/HWID, M;
CREATE	TBC TF MA MC M	CREATE,TBC,TF,MA,MC,M;
GOON	M	GOON, M;
QUEUE	IQ QC IFL	QUEUE (IFL), IQ, QC, BLOCK or BALK (NLBL), SLBLs
TERMINATE	TC or TC	TERMINATE, TC;

Special Node and Routing Symbols

BLOCK	‖	BLOCK
BALK	—•—•—•➤	BALK (NLBL)
Node Label	NLBL	NLBL

† Definition of Codes

TBC	Time Between Creations
TF	Time of First creation
MA	Mark Attribute for creation time
MC	Maximum Creations to be made
M	Maximum branches that an entity can take from a node
IFL	File number

Table 5-3   (continued).

IQ	Initial number of entities in QUEUE
QC	Queue Capacity to hold entities
BLOCK	Queue BLOCKS incoming entities and servers
BALK	Entities BALK from QUEUE node
NLBL	Node LaBeL
SLBL	SELECT node LaBeL
TC	Termination Count to stop simulation
VAR	SLAM II VARiable (see Table 5-1)
VALUE	SLAM II expression for VALUE (see Table 5-2)
TYPE	Statistics TYPE to be collected
ID	IDentifier
H	NCEL/HLOW/HWID
NCEL	Number of interior CELls of a histogram
HLOW	Upper limit of first cell of a histogram
HWID	Cell WIDth for a histogram
QLBL	QUEUE node LaBeL
N	Number of parallel servers
A	Activity number
DUR	DURation of an activity (see Table 5-2)
PROB	PROBability value for selecting an activity
COND	CONDition for selecting an activity

References to nodes are made through node labels NLBLs. When a node label is required, it is placed in a rectangle and appended to the base of the symbol.

Only the first three characters of the statement names and the first four characters of node labels are significant.

5-2.   Perform a manual simulation of the following network for 28 time units by preparing tables similar to those given in Tables 1-2 and 1-3.
(Note: Since entities are initially in the QUEUE nodes, the servers are busy and all entities initially in the system have an attribute 1 value equal to zero.) Compute the utilization of each server, the fraction of time server 1 is blocked, the average number in each queue, and the average time spent in the system by an entity.

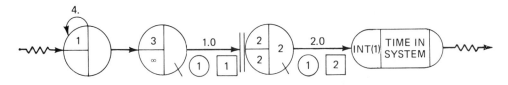

5-3.   Develop a SLAM II network model for the following situation. State any assumptions necessary to model the situation in network form. A machine shop contains two drills, one straightener and a finishing area. Drilling time is normally distributed with a mean of 10 minutes and a standard deviation of one minute. For those parts that need to be straightened, it takes 15 time units on the average and this time is exponentially distributed. To perform the finishing operations on a part takes five minutes and only one part can be finished at a time.

The machine shop processes two types of parts. Type 1 parts arrive every 30 minutes and it takes two minutes for the arriving part to be routed to the drill area. Type 1 parts require the drilling operation, straightening and finishing. Type 2 parts arrive every 20 minutes and require only drilling and finishing. The time to route a Type 2 part from its arrival to the drilling area is 10 minutes. Assume no time delays between drilling, straightening and finishing operations. Assume operators are always available if a machine is available. The network is to be used to obtain information on throughput of parts through the machine shop and utilization of the drills, straightener, and finishing area. Throughput by each part type is also desired. A histogram of the time for each part type to be processed through the machine area is to be obtained. The initial condition of the machine shop is that it is empty.
Embellishments:

(a) Estimate the expected time in the system for each part type under the assumption that there are 20 drills, 10 straighteners and 15 finishing operators. Estimate the variance of the time in the system for each part type.

(b) "Guess" at the expected time in system for each part type in the original model.

(c) Revise your network model to include a probability that finishing operations have to be repeated 10% of the time. How would this change your expected time estimates given in your answer to Embellishment (b).

(d) If a part type 1 has been routed through finishing twice but still requires to be refinished, it must be restraightened. Include this in your model. Estimate the fraction of parts that will need to be restraightened.

5-4.   Prepare the SLAM II input statements for the following network.

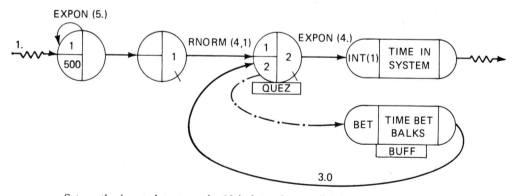

Set up the input data to make 10 independent replications with each run processing the 500 entities completely through the network.

5-5.   Redo the input data for Exercise 5-4 to stop each simulation run at time 2500 or following processing of 500 entities, whichever occurs first. Clear the statistical arrays at time 200 on each run using the MONTR statement.

5-6.   Given the following input statements, draw the equivalent SLAM II network and describe the system.

        GEN,EXERCISE 5-3,INVERT,6/23/78,2;
        LIMITS,1,2,20;
        PRIORITY/1,LVF(2);
        NETWORK;
                CREATE,EXPON(4.0),,1;
                ASSIGN,ATRIB(2)=1.0;
                ACT,,,WAIT;
                CREATE,EXPON(2.0),,1;
                ASSIGN,ATRIB(2)=2.0;
        WAIT   QUEUE(1);
                ACT,EXPON(1.0);
                COLCT,INT(1),TIME IN SYSTEM:
                TERM:
                END NETWORK;
        INIT,0,1000;
                FIN;

5-7.   For the maintenance facility situation involving work stations in series, Example 5-1, analyze the results and redistribute the six storage spaces to work station 1 and work station 2. Test your alternative design using SLAM II. If you were to begin a research and development program for improving the production line, on what quantity would you place your initial efforts?

5-8.   For the maintenance facility situation, Example 5-1, the output from the simulation indicates that the time in workstation 1 is subsumed by the time in workstation 2. Under this assumption, the situation could be modeled as a single workstation with the service time being that of work station 2 which has a queue with a capacity of 7 units. Perform a SLAM II simulation of this situation and compare the results with the original model.

        Embellishment: Perform a mathematical analysis of the single queue, single server system described in this exercise and compare the values obtained with the simulation results.

5-9.   Modify the SLAM II model of the inspection and adjustment stations, Example 5-2, to accommodate the following changes:
        (a) an arrival of television sets to the inspection station involves two television sets to be inspected;
        (b) the adjustor routes 40% of the adjusted sets directly to packing and 60% back to the inspectors;
        (c) by adding a step to the inspection process, it is felt that the probability of sending a set to an adjustor can be decreased to 0.10; the added step takes 5 minues.
        Redraw the network to indicate these changes. For one of the above situations, run the model and analyze the results.

5-10. A new design has been proposed for the television inspection and adjustment situation presented in Example 5-2 so that television sets requiring a third adjustment are sent to a rebuild operation. The rebuild operation is not modeled. For this proposal develop the network. Assume the adjustor spends more time on a television set the second time it is adjusted. (Increase both limits on the uniform distribution by 2). Because of this added time, the probability of requiring a third adjustment is reduced to 0.10.

5.11. For a one server, single queue situation, develop the SLAM II network model when the service time is sample from the probability mass function as shown below.

Probability	Service Time (min.)
.2	4
.3	6
.1	7
.4	10

The interarrival times are exponentially distributed with a mean of 7.75 minutes.

5.19 REFERENCES

1. Pritsker, A. A. B., and P. J. Kiviat, *Simulation With GASP II,* Prentice-Hall, 1969.
2. Pritsker, A. A. B., *Modeling and Analysis Using Q-GERT Networks,* Second Edition, Halsted Press and Pritsker & Associates, Inc. 1979.
3. Schriber, T., *Simulation Using GPSS,* John Wiley, 1974.
4. Standridge, C. R., et al, *TESS FOR SLAM II User's Manual,* Pritsker & Associates, Inc., 1985.

CHAPTER 6

Resources and Gates

6.1   INTRODUCTION

In a network model, an entity is advanced in accordance with the duration of activities. The flow of an entity is regulated by the status of servers. When a service activity is encountered, the entity waits in a queue for the server to become idle. Servers are a particular type of resource that remain stationary, that is, a service activity is only associated with the entity while the entity is flowing through the branch that represents the service activity. Situations arise where an entity requires a resource for a set of activities. SLAM II provides the capability to model this situation through the definition of resource types. For each resource type, the number of units available to be allocated to entities is defined. We refer to the total number of resource units available as the capacity of the resource.

An entity that requires a resource waits for it at an AWAIT node where the number of units of the resource required by the entity is specified. When an entity arrives to an AWAIT node, it proceeds through it to the activity emanating from the node if sufficient units of the resource are available. Otherwise, its flow is halted. A file is associated with the AWAIT node to maintain entities waiting for resources. An entity is removed from the file associated with the AWAIT node when units of the resource required by the entity can be assigned to it. The AWAIT node differs from the QUEUE node in that no service activities follow the AWAIT node.

To allow an entity to acquire a resource currently allocated to an entity that has a lower priority, a PREEMPT node is employed. If a resource cannot be preempted, then the entity waits in a file prescribed at the PREEMPT node in a fashion similar to that for the AWAIT node. PREEMPT nodes can only be used for a resource that has a capacity of one.

Resources are allocated to entities waiting in AWAIT and PREEMPT nodes in a prescribed order. This order is established through the use of a RESOURCE block. Also defined at the RESOURCE block is the initial capacity of the resource type.

When an entity no longer requires the use of a resource, it is routed to a FREE node where a specified number of units of the resource are freed (made available for reallocation). The PREEMPT and AWAIT nodes associated with the resource type are then interrogated to determine if the freed units can be allocated to waiting entities.

The capacity of a resource type can be changed by routing an entity through an ALTER node. ALTER nodes are used to increase or decrease the level of resource availability and can be used to model resource level changes due to machine maintenance, employee breaks, and daily shifts.

In SLAM II, a vehicle for accomplishing the stopping and starting of entity flow is a GATE. Entities can be routed to AWAIT nodes which require that a specified GATE be open before the entity can proceed through the AWAIT node. If the GATE associated with the AWAIT node is closed, the entity waits in a file until the GATE is opened. A GATE is opened when an entity flows through an OPEN node. It can be closed by an entity passing through a CLOSE node. The files in which entities may be waiting for a GATE to be opened are defined at a GATE block. When a gate is opened, all entities waiting at AWAIT nodes for the gate are permitted to pass through the AWAIT node and are routed to the branches emanating from the AWAIT node. For example, a gate can be used to stop the flow of passenger entities in a bus system until a bus entity arrives to an OPEN node. When the passenger entities are loaded on the bus, the bus entity would be routed through a CLOSE node to restrict again the flow of passenger entities onto the bus.

In this chapter, RESOURCES and GATES are described. The modeling of systems using nodes associated with these concepts is presented. Basically, the flow of entities is controlled through the requirements of an entity for units of a resource or for an open gate. The standard branching process for entities presented in Chapter 5 is not changed. Hence, no new ACTIVITY capabilities are required. In fact, only regular activities are needed when modeling systems with RESOURCES and GATES.

6.2   RESOURCE BLOCK

The RESOURCE block is used to identify: the resource name or label, RLBL; the initial resource capacity, that is, number of resource units available, CAP; and the order in which files associated with AWAIT and PREEMPT nodes are to be polled to allocate freed units of the resource to entities. The word "block" is employed instead of "node" because the RESOURCE block has no inputs or outputs as entities do not flow through it. Basically, the RESOURCE block is a definitional vehicle to specify a resource label (RLBL), the available number of units for the resource type, and an allocation procedure for entities waiting for units of the resource. On the network diagram, blocks can be placed together to form a legend. SLAM II assigns numeric codes to each resource name. The resource defined by the first resource block in the NETWORK statements is given a code of 1, the second a code of 2, and so on. The user may specify the resource number, RNUM, directly by using an alternative form for the RESOURCE block.

The RESOURCE, generically referred to as RES, is used in AWAIT, PRE-EMPT, FREE, and ALTER nodes to identify the resource type associated with the nodes. The label RLBL can be any string of characters beginning with an alphanumeric and excluding the special characters [,/( )+–* ';]. However, only the first eight characters are significant. The SLAM II assigned numeric code can also be used to reference a RESOURCE at AWAIT, PREEMPT, FREE and ALTER nodes. In addition, an attribute of an entity arriving to these nodes can carry the numeric code to define the RESOURCE to be acted upon. The initial resource capacity, CAP, is the number of units of the resource that can be allocated at the beginning of a run. During a run, the level of resource capacity can be increased or decreased by entities passing through ALTER nodes. The number of units of a particular resource in use is the number assigned to entities at AWAIT and PRE-EMPT nodes that have not been released at FREE nodes. The SLAM II variable NRUSE(RES) maintains the value of the number of units of resource RES in use. NNRSC(RES) is the value of the number of units of RES currently available. Statistics are automatically collected on resource utilization and availability and are printed as part of the SLAM II summary report.

At the RESOURCE block, file numbers are listed in the order in which the PREEMPT and AWAIT nodes for this resource type are to be polled. The RESOURCE block symbol and alternative resource statements are shown below.

| RLBL | CAP | IFL1 | IFL2 |

```
RESOURCE/RLBL(CAP), IFLs;
or
RESOURCE/RNUM,RLBL(CAP), IFLs;
```

As an example of a RESOURCE block statement, consider a resource with the label MACHINE which has a capacity of 2 and for which it is desired that files 3 and then 7 be polled for entities waiting for a MACHINE. The resource block statement would be written as

RESOURCE/MACHINE(2),3,7;

If this were the first resource block in the list of statements, MACHINE would be identified by SLAM II as resource number 1. For example, the number of units of MACHINE in use could be accessed on a SLAM II network by NRUSE (MACHINE) or NRUSE(1).

If it is desired to specify explicitly that MACHINE be resource 1, then the alternate resource block statement definition could be used as shown below:

RESOURCE/1,MACHINE(2),3,7;

If a specified numeric value is not assigned to a resource then it is assigned a number sequentially in accordance with either the order in which the RESOURCE statements appear in the statement model or the next number if the last specified resource was assigned a number.

6.3   AWAIT NODE

AWAIT nodes are used to store entities waiting for UR units of resource RES or waiting for gate GATE to open. When an entity arrives to an AWAIT node and the units of resource required are available or the GATE is opened, the entity passes directly through the node and is routed according to the M-number prescribed for the node. If the entity has to wait at the node, it is placed in file IFL in accordance with the priority assigned to that file. Regular activities emanate from the AWAIT node.

The symbolism and statement for the AWAIT node are shown below.

```
.AWAIT(IFL/QC),RES/UR,BLOCK or BALK(NLBL),M;
or
 AWAIT(IFL/QC),GATE,BLOCK or BALK(NLBL),M;
```

Normally, RES is specified by a resource label RLBL and GATE by a gate label GLBL. The file number, IFL, queue capacity, QC, and blocking and balking specifications are identical to those used for QUEUE nodes. In particular, IFL can be specified as an attribute of an arriving entity, that is, ATRIB(I) = J,K where I is the attribute number and J through K are the allowable file numbers specified by ATRIB(I). With respect to the file number, AWAIT nodes differ from QUEUE nodes in that the same file number can be associated with more than one AWAIT node. This allows entities to wait in the same file at different AWAIT nodes in the network. The resource, RES, and number of units, UR, can be integers or attribute numbers specified as ATRIB(I). In the latter case, the value of attribute I gives the resource code or number of units required. In Chapter 9, it will be seen that multiple resources can be required at an AWAIT node by specifying RES to be ALLOC(I).

Consider an entity that arrives to the following AWAIT node.

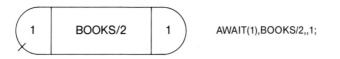

AWAIT(1),BOOKS/2,,1;

The entity that arrives requires two units of the resource BOOKS. If two units of the resource BOOKS are available at the time of arrival of the entity, the two books are allocated to the entity and the entity branches from the AWAIT node in accordance with the M number which is 1. If two books are not available, the entity waits in file 1. No limit is prescribed for the number of entities that can wait in file 1.

The following AWAIT node is the same as the one above except that the number of books required by an arriving entity is prescribed by the value of ATRIB(4). Thus, each entity may require a different number of books before being processed. The resource BOOKS will only be allocated to the first entity in file 1 when BOOKS are available. Thus, if an entity requires 3 BOOKS and it has a higher priority than an entity that requires 1 BOOKS; then the entity that requires 1 BOOKS would wait even though one of the BOOKS was available to be allocated.

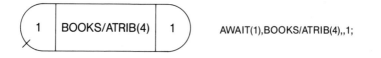

AWAIT(1),BOOKS/ATRIB(4),,1;

The AWAIT node below illustrates that the resource can be specified by an attribute of the arriving entity and that a capacity can be set on the number of entities waiting at an AWAIT node.

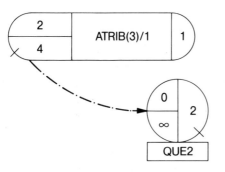

AWAIT(2/4),ATRIB(3)/1,BALK(QUE2),1;

Specifically, an arriving entity requests one unit of the resource defined by the value of ATRIB(3). If the entity has to wait, it waits in file 2. If there are 4 entities waiting in file 2, then the newly arriving entity will balk to QUEUE node QUE2.

In the following node, the file in which an entity waits is described by a range of values given by attribute 2 of the arriving entity. Each arriving entity requests 1 unit of the resource TELEX and the RESOURCE block for TELEX indicates a capacity of 3 with the priority for allocations given first to entities waiting in file 4 and then to entities in file 5 followed by those in file 3. In this situation, an entity that is waiting at this AWAIT node is placed in file 3, 4, or 5 depending on its attribute 2 value. The priority of entities in files 3, 4, and 5 is specified on PRIORITY statements. When 1 unit of TELEX is made available, it will be allocated to the first entity waiting in file 4. If no entities are waiting in file 4, then entities waiting in file 5 will be considered. Similarly, those entities in file 3 will have to wait until all entities in files 4 and 5 are allocated units. A complete description of the allocation and reallocation of resources is given in Chapter 9.

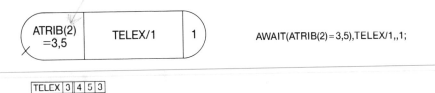

6.4   FREE NODE

FREE nodes are used to release units of a resource type when an entity arrives to the node. Every entity arriving to a FREE node releases UF units of resource type RES where UF and RES are specified values for the FREE node. UF can be a constant or a SLAM II variable. The freed units are then allocated to entities waiting in PREEMPT and AWAIT nodes in the order prescribed by the RESOURCE block. The entity arriving to the FREE node is then routed in accordance with the M-number associated with the FREE node. The symbol and statement for the FREE node are shown below.

The resource RES can be a resource label or an attribute number specified as
ATRIB(I). UF can be a constant or a SLAM II variable.

Consider an entity that arrives to the following FREE node.

At this FREE node, two BOOKS are made available for reallocation when an entity arrives. The reallocation is made to entities in accordance with the list of file
numbers in the RESOURCE block specification. The file numbers are polled in
order and, if entities are waiting, the entities are allocated the BOOKS and they
are placed in the appropriate activity following the AWAIT node. The entity arriving to the FREE node is routed from the FREE node. The routing for this entity
is done before the completion of an activity for any entity that has been reallo-
cated the BOOKS resources.

The FREE node,

causes ATRIB(4) number of BOOKS to be available for reallocation where
ATRIB(4) is the value of the fourth attribute of the arriving entity to the FREE
node.

The FREE node,

frees one unit of the resource defined by the third attribute of an arriving entity.

The reallocation of resource units is a complex process as there may be entities
waiting in different files which require the resource, entities in the same file requiring different units of the resource, or entities in different files requiring different units of the resource. The following describes the procedures used within

SLAM II for reallocating resources that are made available. First, the freed resources are added to the current number of idle resources of the same type. Resources may be idle and entities waiting if an insufficient number of units of the resource are not available. For example, an entity may require 2 units and only 1 is available. A polling of the files associated with the resource type is then initiated. The files are checked in the order listed on the RESOURCE block. For each file, the first entity in the file is polled to determine if sufficient resources are available to satisfy the needs of the entity ranked first. If sufficient resources are not available, the polling continues to the next file. A search for entities in the file is not made as it is assumed that the entity that is ranked first takes precedence over other entities in the file. If resources are sufficient to satisfy the first entity's requirement, that entity is removed from the file and scheduled from the AWAIT node associated with the file. The resources available are then decreased by the amount allocated. The next entity in the file is then polled to see if its resource requirements can be satisfied. The above process continues until either insufficient units of the resource are available to reallocate to entities that are ranked first in each file or there are no further entities waiting in any of the files for the resource type that was freed.

6.5   ILLUSTRATIONS OF THE USE OF RESOURCES

6.5.1   Illustration 6-1.   Resource Usage for Sequential Operations

Consider the situation in which a radio inspector performs an inspection operation and, if the radio requires adjustment, the inspector also performs that operation. Assume that fifteen percent of the radios manufactured require adjustment. In this situation, the inspector can be thought of as a resource that is allocated or assigned to the processing (inspection and adjustment) of the radio. Thus the inspector is not always available to perform another inspection because he may be required to perform the adjustment operation. The modeling of this situation is shown below in both a network and statement form.

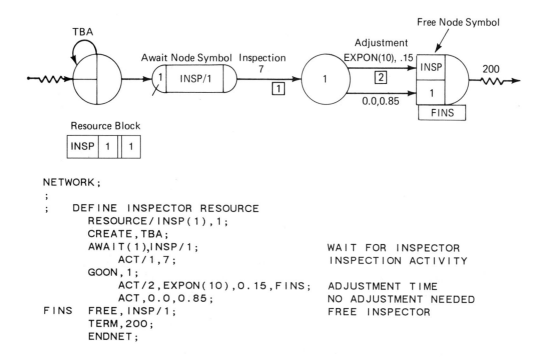

```
NETWORK;
;
;    DEFINE INSPECTOR RESOURCE
        RESOURCE/INSP(1),1;
        CREATE,TBA;
        AWAIT(1),INSP/1;              WAIT FOR INSPECTOR
            ACT/1,7;                  INSPECTION ACTIVITY
        GOON,1;
            ACT/2,EXPON(10),0.15,FINS;  ADJUSTMENT TIME
            ACT,0.0,0.85;             NO ADJUSTMENT NEEDED
FINS    FREE,INSP/1;                  FREE INSPECTOR
        TERM,200;
        ENDNET;
```

In this model, the RESOURCE block indicates that the resource INSP has an initial capacity of one and that it is allocated at an AWAIT node where file 1 is used to maintain waiting entities. After an entity is created, it is routed to the AWAIT node and, if the inspector resource is available, it is processed through ACTIVITY 1. At the GOON node, it is routed either through the adjustment activity (ACTIVITY 2) with probability 0.15 or to an activity that requires zero time with probability 0.85. In either case, the entity flows through the FREE node to make the inspector available to process the next radio waiting in file 1. After freeing the inspector, the radio entity that was inspected and possibly adjusted is terminated at the TERM node. When 200 radios depart the network, one run is completed.

6.5.2   Illustration 6-2.   Single Resource Processing Different Entity Types

Consider the situation in which a professor (PROF) meets with students. Forty percent of the students are classified as type A or B while 60 percent are classified as type C or F. The PROF serves the students on a first-come, first-serve basis but would like to determine the amount of time she spends counseling the two types of students and the amount of time per visit that each type of student spends waiting or being advised. She would also like to have aggregrate statistics on her time spent with students and the time required to counsel a student for each type. The network and statement models for this situation are shown in Figure 6-1 and Figure 6-2.

At the CREATE node, the interarrival time between students is exponentially distributed with a mean of 60. One of the two branches is selected from the CREATE node with 40 percent going to the AWAIT node AWAB and 60 percent going to the AWAIT node AWCF. Both AWAIT nodes reference file 1 so that all students are waiting in file 1 for one unit of the PROF. The RESOURCE block specifies a capacity of 1 for PROF with entities waiting for PROF in file 1. The time to process an A-B type is normally distributed with a mean of 10 and a standard deviation of 10. This is identified as activity 1 on the network. Since negative times are truncated to 0, a fraction of the A-B types will require very little or 0 time (approximately 16 percent). Following servicing, the PROF is available to serve another student if one is waiting. The A-B type who has been counseled is routed to a COLCT node where statistics on the student's time in the system are collected. The entity is then routed to node COMBS where statistics on all students are collected.

A C-F type is routed to AWAIT node AWCF where one unit of the PROF is requested. If the PROF is not available, the student waits in file 1. When the PROF is allocated to the C-F student, the entity is routed over activity 2 which has a counseling time that is uniformly distributed between 30 and 50. Following counseling, one unit of the PROF is freed and statistics are collected on C-F types individually and collectively with the A-B types at node COMBS. Statistics on the PROF's time spent with A-B types are recorded automatically since it is the time that activity 1 is in use. Statistics on the time the PROF spends with C-F types are recorded from the data maintained by SLAM II for activity 2. Resource statistics for PROF will provide information on her total utilization. The number of students waiting and the average waiting time for students will be an output from statistics for file 1. This illustration demonstrates how two classes of entities can be served by a single resource and how statistics by entity class can be obtained. By using the same file number with two AWAIT nodes, a single priority can be applied to a class of entities that require a particular resource for processing.

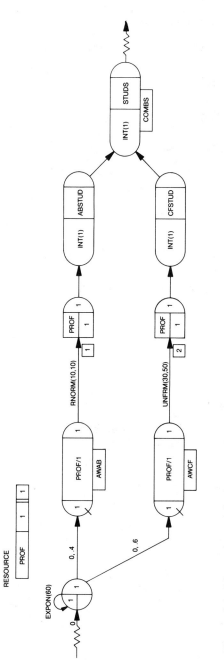

Figure 6-1 Network model of student processing.

```
 1  GEN,PRITSKER,HELP SESSION,9/12/85,1;
 2  LIMITS,1,1,10;
 3  NETWORK;
 4  ;
 5  ;   DEFINE PROFESSOR RESOURCE
 6       RESOURCE/PROF(1),1;
 7       CREATE,EXPON(60,1),,1;          CREATE STUDENTS WANTING HELP
 8       ACT,0,.4,AWAB;                  40% ARE A TO B STUDENTS
 9       ACT,0,.6,AWCF;                  60% ARE C TO F STUDENTS
10  ;
11  ;   PROFESSOR PROCESSES A TO B STUDENTS
12  ;
13  AWAB  AWAIT(1),PROF/1;
14       ACT/1,RNORM(10,10);       A-B STUDENTS
15       FREE,PROF/1;
16       COLCT,INT(1),ABSTUD;
17       ACT,,,COMBS;
18  ;
19  ;   PROFESSOR PROCESSES C TO F STUDENTS
20  ;
21  AWCF  AWAIT(1),PROF/1;
22       ACT/2,UNFRM(30,50);       C-F STUDENTS
23       FREE,PROF/1;
24       COLCT,INT(1),CFSTUD;
25  ;       COLLECT STATISTICS ON ALL STUDENTS
26  COMBS COLCT,INT(1),STUDS;
27       TERM;
28       ENDNETWORK;
29  INIT,0,2400;
30  FIN;
```

Figure 6-2   Statement model of student processing.

6.5.3   Illustration 6-3.   A Flexible Machining System

A flexible manufacturing system (FMS) which performs machining operations on castings was described in Section 4.4. In this illustration, a portion of that system will be modeled which consists of 10 horizontal milling machines which can perform any of three operations. It is desired to evaluate a design in which five of the milling machines are dedicated to performing operation 10, one of the machines is dedicated to perform operation 20 and two of the machines are dedicated to perform operation 30. Two of the 10 milling machines are classified as flexible and the tooling necessary for these machines to perform any of the three operations is made available.

The arrival of castings is scheduled to occur every 22 minutes and the processing times to perform operations 10, 20, and 30 are 120, 40 and 56 minutes respectively. When a casting arrives, it is placed in a queue for one of the four categories of mills described above. Once placed in a queue for a dedicated mill it will not be processed by the flexible mill even if the flexible mill is idle. The network and statement models for this situation are presented in Figure 6-3 and Figure 6-4.

At the CREATE node, castings are inserted into the network every 22 time units and attribute 3 is established as the arrival time. At node SETA, the first attribute of the arriving casting is set with ATRIB(1) defined as the operation number. On the resource statements, the mills dedicated to operation 10 are defined as resource 1, those dedicated to operation 20 as resource 2, those dedicated to operation 30 as resource 3 and the flexible mills are identified as resource 4.

The activities emanating from node SETA determine if a casting should be routed to a dedicated machine or to a flexible mill. It is assumed that a dedicated machine will be used if one is available or if no flexible mill is available. An activity with this condition is shown on the network leading to the ASSIGN node SETM. The availability of a resource is obtained through the SLAM II network variable NNRSC so the joint condition described above is

$$NNRSC(II).GT.0 .OR. NNRSC(MILLF).EQ.0$$

that is, the activity condition either represents that the mill for operation II is available or that the flexible mill, MILLF, is not available. In either case, the casting is assigned to mill II by setting MILL = OPERATION where MILL is equivalenced to ATRIB(2), MILL is used later as a file number and resource number.

If the above condition is not satisfied, then a dedicated machine for operation II is not available and a flexible mill is available. In this case, ATRIB(2) is set to 4 to indicate that the flexible mill is to perform the operation. At AWAIT node AMILL, the casting entity awaits for one unit of the mill as defined by MILL. The casting entity waits in file 1, 2, 3 or 4. When the appropriate mill is freed, the casting entity proceeds through the activity whose duration is specified by PROCESS_TIME which is equivalenced to ARRAY(1,OPERATION). An ARRAY statement (line 3) sets the processing times to 120, 40, and 56.

Following processing by the mill, the mill is freed at a FREE node and the casting entity is routed back to node SETA if an additional operation is to be performed, that is, OPERATION is less than 3. Otherwise, the casting entity has completed all three operations and is routed to a COLCT node where the time the casting was in the FMS system is recorded.

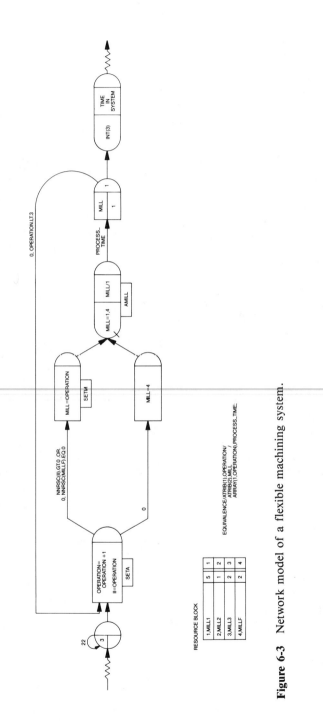

Figure 6-3 Network model of a flexible machining system.

```
 1  GEN,PRITSKER,FMS,2/28/86;
 2  LIMITS,4,3,500;
 3  ARRAY(1,3)/120,40,56;
 4  EQUIVALENCE/ATRIB(1),OPERATION/
 5             ATRIB(2),MILL/
 6             ARRAY(1,OPERATION),PROCESS_TIME;
 7  NETWORK;
 8        RESOURCE/1,MILL1(5),1/2,MILL2(1),2;          DEFINE 10 MILLS
 9        RESOURCE/3,MILL3(2),3/4,MILLF(2),4;          AS 4 RESOURCE TYPES
10        CREATE,22,,3;
11  ;
12  SETA  ASSIGN,OPERATION = OPERATION + 1,
13             II=OPERATION;                           INCREMENT OPERATION
14  ;                                                  NUMBER AND
15  ;                                                  SET II TO PROPOSED MILL
16        ACT,0,NNRSC(II).GT.0.OR.NNRSC(MILLF).EQ.0,SETM; CONDITIONS FOR
17  ;                                                  DEDICATED MILL
18        ACT;
19        ASSIGN,MILL=4;                               ASSIGN FLEXIBLE MILL
20        ACT,,,AMILL;
21  SETM  ASSIGN,MILL = OPERATION;                     ASSIGN DEDICATED MILL
22  AMILL AWAIT(MILL= 1,4),MILL/1;                     WAIT FOR MILL
23        ACT,PROCESS_TIME;                            MILL PROCESSING ACTIVITY
24        FREE,MILL/1,1;                               FREE MILL
25        ACT,0,OPERATION.LT.3,SETA;                   CHECK FOR ANOTHER
26  ;                                                  OPERATION
27        ACT;                                         MILL WORK COMPLETE
28        COLCT,INT(3),TIME IN SYSTEM;                 COLLECT TIME FOR
29  ;                                                  MILL OPERATIONS
30        TERM;
31        END;
32  INIT,0,2400;
33  FIN;
```

Figure 6-4 Statement model of a flexible machining system.

This illustration demonstrates how a complex situation can be reduced in scope in order to get a fast working model of the situation. The scope has been reduced by eliminating portions of the total system such as raw material, lathes, inspection station, and load and unload areas. In addition, the scope of the modeling effort was reduced by making assumptions concerning the routing of castings, that is, the assignment of a casting to a mill made at the time the casting was available to be processed.

6.6   EXAMPLE 6-1.   INVENTORY SYSTEM WITH LOST SALES AND BACKORDERS (2,4)

A large discount house is planning to install a system to control the inventory of a particular radio. The time between demands for a radio is exponentially distributed with a mean time of 0.2 weeks. In the case where customers demand the radio when it is not in stock, 80 percent will go to another nearby discount house to find it, thereby representing lost sales, while the other 20 percent will backorder the radio and wait for the next shipment arrival. The store employs a periodic review-reorder point inventory system where the inventory status is reviewed every four weeks to decide if an order should be placed. The company policy is to order up to the stock control level of 72 radios whenever the inventory position, consisting of the radios in stock plus the radios on order minus the radios on backorder, is found to be less than or equal to the reorder point of 18 radios. The procurement lead time (the time from the placement of an order to its receipt) is constant and requires three weeks.

The objective of this example is to simulate the inventory system for a period of six years (312 weeks) to obtain statistics on the following quantities:

1. number of radios in stock;
2. inventory position;
3. safety stock (radios in stock at order receipt times); and
4. time between lost sales

The initial conditions for the simulation are an inventory position of 72 and no initial backorders. In order to reduce the bias in the statistics due to the initial starting conditions, all the statistics are to be cleared at the end of the first year of the six year simulation period.

Concepts Illustrated. This example illustrates the use of: a RESOURCE block for modeling an inventory level; an AWAIT node for holding backorders; a FREE node for satisfying backorders; the logical .OR. operator for specifying the condition for selecting an activity; the CLEAR option on the MONTR statement for clearing statistics; and the TIMST statement for obtaining time-persistent statistics.

SLAM II Model. The inventory system for this example can be thought of in terms of two separate processes. The first process is the customer arrival process and consists of arriving customers demanding radios. If no radio is available, the customer either backorders a radio or balks to a nearby competitor. The second process is the inventory review through which radios are replenished. This process consists of a review, every four weeks, of the inventory position. If the inventory position is less than or equal to the reorder point, an order is placed. The size of the order is equal to the stock control level minus the inventory position, thus increasing the inventory position to the stock control level. Receipt of the order occurs three weeks later. The radios received are first used to satisfy backorders. Any remaining radios are used to increase the number of radios on-hand.

The two processes described above are modeled in SLAM II as shown in Figure 6-5 by representing the radios on-hand as a resource named RADIO whose capacity is 72 units. The buying and backordering of radios in the customer arrival process can be modeled as entities representing customers arriving to an AWAIT node. Likewise, the replenishment of radios in the inventory review process can be modeled as an entity representing a radio shipment arriving to a FREE node. Thus, the resource RADIO is depleted in the customer arrival process by entities representing customers and replenished in the inventory review process by entities representing radio shipments.

The SLAM II statement model for this example is presented in Figure 6-6. The model employs three XX variables equivalenced to INV_POS, REORDER_PT, and SCL. The INTLC statement (line 8) assigns initial values to these variables of 72, 18, and 72, respectively. The TIMST statement (line 9) causes time-averaged statistics to be collected on the inventory position and the results to be printed using the label INV. POSITION. The RESOURCE block (line 11) is used to define the radios on-hand resource and sets the initial level (availability) to 72. It identifies file 1 as the location of customers awaiting radios. The average number of radios on-hand is equal to the average availability of this resource.

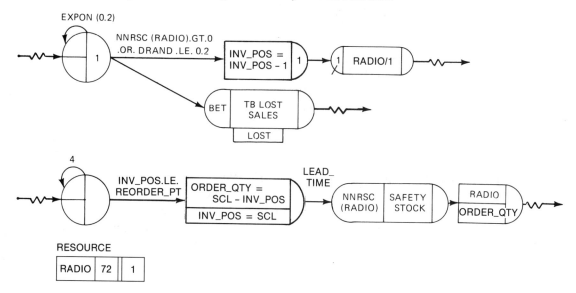

Figure 6-5 Network model for inventory example.

```
 1   GEN,OREILLY,INVENTORY PROBLEM,9/12/83,1,Y,Y,Y/N;
 2   LIMITS,1,2,30;
 3   EQUIVALENCE/XX(1),INV_POS/
 4              XX(2),REORDER_PT/
 5              XX(3),SCL/
 6              ATRIB(1),ORDER_QTY/
 7              3,LEAD_TIME/;
 8   INTLC,INV_POS=72,REORDER_PT=18,SCL=72;
 9   TIMST,INV_POS,INV. POSITION;
10   NETWORK;
11          RESOURCE/RADIO(72),1;
12   ;
13   ;      CUSTOMER ARRIVAL PROCESS
14   ;      ------------------------
15   ;
16          CREATE,EXPON(.2),,,,1;              CREATE ARRIVAL ENTITIES
17          ACT,,NNRSC(RADIO).GT.0.OR.DRAND.LE..2;   CONTINUE IF RADIO AVAIL OR P=.2
18          ACT,,,LOST;                         ELSE BRANCH TO LOST SALE
19          ASSIGN,INV_POS=INV_POS-1;           DECREMENT INVENTORY POSITION
20          AWAIT,RADIO;                        SEIZE A RADIO
21          TERM;                               DEPART THE SYSTEM
22   LOST   COLCT,BET,TB LOST SALES;            COLLECT BET STATS ON LOST SALES
23          TERM;                               DEPART THE SYSTEM
24   ;
25   ;      INVENTORY REVIEW PROCESS
26   ;      ------------------------
27   ;
28          CREATE,4;                           CREATE A REVIEW ENTITY
29          ACT,0,INV_POS.LE.REORDER_PT;        IF POSITION IS BELOW REODER PT.
30          ASSIGN,ORDER_QTY = SCL - INV_POS,
31                 INV_POS = SCL;               ORDER UP TO STOCK CONTROL LEVEL
32          ACT,LEAD_TIME;                      DELAY RECEIPT BY 3 WEEK LEAD_TIME
33          COLCT,NNRSC(RADIO),SAFETY STOCK;    COLLECT STATS ON SAFETY STOCK
34          FREE,RADIO/ORDER_QTY                INCREMENT RADIOS ON HAND
35          TERM;                               END REVIEW
36          END;
37   INIT,0,312;
38   MONTR,CLEAR,52;
39   FIN;
```

Figure 6-6 Statement model of inventory example.

The customer arrival process is modeled by statements 16 through 23. Entities representing customers are generated by the CREATE node with an interarrival time that is exponentially distributed with mean of 0.2 weeks. A maximum of 1 emanating activity is taken at each release of the CREATE statement. The first emanating ACTIVITY (line 17) is taken by entities that represent the non-balking customers; that is, the customers who purchase an available radio or the 20 percent of the customers who backorder a radio when no radio is available. The duration of this ACTIVITY is zero and it is taken conditionally if the current number of available units of RADIO is greater than zero or if the random variable DRAND is less than or equal to 0.2. The non-balking customers arrive at the AS-

SIGN node (line 19) where the inventory position is decremented by 1. The entities then continue to the AWAIT node where they either immediately seize (buy) or wait for one unit of RADIO. Each entity exiting the AWAIT node is destroyed at the TERM statement (line 21) corresponding to the departure of the customer from the system.

The second ACTIVITY (line 18) following the CREATE node has a duration of zero, is selected unconditionally, and ends at the COLCT node (line 22) labeled LOST. Since the CREATE node specifies that at most one emanating ACTIVITY is to be taken at each release, the second ACTIVITY will be taken if and only if the first ACTIVITY is not taken. The entities undertaking this ACTIVITY represent balking customers and are therefore lost sales. These entities are routed by the ACTIVITY to the COLCT node where statistics are collected on the time between lost sales. Each entity is then destroyed at the TERM node.

The inventory review process is modeled by statements 28 through 35. The CREATE node (line 28) creates an entity representing a review every four weeks. The emanating ACTIVITY is taken if the inventory position is less than or equal to the reorder point; otherwise no activity is selected and the review entity is destroyed. At the ASSIGN Node (line 29), the first attribute of the entity, the OR-DER_QTY, is set equal to the stock control level minus the inventory position and then the inventory position is reset to the stock control level. The exiting entry represents a radio shipment with the number of radios in the shipment specified by the ORDER_QTY (ATRIB(1)). The entity next takes the ACTIVITY representing the shipment delay time or lead time. LEAD_TIME is equivalenced to 3 in an EQUIVALENCE statement. At the completion of the ACTIVITY, the entity arrives at the COLCT node (line 33) where statistics are collected on the number of available units of resource RADIO. This value corresponds to the inventory on-hand level at the order receipt time. This quantity is referred to as the safety stock. The entity next moves to the FREE node (line 34) where the number of units of resource RADIO freed is the ORDER_QTY. These radios are then available to the entities representing non-balking customers in the arrival segment of the model. The TERM statement (line 35) then destroys the entity.

The clearing of statistics in order to reduce any bias due to the starting conditions is accomplished by the MONTR statement (line 38) with the CLEAR option. This statement causes all statistical arrays including the file statistics to be cleared at time 52. Therefore, the statistical results for the simulation are based upon values recorded during the last 260 weeks of simulated operation.

This example is easily modified to change the characteristics of the inventory policy and the ordering assumptions. The ease of changing the model, and hence, the simulation analysis, supports the assertion that the original problem formulation for simulation analysis need not be a time consuming process.

Summary of Results. The SLAM II Summary Report for this example is given in Figure 6-7. The report provides statistics on time between lost sales, safety stock, and inventory position which were requested by the user. Automatically obtained

SLAM II SUMMARY REPORT

SIMULATION PROJECT INVENTORY PROBLEM   BY OREILLY

DATE 9/12/1983   RUN NUMBER   1 OF   1

CURRENT TIME   0.3120E+03
STATISTICAL ARRAYS CLEARED AT TIME 0.5200E+02

STATISTICS FOR VARIABLES BASED ON OBSERVATION

	MEAN VALUE	STANDARD DEVIATION	COEFF. OF VARIATION	MINIMUM VALUE	MAXIMUM VALUE	NUMBER OF OBSERVATIONS
TB LOST SALES	0.2069E+01	0.6112E+01	0.2954E+01	0.1923E-02	0.4920E+02	134
SAFETY STOCK	0.3000E+00	0.9234E+00	0.3078E+01	0.0000E+00	0.4000E+01	20

STATISTICS FOR TIME-PERSISTENT VARIABLES

	MEAN VALUE	STANDARD DEVIATION	MINIMUM VALUE	MAXIMUM VALUE	TIME INTERVAL	CURRENT VALUE
INV. POSITION	0.4301E+02	0.1926E+02	-0.3000E+01	0.7200E+02	0.2600E+03	0.4200E+02

FILE STATISTICS

FILE NUMBER	ASSOC NODE LABEL/TYPE	AVERAGE LENGTH	STANDARD DEVIATION	MAXIMUM LENGTH	CURRENT LENGTH	AVERAGE WAITING TIME
1	AWAIT	0.1014	0.5281	5	0	0.0215
2	CALENDAR	2.2308	0.4213	8	2	0.1425

RESOURCE STATISTICS

RESOURCE NUMBER	RESOURCE LABEL	CURRENT CAPACITY	AVERAGE UTILIZATION	STANDARD DEVIATION	MAXIMUM UTILIZATION	CURRENT UTILIZATION
1	RADIO	72	43.3572	19.9391	72	30

RESOURCE NUMBER	RESOURCE LABEL	CURRENT AVAILABLE	AVERAGE AVAILABLE	MINIMUM AVAILABLE	MAXIMUM AVAILABLE
1	RADIO	42	28.6428	0	70

Figure 6-7 SLAM II summary report for inventory example.

are file 1 statistics corresponding to backordered radios and resource statistics corresponding to the average resource utilization or the average number of radios seized. The average number of radios on-hand is the average availability of the RADIO resource and is 28.64. The values from the summary report can be used to compute the average profit for the inventory decision policy employed in the model. An investigation of different parameter settings for this policy can be made by changing the input values of the reorder point, the stock control level, and the time between reviews.

6.7 PREEMPT NODE

The PREEMPT node is a special type of AWAIT node in which an entity can preempt one unit of a resource that has been allocated to some other entity. If the entity using the resource came from an AWAIT node, preemption will always be attempted. The preemption will also be attempted if the priority assigned to the PREEMPT node is greater than the priority of the PREEMPT node from which the entity currently using the resource type came. The symbolism and statement for the PREEMPT node are shown below.

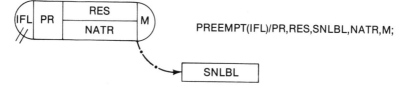

PREEMPT(IFL)/PR,RES,SNLBL,NATR,M;

The definitions of IFL and RES for the PREEMPT node are identical to the specifications for these variables at the AWAIT node and identify the file number and the resource requested at the PREEMPT node. The priority, PR, is specified as LOW(K) or HIGH(K) where K is an attribute number. The incoming entity will attempt to preempt another entity if its Kth attribute gives it a higher priority. A preemption attempt is not satisfied if the resource is currently in use by an entity that: 1) is being processed in a service activity; 2) is in a file; or 3) is performing an activity with an indefinite duration (REL or STOPA). Entities that do not cause a preemption to occur wait for the resource in file IFL.

An entity that is preempted is routed to a node as specified by the send node label SNLBL. The time remaining to process the entity when it is preempted is stored in ATRIB(NATR). If no send node label is specified then the preempted entity is routed to the AWAIT or PREEMPT node at which it was allocated the resource. At that node, it is established as the first entity waiting for the resource. When the resource is reassigned to the preempted entity, its remaining processing time will be used.

As described above, several restrictions are associated with the PREEMPT node. First, preemptions are only allowed for resources having a capacity of 1 unit. Second, an entity holding a resource that currently is in a QUEUE or AWAIT node will not be preempted. Also, if the entity is in a service activity or an activity of indefinite duration, it will not be preempted.

PREEMPT nodes only apply to resources as the concept of preempting a GATE is not meaningful.

The PREEMPT node

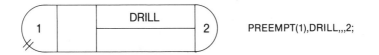

specifies that when an entity arrives to it, the DRILL should be preempted. No ranking is given so the default ranking for preemptions, FIFO, is used. Thus, if an entity had already preempted the DRILL, the current entity would wait in file 1 until the DRILL becomes available. Default values are also used for the send node label, SNLBL, and the attribute for storing the remaining activity time. In this situation, when the preemption occurs, the entity that is preempted will return to the node (AWAIT or PREEMPT) from which it obtained the drill. SLAM II will automatically maintain the remaining processing time for the activity from which it was preempted. The entity that was preempted will be made the first entity in the file of the node to which it is returned and it will start its reprocessing from the activity from which it was preempted. The M-number of 2 specifies that the preemption entity will, after preemption occurs, take at most two branches from the PREEMPT node.

The following PREEMPT node provides values for the SNLBL and NATR fields.

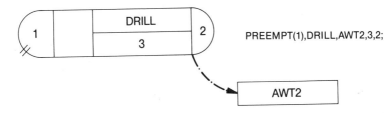

This situation is similar to the one presented above except that the preempted entity will go to a node whose label is AWT2 and the remaining processing time will be stored in ATRIB(3).

To illustrate a ranking for preemptions, consider the following PREEMPT node.

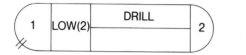

PREEMPT(1)/LOW(2),DRILL,,,2;

In this case, if the entity that arrived to the PREEMPT node has a smaller value of ATRIB(2) than an entity that previously preempted the DRILL, then the new arriving entity will preempt the previous preemptor. Since the entity that is preempted was itself a preempt entity, it will return to the PREEMPT node from which it preempted the DRILL.

The following PREEMPT node shows that the specification of the resource to be preempted can be an attribute value.

PREEMPT(1),ATRIB(4),,,2

In this case, the resource to be preempted is defined by ATRIB(4) of the arriving entity to the PREEMPT node. This capability allows the same PREEMPT node to be used for different machine failures by assigning a numeric resource code to ATRIB(4) and routing all such entities to the PREEMPT node.

6.8   ILLUSTRATION 6-4.   MACHINE BREAKDOWNS

Consider the situation in which packages are to be processed through a scale. The scale encounters failures which stop the processing of packages until the scale is repaired. In this situation, the scale is modeled as a resource and the packages as entities that require one unit of the resource. An entity representing scale status is modeled in a disjoint network. It is delayed by the time to failure, after which the scale status entity arrives to a PREEMPT node. The PREEMPT node stops the weighing of a package by the scale. The package is routed to node HAND where a hand weighing is done. The time to weigh by HAND is twice the remaining process time. The network and statement models for this illustration are shown below.

A repair time for the scale is scheduled from the PREEMPT node. Following the repair time, the scale is freed and the preempted package, if there was one,

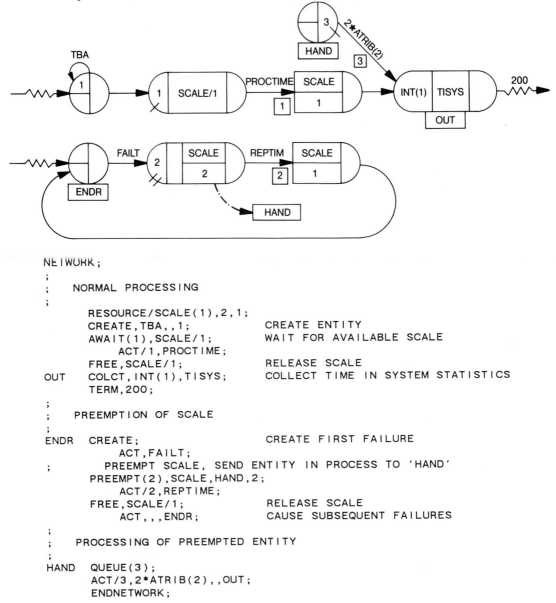

```
NETWORK;
;
;    NORMAL PROCESSING
;
        RESOURCE/SCALE(1),2,1;
        CREATE,TBA,,1;              CREATE ENTITY
        AWAIT(1),SCALE/1;          WAIT FOR AVAILABLE SCALE
            ACT/1,PROCTIME;
        FREE,SCALE/1;             RELEASE SCALE
OUT     COLCT,INT(1),TISYS;       COLLECT TIME IN SYSTEM STATISTICS
        TERM,200;
;
;    PREEMPTION OF SCALE
;
ENDR    CREATE;                   CREATE FIRST FAILURE
            ACT,FAILT;
;         PREEMPT SCALE, SEND ENTITY IN PROCESS TO 'HAND'
        PREEMPT(2),SCALE,HAND,2;
            ACT/2,REPTIME;
        FREE,SCALE/1;             RELEASE SCALE
            ACT,,,ENDR;           CAUSE SUBSEQUENT FAILURES
;
;    PROCESSING OF PREEMPTED ENTITY
;
HAND    QUEUE(3);
        ACT/3,2*ATRIB(2),,OUT;
        ENDNETWORK;
```

can be continued. The next failure of the scale is scheduled by routing the status-of-scale entity back through an activity representing the failure time.

6.9   EXAMPLE 6-2.   A MACHINE TOOL WITH BREAKDOWNS (3)

A schematic diagram of job processing and machine breakdown for a machine tool is given in Figure 6-8. Jobs arrive to a machine tool on the average of one per hour. The distribution of these interarrival times is exponential. During normal operation, the jobs are processed on a first-in, first-out basis. The time to process a job in hours is normally distributed with a mean of 0.5 and a standard deviation of 0.1. In addition to the processing time, there is a set-up time that is uniformly distributed between 0.2 and 0.5 of an hour. Jobs that have been processed by the machine tool are routed to a different section of the shop and are considered to have left the machine tool area.

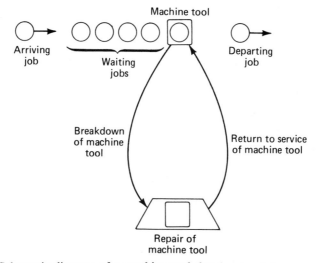

Figure 6-8   Schematic diagram of a machine tool that has breakdowns.

The machine tool experiences breakdowns during which time it can no longer process jobs. The time between breakdowns is normally distributed with a mean of 20 hours and a standard deviation of 2 hours. When a breakdown occurs, the job being processed is removed from the machine tool and is placed at the head of the queue of jobs waiting to be processed. Jobs preempted restart from the point at which they were interrupted.

When the machine tool breaks down, a repair process is initiated which is accomplished in three phases. Each phase is exponentially distributed with a mean of 3/4 of an hour. Since the repair time is the sum of independent and identically distributed exponential random variables, the repair time is Erlang distributed.

The machine tool is to be analyzed for 500 hours to obtain information on the utilization of the machine tool and the time required to process a job. Statistics are to be collected for five simulation runs.

Concepts Illustrated. This example illustrates the concept of preemption of a resource through the use of a PREEMPT node.

SLAM II Model. The machine tool can be considered as a single server. Service involves two operations: job setup and job processing. Since two operations are involved, a resource will be used to model the machine tool. Entities representing jobs will arrive and await the availability of the machine tool if necessary. The jobs will be set up and processed when the machine tool is available and, following processing, will depart the system.

The breakdown of the machine tool will be modeled using a breakdown entity which preempts the machine tool and holds it while a repair operation is performed. The breakdown entity is processed through a disjoint network.

The SLAM II network model of the machine tool processing with breakdowns is shown in Figure 6-9. The corresponding statement model is given in Figure 6-10. Consider first the flow of job entities through the first network segment. Jobs are created at the CREATE node with an exponential time between arrivals having a mean of one hour. The job entities are routed to the AWAIT node. If a TOOL is available they proceed to activity 1. If a TOOL is not available, they wait in file 1. Activity 1 represents the setup time and is uniformly distributed between 0.2 and 0.5 hours. Following setup, the job entity proceeds to activity 2 which represents the machine tool processing operation which is normally distributed with a mean of 0.5 and a standard deviation of 0.1 hours. Following processing, the machine tool is made available by having the job entity pass through a FREE node. The machine tool resource would then be used to process another job if one was waiting in file 1. The job entity proceeds to the COLCT node where time-in-system statistics are computed for the job entity. The job entity is then terminated.

The second network segment illustrates the processing of breakdown entities. The first breakdown is generated at time 20 hours. The CREATE note indicates that only one machine breakdown entity is to be generated. This machine breakdown entity will be recycled in the network segment as machine breakdown times are conditioned upon the time that the machine tool completes repair. Following the creation of the machine breakdown entity it is routed to a PREEMPT node

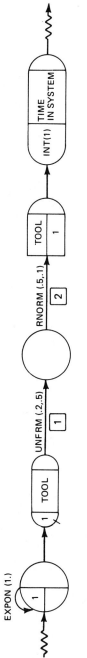

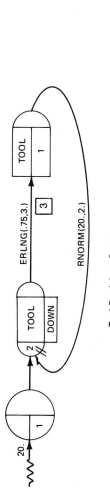

Figure 6-9 SLAM II statement model of a machine tool with breakdowns.

```
 1    GEN,PRITSKER,MACHINE BREAKDOWN,7/5/83,5,,,,NO;
 2    LIMITS,2,1,50;
 3    NETWORK;
 4          RESOURCE/TOOL(1),2,1;
 5          CREATE,EXPON(1.),0,1;                    CREATE ARRIVALS
 6          AWAIT(1),TOOL/1;                          AWAIT THE TOOL
 7          ACT/1,UNFRM(.2,.5);                   SET UP
 8          GOON,1;
 9          ACT/2,RNORM(.5,.1);                   PROCESSING
10          FREE,TOOL/1;                              FREE THE TOOL
11          COLCT,INT(1),TIME IN SYSTEM;          COLLECT STATISTICS
12          TERM;
13    ;
14          CREATE,,20,,1;                        CREATE 1ST BREAKDOWN
15    DOWN  PREEMPT(2),TOOL;                      PREEMPT THE TOOL
16          ACT/3,ERLNG(.75,3.);                  DOWN TIME
17          FREE,TOOL/1;                              FREE THE TOOL
18          ACT,RNORM(20.,2.),,DOWN;              TIME BETWEEN FAILURES
19          END;
20    INIT,0,500;
21    FIN;
```

Figure 6-10   SLAM II network model of a machine tool with breakdowns.

where the resource TOOL is captured. If TOOL was processing a job entity, the entity is interrupted. We will discuss what happens to the interrupted job following the description of this network segment. The machine breakdown entity is then processed by activity 3 which represents the repair time which is Erlang distributed as prescribed. Following repair, the machine tool resource is made available by routing the entity through a FREE node. The next machine breakdown is then scheduled by routing the entity back to the PREEMPT node DOWN through an activity that represents the time between breakdowns which is normally distributed with a mean of 20 and a standard deviation of 2 hours.

Consider now the disposition of the job entity that was preempted. At the PREEMPT node the remaining processing time attribute and the SEND node label were defaulted. The default values for these quantities are: to route the entity back to the AWAIT node where it captured the resource; and to save the remaining processing time. In this case, the job that was preempted is replaced in file 1 since that is the file number associated with the AWAIT node where the job entity captured the resource TOOL. The job entity is made the first entity in file 1. When the resource TOOL is made available, following the completion of the repair activity (activity 3), the job entity that was preempted will be removed from file 1 and will be placed in the activity from which it was preempted. The time to perform the activity will be the remaining service time for the activity. In this case, the job entity can be preempted from either activity 1 or activity 2 and, hence, following repair, it will be reinserted in either activity 1 or activity 2, with the time to perform the activity as the processing time remaining when the job was interrupted.

The control statements shown in Figure 6-10 indicate that five runs are to be made with each run lasting for 500 hours.

Summary of Results. A summary of the output reports from SLAM II is shown in Table 6-1. The results show a high variability in the average time in the system for jobs. This can be attributed to the high variability in both the interarrival times and the service times. A different job arrival sequence was used on each run so that an entirely different 500 hour simulation was performed on each run. The variability of the service times is due to service being the sum of the setup time and the processing time and then sometimes including a machine tool repair time. In 500 hours, 22 machine tool breakdowns are expected, that is, (500-20)/(2.25 + 20).

Table 6-1   Summary of 5 runs for machine breakdown example.

Run Number	Average Time in System	Average Number of Jobs Waiting	Tool Status Percentages		
			In-Use	Idle	In-Repair
1	12.06	12.89	90	1	9
2	7.46	6.78	86	4	10
3	6.39	5.63	86	4	10
4	5.33	4.35	82	7	11
5	9.48	9.68	82	7	11

When a breakdown occurs, all jobs in the system are delayed by the repair time. Based on the results, a large buffer area for holding jobs waiting for processing will be required. Furthermore, if tight due dates are set on jobs, many jobs will be late. Methods should be investigated for better scheduling of job arrivals and for reducing the number of breakdowns.

6.10   ALTER NODE

The ALTER node is used to change the capacity of resource type RES by CC units. CC can be a constant or a SLAM II variable. If CC is positive, the number of available units is increased. If CC is negative, the capacity is decreased. The symbol and statement for the ALTER node are shown below.

When the ALTER node is used to decrease availability (CC is negative), the change is invoked only if a sufficient number of units of the resource are not in use. If this is not the case, the capacity is reduced to the current number in use. Further reductions then occur when resources are freed at FREE nodes. In no case will the capacity of a resource be reduced below zero. Any additional reductions requested when the capacity has been reduced to zero are ignored.

RES can be specified in alphanumeric form by a resource label RLBL previously defined in a resource block. RES can also be specified as an attribute of an arriving entity which defines a resource code to be altered. For example, when an entity arrives to the node

   ALTER,MACHINE/-1;

the capacity of the resource MACHINE is decreased by one unit. If a machine is currently not in use, this change occurs immediately; otherwise the change is made as soon as a MACHINE is freed.

As another example, the node and statement

causes a change in the capacity of the resource corresponding to the value which is carried in ATRIB(2) of an arriving entity. The change in capacity is carried in ATRIB(3) of the arriving entity.

An ALTER node differs from a PREEMPT node in the following respects:

1) An entity arriving to an ALTER node requests a change in the capacity of the resource. The arriving entity is always routed from the ALTER node. An entity that arrives to a PREEMPT node either causes a preemption immediately, in which case the arriving entity is routed from the PREEMPT node, or the entity is queued in the file associated with the PREEMPT node.

2) An entity that arrives to an ALTER node causes the number of available units of the resource, NNRSC, to be decreased by the requested capacity change. Thus, NNRSC can have a negative value which indicates that all units are in use and there are requests to decrease the capacity of the resource.

3) It is not necessary to specify a disposition for the entity which has a resource at an ALTER node since altering a resource's capacity does not stop the processing of any entity. At a PREEMPT node, this is not the case and a send node must be determined.

6.11 EXAMPLE 6-3. PORT OPERATIONS

This problem statement is taken from Schriber (5). A Q-GERT model has also been presented (3). "A port in Africa is used to load tankers with crude oil for overwater shipment. The port has facilities for loading as many as three tankers simultaneously. The tankers, which arrive at the port every 11±7 hours†, are of three different types. The relative frequency of the various types, and their loading time requirements, are as follows:

Type	Relative Frequency	Loading Time, Hours
1	.25	18±2
2	.55	24±3
3	.20	36±4

There is one tug at the port. Tankers of all types require the services of this tug to move into a berth, and later to move out of a berth. When the tug is available, any berthing or deberthing activity takes about one hour. Top priority is given to the berthing activity.

"A shipper is considering bidding on a contract to transport oil from the port to the United Kingdom. He has determined that 5 tankers of a particular type would have to be committed to this task to meet contract specifications. These tankers would require 21±3 hours to load oil at the port. After loading and deberthing, they would travel to the United Kingdom, offload the oil, and return to the port for reloading. Their round-trip travel time, including offloading, is estimated to be 240±24 hours.

"A complicating factor is that the port experiences storms. The time between the onset of storms is exponentially distributed with a mean of 48 hours and a storm lasts 4±2 hours. No tug can start an operation until a storm is over.

---

† All durations given as ranges are uniformly distributed.

"Before the port authorities can commit themselves to accommodating the proposed 5 tankers, the effect of the additional port traffic on the in-port residence time of the current port users must be determined. It is desired to simulate the operation of the port for a one-year period (8640 hours) under the proposed new commitment to measure in-port residence time of the proposed additional tankers, as well as the three types of tankers which already use the port."

Concepts Illustrated. This example illustrates the use of the AWAIT and FREE nodes to model constrained resources. The ALTER node is used to reduce and increase the capacity of a resource during the simulation.

SLAM II Model. In this example, entities representing tankers flow through a network model of the port facilities. The port facilities are constrained by the three berths and one tug. In the previous examples, resources have been modeled as service activities. However, in this example, the tug is required for both the berthing and the deberthing operation and its availability can be altered by storms. In addition, a tanker requires both a berth and a tug before berthing can be undertaken. Therefore, a network model of the port operations can most easily be constructed by using explicit resources to model both the tug and the berths.

The network model for this example is presented in Figure 6-11 and the statement listing is given in Figure 6-12. The explanation of the model will be given in terms of the statement model. The first statement in the network section is the RESOURCE block (line 4) which defines the resource BERTH. The resource BERTH is assigned a capacity of 3 and entities waiting for a BERTH reside in file 1. The resource TUG is defined in statement 5 and has a capacity of one. Entities waiting for the TUG reside in either file 2 or file 3. Recall that the priority for allocating free resources to waiting entities is determined by the order in which these files are listed in the RESOURCE block. Therefore entities waiting in file 2 for a TUG have priority over entities waiting in file 3 for a TUG.

The statement model for this example can be divided into three major segments. The first segment represents the arrival process for the system and consists of statements 9 through 20. The second major segment models the port operations and consists of statements 24 through 45. The last segment models the storm process and includes statements 49 through 54.

The arrival process for this problem is composed of two classes of arrivals. The first arrival class represents the existing tanker traffic consisting of tanker types 1, 2, and 3. These entities are generated by the CREATE node (line 9) and are routed

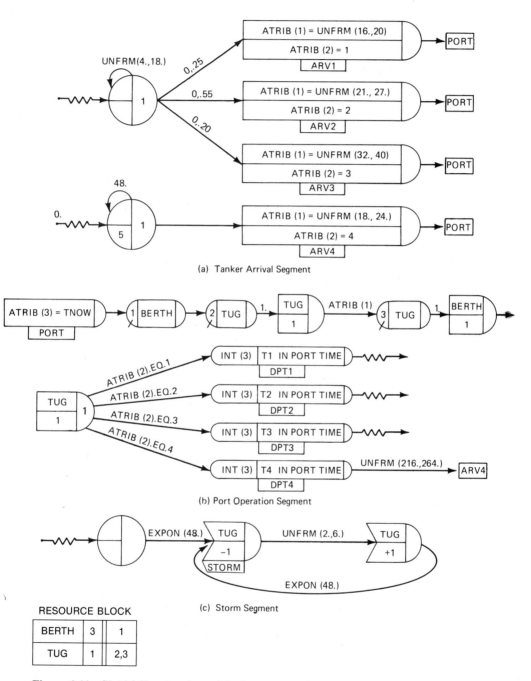

(a) Tanker Arrival Segment

(b) Port Operation Segment

(c) Storm Segment

RESOURCE BLOCK

| BERTH | 3 | 1 |
| TUG | 1 | 2,3 |

Figure 6-11 SLAM II network model of port operations.

```
 1   GEN,PRITSKER,AFRICA PORT,5/21/83,1;
 2   LIMITS,3,3,30;
 3   NETWORK;
 4         RESOURCE/BERTH(3),1;
 5         RESOURCE/TUG(1),2,3;
 6   ;
 7   ; TANKER ARRIVAL SEGMENT
 8   ; ---------------------
 9         CREATE,UNFRM(4.,18.);
10         ACT,,.25,ARV1;
11         ACT,,.55,ARV2;
12         ACT,,.20,ARV3;
13   ARV1  ASSIGN,ATRIB(1)=UNFRM(16.,20.),ATRIB(2)=1;
14         ACT,,,PORT;
15   ARV2  ASSIGN,ATRIB(1)=UNFRM(21.,27.),ATRIB(2)=2;
16         ACT,,,PORT;
17   ARV3  ASSIGN,ATRIB(1)=UNFRM(32.,40.),ATRIB(2)=3;
18         ACT,,,PORT;
19         CREATE,48,0,,5;
20   ARV4  ASSIGN,ATRIB(1)=UNFRM(18.,24.),ATRIB(2)=4;
21   ;
22   ; PORT OPERATION SEGMENT
23   ; ---------------------
24   PORT  ASSIGN,ATRIB(3)=TNOW;
25         AWAIT(1),BERTH/1;              WAIT FOR AN AVAILABLE BERTH
26         AWAIT(2),TUG/1;                WAIT FOR AN AVAILABLE TUG
27         ACT,1;                         TRAVEL TO BERTH
28         FREE,TUG/1;                    RELEASE THE TUG
29         ACT,ATRIB(1);                  TRANSFER THE CARGO
30         AWAIT(3),TUG/1;                WAIT FOR AN AVAILABLE TUG
31         ACT,1;                         TRAVEL TO SEA
32         FREE,BERTH/1;                  RELEASE THE BERTH
33         FREE,TUG/1;                    RELEASE THE TUG
34         ACT,,ATRIB(2).EQ.1,DPT1;
35         ACT,,ATRIB(2).EQ.2,DPT2;
36         ACT,,ATRIB(2).EQ.3,DPT3;
37         ACT,,ATRIB(2).EQ.4,DPT4;
38   DPT1  COLCT,INT(3),T1 IN PORT TIME;
39         TERM;
40   DPT2  COLCT,INT(3),T2 IN PORT TIME;
41         TERM;
42   DPT3  COLCT,INT(3),T3 IN PORT TIME;
43         TERM;
44   DPT4  COLCT,INT(3),T4 IN PORT TIME;
45         ACT,UNFRM(216.,264.),,ARV4;
46   ;
47   ; STORM SEGMENT
48   ; -------------
49         CREATE;
50         ACT,EXPON(48.);
51   STORM ALTER,TUG/-1,1;                REQUEST TUG CAPACITY DECREASE
52         ACT,UNFRM(2.,6.);              DURATION OF THE STORM
53         ALTER,TUG/+1;                  REQUEST TUG CAPACITY INCREASE
54         ACT,EXPON(48.),,STORM;         SCHEDULE THE NEXT STORM
55         END;
56   INIT,0,8640;
57   FIN;
```

Figure 6-12   Statement model for port operations example.

probabilistically by the three emanating ACTIVITY's to either ARV1, ARV2, or ARV3 ASSIGN nodes. At these ASSIGN nodes, ATRIB(1) is set equal to the appropriate loading time and ATRIB(2) is set equal to the appropriate tanker type. Following any of these ASSIGN nodes, the entity is routed to the ASSIGN node labeled PORT.

The second arrival class involves inserting five entities representing the proposed type 4 tankers into the network. The entities are created by the CREATE node (line 19) which generates an entity every 48 time units, with the first entity at time 0, and a maximum of five entities created. At an ASSIGN node (line 20) labeled ARV4, ATRIB(1) is set equal to the loading time and ATRIB(2) is set equal to the tanker type. The entities then continue to the PORT ASSIGN node (line 24).

The second major segment in the model represents the port operations and begins with the ASSIGN statement labeled PORT. Entities arriving to this statement represent tankers arriving to the port. The PORT ASSIGN node records the time of arrival to the port as ATRIB(3) of the entity. The entity then proceeds to the AWAIT node (line 25) where it waits for a BERTH in file 1. Thus, entities which arrive to the AWAIT node when no BERTH is available reside in file 1. When a BERTH is available, the entity continues to the next AWAIT node (line 26) where it waits in file 2 for the TUG. The ACTIVITY (line 27) following this AWAIT node represents the berthing operation and has a duration of one hour. Following berthing, the entity arrives at a FREE node (line 28) which frees one unit of the resource TUG. The ACTIVITY (line 29) that follows represents the tanker loading activity which has a duration of ATRIB(1). Recall that the appropriate loading time for the tanker entity was previously assigned to ATRIB(1) in the arrival segment of the model. Following the loading operation, a tanker requires a tug before the deberthing operaton can begin. The next AWAIT node (line 30) models this requirement by causing the entity to wait in file 3 for a TUG. Since file 3 is listed after file 2 in the RESOURCE block for the TUG, the TUG will be allocated to the deberthing operation only if the TUG is not required for a berthing operation. The ACTIVITY (line 31) with a duration of one hour represents the deberthing operation. When a TUG is finished deberthing, the BERTH (line 32) and the TUG (line 33) are freed.

Statements 34 through 45 represent the tanker departure process from the port. After freeing the TUG, the tanker entity is conditionally branched based on tanker type by the four ACTIVITY's (lines 34-37) to the appropriate departure COLCT node where interval statistics on port residence time are recorded. The entities corresponding to the existing tanker traffic of types 1, 2, and 3 are terminated. The round trip travel time to the United Kingdom and back for tankers of type 4 is represented by the ACTIVITY (line 45) which routes the entity back to the ARV4

ASSIGN node. Therefore the five type 4 tankers continue to cycle through the model until the simulation is terminated after 8640 hours of operations.

The storm segment of the model starts with the creation of a storm entity at a CREATE node (line 49). The first storm is delayed by an exponentially distributed time with a mean of 48 by an ACTIVITY (line 50). The TF option of the CRE-ATE node cannot be used in this case since TF is restricted to be a constant by the SLAM II Processor. At the node with label STORM (line 51), the TUG resource is requested to be altered by −1 units. This decrease in capacity will occur immediately if the tug is not in use or at the end of the tug's current operation. Thus, the tug does not abandon a tanker in stormy waters. The storm duration is uniformly distributed between 2 and 6. This ACTIVITY starts immediately and does not depend on the status of the resource TUG. Following the storm, the TUG resource capacity is increased by 1 at an ALTER node (line 53). The next storm is then scheduled by an ACTIVITY (line 54) and the storm entity is routed back to node STORM. This completes the description of the model.

Summary of Results. The SLAM II Summary Report for this example is shown in Figure 6-13. The first category of statistics is for variables based on observations and consists of the in-port times collected on each tanker type at the COLCT nodes. This is followed by the file statistics for files 1, 2, and 3 which correspond to tankers awaiting a berth, tankers with a berth awaiting a tug for berthing, and loaded tankers awaiting a tug for deberthing, respectively. The last category of statistics for this example is the resource statistics. The results show that on the average of 2.8745 of the 3 BERTH's were utilized and that the TUG was busy 21.48 percent of the time.

The outputs indicate a high utilization of berths and the potential for a large queue of tankers waiting for berths. At one time during the year as many as nine tankers were waiting for a berth. Based on this information, port management should not accept the proposed five new tankers unless an additional berth is constructed or loading times are reduced. The model should be rerun to ascertain the effects of such changes.

S L A M I I S U M M A R Y R E P O R T

SIMULATION PROJECT AFRICA PORT   BY PRITSKER

DATE 5/21/1983   RUN NUMBER 1 OF 1

CURRENT TIME 0.8640E+04
STATISTICAL ARRAYS CLEARED AT TIME 0.0000E+00

STATISTICS FOR VARIABLES BASED ON OBSERVATION

	MEAN VALUE	STANDARD DEVIATION	COEFF. OF VARIATION	MINIMUM VALUE	MAXIMUM VALUE	NUMBER OF OBSERVATIONS
T1 IN PORT TIME	0.3607E+02	0.1452E+02	0.4026E+00	0.1826E+02	0.8634E+02	183
T2 IN PORT TIME	0.4153E+02	0.1540E+02	0.3709E+00	0.2343E+02	0.9896E+02	449
T3 IN PORT TIME	0.5259E+02	0.1399E+02	0.2660E+00	0.3437E+02	0.1019E+03	142
T4 IN PORT TIME	0.3852E+02	0.1403E+02	0.3642E+00	0.2006E+02	0.1009E+03	153

FILE STATISTICS

FILE NUMBER	ASSOC NODE LABEL/TYPE	AVERAGE LENGTH	STANDARD DEVIATION	MAXIMUM LENGTH	CURRENT LENGTH	AVERAGE WAITING TIME
1	AWAIT	1.6095	1.6801	9	2	14.9211
2	AWAIT	0.0315	0.1816	2	0	0.2931
3	AWAIT	0.0305	0.1834	3	0	0.2844
4	CALENDAR	9.0929	0.7388	11	9	6.3102

RESOURCE STATISTICS

RESOURCE NUMBER	RESOURCE LABEL	CURRENT CAPACITY	AVERAGE UTILIZATION	STANDARD DEVIATION	MAXIMUM UTILIZATION	CURRENT UTILIZATION
1	BERTH	3	2.8745	0.3731	3	3
2	TUG	1	0.2148	0.4107	1	1

RESOURCE NUMBER	RESOURCE LABEL	CURRENT AVAILABLE	AVERAGE AVAILABLE	MINIMUM AVAILABLE	MAXIMUM AVAILABLE
1	BERTH	0	0.1255	0	3
2	TUG	0	0.7073	-1	1

Figure 6-13 SLAM II summary report for port operations model.

6.12   GATE BLOCK

A GATE block is used to define the GATE named GLBL, the initial status of the GATE, and the file numbers associated with entities waiting for a gate to be opened at AWAIT nodes. The naming convention for gates is the same as for resources. Also, the SLAM II assigned numeric codes for GATES are made in accordance with the GATE block's location in the network input statements. The GATE on the first GATE statement is given a numeric code of 1, the second is given a code of 2, and so on. GATE blocks are not connected to other nodes and are used only to provide the above definitional information. The symbol and alternative statements for the GATE block are shown below:

| GLBL | OPEN or CLOSE | IFL1 | IFL2 | GATE/GLBL,OPEN or CLOSE,IFLs;

6.13   OPEN NODE

An OPEN node is used to open a GATE with name GLBL or a GATE code specified by an attribute of the arriving entity. Each entity arriving to an OPEN node causes GATE to be opened. When this occurs, all entities waiting for GATE are removed from the files associated with the AWAIT nodes for GATE and are routed in accordance with the M-number of the AWAIT node. The entity that caused GATE to be opened is then routed from the OPEN node. The symbol and statement for the OPEN node are shown below.

GATE M   OPEN,GATE,M;

6.14   CLOSE NODE

A CLOSE node is used to close a GATE with name GLBL or a GATE code specified by an attribute of the arriving entity. An entity ariving to a CLOSE node causes the GATE referenced to be closed. Any entity arriving to an AWAIT node after the GATE is closed will wait for it to be opened. The entity that causes the GATE to be closed at the CLOSE node is routed in accordance with the

M-number associated with the CLOSE node. The symbol and statement for the CLOSE node are shown below.

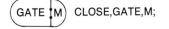 CLOSE,GATE,M;

6.15 ILLUSTRATION 6-5. GATES TO MODEL SHIFTS

Consider the situation in which packages arrive to a post office over a 24-hour period; however, they are only weighed, stamped, and loaded into trucks during the day shift. This can be modeled by having the packages as represented by entities, created and then routed to an AWAIT node that is associated with a gate called DSFT. In a disjoint network, an entity is created that closes the gate after eight hours and then opens it sixteen hours later. The network and statement models for this illustration are shown below. The model involving the processing of the packages through the operations on the day shift has not been detailed.

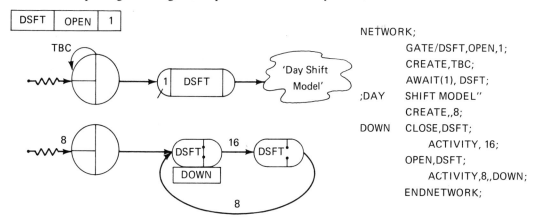

```
NETWORK;
       GATE/DSFT,OPEN,1;
       CREATE,TBC;
       AWAIT(1), DSFT;
;DAY   SHIFT MODEL"
       CREATE,,8;
DOWN   CLOSE,DSFT;
       ACTIVITY, 16;
       OPEN,DSFT;
       ACTIVITY,8,,DOWN;
       ENDNETWORK;
```

6.16 EXAMPLE 6-4. SINGLE-LANE TRAFFIC ANALYSIS

The system to be modeled in this example consists of the traffic flow from two directions along a two-lane road, one lane of which has been closed for 500 meters for repairs (1). Traffic lights have been placed at each end of the closed lane to control the flow of traffic through the repair section. The lights allow traffic to flow for a specified time interval from only one direction. This arrangement is depicted in Figure 6-14. When a light turns green, the waiting cars start and pass the light every two seconds. If a car arrives to a green light when there are no waiting cars, the car passes through the light without delay. The car arrival pattern is exponentially distributed, with an average of 12 seconds between cars from direction 1 and 9 seconds between cars from direction 2. A light cycle consists of green in direction 1, both red, green in direction 2, both red, and then the cycle is repeated. Both lights remain red for 55 seconds to allow the cars in transit to leave the repair section before traffic from the other direction can be initiated.

The objective is to simulate the above system to determine values for the green time for direction 1 and the green time for direction 2 which yield a low average waiting time for all cars.

Concepts Illustrated. This example illustrates the use of the OPEN and CLOSE nodes in conjunction with a gate to control entity flow through a system. Multiple simulation runs are made and activity durations are specified as XX(·) values to facilitate the changing of durations between runs.

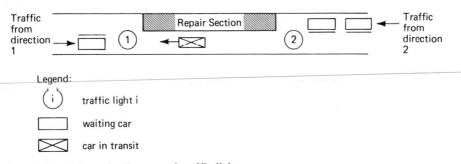

Figure 6-14   Schematic diagram of traffic lights

SLAM II Model. There are three separate processes in this system consisting of: traffic flow from direction 1; traffic flow from direction 2; and information flow representing the traffic light cycle. Each of these processes is modeled by the movement of an entity through a subnetwork. We will model the traffic lights by the gates LIGHT1 and LIGHT2 where an open gate represents a green light and a closed gate represents a red light. To insure that only one car passes through the light at a time, a resource with a capacity of one is employed in conjunction with each gate. These resources are named START1 and START2 corresponding to LIGHT1 and LIGHT2 and represent the starting location before each light. The starting location is seized by each car entity before passing through the light and then freed immediately after it passes the light. In this way only one car can pass through the starting location at a time.

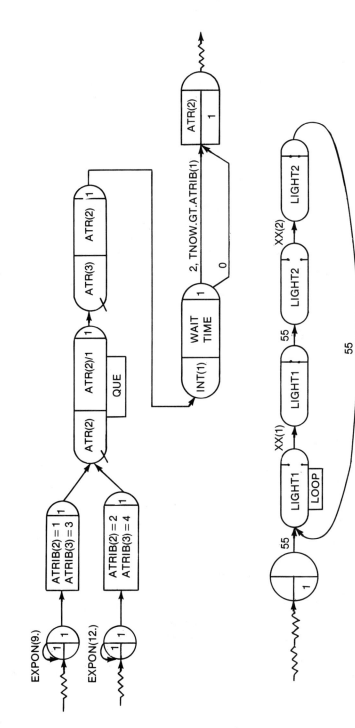

Figure 6-15   SLAM II network model of a traffic light situation.

The network model for this example is depicted in Figure 6-15. Since the decision logic for the traffic flow from both directions is the same, a single network will be used to model both traffic flows with attributes employed to specify the resources and gates required. We will use ATRIB(2) to maintain the resource number and file number associated with the first location before each light. If ATRIB(2) equals 1, the car entity requires the resource START1. If ATRIB(2) equals 2 then resource START2 is required. The numeric values for the resources are defined by the order in which they are specified in the resource block or resource statement. A similar procedure is used for gates, and the GATE LIGHT1 has a numeric code of 1 and the GATE LIGHT2 has a numeric code of 2. Thus we can use ATRIB(2) to indicate the numeric value associated with entities flowing in a particular direction, that is, ATRIB(2)=I where I is the direction of traffic flow and the entities require RESOURCE I and GATE I.

Entities representing cars are created at two CREATE nodes, one for each direction. The time between car arrivals is exponentially distributed, and for direction 1 has a mean of 9 seconds and for direction 2 has a mean of 12 seconds. Following the creation of the entities, ATRIB(2) is set to 1 for direction 1 and set to 2 for direction 2. Entities that wait for a light change will be put in file 3 or 4 and we use ATRIB(3) to indicate these numbers. Thus, entities are assigned an ATRIB(3) value of 3 for direction 1 and 4 for direction 2. Entities from both directions are then routed to the AWAIT node with label QUE where they wait for the START resource as defined by ATRIB(2). Once an entity is allocated the starting location, it proceeds to the next AWAIT node where it waits for the gate defined by ATRIB(2), that is, either LIGHT1 or LIGHT2. If the appropriate light is closed then the entities will wait in file 3 or file 4 in accordance with the value given by ATRIB(3). Note that SLAM II uses the internal information presented in the GATE block that specifies that entities should wait in file 3 if gate number 1 is requested and should wait in file 4 if gate number 2 is requested.

A COLCT node is used to record values of the waiting time of the car at the light and the entity is then routed through one of the two emanating ACTIVITY's since an M-number of 1 is specified. A car that stopped has an arrival time different from the current time, TNOW. The condition specified on the first activity is for those cars that stopped and causes a two second delay for the car to pass the light. Since the M-value of the COLCT node is 1, the second ACTIVITY is taken if and only if the first is not taken. This ACTIVITY models the passage of moving cars that do not incur a delay. The resource defined by ATRIB(2) is then freed and the entity is terminated.

The traffic light segment of the model controls the changes in the traffic lights and consists of a series of OPEN and CLOSE nodes separated by ACTIVITY's. In this segment of the model, resources and gates are referred to by the labels given them in the RESOURCE and GATE blocks. A single entity is entered into the subnetwork by a CREATE node. It is delayed 55 seconds (both lights are red) before opening GATE LIGHT1. The information entity is then delayed by XX(1) seconds before closing GATE LIGHT1. Next, it is delayed by 55 seconds, opens LIGHT2, is delayed XX(2) seconds, closes LIGHT2, and then loops back after a delay of 55 seconds. By specifying the green time for LIGHT1 and LIGHT2 as XX(1) and XX(2) respectively, we can experiment with different values by prescribing new values for XX(1) and XX(2) in INTLC statements.

The SLAM II statement model for this example is given in Figure 6-16 and follows directly from the network model. As specified on the input, three separate runs were made with the following values for XX(1) and XX(2).

Run Number	XX(1)	XX(2)
1	60	45
2	80	60
3	40	30

Summary of Results. The results of primary interest for this problem are the average waiting times for cars in direction 1 and direction 2. The results for the three runs are summarized in the following table. Note that the statistics include the cars that waited for zero time units.

Run Number	Green Times		Average Waiting Time	
	Direction 1	Direction 2	Direction 1	Direction 2
1	60	45	62.35	74.92
2	80	60	66.06	75.79
3	40	30	185.43	185.30

These results indicate that the best combination of green times was obtained from run 1 and suggests additional simulations should be performed with green times in the range of 60 seconds for direction 1 and 45 seconds for direction 2. After further exploration, we should check the statistical stability of the results by making several runs (replications) with different random number seed values for the values of XX(1) and XX(2) that resulted in the smallest average waiting times.

```
1   GEN,OREILLY,TRAFFIC LIGHTS,7/12/83,3,,NO,,NO;
2   LIMITS,4,3,100;
3   INTLC,XX(1)=60,XX(2)=45;
4   NETWORK;
5         RESOURCE/START1,1/START2,2;              DEFINE STARTING PLACES
6         GATES/LIGHT1,CLOSE,3/LIGHT2,CLOSE,4;     DEFINE TRAFFIC LIGHTS
7   ;TRAFFIC LANES
8   ;----------------------
9         CREATE,EXPON(9),,1;                      CREATE ARRIVALS, DIRECTION 1
10        ASSIGN,ATRIB(2)=1,ATRIB(3)=3;            REQUIRES START1, LIGHT1
11        ACT,,,QUE;
12        CREATE,EXPON(12),,1;                     CREATE ARRIVALS, DIRECTION 1
13        ASSIGN,ATRIB(2)=2,ATRIB(3)=4;            REQUIRES START2, LIGHT2
14  QUE   AWAIT(ATRIB(2)=1,2),ATRIB(2);           AWAIT STARTING PLACE
15        AWAIT(ATRIB(3)=3,4),ATRIB(2);           AWAIT GREEN LIGHT
16        COLCT,INT(1),WAIT TIME,,1;              COLLECT WAIT STATISTICS
17        ACT,2,TNOW.GT.ATRIB(1);                 CAR BEGAN STOPPED
18        ACT;                                     CAR BEGAN MOVING
19        FREE,ATRIB(2);                           FREE THE STARTING PLACE
20        TERM;
21  ;TRAFFIC LIGHTS
22  ;--------------
23        CREATE,,,,1;
24        ACT,55;                                  BOTH LIGHTS RED
25  LOOP  OPEN,LIGHT1;                             LIGHT 1 TURNS GREEN
26        ACT,XX(1);                               GREEN TIME 1
27        CLOSE,LIGHT1;                            LIGHT 1 TURNS RED
28        ACT,55;                                  BOTH LIGHTS RED
29        OPEN,LIGHT2;                             LIGHT 2 TURNS GREEN
30        ACT,XX(2);                               GREEN TIME 2
31        CLOSE,LIGHT2;                            LIGHT 2 TURNS RED
32        ACT,55,,LOOP;                            BEGIN NEW CYCLE
33        END;
34  INIT,0,3600;
35  SIMULATE;
```

Figure 6-16 SLAM II statement model of traffic light situation.

6.17   CHAPTER SUMMARY

The symbols and statements associated with resources and gates are summarized in Table 6-2. Five node types are used for modeling with resources and three node types are associated with gates. Resources and gates greatly increase the modeling flexibility available with SLAM II as shown by the illustrations and three examples presented in this chapter. A complete description of each network element presented in this chapter is provided in Appendix B.

Table 6-2 Symbols and statements for resources and gates.†

Name	Symbol	Statement
ALTER	RES / CC / M	ALTER, RES/CC, M;
AWAIT	IFL QC RES/UR or GATE M	AWAIT (IFL/QC), RES/UR or GATE, BLOCK or BALK (NLBL), M;
CLOSE	GATE M	CLOSE, GATE, M;
FREE	RES UF M	FREE, RES/UF, M;
GATE	GLBL OPEN or CLOSE IFL1 IFL2	GATE/GLBL, OPEN or CLOSE, IFLs;
OPEN	GATE M	OPEN, GATE, M;
PREEMPT	IFL PR RES NATR M SNLBL	PREEMPT (IFL)/PR, RES, SNLBL, NATR, M;
RESOURCE	RLBL CAP IFL1 IFL2	RESOURCE/RLBL (CAP), IFLs;

† Definition of codes:

IFL	File number
RES	Generic reference to a RESOURCE type
RLBL	Resource LaBeL as defined at a RESOURCE block
UR	Units Requested
UF	Units to Free

CC	Change Capacity by CC
PR	PRiority specification for preempting
SNLBL	Node to send preempted entity
NATR	ATtRibute number to store remaining processing time
IRC	Initial Resource Capacity (units available)
GATE	Generic reference to a GATE type
GLBL	Gate LaBeL as defined at a GATE block

6.18   EXERCISES

6-1.   Entities are generated at a node whose label is START. The first entity is to be generated at time 5. Thereafter the time between entity arrivals is exponentially distributed with a mean of 2 time units. An unlimited number of entities can be generated at node START. Entities are routed to QUEUE node Q1 if gate G1 is open. The time to reach node Q1 is equal to the capacity of the queue which is 5 minus the current number of entities in Q1. Entities that balk from Q1 leave the system. Initially, there are no entities at Q1 and file 1 is used to store entities waiting at Q1. Two servers process entities waiting at Q1. Processing time of these servers is normally distributed with a mean of 3 and a standard deviation of 1. After service, an entity leaves the system. If an entity's time in the system is greater than 10, gate G1 is closed. Gate G1 is open when an entity spends less than 2 time units in the system or when an arrival finds no one waiting at queue node Q1. It is desired to collect the time between departures for all entities that have arrived to the system. Draw the SLAM II network of the processing of entities as described above.

6-2.   A barber has categorized his customers according to the type of haircut desired. He has determined that the time for a regular haircut is uniformly distributed between 15 and 20 minutes whereas the time for customers who desire a stylized haircut is exponentially distributed with a mean of 20 minutes. The barber has determined that 60 percent of his customers request a hair styling. Assuming that the time between customer arrivals is triangularly distributed with a mode of 20, a minimum of 15 and a maximum of 40, draw a SLAM II network to represent this situation. Include in the network the collection of statistics on the time spent in the system by each type of customer and by both types of customers collectively.

6-3.   Modify the inventory model presented in Example 6-1 to include a variable representing stock-on-hand. Determine whether it is still necessary to use resources to model the inventory situation.

6-4.   A certain machine repair shop consists of a work station where incoming units are repaired and an inspection station where the repaired units are either released from the shop or recycled. The work station has three parallel servers, and the inspection station has one inspector. Units entering this system have interarrival times which are exponentially distributed with a mean of 10.25 time units. The repair time for a unit is

Erlang distributed with mean 22 and variance 242. The "shortest processing time" priority dispatching rule is used at the work station: the unit with the smallest repair time is served first. Repaired units queue up for inspection on a FIFO basis. The inspection of a unit requires 6 time units; the unit is then rejected with a probability p^n, where $p = .15$ and $n =$ the number of times the unit has already been repaired. Rejected units queue up at the work station to be repaired again.

The initial conditions are:

1. Two servers are busy with service completions scheduled for times 1.0 and 1.5, respectively.
2. The first new arrival will occur at time 0.
3. The inspector is idle.

Simulate the operation of this shop for 2000 time units to obtain estimates of the following quantities:

1. Server utilization;
2. Mean, standard deviation, and histogram of total waiting time for each repaired unit;
3. Mean, standard deviation, and histogram of total time spent in the system by each unit;
4. Average number of units in the system;
5. Mean, standard deviation, and histogram of number of repair cycles required of a unit before it leaves the shop.

Embellishments:

(a) Modify the priority dispatching rule used at the work station so that recycled items are processed ahead of new arrivals. Among recycled units priority is based on time spent in the system.
(b) Let the repair time be uniformly distributed in the range 0-48.
(c) Modify the original problem so that all units arriving before time 2000 are processed, and statistics concerning these items should be included in the overall statistics for the simulation.
(d) Evaluate the effect of a lognormally distributed inspection time with mean 6 and standard deviation 1.5.

6-5.   Perform an analysis of the situation presented in Illustration 6-3 to determine the number of dedicated mills required to produce 173 castings in 80 hours.

6-6.   For the flexible milling system described in Illustration 6-3, assume that the use of a flexible mill to perform an operation increases the operation time by 10 percent. Illustrate the modeling changes required to model this situation.

Embellishment: Instead of an increase in processing time when an operation is performed by a flexible mill, add a setup time to the processing time for a flexible mill if the operation number performed by the flexible mill changes. The setup time for changing a flexible mill from operation 10 to 20 is five minutes, from 20 to 30 is eight minutes, from 10 to 30 is seven minutes, from 20 to 10 is ten minutes, from 30 to 10 is eight minutes, and from 30 to 20 is six minutes. Make the modeling changes required to include this setup time for the flexible mills.

6-7.   Cargo arrives at an air terminal in unit loads at the rate of two unit loads per minute. At the freight terminal there is no fixed schedule, and planes take off as soon as they can be loaded to capacity. Two types of planes are available for transporting cargo. There are three planes with a capacity of 80 unit loads and two planes with a capacity of 140 unit loads. The round trip time for any plane is normally distributed with a mean of 3 hours, a standard deviation of 1 hour, and minimum and maximum times of 2 and 4 hours, respectively. The loading policy of the terminal manager is to employ smaller planes whenever possible. Only when 140 unit loads are available will a plane of type 2 be employed. Develop a SLAM II network to model this system to estimate the number of unit loads waiting and the utilization of the two types of planes over a 100 hour period. Assume at first that the loading time of planes is negligible. Embellish the model to include a one minute per unit loading time.

6-8.   Change the model presented in Example 6-2 so that the utilization of the machine tool does not include the repair time of the machine tool.

6-9.   A machine tool processes two different types of parts. The time between arrivals of Type 1 parts is triangularly distributed with a mode of 30 minutes, a minimum of 20 minutes, and a maximum of 50 minutes. The interarrival time of Type 2 parts is a sample from a triangular distribution with a mode of 50 minutes, a minimum of 35 minutes, and a maximum of 60 minutes. Processing time for Type 1 parts is exponentially distributed with a mean of 20 minutes. For Type 2 parts processing time is a sample from a uniform distribution with a minimum of 15 minutes and a maximum of 20 minutes. Processing time includes an inspection of the completed part. Fifteen percent of the parts fail inspection and return to the end of the queue of parts awaiting processing. Assume that parts which fail inspection have a rework time equal to 90% of the previous processing time. Develop a SLAM II network to collect statistics on the time spent in the system by a part and the utilization of the machines. Simulate this system for 2400 minutes.

Embellishments:

(a) Include a downtime of one-half hour after 4 simulated hours, recurring every 8 hours. This allows for preventive maintenance. Assume that preventive maintenance is performed after the completion of the current job and that the machine is always placed back into service at the end of the scheduled maintenance period.

(b) Any part which fails inspection more than once is scrapped. Collect statistics on the time between scraps. Run this model and identify the effect on throughput and in-process inventory.

6.19   REFERENCES

1. Bobillier, P. A., B. C. Kahan and A. R. Probst, *Simulation with GPSS and GPSS V,* Prentice-Hall, 1976.
2. Pritsker, A. A. B., *The GASP IV Simulation Language,* John Wiley, 1974.
3. Pritsker, A. A. B., *Modeling and Analysis Using Q-Gert Networks (Second Edition),* Halsted Press and Pritsker & Associates, Inc., 1979.
4. Pritsker, A. A. B. and C. E. Sigal, *Management Decision Making: A Network Simulation Approach,* Prentice-Hall, 1983.
5. Schriber, T., *Simulation Using GPSS,* John Wiley, 1974.

CHAPTER 7

Logic and Decision Nodes

7.1   INTRODUCTION

Many systems to be modeled involve logical operations and complex rules for deciding among sets of alternatives. As examples of the type of logic and decision processes that need to be modeled, consider the following rules and situations:

- Wait until three people arrive before sending a bus to the airport.
- Do not start a military engagement until confirming intelligence signals from two different sources are received.
- A pallet requires 24 cartons to be loaded before it is to be moved.

- Route a job to the shortest queue.
- Select Professor A's course unless there are over 80 students in it in which case select Professor B's course.
- Assign a patient an examining room only after the patient's insurance has been verified and the patient's records have been received.
- Start an assembly operation after receiving at least one nut, one bolt, and one washer.
- Select the next job to be done from different types of jobs based on the longest waiting time of the first job of each type.
- Start the car after the doors have been closed, seat buckles fastened, and the key is inserted.

As can be seen from the above list, the modeling of diverse systems involves extensive logic and procedural specifications. SLAM II provides the capability to model such situations through the use of logic and decision nodes.

There are five nodes in SLAM II which perform logic and decision operations. The ACCUMULATE node is used to accumulate a specified number of entities. The BATCH node generalizes the concept of the ACCUMULATE node and allows the identity of each entity put into a batch to be maintained. The UNBATCH node causes individual entities of a batch to be routed into the network. The MATCH node is used to identify entities that have a common characteristic and to cause entities to wait until a prescribed set of entities with the common characteristic have reached the MATCH node. The SELECT node is used for routing purposes. It provides rules for selecting a QUEUE node from a set of parallel QUEUE nodes. It also provides decision logic to select an entity from a set of parallel QUEUE nodes when a server becomes available. A third use of the SELECT node is to select a service activity from among a set of available service activities when an entity arrives to a QUEUE node. One of the queue selection rules for the SELECT node is ASSEMBLY which provides for a selection based on the availability of entities from a set of different QUEUE nodes.

The logic and decision nodes described in this chapter perform operations on an entity or a set of entities. This differs from the branching logic on an activity which is applied to a single entity to determine its route through a network.

7.2   ACCUMULATE NODE

The ACCUMULATE node or ACCUM node accumulates entities until a prescribed number is reached. When the number is achieved, the node is released.

The release of an ACCUMULATE node causes branching from the node to be initiated. At the ACCUMULATE node, a release specification is required. This specification involves the number of incoming entities needed to release the node for the first time (FR), the number required for subsequent releases (SR), and a rule for deciding which entity's attributes to save when more than one incoming entity is required to release the node (SAVE). The possible rule specifications are:

1. Save the attributes of the first entity arriving to the node (FIRST);
2. Save the attributes of the entity that causes the release of the node (LAST);
3. Save the attributes of the incoming entity that has the highest value of attribute I (HIGH(I));
4. Save the attributes of the incoming entity that has the lowest value of attribute I (LOW(I));
5. Create a new entity whose attributes are equal to the sum of the attributes of all incoming entities (SUM);
6. Create a new entity whose attributes are equal to the product of the incoming attributes of all incoming entities (MULT).

As an example of the release mechanism specification, consider that two incoming entities are required to release the node for the first time and only one incoming entity is required to release the node on subsequent times. On the first release, it is desired to save the attributes of the second arriving entity. The specification of the release mechanism for this situation is: 2,1,LAST. The specification of the release mechanism is given by FR,SR, and SAVE where:

$$FR = \text{First release requirement;}$$
$$SR = \text{Subsequent release requirement; and}$$
$$SAVE = \text{Rule for saving attributes when entities are accumulated.}$$

The symbol and statement for the ACCUMULATE node are:

ACCUMULATE,FR,SR,SAVE,M;

The above example would be modeled as:

ACC,2,1,LAST;

where the default value of $M = \infty$ is used. If the SAVE criterion was specified as SUM, then the sum of the attributes of the first two entities arriving would be maintained as the attributes of the entity routed from the node. By using this criterion and by specifying zero attribute values for selected attributes for each entity, a mixture of attribute values can be obtained.

The ACCUMULATE node is used extensively in project planning networks (PERT,CPM) where multiple activities must be completed prior to the start of additional activities. The SAVE criterion can be used to maintain information about entities that cause the ACCUMULATE node to be released.

FR and SR may also be specified by an attribute of the first arriving entity following a release of the accumulate node. For example,

    ACCUM,2,ATRIB(1),SUM,1;

specifies that 2 entities cause the first release and that subsequent releases require ATRIB(1) entity arrivals where ATRIB(1) of the first of these arrivals provides the SR value.

7.3   EXAMPLE 7-1.   ANALYSIS OF A PERT-TYPE NETWORK

PERT is a technique for evaluating and reviewing a project consisting of interdependent activities (4). A number of books have been written that describe PERT modeling and analysis procedures (1, 5, 8). A PERT network is a graphical illustration of the relations between the activities of a program.

A PERT network model of a repair and retrofit project (7) is shown in Figure 7-1 and activity descriptions are given in Table 7-1. All activity times will be assumed to be triangularly distributed. For ease of description, activities have been aggregated. The activities relate to power units, instrumentation, and a new assembly and involve standard types of operations.

In the following description of the project, activity numbers are given in parentheses. At the beginning of the project, three parallel activities can be performed that involve: the disassembly of power units and instrumentation (1); the installation of a new assembly (2); and the preparation for a retrofit check (3). Cleaning, inspecting, and repairing the power units (4) and calibrating the instrumentation (5) can be done only after the power units and instrumentation have been disassembled. Thus, activities 4 and 5 must follow activity 1 in the network. Following the installation of the new assembly (2) and after the instruments have been calibrated (5), a check of interfaces (6) and a check of the new assembly (7) can be

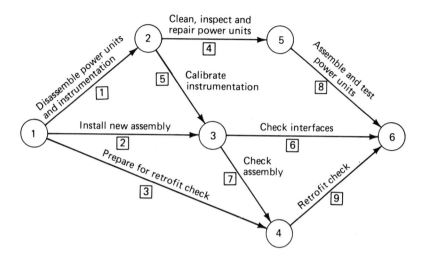

Figure 7-1. PERT network model of a retrofit project.

made. The retrofit check (9) can be made after the assembly is checked (7) and the preparation for the retrofit check (3) have been completed. The assembly and test of power units (8) can be performed following the cleaning and maintenance of power units (4). The project is considered completed when all nine activities are completed. Since activities 6, 8, and 9 require the other activities to precede them, their completion signifies the end of the project. This is indicated on the network by having activities 6, 8, and 9 incident to node 6, the sink node for the project. The objective of this example is to illustrate the procedures for using SLAM II to model and simulate project planning networks.

Table 7-1 Description of retrofit project activities.

Activity Number	Description	Mode	Minimum	Maximum	Average
1	Disassemble power units and instrumentation	3	1	5	3
2	Install new assembly	6	3	9	6
3	Prepare for retrofit check	13	10	19	14
4	Clean, inspect, and repair power units	9	3	12	8
5	Calibrate instrumentation	3	1	8	4
6	Check interfaces	9	8	16	11
7	Check assembly	7	4	13	8
8	Assemble and test power units	6	3	9	6
9	Retrofit check	3	1	8	4

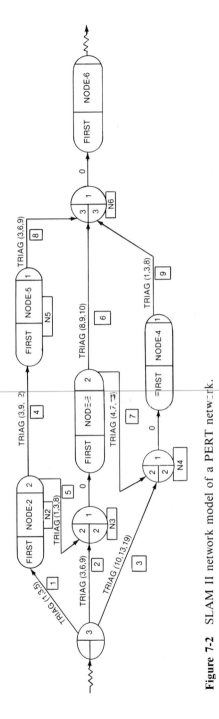

Figure 7-2 SLAM II network model of a PERT network.

Concepts Illustrated. Multiple runs are illustrated by this example. The ACCU-MULATE node is used to model activity precedence relations. The suppression of intermediate reports is accomplished by setting variables on the GEN statement. Activity start times are collected over multiple runs using a FIRST specification at COLCT nodes.

SLAM II Model. The SLAM II network model corresponding to the PERT network is shown in Figure 7-2. The SLAM II network is similar to the PERT network with the addition of: 1) the number of first and subsequent releases (equal to the number of incoming branches); and 2) a specification that FIRST statistics are to be collected after a node is released. When a single incoming activity releases the node, a COLCT node can be used directly. When more than one activity is required, an ACCUM node is also required.

In this SLAM II model, the CREATE node is only used to generate a single entity and, hence, no further specification is required (all default values are appropriate). Following the collection of statistics at the COLCT node after node N6, a run is completed. Four hundred runs are made as specified on the GEN statement. Statistics are not reset at the end of each run because the input for JJCLR on the INIT statement was NO.

The SLAM II statement model for this example is given in Figure 7-3. The coding of the statement model follows directly from the network model.

```
 1   GEN,OREILLY,PERT NETWORK,3/15/1982,400,,NO,,NO,YES/400;
 2   ;
 3   ;     PERFORM 400 ITERATIONS
 4   ;
 5   LIM,,1,700;
 6   NETWORK;
 7         CREATE;
 8         ACT,TRIAG(1.,3.,5.),,N2;        DISASSEMBLE PWR UNITS & INSTRU
 9         ACT,TRIAG(3.,6.,9.),,N3;        INSTALL NEW ASSEMBLY
10         ACT,TRIAG(10.,13.,19.),,N4;  PREPARE RETROFIT CHECK
11   N2    COLCT,FIRST,NODE 2,10,0.0,0.5;
12         ACT,TRIAG(3.,9.,12.),,N5;       CLEAN, INSPECT & REPAIR PWR UNITS
13         ACT,TRIAG(1.,3.,8.),,N3;        CALIBRATE INSTRU
14   N3    ACCUM,2,2;
15         COLCT,FIRST,NODE 3,20,3,0.5;
16         ACT,TRIAG(8.,9.,16.),,N6;       CHECK INTERFACES
17         ACT,TRIAG(4.,7.,13.),,N4;       CHECK ASSEMBLY
18   N4    ACCUM,2,2;
19         COLCT,FIRST,NODE 4,20,10.,0.5;
20         ACT,TRIAG(1.,3.,8.),,N6;        RETROFIT CHECK
21   N5    COLCT,FIRST,NODE 5,20,12.,0.5;
22         ACT,TRIAG(3.,6.,9.),,N6;        ASSEMBLY & TEST PWR UNITS
23   N6    ACCUM,3,3;
24         COLCT,FIRST,PROJ. COMPLETION,20,15.,0.5;
25         TERM;
26         ENDNETWORK;
27   INIT,,,NO;
28   FIN;
```

Figure 7-3 SLAM II statement model for PERT Network example.

Summary of Results. The final summary report for 400 independent simulations of the network is shown in Figure 7-4. The average time to complete the project is 20.78 time units with a standard deviation of 2.14 time units. By the central limit theorem, the average project duration is approximately normally distributed.

S L A M I I S U M M A R Y R E P O R T

SIMULATION PROJECT PERT NETWORK	BY OREILLY
DATE 3/15/1982	RUN NUMBER 400 OF 400

CURRENT TIME   0.2172E+02
STATISTICAL ARRAYS CLEARED AT TIME   0.0000E+00

STATISTICS FOR VARIABLES BASED ON OBSERVATION

	MEAN VALUE	STANDARD DEVIATION	COEFF. OF VARIATION	MINIMUM VALUE	MAXIMUM VALUE	NUMBER OF OBSERVATIONS
NODE 2	0.2992E+01	0.7788E+00	0.2603E+00	0.1270E+01	0.4851E+01	400
NODE 3	0.7451E+01	0.1300E+01	0.1744E+00	0.4456E+01	0.1198E+02	400
NODE 4	0.1596E+02	0.2037E+01	0.1276E+00	0.1182E+02	0.2245E+02	400
NODE 5	0.1102E+02	0.2041E+01	0.1852E+00	0.5562E+01	0.1571E+02	400
PROJ. COMPLETION	0.2078E+02	0.2137E+01	0.1028E+00	0.1577E+02	0.2697E+02	400

Figure 7-4 SLAM II summary report for 400 simulations of retrofit project.

Since each of the 400 observations of the project completion time are performed in an independent manner, an estimate of the standard deviation of the average project completion time is 2.14 divided by 20 or 0.107. Following the discussion presented in Section 2.11.1, a 95% confidence interval on the average project completion time is $20.78 \pm 1.96*0.107$ where 1.96 is the critical value of the normal distribution corresponding to a 95% confidence. The critical value for the normal distribution is used even though the estimate for the standard deviation is employed since the number of observations is large. The limits for the confidence interval are (20.57, 20.99) which indicates that the mean of the network model is expected to be in a narrow range. Specifically, the 95% confidence interval statement says that only one out of twenty confidence intervals calculated in the above manner would not contain the mean duration of the project network model.

The calculation of a tolerance interval is based on the material presented in Section 2.11.2. For this example, the assumptions required are satisfied. We compute with 95% confidence that at least 95% of the population of *average* completion times generated from the model should be in the tolerance interval (20.56, 21.00). For this case, the chi-square critical value is 353.75 for $\delta = .05$ with 399 degrees of freedom.

We can also compute that the probability is at least $0.9025 = (.95)(.95)$ that a sample of the project completion time generated from the model will be in the tolerance interval (16.33, 25.23).

The average values for nodes 2, 3, 4, and 5 provide estimates of the average starting times for activities emanating from these nodes. Additional information concerning all nodes is available on the histograms obtained. The histogram for project completion is shown in Figure 7-5. From Figure 7-5, estimates of the probability that the project will be completed by a certain time can be made. Thus, it is estimated that the probability of the project being completed by 19 time units is 0.210; hence, the probability of the project taking more than 19 time units is 0.790. This provides an indication of the gross nature of PERT assumptions since the expected completion time as estimated using PERT techniques is 19 time units (3).

```
                              **HISTOGRAM NUMBER  5**

                                 PROJ. COMPLETION


  OBSV     RELA     CUML      UPPER
  FREQ     FREQ     FREQ    CELL LIMIT     0      20      40      60      80     100
                                           +   +    +    +    +    +    +    +    +    +
    0     0.000    0.000    0.1500E+02     +                                         +
    0     0.000    0.000    0.1550E+02     +                                         +
    1     0.002    0.002    0.1600E+02     +                                         +
    3     0.007    0.010    0.1650E+02     +C                                        +
    8     0.020    0.030    0.1700E+02     +*C                                       +
    6     0.015    0.045    0.1750E+02     +*C                                       +
   17     0.043    0.087    0.1800E+02     +** C                                     +
   27     0.068    0.155    0.1850E+02     +***    C                                 +
   22     0.055    0.210    0.1900E+02     +***      C                               +
   35     0.087    0.298    0.1950E+02     +****       C                             +
   36     0.090    0.388    0.2000E+02     +*****        C                           +
   36     0.090    0.478    0.2050E+02     +*****           C                        +
   30     0.075    0.553    0.2100E+02     +****              C                      +
   33     0.083    0.635    0.2150E+02     +****                C                    +
   29     0.072    0.707    0.2200E+02     +****                  C                  +
   33     0.083    0.790    0.2250E+02     +****                     C               +
   21     0.052    0.842    0.2300E+02     +***                        C             +
   20     0.050    0.892    0.2350E+02     +***                          C           +
   12     0.030    0.922    0.2400E+02     +**                            C          +
   13     0.032    0.955    0.2450E+02     +**                             C +
    4     0.010    0.965    0.2500E+02     +*                              C +
   14     0.035    1.000       INF         +**                                     C
  ___                                       +   +    +    +    +    +    +    +    +    +
  400                                       0      20      40      60      80     100
```

Interpretive Statements

1. Probability of project completion by 19 time units = 0.210
2. Probability of project taking more than 19 time units = 1 − 0.210 = 0.790
3. Probability of project taking more than 24 time units = 1 − 0.922 = 0.078

Figure 7-5 Histogram of retrofit project completion time.

7.4 BATCH NODE

The BATCH node is used to accumulate entities to a specified level and then to release a single entity which represents the batch.

At a BATCH node, one or more batches of entities may be accumulated with the option to later unbatch and restore the individual members of the batch. This capability is useful for modeling pallets and transfer cars that accumulate a full load before moving.

The node is released when the sum of the values of an attribute for all batch members is greater than or equal to a threshold. The attribute number carrying the values to be summed is given the name NATRS. The threshold can be a constant value or the value carried in the ith attribute of the first entity of the batch. The batch may also be released upon the arrival of an entity with a negative value for a specified attribute, NATRB. This allows the overriding of the threshold requirement. For example, if the threshold is 10 and the sum of the ATRIB(NATRS) values is 7 for five entities waiting, then an entity arrival with ATRIB(NATRB)<0 releases a batch of 6 entities.

The entity released from the BATCH node has attributes which are a combination of the member entities. The combination of attributes is specified by a SAVE criterion. As part of the SAVE criterion, specific attributes of the batched entity may be defined to be the sum of the attributes of all individual entities forming the batch. A RETAIN field allows the modeler to save all the individual entities and their attributes that form the batch. If RETAIN is specified as ALL(NATRR), the individual entities are retained by SLAM II and ATRIB (NATRR) is set by SLAM II to be an internal reference to the individual entities. The modeler can then retrieve these entities by sending the batched entity through an UNBATCH node.

A BATCH node can also be used to simultaneously sort groups of entities into multiple batches. The number of batches simultaneously being accumulated for a BATCH node is specified by NBATCH. An entity is placed in a batch (sorted) based on the NATRBth attribute value of the arriving entity. This is the same attribute value that is used to release a batch when it is negative. Thus, the negative of the number of a particular batch to be released should be assigned to ATRIB (NATRB) for this purpose. A maximum of M activities are initiated at each release of a BATCH node.

As can be seen from the above description, the BATCH node performs many functions and is quite complex. The characteristics included with the BATCH node have been those found to be necessary when solving applied manufacturing and industrial problems.

The symbol and statement for the BATCH Node are shown below.

BATCH,NBATCH/NATRB,THRESH,NATRS,SAVE,RETAIN,M;

NBATCH is the total number of batches that will be accumulated concurrently at the BATCH node. NATRB is the number of the attribute that specifies the batch for the arriving entity, that is, the value of ATRIB(NATRB) is to be the same for entities in a batch. A secondary use for ATRIB(NATRB) is to cause the batch to be released when the negative of the batch number is given to the value of this attribute. For this case, the arriving entity is included in the batch. THRESH is the release threshold and can be a constant or specified as an attribute of the first arriving entity. Thus, if THRESH is specified as ATRIB(I), the Ith attribute of the first entity in a batch defines the threshold.

NATRS is the number of the attribute which contains the value to be summed. Thus, for entities arriving to the BATCH node which have the same value for ATRIB(NATRB), a sum is maintained of the values of ATRIB(NATRS). When this sum is greater than or equal to THRESH, a batched entity is formed and released from the BATCH node.

SAVE is used to specify a criterion for defining the attributes of the batched entity. The criterion specifies which entity in the batch should be used as a basis to define the attributes of the batched entity. The options for the criterion are:

FIRST entity included in the batch;
LAST entity included in the batch;
Entity with the lowest value of attribute I, LOW(I); and,
Entity with the highest value of attribute I, HIGH(I).

In addition to specifying this criterion, a list of attribute numbers can be given. For each of the attribute numbers in the list, the sum of the attribute values of each entity included in the batch will be obtained, and the sum will be used as the value of the corresponding attribute of the batched entity. For example, FIRST/3,5 specifies that attribute 3 of the batched entity should be the sum of the attribute 3 values of every entity included in the batch and attribute 5 of the batched entity should be the sum of the values of attribute 5 of each entity included in the batch. All other attribute values for the batched entity would be taken from the first entity making up the batch.

RETAIN indicates whether the individual entities included in the batch should be maintained for future use. The specification, ALL(NATRR), is used for this purpose where NATRR is an internal reference maintained in ATRIB(NATRR) by SLAM II to enable the individual entities to be accessed at an UNBATCH node. If it is not necessary to retain the individual entities, then the field can be specified as NONE. NONE is the default for this field.

Because of the complexity of the BATCH node, detailed examples of its use are given below. Let us define a BATCH node that maintains 5 batches where an entity's batch type is stored as attribute 2. A batch is formed when the sum of the values of attribute 3 for entities of the same type has a value which is 100 or greater. The individual entities making up the batch are not needed and the entity representing the batch should have the attributes of the first arriving entity of a batch except for attribute 4 which should be the sum of the values of attribute 4 of each individual entity. The BATCH node and statement that accomplishes the above is

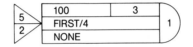

BATCH,5/2,100,3,FIRST/4,NONE,1;

This statement defines a BATCH node which will maintain a maximum of five separate batches as sorted on the value of ATRIB(2). When the sum of the values of ATRIB(3) for a batch reaches or exceeds 100, a batched entity will be released. This batched entity has the attributes of the FIRST entity forming the batch with the exception of ATRIB(4), which is the sum of all the ATRIB(4) values of entities in the batch. The attributes of the entities making up the batch are not retained since RETAIN is specified as NONE. Since the M value is 1, at most one activity will be initiated with each release of a batch from this node.

The BATCH node and statement

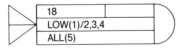

BATCH,,18,,LOW(1)/2,3,4,ALL(5);

maintains one batch (the default value for NBATCH). It is released when 18 entities arrive (the default for NATRS is to count each entity). The emanating entity will carry the attributes of the entity arriving with the lowest value of ATRIB(1). However, the values of ATRIB(2), ATRIB(3), and ATRIB(4) of the batched entity are the sum of the values of the corresponding attributes for all 18 entities in the batch. All of the original entities will be saved for later unbatching. The reference to the original entities is maintained in ATRIB(5) of the batched entity.

As a third example of the BATCH node, consider

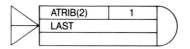

BATCH,,ATRIB(2),1,LAST;

which accumulates a batch until the sum of ATRIB(1) for all entities is greater than or equal to the value of ATRIB(2) of the first entity included in the batch. The batched entity will carry the attributes of the last entity included in the batch. The attributes of the other entities included in the batch will not be saved.

7.5 UNBATCH NODE

The UNBATCH node is used to put the entities of a batch back in the network or to split an entity into multiple entities. The symbol and statement for the UNBATCH node are shown below

          UNBATCH,NATRR,M;

where NATRR is the number of the attribute specifying whether each individual entity of the batch is to be routed from the UNBATCH node, or a specified number of entities identical to the arriving entity is to be routed into the network. In the latter case, the UNBATCH node operates in the same fashion as a GOON node if ATRIB(NATRR) is 1. If ATRIB(NATRR) was set by SLAM II at a

BATCH node, each of the individual entities of the batch is released from the UNBATCH node, and the arriving entity to the UNBATCH node is destroyed. When the entities of a batch are routed from the UNBATCH node, they are routed in the reverse order of their arrival to the BATCH node. If it is desired to maintain the order in which they arrived to the BATCH node then a zero duration activity or a node-to-node transfer (an implied zero time activity) should be used following the UNBATCH node.

If ATRIB(NATRR) is set by the modeler, it defines the number of identical entities to be released from the UNBATCH node. All attributes of such entities are set equal to the attributes of the arriving entity to the UNBATCH node.

Note that the UNBATCH node can insert a large number of entities into the network. For example, if ATRIB(NATRR) is set at an ASSIGN node by the user to be 50 and the M value for the UNBATCH node is 5, then up to 5 entities will be inserted into the network for each of the 50 entities to be split from the batched entity. Thus, up to 250 entities could be inserted into the network at this UNBATCH node.

Consider the following UNBATCH node.

    UNBATCH,3,1;

This UNBATCH node uses the value in the incoming entity's attribute 3 to define the entities to be released from the UNBATCH node. If ATRIB(3) is set by SLAM II, the individual entities of the batch are restored. If ATRIB(3) is set by the user, then ATRIB(3) identical entities are routed from the UNBATCH node. If ATRIB(3) is zero then the arriving entity is routed from the UNBATCH node. At most, one activity branch is activated for each entity.

7.6   ILLUSTRATION 7-1.   BATCHING AND UNBATCHING

Consider the situation shown in Figure 7-6 involving metal parts as they complete a series of machining operations on one of two production lines. The parts are automatically loaded into one of five racks and processed through a washing unit to remove machine oil and dirt. Six parts of the same type produced on the same line are required to fill each rack before the rack enters the washer. After washing, parts are unloaded from the rack and are routed to other stations that are outside the model described in this illustration. The network model of this subsystem is shown in Figure 7-7. The statement model is shown in Figure 7-8.

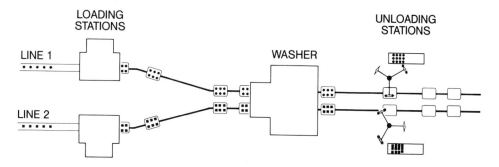

Figure 7-6 Schematic of batching and unbatching illustration.

The parts in this model arrive from one of two CREATE nodes with interarrival times specified by a uniform distribution between 0.3 time units and 0.7 time units. The entity representing a part is then assigned an ATRIB(1) value of 1 or 2 to identify the production line on which it was produced. A wash time for the part is then assigned to ATRIB(4). Entities from both lines are routed to the BATCH node, RACK, which sorts the entities into two groups according to the value in ATRIB(1). Note that the loading time into the rack is included on the activities prior to the BATCH node. Each entity counts as one part from a particular line toward the release threshold of 6 parts needed to fill the rack. When the sixth part from one of the lines arrives, a representative or batched entity is released from the node which carries a reference in ATRIB(2) to the individual parts in the rack. The attributes of the batched entity are those of the LAST part entering the rack except for attribute 4 which is the sum of the wash times. The rack entity then enters a QUEUE node prior to being serviced in the washer which has a queue capacity of 10 racks. The washer can wash 5 racks at the same time and the wash time is taken as the fourth attribute of the rack entity which is the sum of the wash times of each part. After washing, the 6 entities are reestablished in the network at the UNBATCH node, UNRK, and the batched entity is automatically destroyed. Each part entity is reestablished through the ATRIB(2) value of the batched entity as was specified at the BATCH node by ALL(2). The part entities

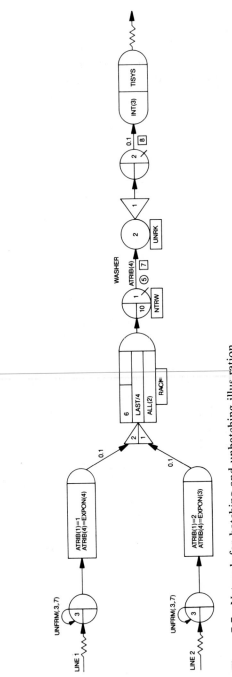

Figure 7-7 Network for batching and unbatching illustration.

```
NETWORK;
;
; CREATE ENTITIES FROM LINE 1
;
        CREATE,UNFRM(.3,.7),,3;
        ASSIGN,ATRIB(1)=1,ATRIB(4)=EXPON(4);
        ACT,0.1,,RACK;
;
; CREATE ENTITIES FROM LINE 2
;
        CREATE,UNFRM(.3,.7),,3;
        ASSIGN,ATRIB(1)=2,ATRIB(4)=EXPON(3);
        ACT,0.1;
;
; COMBINE 6 ENTITIES FROM THE SAME LINE
;
 RACK BATCH,2/1,6,,LAST/4,ALL(2),1;
 NTRW QUEUE(1),,10;
        ACT(5)/7,ATRIB(4);       WASH ACTIVITY
 UNRK UNBATCH,2,1;               BRING ENTITIES BACK
;
;        PROCESS INDIVIDUAL ENTITIES
        QUEUE(2);
        ACT/8,0.1;
        COLCT,INT(3),TISYS;
        TERMINATE;
        ENDNETWORK;
```

Figure 7-8 Statement model for batching and unbatching illustration.

exit the UNBATCH node in the reverse order of arrival to the BATCH node but are placed in their original order because of the zero time activity to the QUEUE node. In this model of the subsystem, the time through the subnetwork for each part is collected at the COLCT node since ATRIB(3) is the mark time established at the CREATE nodes for each part type.

7.7 SELECT NODE

SELECT nodes are points in the network where a decision regarding the routing of an entity is to be made and the decision concerns either QUEUE nodes or servers or both. To accomplish the routing at the SELECT node, the modeler chooses a *queue selection rule (QSR)* and/or a *server selection rule (SSR)*. The rule establishes the decision process by which SLAM II will route entities when a decision point is reached. The decision points in the SLAM II network occur at the following times:

1. An entity is to be routed to one of a set of parallel queues;
2. A service activity has been completed and parallel queues exist that have entities waiting for the service activity. In this situation, a SELECT node is used to decide from which QUEUE node an entity should be taken; and
3. An entity is to be routed to one of a set of non-identical idle servers.

The SELECT symbol is more complex than other SLAM II symbols in that the decision at the node can involve both a "looking ahead" and a "looking behind" capability. A look ahead capability is necessary to route entities to one of a set of parallel queues and to select from a set of parallel servers. These are decision types

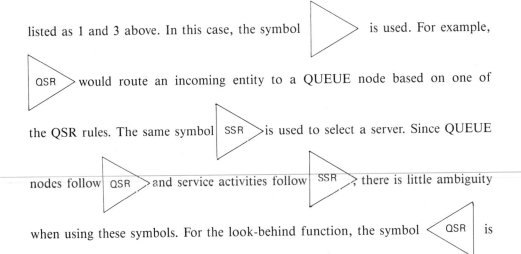

listed as 1 and 3 above. In this case, the symbol ▷ is used. For example, QSR▷ would route an incoming entity to a QUEUE node based on one of the QSR rules. The same symbol SSR▷ is used to select a server. Since QUEUE nodes follow QSR▷ and service activities follow SSR▷ there is little ambiguity when using these symbols. For the look-behind function, the symbol ◁QSR is

used where QUEUE nodes would precede the symbol and the SELECT node would perform the function listed as 2 above. A single SELECT node can perform both the look-behind and look-ahead functions in which case we use the symbol

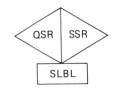

where the SELECT node label, SLBL, has been attached. In all cases, we employ the same statement type which is

SLBL SELECT,QSR,SSR,BLOCK or BALK(NLBL),QLBLs;

When the SELECT node is used for a single purpose, the field in the QSR or SSR rule not required should be defaulted. In the above statement, the QLBLs are the QUEUE node labels associated with the QSR rule. The QUEUE nodes could be before or after the SELECT node.

Five observations regarding the SELECT node are:

1. QUEUE nodes cannot be on both sides of a given SELECT node.
2. If service activities follow a SELECT node then QUEUE nodes must precede the SELECT node to hold entities when all the service activities are ongoing.
3. Balking and blocking occur at a SELECT node when all following QUEUE nodes are at their capacity and the BALK or BLOCK option is prescribed. The symbolism for this is shown below.

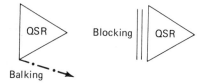

4. Whenever the look-behind capability is required, the preceding QUEUE nodes must refer to the SELECT node to transfer arriving entities.
5. A SELECT node always requires a node label.

A list of queue selection rules (QSR) available for specification at the SELECT node is given in Table 7-2. The list of server selection rules (SSR) is given in Table 7-3. Illustrations of the use of SELECT nodes follow.

1. Route entities to QUE1 or QUE2 based on the smallest number in queue (SNQ).

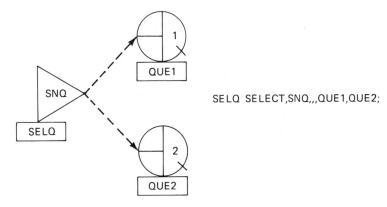

SELQ SELECT,SNQ,,,QUE1,QUE2;

This network segment could model the routing of customer entities to parallel queues before two airline ticket sellers.

2. Select a server from servers 1, 2, and 3 to process entities waiting in queue, WAIT. It is preferred to use server 1 to server 2 and server 2 to server 3, that is, a preferred order for selecting servers is to be used.

Table 7-2   Priority rules associated with SELECT nodes for selecting from a set of queues.

Code	Definition
POR	Priority given in a preferred order.
CYC	Cyclic Priority—transfer to first available QUEUE node starting from the last QUEUE node that was selected.
RAN	Random Priority—assign an euqal probability to each QUEUE node that has an entity in it.
LAV	Priority given to the QUEUE node which has had the largest average number of entities in it to date.
SAV	Priority is given to the QUEUE node which has had the smallest average number of entities in it to date.
LWF	Priority is given to the QUEUE node for which the waiting time of its first entity is the longest.
SWF	Priority is given to the QUEUE node for which the waiting time of its first entity is the shortest.
LNQ	Priority is given to the QUEUE node which has the current largest number of entities in the queue.
SNQ	Priority is given to the QUEUE node which has the current smallest number of entities in the queue.
LRC	Priority is given to the QUEUE node which has the largest remaining unused capacity.
SRC	Priority is given to the QUEUE node which has the smallest remaining unused capacity.
ASM	Assembly mode option—all incoming queues must contribute one entity before a processor may begin service (this can be used to provide an "AND" logic operation).
NQS(N)	User written function to select a QUEUE node. N is an integer to differentiate between the use of NQS at different SELECT nodes (see Chapter 9).

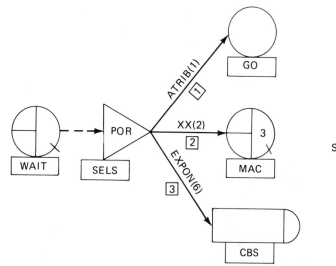

```
SELS   SELECT,,POR,,WAIT;
          ACT/1,ATRIB(1),,GO;
          ACT/2,XX(2),,MAC;
          ACT/3,EXPON(6.),,CBS;
```

This network segment could represent three machines that can be used to process jobs waiting in QUEUE node WAIT. The processing time and the routing after processing is modeled as being machine dependent.

3. Illustrations 1 and 2 are combined below so that SELECT node SELS takes entities from QUEUE nodes QUE1 and QUE2 (rather than QUEUE node WAIT). A cyclic queue selection rule is used at SELECT node SELS.

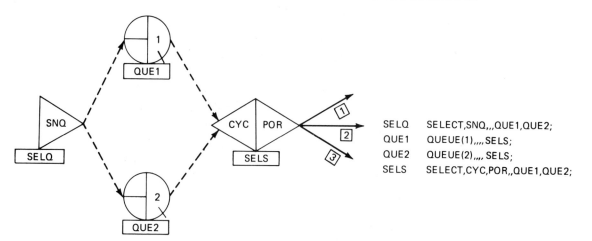

```
SELQ   SELECT,SNQ,,,QUE1,QUE2;
QUE1   QUEUE(1),,,,SELS;
QUE2   QUEUE(2),,,,SELS;
SELS   SELECT,CYC,POR,,QUE1,QUE2;
```

In this network model, entities are routed by SELECT node SELQ to either QUE1 or QUE2 depending on which QUEUE node has fewer entities in it at the time of routing. If the queues have an equal number of entities in them, then QUE1 is selected as it is listed first. When an entity arrives to either QUEUE node and a server (activity 1, 2, or 3) is not busy, the entity is routed to the free server. If more than one server is free, the POR server selection rule associated with SELECT node SELS will select the first free server activity listed after the SELS SELECT node statement. Thus, the POR server selection rule gives priority to servers in the order they are listed in the statement model.

Table 7-3   Priority rules associated with SELECT nodes for selecting from a set of servers.

Code	Definition
POR	Select from free servers in a preferred order.
CYC	Select servers in a cyclic manner. That is, select the next free server starting with the last server selected.
LBT	Select the server that has the largest amount of usage (busy time) to date.
SBT	Select the server which has the smallest amount of usage (busy time) to date.
LIT	Select the server who has been idle for the longest period of time.
SIT	Select the server who has been idle for the shortest period of time.
RAN	Select randomly from free servers according to preassigned probabilities.
NSS(N)	User written function to select a server. N is an integer to differentiate between the use of NSS at different SELECT nodes (see Chapter 9).

When a server becomes free and entities are waiting in both QUE1 and QUE2, SELECT node SELS uses the CYCLIC rule and takes an entity from the QUEUE node that was not selected when the last entity was routed to a server.

4. Server 1 can process units stored in QUEUE nodes Q1 or Q2. Server 1 uses a preferred order for selecting entities from Q1 and Q2. Server 2 can only serve entities in Q2. If an entity arrives to Q2 when both servers are free, the use of server 2 is preferred. The network for this situation is shown below. In this network model segment, when activity 1 completes service on an entity, SELECT node S1 interrogates Q1 first, and only if it is empty will Q2 be interrogated. This order is specified by the order in which the QUEUE nodes are listed on the S1 SELECT node statement. Entities arriving to Q2

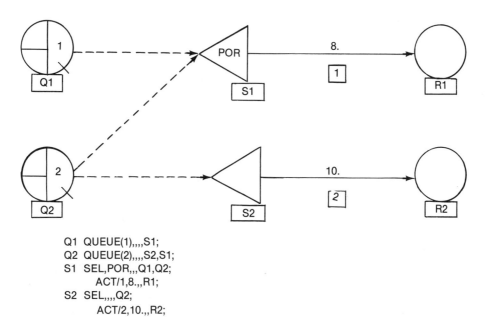

```
Q1  QUEUE(1),,,,S1;
Q2  QUEUE(2),,,,S2,S1;
S1  SEL,POR,,,Q1,Q2;
        ACT/1,8.,,R1;
S2  SEL,,,,Q2;
        ACT/2,10.,,R2;
```

seek service first through S2 then S1 (see order of listing SELECT node labels). Entities waiting at Q2 can be selected by either S2 or S1 when activities
2 or 1 complete a service activity as Q2 is in the QUEUE node list for both
S1 and S2.

7.7.1   ASSEMBLY Queue Selection Rule

One of the queue selection rules listed in Table 7-3 is the ASM or assembly rule.
This rule differs from the other rules in that it involves the combining of two or
more entities into an assembled entity. In this case the selection process requires
that at least one entity be in each QUEUE node before any entity will be routed to
a service activity. An air freight example of this assembly procedure is the requirement for both an aircraft entity and a cargo entity to be available before aircraft
loading can begin. A network segment is shown for this situation where aircraft
entities wait at QUEUE node ACFT, cargo entities wait at QUEUE node CARG,
and aircraft loading is modeled as activity 3.

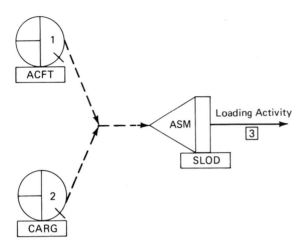

SLOD   SELECT,ASM,,,ACFT,CARG;

SELECT node SLOD does not route an entity to service activity 3 until there is one entity in each QUEUE node. At the SELECT node, ASM is prescribed as the QSR procedure. In addition, the dashed lines preceding the SELECT node have been joined prior to the SELECT node to indicate that an entity is required at each QUEUE node before any entity is sent to the server. A SELECT node that employs the ASM queue selection rule can assemble entities from two or more queues.

When entities are assembled by a SELECT node, it may be desirable to specify a save attribute criterion, SAVE, by which the appropriate attributes of one of the entities to be assembled can be maintained. The concept is similar to that presented earlier for ACCUMULATE nodes. The SAVE criterion is specified in the same manner as was done for the ACCUMULATE node and the SELECT node symbol is expanded to accommodate this specification. The SAVE criterion can be based on HIGH(I), LOW(I), SUM, or MULT.

If no SAVE criterion is specified, then the attributes of the entity in the first queue node listed in the select statement will be assigned to the assembled entity. The symbolism and statement for the ASSEMBLY queue selection rule are shown below. Note that the SAVE criterion is specified after ASM in the statement and that the delimiter is a slash.

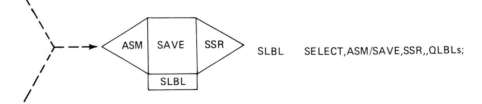

SLBL   SELECT,ASM/SAVE,SSR,,QLBLs;

As an example, consider the aircraft and cargo example presented previously. Suppose we desired to save the attributes of the aircraft. If aircraft entities are identified by a value of 1 in attribute 3 and cargo entities have a value of 2 in attribute 3, then specifying that the attributes of entities having a low value of attribute 3 will cause the aircraft entity's attributes to be saved. The statement for this case would be:

SLOD   SELECT,ASM/LOW(3),,,ACFT,CARG;

7.8   EXAMPLE 7-2.   A TRUCK HAULING SITUATION

The system to be modeled in this example consists of one bulldozer, four trucks, and two man-machine loaders (2,7). The bulldozer stockpiles material for the loaders. Two piles of material must be stocked prior to the initiation of any load operation. The time for the bulldozer to stockpile material is Erlang distributed and consists of the sum of two exponential variables each with a mean of 4. (This corresponds to an Erlang variable with a mean of 8 and a variance of 32.) In addition to this material, a loader and an unloaded truck must be available before the loading operations can begin. Loading time is exponentially distributed with a mean time of 14 minutes for server 1 and 12 minutes for server 2.

After a truck is loaded, it is hauled, then dumped and must be returned before the truck is available for further loading. Hauling time is normally distributed. When loaded, the average hauling time is 22 minutes. When unloaded, the average time is 18 minutes. In both cases, the standard deviation is 3 minutes. Dumping time is uniformly distributed between 2 and 8 minutes. Following a loading operation, the loader must rest for a 5 minute period before he is available to begin loading again. A schematic diagram of the system is shown in Figure 7-9. The system is to be analyzed for 8 hours and all operations in progress at the end of 8 hours should be completed before terminating the operations for a run.

Concepts Illustrated. This example illustrates the use of the SELECT node for routing entities from multiple QUEUE nodes to multiple service ACTIVITY's. Additional concepts illustrated by this example include the use of the ACCUM node for the combining entities, the use of conditional branching for testing system status, the representation of several entity types by entities within the same model, and the ending of a simulation by completing processing of all entities in the system.

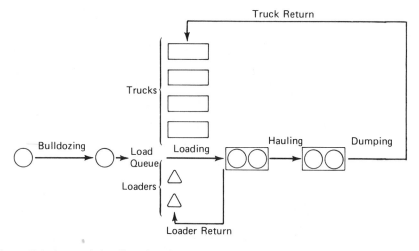

Figure 7-9 A truck hauling situation.

SLAM II Model. In this example, there are four distinct entities to be modeled. The first entity type is the pile of material created by the bulldozing operation. Since two piles must be combined to make one load, the entities representing piles of material must be combined two at a time to create a new entity representing a load. This accumulation of entities can be modeled with an ACCUM node that requires that two entities be combined for the first release (FR–2), and two entities be combined for subsequent releases (SR=2). No attributes are used in this example, and, hence, no SAVE rule is required at the ACCUM node. Before the loading operation can begin, in addition to a load, there must also be an available truck and loader, each represented by separate entities. The loading operation can be performed by either of two non-identical servers.

This process can be modeled by employing separate QUEUE nodes for the trucks, loads, and loaders, in conjunction with a following SELECT node with two emanating servers. By specifying the ASM (assembly) and LIT (longest idle time) options for the queue and server selection rules, respectively, an entity is required in each queue before a service can be initiated, and available servers are selected based on longest idle time for each server. Following a loading operation, the entity is split and two entities are routed. One represents the truck and the other the loader. The entity representing the loader is delayed 5 minutes before being available to begin loading again. The consecutive activities of hauling, dumping, and return trip can be represented by serial ACTIVITY's with the last ACTIVITY returning the truck entity to the QUEUE node representing available trucks. The SLAM II network model of the system is depicted in Figure 7-10.

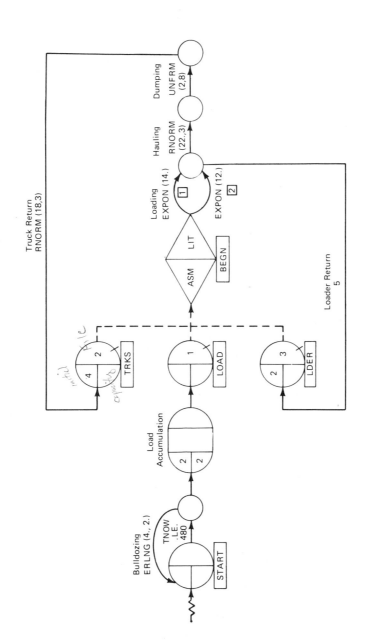

Figure 7-10 SLAM II network model of truck hauling system.

```
1    GEN,PRITSKER,TRUCK HAULING,7/12/83,1,,,Y/N;
2    LIMITS,3,1,50;
3    NETWORK;
4    START CREATE;                                    CREATE LOAD TRANSACTIONS
5          ACT,,TNOW.LE.480;                          STOP IF AFTER 8 HOURS
6          GOON;                                      ELSE
7          ACT,ERLNG(4.,2.),,START;                   BRANCH BACK TO START
8          ACT;                                       AND CONTINUE
9          ACCUM,2,2;                                 ACCUMULATE TWO PILES
10   LOAD  QUEUE(1),,,,BEGN;                          QUEUE OF LOADS
11   TRKS  QUEUE(2),4,,,BEGN;                         QUEUE OF TRUCKS
12   LDER  QUEUE(3),2,,,BEGN;                         QUEUE OF LOADERS
13   BEGN  SELECT,ASM,LIT,,LOAD,TRKS,LDER;            ASM OF LOAD,TRKS, AND LDER
14         ACT/1,EXPON(14.);                          LOADER1 TIME
15         ACT/2,EXPON(12.);                          LOADER2 TIME
16         GOON;
17         ACT,5,,LDER;                               LOADER RESTING TIME
18         ACT,RNORM(22.,3.);                         TRUCK HAULING TIME
19         GOON;
20         ACT,UNFRM(2.,8.);                          TRUCK DUMPING TIME
21         GOON;
22         ACT,RNORM(18.,3.),,TRKS;                   TRUCK RETURN TIME
23         END
24   FIN;
```

Figure 7-11 SLAM II statement listing for truck hauling example.

The SLAM II simulation statement listing for this example is shown in Figure 7-11. The network description begins with the CREATE statement (line 4) which creates the first entity at time 0. Thereafter, the time between entities is specified according to the Erlang distribution. The parameters for the Erlang distribution are: each exponential sample has a mean of 4 and there are to be 2 exponential samples. The entity continues through the conditional ACTIVITY (line 5) if TNOW is less than 480; otherwise the activity is not released and the entity is destroyed. This condition stops the creation of piles by the bulldozer after 480 minutes of operation. Since there is no INIT statement, an ending time for the simulation is not specified. In this situation, the run will end when all operations in progress at time 480 are completed since the arrival process was halted. SLAM II ends a run when the executive routine attempts to advance time and no events exist on the event calendar.

Entities representing piles arrive at the ACCUM statement (line 9), where they are combined in pairs to form load transactions and continue to the LOAD QUEUE statement (line 10). The LOAD QUEUE is related to the TRKS QUEUE (line 11) and the LDER QUEUE (line 12), all of which have the statement labeled BEGN as a following SELECT node. The TRKS QUEUE represents the queue of waiting trucks which are stored in file. 2. There are initially four trucks in the queue. The LDER QUEUE represents the queue of waiting loaders which are stored in file 3. The queue is initialized to have two loaders in the queue at the beginning of the simulation. The SELECT statement labeled BEGN (line 13) employs the ASM queue selection rule, LIT server selection rule, has neither balking nor blocking, and selects from the preceding QUEUE's labeled LOAD, TRKS, and LDER. Following the SELECT statement are two non-identical servers. The first server (line 14) representing loader 1 is assigned an activity number of 1 and has a service time which is exponentially distributed with a mean of 14 minutes. The second server (line 15) representing loader 2 is assigned an activity number of 2 and has a service time which is exponentially distributed with a mean of 12 minutes.

At the end of the loading operation, the entity arrives at the GOON statement (line 16). Following the GOON statement are two regular ACTIVITY's. The first ACTIVITY (line 17) representing the loader resting time has a duration of five minutes and routes the entity to the statement labeled LDER. The entity proceeding through the second ACTIVITY statement (line 18), representing the truck hauling operation, continues to a GOON statement (line 19). Next, the entity continues through the ACTIVITY statement (line 20) representing the dumping operation which ends at a GOON statement (line 21). The last ACTIVITY statement (line 22) models the return trip of the truck and routes the entity to the statement labeled TRKS.

In summarizing the SLAM II model for this example, it is interesting to note the representation of loaders as both service ACTIVITY's and entities queueing up at the LDER QUEUE. This dual representation is used to include the resting time for the loader. If the resting time requirement were omitted from the problem statement, the loaders need only be modeled by the service activities, and thus statements 12 and 17 could be deleted from the model.

Summary of Results. The SLAM II Summary Report for this example is shown in Figure 7-12. The first category of statistics is the file statistics for files 1, 2, and 3, and correspond to the queue of loads waiting for service, the queue of idle trucks, and the queue of available loaders, respectively. The second category of statistics is the service activity statistics representing the two loaders. A count of entities completing the two activities indicates that a total of 29 loads were processed through the system. This small number of observations suggest that considerable variation in both queue lengths and service activity utilizations can be expected between simulation runs employing different seed values.

```
                         S L A M   I I   S U M M A R Y   R E P O R T

              SIMULATION PROJECT TRUCK HAULING          BY PRITSKER

              DATE  7/12/1983                           RUN NUMBER    1 OF    1


              CURRENT TIME   0.4988E+03
              STATISTICAL ARRAYS CLEARED AT TIME  0.0000E+00


                  **FILE STATISTICS**

 FILE     ASSOC NODE      AVERAGE     STANDARD     MAXIMUM    CURRENT    AVERAGE
NUMBER   LABEL/TYPE       LENGTH      DEVIATION    LENGTH     LENGTH     WAITING TIME

   1     LOAD QUEUE       0.6629      0.9595          3          0       11.4023
   2     TRKS QUEUE       0.8554      1.1152          4          4       12.9296
   3     LDER QUEUE       1.1249      0.7326          2          2       18.1005
   4         CALENDAR     4.3987      1.3419          8          0        6.7930


                  **SERVICE ACTIVITY STATISTICS**

ACTIVITY   START NODE OR    SERVER    AVERAGE      STANDARD    CURRENT      AVERAGE    MAXIMUM IDLE   MAXIMUM BUSY    ENTITY
 INDEX     ACTIVITY LABEL   CAPACITY  UTILIZATION  DEVIATION   UTILIZATION  BLOCKAGE   TIME/SERVERS   TIME/SERVERS    COUNT

   1       LOADER1 TIME        1       0.3014      0.4589         0         0.0000       62.0796        38.5929         16
   2       LOADER2 TIME        1       0.2830      0.4505         0         0.0000       63.7792        36.2041         13
```

Figure 7-12   SLAM II summary report for truck hauling example.

7.9   MATCH NODE

MATCH nodes in SLAM II are nodes that match entities residing in specified QUEUE nodes that have equal values of a specified attribute. When each QUEUE node preceding a MATCH node has an entity with the specified common attribute value, the MATCH node removes each entity from the corresponding QUEUE node and routes it to a node associated with the QUEUE node. Thus, each entity is routed individually. The symbol and statement for the MATCH node are:

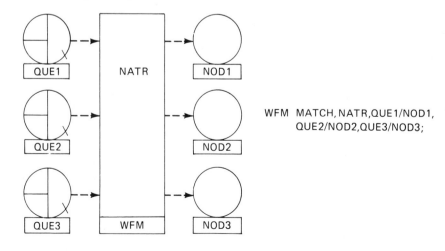

```
WFM  MATCH,NATR,QUE1/NOD1,
     QUE2/NOD2,QUE3/NOD3;
```

Note that for the MATCH node, there are nodes on both sides of the symbol and a node-to-node transfer is made when a match occurs. If there is no route node specified for one of the QUEUE nodes, the entity in that QUEUE node is destroyed after a match is made. The attribute number on which the match is based is specified within the MATCH node symbol. Only QUEUE nodes can precede MATCH nodes and the initial number in the QUEUE node must be zero. Illustrations of the use of MATCH nodes are shown below.

1. Hold entities in files 1 and 2 at QUEUE nodes TYP1 and TYP2, until there is an entity in each QUEUE node that has an attribute 3 value that is the same. Route both entities to ACCUMULATE node MAA and save the attribute set of the entity whose attribute 2 value is the largest.

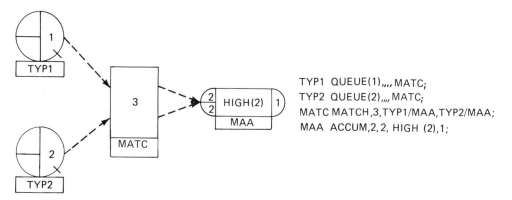

```
TYP1 QUEUE(1),,,,MATC;
TYP2 QUEUE(2),,,,MATC;
MATC MATCH,3,TYP1/MAA,TYP2/MAA;
MAA ACCUM,2,2, HIGH (2),1;
```

This model segment could be used to represent an aircraft and crew where only a particular crew can be used with a given aircraft.

2. Hold a patient entity until his health records arrive. Route the patient to the queue before the doctor's office when both the patient and his records are available. Destroy the record entity. Patient identification is maintained as attribute 1.

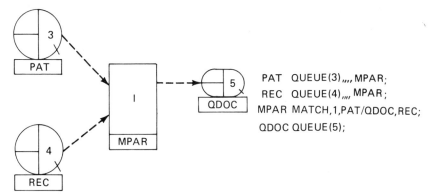

```
PAT  QUEUE(3),,,,MPAR;
REC  QUEUE(4),,,,MPAR;
MPAR MATCH,1,PAT/QDOC,REC;
QDOC QUEUE(5);
```

7.10   ILLUSTRATION 7-2.   A MAIL ORDER CLOTHES MANUFACTURER†

A clothes manufacturer supplies shirts and pants to a centralized distributor. The time to produce one dozen shirts is triangularly distributed with a minimum value of 11, maximum value of 22 and a modal value of 16. The time to produce one dozen pair of pants is also triangularly distributed with a minimum of 11, maximum of 22 and a most likely value of 19. After manufacturing the garments, they are transported by carts to packaging and the time for this transport is 7 minutes.

The time between the receipt of orders from the distributor is exponentially distributed with a mean of 30. Each order is made up of a combination of shirts and pants. From past purchases, the size of an order in terms of the number of dozen of pants and shirts is randomly distributed. The probability that an order is for 1, 2, 3, or 4 dozen of a product (either shirts or pants) is 0.6, 0.2, 0.15, and 0.05, respectively. Orders are filled as quickly as possible. When an order is filled, it is routed by a cart to a loading dock where it is packaged with other orders in a crate. A crate has a capacity of 112 cubic feet. Shirts are packaged in an 8 foot cubic box and pants in a 12 foot cubic box. A crate is considered loaded as soon as 100 cubic feet of its space or more has been filled. An evaluation of this rule is desired.

In order to buy insurance on the merchandise in a shipment, the dollar value of a crate and the number of crates shipped per week are desired. The dollar value for a dozen shirts is $60 and for a dozen pants is $102. Statistics are desired over five 8 hour days on the average time between the filling of a crate and the dollar value of the goods in a crate.

The SLAM II network model of the mail order clothes manufacturer is shown in Figure 7-13. In this model, clothes are manufactured throughout the day. In Figure 7-13, production of shirts and pants is modeled by two CREATE nodes with the time between creation as the production time for a dozen of the product. Each CREATE node is followed by an activity whose duration is 7 minutes representing the transport time. At QUEUE node Q2, entities representing a dozen shirts are stored in file 2. At QUEUE node Q4, entities representing a dozen pants are stored. These queues represent the inventory of shirts and pants.

When an order arrives, an entity representing a demand for one dozen shirts is placed in QUEUE node Q1. SELECT node ASM1 is used to assemble an order for a dozen shirts with an entity in node Q2 representing an inventory of a dozen shirts. The attributes of the order for a dozen shirts are saved by specifying that the entity with a high value of attribute 5 specifies which of the entity's attributes are to be saved.

---

†This illustration was designed and developed by Carole Vasek of Pritsker & Associates, Inc.

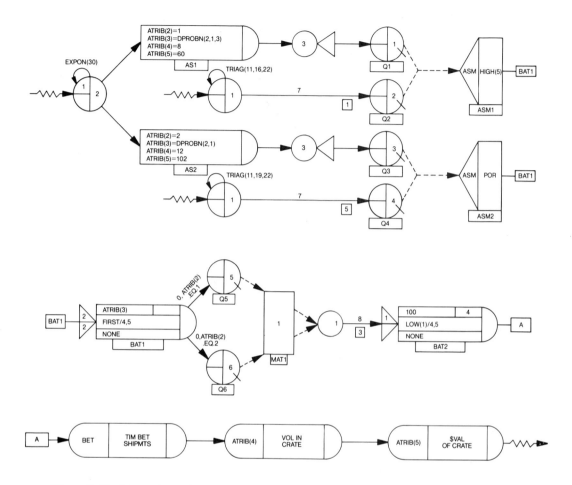

Figure 7-13 Network model of a mail order clothes manufacturer.

The new entity is routed to BATCH node BAT1 where the entities are batched until the number of dozen of shirts in an order is reached. This counting of entities is modeled by specifying ATRIB(3) as the threshold value and not specifying an attribute defining an amount to batch toward the threshold value.

At ASSIGN node AS1, ATRIB(3) is assigned a value which represents the number of dozens of shirts requested on an order. Also assigned at node AS1 is a value of 1 to ATRIB(2) to indicate the product type (shirts), a value of 8 to ATRIB(4) which is the volume of a shirt box, and a value of $60 to ATRIB(5) which is the value of 1 dozen shirts. To summarize, the definitions of the attributes are given below.

Attribute	*Definition*
1	Mark time of a receipt of an order.
2	Product type: 1 for shirts
	2 for pants
3	The number of dozens of product in the order.
4	Volume of the product box.
5	The value of the product.

The receipt of an order for products is generated at a CREATE node with an exponentially distributed time between creations with a mean of 30. Attribute 1 is marked at the CREATE node and an M value of 2 is prescribed. This splits the order into a shirt entity and a pants entity. The shirt entity is sent to ASSIGN node AS1 and the pants entity is sent to ASSIGN node AS2. We will describe the routing of the shirts entity. From the ASSIGN node AS1, it is routed to an UN-BATCH node where the entity is split into entities in accordance with the value of attribute 3. If four dozen shirts were requested on the order, the entity is split into four entities each of which is routed to QUEUE node Q1. A similar network section models the processing of an order entity for pants.

After a shirt order is assembled with shirts from inventory, the entity is routed to BATCH node BAT1 where shirt entities are combined into an entity representing all the shirts on a particular order. This is accomplished by batching on attribute 2 until a threshold defined by ATRIB(3) is reached. ATRIB(3) of the first entity of a new batch provides the number of dozens of shirts in the order. Two different batches are put together at node BAT1 as represented by the 2 given for the NBATCH parameter. The save criterion specifies that the attributes of the batched entity are to be those of the first arriving entity except for attributes 4 and 5 whose values are the sum of these attributes of all entities included in the batch. This causes attribute 4 of the batched entity to be the sum of the volumes of the boxes and attribute 5 to be the dollar value of the product for the order.

When a batched entity representing the shirts for an order has been created, it is routed to QUEUE node Q5. Similarly, a batched entity representing the pants for the order is routed to QUEUE node Q6. MATCH node MAT1 causes the entities to wait until both the batched pants entity and the batched shirts entity from the same order are in the appropriate QUEUE nodes. When this occurs, they are both routed to GOON node GO1 and, after a transportation delay, of eight minutes are sent to BATCH node BAT2 where they are put into a crate. The batching continues until the volume attribute, ATRIB(4), exceeds the crate threshold of 100. In this way, a batched entity with the attributes of the entity with the lowest mark time is created. The value of all products in the crate is given by attribute 5 and the volume of all boxes by attribute 4. Statistics are then collected at three sequential COLCT nodes on the time between shipments, the dollar value of a crate, and the volume of boxes that have been put in a crate.

This model illustrates the use of the different types of SLAM II decision and logic nodes. A statement model corresponding to the network model of Figure 7-13 is shown in Figure 7-14.

```
1   GEN,VASEK,CLOTHES MANUFACTURER,10/14/85;
2   LIMITS,6,6,175;
3   ;
4   ;   DEFINE PROBALISTIC ASSIGNMENT OF NUMBER OF DOZENS IN
5   ;            EACH ORDER
6   ;
7   ARRAY(1,4)/ 1, 2, 3, 4;
8   ARRAY(2,4)/.6,.8,.95,1.0;
9   NETWORK;
10  ;
11  ;   CREATE ORDERS
12  ;
13      CREATE,EXPON(30),,1,,2;
14      ACT,,,AS1;
15      ACT,,,AS2;
16  AS1 ASSIGN,ATRIB(2)=1,ATRIB(3)=DPROBN(2,1),ATRIB(4)=8,
17      ATRIB(5)=60;
18      UNBATCH,3;            SPLIT INTO ENTITIES FOR 1 DOZEN
19  Q1  QUE(1),,,,ASM1;
20  ;
21  ;   MANUFACTURE SHIRTS
22  ;
23      CREATE,TRIAG(11.,16.,22.),,,,1;
24      ACT/1,7;             MOV1 SHRTS
25  Q2  QUE(2),,,,ASM1;
26  ;
27  ;   ASSEMBLY A DOZEN SHIRTS WITH AN ORDER
28  ;
29  ASM1 SELECT,ASM/HIGH(5),,,Q1,Q2;
30      ACT,,,BAT1;
31  ;
32  ;   ORDERS FOR PANTS
33  ;
34  AS2 ASSIGN,ATRIB(2)=2,ATRIB(3)=DPROBN(2,1),ATRIB(4)=12,ATRIB(5)=102;
35      UNBATCH,3;           SPLIT INTO ENTITIES FOR 1 DOZEN
36  Q3  QUE(3),,,,ASM2;
37  ;
38  ;   MANUFACTURE PANTS
39  ;
40      CREATE,TRIAG(11.,19.,22.),,,,1;
41      ACT/5,7;             MOV1 PNTS
42  Q4  QUE(4),,,,ASM2;
43  ;
44  ;   ASSEMBLY A DOZEN PANTS WITH AN ORDER
45  ;
46  ASM2 SELECT,ASM,,,Q3,Q4;
```

Figure 7-14   Statement model of a mail order clothes manufacturer.

```
47          ACT;
48    ;
49    ;    BATCH NO. OF DOZ. REQUIRED FOR ORDER
50    ;
51  'BAT1 BATCH,2/2,ATRIB(3),,FIRST/4,5,,1;   BATCH SHIRTS AND PANTS SEPARATELY
52          ACT,,ATRIB(2).EQ.1,Q5;
53          ACT,,ATRIB(2).EQ.2,Q6;
54    Q5    QUE(5),,,,MAT1;
55    Q6    QUE(6),,,,MAT1;
56    ;
57    ;    MATCH THE SHIRT AND PANTS ENTITIES
58    ;
59   'MAT1 MATCH,1,Q5/G01,Q6/G01;
60    G01   GOON,1;
61          ACT/3,8,,BAT2;    TRNSPORT TIM
62    ;
63    ;    BATCH THE SHIRTS AND PANTS INTO A CRATE ENTITY
64    ;
65   'BAT2 BATCH,,100,4,LOW(1)/4,5;
66    ;
67    ;    COLLECT STATISTICS ON TIME BETWEEN SHIPMENTS, THE VOLUME
68    ;       IN EACH CRATE AND THE $ VALUE OF EACH CRATE
69    ;
70          COLCT,BET,TIME BET SHIPMT;
71          COLCT,ATRIB(4),VOL IN CRATE;
72          COLCT,ATRIB(5),$VAL OF CRATE;
73          TERM;
74          END;
75  INIT,0,2400;
76  FIN;
```

Figure 7-14 (continued).

7.11 SUMMARY OF SYMBOLS AND STATEMENTS

Table 7-3 presents the logic and decision nodes of SLAM II. The decision and logic nodes are used in conjunction with activities to model the grouping and control of entities. The SELECT node performs a decision function which routes entities to queues, from queues, and to non-identical servers based on heuristic rules.

Table 7-3 Logic and decision nodes of SLAM II networks†

Name	Symbol	Statement
ACCUMULATE		ACCUMULATE, FR, SR, SAVE, M;
BATCH		BATCH, NBATCH/NATRB, THRESH, NATRS, SAVE, RETAIN, M;
MATCH		MLBL MATCH, NATR, QLBL/NLBL, repeats . . .;
SELECT		SELECT, QSR, SSR, BLOCK or BALK (NLBL), QLBLs;
SELECT VARIATIONS		
ASSEMBLY at SELECT nodes		SELECT, ASM/SAVE, SSR,, QLBLs;
UNBATCH		UNBATCH, NATRR, M;

---

† Definitions of Codes

FR   First Release requirement to release node
SR   Subsequent Release requirement to release node

SAVE	Criterion for determining which attributes to SAVE
M	Maximum number of branches from a node along which an entity can be routed.
NBATCH	Number of different batches accumulated concurrently at a BATCH node
NATRB	Number of an ATtRibute which identifies Batch membership
THRESH	THRESHold value for specifying the size of a batch
NATRS	Number of an ATtRibute on which a Sum of values is to be kept
RETAIN	Specification for RETAINing individual entities included in a batch
NATR	Number of an ATtRibute
QLBL	Queue node LaBeL
NLBL	Node LaBeL
QSR	Queue Selection Rule (See Table 7-1)
SSR	Server Selection Rule (See Table 7-2)
BLOCK	SELECT node BLOCKs incoming entities and servers
BALK	Entities BALK from SELECT node
ASM	ASseMbly of entities at SELECT node
NATRR	Number of an ATtRibute for Referencing individual entities of a batch or for splitting entities from an UNBATCH node

7.12 EXERCISES

7-1. In SLAM II, there are many ways to halt the flow of an entity through a network. Describe the node type or procedures you would use to stop and start the flow of an entity in each of the following situations:

(a) An entity called dinner cannot be served until the steak, potatoes, and salad are prepared.

(b) An entity called steak cannot be inserted into the oven until five minutes after the oven is turned on.

(c) The dessert, apple pie and ice cream, cannot be served until four people have completed eating.

(d) One of the diners requires that the steak be medium rare and that the salad dressing be bleu cheese.

(e) The eating-of-dinner activity is not started until wine is poured for all diners.

7-2. At a drive-in bank where there is only one teller, there is space for five waiting cars. If a customer arrives when the waiting line is full, the customer drives around the block and tries to join the waiting line again. The interarrival time between customer arrivals is exponentially distributed with a mean of 10. The time to drive around the block is normally distributed with a mean of 2 and a standard deviation of 0.5. The teller service time is uniformly distributed between and 6 and 12. When a customer arrives and can join the queue, it takes a negligible amount of time to become a member of the queue. Initially, no customers are waiting to be served and the teller is idle. Draw the SLAM II network associated with this situation which collects statistics on the customer's time in the queue, time in the system, and time between balks.

Embellishment: For this banking situation, cars depart from the teller into a street. The amount of time for a car to find a gap large enough to depart into the street is exponentially distributed with a mean of 3. The design of the drive-in bank parking lot only allows five cars to be waiting to enter the street. Modify your network to include this new feature.

7-3. A server is stationed by a conveyor belt and the server can take items off the conveyor belt only if he is idle. Items arrive to the conveyor belt with the time between arrivals a constant 10 time units. Once the item is placed on the conveyor belt, it takes three time units for it to reach the service station. If the server is busy, the item continues on the conveyor belt and returns to the server in 9 time units. Service for the item is exponentially distributed with a mean of 2.5. When the server finishes working on an item, he places it on a second conveyor belt to be processed by a second server. The item spends five time units on the second conveyor belt before arriving at the second server. If the second server is busy, the item stays on the second conveyor belt for 12 time units before it is returned to the second server. The service time of the second server is normally distributed with a mean of 2.0 and a standard deviation of 1. After being served by the second server, the item departs the system. Draw a SLAM II network of this situation that collects information on the amount of time an item spends in the system and the number of items on each conveyor belt.

7-4. Convert a PERT network with which you are familiar into a SLAM II network representation. For this SLAM II network, presume that there is a probability that some activities in the network will fail which would cause project failure. Redraw the SLAM II network to represent this situation.

7-5. Describe how an ACCUMULATE node of a SLAM II network can be used to represent the following logic operations: all preceding activities must be completed before successor activities can be started; any one of the preceding activities must be completed before the activity can be started; and three out of five of the preceding activities must be completed before the activity can be started (or, in general, a majority voting type of logic).

7-6. At an airline terminal, five ticket agents are employed and current practice is to allow queues to form before each agent. Time between arrivals to the agents is exponentially distributed with a mean of 5 minutes. Customers join the shortest queue at the time of their arrival. The service time for the ticket agents is uniformly distributed between 0.5 and 1.5 minutes. The queues of the ticket agents are not allowed to exceed two customers each. If the queues of all ticket agents are full, the customer goes directly to his gate to be served by a stewardess. Develop the SLAM II network from which the total time a customer spends at the ticket agent windows, the utilization of the ticket agents, and the number of customers per minute that cannot gain service from the ticket agents can be determined.

Embellishments: (a) The airline has decided to change the procedures involved in processing customers by the ticket agents. A single line is formed and the customers are routed to the ticket agent that becomes free next. A tenth of a minute service time is added to the processing time of each ticket agent. Space available in the single line for waiting customers is ten. Develop the SLAM II network for this revised situation.

(b) It has been found that a subset of the customers purchasing tickets are taking a long period of time. By segregating ticket holders from non-ticket holders, improvements can be made in the processing of customers. To accomplish this segregation, four ticket agents are used for checking in customers and one agent is used for purchases. The time to check in a person is uniformly distributed between 0.2 and 1 minute and the time to purchase a ticket is exponentially distributed with a mean of 5 minutes. Assuming that 15 percent of the customers will be purchasing tickets, develop the SLAM II network for this situation. The time between all customer arrivals is exponentially distributed with a mean of 5 minutes.

7-7.   For the model of a Mail Order Clothes Manufacturer given in Illustration 7-2, revise the model to include size as part of the order. For shirts, 30% are for size 15; 50% are for size 16; and 20% are for size 17. In addition, pants come in sizes 30, 32, 34 and 36. The probabilities of an order being for these sizes are 0.10, 0.20, 0.40, and 0.30, respectively. Determine a production schedule or procedure for producing shirts and pants considering that it takes 2 minutes to setup to change to a new size. Establish an objective for determining a good production schedule.

7-8.   In the model of Illustration 7-2 of the Mail Order Clothes Manufacturer, show how resources can be used to represent the production of shirts and pants to avoid the splitting of an order into entities representing 1 dozen units of product.

7-9.   There are three stations on an assembly line and the service time at each station is exponentially distributed with a mean of 10. Items flow down the assembly line from server 1 to server 2 to server 3. A new unit is provided to server 1 every 15 time units. If any server has not completed processing its current unit within 15 minutes, the unit is diverted to one of two off-line servers who complete the remaining operations on the job diverted from the assembly line. One time unit is added to the remaining time of the operation that was not completed. Any following operations not performed are done so by the off-line servers in an exponentially distributed time with a mean of 16. Draw the SLAM II network to obtain statistics on the utilization of all servers, and the fraction of items diverted from each operation.

Embellishments: (a) Assume the assembly line is paced and that the movement of units can only occur at multiples of 15 minutes.

(b) Allow one unit to be stored between each assembly line server.

(c) If a server is available, route units back to the assembly line from the off-line servers.

7-10.   Simulate the activities of the PERT network described below 400 times. Compute statistics and prepare a histogram on the time to reach each node of the network. Compare the results with the PERT calculations for the network.

Embellishment: (a) Based on the SLAM II simulation of a PERT network, develop a schedule of early start times and late start times for the activities in the network.

Activity Number	Start Node	End Node	Distribution Type	Mean	Variance
1	1	2	Lognormal	10	4.00
2	1	3	Exponential	6	36.00
3	2	4	Uniform	7	3.00
4	2	9	Gamma	14	21.00
5	3	2	Exponential	3	9.00
6	3	4	Uniform	13	5.33
7	3	6	Normal	5	1.00
8	4	9	Erlang	8	32.00
9	6	5	Constant	7	0.00
10	6	7	Normal	4	2.16
11	6	7	Normal	4	3.00
12	5	4	Normal	2	1.20
13	5	8	Normal	6	10.40
14	7	8	Normal	8	26.40
15	8	9	Normal	5	2.00

7-11. Consider a banking system involving two inside tellers and two drive-in tellers. Arrivals to the banking system are either for the drive-in tellers or for the inside tellers. The time between arrivals to the drive-in tellers is exponentially distributed with a mean of 0.75 minutes. The drive-in tellers have limited waiting space. Queueing space is available for only three cars waiting for the first teller and four cars waiting for the second teller. The first drive-in teller service time is normally distributed with a mean of 0.5 minutes and standard deviation of 0.25 minutes. The second drive-in teller service time is uniformly distributed between 0.2 and 1.0 minutes. If a car arrives when the queues of both drive-in tellers are full, the customer balks and seeks service from one of the inside bank tellers. However, the inside bank system opens one hour after the drive-in bank. Customers who directly seek the services of the inside tellers arrive through a different arrival process with the time between arrivals exponentially distributed with a mean of 0.5 minutes. However, they join the same queue as the balkers from the drive-in portion. A single queue is used for both inside tellers. A maximum of seven customers can wait in this single queue. Customers who arrive when there are seven in the inside queue balk and do not seek banking service. The service times for the two inside tellers are triangularly distributed between 0.1 and 1.2 minutes with a mode of 0.4 minutes. Simulate the operation of the bank for an 8 hour period.

7-12. Modify the bank teller operations described in Exercise 7-11 to model a credit inquiry on selected non-drive-in bank customers. A credit inquiry is performed by the bank manager on new customers. Ten percent of the non-drive-in customers are in this category. The bank manager obtains the necessary information on the customer and initiates the inquiry which takes between 2 and 5 minutes, uniformly distributed. The time for a credit inquiry is exponentially distributed with a mean of 5 minutes during which time the customer waits in a separate room. The manager processes other customers during the time the credit inquiry is being performed, and there is no limit to the number of simultaneous credit inquiries that can be done. When the credit inquiry is completed, the customer for which the credit inquiry was made is served again by the manager and is given preference over customers who have not seen the manager. The manager completes any information gathering task before he issues the credit inquiry which takes 1 minute. Five percent of the credit inquiries result in a negative response and the customer is not routed to the tellers. The time to give a negative response is exponential with a mean of 10 minutes.

Embellishment: Model the manager in the case where the rule is that two inquiries are made for each customer requiring an inquiry. Each inquiry has a 0.05 negative response probability.

7-13.   Kits consist of two parts, Part A and Part B. A kit arrives to a cleaning and assembly station every 10 minutes. The station attempts to process each kit in 60 minutes. Functions of the station are to disassemble the kit into its two parts which takes eight minutes on the average and is exponentialy distributed. Each part is then sent to a cleaning station. The time to clean Part A is normally distributed with a mean of 12 and a standard deviation of 2. There is only one cleaning machine for Part A. The time to clean Part B is uniformly distributed between 20 and 28 minutes and there are two cleaning machines for Part B. Inspection and assembly are done at the same time. There are two expert assemblers and one trainee. From past analysis it is known that Part B holds up the reassembling of the kit, and a procedure that specifies that if Part type B is within 10 minutes of the due date it should be assembled by one of the expert assemblers. When a choice exists between an expert assembler and the trainee, the expert assembler is used. The time to perform the assembly operation is triangularly distributed for the most expert of the assemblers with a modal value of 25 minutes, minimum of 16 minutes and a maximum of 40 minutes. The second expert assembler takes 1.1 times the time of the first expert assembler. When a choice exists, the faster assembler is used. The trainee assembler takes 1.25 of the time of the most expert assembler. Assume that Part A and Part B do not have to be reassembled from the same kit in which they arrived.

Embellishment: Change the model to require that each Part A is assembled with the Part B of the kit from which they were disassembled.

7-14.   An indexed rotary table serves eight machining centers. A part enters the table for Operation 1, then is rotated in turn through all eight operations before being unloaded. The table can rotate only when all eight machines have finished their current operation. Suppose that processing time on each machine is uniformly distributed between 3 and 6 minutes, and that indexing (rotation) takes one minute. Model the rotary table to determine the utilization of each machine center in a 50-part production run. Estimate the length of the production run. Assume that the simulation starts at time 8 minutes with each machine center ready to work on an operation.

Embellishment: Build two different models of the rotary table.

7-15.   Develop a 99.7 percent confidence interval for the project completion time in the PERT model, Example 7-1. Compute an interval such that with 90 percent confidence at least 80 percent of the population of average completion times are in the interval.

7.13 REFERENCES

1. Archibald, R. D. and R. L. Villoria, *Network-Based Management Systems (PERT/CPM),* John Wiley, 1968.
2. Halpin, D. W. and W. W. Happ, "Digital Simulation of Equipment Allocation for Corps of Engineering Construction Planning," U.S. Army, CERL, Champaign, Illinois, 1971.
3. MacCrimmon, K. R. and C. A. Ryavec, "An Analytical Study of the PERT Assumptions," *Operations Research,* Vol. 12, 1964, pp. 16-38.
4. Malcolm, D. G., J. H. Rosenbloom, C. E. Clark and W. Frazer, "Application of a Technique for Research and Development Program Evaluation," *Operations Research,* Vol. 7, 1959, pp. 616-669.
5. Moder, J. J. and C. R. Phillips, *Project Management with CPM and PERT (Second Edition),* Van Nostrand-Reinhold, 1970.
6. Pritsker, A. A. B., *The GASP IV Simulation Language,* John Wiley, 1974.
7. Pritsker, A. A. B., *Modeling and Analysis Using Q-Gert Networks (Second Edition),* Halsted Press and Pritsker & Associates, Inc., 1979.
8. Weist, J. and F. Levy, *Management Guide to PERT-CPM,* Prentice-Hall, 1969.

CHAPTER 8

SLAM II Processor, Inputs and Outputs

8.1   INTRODUCTION

In Chapters 5, 6, and 7, SLAM II was presented both as a graphical framework for conceptualizing network models and as a language for describing network models in statement form. SLAM II is also a computer program which interprets and executes the statement equivalent of a network to act out or simulate a model of the real system. In this chapter, we discuss the use of the SLAM II simulation program to simulate network models. We begin by presenting an overview of the network simulation procedure employed by the SLAM II processor. This is followed by a detailed description of the input statements used in constructing network models, the control statements used when performing experiments with the SLAM II processor, and the standard output reports obtained from a SLAM II simulation. A SLAM II Quick Reference Manual is available that covers material from this chapter and the appendices in a more direct manner (1). It also provides information on SLAM II COMMON blocks and dimensions.

8.2   NETWORK ANALYSIS

The network modeling approach consists of modeling a system as a set of entities which flow through a network of nodes and activities. As entities flow through a network, they occupy servers; advance time; await, seize and free resources; open and close gates; queue up in files; change variable values; and, in general, cause changes in the state of the system. A fundamental observation, which forms the basis for the network simulation approach employed by SLAM II, is that these changes can only occur at the time of arrival of an entity to a node. The SLAM II processor generates a complete portrayal of the changes in state of a network model by processing in a time-ordered sequence the events representing the arrival of an entity to a node.

The mechanism employed for maintaining the time-ordered sequence of entity arrival events is the event calendar. The event calendar consists of a list of entity arrival events, each characterized by an "event time" and an "end node". The event time specifies the time at which the entity arrival is to occur. The end node specifies the node to which the entity is to arrive. The events on the event calendar are ranked low-value-first (LVF) based on their event time.

The next-event processing logic employed by SLAM II for simulating networks is depicted in Figure 8-1. The processor begins by interpreting the SLAM II state-

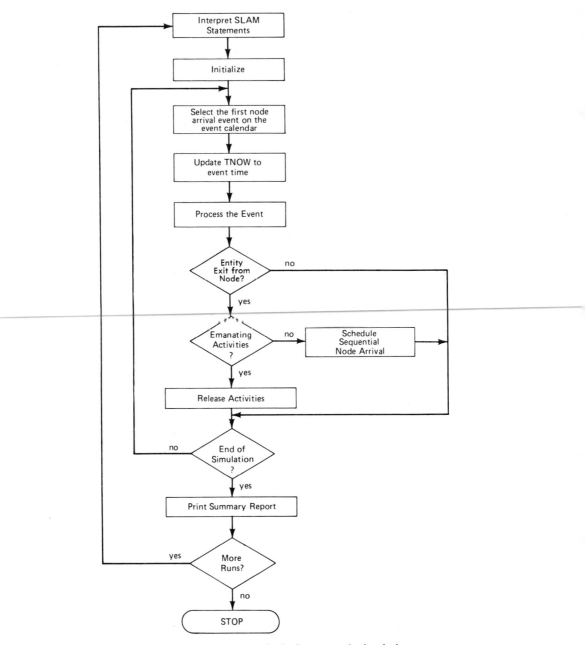

Figure 8-1 SLAM II next event processing logic for network simulations.

ments. This is followed by an initialization phase which is completed prior to the start of the simulation. During this initialization phase, the processor places on the event calendar an entity arrival event to occur at each CREATE node at the time of the first release of the node. Also, entities initially in QUEUE nodes are created and end-of-service activity events are scheduled where appropriate. Therefore, the event calendar will initially contain one arrival event corresponding to each CREATE node and one end-of-service event for each busy server in the network.

The execution phase of a simulation begins by selecting the first event on the calendar. The processor advances the current simulated time, TNOW, to the event time corresponding to this event. It processes the event by performing all appropriate actions based on the decision logic associated with the node type to which the entity is arriving. For example, if the entity is arriving to an AWAIT node, the decision logic involved with realizing the event consists of testing to determine if the required level of resource is available; if yes, the entity seizes the desired units of resource and exits the node; otherwise, the entity is placed in the specified file and awaits the required resource. Although the decision logic is different for each node, the logic will result in one of three possible outcomes for the arriving entity:

1. The entity will be routed to another node;
2. The entity will be destroyed at the node; or
3. The entity will be delayed at the node based on the state of the system.

The routing of an entity from a node involves a test for activities emanating from the node. If there are no emanating activities, then the entity is scheduled to arrive at the sequential node at the current time, TNOW. If there is no sequential node, the entity is destroyed. When there are emanating activities, as many as M activities are selected in accordance with the probability or conditions associated with each activity where M is the M-number associated with the node. If an activity is selected, the entity or its duplicate is routed to the end node of the activity at the current time, TNOW, plus the duration of the activity.

After all events have been scheduled, the SLAM II processor tests for one of the following end-of-simulation conditions:

1. TNOW is greater than or equal to the user specified ending time of the simulation;
2. There are no events on the event calendar; or
3. A TERMINATE node has been released.

If none of the end of simulation conditions are satisfied, the SLAM II processor selects the next arrival event on the event calendar and continues. When a simulation run is ended, statistics are calculated and the SLAM II Summary Report is printed. A test is then made to determine if additional simulation runs are to be executed. If more runs remain, the next run is initiated. If all simulation runs have been completed, execution by the SLAM II processor is stopped and control is returned to the operating system.

The next event logic described above is well defined except when two or more arrival events are scheduled to occur at the same instant in time. To illustrate the problem, consider the case where we have a gate named DOOR which is currently open, and entities are scheduled to arrive at precisely the same time to the following statements:

Statement 1	*Statement 2*
AWAIT, DOOR;	CLOSE, DOOR;

If statement 1 is processed before statement 2, the gate DOOR will be open and the entity will exit the AWAIT node. However, if statement 2 is processed before statement 1, the transaction arriving at statement 1 will be delayed at the AWAIT node because the gate DOOR was just closed. Therefore, the order in which simultaneous events are processed can affect the results of the simulation.†

In describing the node arrival tie-breaking rules employed by SLAM II, it is convenient to classify all events on the event calendar as either "current events" or "future events." Current events are those events which are scheduled to occur at the current simulation time (TNOW), whereas future events are those events which are scheduled to occur at some simulated time in the future (not at TNOW). The reason for conceptually distinguishing between current and future events is that different tie-breaking procedures are employed depending upon whether the event is initially scheduled as a current event or as a future event.

The scheduling of an arrival event as a current event results from scheduling either a sequential node arrival or an end of activity node arrival where the activity has a duration of zero. In either case, the time at which the event is to occur is the current simulation time (TNOW). The scheduling procedure employed by SLAM II is always to place current events at the top of the event calendar, regardless of other events on the calendar. Therefore, current events are scheduled onto the calendar on a last-in, first-out (LIFO) basis. In the case where several zero duration activities emanate from a single node, the arrival events corresponding to the end of the activities are placed at the front of the calendar in the order that

---

† This difficulty is not as severe as in some simulation languages where an integer-valued clock time is used.

the activities appear in the statement model; that is, the first event would correspond to the first zero duration activity, the second event would correspond to the second zero duration activity, and so on.

Since the only mechanism for advancing time in a SLAM II network model is the activity, the scheduling of a future event necessarily corresponds to the arrival event resulting from an end of activity of nonzero duration. The tie-breaking procedure for scheduling future events employs a secondary ranking procedure which is defaulted to first-in, first-out (FIFO). The secondary ranking can be specified by the modeler (see Section 8.5.3) to be FIFO; LIFO; LVF(NATR), low-value-first based on attribute NATR; or HVF(NATR), high-value-first based on attribute NATR. In the case of the LVF or HVF secondary ranking, ties can still exist and they are broken using the FIFO rule.

At this point, the reader may be questioning the rationale of scheduling current events using LIFO as the tie-breaking rule while scheduling future events using a user specified tie-breaking rule. Why not just use one rule and avoid distinguishing between current and future events? The reason is that the LIFO rule is not only considerably more efficient in terms of computer execution time for scheduling current events, but it moves one entity at a time as far through the network as possible until it is either destroyed, delayed by the status of the system, or encounters a time delay. A FIFO rule would advance one entity one node, then advance the next entity one node, and so on. The advantage of the LIFO flow pattern will become apparent when the reader encounters traces of entity flow (the TRACE option is discussed in Section 8.5.9).

8.3 GENERAL FORMAT CONVENTIONS

Input statements are read by the SLAM II processor from a terminal or its equivalent. Each input statement is uniquely defined by the first three letters of the statement name. Each field in the statement is separated by a comma [,], slash [/], or offset in parentheses [( )]. The appropriate field separator is dependent upon the specific statement type and field entry and is specified in the description of each statement type. Basically, slashes and parentheses are used for optional specifications and are optional also. Commas separate fields that are not optional. The special characters [+−*/] are used to denote arithmetic operations and the semi-colon character [;] is used to denote the end of a statement. *These characters should not be employed within network statement labels or within user defined alphanumeric names.*

A field in a statement can be defaulted by simply omitting the entry while in-cluding the terminator. For example, an activity with zero duration, no branching condition, and an end node labeled EXIT can be specified by defaulting the dura-tion and condition fields as follows:

ACTIVITY,,,EXIT;

Also, if a statement is terminated with remaining fields unspecified, the remaining fields take on their default values. For example, an ACTIVITY of duration 1, un-conditionally taken, with an end node as the sequential node can be specified by defaulting the last two fields as follows:

ACTIVITY, 1;

The end node for this activity is assumed to be the next node statement encoun-tered.

Input statements are read using a free format which permits a statement to be spaced across a line or over several lines. One restriction is that a field may not be split between lines. A continuation of a statement to the next line is assumed if the last non-blank character of the statement is either a [,/(+−*)] or the last period of a logical operator such as [.AND.]. If the last non blank character is any other character, an end of statement is assumed. However, the preferred method for end-ing a statement is the explicit use of the statement terminator [;] which permits the inclusion of comments following the terminator. All blanks are ignored, except within alphanumeric fields, and therefore can be freely employed to improve the readability of statements. For example, the following ACTIVITY statement:

ACTIVITY,10,ATRIB(1).EQ.1.AND.TNOW.LT.100,LOOP;

can be spaced over three lines as follows:

ACTIVITY,10,
    ATRIB(1).EQ.1.AND.
    TNOW.LT.100,   LOOP;

However, it should be noted that the three lines would require a longer processing time that that required for the single line.

Numeric data can be entered as whole numbers (integers) or numbers with a fractional part (decimal numbers), and may be signed or unsigned. In addition, extremely large or extremely small numbers can be entered in scientific notation using an E format. For example, the number ten can be entered as 10 or 10. or +10. or 1.E + 1 or 1E1 or 100E − 1. If a decimal number is entered in a field specified to be integer, the fractional part is dropped. Likewise if an integer is entered in a field specified to be decimal, its decimal equivalent is used. Therefore, the SLAM II input processor does not distinguish between 1. and 1 regardless of the field type specified.

Alphanumeric fields can be inputted as any string of characters (including numerals) which begins with an alphabetic character *but does not contain the special characters* [,/( )+−*;]. Blank characters are significant within alphanumeric fields; hence L 1 is not considered the same as L1. Since fields cannot be split between lines, an alphanumeric field is limited to a maximum of 80 characters. However, depending upon the field being entered, only the first 3, 4, 8, 16, or 20 characters are read. Examples of valid alphanumeric fields are BARBER or CRANE or S927 or K?1. or J. DOE.

8.4 NETWORK STATEMENT FORMATS

Network statements are prepared beginning in column 7 or after, with columns 1 through 5 reserved for node labels. Although node labels can appear anywhere in the first five spaces, *only the first four characters of a label are significant.* This means that the label LANE1 would not be distinguished from the label LANE2.

The network statements must be preceded by a NETWORK statement and followed by an ENDNETWORK statement. The NETWORK statement consists of the characters NET entered anywhere on a line and denotes to the SLAM II processor that the lines to follow are network statements. The NETWORK statement has two additional fields:

NETWORK,option,device;

which provide a means to store a decoded network for future loading. To store a decoded network, the word SAVE is put in the option field and a logical unit number (device or tape number) is given in the device field. Thus, the statement

NETWORK,SAVE,3;

causes the decoded network to be written in binary form to logical unit number 3. In future uses of the network, the statement

NETWORK,LOAD,3;

causes the decoded network to be used. This saves the computer time necessary to decode a large set of network statements. When the option is specified as LOAD, no network statements follow the NETWORK statement and no ENDNETWORK statement is used. The defaults for the option and device fields assume that a network is to be decoded. The ENDNETWORK statement consists of the characters END typed on a line beginning in or after column 7 and denotes an end to all network statements. A list of network statements is presented in Table 8-1 which includes the default values for each field.

Some of the features of the network input statements require further explanation. The overall design is intended to increase the readability of the statements without encumbering the user with extraneous information requirements. This goal led to the use of four delimiters to separate values:

1. Commas are used to separate fields.
2. Slashes (virgules) are used to allow an optional or normally defaulted value to be contained as a second value in a field. Slashes may also be used to indicate that a set of fields is to be repeated.
3. Parentheses are used to indicate a capacity or associated file.
4. A semicolon is used to end a statement.

With these delimiters, efficient input statement preparation has been obtained. The examples demonstrate this point. As with any simulation language, it does require using the language to feel comfortable with the wide range of alternatives.

Another aspect of the input statements is that all node types have been given verbs as names. This corresponds to the modeling approach which requires decisions and logical functions to be performed at the nodes of the network.

8.5   SIMULATION CONTROL STATEMENTS

In this section, we describe additional statement types which are used in writing SLAM II simulation programs. In contrast to network input statements which begin in or after column 7, the control statements can start in any column. Typi-

Table 8-1 SLAM II network statement types.

Statement Form	Statement Defaults(ND = no default)
Nodes	
ACCUM,FR,SR,SAVE,M;	ACCUM,1,1,LAST,∞;
ALTER,RES/CC,M;	ALTER,ND/ND,∞;
ASSIGN,VAR = value,VAR = value,...,M;	ASSIGN,ND = ND,ND = ND,...,∞;
AWAIT(IFL/QC),RES/UR or GATE,BLOCK or BALK(NLBL),M;	AWAIT(first IFL in RLBL's or GLBL's list/∞),ND/1,none,∞;
BATCH,NBATCH/NATRB,THRESH, NATRS,SAVE,RETAIN,M;	BATCH,1/none,ND, entity count,LAST,NONE,∞;
CLOSE,GATE,M;	CLOSE,ND,∞;
COLCT(N),TYPE or VARIABLE, ID,NCEL/HLOW/HWID,M;	COLCT(ordered),ND,blanks,no histogram/0./1
CREATE,TBC,TF,MA,MC,M;	CREATE,∞,0,no marking,∞,∞;
DETECT,XVAR,XDIR,VALUE,TOL,M;	DETECT,ND,ND,ND,0,∞;
ENTER,NUM,M;	ENTER,ND,∞;
EVENT,JEVNT,M;	EVENT,ND,∞;
FREE,RES/UF,M;	FREE,ND/1,∞;
GOON,M;	GOON,∞;
MATCH,NATR,QLBL/NLBL,...,M;	MATCH,ND,ND/no routing,ND/no routing,...,∞,
OPEN,GATE,M;	OPEN,ND,∞;
PREEMPT(IFL)/PR,RES,SNLBL, NATR,M;	PREEMPT(first IFL in RLBL's list)/no priority,ND,AWAIT node where transaction seized resource, none,∞;
QUEUE(IFL),IQ,QC,BLOCK or BALK(NLBL),SLBLs;	QUEUE(ND),0,∞, none,none,none;
SELECT,QSR/SAVE,SSR,BLOCK or BALK (NLBL),QLBLs;	SELECT,POR/none,POR,none,ND;
TERMINATE,TC;	TERMINATE,∞;
UNBATCH,NATRR,M;	UNBATCH,ND,∞;
Blocks	
GATE/GLBL,OPEN or CLOSE,IFLs/repeats;	GATE/ND,OPEN,ND/repeats;
RESOURCE/RLBL(IRC),IFLs/repeats;	RESOURCE/ND(1),ND/repeats;
Regular Activity	
ACTIVITY/A,duration,PROB or COND,NLBL; ID	ACTIVITY/no ACT number,0.0,take ACT,ND; blank
Service Activity	
ACTIVITY(N)/A, duration,PROB, NLBL; ID	ACTIVITY(1)/no ACT number,0.0,1.0,ND; blank

cally, they are started in Column 1. A list of the statement types is presented in Table 8-2 in alphabetical order.

The control statement types related to network processing are described in the order of their most frequent use. The EQUIVALENCE and ARRAY statements were described in Chapter 5. The CONTROL statements CONTINUOUS, SEVNT and STAT are described in later chapters. The STAT statement is described in Chapter 11 where it is used in conjunction with calls to subroutine COLCT to obtain statistics on observations. The combination of subroutine COLCT and the STAT statement provides the same capabilities as the COLCT node. The CONTINUOUS statement is used to define continuous variables and their evaluation procedures. The statement is described in Chapter 10. The SEVNT statement is used to specify state events which are events specified as the crossing of a threshold value by a variable. This statement is described in Chapter 13.

8.5.1   GEN Statement

The GEN statement provides general information about a simulation in the format shown below.

GEN,NAME,PROJECT,MONTH/DAY/YEAR,NNRNS,ILIST, IECHO,
    IXQT/IWARN,IPIRH,ISMRY/FSN,IO;

The GEN statement must be the first statement in any SLAM II simulation program. Included on the GEN statement are: the analyst's name, a project identifier, date, number of simulation runs, and report options. The NAME and PROJECT are both alphanumeric fields with 20 characters of significance and are used for output reports to identify the analyst and the project. Recall that blanks are significant within alphanumeric fields. The MONTH, DAY, and YEAR are entered as integers separated by slashes. The SLAM II variable NNRNS is entered as an integer, has a default value of 1, and denotes the number of simulation runs to be made. The next six fields are normally specified as YES and correspond to the following options:

ILIST   If yes, a numbered listing of all input statements is printed including error messages if any; otherwise the listing is omitted.

IECHO   If yes, an echo summary report is printed; otherwise the report is omitted.

Table 8-2 SLAM II control statements.

Statement Form

ARRAY(IROW,NELEMENTS)/initial values;
CONTINUOUS,NNEQD,NNEQS,DTMIN,DTMAX,DTSAV,W or F or
  N,AAERR,RRERR;
ENTRY/IFL,ATRIB(1),ATRIB(2),...,ATRIB(MATR)/repeats;
EQUIVALENCE/SLAM II variable,name/repeats;
FIN;
GEN,NAME,PROJECT,MO/DAY/YEAR/,NNRNS,ILIST,IECHO,IXQT/IWARN,
  IPIRH,ISMRY/FSN,IO;
INITIALIZE, TTBEG,TTFIN,JJCLR/NCCLR,JJVAR,JJFIL;
INTLC,VAR = value,repeats;
LIMITS,MFIL,MATR,MNTRY;
MONTR,option,TFRST,TSEC,variables;
NETWORK,SAVE or LOAD,device;
PRIORITY/IFL,ranking/repeats;
RECORD(IPLOT),INDVAR,ID,ITAPE,P or T or B,DTPLT,TTSRT,TTEND,KKEVT;
SEEDS,ISEED(IS)/R,repeats;
SEVNT,JEVNT,XVAR,XDIR,VALUE,TOL;
SIMULATE;
STAT,ICLCT,ID,NCEL/HLOW/HWID;
TIMST,VAR,ID;
VAR,DEPVAR,SYMBL,ID,LOORD,HIORD;

IXQT	If yes, execution is attempted if no input errors are detected. If specified as no, execution is not attempted.
IWARN	If yes, a warning message is printed when an entity is destroyed before reaching a TERMINATE node; otherwise the printing is omitted.
IPIRH	If yes, the heading INTERMEDIATE RESULTS is printed prior to execution of each simulation run; otherwise the printing is omitted.
ISMRY	If yes, the SLAM II Summary Report is printed in accordance with the next field specification, FSN; otherwise the report is suppressed.

If the summary report option is answered as YES, a report frequency field, FSN, is to be specified as: F, after the first run only; S, after the first and last runs; or N, an integer, specifying after every Nth run. The last field, IO, specifies the number of columns to be used for output reports. The options are 72 or 132 columns which are typically used for terminal and line printer outputs respectively.

8.5.2   LIMITS Statement

The format of the LIMITS statement is shown below.

LIMITS,MFIL,MATR,MNTRY;

The second statement in a SLAM II simulation program is the LIMITS statement. The LIMITS statement is used to specify integer limits on the largest file number used (MFIL), the largest number of attributes per entity (MATR), and the maximum number of concurrent entries in all files (MNTRY). Normally an estimate of MNTRY must be made, as it is the total number of entities that can exist in the model at one time. We recommend the judicious use of a safety factor when estimating MNTRY.

8.5.3   PRIORITY Statement

The format of the PRIORITY statement is shown below.

PRIORITY/IFILE, ranking/repeats;

The PRIORITY statement is used to specify the criterion for ranking entities within a file. There are four possible specifications for the criterion:

FIFO   Entries are ranked based on their order of insertion in the file with early insertion given priority. This is a first-in, first-out ranking criterion.

LIFO   Entries are ranked based on their time of insertion with late insertions given priority. This is a last-in, first-out ranking criterion.

HVF(N)   The entries are ranked high-value-first based on the value of the Nth attribute.

LVF(N)   The entries are ranked low-value-first based on the value of the Nth attribute.

The default for the criterion for all files is FIFO, therefore a PRIORITY statement for a file need be included only if the file is ranked LIFO, HVF(N), or LVF(N). A file ranking is specified by entering the file number IFILE followed by a comma and the file ranking. The rankings for different files can be specified on a single PRIORITY statement by separating the inputs with slashes.

The PRIORITY statement can also be used to specify the secondary ranking procedure for breaking ties between simultaneous node arrivals which are scheduled as future events by specifying the file value, IFILE, with the alphanumeric characters NCLNR denoting the event calendar. The event code can be used as the attribute specification by inserting the characters JEVNT for the attribute number. The following statement specifies that file 3 is to be ranked on a LIFO basis and the tie-breaking rule for events is high-value-first based on attribute 4:

PRIORITY/3,LIFO/NCLNR,HVF(4);

If the secondary ranking for the event calendar is to be low-value-first based on the event code, the statement would be:

PRIORITY/3,LIFO/NCLNR,LVF(JEVNT);

8.5.4   TIMST Statement

The format for the TIMST statement is shown below.

TIMST,VAR,ID,NCEL/HLOW/HWID;

The TIMST statement is normally employed to initiate the automatic collection of time-persistent statistics on the global variable XX(N), or on the state or derivative variables SS(N) and DD(N). To employ the TIMST statement to initiate statistics on the indexed variable XX(N) where N is an integer, the user simply enters XX(N) in the variable field followed by a comma and an alphanumeric identifier (ID) which is to be used in displaying the statistics in the SLAM II Summary Report. The first 16 characters of the identifier are significant. An example of the TIMST statement to collect statistics on the global variable XX(1) defined to be the number of entities currently in the network is shown below.

TIMST,XX(1),NUMBER IN SYSTEM;

Histograms can also be obtained on the fraction of time that the variable VAR was within a range of values. The number of cells in the histogram and the width of each cell is specified by NCEL, HLOW and HWID as discussed in Section 5.12. Any SLAM II time-persistent variable may be used for the VAR field. The statement TIMST, NNQ(1), QUEUE LENGTH, 10/0/1; specifies that statistics and a histogram be obtained with the histogram presenting the amount of time the queue length is 0, 1, 2, $\cdots$.

8.5.5 SEEDS Statement

The format for the SEEDS statement is shown below.

SEEDS,ISEED(IS)/R, repeats;

The purpose of the SEEDS statement is to permit the user to specify the starting unnormalized random number seed for any of the 10 random number streams available within SLAM II and to control the reinitialization of streams for multiple simulation runs. The seeds are entered as integers with the stream number of the seed given in parentheses. If the stream number is not specified, then stream numbers are assigned based on the position of the seed. The first seed is for stream 1, the second for stream 2, and so on. The reinitialization of each stream is controlled by specifying YES or NO as a subfield following a slash immediately after the seed value and stream number. If the subfield is not included, the default case is assumed and the seed values are not reinitialized. If the SEEDS input statement is not included, the SLAM II processor uses default seed values. An example of the SEEDS input statement is shown below:

SEEDS,9375295(1)/YES,0(2)/YES,6315779(9),2734681;

This statement initializes the seed for stream 1 to 9375295 and specifies that this value be used as the first value for each run. The zero specifies that the default value should be used for stream 2 and that it should be the first value for each run. The seed value for stream 9 is 6315779 and it is not to be reinitialized on subsequent runs. The seed value for stream 10 is 2734681 and is not to be reinitialized. If antithetic random numbers are desired for a run then a negative seed value is used. The sequence of random numbers generated will be the complement of the numbers generated from the use of a positive value of the seed. The function XRN (IS) can be used to access the value of the last random number generated from stream IS.

8.5.6   INTLC Statement

The format for the INTLC statement is shown below.

INTLC,VAR = value,repeats;

The INTLC statement is used to assign initial values to the SLAM II variables XX(N), SS(N), or DD(N) where N is an integer. The initial value can be specified on the statement simply by separating each assignment by a comma. An example is:

INTLC,XX(1)=0, XX(2)=3.0;

8.5.7   INITIALIZE Statement

The format for the INITIALIZE statement is shown below.

INITIALIZE,TTBEG,TTFIN,JJCLR/NCCLR,JJVAR,JJFIL;

The INITIALIZE statement is used to specify the beginning time (TTBEG) and ending time (TTFIN) for a simulation, and initialization options for clearing statistics, initializing variables, and initializing files. The last three fields are specified as YES or NO and are normally defaulted to YES. If JJCLR is specified as YES, NCCLR specifies the number of the collect variable up to which clearing is to be performed. If JJCLR is specified as NO, NCCLR specifies the collect variable number up to which clearing is not to be performed. The default value for NCCLR is that JJCLR applies to all collect variables. If JJVAR is specified as YES, TNOW is initialized to time TTBEG, and the variables XX(N), SS(N), and DD(N) are initialized to their starting values before each simulation run. If the field is specified as NO, the initializations are not performed. If JJFIL is specified as YES, the filing system is initialized before beginning each simulation run, otherwise it is not. The initialization of the filing system causes all file statistics to be cleared, removes all entries from the files, and places any initial QUEUE node entities into appropriate files.

8.5.8   ENTRY Statement

The format for the ENTRY statement is shown below.

ENTRY/IFILE,ATRIB(1),ATRIB(2),. . . ,ATRIB(MATR)/repeats;

The ENTRY statement is used to place initial entries into files. An entry is specified by entering the file number (IFILE) followed by the attributes of the entry separated by commas. The slash is used to denote the beginning specification of a new entry and causes any unspecified attribute values for the last entry to default to zero. If file IFILE is associated with a QUEUE or AWAIT node, the entry is processed as an entity arrival to the node at time TTBEG (beginning time of the simulation). In the case where multiple AWAIT nodes employ the same file, the arrival is processed at the first AWAIT node listed in the network model. An example of an ENTRY statement is shown below.

ENTRY/1,7.0,3.2/3,10.,5.5;

This statement inserts an entity into file 1 with attribute 1 equal to 7 and attribute 2 equal to 3.2, and an entity into file 3 with attributes 10 and 5.5. The ENTRY statement applies only to a single run and must be repeated when used with multiple runs.

8.5.9   MONTR Statement

The format for the MONTR statement is shown below.

MONTR,Option,TFRST,TSEC,Variables;

The MONTR statement is used to monitor selected intermediate simulation results. The MONTR statement can also be used to clear statistical arrays after a "warm up" period in order to reduce any bias that is due to initial starting conditions. The values on the MONTR statement consist of the MONTR option, the time for the first execution of the option (TFRST), and a time for the successive executions or the completion of the option (TSEC). The times TFRST and TSEC default to TTBEG and infinity, respectively. If TSEC is defaulted, the MONTR option is executed only at time TFRST. However, if TSEC is specified, the MONTR option is executed at time TFRST and, except for the TRACE option, every TSEC time units thereafter. For the TRACE option, TSEC specifies the stopping time for the TRACE. There are five MONTR options available and they are listed below:

SUMRY   Causes a SLAM II Summary Report to be printed.
FILES   Causes a listing of all entries in the files to be printed.
STATES   Causes the continuous variables SS(N) and DD(N) to be printed.
CLEAR   Causes all statistical arrays, including the file statistics, to be cleared.
TRACE   Causes the starting and stopping of detailed tracing of each entity as it moves through the network. The trace will start at time TFRST and end at TSEC. A list of nodes can be provided with the TRACE option as follows: TRACE (nodelist). This causes an output only when one of the nodes in the list is processed. Also for the TRACE option, a list of variables to print with the trace may be defined. Attributes may be requested by specifying an index value, I, or by name, ATRIB(I). (XX variables may also be requested by a negative index.)

Below are examples of the MONTR statement:

Statement	*Description*
MONTR,SUMRY,200.,100.;	Summary report at TNOW = 200 and every 100 time units thereafter
MONTR,CLEAR,500;	Clear STATISTICS at TNOW = 500
MONTR,TRACE,0,150,2,3,7;	Start TRACE at TNOW = 0, stop TRACE at TNOW = 150. Print attributes 2, 3, and 7 on TRACE OUTPUT
MONTR,TRACE(RIGH,LEFT), 0,10,ATRIB(1),NNQ(1), NNRSC(2);	From time 0 to 10, trace events associated with the nodes labeled RIGH and LEFT. Print the values of the first attribute of each entity, the number in file 1, and the availability of resource 2 with each trace message.

MONTR statements apply only to a single run and must be restated for each run.

8.5.10   RECORD Input Statement

The RECORD input statement provides general information concerning the values to be recorded at save times. This includes an explicit specification for a plot number, IPLOT, or an implicit value determined by the order in which the RECORD statement appears. Also provided on the RECORD statement are definitions for the independent variable, the storage medium, and detailed specifications concerning the type and time interval for the output reports.

The fields of the RECORD statement are:

RECORD(IPLOT),INDVAR,ID,ITAPE,P or T or B,DTPLT,TTSRT,TTEND,KKEVT;

with the definitions of the variables listed below.

Variable	Definition	Default
IPLOT	PLOT number for user calls to subroutine GPLOT (IPLOT)	Optional
INDVAR	The name of the variable which is to serve as the independent variable for the table or plot corresponding to this RECORD statement. Possible variables are: TNOW, SS(·), DD(·), XX(·), NNACT(·), NNCNT(·), NNGAT(·), NNRSC(·), NRUSE(·), NNQ(·), II, ARRAY(I,J), or USERF(·).	Required
ID	An alphanumeric identifier for the independent variable. The first 16 characters of the identifier are significant.	Blanks
ITAPE	Peripheral device number on which the recorded variables are to be stored. If ITAPE is set to 0, the recorded variables are stored in NSET/QSET.	0
P or T or B	A description of the output format for the recorded information. A P indicates a plot is to be printed, a T indicates that a table is to be printed, and a B indicates that both a plot and a table are to be printed.	P
DTPLT	The printing increment of the independent variable between successive plot points. For tables, DTPLT is not employed and all recorded values are printed in the table.	5.0

TTSRT	The time at which the recording of values is to be initiated.	TTBEG
TTEND	The time at which the recording of values is to be halted.	TTFIN
KKEVT	A YES or NO field to indicate if values of variables before and after event times are to be recorded.	YES

For each RECORD statement included, there is one independent variable and a set of dependent variables. The specification of the dependent variables is given immediately following the RECORD statement in a manner to be illustrated in Section 8.5.11. The frequency with which data are recorded for a plot is normally specified by the DTSAV value on a CONTINUOUS statement. The user may override this specification by assigning a plot number, IPLOT, on the RECORD statement. When IPLOT is specified, the user records values for the plot by calling subroutine GPLOT(IPLOT). Subroutine GPLOT accesses the values specified on appropriate VAR statements and stores the values for future plots and tables. RECORD statements without plot numbers must follow RECORD statements with plot numbers in the input statement sequence. The independent variable defined on the RECORD statement is used to specify the first column of a table or the abscissa for a plot. When the value of the variable ITAPE is set to zero, space in the NSET/QSET array is used for storing the independent variable and each dependent variable. The number of values to be stored is the product of the number of recordings and the number of dependent variables plus one.

When no CONTINUOUS statement is used, the variable DTPLT is used to define the recording frequency, DTSAV. In such a case, DTSAV is set to the largest DTPLT value.

If KKEVT = YES, values are recorded before and after each event of the simulation. This double recording is to insure that any changes made in event routines are recorded. To suppress recordings at event times, the value of KKEVT should be set to NO.

When ITAPE is set greater than zero, peripheral device numbers must be specified for storing the values to be recorded. This should be done through the job control language or program cards as required. If a peripheral device is used, up to ten plots and/or tables with as many as ten variables per plot may be prepared. When the storage medium is core, a check is made to see if sufficient space has been allocated to the array QSET. If there is not sufficient space, the simulation continues but a message indicating that the complete plot could not be obtained due to insufficient space is printed at the end of the plot.

8.5.11 VAR Input Statement

The VAR statement is used in conjunction with the RECORD statement to define the dependent variables that are to be recorded for each value of the independent variable. The form for the VAR statement is:

VAR,DEPVAR,SYMBL,ID,LOORD,HIORD;

The definitions of the fields for the VAR statement are shown below.

Variable	Definition	Default
DEPVAR	The name of the dependent variable defined by the statement. The dependent variable can be TNOW, SS($\cdot$), DD($\cdot$), XX($\cdot$), NNACT($\cdot$), NNCNT($\cdot$), NNGAT($\cdot$), NNRSC($\cdot$), NRUSE($\cdot$), NNQ($\cdot$), II, ARRAY(I,J), or USERF($\cdot$).	Required
SYMBL	The symbol used on the plot to identify the dependent variable. If only a table is requested, this field can be defaulted.	Blank
ID	An alphanumeric label of up to 16 characters to identify the dependent variable.	Blanks
LOORD	The low ordinate specification that defines the left-hand scale for plots. This can be a value; the minimum observed, MIN; or MIN rounded down to a nearest multiple of IVAL, MIN(IVAL).	MIN
HIORD	The high ordinate value that defines the right-hand scale for plots. This can be a value; the maximum observed, MAX: or MAX rounded up to a nearest multiple of IVAL, MAX(IVAL).	MAX

The specification of the low and high ordinate values has been made flexible to enable the obtaining of properly scaled plot information. Specific values can be assigned for the low and the high ordinate fields. In the low ordinate field, MIN can be specified to indicate that the minimum value obtained during a simulation run should be prescribed for the left-hand scale. If MIN(IVAL) is specified, the minimum value observed during the simulation run is rounded down to the nearest multiple of IVAL and this value is used as the left-hand scale. For example, if the lowest value observed was 547 and the specification MIN(100) was made, the low ordinate would be set to 500.

For the high ordinate value, the specification MAX would cause the maximum value observed to be used for the right-hand scale. The specification MAX(IVAL) would cause the right-hand scale to be set to the nearest multiple of IVAL above the maximum value observed. Thus, if the maximum value observed was 2317 and the specification for the high ordinate was MAX(1000), the right-hand scale would be set at 3000.

The number of VAR statements following a RECORD statement specifies the number of dependent variables associated with each recording of values. *The dependent variables associated with a recording must be specified immediately after the RECORD statement.* The symbol character used for the dependent variable may be any character in the FORTRAN character set except the special characters [,/( )+-*;].

8.5.12 SIMULATE Statement

The SIMULATE statement consists of a single field as shown below.

SIMULATE;

The SIMULATE statement is used when making multiple simulation runs. One simulation run is executed for the statements preceding the SIMULATE statement. Following each SIMULATE statement, the user can insert any updates such as new random number seeds using the SEEDS statement, new ENTRY statements, or new initial values for the XX(N) variables using the INTLC statements. If only one simulation run is being made (NNRNS=1), or multiple runs are being made (NNRNS>1) with the same starting conditions, the SIMULATE statement is not required. Recall, however, that ENTRY and MONTR statements apply for only one run.

8.5.13 FIN Statement

The FIN statement consists of a single field as shown below.

FIN;

The FIN statement denotes the end to all SLAM II input statements. The FIN statement causes the execution of all remaining simulation runs. This completes the description of network input statements and statements for executing network models. The description of other input statements that relate to discrete, continuous, and combined models is given in later chapters.

8.6   PROGRAM SETUP FOR NETWORK MODELS

The statements in a simulation program other than the network statements can be input in any order with the following exceptions:

1. The GEN statement must be the first statement in the deck, the LIMITS statement must be the second statement in the deck, and the FIN statement must be the last SLAM II statement.
2. The network description statements must be immediately preceded by the NETWORK statement and immediately followed by the ENDNETWORK statement.
3. The INITIALIZE statement (if used) must precede ENTRY and MONTR statements (if any).
4. A MONTR statement with the TRACE option which includes a nodelist must follow the ENDNETWORK statement.

8.7   SLAM II OUTPUT REPORTS

The purpose of this section is to describe the output reports which are generated by the SLAM II processor. The output reports include the input listing, echo report, trace report, and SLAM II Summary Report. A description of each report follows.

8.7.1   Statement Listing and Input Error Messages

The SLAM II processor interprets each input statement and performs extensive checks for possible input errors. If the variable ILIST on the GEN statement is specified as YES or defaulted, the processor prints out a listing of the input statements. Each statement is assigned a line number and if an input error is detected an error message is printed immediately following the statement where the error occurred. The following types of input error messages will result from incorrect input statements. Underlines presented below represent values to be inserted by the SLAM II processor.

FIELD NUMBER _____ MUST NOT BE DEFAULTED.

FIELD NUMBER _____ MUST NOT BE NUMERIC.

FIELD NUMBER _____ MUST NOT BE ALPHABETIC.

Each item of data in a statement counts as a field. Fields are counted left to right, with the first field for a node being its label (even if the label is omitted). For example, the following statement with field numbers shown circled over the statement would result in the message FIELD NUMBER 5 MUST NOT BE ALPHABETIC.

The incorrect field in this contains "XX" which is not a valid entry for the maximum number allowed in a queue.

ILLEGAL USE OF MATCH NODE: The correspondence between QUEUE nodes and associated SELECT or MATCH nodes could not be resolved. References to associated nodes have been omitted from the input statements.

INCORRECT STATEMENT SEQUENCE: This message is caused by an illogical or incorrect statement sequence. Examples are an ACTIVITY following a TERM node, a RESOURCE definition after an AWAIT node statement, or a STAT statement following a network.

INCORRECT VARIABLE/NUMERIC INPUT ENDING IN FIELD _____. Caused by an error found in a compound field. For example, the statement

would result in this error for field 5.

INITIAL CREATE RELEASE BEFORE TTBEG: This error condition is detected during initialization while scheduling initial CREATE node releases. A

CREATE node may not be released before the beginning time of the simulation.

_____ *IS AN ILLEGAL OR UNRECOGNIZED NAME FOR FIELD* _____.
Caused by an unrecognizable alphabetic field. For example, the statement

    QUEUE(1);

would result in the message

    QUEUE IS AN ILLEGAL OR UNRECOGNIZED NAME FOR FIELD 2.

Recall that the label field is the first field on node input statements.

MONTR REQUIRES MATR GE 2: The MONTR statement requires two attributes in order to schedule the specified option. Change the LIMITS statement to allow for at least two attributes.

MORE THAN 50 FIELDS OR UNRECOGNIZED FIELD TYPE: Caused by the failure to parse an input field, either because the statement is too long or because an alphabetic character was found in a field beginning with a number. A common cause of this error is typing an O instead of a zero.

MULTIPLE DEFINED FILE NUMBER: Caused by assigning a file number to more than one QUEUE node or to a QUEUE and an AWAIT node.

THE LABEL __ IS REDUNDANT: Caused by assigning the same label to two or more nodes.

THE VALUE _____ IS OUT OF RANGE: Caused by an index out of range, such as referencing ATRIB(6) when only 5 attributes were specified on the LIMITS statement. The number found to be out of range is included in the error message.

UNRESOLVED LABEL: A list is given of non-existent nodes to which a reference was made.

VARIABLE OR STATISTICAL ARRAY BOUNDS ARE EXCEEDED: Caused by defining more occurrences of a modeling element than allowed. For instance, without redimensioning, up to 50 sets of time-persistent statistics are allowed in a SLAM II model. The 51st TIMST statement encountered would result in this message.

DIMENSION OF NSET/QSET EXCEEDED: Caused by specifying more information than can be saved in the slam storage array NSET. This message may be encountered following a LIMITS statement in which case the filing space required for the number of entries and attributes specified would overflow the storage provided. The error may also occur during the decoding of network statements in which case the amount of space left after providing for the filing system is not sufficient to store the network description. To recover from this error, restructure the file storage set aside by the LIMITS statement or increase the dimensions of NSET and QSET in the MAIN program.

All input errors are treated as fatal errors in SLAM II; that is, no execution is attempted if one or more input errors are detected.

8.7.2 Echo Report

The SLAM II Echo Report provides a summary of the simulation model as interpreted by the SLAM II processor. This report is particularly useful during the debugging and verification phases of the simulation model development process. An illustration of this report is given in Chapter 13, Figure 13-10.

8.7.3 Trace Report

The Trace Report is initiated by the MONTR statements using the TRACE option and causes a report summarizing each entity arrival event to be printed during execution of the simulation. The Trace Report generates a detailed account of the progress of a simulation by printing for each entity arrival event, the event time, the node label and type to which the entity is arriving, and the attributes of the arriving entity. In addition, a summary of all regular activities which emanate from the node is printed denoting if the activity was scheduled, the duration of the activity, and the end node of the activity. Illustrations of trace reports are given in Figures 5-9 and 12-1.

8.7.4 SLAM II Summary Report

The SLAM II Summary Report displays the statistical results for the simulation and is automatically printed at the end of each simulation run. The report consists of a general section followed by the statistical results for the simulation categorized by type. The output statistics provided by the report are defined in Figure 8-2 and Table 8-3. The first category of statistics is for variables based on discrete observations and includes the statistics collected within network models by the COLCT statement. The second category of statistics is for the $XX(\cdot)$ variables over time. This is followed by statistics on all user files. The next two categories

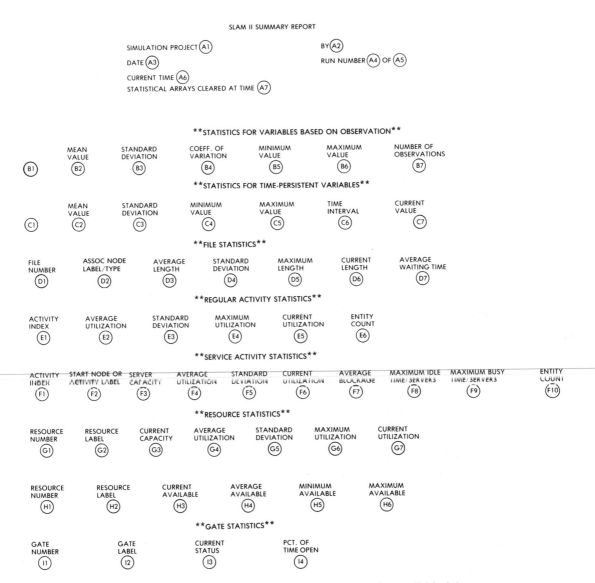

Figure 8-2 Definition of SLAM II output statistics corresponding to Table 8-3.

correspond to statistics collected on regular and service activities, respectively. The last category of statistics is for resource and gate statistics and is followed by the printout of histograms. A SLAM II Summary Report includes only those categories of statistics which are applicable to the particular simulation, and therefore may include none, some or all of the above categories.

Table 8-3 Definition of output statistics corresponding to Figure 8-2.

General Section

(A1)   The first 20 characters of the project title

(A2)   The first 20 characters of the analyst name

(A3)   The MONTH/DAY/YEAR

(A4)   The number of the simulation run

(A5)   The number of simulation runs to be made

(A6)   The current value of TNOW

(A7)   Time at which all statistical arrays were last cleared

Statistics for Variables Based on Observation

(B1)   The first 16 characters of the statistics label

(B2)   The arithmetic mean of the observations

(B3)   The standard deviation of the observations

(B4)   The coefficient of variation (standard deviation/mean)

(B5)   The minimum value over all observations

(B6)   The maximum value over all observations

(B7)   The number of observations

Statistics for Time-Persistent Variables

(C1)   The label for the variable

Table 8-3 (Continued)

(C2) The average value of the variable over time

(C3) The standard deviation over time

(C4) The minimum value of the variable over time

(C5) The maximum value of the variable over time

(C6) The time interval over which the statistics are accumulated

(C7) The current value of the variable

File Statistics

(D1) The file number or event calendar

(D2) The node type (if any) associated with the file

(D3) The average number of entities in the file over time

(D4) The standard deviation of the number of entities in the file over time

(D5) The maximum number of entities in the file at any one time

(D6) The current number of entities in the file

(D7) The average waiting time of all entities that arrived to the file including those that did not wait

Regular Activity Statistics

(E1) The activity index number for the activity

(E2) The average number of entities in the activity

Table 8-3 (Continued)

(E3) The standard deviation of the number of entities in the activity

(E4) The maximum number of entities in the activity at any one time

(E5) The number of entities currently in the activity

(E6) The number of entities which have completed the activity

Service Activity Statistics

(F1) The activity number for the activity. A zero if a number is not assigned. Statistics are listed in the same order as the input statements.

(F2) The label and type (QUEUE or SELECT) of the start node of the activity.

(F3) The number of parallel identical servers represented by the activity.

(F4) The average number of entities in service over time. If the capacity of the server is 1, this corresponds to the fraction of time the server is busy.

(F5) The standard deviation of the number of entities in service over time.

(F6) The current number of entities in service.

(F7) The average number of servers blocked over time. If the capacity of the server is 1, this corresponds to the fraction of time blocked.

Table 8-3   (Continued)

 If the capacity of the server is 1, this value specifies the maximum idle time of the server. If the capacity of the server is greater than 1, this value specifies the maximum number of idle servers.

 Maximum busy time for a single server or maximum number of busy servers.

(F10) If the service activity is assigned an activity index number, the number of entities completing service; otherwise no value is printed.

Resource Utilization Statistics

(G1) The resource number assigned by the processor

(G2) The resource label as specified on the RESOURCE block

(G3) The current capacity of the resource

(G4) The average utilization of the resource over time

(G5) The standard deviation of the resource utilization over time

(G6) The maximum number of units of resource utilized at any one time

(G7) The current number of units of resource in use

Resource Availability Statistics

 The resource number assigned by the processor

 The resource label as specified on the RESOURCE block

Table 8-3   (Continued)

 The current number of units of resource available

 The average availability of the resource over time

(H5) The minimum number of units of resource available at any one time

(H6) The maximum number of units of resource available at any one time

<u>Gate Statistics</u>

(I1) The gate number assigned by the processor

(I2) The gate label as specified on the GATE block

(I3) The current status of the gate

(I4) The percentage of time the gate was opened

8.7.5   Plot and Table Reports

The output of a plot or table is part of the SLAM II Summary Report. For convenience, they are described separately since either can be obtained using calls to subroutine PRNTP as described in Section 12.7. For a table, all values recorded are printed with the independent variable given in the first column. The output of a plot progresses line by line with each line consisting of three components. The first is the value of the independent variable associated with that line. The second component contains 101 print positions corresponding to the plot points. The third component of a plotted line is a duplicates column that indicates any instances in which two different plot symbols occupy the same plot position. If this happens, the first symbol processed is plotted and the duplicates column contains a symbol pair showing the plotted symbol followed by the symbol that should occupy the same plot position. As many as five duplicates may be printed per plot line. If there are more, the first five are printed followed by an asterisk. The symbols used on a plot are prescribed on a VAR input statement. Illustrations of plots and tables are given throughout the book.

8.8   CHAPTER SUMMARY

This chapter summarizes the input statements and output reports associated with SLAM II network models. Over 95 percent of all inputs and outputs related to SLAM II are presented in this chapter. The overall processing logic for simulating SLAM II networks is also described in this chapter.

8.9   EXERCISES

8-1.   Obtain a trace for Example 6-2 from which you can determine if the operation time of jobs preempted is equal to the remaining processing time on the operation from which they were preempted.

8-2.   In the following input statements, detect at least eight errors.

```
GEN,EXERCISE 8-2,ERRORS,JULY/23/1986,2;
LIMITS,2,1,0;
PRIORITY / 3,HVT(2);
INTLC,XX(1)=2.0;
NETWORK;
      CREATE,UNIFORM(XX(1),10.);
      AWAIT(1),TELLER;
      ACT,EXPON (4.0);
      FREE, TELLER;
      TERM;
INIT,0,100;
SIMULATE;
INTLC,XX(1)=4.0;
FIN;
```

8-3.   Prepare the input data for Illustration 5-2 in Section 5.10.2 to make 5 runs with each run lasting 1000 time units. On the second run, the mean and standard deviation for the normal distribution should be changed to 0.0 and 2.0. No additional changes are required on the third run. On the fourth run, the ranking at QUEUE node QOFS should be changed to last-in, first-out. No additional changes are required for run number 5.

8-4.   For the thief of Baghdad problem given in Exercise 2-7 draw the network and prepare the data input for 1000 runs using the following information: $p_F = 0.3$, $p_S = 0.2$, $P_L = 0.5$ and the time in the short tunnel is exponentially distributed with a mean of 3, and the time in the long tunnel is lognormally distributed with a mean of 6 and a standard deviation of 2.

Suppose the thief's remaining time to live is normally distributed with a mean of 10 and a standard deviation of 2. Redraw the network and redo the data input in order to ascertain the probability that the thief reaches freedom before he dies based on 1000 simulations of the network.

8-5.   For the following network, describe the chronological sequence of node arrival events if the secondary ranking for future events if FIFO.

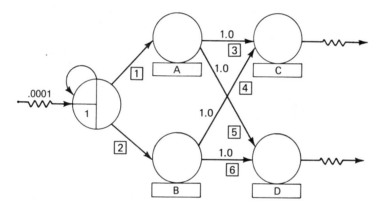

Embellishment: Repeat the exercise if the secondary ranking for future events is LIFO.

8-6.   Prepare the SLAM II input statements for Exercise 8-5 to make two runs. The secondary ranking for future events should be FIFO on the first run and LIFO on the second run. Obtain a TRACE report on both runs.

8-7.   Perform a simulation of a queueing situation using SLAM II and discuss the meaning of each output statistic that is given on the SLAM II Summary Report.

8-8.   Prepare a data worksheet for the input to the simulation program for which there are 2 files with a maximum number of 25 entries in all files and a maximum of 4 attributes for any entry. File 1 is a high-value-first file with priority based on attribute 4. Statistics on 5 variables are to be collected in subroutine COLCT and the labels for these variables are HT, WT, TOL, SIZE, and GRADE. Time-persistent statistics are to be recorded on SS(1) and SS(2) and labeled as BAL and POL, respectively. A histogram is to be collected on the variable GRADE, and it is desired to have 15 cells in the histogram, not including the end cells. The lower limit for the histogram is 10 and a cell width of 5 is desired. There are three equations written in terms of the DD(·) variables and two written in terms of the SS(·) variables. If accuracy cannot be met or tolerance on the state-event conditions cannot be satisfied, a warning is to be printed but the simulation is to be continued. The absolute error and relative error for accuracy computation are specified as $1.E - 4$ and $1.E - 5$, respectively. The minimum step size allowed is .001 and the maximum step size permitted is 1.0. The values of the state variables are to be recorded at the end of each step and at event times. One plot is to be obtained using core storage and 5000 words of core are to be allocated for the file. The independent variable of the plot will be TIME and there will be 3 dependent variables for the plot. Plot points should be obtained every 2 time units and both a table and a plot are desired. The minimum and maximum values obtained from the simulation are to be rounded to the nearest integer and used for scaling. The symbols X, Y, and Z are to be plotted for the variables SS(3), SS(4), and SS(5) and the labels for these variables are XPOS, YPOS, and ZPOS, respectively.

The beginning time for the simulation is 10 and the simulation will be completed by an end-of-simulation event. Statistical storage areas are to be cleared. Only one random number stream is to be used and the initial random number seed is 567471923. The filing array is to be initialized and an event of type 2 is to occur at time 11. An end-of-simulation event with event code 3 is to occur at time 225. An entry with ATRIB(1)=3 and ATRIB(2)=5 is to be stored in file 2. Events occurring between time 50 and 100 are to be monitored. Summary reports are desired, starting at time 100 and for every 25 time units thereafter, until the end of the simulation. Only one simulation run is to be made.

8.10   REFERENCES

1. O'Reilly, J. J., *SLAM II® Quick Reference Manual,* Pritsker & Associates, 1984.

CHAPTER 9

Network Modeling With User-Written Inserts

9.1   INTRODUCTION

In this chapter, we describe the procedures for constructing network models with user-written FORTRAN inserts. We begin by describing the EVENT and EN-TER nodes which provide the key interface points between the network model and user code. The support subprograms of SLAM II for file manipulations, setting attribute values, and activity durations are presented. We next describe a set of SLAM II provided subprograms which allow the modeler to change the status of the network elements by allocating and freeing units of resources, altering resource capacities, opening and closing gates, stopping activities, and specifying selection rules from user-written subprograms.

9.2   EVENT NODE

The symbol and the input statement for the EVENT node are shown below.

EVENI, JEVNT, M;

The EVENT node is included in the network model to interface the network portion of a model with event coding written by the modeler. The EVENT node causes subroutine EVENT(JEVNT) to be called every time an entity arrives to the EVENT node. The value of JEVNT specifies the event code to be executed, and M specifies the maximum number of emanating activities to be taken following the processing of the EVENT node. Since the logic associated with the EVENT node is coded by the modeler, complete modeling flexibility is obtained. Thus, if the modeler is faced with an operation for which a standard network node is not provided, the modeler can employ the EVENT node to perform the specialized logic required.

The statement format for the EVENT node consists of the node name EVENT beginning in or after column seven, followed by the event code JEVNT and the M value separated by commas.

The procedure for coding an event created by an EVENT node is identical to the procedure for coding any FORTRAN subroutine. Subroutine EVENT maps the event code JEVNT onto the appropriate event subroutine containing the coding for the event logic. In coding the event logic, the modeler has access to the SLAM II provided subprograms for performing commonly encountered functions such as random sampling, file manipulation, and data collection.

When an entity arrives to an EVENT node, the SLAM II processor loads the attributes of the arriving entity into the ATRIB vector prior to the call to subroutine EVENT(JEVNT).

The general form for subroutine EVENT is shown below.

```
   SUBROUTINE EVENT(I)
   GO TO (1,2),I
 1 Code for event node 1
   RETURN
 2 Code for event node 2
   RETURN
   END
```

The argument I is the event code that is defined for the EVENT node, that is, JEVNT. This code is used to determine which node has had an entity arrive to it. The EVENT subroutine is written by the user and normally consists of a computed GOTO statement indexed on the event code I causing a transfer to the appropriate statement where the logic associated with the EVENT node is written. In many cases, the logic for the EVENT node is written in a separate subroutine to facilitate the identification of the logic and the documentation of the code associated with an EVENT node. An example of subroutine EVENT(I) is shown below for a simulation model with two EVENT nodes consisting of a lunch break coded in subroutine LUNCH and assigned event code 1, and an end-of-lunch break coded in subroutine ELUNCH and assigned event code 2.

```
   SUBROUTINE EVENT(I)
   GO TO (1,2),I
 1 CALL LUNCH
   RETURN
 2 CALL ELUNCH
   RETURN
   END
```

Following the return from subroutine EVENT(JEVNT), the SLAM II processor assigns the values in the ATRIB array as the attributes of the entity exiting from the EVENT node. In this way, the values of the attributes are passed to the modeler in the SLAM II labeled COMMON block SCOM1 which is shown below.

COMMON/SCOM1/ATRIB(100),DD(100),DDL(100),DTNOW,II,MFA,MSTOP,NCLNR
1,NCRDR,NPRNT,NNRUN,NNSET,NTAPE,SS(100),SSL(100),TNEXT,TNOW,XX(100)

Thus, the modeler can make assignments to the attributes of an entity which passes through the EVENT node. The use of the ATRIB buffer should be reserved for this purpose when processing an event initiated by an entity arrival to an EVENT node.

9.2.1   Accessing Network Values

The logic at an EVENT node may involve testing the status or changing the value of variables associated with the network portion of the model. In subroutine EVENT, values of network related variables can be obtained using the SLAM II functions listed in Table 9-1.

Table 9-1   Functions for accessing the status of an activity, gate, or resource.

Function	Definition
NNACT(I)	Number of active entities in activity I at current time
NNCNT(I)	The number of entities that have completed activity I
NNGAT(I)	Status of gate number I at current time: $0 \rightarrow$ open, $1 \rightarrow$ closed
NRUSE(I)	Current number of resource type I in use
NNRSC(I)	Current number of resource type I available

On a SLAM II network, ARRAY is a two dimensional table for global values. The values in ARRAY can be accessed and changed in event routines by using SLAM II provided subprograms. Table 9-2 provides the definitions for function GETARY and subroutines PUTARY and SETARY for storing and retrieving values from ARRAY.

Table 9-2 Subprograms for storing and retrieving values from the SLAM II global table ARRAY.

FUNCTION GETARY(IR,IC)   Returns the value of row IR, column IC from the global table, ARRAY.

SUBROUTINE PUTARY(IR,IC,VAL)   Sets the value of row IR, column IC of the global table, ARRAY, to VAL.

SUBROUTINE SETARY(IR,VALUE)   Sets the values of row IR of the global table to the values passed in the vector VALUE. Note that the number of elements in VALUE should correspond to the number of elements in row IR.

The statement X2=GETARY(3,4) sets X2 equal to the value of the fourth cell in the third row of ARRAY. Internal to SLAM II, the values in ARRAY are stored in a one dimensional form in QSET. Hence, there is no access to a variable within SLAM II by the name ARRAY. To put the value 10 into the seventh cell of row 1, the following statement would be used.

CALL PUTARY(1,7,10.0)

If it is desired to set all the values in row 2 of ARRAY and there are three cells in this row then the following statements should be used.

DIMENSION VALUE(3)
VALUE(1) = 100.
VALUE(2) = 200.
VALUE(3) = 300.
CALL SETARY(2,VALUE)

9.3 ENTER NODE

The ENTER node is provided to permit the modeler to insert selectively an entity into the network from a user-written subroutine. The symbol and the input statement for the ENTER node are shown on the next page.

ENTER, NUM, M;

Each ENTER node has a unique user-assigned integer code NUM and a M value which specifies the maximum number of emanating activities to be taken at each release of the node. The ENTER node is released following a return from a user written routine in which a call is made to subroutine ENTER(NUM,A) where NUM is the numeric code of the ENTER node being released and A is the name of the array containing the attributes of the entity to be inserted into the network at the ENTER node NUM. The ENTER node can also be released by entity arrivals to the node.

The statement format for the ENTER node consists of the node name ENTER beginning in or after column seven followed by the numeric code NUM and the M value separated by commas.

9.4   FILE MANIPULATIONS

A file provides the mechanism for storing the attributes of an entity in a prescribed ranking with respect to other entities in the file. In network models, files are used to maintain entities waiting at QUEUE, AWAIT, and PREEMPT nodes.

Associated with a file is a ranking criterion which specifies the procedure for ordering entities within the file. Thus, each entity in a file has a rank which specifies its position in the file relative to the other members in the file. A rank of one denotes that the entity is the first entry in the file. Possible ranking criterion for files are: first-in, first-out (FIFO); last-in, first-out (LIFO); high-value-first (HVF) based on an attribute value; and low-value-first (LVF) based on an attribute value. The ranking criterion for entries in a file is assumed to be FIFO unless otherwise specified using the PRIORITY statement described in Chapter 8.

In SLAM II, files are distinguished by integer numbers assigned to the files by the user. SLAM II automatically collects statistics on each file and provides the function NNQ(IFILE) which returns the number of entries in file IFILE. For example, NNQ(2) denotes the number of entries in file 2.

The modeler can, when referencing files associated with network nodes, employ any of the file manipulation routines which are available in discrete event simulations.

9.4.1   Subroutine FILEM(IFILE,A)

Subroutine FILEM(IFILE,A) files an entry with attributes specified in the buffer array A into file IFILE. The entry's rank in the file is determined by SLAM II based upon the priority specified by the user. As noted earlier, if no priority is specified then a first-in, first-out ranking is assumed. The attributes of the entry are stored with the entry and are returned when the entry is removed from the file. As an example of the use of subroutine FILEM(IFILE,A), the following statements cause an entry with its first attribute set equal to the current simulation time to be inserted into file 1.

        A(1) = TNOW
        CALL FILEM(1,A)

If IFILE is associated with a network node, the SLAM II processor does not directly insert the entity into the file, but processes the entity as an arrival to the node immediately following the return from the routine. It will be the next event processed unless other calls to subroutine FILEM are made in the event routine. For example, assuming that file 1 is associated with a QUEUE node in the network model, the above two statements would schedule an entity arrival to occur at the QUEUE node with the first attribute of the entity equal to TNOW. Upon return to the SLAM II processor, the entity arrival to the QUEUE node with file 1 would be processed with the arriving entity either going directly into service if a server is idle, balking if the queue is full, or being inserted into file 1 to wait for a server. Note, however, that the value of the number of entries in file 1, obtained from NNQ(1), would not change until after the return to the SLAM II processor.†

9.4.2   Subroutine RMOVE(NRANK,IFILE,A)

Subroutine RMOVE(NRANK,IFILE,A) removes an entry with rank NRANK from file IFILE and places its attributes into the attribute buffer array A. As an example of the use of subroutine RMOVE, the following statement removes the last entry in file 1 and places its attributes in the array A.

---

† SLAM II provides Subroutine FFILE(IFILE,A) which directly places the entity in file IFILE. No check is made on the status of following service activities.

CALL RMOVE(NNQ(1),1,A)

The dimension of the user-defined array A should be set greater than or equal to the maximum number of attributes per entry, MATR, plus 2.

As a second example, the following statement causes the second entry in file 3 to be removed and its attributes to be placed in the array ATRIB.

CALL RMOVE(2,3,ATRIB)

Since the attributes of the entity that was the second entry in file 3 are in the buffer vector ATRIB, it is these attributes that are returned to the network. If branching is performed from the EVENT node then an entity with the attributes as defined in ATRIB is routed over the activities emanating from the EVENT node. Only when this is desired should ATRIB be used as the third argument to subroutine RMOVE.

9.5   USER FUNCTION USERF

To permit a modeler, the flexibility to use FORTRAN code throughout a network, SLAM II provides the function USERF(IFN). The argument IFN is a code established by the user and is referred to as a user function number. The name USERF can be used in a network in locations that allow a SLAM II variable to be an option. It is most frequently used in the following two situations:

1. An entity passes through an ASSIGN node and one of the assignments is to be made in USERF; and
2. A duration for an activity is specified as USERF.

The function USERF allows a modeler to make programming inserts into the network model. In the programming insert, the user can employ all FORTRAN coding procedures and the SLAM II subprograms that will be described in Chapters 11 and 12 with the restriction that file operations associated with the event calendar are not permitted. Specifically, no calls should be made to subroutines SCHDL, OPEN, FREE, or ENTER. Subroutine RMOVE and then FILEM or ULINK and then LINK may only be used in pairs. When it is desired to manipulate files, an EVENT node must be employed.

As an example of the uses of function USERF, consider the following single-server, single-queue network.

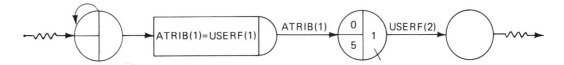

In this network, user function 1 is used to assign a value to attribute 1 at the ASSIGN node, and user function 2 is used to specify the duration of the service time of activity 1. ATRIB(1) is the duration of the activity representing the entity's travel to the QUEUE node. In this example, we will make this time a function of the number of remaining spaces in the QUEUE node, that is, 5−NNQ(1). If ATRIB(1) was used only as the duration of the activity then user function 1 could have been specified for the activity. Here we illustrate a concept and demonstrate the use of function USERF to assign an attribute value. If future decisions were made based on the value of ATRIB(1) then the model as depicted would be required.

The service time for activity 1 will also be made a function of the number of entities waiting for service. The general form for writing function USERF(IFN) is shown below.

```
    FUNCTION USERF(IFN)
    GO TO (1,2),IFN
  1 Set USERF as the time to travel to QUEUE node
    RETURN
  2 Set USERF as service time to be a function of number in queue
    RETURN
    END
```

This example illustrates how IFN, the user function number, is decoded to allow different user functions to be employed throughout the network model.

For this example, user function 1 only requires USERF to be set equal to 5−NNQ(1). When a return is made, SLAM II assigns the value of USERF to ATRIB(1). For user function 2, we require knowledge of how service time varies as a function of the number of entities in the QUEUE node. For illustrative purposes, we will assume an exponential service time whose mean decreases as the number in the queue is increased to a value of 3. When there are more than 3 in the queue, the mean service time increases. A table of mean service time as a function of the number in the queue is shown below.

Number in Queue	Mean Service Time
0	10.
1	9.
2	8.
3	7.
4	9.
5	10.

The specific coding for function USERF(IFN) for this situation would be:

```
      FUNCTION USERF(IFN)
      DIMENSION ST(6)
      DATA ST/10.,9.8.,7.,9.,10./
      GO TO (1,2),IFN
    1 USERF = 5 - NNQ(1)
      RETURN
    2 XNINQ = NNQ(1)
      AVEST = GTABL(ST,XNINQ,0.0,5.0,1.0)
      USERF = EXPON(AVEST,1)
      RETURN
      END)
```

The first two statements of the function dimension the array ST and establish values for ST through a DATA statement. Next, the function number is decoded. At statement 1, USERF is established as the value to be assigned when user function 1 is invoked. At statement 2, XNINQ is established as the current number in file 1. The average service time is determined through the use of function GTABL which is the SLAM II table look-up function described in Chapter 12, Section 12.9.4. The arguments to function GTABL are the dependent variable values, the value of the independent variable for which the table look-up is desired, the independent variable corresponding to the first value in the ST array, the value of the independent variable corresponding to the last value of the ST array, and the increment between the values of the independent variables corresponding to the values of the dependent variables given in the array ST. The value of USERF is then set equal to a sample from an exponential distribution using AVEST as the average service time and employing stream number 1. Note that the stream number must be given in function EXPON since direct coding is being employed. Also note that the arguments to function GTABL must all be real values to correspond to the variable types used in function GTABL.

This example illustrates the flexibility available by allowing program inserts to be made in network models. Function USERF can be made as complex as necessary to meet the modeling requirements of the user.

9.6   EXAMPLE 9-1.   DRIVE-IN BANK WITH JOCKEYING

A drive-in bank has two windows, each manned by a teller and each with a separate drive-in lane. The drive-in lanes are adjacent. From previous observations, it has been determined that the time interval between customer arrivals during rush hours is exponentially distributed with a mean time between arrivals of 0.5 time units. Congestion occurs only during rush hours, and only this period is to be analyzed. The service time is normally distributed for each teller with a mean service time of one time unit and a standard deviation of 0.3 time units. It has also been shown that customers have a preference for lane 1 if neither teller is busy or if the waiting lines are equal. At all other times, however, a customer chooses the shortest line. After entering the system, a customer does not leave until served. However, the last customer in a lane may change lanes if there is a difference of two customers between the two lanes. Because of parking space limitations only three cars can wait in each lane. These cars, plus the car of the customer being serviced by each teller, allow a maximum of eight cars in the system. If the system is full when a customer arrives, the customer balks and is lost to the system.

The initial conditions are as follows:

1. Both drive-in tellers are busy. The initial service time for each teller is normally distributed with a mean of one time unit and a standard deviation of 0.3 time units.
2. The first customer is scheduled to arrive at 0.1 time units.
3. Two customers are waiting in each queue.

The objective is to develop a simulation model that can be used to analyze the banking situation in terms of the following statistics:

1. Teller utilization $= \dfrac{\text{total time performing service}}{\text{total simulation time}}$
2. Time-integrated average number of customers in the system.
3. Time between departures from the drive-in windows.
4. Average time a customer is in the system.
5. Average number of customers in each queue.

6. Percent of arriving customers who balk.
7. Number of times cars jockey.

The system is to be simulated for 1000 time units.

Concepts Illustrated. This example illustrates user written code to augment the modeling concepts in a network model. Two key concepts which are specifically illustrated by this example are: 1) the use of the EVENT node to model jockeying; and 2) the referencing of a common file number between the network and discrete event portions of a model. In addition, this example illustrates balking from a SELECT node.

SLAM II Model. The drive-in bank example is amenable to modeling with network concepts with the exception of the jockeying of cars between lanes when the lanes differ by two cars or more. Therefore, we will represent the drive-in bank system as a network model with the jockeying of cars modeled using an EVENT note. Since jockeying can occur only when a teller completes service on a customer, the EVENT node for processing the jockeying of cars between lanes will be executed following the end-of-service for each entity in the system.

The network model for this example is depicted in Figure 9-1. Arriving cars are created by the CREATE node which marks current time as ATRIB(1). The first entity is created at time 0.1, and the time between car arrivals is exponentially distributed with a mean of 0.5 time units. The SELECT node routes each entity to either the LEFT or RIGHT QUEUE node based on the smallest number in the queue rule (SNQ). The LEFT QUEUE node initially contains two entities and permits a maximum of three waiting entities which are stored in file 1. The LEFT QUEUE is followed by an ACTIVITY representing teller 1 which is prescribed as activity number 1. The duration of the activity is normally distributed with a mean of 1.0 and a standard deviation of 0.3. The ACTIVITY routes the entity to an EVENT node whose event code is 1. The RIGHT QUEUE node also initially contains two entities and permits a maximum of three waiting entities which are stored in file 2. The ACTIVITY following the RIGHT QUEUE node is assigned activity number 2, is normally distributed, and routes the entities to the EVENT node with event code 2. This ACTIVITY represents the service by teller 2.

At the EVENT nodes, SLAM II calls the user-coded subroutine EVENT (JEVNT) with JEVNT set equal to 1 or 2 to process the discrete event logic associated with the node. The jockeying event logic is coded directly in subroutine EVENT(JEVNT) as depicted in Figure 9-2.

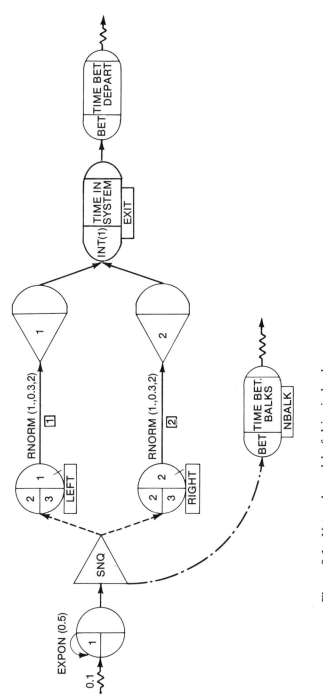

Figure 9-1   Network model of drive-in bank.

```
      SUBROUTINE EVENT(I)
      DIMENSION A(10)
      NL1=NNQ(1)+NNACT(1)
      NL2=NNQ(2)+NNACT(2)
      GO TO (1,2),I
C*****IF THE NUMBER IN LANE 2 EXCEEDS LANE 1 BY 2
    1 IF (NL2.LT.NL1+2) RETURN
C*****THEN JOCKEY FROM 2 TO 1
      CALL RMOVE(NNQ(2),2,A)
      CALL FILEM(1,A)
      RETURN
C*****IF THE NUMBER IN LANE 1 EXCEEDS LANE 2 BY 2
    2 IF(NL1.LT.NL2+2) RETURN
C*****THEN JOCKEY FROM 1 TO 2
      CALL RMOVE(NNQ(1),1,A)
      CALL FILEM(2,A)
      RETURN
      END
```

Figure 9-2 Subroutine EVENT for model of the drive-in bank.

In subroutine EVENT(JEVNT), the variables NL1 and NL2 are computed as the number of cars in lane 1 (LEFT) and lane 2 (RIGHT), respectively. Note that the number in each lane is calculated to include the customer in service, if any. At statement 1, jockeying from the RIGHT lane to the LEFT lane is investigated. If the number of cars in the RIGHT lane exceeds the number in the LEFT lane by two, then the last entity in file 2 (whose rank is NNQ(2)) is removed from the RIGHT QUEUE node by a call to RMOVE with IFILE = 2 and is scheduled to arrive at the LEFT QUEUE node at the current simulated time by a call to subroutine FILEM with IFILE = 1. This completes the code for jockeying and is followed by a RETURN statement.

At statement 2, if the number of cars in the LEFT lane exceeds the number in the RIGHT lane by two, the reverse jockeying procedure is executed. If the lanes do not differ by two cars, then a return from subroutine EVENT is made without a car jockeying.

Following execution of the EVENT node, each entity (car) continues to the COLCT node EXIT where interval (INT) statistics are collected using the mark time in ATRIB(1) as a reference. The results are displayed using the identifier TIME IN SYSTEM. A second COLCT node collects statistics on the time between departures and displays the results using the identifier TIME BET. DEPART. The entities are then terminated. The input statements for the model are shown in Figure 9-3.

The control statements for this model include GEN, LIMITS, TIMST, SEEDS, INIT and FIN statements. The GEN, LIMITS, INIT and FIN statement are in standard form. The TIMST statement is employed to obtain statistics on the num-

```
 1  GEN,OREILLY,DRIVE IN BANK,4/12/82,1;
 2  LIMITS,2,1,75;
 3  TIMST,USERF(1),NO. OF CUST;               USER COMPUTED STATISTICS
 4  SEEDS,4367651(1),6121137(2);
 5  NETWORK;
 6        CREATE,EXPON(.5,1),,.1,1;           CREATE CUSTOMERS
 7        SELECT,SNQ,,BALK(NBALK),LEFT,RIGHT; PLACE CUSTOMER IN SHORTEST Q
 8  LEFT  QUEUE(1),2,3;
 9        ACT/1,RNORM(1,.3,2);                TELLER 1
10        EVENT,1;                            CHECK FOR JOCKEYING INTO LEFT QUEUE
11        ACT,,,EXIT;
12  RIGHT QUEUE(2),2,3;
13        ACT/2,RNORM(1,.3,2);                TELLER 2
14        EVENT,2;                            CHECK FOR JOCKEYING INTO RIGHT QUEUE
15  EXIT  COLCT,INT(1),TIME IN SYSTEM;        COLLECT STATISTICS
16        COLCT,BET,TIME BET. DEPART;
17        TERM;
18  NBALK COLCT,BET,TIME BET. BALKS;
19        TERM;
20        END;
21  INIT,0,1000;
22  FIN;
```

Figure 9-3 Input statements for model of the drive-in bank.

ber of customers in the drive-in bank system. Since there is not a SLAM II variable that represents this value, a user function, USERF(1), is written to compute
the value. Time persistent statistics are collected on the computed value by using
the TIMST control statement.

   TIMST,USERF(1),NO. OF CUST;

This statement specifies that user function 1 is to return a value as USERF on
which time persistent statistics are to be computed. The coding for function
USERF(I) is shown in Figure 9-4 where USERF is set equal to the sum of the
number in Teller 1's subsystem (NNQ(1)+NNACT(1)) and the number in Teller 2's
subsystem (NNQ(2)+NNACT(2)).

   The SEEDS control statement sets the initial seed value for stream number 1 to
be 4367651 and the initial seed value for stream number 2 to be 6121137.

```
      FUNCTION USERF(I)
C*****
C  CALCULATES THE TOTAL NUMBER OF CUSTOMERS IN THE SYSTEM
C*****
      USERF=NNQ(1)+NNACT(1)+NNQ(2)+NNACT(2)
      RETURN
      END
```

Figure 9-4 Function USERF for model of the drive-in bank.

Summary of Results. The results for this example are summarized by the SLAM II Summary Report shown in Figure 9-5. Statistics for variables based on observations are given first with the identifiers specified at the COLCT nodes. For this example, observation statistics were collected on the time in the system for each customer and the time between departures of customers. From the summary report, it is seen that 1854 customers were served and the average time in the system was 2.317 minutes. Also customers leave the bank every 0.54 minutes on the average.

S L A M   I I   S U M M A R Y   R E P O R T

SIMULATION PROJECT DRIVE IN BANK   BY OREILLY

DATE 4/12/1982   RUN NUMBER   1 OF   1

CURRENT TIME   0.1000E+04
STATISTICAL ARRAYS CLEARED AT TIME 0.0000E+00

STATISTICS FOR VARIABLES BASED ON OBSERVATION

	MEAN VALUE	STANDARD DEVIATION	COEFF. OF VARIATION	MINIMUM VALUE	MAXIMUM VALUE	NUMBER OF OBSERVATIONS
TIME IN SYSTEM	0.2317E+01	0.1018E+01	0.4393E+00	0.3039E+00	0.5068E+01	1854
TIME BET. DEPART	0.5388E+00	0.3610E+00	0.6700E+00	0.9155E-04	0.3004E+01	1853
TIME BET. BALKS	0.6982E+01	0.1315E+02	0.1884E+01	0.1413E-01	0.7082E+02	142

STATISTICS FOR TIME-PERSISTENT VARIABLES

	MEAN VALUE	STANDARD DEVIATION	MINIMUM VALUE	MAXIMUM VALUE	TIME INTERVAL	CURRENT VALUE
NO. OF CUST	0.4312E+01	0.2154E+01	0.0000E+00	0.8000E+01	0.1000E+04	0.7000E+01

FILE STATISTICS

FILE NUMBER	ASSOC NODE LABEL/TYPE	AVERAGE LENGTH	STANDARD DEVIATION	MAXIMUM LENGTH	CURRENT LENGTH	AVERAGE WAITING TIME
1	LEFT QUEUE	1.3105	1.0327	3	3	1.1078
2	RIGH QUEUE	1.1476	0.9926	3	2	1.2313
3	CALENDAR	2.8534	0.4323	5	3	0.3366

SERVICE ACTIVITY STATISTICS

ACTIVITY INDEX	START NODE OR ACTIVITY LABEL	SERVER CAPACITY	AVERAGE UTILIZATION	STANDARD DEVIATION	CURRENT UTILIZATION	AVERAGE BLOCKAGE	MAXIMUM IDLE TIME/SERVERS	MAXIMUM BUSY TIME/SERVERS	ENTITY COUNT
1	TELLER 1	1	0.9394	0.2385	1	0.0000	2.2024	84.8427	942
2	TELLER 2	1	0.9140	0.2804	1	0.0000	4.0954	71.4718	912

Figure 9-5 SLAM II summary report for drive-in bank.

The second category of statistics for this example is for time-persistent variables. The results show the average number of customers in the system is 4.312.

The third category of statistics relate to the number of customers waiting in line for teller 1 and teller 2, respectively. The results show that there was an average of 1.3105 customers waiting in file 1 for service by teller 1 and an average of 1.1476 customers waiting in file 2 for service by teller 2. The final category of statistics for this example is service activity statistics. The results show that teller 1 was busy 93.94 percent of the time while teller 2 was busy 91.40 percent of the time. The slightly longer length of the waiting line and higher utilization for teller 1 as compared to teller 2 is because of the preference by customers for teller 1 when both waiting lines are equal. This information could be used by bank management for assigning tellers to windows.

9.7   SAMPLING FROM PROBABILITY DISTRIBUTIONS

An important aspect of many systems of interest is the stochastic nature of one or more elements in the system. For example, in queueing systems the arrival times and service times are usually not known with certainty but are depicted as random variables. Therefore, in order to build models of such systems, the simulation programmer needs to be able to sample from commonly encountered distributions such as the exponential, normal, and beta as well as from user-defined distributions. SLAM II provides FORTRAN function statements for this purpose. A list of such functions is summarized in Table 9-1. Note that the functional names are the same as those employed in network models except for DPROB and the parameter specifications are identical to those specified in Chapter 5. The reader should note that these are FORTRAN function subprograms, and all arguments must be specified with the correct type of FORTRAN variable (integer or real). In particular, the stream parameter cannot be defaulted as in the case of network statements.

FUNCTION DPROB allows the user to obtain a sample from a user-defined discrete distribution (a probability mass function). The user must convert the probability mass function into a cumulative distribution. Two user arrays are required to describe this distribution: the first for the potential values of the random variable and the second for the associated cumulative probabilities. The arguments of FUNCTION DPROB are: the user array name for the cumulative probabilities (CPROB), the user array name for the associated values (VALUE), the number of

Table 9-1 Function statements to obtain random samples.

Function	Descriptions†
DRAND(IS)	A psuedo-random number obtained from random number stream IS
XRN(IS)	Last random number generated from stream IS
EXPON (XMN,IS)	A sample from an exponential distribution with mean XMN using random number stream IS
UNFRM(ULO,UHI,IS)	A sample from a uniform distribution in the interval ULO to UHI using random number stream IS
WEIBL(BETA,ALPHA,IS)	A sample from a Weibull destribution with shape parameter ALPHA and scale parameter BETA using random number stream IS
TRIAG (XLO,XMODE,XHI,IS)	A sample from a triangular distrubution in the interval XLO to XHI with mode XMODE using random number stream IS
RNORM (XMN,STD,IS)	A sample from a normal distribution with mean XMN and standard deviation STD using random number stream IS
RLOGN (XMN,STD,IS)	A sample from a lognormal distribu;tion with mean XMN and standard deviation STD using random number stream IS
ERLNG (EMN,XK,IS)	A sample from an Erlang distribution which is the sum of XK exponential samples each with mean EMN using random number stream IS
GAMA (BETA,ALPHA,IS)	A sample from a gamma distribution with parameters BETA and ALPHA using random number stream IS
BETA (THETA,PHI,IS)	A sample from a beta distribution with parameters THETA and PHI using random number stream IS
NPSSN (XMN,IS)	A sample from a Poisson distribution with mean XMN using random number stream IS
DPROB (CPROB,VALUE,NVAL,IS)	A sample from a user-defined discrete probability function with cumulative probabilities and associated values specified in arrays CPROB AND VALUE, with NVAL values using random stream IS.

† See Chapter 18 for the equations of the distribution functions associated with the random variables and Chapter 5 for the definitions of the arguments and their calculation.

values (NVAL) contained in CPROB and VALUE, and the random stream number (IS).

As an example of the use of FUNCTION DPROB, consider the following probability mass function for the random variable Z:

Z	$f(Z)$
3	.1
4	.4
5	.5

First, the probability mass function is converted to a cumulative distribution. We define the cumulative distribution using arrays $Z(I)$ and $FZ(I)$ as follows:

I	$Z(I)$	$FZ(I)$
1	3	.1
2	4	.5
3	5	1.0

These values can be conveniently assigned to the arrays by a FORTRAN DATA statement in the subroutine calling FUNCTION DPROB or can be read from input records. The following statement draws a sample, ZSAMP, from the desired probability mass function using random stream number 1.

ZSAMP = DPROB(FZ,Z,3,l)

Function DPROB differs from the network discrete probability function, DPROBN, in that for DPROBN, rows of the SLAM II variable ARRAY must be specified as the first two arguments.

9.8 USER-WRITTEN INITIALIZATION AND POST-RUN PROCESSING ROUTINES: SUBROUTINES INTLC AND OTPUT

At the beginning of each simulation run, the SLAM II processor calls subroutine INTLC to enable the modeler to set initial conditions and to schedule initial events. All the subroutines and functions included within SLAM II can be used at this time. Subroutine OTPUT is called at the end of each simulation run and is used for end-of-run processing such as clearing out files and printing special results for the simulation run. SLAM II provides many subroutines and functions to access all variables and statistics collected during a run. These subroutines are described in detail in Chapters 11 and 12.

9.9   CHANGING THE STATUS OF A RESOURCE

SLAM II provides the modeler with subroutines which free a specified number of units of a resource, alter the capacity of a resource by a specified amount, and support complex resource allocation decisions. Before describing these routines, the method for referencing a resource needs to be established.

In network statements, resources are referenced by either the resource name or number. In the resource subroutines, a resource must be referenced by its number. Recall from Chapter 5 that if a resource number is not specified, the SLAM II processor automatically numbers the resources in the order in which the RESOURCE blocks are inputted to the SLAM II processor. The first resource is assigned number 1, the second resource is assigned number 2, and so on. For example, consider the following set of RESOURCE blocks included in the network portion of a model.

    RESOURCE/TUG(2), 1, 2/BERTH(3), 3;
    RESOURCE/CREW, 4;

The processor would assign the TUGs as resource number 1, the BERTHs as resource number 2, and the CREW as resource number 3. When changing the status of a resource from within a discrete event, the resource number is included as an argument to the appropriate subroutine to distinguish between resource types.

9.9.1   Freeing Resources: Subroutine FREE

Subroutine FREE(IR,N) releases N units of resource number IR. The freed units of the resource are made available to waiting entities according to the order of the file numbers specified in the RESOURCE statement included in the network model. For the RESOURCE blocks depicted above, the statement

CALL FREE(1,2)

would release 2 units of the resource TUG. These tugs are made available to entities waiting in file 1. If both tugs are not used, then file 2 would be interrogated to determine if an entity was waiting for the use of a tug. The execution of the statement CALL FREE(1,2) is identical to the execution of the following network statement.

FREE, TUG/2;

9.9.2   Altering Resource Capacities: Subroutine ALTER

Subroutine ALTER(IR,N) changes the capacity of resource number IR by N units. In the case where the capacity of the resource is decreased below current utilization, the excess capacity is destroyed as it becomes freed. The capacity can be reduced to a minimum of zero with additional reduction requests having no effect. Assuming the RESOURCE blocks depicted earlier, the following statement would reduce the number of tugs by 1.

CALL ALTER(1,−1)

This statement produces the same effect as the execution of the following network statement.

ALTER,TUG/−1;

9.10   USER WRITTEN RESOURCE ALLOCATION PROCEDURES: SUBROUTINES ALLOC AND SEIZE

AWAIT nodes are used to store entities waiting for resources. When an AWAIT node description includes a resource name, an entity cannot pass through the

AWAIT node until a specified number of units of that resource are available. When this condition occurs, the proper number of resources of that type are "seized" (set busy) and the entity proceeds from the AWAIT node.

Instead of naming a specific resource type, the modeler may specify ALLOC(I) in the field normally used for the resource label. (Note that "ALLOC" is no longer an acceptable resource or gate label.) With this specification, subroutine ALLOC(I,IFLAG) is called by SLAM II when an entity arrives to the AWAIT node whose resource is specified as ALLOC(I). Subroutine ALLOC is also called when the file associated with that AWAIT node is polled as the result of a newly available or freed resource.

Subroutine ALLOC is coded to determine which, if any, entity in the file can proceed. In coding Subroutine ALLOC, all FORTRAN coding procedures and SLAM II subprograms can be used with the restrictions that no operations on the event file are allowed and subroutines FREE and ALTER may not be called from subroutine ALLOC. SLAM II loads the ATRIB buffer with the attributes of the first entity in the AWAIT node file prior to calling ALLOC.

There are two arguments for subroutine ALLOC:

1. A user code, I, to differentiate calls to subroutine ALLOC from different AWAIT nodes; and
2. A flag, IFLAG, set by the user to inform SLAM II of the entity to which an allocation has been made. If IFLAG is set to zero, no allocation has been made.

In subroutine ALLOC, if it is determined that an entity should proceed from the AWAIT node, the user will seize the appropriate resources (discussed below). IFLAG should be set to plus or minus the rank of the entity to be removed from the AWAIT node file for processing in the network. A positive IFLAG value indicates that the ATRIB buffer has been set in ALLOC to the attribute values to be associated with the entity removed from the file. If IFLAG is set to the negative of the rank, the entity is removed and no change is to be made to its attributes. If no allocation is possible, IFLAG should be set to 0.

To summarize, the SLAM II user codes subroutine ALLOC(I,IFLAG) in a manner similar to function USERF(IFN). The arguments for subroutine ALLOC are listed below:

I   Input argument to the subroutine to provide a code to reference a particular allocate function.

IFLAG   Output argument set by the user to inform SLAM II of actions to be taken. If IFLAG is zero, no allocation of resources was made. If

IFLAG is a positive integer, it is the rank of the entity to be removed from the AWAIT node file. The entity's attributes are given by the buffer ATRIB. If IFLAG is a negative integer, it is the negative of the rank of the entity to be removed. The attributes of the entity should not be changed.

When the resources to be allocated at an AWAIT node are determined by the user in subroutine ALLOC, the required number of units of the required resource is seized through a call to subroutine SEIZE(IR,NU) where IR is the numeric code for the resource type and NU is the number of units to be seized. Subroutine SEIZE then sets NU units of resource type IR busy and updates the statistics for that resource. For example, the statement CALL SEIZE(3,2) sets busy 2 units of resource type 3. Before invoking subroutine SEIZE, the user must test to determine if 2 units of resource type 3 are available. Through the use of subroutine ALLOC and SEIZE, complex resource allocation rules can be included in simulation models.

9.10.1 Illustration 9-1. Joint and Alternative Resource Requirements

Consider a situation in which there are three operators who are modeled as resources. The resource names are OP1, OP2, and OP3. The resource blocks for defining these operators and the files at which they are allocated are shown below.

```
RESOURCE/OP1(1),1,3;
RESOURCE/OP2(1),2,4;
RESOURCE/OP3(1),1,3;
```

From the resource block definition we see there is one unit of each resource and that resources OP1 and OP3 are allocated at files 1 and 3. Resource OP2 is allocated at files 2 and 4. Resource OP2 will not be used in this illustration. The numeric value assigned by SLAM II to resource OP1 is 1 and the numeric value assigned to resource OP3 is 3.

In the network model, suppose we require that activity 1 be performed by both OP1 and OP3. Because this is a complex resource allocation involving the joint allocation of two resources, subroutine ALLOC must be used. The network statements for the AWAIT node and ACTIVITY are listed below.

```
AWAIT(1),ALLOC(1);
ACT/1,10;
```

These network statements show that an entity arriving to the AWAIT node will pass through file 1 to ACTIVITY 1 if allocation rule 1 indicates that resources are available to work on the arriving entity. The code for allocation rule 1 is shown below.

```
            IFLAG = 0
            GO TO(1,2),I
C****
C****   ALLOCATION RULE 1 - SEIZE RESOURCES 1 AND 3
C****
      1   IF (NNRSC(1).LE.0.OR.NNRSC(3).LE.0) RETURN
            CALL SEIZE (1,1)
            CALL SEIZE(3,1)
            IFLAG=-1
            RETURN
```

In the above code, IFLAG is initially set to 0 indicating that no allocation could be made. Next, a test on the allocation rule code is made as we will have two allocation codes associated with this illustration. For I = 1 a transfer is made to statement 1. At statement 1 if either resource 1 (OP1) or resource 3 (OP3) is not available, no allocation can be made. NNRSC(I) is the current number of resources available for resource I. If the condition in statement 1 is not satisfied then both resources are available. Note that NNRSC(I) can be less than 0 if requests to decrease the resource capacity are pending. Subroutine SEIZE is then called twice to assign resource 1 and resource 3. The variable IFLAG is set equal to -1 to inform SLAM II that the first entity in file 1 should be routed from the AWAIT node. The minus sign indicates that the entity's attributes should not be changed.

Consider now a second allocation function represented by the following statements.

```
    AWAIT(3/2),ALLOC(2),BALK(OUT);
      ACT/5,22;
```

The above code indicates that arriving entities to the AWAIT node will wait in file 3 which has a queue capacity of 2. If an entity arrives with two units in file 3 it will balk to a node labeled OUT. When a entity arrives to file 3 and it is empty, subroutine ALLOC is invoked with a request to use allocation rule 2. Allocation rule 2 will be coded to have either resource 1 or resource 3 assigned to an entity.

Subroutine ALLOC, which includes both allocation rule 1 and allocation rule 2, is shown below. For rule 2, subroutine ALLOC will be entered with I=2. IFLAG will again be set to zero and a transfer will be made to statement 2. At statement 2, a check is made to see if either resource is available. If neither is available, a return is made to SLAM II with IFLAG=0 to indicate that the entity should be filed in file 3.

```
      SUBROUTINE ALLOC(I,IFLAG)
      COMMON/SCOM1/ATRIB(100),DD(100),DDL(100),DTNOW,II,MFA,MSTOP,NCLNR
     1,NCRDR,NPRNT,NNRUN,NNSET,NTAPE,SS(100),SSL(100),TNEXT,TNOW,XX(100)
      IFLAG=0
      GO TO (1,2),I
C
C     ALLOCATION RULE 1 — SEIZE RESOURCES 1 AND 3
C
    1 IF(NNRSC(1).LE.0.OR.NNRSC(3).LE.0) RETURN
      CALL SEIZE(1,1)
      CALL SEIZE(3,1)
      IFLAG= −1
      RETURN
C
C     ALLOCATION RULE 2 — SEIZE RESOURCE 1 OR 3
C     SAVE SELECTED RESOURCE IN ATRIB(1)
C
    2 IF (NNRSC(1).LE.0.AND.NNRSC(3).LE.0) RETURN
      IF (NNRSC(1).GT.0)THEN
        CALL SEIZE (1,1)
        ATRIB(1)=1
        IFLAG=1
        RETURN
      ELSE
        CALL SEIZE(3,1)
        ATRIB(1)=3
        IFLAG=1
        RETURN
      ENDIF
      CALL ERROR (3)
      RETURN
      END
```

Next, a test is made to see if resource 1 is available. If it is, a call to subroutine SEIZE is made to seize one unit of resource 1. ATRIB(1) is set to 1 to indicate that the current entity is to be processed by resource 1. IFLAG is then set equal to 1 which specifies that the first entity in file 3 is to be removed from the file and, since IFLAG is positive, that the attribute buffer is to be changed. Before subroutine ALLOC is called, SLAM II resets the ATRIB buffer to the attributes of the first entity in the AWAIT file. Thus, it is not necessary to copy the other attributes of the entity when prescribing a new set of attributes. If, however, an entity other than the first entity is requested to be removed from the AWAIT file, the attributes should be copied into the ATRIB buffer if any attributes of the entity are to be changed.

The remainder of the code for allocation rule 2 performs the same functions when resource 3 is available and resource 1 is not. Note that a preferred choice for resource 1 over resource 3 when both are available is coded directly into allocation rule 2.

9.10.2   Illustration 9-2.   Port Operations Revisited

In Example 6-3, port operations were modeled in which there were three berths and a single tug available to process tankers. In that model, berths were assigned to tankers and then, if available, a tug was assigned to the tanker which already had a berth allocated to it. By first assigning berths and tugs to a tanker, the utilization of the resource BERTH included the time that a tanker is waiting for a TUG. For that model, the definition of utilization for a berth is the fraction of time that a tanker is assigned to it. More realistically, berth utilization should be the fraction of time that the tanker was in the berth or in the process of being berthed. To obtain this resource statistic, it is necessary to assign jointly the BERTH and the TUG to a tanker. This can be accomplished in SLAM II by using the joint allocation procedure described in Illustration 9-1 above. The portion of the network model showing the AWAIT node with allocation rule 1 is given below.

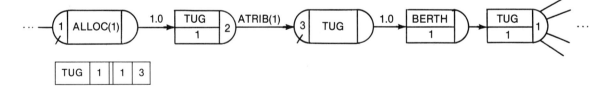

In Example 6-3, it was also assumed that the TUG would always attempt to berth a tanker before initiating a deberthing operation. This order for allocating a TUG was specified through the resource block where the TUG was to be allocated to entities waiting in file 2 before being allocated to entities waiting in file 3.

For this illustration, we assume that a TUG should be allocated to the deberthing activity if it is in port and a tanker is ready to leave. Thus, if a tanker is waiting for deberthing and the TUG has just finished a berthing operation, the

TUG should be allocated to the tanker waiting in file 3. To model this procedure, we must maintain information on the TUG's last operation. Let XX(1) be equal to 1 if the TUG is performing a berthing operation, and let XX(1) be 0 if the TUG is idle or performing a deberthing operation.

When the TUG is freed, entities in file 1 will be considered for allocation (see resource block for TUG). From the revised network, allocation rule 1 will be used for making this decision. In ALLOC, we will not consider a berthing operation if the TUG has just finished a berthing operation (XX(1)=1) and there is a tanker waiting for deberthing (NNQ(3).GT.0).

The test on these conditions are the second and third executable statements in ALLOC (see below). If the TUG is not allocated for berthing, then XX(1) is set to zero, and a return is made. After the return is made, the tug is considered for processing entities in file 3 as specified by the RESOURCE block. In this case, the tug is allocated by SLAM II since an entity is waiting (NNQ(3).GT.0).

```
      SUBROUTINE ALLOC(I,FLAG)
      COMMON/SCOM1/ATRIB(100),DD(100),DDL(100),DTNOW,II,MFA,MSTOP,NCLNR
     1,NCRDR,NPRNT,NNRUN,NNSET,NTAPE,SS(100),SSL(100),TNEXT,TNOW,XX(100)
      IFLAG=0
C*****
C     CALLED TO CHECK FOR ALLOCATION OF TUG AND BERTH FOR BERTHING
C     ALLOCATE IF A) TUG WAS LAST DEBERTHING, XX(1) = 0., OR
C                  B) TUG WAS BERTHING BUT NO SHIPS ARE REASY TO DEBERTH
C*****
      IF ( XX(1) .EQ. 0.) GO TO 10
      IF ( NNQ(3).EQ. 0.) GO TO 10
      RETURN
C*****
C     BERTHING WILL OCCUR IF BOTH TUG AND BERTH ARE AVAILABLE
C*****
   10 IF ( NNRSC(1) .LE. 0 .OR. NNRSC(2) .LE. 0 ) THEN
         XX(1) = 0.
         RETURN
      ENDIF
      CALL SEIZE(1,1)
      CALL SEIZE(2,1)
      IFLAG = -1
      XX(1) = 1
      RETURN
      END
```

The coding for subroutine ALLOC shows that both the BERTH and TUG resources are seized in ALLOC when a berthing activity is to be performed. Next, IFLAG is set to −1 to communicate to SLAM II that the tanker entity should be removed from file 1 without changing its attribute values. XX(1) is then set to 1 to indicate that the tug is performing a berthing activity.

9.10.3   Illustration 9-3.   Assembling Resources in the Truck Hauling Situation

This illustration demonstrates how subroutine ALLOC can be used to provide an assembly capability for resources. The truck hauling situation modeled in Example 7-2 is used to demonstrate this capability. In that example, a SELECT node with the assembly (ASM) queue selection rule is used to assemble a truck, a load and a loader. The use of a loader as both a service activity and as an entity to be assembled was required to model the situation in Chapter 7.

In the network shown below, an AWAIT node is used to queue loads that wait for both a truck and a loader. Trucks and loaders are modeled as resources with the labels TRKS and LDER which have resource numbers 1 and 2 respectively. In subroutine ALLOC, a check is made to determine if both a TRKS and LDER are available. If available, they are allocated and a load continues through the network. Subroutine ALLOC is shown below.

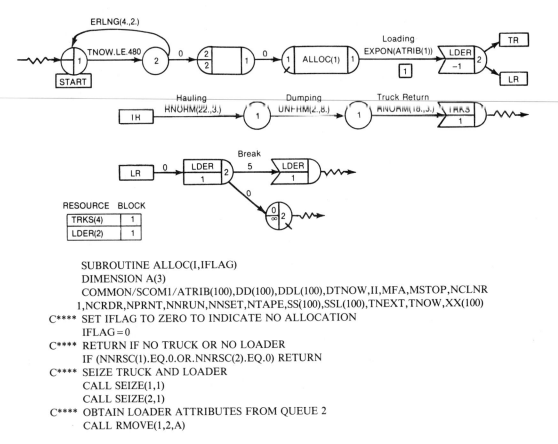

```
          SUBROUTINE ALLOC(I,IFLAG)
          DIMENSION A(3)
          COMMON/SCOM1/ATRIB(100),DD(100),DDL(100),DTNOW,II,MFA,MSTOP,NCLNR
         1,NCRDR,NPRNT,NNRUN,NNSET,NTAPE,SS(100),SSL(100),TNEXT,TNOW,XX(100)
    C**** SET IFLAG TO ZERO TO INDICATE NO ALLOCATION
          IFLAG=0
    C**** RETURN IF NO TRUCK OR NO LOADER
          IF (NNRSC(1).EQ.0.OR.NNRSC(2).EQ.0) RETURN
    C**** SEIZE TRUCK AND LOADER
          CALL SEIZE(1,1)
          CALL SEIZE(2,1)
    C**** OBTAIN LOADER ATTRIBUTES FROM QUEUE 2
          CALL RMOVE(1,2,A)
```

```
C**** MODIFY ATRIB BUFFER FOR FIRST ATTRIBUTE ONLY
      ATRIB(1) = A(1)
      IFLAG = 1
      RETURN
      END
```

In ALLOC, the communication variable, IFLAG, is set to zero so that any return from subroutine ALLOC before IFLAG is reset will indicate to SLAM II that no allocation was made. Next, if either NNRSC(1) or NNRSC(2) is zero, no allocation can be made. Otherwise, one unit of TRKS is seized and one unit of LDER is seized. The first entity in file 2 (see QUEUE node on network) is then removed as it contains the attributes of the loader. In particular, attribute 1 contains the mean loading time for activity 1. In the call to subroutine RMOVE, the variable A is used for storing attribute values. The attribute buffer ATRIB is not altered by this call to subroutine RMOVE. Next, ATRIB(1) is set equal to A(1) to have the first attribute of the load entity represent the mean loading time. In this example, other attributes are not used but if they were this would permit the attributes of the load entity to be unchanged except for the first value. IFLAG is then set to 1 to indicate to SLAM II that an allocation has been made and that a load entity should be removed from file 1.

In this model, QUEUE node 2 is used to store two loader entities. The following entry statement is used to put the attributes of the two loaders into file 2.

    ENTRY/2,14./2,12.;

With this ENTRY statement, two entities will be put in file 2 with the first entity having an attribute 1 value of 14 and the second entity having an attribute 1 value of 12. No priority statement is used in this model and, hence, the entities in file 2 are ranked on a first-in-first-out basis. This ensures that the loader who is idle the longest is used to process the next LOAD.

To model the five minute break for each LDER, an ALTER node is used following the loading activity to change the capacity of resource LDER by −1. This starts the loader's break. Following this ALTER node, two entities are generated representing a load entity with a truck and a loader entity. The former entity is sent over hauling, dumping and truck return activities after which a TRKS resource is freed.

From the ALTER node, the loader entity is routed to a FREE node which frees one unit of the LDER resource. By processing the ALTER node and then the FREE node, we guarantee that the freed LDER will not be reassigned before the LDER capacity is reduced because of the break. The five minute break is modeled by an activity which is followed by an ALTER node which alters the capacity of resource LDER by +1. This puts the loader back in operation and SLAM II will try to allocate the loader to loads waiting in file 1 through a call to subroutine ALLOC.

The procedure of using an ALTER node in series with a FREE node and then another ALTER node is required to correctly estimate the utilization of the loader. During the five minute break time, the loader is not considered utilized as the capacity of the resource has been reduced by the one loader on break. The utilization of the loader will then be computed as the ratio of the time the loaders were involved in activity 1 divided by twice the total simulation time period minus the time spent by loaders on breaks.

Following the FREE node, an entity is routed to the QUEUE node associated with file 2. The loader is put in file 2 during the time period that he is not loading and during the time that he is on a break. Entities are removed from file 2 from subroutine ALLOC and, hence, there is no need for an activity emanating from the QUEUE node.

9.11 CHANGING THE STATUS OF A GATE

SLAM II provides the capability to open and close gates using FORTRAN statements. A gate must be referenced by its gate number. The gate number is automatically assigned by SLAM II to each gate in the order that the GATE blocks are inputted to the SLAM II processor. For example, consider the following set of GATE blocks included in the network portion of a model:

```
GATE/DOOR1,OPEN,4;
GATE/DOOR2,CLOSED,5;
```

The processor would assign DOOR1 as gate number 1 and DOOR2 as gate number 2.

9.11.1   Opening a Gate: Subroutine OPEN

Subroutine OPEN(IG) opens gate number IG and releases all waiting entities in the AWAIT files specified in the GATE block. For the GATE blocks depicted above, the following statement would open DOOR2:

CALL OPEN(2)

This statement is equivalent to the following network statement.

OPEN,DOOR2;

9.11.2   Closing a Gate: Subroutine CLOSX

Subroutine CLOSX(IG) closes gate number IG. For the GATE blocks depicted above, the following statement would close DOOR1.

CALL CLOSX(1)

This statement is equivalent to the following network statment.

CLOSE,DOOR1;

9.12   USER-WRITTEN QUEUE SELECTION: FUNCTION NQS

In a network model, SELECT nodes are used to route an entity to one of a set of parallel QUEUE nodes or to select an entity from a set of parallel queues for processing by a server. The SLAM II standard queue selection rules are presented in Table 7-2. A user-coded queue selection option is available to model complex selection logic. By specifying NQS(N) as the queue selection rule for a SELECT node, the user-written function NQS is called with argument N whenever a queue selection from the SELECT node is needed. The SLAM II user writes function NQS to execute the desired queue selection logic and returns the file number of the selected QUEUE as the assigned value to NQS. When queue selection is done by the user, it is the user's responsibility to insure that a feasible choice is made, that is, an entity must be able to be routed to the QUEUE node or taken from the QUEUE node. Whenever there are no feasible choices, the user should set NQS to zero to indicate that no QUEUE node was selected.

The network segment shown on the next page and its corresponding statement model shown below illustrate how to code function NQS. At SELECT node SSS, arriving entities are to be routed to QUEUE nodes Q1, Q2, and Q3 based on the value of XX(1). In a section of the model not shown, the modeler has specified the value of XX(1) to be 1, 2, or 3. Queue selection function 1 will be used to make this selection and NQS(1) is specified as the queue selection rule for SELECT node SSS. In function NQS(N) also shown on the next page, a transfer to statement 1 is made when the queue selection code number is 1. At statement 1, the value of NQS is set equal to XX(1) to indicate that the QUEUE node with the file number XX(1) is to be selected for the arriving entity. In this situation, no check is made on the feasibility of routing the arriving entity to the prescribed QUEUE node as all three QUEUE nodes have unlimited capacity.

At SELECT node SEL, queue selection function 2 is prescribed for selecting entities from Q1 or Q2 when either activity 1, 2, or 3 is completed. The coding for queue selection function 2 is shown in function NQS starting at statement 2. First it is determined whether an entity is waiting for the service activities following SELECT node SEL. Thus, if the number in Q1 and the number in Q2 are 0, a

```
NETWORK;
        •
        •
        •
SSS    SELECT,NQS(1),,,Q1,Q2,Q3;
Q1     QUEUE(1),0,,,SEL;
Q2     QUEUE(2),0,,,SEL;
Q3     QUEUE(3),0;
       ACT/4,20,,N1;              N1 NOT SHOWN
SEL    SELECT,NQS(2),NSS(1),,Q1,Q2;
       ACT/1,10;
       ACT/2,12;
       ACT/3,14;
        •
        •
        •
```

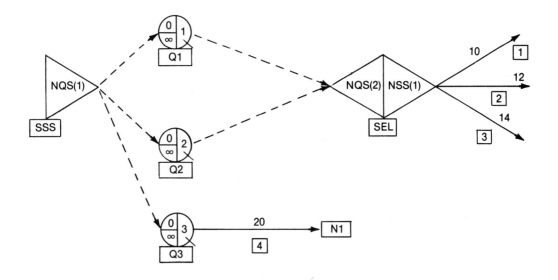

```
            FUNCTION NQS(N)
            COMMON/SCOM1/ATRIB(100),DD(100),DDL(100),DTNOW,II,MFA,MSTOP,NCLNR
           1,NCRDR,NPRNT,NNRUN,NNSET,NTAPE,SS(100),SSL(100),TNEXT,TNOW,XX(100)
            GO TO (1,2),N
C****  SELECT QUEUE BASED ON VALUE OF XX(1)
       1 NQS = XX(1)
            RETURN
       2 NQS = 0
C****  TEST FEASIBILITY — IS AN ENTITY WAITING FOR SERVICE?
            IF (NNQ(1).EQ.0.AND. NNQ(2).EQ.0) RETURN
            IF(NNQ(1).EQ.0) THEN
              NQS = 2
              RETURN
            ELSE
              IF(NNQ(2).EQ.0) THEN
                 NQS = 1
                 RETURN
                   ENDIF
            ENDIF
C****  SELECT QUEUE FOR WHICH SUM OF ATRIB(3) VALUES
C****  FOR ALL ENTITIES IS THE LARGEST
            IF(SUMQ(3,1).GE.SUMQ(3,2))THEN
              NQS = 1
                ELSE
              NQS = 2
                ENDIF
            RETURN
            END
```

value of 0 is assigned to NQS to indicate to SLAM II that the server should be set to idle status. Next, if no entity exists in either queue, the other queue is selected. When entities are available in both queues, the queue selection rule is to select an entity from the queue for which the sum of all attribute 3 values in that queue is larger. This is accomplished by using the SLAM II function SUMQ (described in Chapter 12) and comparing the values obtained.

As seen from the above, the coding of function NQS permits complex queue selection rules to be included in a SLAM II network.

9.13   USER-WRITTEN SERVER SELECTION: FUNCTION NSS

In a SLAM II network when more than one service activity follows a queue or set of queues and the service activities are not identical, a SELECT node is necessary to make a selection from among the non-identical servers. Standard server selection rules are provided by SLAM II and are listed in Table 7-3. A user coded server selection rule can be associated with a SELECT node by specifying NSS(N) in the server selection field of the SELECT node where N is a numeric index which allows the user to differentiate between user-written selection rules employed at different SELECT nodes in the network.

Function NSS is called when an entity arrives to an empty QUEUE node associated with the SELECT node whose server selection rule is NSS. In function NSS, the index is used to branch to the appropriate portion of the code for the SELECT node. In function NSS, an activity number is to be assigned to NSS to indicate the service activity selected. The service activity selected must have an available server to process the newly arriving entity. If all servers following the SELECT node are busy, NSS should be returned as zero in which case the newly arriving entity will be placed in the QUEUE node to which it arrived. The service activity numbers following the SELECT node are known to the user as they are associated with the activity statements following the SELECT node in the network input statements.

When a value of NSS is returned, SLAM II takes the appropriate action of either putting the arriving entity into a QUEUE node or making the selected service activity busy by scheduling the start of the activity.

The example used in the section on function NQS will be continued in this section. The appropriate portions of the statement model to illustrate function NSS are:

```
SEL   SELECT,NQS(2),NSS(1),,Q1,Q2;
         ACT/1,10;
         ACT/2,12;
         ACT/3,14;
```

When an entity arrives to either QUEUE node Q1 or Q2 and there are no entities waiting at the time of arrival, function NSS with an argument of 1 is called in order to determine if server 1, 2, or 3 can process the arriving entity. If all three servers are busy, then the arriving entity is put into the appropriate file. If only one server is free, that server is selected for processing the entity. If more than one server is free, a selection rule is coded in function NSS. The coding for function NSS is shown below.

```
         FUNCTION NSS(N)
         DIMENSION NSER(3)
         GO TO (1),N
    1    NSS = 0
C****    SET NUMBER AVAILABLE TO 0
         J = 0
         DO 10 I = 1,3
         IF (NNACT(I).NE.0) GO TO 10
C****    SET FIRST AVAILABLE SERVER NUMBER TO NSER(J)
         J = J + 1
         NSER(J) = I
   10    CONTINUE
C****    IF NO SERVER AVAILABLE, RETURN WITH NSS = 0
         IF (J.EQ.0) RETURN
         XJ = J + 1
C****    RANDOMLY SET A SERVER RANK BETWEEN 1 AND J
         NR = UNFRM(1.,XJ,1)
C****    GET SERVER NUMBER FROM SERVER RANK VECTOR
         NSS = NSER(NR)
         RETURN
         END
```

In NSS, only one server selection function is indicated and only the code is given for server selection function 1. At statement 1, the selected server number NSS is set to 0 to indicate that if an early return is made before NSS is set, a server could not be selected. In the middle part of the function, each service activity status is tested and if a server is available, the service activity number is stored in the vector NSER(J) where J is an index for available servers. If no servers are available, J is zero and a return is made with NSS = 0.

If one or more servers are available then an integer is selected from 1 to the number of available servers. Note that XJ is set equal to J + 1 and a number drawn between 1 and XJ. By using the integer variable NR, these values will be truncated down to the next lowest integer so that an integer from 1 to J will be obtained as the value of NR. In this way, each integer has an equal likelihood of being selected. The selected server activity number, NSS, is then obtained from the NRth location of the vector NSER.

The code shown for function NSS is equivalent to selecting randomly from among the available service activities. It is equivalent to the standard priority rule RAN with an equal probability assigned to each of the servers following the SELECT node.

9.14   STOPPING AN ACTIVITY: SUBROUTINE STOPA

In complex systems, the length of a specific activity may not be known a priori but may depend upon the dynamics of the system. For example, in a queueing system the service rate may be a function of the number of entities waiting for the server and thus may change over time. Therefore the duration of the activity is unknown at the start time of the activity and is affected by future arrivals. One way of modeling an indefinite activity duration using network concepts is to specify the activity duration as keyed to STOPA(NTC). This allows the modeler to stop selectively a specific entity undergoing an activity without stopping the other entities undergoing the same activity.

To stop an activity for a network entity from user-written code, the modeler must specify the duration of the activity in the network model as STOPA(NTC) where NTC is a positive integer which is user-assigned as an entity code to distinguish the entity from other entities in the same activity or elsewhere in the network. If NTC is specified as a real value it is truncated to the nearest integer. The value of NTC can be specified as a number, a SLAM II variable, or a SLAM II random variable. By specifying the entity code as an attribute of the entity or as a random variable, the modeler can assign different entity codes to entities within the same activity. The activity statement

ACT,STOPA(1);

specifies that all entities in the activity are to be assigned entity code 1. The statement

ACT,STOPA(ATRIB(3));

specifies that the third attribute of each entity is to be assigned as the entity code.

The mechanism for stopping the activity for a network entity from FORTRAN code is to call subroutine STOPA(NTC). A call to subroutine STOPA(NTC) causes an end of activity to occur for every entity with entity code NTC that is being processed by an activity whose duration is specified as STOPA. For example, execution of the statement CALL STOPA(1) causes all activities to be completed whose duration was specified as STOPA(1).

Since the execution of the call STOPA(NTC) statement causes the end of an activity to occur for each entity with entity code NTC, there may be none, one, or several activities ended by a call to STOPA(NTC). For each activity that is ended, the end-of-activity event for the entity in the activity is placed at the top of the event calendar to be processed immediately following the return from the discrete event. If more than one activity is stopped in this manner, then the end-of-activity event for each entity is processed following the return to SLAM II in the order in which the activities were started. If no entities are currently keyed to STOPA(NTC), then execution of a call to STOPA(NTC) has no effect.

9.15   EXAMPLE 9-2.   PSYCHIATRIC WARD

Clients of a psychiatric ward arrive at the rate of two per day. Each client is given a test and the test scores are uniformly distributed between 30 and 44. When the ward is full, clients are not admitted if their score is greater than 41. The ward has space for 25 patients. A patient's test score is estimated to change daily in a uniform manner in the range from −0.2 to 1.2. When a patient's test score reaches 48, the patient is discharged from the ward. If a potential patient arrives to the ward and the ward is full, a current patient will be bumped from the ward if a test score of 47 or higher has been achieved. Initially there are 18 patients in the ward and their test scores range from 30 to 44. The objective is to simulate the operation of the ward for 1000 days to determine the average time in the system for each patient, the ward utilization, the number of clients balking, and number of patients bumped.

Concepts Illustrated. This example illustrates the use of the EVENT node, subroutine EVENT, ENTER node, subroutine ENTER, the STOPA duration specification, and subroutine STOPA. This example also illustrates the use of the function USERF and subroutine INTLC.

SLAM II Model. In this problem, the discharge from the system for each patient in the ward cannot be scheduled in advance. The test scores for each patient and the ranking of patients based on test scores change daily. This process can be modeled using SLAM II by representing patients in the ward as entities in an ACTIVITY with a duration specified as STOPA(II) where II denotes the space or bed number in the ward. The test scores for patients in each of the twenty-five spaces is maintained in the SLAM II variable XX(II). If space number II is not occupied, then XX(II) is set equal to 0.

The network portion of the model is depicted in Figure 9-6. Entities which represent clients seeking admittance to the ward are created at the CREATE node with the current time marked as ATRIB(1) and the time between entities exponentially distributed with mean of 0.5 day. The entities proceed to the ASSIGN node where the initial test score is assigned as ATRIB(2) of the entity. The M-number for the ASSIGN node is specified as 1, thus a maximum of one of the three emanating ACTIVITY's will be taken. If the ward is full (the number of active entities in the ACTIVITY with index number 1 is 25) and the test score for this client (ATRIB(2)) is greater than 41, then the entity is routed to the COLCT node labeled BALK. If this ACTIVITY is not taken and the ward is full, then the entity is routed to the EVENT node labeled BUMP. If neither of these ACTIVITY's is taken, then the entity is routed to the ASSIGN node labeled WARD.

Consider the entities which arrive to the COLCT node BALK. At each entity arrival to the node, statistics are collected on the time between entity arrivals with the results displayed using the identifier TIME BET. BALKS. The entities are then terminated from the system. No terminate count is used so that the entity is destroyed, but no end of run will occur because of it.

Next, consider the entities which are routed to the EVENT node BUMP, which represents a client attempting to bump one of the patients currently on the ward. Determining if a patient can be bumped is a complex decision. Based on the current test scores of the patients in the ward, a new client is not admitted or a patient is bumped (discharged). In the former case, the client entity is assigned an ATRIB(2) value equal to 48. A return to the EVENT node then causes the entity to be routed to the node BALK based on the condition that ATRIB(2).GE.48 on the activity emanating from node BALK. In the latter case, the client entity enters into the network through ENTER node 2 to arrive at the ASSIGN node WARD. Before describing the code for event 1, we will complete our description of the network portion of the model.

The entities, which are routed to the ASSIGN node WARD, represent clients who are to be assigned an available space in the ward. At the ASSIGN node, the index variable II is set equal to the value computed in the user-coded function

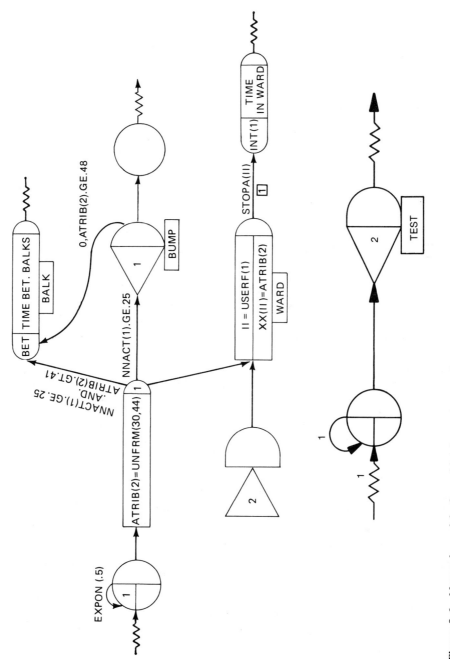

Figure 9-6 Network model of psychiatric ward example.

```
      FUNCTION USERF(I)
      COMMON/SCOM1/ ATRIB(100),DD(100),DDL(100),DTNOW,II,MFA,MSTOP,NCLNR
     1,NCRDR,NPRNT,NNRUN,NNSET,NTAPE,SS(100),SSL(100),TNEXT,TNOW,XX(100)
      DO 10 J=1,25
C
C *** SET THE USERF EQUAL TO EMPTY BED NUMBER
C
      IF(XX(J).GT.O.) GO TO 10
      USERF=J
      GO TO 20
   10 CONTINUE
C
C *** IF AN EMPTY BED IS NOT FOUND -- CALL ERROR.
C
      CALL ERROR(1)
   20 RETURN
      END
```

Figure 9-7   Function USERF(I) for psychiatric ward example.

USERF(I) with I=1. The coding for function USERF is shown in Figure 9-7 and it returns the value of USERF as the smallest index J for which XX(J) is equal to zero. An XX(J)=0 indicates that space J is available. Since entities should only arrive to the ASSIGN node WARD when at least one space is available, a call is made to subroutine ERROR to cause an error exit from SLAM.II if all spaces are full. Following the assignment of the first available space number to II, we set XX(II) to the initial test score for this patient.

The ACTIVITY following the ASSIGN node WARD represents the duration for which the patient remains in the ward. The duration for the ACTIVITY is specified to be STOPA(II) denoting that the duration is to be terminated by a call to subroutine STOPA(NTC) with NTC equal to space number II of the patient who is to leave the ward. In this manner, each patient can be selectively discharged from the ward by specifying NTC as the space number for the patient. Following discharge from the ward, interval statistics on the time in system are collected at the COLCT node using the mark time in ATRIB(1) as a reference time. The identifier is TIME IN WARD. The entity is then terminated. A disjoint network is used to create an entity every day which arrives at EVENT node TEST that initiates the testing event for the model. The input statements for this example are shown in Figure 9-8.

There are two events that are coded outside of the network model. The first event is initiated by an entity arrival to the EVENT node BUMP and is coded in subroutine BUMP. The second event processes the daily changes in the patients' test scores and discharges patients as necessary. The logic for this event is coded in subroutine TEST. Subroutine EVENT(I) which maps the event code onto the appropriate event subroutine is shown in Figure 9-9.

```
 1  GEN,OREILLY,PSYCHIATRIC WARD,12/20/83,1;
 2  LIMITS,0,2,150;
 3  NETWORK;
 4  ;
 5        CREATE,1,1;              DAILY RETEST OF OCCUPIED BEDS
 6  TEST  EVENT,2;
 7        TERM;
 8  ;
 9        CREATE,EXPON(.5),,1;              PATIENT ARRIVALS
10  ADMT  ASSIGN,ATRIB(2)=UNFRM(30,44),1;  ASSIGN PATIENT SCORE
11        ACT,,NNACT(1).GE.25.AND.
12                ATRIB(2).GT.41,BALK;     BED NOT AVAILABLE
13        ACT,,NNACT(1).GE.25,BUMP;        TRY TO BUMP
14        ACT,,,WARD;                      BED AVAILABLE
15  BALK  COLCT,BET,TIME BET. BALKS;       COLLECT BALK STATISTICS
16        TERM;
17  ;
18  ;       IS A PATIENT BUMPABLE?
19  BUMP  EVENT,1;
20        ACT,0,ATRIB(2).GE.48,BALK;             BUMPED PATIENT STATS
21        ACT;
22        TERM;
23  ;
24  ;       ASSIGN PATIENT A BED
25        ENTER,2;
26  WARD  ASSIGN,II=USERF(1),XX(II)=ATRIB(2);
27        ACT/1,STOPA(II);
28        COLCT,INT(1),TIME IN WARD;       WAIT FOR CALL TO STOPA
29        TERM;
30        ENDNETWORK;
31  INITIALIZE,0,500;
32  FIN;
```

Figure 9-8 Statement model for psychiatric ward example.

```
        SUBROUTINE EVENT(I)
        GO TO (1,2),I
C *** EVENT CODE 1---SUBROUTINE BUMP
      1 CALL BUMP
        RETURN
C *** EVENT CODE 2---SUBROUTINE TEST
      2 CALL TEST
        RETURN
        END
```

Figure 9-9 Subroutine EVENT for psychiatric ward example.

Subroutine BUMP contains the processing logic for the EVENT node BUMP and is depicted in Figure 9-10. Subroutine BUMP is only called when a new client arrives. Initially, the variable SCORE is set to 47, and the variable JJ is set to 0. The DO loop then searches the current test scores of patients and sets JJ equal to the index of the patient with the largest test score greater than 47. If no patient has a score greater than 47, a branch is made to statement 20 where ATRIB(2) is set to 48 to represent the departure of the new client. A return from subroutine BUMP is then made. If a patient can be bumped, an entity is entered into the network at ENTER node 2 and the bumped patient is discharged from the ward by setting the test score for space JJ to 0, that is, XX(JJ)=0.0, and calling subroutine STOPA(JJ). A return from subroutine BUMP is then made. Note that in the latter case, the order in which the subroutine calls are executed is important. The call to subroutine ENTER(2,ATRIB) is executed first and causes an entity arrival to the ENTER node to be scheduled at the top of the calendar. This is followed by the call to subroutine STOPA(JJ) which causes the end-of-activity event for the patient with entity code JJ to be scheduled at the top of the event calendar. Therefore upon the return to SLAM II, the end-of-activity event will be processed prior to the entity arrival to ENTER node 2.

The coding for subroutine TEST which processes the daily changes in each patient's test score is depicted in Figure 9-11. The processing is performed in the DO

```
      SUBROUTINE BUMP
      COMMON/SCOM1/ ATRIB(100),DD(100),DDL(100),DTNOW,II,MFA,MSTOP,NCLNR
     1,NCRDR,NPRNT,NNRUN,NNSET,NTAPE,SS(100),SSL(100),TNEXT,TNOW,XX(100)
      SCORE=47.
      JJ=0
C
C *** CHECK TO SEE IF ANY PATIENTS IN THE WARD HAVE SCORES
C *** GREATER THAN 47. IF SO DETERMINE WHICH PATIENT HAS
C *** THE HIGHEST SCORE.
C
      DO 10 J=1,25
      IF(XX(J).LE.SCORE) GO TO 10
      SCORE=XX(J)
      JJ=J
   10 CONTINUE
      IF(JJ.EQ.0) GO TO 20
C
C *** ROUTE THE NEW PATIENT THROUGH THE ENTER NODE AND INTO THE
C *** WARD.  CALL STOPA FOR THE BUMPED PATIENT.
C
      CALL ENTER(2,ATRIB)
      XX(JJ)=0.
      CALL STOPA(JJ)
      RETURN
C
C *** A PATIENT ELIGIBLE FOR BUMPING WAS NOT FOUND.  PLACE
C *** THE NEW PATIENT INTO THE NETWORK AND MAKE HIM BALK.
C
   20 ATRIB(2)=48.
      RETURN
      END
```

Figure 9-10 Subroutine BUMP for psychiatric ward example.

```
      SUBROUTINE TEST
      COMMON/SCOM1/ ATRIB(100),DD(100),DDL(100),DTNOW,II,MFA,MSTOP,NCLNR
     1,NCRDR,NPRNT,NNRUN,NNSET,NTAPE,SS(100),SSL(100),TNEXT,TNOW,XX(100)
C
C *** DAILY UPDATE PATIENT SCORES AND SEE IF THE PATIENT IS
C *** WELL ENOUGH (SCORE AT LEAST EQUAL TO 48) TO LEAVE.
C *** IF PATIENT IS WELL ENOUGH TO LEAVE CALL STOPA WITH
C *** THE PATIENT'S BED NUMBER
C
      DO 10 J=1,25
      IF(XX(J).EQ.0.) GO TO 10
      XX(J)=XX(J)+UNFRM(-.2,1.2,1)
      IF(XX(J).LT.48.) GO TO 10
      CALL STOPA(J)
      XX(J)=0.
   10 CONTINUE
      RETURN
      END
```

Figure 9-11 Subroutine TEST for psychiatric ward example.

loop which changes each patient's score by adding a uniform sample between –0.2 and 1.2 using stream 1 to their current score. Those patients with test scores exceeding 48.0 are discharged from the ward by calling STOPA(J) and setting XX(J)=0.0 where J is the space number occupied by a discharged patient.

The initial conditions for the simulation are established in subroutine INTLC which is shown in Figure 9-12. This subroutine inserts the initial 18 patients into the network through ENTER node 2.

```
      SUBROUTINE INTLC
      COMMON/SCOM1/ATRIB(100),DD(100),DDL(100),DTNOW,II,MFA,MSTOP,NCLNR
     1,NCRDR,NPRNT,NNRUN,NNSET,NTAPE,SS(100),SSL(100),TNEXT,TNOW,XX(100)
      ATRIB(1)=TNOW
C
C *** ASSIGN PATIENT SCORES AND PUT 18 PATIENTS IN THE SYSTEM
C
      DO 10 J=1,18
      ATRIB(2)=UNFRM(30.,44.,1)
   10 CALL ENTER(2,ATRIB)
      RETURN
      END
```

Figure 9-12 Subroutine INTLC for psychiatric ward example.

Summary of Results. The SLAM II Summary Report shown in Figure 9-13 displays statistics for the time in the ward for each patient, the time between balks from the system, and the utilization of the twenty-five spaces in the ward. The results show that on the average 24.8 of the 25 available spaces in the ward are utilized and that over half of the patients seeking admittance to the ward balked from the

```
                              S L A M   I I   S U M M A R Y   R E P O R T

                  SIMULATION PROJECT PSYCHIATRIC WARD           BY OREILLY

                  DATE 12/20/1983                               RUN NUMBER    1 OF    1


                  CURRENT TIME   0.5000E+03
                  STATISTICAL ARRAYS CLEARED AT TIME   0.0000E+00


                  **STATISTICS FOR VARIABLES BASED ON OBSERVATION**

                    MEAN         STANDARD     COEFF. OF    MINIMUM      MAXIMUM      NUMBER OF
                    VALUE        DEVIATION    VARIATION    VALUE        VALUE        OBSERVATIONS

TIME BET. BALKS     0.8923E+00   0.1286E+01   0.1441E+01   0.6104E-04   0.1190E+02   556
TIME IN WARD        0.2298E+02   0.8004E+01   0.3484E+00   0.5112E+01   0.4506E+02   526


                     **REGULAR ACTIVITY STATISTICS**

ACTIVITY            AVERAGE       STANDARD     MAXIMUM CURRENT   ENTITY
INDEX/LABEL         UTILIZATION   DEVIATION    UTIL    UTIL      COUNT

   1 IN WARD        24.8383       0.5928       25      25        526
```

Figure 9-13   SLAM II summary report for psychiatric ward example.

system as the result of the lack of space on the ward. These results clearly indicate the need for additional space for patients. Additional runs should be made with increased spaces available to determine the number of spaces that should be added. Cost information should also be added to the model to access the worth of proposed new additions. If space cannot be added then research on new patient-improvement procedures and on the criterion for discharge should be initiated.

9.16   CHAPTER SUMMARY

This chapter has provided many capabilities for inserting user-written logic directly into a network model. These user-written inserts increase the flexibility of the network-oriented world view. The EVENT node provides a general capability to include in the network a node to perform logic functions and attribute assignments limited only by the programming capabilities of the modeler. The ENTER node allows entities to be inserted in the network based on logic written in event routines.

A discussion of filing and removing entities waiting at QUEUE nodes or AWAIT nodes is given. A large complement of subprograms is described for extending the logical capabilities provided by SLAM II's routing, selection, and resource allocation methods. In addition, advance procedures for allocating resources to entities and for stopping activities based on external conditions is described. An example illustrating the jockeying of entities from one queue to another is presented. A second example illustrates the activity scanning features associated with the STOPA specification as well as user-written modeling features associated with EVENT, ENTER, and USERF subprograms.

9.17   EXERCISES

9-1.   In Example 6-2, add the feature that if there are more than three jobs to be processed by the machine tool when it breaks down, all jobs except the last three to arrive are routed to a subcontractor. The job in progress is also routed to the subcontractor.

9-2.   For Example 6-2, redevelop the model to include the possibility that the repairman process breaks down and a delay of three hours is incurred in order to get a spare part for the repair process. The time between repair process breakdowns is exponentially distributed with a mean of 100 hours. If the repair breakdown occurs when the repair process is not active, no action is taken.

9-3.   Model and analyze the admitting process of a hospital as described below. The following three types of patients are processed by the admitting function:

Type 1.   Those patients who are admitted and have previously completed their pre-admission forms and tests;

Type 2.   Those patients who seek admission but have not completed pre-admission; and

Type 3.   Those who are only coming in for pre-admission testing and information gathering.

Service times in the admitting office vary according to patient type as given below.

Patient Types and Service Times

Patient Type	Relative Frequency	Mean Time to Admit
1	0.90 before 10:00 A.M.	15 minutes
	0.50 after 10:00 A.M.	
2	0.10 always	40 minutes
3	0 before 10:00 A.M.	30 minutes
	0.40 after 10:00 A.M.	

Note: All of the above times are normally distributed
with $\sigma = 0.1\mu$ (min. $= 0.0$).

On arrival to admitting, a person waits in line if the two admitting officers are busy. When idle, an admitting officer selects a patient who is to be admitted before those who are only to be pre-admitted. In addition, Type 1 patients are given highest priority. After filling out various forms in the admitting office, Type 1 patients are taken to their floor by an orderly while Type 2 and 3 patients walk to the laboratory for blood and urine tests. Three orderlies are available to escort patients to the nursing units. Patients are *not* allowed to go to their floor by themselves as a matter of policy. If all orderlies are busy, patients wait in the lobby. Once patients have been escorted to a floor, they are considered beyond the admitting process. It takes the orderly 3 time units to return to the admitting room. Those patients who must go to the lab are always ambulatory, and as a result require no escorts. After arriving in the lab, they wait in line at the registration desk. After registration, they go to the lab waiting room until they are called on by one of two lab technicians. After the samples are drawn, they walk back to the admitting office if they are to be admitted or leave if only preadmission has been scheduled. Upon return to admitting, they are processed as normal Type 1 patients. The admitting office is open from 7:00 A.M. until 5:00 P.M. However, no pre-admissions (Type 3) are scheduled until 10:00 A.M. because of the heavy morning workload in the lab. At 4:00 P.M., incoming admissions are sent to the outpatient desk for processing. However, Type 2 patients returning from the lab are accepted until 5:00 P.M. which is the time both admitting officers go home and the office is closed. Analyze the above system for 10 days. It is of interest to determine the time in the system, that is, the time from arrival until on a floor (Type 1 and 2) or exit from the lab (Type 3). Also, determine the time between arrivals to the laboratory. Assume all patient queues are infinite and FIFO ranked except where noted. Activty times are specified below.

Activity Times (all times in minutes)

Explanation	Distribution: Parameters
time between arrivals to admitting office, t_1	exponential: mean = 15
travel time between admitting and floor, t_2	uniform: min = 3, max = 8
travel time between admitting and lab or lab and admitting, t_3	uniform: min = 2, max = 5
service time at lab registration desk, t_4	Erlang-3: mean = 4.5, k = 3
time spent drawing lab specimen, t_5	Erlang-2: mean = 5, k = 2
time for orderly to return from floor to admitting desk, t_6	constant: 3

9-4. There are three stations on an assembly line and the service time at each station is exponentially distributed with a mean of 10. Items flow down the assembly line from server 1 to server 2 to server 3. A new unit is provided to server 1 every 15 time units. If any server has not completed processing its current unit within 15 minutes, the unit is diverted to one of two off-line servers who complete the remaining operations on the job diverted from the assembly line. One time unit is added to the remaining time of the operation that was not completed. Any following operations not performed are done so by the off-line servers in an exponentially distributed time with a mean of 16. Obtain statistics on the utilization of all servers, and the fraction of items diverted from each operation.

Embellishments: (a) Assume the assembly line is paced and that the movement of units can only occur at multiples of 15 minutes.

(b) Allow one unit to be stored between each assembly line server.

(c) If a server is available, route units back to the assembly line from the off-line servers.

9-5. A conveyor system involves five servers stationed along a conveyor belt. Items to be processed by the servers arrive at the first server at a constant rate of four per minute. Service time for each server is 1 minute, and is exponentially distributed. No storage is provided before each server; therefore, the server must be idle if he is to remove the item from the conveyor belt. If the first server is idle, the item is processed by that server. At the end-of-service time, the item is removed from the system. If the first server is busy when the item arrives, it continues down the conveyor belt until it arrives at the second server. The delay time between servers is 1 minute. If an item encounters a situation in which all servers are busy, it is recycled to the first server with a time delay of 5 minutes. Simulate the above conveyor system for 100 time units to determine statistics concerning the time spent in the system by an item, the percentage of time each server is busy, and the number of items in the conveyor system.

Embellishments: (a) Repeat the simulation with a time delay of 2 minutes between servers. Is there an effect on the utilization of the servers because of a change in the time delay between servers, that is, the speed of the conveyor?

(b) Evaluate the situation in which the last server has sufficient space for storage so that all items passing servers 1, 2, 3, and 4 are processed by server 5. Simulate this situation.

(c) Assess the increased performance obtained by allowing a one item buffer before each server. Based on the results of this study, specify how you would allocate ten buffer spaces to the five servers.

(d) Discuss how you would evaluate the tradeoffs involved between reducing the number of servers in the conveyor system versus increasing the buffer size associated with each server.

9-6. Let the storage space before each server for the conveyor system described in Exercise 9-5 be two units. When storage exists before each server of a conveyor system, decisions regarding the removal of items must be established. Propose decision rules for determining whether items should be removed from the conveyor belt and placed in storage before a particular server. Simulate the decision rules to obtain the statistics requested in Exercise 9-5.

9-7. Simulate the following resource constrained PERT network (1) to evaluate a set of dispatching rules for determining the order in which to perform activities when the resources available are insufficient to perform all the activities that have been released. The following statistics are to be recorded based on 100 simulations of the network for each dispatching rule.

1. Average time at which each node is realized.
2. Minimum time at which each node is realized.
3. Maximum time at which each node is realized.
4. Standard deviation of the time each node is realized.
5. Histograms of the time to realize nodes.

6. Percentage of the simulation runs in which an activity was on a critical path (a criticality index).
7. Network completion time distribution.
8. Utilization of each resource type.

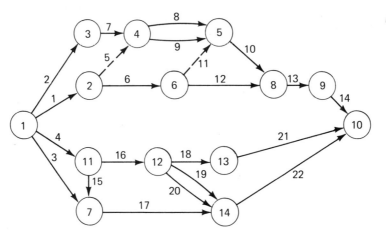

Parameters of Activities of the Network for Exercise 9-7

Activity Number	Distribution	Mean Time	Variance	Requirement for Resource Type			
				1	2	3	4
1	Lognormal	7	2	2	0	1	3
2	Constant	3	0	2	0	1	1
3	Constant	10	0	0	2	0	0
4	Constant	5	0	1	1	1	1
5	Dummy	0	0	0	0	0	0
6	Constant	5	0	2	0	3	2
7	Uniform	3	$1/3$	1	0	2	3
8	Gamma	8	6	3	0	0	1
9	Constant	3	0	1	0	0	0
10	Constant	5	0	2	0	1	3
11	Dummy	0	0	0	0	0	0
12	Constant	2	0	2	1	2	2
13	Constant	6	0	1	1	0	0
14	Constant	10	0	0	0	0	0
15	Exponential	10	0	1	1	1	0
16	Constant	3	0	1	1	2	0
17	Constant	5	0	0	1	1	1
18	Constant	2	0	0	1	0	3
19	Normal	5	4	1	0	0	2
20	Constant	1	0	1	1	0	0
21	Constant	15	0	0	1	2	0
22	Constant	5	0	0	1	0	0

Resource Availability and Types

Type	Description	Availability
1	Systems analyst	3
2	Marketing personnel	2
3	Maintenance personnel	3
4	Engineering personnel	3

9-8. A multiprocessor computing system is composed of two processors (CPUs) sharing a common memory (CM) of 131 pages, four disk drives, each of which can be accessed by either processor, and a single data channel. A schematic diagram of the system is shown below.

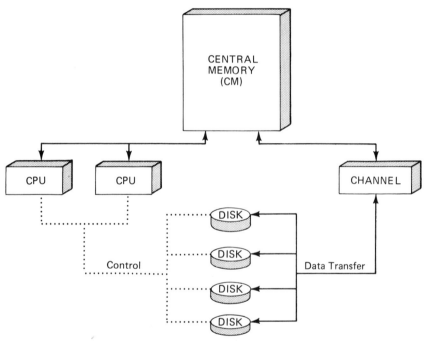

Schematic Diagram of a Multiprocessor Computing System

Jobs arrive to the system at an average rate of 12 jobs per minute in accordance with a Poisson distribution. Total CPU time for a job follows a normal distribution with a mean of 10 seconds and a standard deviation of 3 seconds. The CPU processing time consists of bursts, each of which is followed by an I/O requirement. Each burst follows a negative exponential distribution whose mean is equal to the reciprocal of the average I/O rate of a job. The average I/O rate per job varies uniformly from 2 to 10. I/O operations are assigned to a specific disk upon arrival.

Jobs arriving to the system are assigned a priority which is inversely related to their memory requirements. The CM requirements for a job are distributed uniformly between 20 and 60 pages. Once memory has been allocated to a job it begins execution on any available CPU. Upon issuing an I/O request, the job can continue using the CPU as long as only one I/O request is outstanding. Thus, if an I/O request is made and the job has an I/O request pending, the CPU is relinquished and the I/O request is queued. Following completion of a nonpending I/O request, CPU operations on the job can be reinitiated if a CPU is available.

After a CPU burst, an I/O request is automatically made on the disk assigned to the job, that is, direct access to the disks from the CPUs is made. The seek time to locate the proper position on any disk is assumed to be uniformly distributed between 0. and .075 seconds. Only one seek operation per disk can be performed at a time. Following the seek operation, a data transfer through the data channel is made. The transfer time is equal to .001*(2.5+U) where U is uniformly distributed between 0 and 25. After the transfer, the I/O request is considered satisfied.

Simulate the computing system to determine the residence time of jobs processed by the system, and the utilization statistics for the four disks, the input/output channel, and the two processors (CPUs). Also, the average use of memory is to be obtained. Statistics on the number of jobs waiting for a resource and the waiting time of jobs are also desired. The simulation is to be run for 12,000 seconds.

9-9. A machine tool processes two different types of parts. The time between arrivals of Type 1 parts is triangularly distributed with a mode of 30 minutes, a minimum of 20 minutes, and a maximum of 50 minutes. The interarrival time of Type 2 parts is a sample from a triangular distribution with a mode of 50 minutes, a minimum of 35 minutes, and a maximum of 60 minutes. Processing time for Type 1 parts is exponentially distributed with a mean of 20 minutes. For Type 2 parts processing time is a sample from a uniform distribution with a minimum of 15 minutes and a maximum of 20 minutes. Processing time includes an inspection of the completed part. Fifteen percent of the parts fail inspection and return to the end of the queue of parts awaiting processing. Assume that parts which fail inspection have a rework time equal to 90% of the previous processing time.

Include in the model a cart that transports parts from arrival to the processing area. Once the parts have passed inspection the same cart transports the finished product to the packaging area. No cart is required if a part requires reprocessing. The cart transports four parts at a time to the processing area. All transport times are 5 minutes. A packaged product includes one of each type of part. The packaging time is exponentially distributed for each of the two products with a mean of 20 minutes. There is only one cart and it is freed immediately after transporting the products. In addition, if more than three parts are waiting for the machine at a scheduled maintenance time, remove all but the first 3 parts for subcontracting. Make the probability of failing inspection 10 percent for Type 1 parts and 13% for Type 2 parts. Model and analyze this situation for 2400 minutes collecting utilization statistics and time spent in the system by a part.

Embellishment: Using the following data, replace the constant 5-minute travel time with the actual time that it takes for the cart to travel between stations. Assume that the cart remains at the station where it was last stopped until it is needed elsewhere. The cart travels at the rate of 100 feet per minute. The distances between the arrival and the machining areas is 45 feet and between the machining and packaging areas is 55 feet.

9.18 REFERENCES

1. Pritsker, A. A. B., *The GASP IV Simulation Language,* John Wiley, 1974.
2. Pritsker, A. A. B., *Modeling and Analysis Using Q-GERT Networks,* Halsted Press and Pritsker & Associates, (Second Edition), 1979.

CHAPTER 10

Network Modeling With Continuous Variables

10.1   INTRODUCTION

Models are often classified as either discrete change, continuous change or combined discrete-continuous change. In discrete change models, the state of the model changes discretely at isolated points in time called event times. In a network model, the event times are imbedded in the network description and are equivalent to times at which an entity arrives to a node.

Continuous change models are characterized by variables defined in terms of equations that describe their behavior over time. Continuous models are built in SLAM II by describing the equations for the state variables in a user-written subroutine called STATE and by having SLAM II call subroutine STATE whenever updated values of the state variables are needed.

In this chapter we introduce the concept of state variables in order to illustrate how network models can be developed in conjunction with variables whose values

342

change continuously with time. The defining equations for state variables are written in subroutine STATE. Discrete changes can be made to state variables at ASSIGN nodes in the network. A state variable can be monitored to detect when its value crosses a threshold. Such crossings are called state events, and when this occurs, an entity can be placed in the network to initiate any changes that are to take place. In this chapter, the concepts associated with state variables and state events are introduced.

10.2   STATE VARIABLES

A state variable has a value which changes, primarily, in accordance with a defining equation. The value of a state variable does not stay constant between event times but changes in accordance with the equation that describes its behavior over time. Examples of state variables are: the level of oil in a storage tank; the position of a projectile; the height of a cab on an overhead crane; and the probability that there are no items in the system at a particular time.

The SLAM II variable SS(I) is used to represent state variable I. The derivative of state variable I is defined by the SLAM II variable DD(I), that is,

DD(I) = dSS(I)/dt

SLAM II solves for the values of state variables over time by taking steps in simulated time. The state variable value at each immediately preceding step is maintained as the SLAM II variable SSL(I). With these SLAM II variables, it is possible to write a difference equation for state variable I in subroutine STATE in the following fashion:

SS(I) = SSL(I) + DTNOW*RATE(I)

where DTNOW is the step size and RATE(I) is a user variable that specifies the rate of change of state variable I during DTNOW. This equation states that the value of state variable I at time TNOW is equal to the last value of state variable I plus the amount of change that would occur over time DTNOW. The value of RATE(I) could be specified as a function of other system variables. The SLAM II executive calls subroutine STATE every time a step is to be made so that the updating of time is implicit in the equation given above for SS(I).

An alternative method for defining SS(I) is to define its derivative in subroutine STATE. In this case the derivative is integrated by SLAM II to obtain the value of SS(I) at each step. Hence, an equivalent formulation for the above difference equation would be the following statement in subroutine STATE:

DD(I) = RATE(I)

The user may define SS(I) by either an algebraic expression, a difference equation, or a DD(I) equation. However only one method for each variable is allowed.

More formally, the SLAM II variables SS(I), SSL(I), DD(I), and DDL(I) are used to define the state variables and their derivatives at times TNOW and TTLAS, where

    TNOW = time at which values of the state variables are being computed.
    TTLAS = time at the beginning of the current step (the time at which the values for the state variables were last accepted).
  DTNOW = TNOW–TTLAS.
     SS(I) = value of state variable I at time TNOW.
   SSL(I) = value of state variable I at time TTLAS.
   DD(I) = value of the derivative of state variable I at time TNOW.
  DDL(I) = value of the derivative of state variable I at time TTLAS.

As mentioned above, to define a state variable by a differential equation, the user would write a defining equation in terms of DD(·). For example, if the differential equation for state variable j is given by

$$dy_j/dt = Ay_j + B$$

the corresponding SLAM II statement is

$$DD(J) = A*SS(J) + B$$

where SS(J) represents y_j.

In this situation, the SLAM II processor calls subroutine STATE many times within a step in order to obtain estimates of the derivatives (DD(J)) within the step. These estimates are used to compute SS(J) at TNOW from the equation

$$SS(J) = SSL(J) + \int_{TTLAS}^{TNOW} DD(J)\ dt$$

where TTLAS is the time at the beginning of the step and corresponds to the time at which SSL(J) was computed. The integration involved in the above equation is performed by a Runge-Kutta-Fehlberg (RKF) numerical integration algorithm in which a user-prescribed single-step accuracy specification is maintained (1,5,6). If the accuracy is not maintained, the step size is reduced and the integration recalculated as described in Chapter 13. This process is repeated if necessary until a user-specified minimum step size, DTMIN, is encountered.

10.3   CODING SUBROUTINE STATE

Subroutine STATE is written to compute the current value of each state variable or its derivative. SLAM II permits state variables to be defined by state equations or derivative equations in subroutine STATE. Subroutine STATE is frequently called, especially if there are active derivative equations, and therefore should contain only essential code. It is most efficient if the state variables are numbered sequentially.

The state storage array consists of four one-dimensional arrays. The vectors SS(·) and DD(·) contain values associated with time TNOW. The vectors SSL(·) and DDL(·) contain values associated with time TTLAS, the most current value of simulated time for which model status has been completely updated. When the model status has been updated to TNOW, TTLAS is reset to TNOW.

The problem-specific definition of the state storage array is determined by the user. There are several policies regarding the writing of state equations that must be followed. The user inputs the variable NNEQD which is defined as the largest subscript used in a derivative equation. Therefore, equations defining the rate of change of state variables, that is, equations for DD(I), must satisfy the expression $I \leq NNEQD$. Equations defining SS(I) must satisfy the expression $NNEQD + 1 \leq I \leq NNEQT$ where NNEQT is the largest subscript I in the defining equations for SS(I). Thus, if there is an equation defining DD(M) and another defining SS(N), then $M \leq NNEQD$; $M < N$; and $NNEQD + 1 \leq N \leq NNEQT$. NNEQS = NNEQT − NNEQD is the largest possible number of equations written in terms of SS(I). Since it is most efficient to have the DD(·) numbered sequentially, NNEQD is often referred to as the number of defining equations written for DD(·) variables. Similarly, NNEQS is often referred to as the number of defining equations written for SS(·) variables. NNEQD, NNEQS, or both can be zero.

The above numbering policy can be summarized by the following four cases:

Case 1. NNEQD = 0; NNEQS = 0. No continuous variables are included in the simulation, that is, a discrete or network simulation is to be performed.

Case 2. NNEQD > 0; NNEQS = 0. All state variables are defined by differential equations written for DD(I). SLAM II uses an integration algorithm to compute SS(I). Multiple calls to subroutine STATE are made to evaluate DD(I) for use in the numerical integration algorithm.

Case 3. NNEQD = 0; NNEQS > 0. All state variables are defined by algebraic or difference equations written for SS(I). The user must compute SS(I) in subroutine STATE. In this case, the variable DD(I) is not used by SLAM II.

Case 4. NNEQD > 0; NNEQS > 0. The first NNEQD state variables are defined by differential equations, and the next NNEQS variables are defined by algebraic or difference equations. The evaluation procedure is a combination of cases (2) and (3) above.

In subroutine STATE, the order in which the equations are written is left to the user, that is, a statement defining SS(5) can precede one defining DD(3). Because SLAM II does not change the execution sequence of state equations, correct sequencing of state and derivative equations is the responsibility of the user. If the defining equations for DD(·) do not involve other DD(·) variables, any order is permitted and the integration procedure simultaneously solves for all DD(·) and corresponding SS(·) variables. Thus DD(1) = A * SS(1) + B * SS(2) and DD(2) = C * SS(1) + F * SS(2) can be written in either order, and the values obtained for DD(1), DD(2), SS(1), and SS(2) will be the same.

If the defining equations for DD(·) do involve other DD(·) variables, ordering becomes important. For example, if DD(3) = f(DD(5)), the equation for DD(5) must precede the equation for DD(3) according to standard FORTRAN conventions. If there are simultaneous equations involving DD(·) variables, the user must develop an algorithm for solving the set of equations. This is also the case when a set of simultaneous algebraic state equations is written.

In subroutine STATE, the equations for DD(·) and SS(·) can be written in a variety of forms. The equations†

DD(M) = RATE

and

SS(M) = SSL(M) + DTNOW*RATE

---

† Equations that define both DD(M) and SS(M) are permitted only if M > NNEQD. In this case, DD(M) is *not* related to SS(M) by SLAM II and is processed as if it were an SS(·) variable.

are essentially equivalent (SLAM II sets DTNOW = TNOW–TTLAS). When an equation for DD(M) is written, values of SS(M) are obtained through the integration routine contained within SLAM II. Values of DDL(M) and SSL(M) are automatically maintained. The step size is automatically determined to meet specified accuracy requirements on the computation of SS(M) and tolerances on state-event occurrences. When the equation is written for SS(M), only SSL(M) is maintained by SLAM II. Accuracy requirements are not specified on SS(M), and the step size DTNOW for updating SS(M) is maintained at the maximum value specified by the user unless there is an intervening event. DTNOW is automatically reduced if there are intervening time-events, state-events, or a need to record a value of SS(M) for eventual communication (output).

The form of the equations for DD(·) and SS(·) is limited only by FORTRAN statement types. Thus, the user has a great deal of flexibility in defining the state variables for the model. In fact, the user can make the description conditional on time or on any model variable. State- or time-events could be used to trigger the change by resetting the rate values or by setting an indicator at the time of the event.

10.4 CONTINUOUS INPUT STATEMENT

When state variables are included in a SLAM II model, it is necessary to communicate to the SLAM II processor: the number and type of state variables; limits on the step size to be taken when updating values of state variables; the recording interval desired for state variables; and whether messages associated with the accuracy of state variable evaluation are desired. The information about these quantities is given on the CONTINUOUS statement, and a description of the fields for this statement is now provided.

The largest subscript for a differential equation is specified by NNEQD. For state variables defined by difference or algebraic equations, the largest subscript is NNEQD+NNEQS. The user may specify the largest step size to be taken, DTMAX, and the smallest step size desired, DTMIN. These values control the step size set by SLAM II in meeting tolerance and accuracy requirements. No step will be taken smaller than DTMIN in order to achieve accuracy requirements or to detect a state event. The recording frequency specifies that values of the state variables are desired every DTSAV time units. Information on the error limits for numerical integration is specified by a relative error allowed (a fraction of the state variable value) and an absolute error allowed.

The SLAM II input statement to specify values for models with state variables is

CONTINUOUS,NNEQD,NNEQS,DTMIN,DTMAX,DTSAV,W or F or N,AAERR,RRERR;

The definitions of the variables on the CONTINUOUS statement are presented below along with their default values.

Variable	Definition	Default
NNEQD	Largest subscript for DD($\cdot$) when defined by a derivative equation. (For efficiency, NNEQD should be the number of derivative equations.)	0
NNEQS	Number of state variables that can be defined by state equations.	0
DTMIN	Minimum allowable step size; not used if NNEQD + NNEQS = 0.	0.01*DTMAX
DTMAX	Maximum allowable step size; not used if NNEQD + NNEQS = 0.	DTSAV
DTSAV	The frequency at which data values are recorded for RECORD statements. If DTSAV < 0, recording done at event times only. If DTSAV = 0, recording done at the end of each step. If DTSAV > 0, recording done every DTSAV time units and at each event time. When user plots are specified on a RECORD statement, the calls to GPLOT control the frequency at which data values are recorded.	High Value†
W or F or N	Indicates type of error check in Runge-Kutta integration or in state-event crossing detection when a step size smaller that DTMIN is required. If F is specified, then a *fatal* error occurs. If W is specified, then a *warning*	W

---

† High value is prescribed by the SLAM II variable HIVAL. The default for DTSAV if no CONTINUOUS statement is used is the largest DTPLT value on any RECORD statement.

message is printed before proceeding. If N is specified, then execution proceeds with *no* warning message given.

AAERR	Absolute local truncation error allowed in Runge-Kutta integration. Used with RRERR to control accuracy; not used when NNEQD=0.	0.00001
RRERR	Relative local truncation error allowed in Runge-Kutta integration; used with AAERR to control accuracy.	0.00001

The numerical integration accuracy is controlled by the specification of AAERR and RRERR. The RKF algorithm estimates the single-step error for each variable defined by a differential equation. The Ith error estimate is compared to TERR where

$$TERR(I) = AAERR + ABS(SS(I))*RRERR$$

If the error estimate is less than or equal to TERR(I) for each I, the values of SS(I) are accepted. If not, the step size is reduced and the integration algorithm is reapplied. The default values of AAERR and RRERR are stringent and a significant reduction in running times can be achieved by liberalizing these values subject, of course, to the accuracy requirements of the simulation model.

10.5 STATE EVENTS

A state event is defined as a point in time at which a variable crosses a threshold. When a state event occurs, changes to the model and the variables in the model can be made. In this chapter, when a state event occurs, we will only consider the introduction of an entity into a network to initiate the desired status changes due to the occurrence of the state event. Note that state events can be caused by either state variables or network variables crossing a threshold.

A state event is defined in terms of a crossing variable, a direction of crossing and a value that describes the threshold. In addition, a tolerance for the crossing is prescribed. Illustrations of state events are shown in Figure 10-1. In Figure

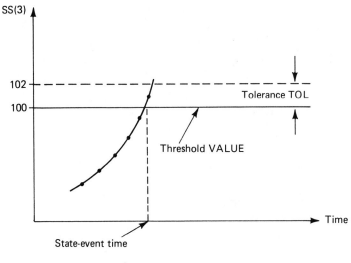

(a) Positive Crossing

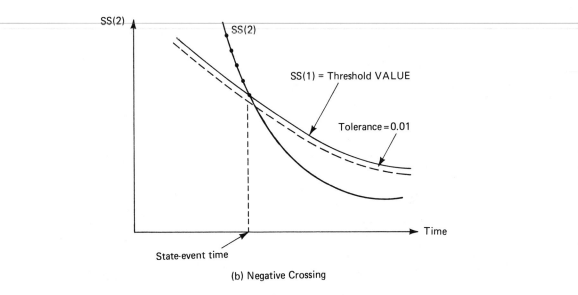

(b) Negative Crossing

Figure 10-1 Illustrations of positive and negative crossings defining state events.

10-1(a), a positive crossing of state variable SS(3) of the threshold value 100 is shown. The crossing occurs when the value of SS(3) is below 100 at the end of one step (at time TTLAS) and equal to or above the value 100 at the end of the step (at time TNOW). In the figure, a tolerance TOL of 2 is shown. If the value of SS(3) is greater than 102 at the end of the step, then a crossing out of tolerance is indicated. In such a case, SLAM II would try to reduce the step size, DTNOW, and recalculate the value of the state variable for this shortened step. When this is done, a new value for TNOW is determined since the value given to DTNOW by SLAM II is smaller. Step size reductions will only be made as long as DTNOW is greater than DTMIN.

In Figure 10-1(b), an illustration of a negative crossing defining a state event is given. In this figure, the state event is defined as SS(2) crossing SS(1) in the negative direction with a tolerance of 0.01. The crossing is defined when the value of SS(2) is above SS(1) at the beginning of the step and equal to or below SS(1) at the end of the step. The concept of being within or out of tolerance for negative crossings is similar to that described for positive crossings.

State events can be defined for the following crossing variables: SS, DD, a SLAM II network variable, and USERF(I).

10.6 DETECT NODE

A DETECT node is used in a network to describe the conditions for a state event. When the conditions for the state event are realized, the DETECT node is released and an entity with all attribute values equal to 0 is routed from the DETECT node in accordance with the M number of the DETECT node. The symbol for the DETECT node is given below.

The DETECT node provides the primary interface between the continuous and network portions of a model. When a DETECT node is included in the network model, it is released whenever the crossing variable, XVAR, crosses a threshold VALUE in the direction specified by XDIR. The value of TOL specifies an interval beyond the threshold value for which a detection of a crossing is considered within tolerance. If a crossing occurs beyond the allowable tolerance, the SLAM II processor reduces the step size until the crossing is within tolerance or until the step size is reduced to the user prescribed minimum step size, DTMIN.

The DETECT node is also released whenever an entity arrives to it. A maximum of M emanating activities are initiated at each release. The statement format for the DETECT node is:

DETECT, XVAR, XDIR, VALUE, TOL, M;

where
XVAR specifies the crossing variable and can be any SLAM II status variable
XDIR is the crossing direction and can be specified as X, XP, or XN where
  X → either direction; XP → positive direction and XN → negative direction
VALUE is a SLAM II status variable or a numeric value
TOL is a numeric value which specifies the tolerance within which the crossing
  is to be detected.

One common use of the DETECT node is to specify or key the duration of an activity to the release of the DETECT node. In this way, the time an entity spends in an activity can be keyed to the time when a continuous state variable achieves a specified condition. This is illustrated by the following queueing model where the ACTIVITY completes service for the entity whenever the value of SS(1) crosses in the negative direction the value of SS(2) with a prescribed tolerance of 0.01. In this situation, the ACTIVITY could be the unloading of a tanker where SS(1) is the amount in the tanker and SS(2) is the amount to be left in the tanker after unloading.

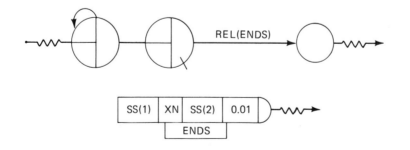

The above network segment operates in the following fashion. When SS(1) crosses SS(2) in a negative direction, node ENDS is released and an entity is created. Recall that SS(1) and SS(2) are varying in accordance with equations written in subroutine STATE. The entity is immediately terminated since a TERM symbol follows the DETECT node. The release of the DETECT node ENDS causes the release(REL) of any entities in the activity following the QUEUE node whose duration is specified by REL(ENDS). Thus any entities being processed by this service activity are completed when the state event as represented by the DETECT node ENDS occurs. As discussed in Chapter 5, the duration specification REL(ENDS) only applies if an entity is currently engaged in the ACTIVITY. If DETECT node ENDS detects a crossing and no entity is being processed by the ACTIVITY, or any other activity keyed to ENDS, then the crossing has no effect.

As another illustration of the DETECT node, consider the following network segments.

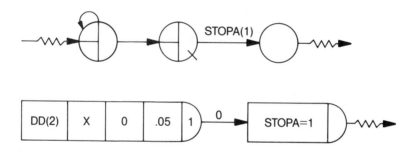

These network segments are similar to the ones shown above with the exception that the service duration is now specified as STOPA(1) to indicate that entities are to leave the service activity when an assignment of 1 is made to STOPA.

A DETECT node is used to monitor the value of DD(2), the derivative of state variable 2. When the derivative crosses 0 in either direction with a tolerance of .05, the DETECT node is released. Recall that a derivative crossing 0 indicates a maximum, minimum or point of inflection for the state variable. Thus, when one of these conditions occurs, an entity is put into the network following the DETECT node and, in the above situation, is routed to the ASSIGN node where STOPA is set equal to 1. When this assignment is made, service on the entities in the service activity would be completed. At this time, entities waiting in the QUEUE node would start service. These activities would not be completed until the next assignment of the value 1 to STOPA.

Through the use of STOPA(I), selective entities can be moved out of an activity by prescribing that an entity's duration be keyed to the value of I. For example the activity specification STOPA(ATRIB(1)) would cause only entities with ATRIB(1)=1 to complete the activity when STOPA is assigned a value of 1 at the ASSIGN node.

10.7   EXAMPLE 10-1. TANKER-REFINERY OPERATIONS (4)

A fleet of 15 tankers carries crude oil from Valdez, Alaska to an unloading dock near Seattle, Washington. It is assumed that all tankers can be loaded simultaneously in Valdez, if necessary. In Seattle, there is only one unloading dock, which supplies a storage tank that feeds a refinery through a pipeline. The storage tank receives crude from a tanker at the dock at a constant rate of 300 tb/day†. The storage tank supplies crude to the refinery continuously at a constant rate of 150 tb/day. The unloading dock is open from the hours of 6 a.m. to 12 p.m. Safety considerations require the stopping of unloading of the crude when the dock is shut down. The completion of the tanker unloading occurs when the amount of crude remaining in the tanker is less than 7.5 tb.

The storage tank has a capacity of 2000 tb. When it is full, unloading is halted until the amount in the tank decreases to 80 percent of capacity. When the storage tank is nearly empty (less than 5 tb), supply to the refinery is halted until 50 tb is reached to avoid the possibility of frequent refinery start-ups and shut-downs. The characteristics associated with the tankers are listed below.

1. Nominal carrying capacity is 150 tb.
2. Travel time loaded is normally distributed with a mean of 5.0 days and standard deviation of 1.5 days.
3. Travel time unloaded is normally distributed with a mean of 4.0 days and a standard deviation of 1 day.
4. Time to load is uniformly distributed in the interval 2.9 to 3.1 days.

The initial conditions for the simulation are that the storage tank is half full and the tankers are to arrive at their loading points at ½ day intervals, starting with the first at time 0.

The objective of this example is to simulate the above systems for 365 days to obtain estimates of the following quantities:

1. Unloading dock utilization.
2. Loading dock utilization.
3. Time refinery has a crude input available.
4. Amount of crude in the storage tank.
5. Tanker round trip time.
6. Tanker waiting time.
7. Number of tankers waiting for unloading.

---

† tb = thousand barrels

Concepts Illustrated. This example illustrates the use of the DETECT node for modeling state-events in combined network-continuous simulation models. The use of a single state variable to represent the amount of crude oil to be unloaded simplifies the system state description. Since all tankers are scheduled to arrive to Valdez, abnormal initial conditions exist. MONTR statements are used to CLEAR statistics on days 65 and 165, and to obtain a summary report over the period from day 65 to day 165.

SLAM II Model.   The tanker problem is simulated using a combined network-continuous model. The continuous variables are used to represent the level of crude oil in the tanker being unloaded and in the storage tank. The network is used to model the movement of tankers through the system and the interactions between the continuous and discrete elements of the system.

Two state variables are used in this simulation: SS(1), the amount of crude in a tanker at the unloading dock; and SS(2), the amount of crude in the storage tank. The state variable SS(1) represents the amount of crude available to be unloaded. SS(1) will be zero when no tanker is in the unloading dock; otherwise it will be equal to the amount of crude in a tanker that is in the unloading dock.

When a tanker leaves the unloading dock, SS(1) either becomes zero or is set equal to the amount of crude in the next waiting tanker to be unloaded. By defining SS(1) in this manner, a separate state variable for the amount of crude in each tanker need not be defined.

There are three XX variables which are used in the simulation to control the flow of crude between the unloading tanker and the refinery. Each of these variables represents a valve which is open when equal to 1.0 and closed when equal to 0. XX(1) is used to represent the dock input valve and is open between the dock operating hours of 6 a.m. to 12 p.m., and is closed otherwise. XX(2) is used to model the storage tank input valve and is closed whenever the storage tank crude level, SS(2), reaches the tank capacity of 2000 tb. It is reopened when the level of crude decreases to 1600 tb. XX(3) is used to represent the storage tank output valve and is closed whenever the storage tank crude level has decreased to less than 5 tb, thereby halting the flow to the refinery. XX(3) is reset to open when the crude level in the storage tank has increased to 50 tb, thereby restoring the flow of crude to the refinery. A schematic diagram depicting the arrangement of the three valves is provided in Figure 10-2. Note that XX(1) can be opened and closed by scheduling a time-event whereas XX(2) and XX(3) require the concept of a state-event.

The equations describing the state variables SS(1) and SS(2) are coded in subroutine STATE shown in Figure 10-3. The variable RATIN represents the flow rate

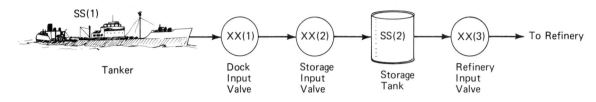

Figure 10-2 Crude oil flow from tanker to refinery.

of crude into the storage tank. It is set to zero if XX(1), XX(2), or SS(1) is zero, and is set equal to 300 otherwise. The variable RATOUT, representing the flow rate of crude from the storage tank to the refinery, equals 150 if XX(3) = 1. and equals 0 if XX(3) = 0. Equations for state variables SS(1) and SS(2) are written as difference equations in terms of RATIN and RATOUT. In this case, we are integrating the state equations explicitly in subroutine STATE. An alternative would be to code subroutine STATE in terms of the derivatives of the state variables as follows:

DD(1)= –RATIN
DD(2)= RATIN –RATOUT

In this case, the equations would be integrated by SLAM II using the Runge-Kutta-Fehlberg integration algorithm to determine SS(1) and SS(2).

```
      SUBROUTINE STATE
      COMMON/SCOM1/ ATRIB(100),DD(100),DDL(100),DTNOW,II,MFA,MSTOP,NCLNR
     1,NCRDR,NPRNT,NNRUN,NNSET,NTAPE,SS(100),SSL(100),TNEXT,TNOW,XX(100)
C****RATIN = O IF DOCK OR STORAGE INPUT CLOSED OR NO WAITING TANKER - ELSE 300
      RATIN=300.
      IF(XX(1)*XX(2)*SS(1).EQ.0.0) RATIN=0.
C****RATOUT=O IF REFINERY INPUT OFF - ELSE 150
      RATOUT=150.*XX(3)
      SS(1)=SSL(1)-DTNOW*RATIN
      SS(2)=SSL(2)+DTNOW*(RATIN-RATOUT)
      RETURN
      END
```

Figure 10-3 Subroutine state for tanker example.

The state equations for this example appear relatively simple but are deceptive since RATIN and RATOUT have different values during the simulation because of the status of the system. These equations could be made more complex if RATIN or RATOUT were functions of the type of crude or the level of crude in the storage tank and tanker. Since these aspects of the system do not add to the organizational aspects of the model, they are not included.

The network model for this example can be viewed as three subprocesses consisting of the tanker flow through the system, the start-up and shut-down of dock operations, and the state-events. Each of these subprocesses is modeled as a separate disjoint network.

The network for the tanker flow subprocess is depicted in Figure 10-4. The initial 15 tankers are created by the CREATE node at 0.50 day intervals, beginning with the first at time 0. The tankers proceed to the ASSIGN node labeled VLDZ where their arrival time to Valdez is marked as ATRIB(1). The tankers then undertake the loading activity which is represented by ACTIVITY 1. The trip from Valdez to Seattle is modeled by ACTIVITY 2. Upon completion of ACTIVITY 2, the tankers then wait in file 1 at the AWAIT node for the resource DOCK. A single unit of resource DOCK is available as specified by the resource block. When the DOCK becomes available, the state variable SS(1) is set to 150 at the ASSIGN node indicating that there is 150 tb of crude available for unloading. The tanker then undergoes ACTIVITY 3 which represents the unloading activity. This ACTIVITY is completed at the next release of the node labeled ENDU. The node labeled ENDU is a DETECT node which is released when SS(1) crosses, in the negative direction, the threshold value of 7.5 which indicates that the state-event "end-of-unloading" has occurred.

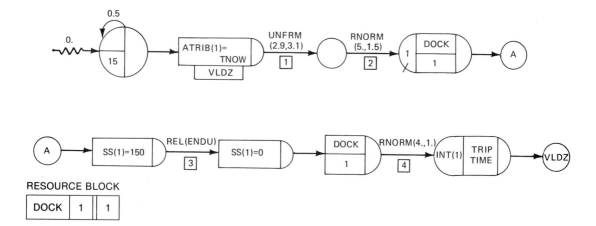

Figure 10-4 Tanker flow subprocess for tanker example.

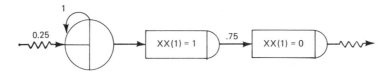

Figure 10-5 Shift start-up/shut-down subprocess for tanker example.

At the completion of unloading, the tanker entity is routed to the ASSIGN node where SS(1) is set to 0, and then releases the DOCK at the FREE node. The return trip to Valdez is modeled by ACTIVITY number 4. At the COLCT node, statistics are collected on the round trip time for the tanker which is then routed to the ASSIGN node labeled VLDZ to repeat the cycle through the network.

The network for the shift start-up and shut-down subprocess is depicted in Figure 10-5. The CREATE node inserts an entity into the network beginning at time 0.25 days (6 a.m.), and then daily thereafter. At the ASSIGN node, the dock input valve is opened by setting XX(1) equal to 1. The dock status remains open during the 0.75 days required for the entity to transverse the ACTIVITY before being closed at the ASSIGN node where XX(1) is reset to 0. The dock status remains closed until the next entity is inserted into the network at 6 a.m. the next day.

There are five possible conditions that could result in a state-event and these are listed below.

Condition	State-Event
The level of crude in the unloading tanker, SS(1), has decreased to 7.5.	Tanker unloading is completed
The level of crude in the storage tank, SS(2), has decreased to 5 tb.	Stop supply to refinery by setting XX(3) = 0
The level of crude in the storage tank has increased to 50 tb.	Start supplying refinery by setting XX(3) = 1
The level of crude in the storage tank has reached its capacity of 2000 tb.	Close input to the storage tank by setting XX(2) = 0
The level of crude in the storage tank has decreased to 1600 tb.	Open input to the storage tank by setting XX(2) = 1

These five state-events are modeled by the five subnetworks depicted in Figure 10-6. The first subnetwork is used to detect the end of unloading state-event and causes the completion of ACTIVITY 3 whose duration is keyed to the release of the node labeled ENDU. The other four subnetworks are used to detect and proc-

ess state-events which cause the opening and closing of the storage tank and refinery input valves. The tolerance for each state-event is set at 5. The value prescribed

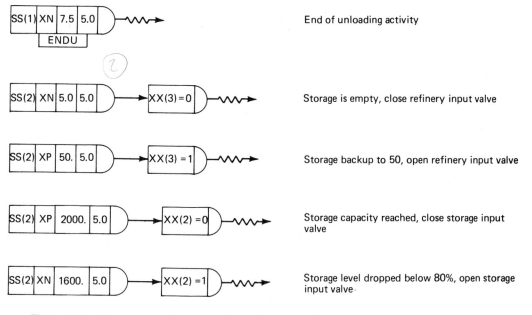

SS(1)	XN	7.5	5.0	End of unloading activity
	ENDU			

SS(2) XN 5.0 5.0 → XX(3)=0   Storage is empty, close refinery input valve

SS(2) XP 50. 5.0 → XX(3)=1   Storage backup to 50, open refinery input valve

SS(2) XP 2000. 5.0 → XX(2)=0   Storage capacity reached, close storage input valve

SS(2) XN 1600. 5.0 → XX(2)=1   Storage level dropped below 80%, open storage input valve

Figure 10-6   State event subprocesses for tanker example.

for a tolerance is set according to the accuracy with which a state-event should be
detected. The value of the tolerance should also consider the value given to
DTMIN and the maximum rate of change of the state variable. In this example,
DTMIN = 0.0025 days and the maximum rate is 300 tb/day, hence tolerances of
0.75 tb or greater should enable detection of state-events within tolerance.

The input statements for this example are depicted in Figure 10-7. In addition to
the network statements, the necessary control statements are included to obtain: a
plot of the crude level in an unloading tanker and the level in the storage tank;
and time-persistent statistics on the refinery input availability and the average
crude level in the storage tank. The INITIALIZE statement specifies that the
model is to be simulated for 365 days. MONTR statements with the CLEAR option are used to clear statistics at time 65 and 165.

Summary of results. The SLAM II Summary Report for this example is given in
Figure 10-8. As can be seen from the output statistics, the refinery is operated 100
percent of the time from day 165 to day 365. This high percentage of refinery utilization occurs at the expense of the tankers which wait on the average 1.52 days

```
 1  GEN,PRITSKER,TANKER FLEET,5/7/81,1;
 2  LIMITS,1,2,100;
 3  TIMST,XX(3),REFN INPUT AVAIL;
 4  CONT,0,2,.0025,.25,.25;
 5  RECORD,TNOW,DAYS,0,P,.25;
 6  VAR,SS(1),T,TANKER  LEVEL,0,300;
 7  VAR,SS(2),S,STORAGE LEVEL,0,2000;
 8  TIMST,SS(2),STORAGE LEVEL;
 9  INTLC,SS(2)=1000,XX(2)=1,XX(3)=1,XX(1)=0;
10  NETWORK;
11  ;
12  ;TANKER FLOW SUBPROCESS
13  ;---------------------
14        RESOURCE/DOCK(1),1;
15        CREATE,.5,0,,15;               CREATE INITIAL ARRIVALS
16  VLDZ  ASSIGN,ATRIB(1)=TNOW;          MARK ARRIVAL TIME TO VALDEZ
17        ACT/1,UNFRM(2.9,3.1);          LOADING
18        GOON;                          END OF LOADING
19        ACT/2,RNORM(5.,1.5);           TO SEATTLE
20        AWAIT(1),DOCK/1;               AWAIT THE DOCK
21        ASSIGN,SS(1)=150;              RESET TANKER CRUDE LEVEL
22        ACT/3,REL(ENDU);               UNLOADING
23        ASSIGN,SS(1)=0;                SET TANKER CRUDE LEVEL TO 0
24        FREE,DOCK;                     FREE THE DOCK
25        ACT/4,RNORM(4,1);              RETURN TRIP TO VALDEZ
26        COLCT,INT(1),TRIP TIME;        COLLECT STATISTICS
27        ACT,,,VLDZ;                    BRANCH TO VLDZ
28  ;
29  ;SHIFT START UP/SHUT DOWN SUBPROCESS
30  ;----------------------------------
31        CREATE,1.,,25;
32        ASSIGN,XX(1)=1;                BEGIN SHIFT AT 6 A.M.
33        ACT,.75;                       CONTINUE FOR 3/4 DAYS
34        ASSIGN,XX(1)=0;                CLOSE SHIFT AT 6 P.M.
35        TERM;
36  ;
37  ;STATE EVENT SUBPROCESSES
38  ;-----------------------
39  ENDU  DETECT,SS(1),XN,7.5,5;         END OF UNLOADING ACTIVITY
40        TERM;
41        DETECT,SS(2),XN,5,5;           STORAGE IS EMPTY
42        ASSIGN,XX(3)=0;                CLOSE REFINERY INPUT SWITCH
43        TERM;
44        DETECT,SS(2),XP,50,5;          STORAGE BACK UP TO 50
45        ASSIGN,XX(3)=1;                OPEN REFINERY INPUT SWITCH
46        TERM;
47        DETECT,SS(2),XP,2000,5;        STORAGE CAPACITY REACHED
48        ASSIGN,XX(2)=0;                CLOSE STORAGE INPUT SWITCH
49        TERM;
50        DETECT,SS(2),XN,1600,5;        STORAGE DROPPED BELOW 80
51        ASSIGN,XX(2)=1;                OPEN STORAGE INPUT SWITCH
52        TERM;
53        ENDNETWORK;
54  ;
55  INITIALIZE,0,365;
56  MONTR,CLEAR,65;
57  MONTR,SUMRY,165;
58  MONTR,CLEAR,165;
59  FIN;
```

Figure 10-7 SLAM II input statements for tanker fleet model.

for the unloading dock. This is further illustrated by the file statistics which indicate that the average number waiting for the unloading dock is 1.58, and that as many as 5 tankers were waiting at one time. The resource statistics indicate that there was a tanker in the unloading dock 95 percent of the time. This statistic can also be obtained from the statistics for ACTIVITY 3. Also available from the activity statistics is the average number of tankers being unloaded as this quantity is the average utilization of activity 1. From the output, it is seen that approximately 3.16 tankers are being loaded and the maximum number of tankers loaded concurrently is 7.

```
                 S L A M   I I   S U M M A R Y   R E P O R T

        SIMULATION PROJECT TANKER FLEET              BY PRITSKER

        DATE  5/ 7/1981                              RUN NUMBER   1 OF   1


        CURRENT TIME   0.3650E+03
        STATISTICAL ARRAYS CLEARED AT TIME  0.1650E+03


            **STATISTICS FOR VARIABLES BASED ON OBSERVATION**

                MEAN        STANDARD     COEFF. OF    MINIMUM     MAXIMUM     NUMBER OF
                VALUE       DEVIATION    VARIATION    VALUE       VALUE       OBSERVATIONS

   TRIP TIME    0.1436E+02  0.2045E+01   0.1425E+00   0.9547E+01  0.1999E+02  211


            **STATISTICS FOR TIME-PERSISTENT VARIABLES**

                MEAN        STANDARD     MINIMUM      MAXIMUM     TIME        CURRENT
                VALUE       DEVIATION    VALUE        VALUE       INTERVAL    VALUE

REFN INPUT AVAIL 0.1000E+01 0.0000E+00   0.1000E+01   0.1000E+01  0.2000E+03  0.1000E+01
STORAGE LEVEL    0.1807E+04 0.1085E+03   0.1592E+04   0.2004E+04  0.2000E+03  0.1661E+04


                      **FILE STATISTICS**

   FILE   ASSOC NODE   AVERAGE    STANDARD     MAXIMUM  CURRENT   AVERAGE
   NUMBER LABEL/TYPE   LENGTH     DEVIATION    LENGTH   LENGTH    WAITING TIME

     1     AWAIT        1.5759     1.2173        5        2        1.5080
     2     CALENDAR    14.2213     1.3584       18       13        1.2197


                  **REGULAR ACTIVITY STATISTICS**

   ACTIVITY       AVERAGE      STANDARD    MAXIMUM CURRENT  ENTITY
   INDEX/LABEL    UTILIZATION  DEVIATION   UTIL    UTIL     COUNT

     1 LOADING      3.1566      1.5206        7       5       208
     2 TO SEATTLE   5.1803      1.4911       10       4       208
     3 UNLOADING    0.9529      0.2119        1       1       207
     4 RETURN TRIP  4.1344      1.5377        8       3       211
```

Figure 10-8 SLAM II summary report for tanker fleet model.

RESOURCE STATISTICS

RESOURCE NUMBER	RESOURCE LABEL	CURRENT CAPACITY	AVERAGE UTILIZATION	STANDARD DEVIATION	MAXIMUM UTILIZATION	CURRENT UTILIZATION
1	DOCK	1	0.9529	0.2119	1	1

RESOURCE NUMBER	RESOURCE LABEL	CURRENT AVAILABLE	AVERAGE AVAILABLE	MINIMUM AVAILABLE	MAXIMUM AVAILABLE
1	DOCK	0	0.0471	0	1

STATE AND DERIVATIVE VARIABLES

(I)	SS(I)	DD(I)
1	0.1378E+02	0.0000E+00
2	0.1661E+04	0.0000E+00

Figure 10-8   (continued).

A plot of the state variables during the initial and middle portions of the simulation is depicted in Figure 10-9. At the start of the simulation, all the tankers were scheduled to arrive at the loading dock at 0.50 day intervals. Thus, there were no tankers to be unloaded. This is represented on the plot by the letter T which is at the zero point. Note that the scales for each variable are different and that T is plotted from 0 to 300 whereas S, the level in the storage tank, is plotted from 1 to 2000. Cursors are provided at the 0, 25, 50, 75, and 100 percent levels.

Since there are no tankers in the unloading dock initially, the amount in the storage tank is depleted by the amount being sent to the refinery. From the plot, this depletion continues until the first tanker arrives and begins unloading its crude. Since the input rate for the storage tank is greater than its output rate, the amount of crude in the storage tank increases momentarily as the tanker is being unloaded. Other tankers arrive and provide sufficient crude to replenish the storage tank after the initial depletion period.

As can be seen, the plot illustrates the combined discrete-continuous nature of the simulation. The second part of the plot illustrates the steady-state behavior of the system. The amount of crude in the storage tank oscillates between 1600 and 2000, and tankers are in the unloading dock waiting for storage space to become available to unload their crude.

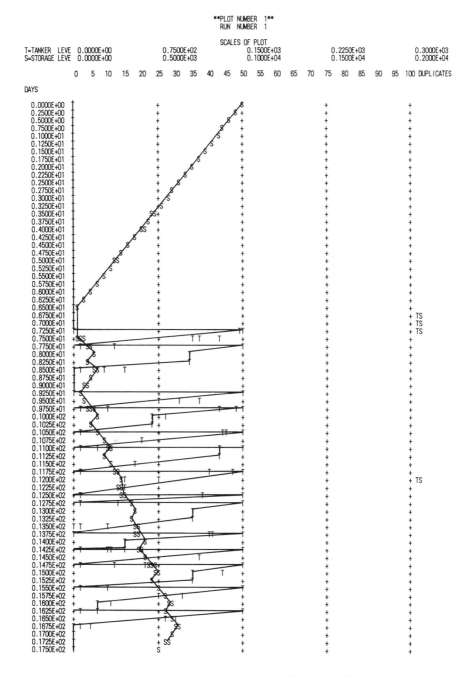

Figure 10-9 Plot of tanker level and storage tank level in tanker fleet model.

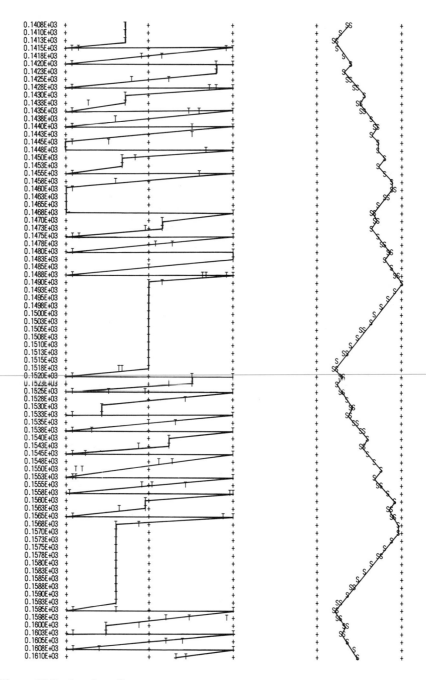

Figure 10-9 (continued).

10.8   EXAMPLE 10-2. A CHEMICAL REACTION PROCESS(2,3,4)

A hydrogenation reaction is conducted in four reactors operating in parallel. Each reactor may be started, stopped, discharged, or cleaned independently of the others. A compressor with a constant molal flow rate provides a supply of hydrogen gas to the reactors through a surge pressure tank and individual valves for each reactor. The valve connecting each reactor to the surge tank is adjusted by controls that make the effective pressure in each reactor the minimum of surge tank pressure and critical pressure (100 psia). Figure 10-10 is a schematic representation of the compressor, surge tank, valves, and reactors.

Initially the surge tank pressure is equal to 500 psia. Each reactor is charged with a fresh batch of reactant and the four reactors are scheduled to be turned on at half hour intervals beginning with reactor 1 at time 0. As a reaction proceeds, the concentration of the reactant (and the demand for hydrogen for that reactor) decreases. As long as surge tank pressure remains above the critical pressure of 100 psia, the decrease in concentration is exponential. The concentration of each reactant is monitored until it decreases to 10% of its initial value at which time the reactor is considered to have completed a batch. At this time the reactor is turned off, discharged, cleaned, and recharged. The time to discharge a reactor is exponentially distributed with a mean time of one hour. The time to clean and recharge a reactor is normally distributed with a mean of one hour and a standard deviation of one-half hour.

The operating policy for the system prescribes that the last reactor started will be immediately turned off whenever surge tank pressure falls below the critical

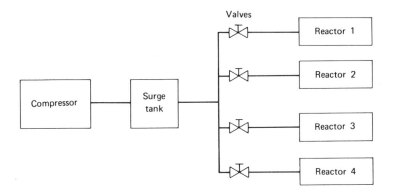

Figure 10-10   Schematic diagram for reactor example.

value of 100 psia. All other reactors that are on at that time will continue, but no reactor will be started if surge tank pressure is below a nominal pressure of 150 psia.

The objective for this example is to model and analyze a chemical reaction process that involves a continuous batch process subject to controls imposed by surge tank pressure. In addition to a plot of the operation of the process, statistics on the maintenance time, the number of reactors on, and the surge tank pressure are to be obtained.

Concepts Illustrated. A network model is used to portray the status of a reactor when it is not processing a batch. While processing a batch, the concentration level in the reactor is described by a set of first order differential equations. SLAM II's numerical integration algorithm is used to obtain the values of the concentration levels. The status of a reactor is maintained on the network in the SLAM II variable ARRAY. In the equations describing the rate of change of concentration, function GETARY is employed to access the status of a reactor. DETECT nodes are used to monitor the continuous variables which can initiate state changes.

SLAM Model. The SLAM II model of the reactor system consists of three parts: 1) equations describing the change of concentration level in each of the reactors; 2) reactor entities moving through activities that represent different states for the reactor; and 3) DETECT nodes for monitoring and detecting when reactor concentration and surge tank pressure cross thresholds.

The definitions of the parameters and variables of the model are given in Table 10-1. The basic equation for the rate of change in concentration of reactor I are

$$\frac{dSS(I)}{dt} = -RK(I)*PEFF*SS(I), \qquad I = 1,4$$

where RK(I) is the reaction constant for reaction I and PEFF is the effective surge tank pressure. The SLAM II coding of this differential equation is directly obtained by substituting DD(I) for dSS(I)/dt. SS(5) is the state variable used to represent surge tank pressure. Using the definitions provided by Table 10-1, the model's state equations describing the chemical reactor system are:

$$\text{PEFF} = \text{Minimum of (PCRIT,SS(5))}$$

$$\text{DD(I)} = \begin{cases} -\text{RK(1)*SS(I)*PEFF} & \text{if reactor I is on} \\ 0 & \text{if reactor I is off} \end{cases}$$

$$\text{F(I)} = -\text{DD(I)*V(I)}$$

$$\text{SUMF} = \text{F(1)+F(2)+F(3)+F(4)}$$

$$\text{DD(5)} = \text{(RR*TEMP/VS)*((FC*SUMFO)-SUMF)}$$

$$= \text{RTV*(FCOMP-SUMF)}$$

Table 10-1 Parameters and variables for chemical reaction model.

Model Parameters	Initial Value	Definition
RK(1)	0.03466	Reaction constant for product 1
RK(2)	0.00866	Reaction constant for product 2
RK(3)	0.01155	Reaction constant for product 3
RK(4)	0.00770	Reaction constant for product 4
V(1)	10.0	Volume of reactor 1
V(2)	15.0	Volume of reactor 2
V(3)	20.0	Volume of reactor 3
V(4)	25.0	Volume of reactor 4
VS	50.0	Volume of surge tank
RR	10.73	Gas constant
TEMP	550.0	System operating temperature
FC	0.19	Flow constant of compressor; FC ★ SUMFO is compressor molal flow rate
PNOM	150.0	Nominal pressure
PCRIT	100.0	Critical pressure
PEFF	—	Minimum of critical pressure and surge tank pressure
SUMFO	Calculated	Maximum possible molal flow of hydrogen to all reactors
RTV	RR ★ TEMP/VS	Composite reaction constant
FCOMP	FC ★ SUMFO	Compressor molal flow rate
Model Variables		
SS(1)	XX(1)=0.1	Concentration of reactor 1
SS(2)	XX(2)=0.4	Concentration of reactor 2
SS(3)	XX(3)=0.2	Concentration of reactor 3
SS(4)	XX(4)=0.5	Concentration of reactor 4
SS(5)	XX(5)=500.0	Surge tank pressure
DD(I),I=1,4	0.0	Derivative of SS(I) with respect to time
F(I),I=1,4	0.0	Molal flow rate of hydrogen to reactor I
SUMF	0.0	Total molal flow of hydrogen to all reactors
XX(5)	0.0	Number of reactors on
ARRAY(1,I), I=1,4	0.0	A flag indicating whether reactor I is on (1.0) or off (0.0)

The coding of these equations is performed in subroutine STATE which is shown in Figure 10-11. As can be seen, there is a direct translation of the equations defined for the reactors to the FORTRAN code shown in subroutine STATE. In the statement for DD(I), the variable STATUS is employed. When the reactor is idle, STATUS is 0 and when the reactor is working STATUS is 1. In this way, the rate of change of concentration level is made 0 when the reactor is idle. In the network model, the SLAM global array, ARRAY(1,I), is used to maintain the status of reactor I. In subroutine STATE, the value for the status of reactor I is obtained by using the SLAM function GETARY(1,I).

```
      SUBROUTINE STATE
      DIMENSION RK(4),F(4),V(4)
      COMMON/SCOM1/ ATRIB(100),DD(100),DDL(100),DTNOW,II,MFA,MSTOP,NCLNR
     1,NCRDR,NPRNT,NNRUN,NNSET,NTAPE,SS(100),SSL(100),TNEXT,TNOW,XX(100)
      EQUIVALENCE (PCRIT,XX(8)),  (FCOMP,XX(9)),  (RTV,XX(10))
      EQUIVALENCE (RK(1),XX(21)),  (V(1),XX(25))
      SUMF = 0
      PEFF = AMIN1(SS(5),PCRIT)
C
C *** DEFINE THE DIFFERENTIAL EQUATIONS FOR CONCENTRATE IN REACTORS 1 - 4
C
      DO 10 I=1,4
        STATUS = GETARY(1,I)
        DD(I)=-RK(I)*SS(I)*STATUS*PEFF
        F(I) =-DD(I)*V(I)
C**** SET TOTAL MOLAL FLOW FOR ALL REACTORS
        SUMF = SUMF+F(I)
10      CONTINUE
C
C *** SET SURGE TANK PRESSURE RATE
C
      DD(5)=RTV*(FCOMP-SUMF)
      RETURN
      END
```

Figure 10-11 Subroutine STATE for chemical reactor process.

The network model depicting reactor status changes is shown in Figure 10-12. In the legend block for this example, a gate PRESSURE is defined. This gate is used to prohibit reactors from being started until the surge pressure is above the nominal value of 150 psia. Reactor entities waiting for the gate PRESSURE to open are placed in file 1. The remainder of the legend block consists of equivalences between SLAM II variables and names employed on the network. Attribute 1 of the reactor entity is given the name REACTOR. DISCHARGE is a sample from an exponential distribution with a mean of 1. C_AND_R represents the time to clean and recharge the reactor and is a sample from a normal distribution with mean 1 and standard deviation 0.5. PNOM, the nominal pressure, is equvalenced to XX(7) and PCRIT, the critical pressure, is equivalenced to XX(8).

At a CREATE node, four reactor entities are generated at half-day intervals which are routed to an ASSIGN node where a reactor number is assigned from the count of the number of entities passing over activity 10. This assigns reactor numbers of 1, 2, 3 and 4 to the four entities created. The reactor entity is then sent to GOON node START which routes the entity to the AWAIT node LOWP if the surge pressure is less than the nominal pressure. If this is not the case, the status of the reactor is changed to busy and the number of reactors working, XX(5), is increased by one. The reactor number is recorded in XX(6) to identify the last reactor started. When reactor I is busy, its concentration level is modeled by a continuous variable SS(I) which is defined in accordance with the equations in subroutine STATE. Thus, the reactor entity is no longer required in the network so it is terminated following the ASSIGN node at which the appropriate status changes are made.

The reactor entity will be recreated when its concentration level decreases below its prescribed batch-ending threshold. For each reactor, the value of its concentration level crossing this threshold value is modeled at a DETECT node as shown in Figure 10-12. When the crossing is detected, a reactor entity is created and routed to an ASSIGN node where the reactor number is assigned. At ASSIGN node ENDR, the status of the reactor is changed to idle and the number of reactors busy is decreased by 1. The initial concentration of the reactor is reset by the statement SS(II) = XX(II) where II is the reactor number. The variable II is required at this point since the argument for an SS or XX variable must be an integer. To record the time the reactor starts its down period, ATRIB(2) is set equal to TNOW. Activities 2 and 3 model the discharge and clean and repair activities. Statistics on the discharge and clean and repair time are then collected at the COLCT node using the INT(2) specification. The reactor entity is then routed to the GOON node START and the process is reapplied.

A DETECT node is used to monitor the surge pressure. When the surge pressure crosses the critical pressure, PCRIT, in the negative direction, an entity is routed to an ASSIGN node where the last reactor started, XX(6), is made idle. The entity created at the DETECT node is then used to represent the reactor made idle by routing it to the AWAIT node LOWP. The last DETECT node shown in Figure 10-12 is used to determine when the surge pressure crosses the nominal pressure in a positive direction. This state event opens the gate PRESSURE which allows all reactor entities waiting in file 1 to be routed to the GOON node START. The gate PRESSURE is then closed so that new arrivals to the AWAIT node LOWP are forced to wait in file 1.

This completes the description of the network model for the chemical reactor process. In Figure 10-13, a complete statement model is provided. The

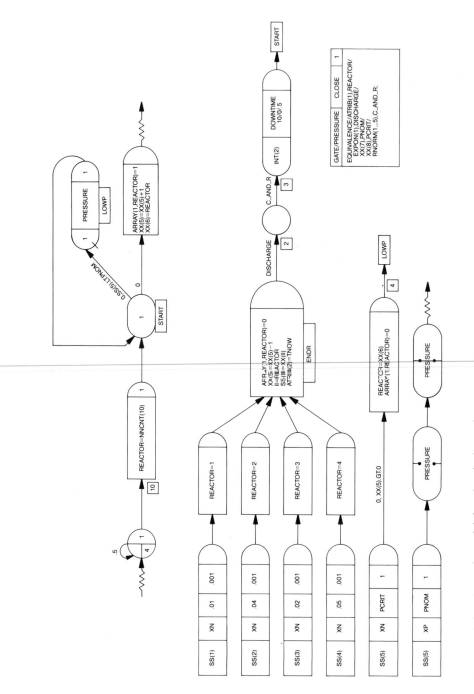

Figure 10-12 Network model for chemical reaction process.

```
 1  GEN,PRITSKER,CHEM REACTOR,1/9/86;
 2  LIMITS,1,2,50;
 3  TIMST,XX(5),NO. BUSY REACT;
 4  TIMST,SS(5),SURGE PRESSURE;
 5  CONTINUOUS,5,0,.001,.1,.1;
 6  ARRAY(1,4)/0,0,0,0;                             REACTOR STATUS
 7  EQUIVALENCE/ATRIB(1),REACTOR/EXPON(1),DISCHARGE;
 8  EQUIVALENCE/XX(7),PNOM/XX(8),PCRIT/RNORM(1,.5),C_AND_R;
 9  ;
10  ;   XX(I) = INITIAL CONCENTRATION FOR REACTOR I, I=1,4
11  ;   XX(5) = NUMBER OF BUSY REACTORS
12  ;   XX(6) = LAST REACTOR STARTED, XX(7) = NOMINAL PRESSURE
13  ;   XX(8) = CRITICAL PRESSURE, XX(9) = FCOMP, XX(10)=RTV
14  ;   XX(21:24) = REACTION CONSTANTS 1-4
15  ;   XX(25:28) = VOLUME OF REACTORS 1-4
16  ;
17  INTLC,SS(1)=0.1,      SS(2)=0.4,      SS(3)=0.2,      SS(4)=0.5, SS(5)=500;
18  INTLC,XX(1)=0.1,      XX(2)=0.4,      XX(3)=0.2,      XX(4)=0.5;
19  INTLC,XX(7)=150,      XX(8)=100,      XX(9)=4.352,    XX(10)=118.03;
20  INTLC,XX(21)=0.03466, XX(22)=0.00866, XX(23)=0.01155, XX(24)=0.00770;
21  INTLC,XX(25)=10,      XX(26)=15,      XX(27)=20,      XX(28)=25;
22  ;
23  RECORD,TNOW,TIME,0,P,.1,0,20;
24      VAR,SS(1),1,CONC 1,0,.5;
25      VAR,SS(2),2,CONC 2,0,.5;
26      VAR,SS(3),3,CONC 3,0,.5;
27      VAR,SS(4),4,CONC 4,0,.5;
28      VAR,SS(5),P,PRESSURE,0,1000;
29      VAR,PNOM ,!,PNOM,0,1000;
30      VAR,PCRIT,!,PCRIT,0,1000;
31  INIT,0,300;
32  ;
33  NETWORK;
34          GATE/PRESSURE,CLOSED,1;
35          CREATE,.5,,,4;                          CREATE INITIAL BATCHES.
36            ACT/10;
37          ASSIGN,REACTOR=NNCNT(10);               ASSIGN REACTOR NUMBER.
38  START GOON,1;                                   BEGIN A BATCH.
39            ACT,,SS(5).LT.PNOM,LOWP;              IF PRESSURE.LT.PNOM, GO TO LOWP
40            ACT;                                  ELSE
41          ASSIGN,ARRAY(1,REACTOR)=1,
42                XX(5)=XX(5)+1,
43                XX(6)=REACTOR;                    TURN ON REACTOR.
44          TERM;
45  LOWP  AWAIT(1),PRESSURE;                        WAIT HERE IF PRESSURE IS TOO LOW
46          ACT,,,START;                            TO START A BATCH.
47  ;
48          DETECT,SS(1),XN,.01,.001;               STOP REACTOR 1: BATCH COMPLETE
49          ASSIGN,REACTOR=1;
50            ACT,,,ENDR;
51          DETECT,SS(2),XN,.04,.001;               STOP REACTOR 2: BATCH COMPLETE
52          ASSIGN,REACTOR=2;
53            ACT,,,ENDR;
```

Figure 10-13 Statement model for chemical reaction process.

```
54          DETECT,SS(3),XN,.02,.001;           STOP REACTOR 3: BATCH COMPLETE
55          ASSIGN,REACTOR=3;
56            ACT,,,ENDR;
57          DETECT,SS(4),XN,.05,.001;           STOP REACTOR 4: BATCH COMPLETE
58          ASSIGN,REACTOR=4;
59   ENDR   ASSIGN,ARRAY(1,REACTOR)=0,
60                 XX(5)=XX(5)-1,
61                 II=REACTOR, SS(II)=XX(II),ATRIB(2)=TNOW;
62            ACT/2,DISCHARGE;                  DISCHARGE
63          GOON;
64            ACT/3,C_AND_R;                    CLEAN&RECHRG
65          COLCT,INT(2),DOWNTIME,10/0/.5;
66            ACT,,,START;
67   ;
68          DETECT,SS(5),XN,PCRIT,1;            STOP LAST REACTOR STARTED IF
69            ACT,,XX(5).GT.0;                  SURGE TANK PRESSURE FALLS
70   ;                                          BELOW CRITICAL PRESSURE.
71          ASSIGN,REACTOR=XX(6),
72                 ARRAY(1,REACTOR)=0,XX(5)=XX(5)-1;
73          ACT/4,,,LOWP;                       INTERRUPT
74   ;
75          DETECT,SS(5),XP,PNOM,1;             ALLOW REACTORS TO PROCEED WHEN
76          OPEN,PRESSURE;                      SURGE TANK PRESSURE RETURNS TO
77          CLOSE,PRESSURE;                     NOMINAL PRESSURE.
78          TERM;
79          END;
80   FIN;
```

Figure 10-13 (continued).

CONTINUOUS statement shows that there are five differential equations and no state equations. The minimum step size is 0.001 days and the maximum step size is 0.1 days. The ARRAY statement establishes the status of each reactor as idle at the beginning of the simulation. The EQUIVALENCE statement allows for the use of names in the network model. The INTLC statement initializes the variables associated with the model. The concentration of the four reactors and the system pressure are plotted through the use of RECORD and VAR statements. The nominal and critical pressure levels will also be plotted. The INIT statement indicates that a 300 day simulation run is desired. The network statements given in Figure 10-13 describe the flow of entities as presented in Figure 10-12.

Summary of Results. The SLAM II Summary Report for this example is given in Figure 10-14. The statistics on downtime show that 281 maintenance operations were completed in the 300 hour simulation. For the 281 observations, the average downtime is 1.9 hours which is not statistically different from the expected value of 2 hours. The statistics for time persistent variables indicate that the average number of busy reactors is 1.8 and the average surge pressure is 246.6 psia.

SLAM II SUMMARY REPORT

SIMULATION PROJECT CHEM REACTOR   BY PRITSKER

DATE 1/ 9/1986   RUN NUMBER   1 OF   1

CURRENT TIME   0.3000E+03
STATISTICAL ARRAYS CLEARED AT TIME 0.0000E+00

STATISTICS FOR VARIABLES BASED ON OBSERVATION

	MEAN VALUE	STANDARD DEVIATION	COEFF. OF VARIATION	MINIMUM VALUE	MAXIMUM VALUE	NUMBER OF OBSERVATIONS
DOWNTIME	0.1909E+01	0.8934E+00	0.4680E+00	0.1199E+00	0.5639E+01	281

STATISTICS FOR TIME-PERSISTENT VARIABLES

	MEAN VALUE	STANDARD DEVIATION	MINIMUM VALUE	MAXIMUM VALUE	TIME INTERVAL	CURRENT VALUE
NO. BUSY REACT	0.1835E+01	0.7816E+00	0.0000E+00	0.4000E+01	0.3000E+03	0.2000E+01
SURGE PRESSURE	0.2466E+03	0.1962E+03	0.4945E+02	0.1268E+04	0.3000E+03	0.1313E+03

FILE STATISTICS

FILE NUMBER	ASSOC NODE LABEL/TYPE	AVERAGE LENGTH	STANDARD DEVIATION	MAXIMUM LENGTH	CURRENT LENGTH	AVERAGE WAITING TIME
1	LOWP AWAIT	0.3544	0.5694	3	0	0.4074
2	CALENDAR	1.8056	0.9144	5	2	0.1530

REGULAR ACTIVITY STATISTICS

ACTIVITY INDEX/LABEL	AVERAGE UTILIZATION	STANDARD DEVIATION	MAXIMUM UTIL	CURRENT UTIL	ENTITY COUNT
2 DISCHARGE	0.9148	0.8003	3	2	281
3 CLEAN&RECHRG	0.8859	0.8008	4	0	281
4 INTERRUPT	0.0000	0.0000	1	0	186
10	0.0000	0.0000	1	0	4

GATE STATISTICS

GATE NUMBER	GATE LABEL	CURRENT STATUS	PCT. OF TIME OPEN
1	PRESSURE	CLOSED	0.0000

STATE AND DERIVATIVE VARIABLES

(I)	SS(I)	DD(I)
1	0.1000E+00	0.0000E+00
2	0.4000E+00	0.0000E+00
3	0.1567E+00	-0.1809E+00
4	0.1255E+00	-0.9660E-01
5	0.1313E+03	-0.1985E+03

Figure 10-14 SLAM II Summary Report for chemical reaction model.

From the file statistics, it is seen that 0.354 reactors are waiting for surge pressure to build up and that the average waiting time for a reactor is approximately 0.4 hours. At the end of the simulation, there were no reactors waiting but during the simulation as many as 3 reactors are waiting for the gate PRESSURE to open.

The activity statistics indicate a high utilization for the discharge and clean and repair activities. The average number of reactors in the discharge operation is 0.91 and the average in cleaning and repair is 0.89. Since there is no resource restriction for discharge and clean and repair operations, the reactors could proceed in these activities in parallel. In fact, 3 reactors were discharged simultaneously during the simulation run and 4 reactors were cleaned and repaired simultaneously. If a restriction is placed on the number of concurrent cleaning and repair activities, then a decrease in the number of batches processed, 283, could be expected. The number of batches processed is the sum of the number of discharge operations plus the number of reactors currently being discharged and cleaned. Activity 4 in the network represents the detection of a surge pressure decreasing below the critical value which causes an interruption in the processing of a reactor. In this simulation, 186 entities traversed activity 4. This represents a large number of interruptions of the chemical reaction process and indicates that further design may be needed to eliminate some of the starting and stopping of reactors.

By dividing the average number of reactors in each state by the number of reactors in the system, we can obtain the percentage of time that reactors are in each possible state. This is shown below.

Possible State	Percent of Reactors in the State
Operating	46
Waiting	09
Discharging	23
Cleaning and Recharging	22

With this type of information, a detailed assessment of the operation of the four reactors coupled through the surge tank can be made.

As expected, statistics on the gate PRESSURE show the percent of time opened as zero because the gate is closed immediately after it is opened. Figure 10-14 also provides the state and derivative variable values at the end of the simulation. In the model, the concentration level for a reactor is set to its initial concentration level while the reactor is being discharged and cleaned. This is done for plot display purposes. Thus, at the end of the simulation, reactors 1 and 2 have concentration levels at their initial amount. Reactors 3 and 4 are processing batches as indicated by the negative DD(·) values. Surge pressure is 131.3 psia and decreasing at a rate of 198.5 psia/hr. Thus, we can expect the surge pressure to become critical in approximately 0.15 hours.

Figure 10-15 is a plot of the concentration levels for each reactor and the surge tank. The heading lists the user-specified plot symbol and associated identifier as well as the scale for each variable to be plotted. Thus, the plot symbol "P" represents the system pressure on a scale ranging from 0 to 1000 psia. From the plot, it is seen that only reactor 1 is on initially and that pressure rose rapidly until reactor 2 is turned on at time 0.5. Reactor 1 is turned off because of a batch completion at about 0.7 hours. For display purposes, the concentration levels are plotted at their initial values while the reactors are off. Beginning with the start of reactor 3 at time 1.0, pressure fell rapidly. There is an obvious change in rate in the pressure curve at time 1.5 when reactor 4 is started. Pressure first went critical at about 2.0 hours causing reactor 4 to be stopped since it was the last one started. At time 2.3, pressure increases above nominal and reactor 4 is restarted which quickly drives the pressure down again to the critical level at time 2.4. From Figure 10-15, it is seen how the controls and operating policies affect the batch time and batch processing. In addition, it shows that the length of the downtime can have a significant impact on the utilization of the reactors.

To graphically see the variation of downtime, Figure 10-16 shows the histogram for downtime. This histogram shows a range of downtimes from a half hour to a downtime greater than five hours. The histogram illustrates that the downtime has a bell shaped curve as to be expected as downtime is the sum of two random variables, one of which is normally distributed and the other exponentially distributed. Since the mean value of both distributions is one, the expected downtime is 2.0 hours with a variance equal to the sum of the variances of the individual components.

```
                          **HISTOGRAM NUMBER  1**

                               DOWNTIME


     OBSV    RELA    CUML      UPPER
     FREQ    FREQ    FREQ    CELL LIMIT    0        20        40        60        80        100
                                          +     +     +     +     +     +     +     +     +     +
        0   0.000   0.000   0.0000E+00    +                                                   +
        9   0.032   0.032   0.5000E+00    +**                                                 +
       29   0.103   0.135   0.1000E+01    +***** C                                            +
       58   0.206   0.342   0.1500E+01    +*********         C                                +
       73   0.260   0.601   0.2000E+01    +**************              C                       +
       49   0.174   0.776   0.2500E+01    +**********                            C            +
       30   0.107   0.883   0.3000E+01    +*****                                       C      +
       20   0.071   0.954   0.3500E+01    +****                                            C  +
        7   0.025   0.979   0.4000E+01    +*                                               C+
        3   0.011   0.989   0.4500E+01    +*                                               C+
        0   0.000   0.989   0.5000E+01    +                                                C+
        3   0.011   1.000      INF        +*                                                C
      ---                                 +     +     +     +     +     +     +     +     +     +
      281                                 0        20        40        60        80        100
```

Figure 10-16 Histogram of downtime of reactors.

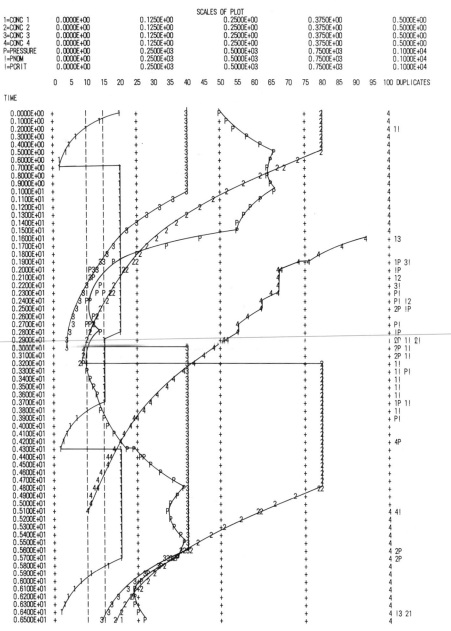

Figure 10-15 Plot of state variables for chemical reaction process.

This example further illustrates the concepts in a combined network-continuous simulation. The summary reports and plots demonstrate the different types of information that can be obtained from a simulation of a combined model.

10.9   CHAPTER SUMMARY

In this chapter the concept of a state variable is introduced and the procedures for writing differential and difference equations in subroutine STATE are described. The interface between a network model and state events is through the DETECT node. When a DETECT node establishes that a state event has occurred, the DETECT node is released and an entity is put into the network. A tanker refinery model is presented in which the flow of the tanker through operations is modeled in network form and the amount of crude oil in a tanker and storage tank is characterized by state variables. A second example involving the modeling of chemical reactors is presented. Pressure is modeled as a continuous variable and discharge, maintenance and recharge operations are modeled using network concepts. The framework provided by SLAM II to develop these combined network-continuous models involves only a few new constructs.

10.10   EXERCISES

10-1.   For the tanker refinery problem, assume that the rate of input to the storage tank decreases exponentially from the nominal value by the factor $1-e^{-x/5}$, where $x=SS(1)/150$. Determine the effects on the system due to this change in flow rate.

10-2.   Develop equations that make the flow rate from tanker to storage tank a function of the level of crude in the tanker, the level of crude in the storage tank, and the viscosity of the crude in the tanker. Develop the code for subroutine STATE to simulate the developed flow equations.

10-3.   It has been proposed that offshore unloading docks be built for unloading tankers at Seattle. Three such docks have been proposed, each of which can process at a rate two-thirds that of the current dock. Compare system operation between the three offshore unloading docks operating on only 1 shift versus the current unloading dock operating on a three-shift basis.

10-4.   For the problem stated in Example 10-1, an elaboration of the refinery is to be made. Consider that the refinery consists of four phases, the first of which is processing. Crude is taken from the initial storage tank and processed in a processing unit at a rate of 150 tb/day. After processing, the material is stored in an intermediate storage tank (phase 2), after which it goes through a filtering unit (phase 3). The rate of filtering is dependent on the conditions of the filter. A filter has an expected life of 30 days. Initially, when the filter is new, the filter rate is 200 tb/day. The rate of flow through the filter decreases linearly until it is 100 tb/day after 30 days. The time to replace a filter is exponentially distributed with a mean of 0.5 day. The fourth phase of the refinery is a finished product storage tank. The finished product is removed from the tank according to the demand for the product, which is cyclic over an approximately 90-day period with a mean of 150 tb/day. All other characteristics of the system are identical to those presented in Example 10-1, except that the unloading dock works on a 24-hour schedule. Simulate the above system for 200 days, using different filter replacement policies to obtain the output requested in Example 10-1 and the following quantities: processor utilization, filter utilization, amount in intermediate storage tank, amount in final storage tank, and percentage of time that demand is lost. Initial conditions for the simulation are that the raw material storage tank is 50% full, the intermediate tank is 60% full, and the final storage tank is 40% full. The capacity of each storage tank is 2000 tb.

10-5. In the tanker problem discuss the changes necessary to model the following embellishments:

(a) Introduce a new super tanker into the fleet arriving empty at Valdez, Alaska on day 70. The new tanker has a capacity of 450 tb. The travel times for the super tanker are distributed according to the triangular distribution with a mode equal to the mean of the regular tankers. All other characteristics for the super tanker remain the same. Assume minimum and maximum travel times of 0.9 and 1.2 of the modal value, respectively.

(b) Retire the first three tankers that complete a round trip following the introduction of the new super tanker.

(c) Cause the super tanker to have priority over other tankers waiting for unloading at Seattle.

(d) For the super tanker, unloading rate is a function of the level of crude in the super tanker, that is, $dx/dt = -x$ where x is the amount of crude in the super tanker in thousands of barrels (tb).

10-6. For the chemical reaction process described in Example 10-2, assume that the conditions under which reactors may be started when the surge tank pressure exceeds 150 psi are changed to allow only one reactor to start for each increase of 50 psi above the critical pressure. In addition, suppose that it is required to have a maintenance man available to clean the reactor. Rewrite the program under the condition that only one maintenance man is available and reactors requiring cleaning and recharging must wait if the maintenance man is busy. Determine the effect on throughput of the new control and requirement to have a maintenance man perform the cleaning and recharging.

10-7. For Example 10-2, solve the differential equations in terms of the initial conditions and rewrite the simulation program to use the derived solution.

10-8. A machine tool processes two different types of parts. The time between arrivals of Type 1 parts is triangularly distributed with a mode of 30 minutes, a minimum of

20 minutes, and a maximum of 50 minutes. The interarrival time of Type 2 parts is a sample from a triangular distribution with a mode of 50 minutes, a minimum of 35 minutes, and a maximum of 60 minutes. Processing time for Type 1 parts is exponentially distributed with a mean of 20 minutes. For Type 2 parts processing time is a sample from a uniform distribution with a minimum of 15 minutes and a maximum of 20 minutes. Processing time includes an inspection of the completed part. Fifteen percent of the parts fail inspection and return to the end of the queue of parts awaiting processing. Assume that parts which fail inspection have a rework time equal to 90% of the previous processing time. The machine tool tends to breakdown after 8 to 8½ hours of use. That is, the rate of wear on the tool is dependent on the status of the machining operation (zero if idle, one if operating). When a breakdown occurs, processing stops and the part currently being processed returns to the end of the queue to be reworked. Repair time is normally distributed with a mean of 30 minutes and a standard deviation of 8 minutes. If the number of parts awaiting processing reaches four units, the processing time is decreased by 10%. (The machine operator works faster in an attempt to catch up.) In this event, the failure rate increases by 10%. When the number in the queue decreases to one unit, processing time and failure rate return to normal levels. Develop a SLAM II model to represent this situation, collecting statistics on the time spent in the system by a part. Simulate the system for 2400 minutes.

10.11   REFERENCES

1. Fehlberg, E., "Low-Order Classical Runge-Kutta Formulas with Step-Size Control and Their Application to Some Heat Transfer Problems," NASA Report TR R-315, Huntsville, Alabama, April 15, 1969.
2. Hurst, N. R., GASP IV: A Combined Continuous/Discrete FORTRAN based Simulation Language, Unpublished Ph.D. Thesis, Purdue University, Indiana, 1973.
3. Hurst, N. R. and A. A. B. Pritsker, "Simulation of a Chemical Reaction Process Using GASP IV," *SIMULATION,* Vol. 21, 1973, pp. 71-75.
4. Pritsker, A. A. B., *The GASP IV Simulation Language,* John Wiley, 1974.
5. Shampine, L. F., and R. C. Allen, Jr., Numerical Computing: An Introduction, W. B. Saunders, 1973.
6. Shampine, L. F. et al., "Solving Non-Stiff Ordinary Differential Equations—The State of the Art," *SIAM Review,* Vol. 18, 1976, pp. 376-411.

CHAPTER 11

Discrete Event Modeling and Simulation

11.1   INTRODUCTION

The network orientation presented in the previous chapters is a valuable approach to modeling a large class of systems. However, for some systems the network orientation lacks the flexibility needed to model the system under study. For systems requiring flexibility beyond that afforded by the network orientation, the discrete event orientation provides a useful approach to simulation modeling. In this chapter, we describe and illustrate the basic concepts and procedures employed in constructing discrete event simulation models using SLAM II. The discussion of advanced discrete event concepts is deferred until Chapter 12.

11.2   DISCRETE EVENT ORIENTATION

In this section, we review the concepts of discrete event modeling described in Chapter 3. The world view embodied in a discrete event orientation consists of modeling a system by describing the changes that occur in the system at discrete points in time. An isolated point in time where the state of the system may change is called an "event time" and the associated logic for processing the changes in state is called an "event." A discrete event model of a system is constructed by defining the event types that can occur and then modeling the logic associated with each event type. A dynamic portrayal of the system is produced by causing the changes in states according to the logic of each event in a time-ordered sequence.

The state of a system in a discrete event model is similar to that of a network model and is represented by variables and by entities which have attributes and which belong to files. The state of the model is initialized by specifying initial values for the variables employed in the simulation, by creating the initial entities, and by the initial scheduling of events. During execution of the simulation, the model moves from state to state as entities engage in activities. In discrete event simulation, system status changes only occur at the beginning of an activity when something is started or at the end of the activity when something is terminated. Events are used to model the start and completion of activities.

The concept of an event which takes place instantaneously at a point in time and either starts or ends an activity is a crucial one. This relationship is depicted in Figure 11-1. Time does not advance within an event and the system behavior is simulated by state changes that occur as events happen.

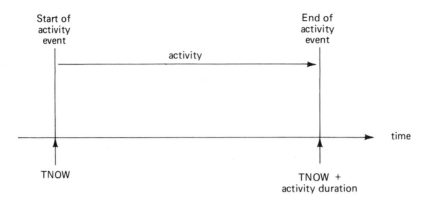

Figure 11-1   Relationship between activities and events.

When an event occurs, the state of the model can change in four ways: by altering the value of one or more variables; by altering the number of entities present; by altering the values assigned to one or more attributes of an entity; or by altering the relationships that exist among entities through file manipulations. Methods are available within SLAM II for accomplishing each of these changes. Note than an event can occur in which a decision is made not to change the state of the model. Schruben has developed an interesting graphic methodology for portraying events and the conditions which define the relations between events (6).

Events are scheduled to occur at a prescribed time during the simulation. Events have attributes and are maintained in chronological order in a file. For example, when scheduling an end-of-service event, the attributes of the customer undertaking service are part of the event and are then made available at the time of the end-of-service event processing. Thus, if an entity undertakes a series of activities with the end of each activity represented by an event, the attributes of the entity will be associated with each event that is processed.

11.2.1   Event Model of a Queueing Situation

To illustrate the concepts involved in discrete event modeling, consider a bank with one teller where it is desired to obtain statistics on the time each customer spends in the system. Customers arrive to the system, wait for the teller, undergo service, and depart the system. For simplicity, it will be assumed that there is no time delay between the time the service ends for one customer and begins for the next waiting customer, if any. The states of the system will be measured by the number of customers in the system and the status of the teller. Two event types can be used to model the changes in system state: a customer arrival event, and an end-of-service event. As a modeler, one must decide on the events that model the system. Here we assume all significant changes in system status can occur only at the arrival time of a customer or at the time that a service ends. In this model, it is assumed that the state of the system does not change between these event times.

There is one activity in the model of this system consisting of the service activity. The service activity for a customer begins at either the time of arrival of the customer or at the time that the teller completes service for another customer. Therefore, the starting time for the service activity can be either in the customer arrival event or in the end-of-service event. The ending time for the service activity is in the end-of-service event.

In constructing the event logic for this example, we employ the variable BUSY to describe the status of the teller where a value of 1 denotes busy and a value of 0 denotes idle. Customers are represented as entities with one attribute denoting the arrival time of the customer. This attribute is used in the model for collecting statistics on the time in the system for each customer in a fashion similar to the mark time employed in network modeling. The model employs a file ranked first-in, first-out (FIFO) for storing entities representing customers waiting for service when the teller is busy. The state of the model at any instant in time is defined by the value of the variable BUSY, location of the customer, the entities, and attribute values of the entities.

The initialization logic for this example is depicted in Figure 11-2. The variable

Figure 11-2   Initialization logic for bank teller problem.

BUSY is set equal to 0 to indicate that the teller is initially idle. The arrival event corresponding to the first customer is scheduled to occur at time 0. Thus, the initial status of the system is empty and idle with the first customer arrival scheduled at time 0.

The logic for the customer arrival event is depicted in Figure 11-3. The first action which is performed is the scheduling of the next arrival to occur at the current time plus the time between arrivals. Thus, each arrival will cause another ar-

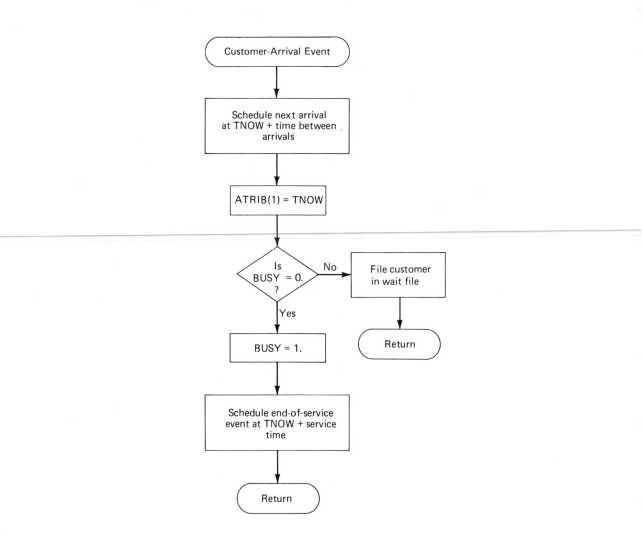

Figure 11-3 Customer arrival event logic for bank teller problem.

rival to occur at some later time during the simulation. In this way only one arrival event is scheduled to occur at any one time; however, a complete sequence of arrivals is generated.† The first attribute of the arrival entity associated with the event is then set equal to TNOW to mark the arrival time of the customer to the system. A test is then made on the status of the teller to determine if the service activity can begin or if the entity representing the customer must be placed in a file to wait for the teller. If service can begin, the variable BUSY is set equal to one, the end-of-service event for the current customer is scheduled and the event processing is completed. Otherwise, the entity and its attributes are placed in the file representing waiting customers and the event processing is completed.

At each end-of-service event, statistics are collected on the time in the system for the customer completing service. The first waiting customer, if any, is removed from the wait file and placed into service. The logic for the end-of-service event is depicted in Figure 11-4. The variable TSYS, corresponding to the time in the system for this customer, is computed as TNOW – ATRIB(1). Since ATRIB(1) was previously assigned the arrival time of the customer, TSYS is the elapsed time between the arrival and departure of the customer on which service was completed. A test is then made to determine if there are customers waiting for service. If not, the status of the teller is made idle by setting BUSY equal to 0 and the event processing is completed. Otherwise, the first customer entity is removed from the wait file. Processing of the customer is modeled by scheduling an end-of-service event to occur at the current time plus the service time.

This simple example illustrates the basic concepts of discrete event simulation modeling. Variables and entities with their attributes and file memberships make up the static structure of a simulation model. They describe the state of the system but not how the system operates. The events specify the logic which controls the changes that occur at specific instants of time. The dynamic behavior is then obtained by sequentially processing events and recording status values at event times. Since status is constant between events, the complete dynamic system behavior is obtained in this manner.

This example illustrates several important types of functions which must be performed in simulating discrete models. The primary functional requirement in discrete event simulation involves scheduling events, placing events in chronological order, and advancing time. Other functional requirements which are illustrated by this example include mechanisms for file manipulations, statistics collecting and

---

† The next arrival event could also be read from an input device which contained an "historical" sequence of events. In this case, we refer to the simulation model as being "trace driven." Of course, the historical sequence could have been generated by a separate simulation model.

reporting, and random sample generation. In Section 11-3, we describe the use of SLAM II for performing these and other functions in discrete event simulation modeling. First, we discuss how SLAM II can be used to study inventory systems.

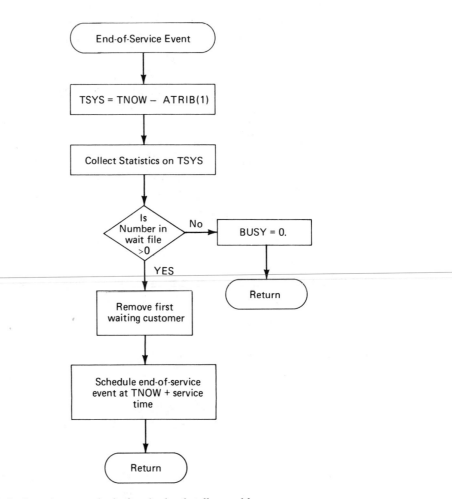

Figure 11-4 End-of-service event logic for the bank teller problem.

11.2.2 Event Modeling of Inventory Systems.

To illustrate the generality of the event modeling approach, we now consider the analysis of inventory systems and indicate the event modeling procedures. All inventory systems involve the storage of particular items for future sale or use. Demands are imposed on an inventory by customers seeking supplies of an individual item. A company must establish an inventory policy that specifies: when an order for additional items should be placed (or manufactured); and how many of the items should be ordered at each order time. When the items stored in inventory are discrete units, simulation of the inventory system using the next event philosophy is quite natural. The primary system variables are the number of units in storage, called the *inventory-on-hand,* and the number of units that are theoretically on the books of the company, called the *inventory position.* The inventory position is equal to the inventory-on-hand, plus the number of units on order (the number due-in), minus the number of units that have been backordered by customers (the number due-out). The inventory-on-hand represents the number of units available to be sold to customers. The inventory position represents the number of units that is or will be available to the company for potential sale. Inventory decisions regarding when to place an order and how much to order are normally based on the inventory position. Other attributes associated with inventory situations are: the number of units sold, the number of orders placed, the number of lost sales and/or the number of backorders, and the number of times the inventory position is reviewed to make a decision whether to place an order.

The events associated with an inventory situation at which these attributes could change their value are: a demand by a customer for the items in inventory, a review of the status of the inventory position to determine if an order should be placed, the receipt of goods from the distributor to be placed in inventory, and the satisfying of the customers' demands for the item. Each of these events could be modeled as a distinct event. However, several of the above will occur at the same point in time. When this happens, it is efficient to combine the events into a single event. For example, the satisfying of a customer demand would normally take place either when the customer demand is placed or when a shipment of units is received from the distributor. When this is done, the modeler assumes that the time delay in delivering a unit to a customer has an insignificant effect on the expected cost or profit associated with the inventory situation.

Customer demand events are normally modeled by specifying the time between customer demands. In this way, each demand event can be used to specify the time of the next demand. Then only the first demand event, which starts the demand process, and the time between demands would need to be specified.

Two types of review procedures to determine if an order should be placed are common in industry. A *periodic review procedure* specifies that a review should be performed at equally spaced points in time, for example, every month. The other review procedure involves the examination of the inventory position every time it decreases. This type of review is called *transaction reporting* and involves keeping a running log of the inventory position. When a transaction reporting review procedure is being studied, no separate review event is required since the inventory position can only be decreased when a customer demand occurs. Thus, the review procedure can be incorporated into the customer demand event.

The event involving the receipt of units from the distributor does not change the inventory position since the inventory position was increased when the units were ordered. However, the inventory-on-hand and the number of backorders could change when the receipt event occurs. The time at which the receipt event occurs is based on the time at which it was decided to place an order (at a review time) plus the time for processing the order and shipping the units. The time from the placing of the order to the receipt of the units is called the *lead time*.

The above general discussion of events that can occur in an inventory situation provides a general framework for simulating any inventory system. To obtain a simulation using SLAM II, it would only be necessary to write subroutines for each of the events required by the specific inventory situation. The points in time at which the individual events occur would automatically be handled by SLAM II using the time between events as specified by the user in the appropriate event routines. The user would schedule the next demand event (in a subroutine representing all demand events) and the receipt of units, if it was decided that an order was to be placed.

Of significant importance in this modeling of systems is the absence of the need to specify the details about the interdemand time distribution, the lead time distribution, the number of units requested by a customer at a demand point, and so forth, when discussing the overall procedure for simulating an inventory system. This information is only important when simulating a specific system. The SLAM II organizational structure for discrete event simulation allows the modeling of systems prior to data collection and, in fact, helps to specify the data required to analyze a system.

11.3   THE DISCRETE EVENT FRAMEWORK OF SLAM II

To simulate a discrete event model of a system using SLAM II, the analyst codes each discrete event as a FORTRAN subroutine. To assist the analyst in this task, SLAM II provides a set of FORTRAN subprograms for performing all commonly encountered functions such as event scheduling, statistics collection, and random sample generation. The advancing of simulated time (TNOW) and the order in which the event routines are processed are controlled by the SLAM II executive program. Thus, SLAM II relieves the simulation modeler of the task of sequencing events in their proper chronological order.

Each event subroutine is assigned a positive integer numeric code called the *event code,* in the same fashion as the event code defined at an EVENT node. The event code is mapped onto a call to the appropriate event subroutine by subroutine EVENT(I) where the argument I is the event code. This subroutine is written by the user as was discussed in Chapter 9 and consists of a computed GO TO statement indexed on I causing a transfer to the appropriate event subroutine call followed by a return. An example of the EVENT(I) subroutine for a simulation model with two events consisting of an arrival event coded in subroutine ARVL and assigned event code 1, and an end-of-service event coded in subroutine ENDSV and assigned event code 2 is depicted in Figure 11-5.

```
            SUBROUTINE EVENT(I)
            GO TO (1,2),I
     C
     C *** DEFINE EVENT CODE 1 AS SUBROUTINE 'ARVL'
     C
        1   CALL ARVL
            RETURN
     C
     C *** DEFINE EVENT CODE 2 AS SUBROUTINE 'ENDSV'
     C
        2   CALL ENDSV
            RETURN
            END
```

Figure 11-5   Sample subroutine EVENT(I).

The executive control for a discrete event simulation is provided by subroutine SLAM which is called from a user-written *main* program. The reason for making SLAM a subroutine and not the main program is to allow the user to dimension the SLAM II storage arrays NSET and QSET in the main program without the need to recompile the SLAM II executive control program. The array QSET is in unlabeled COMMON and is equivalenced to the array NSET which is prescribed to have the same dimension. This allows both integer and real data to be stored within a single contiguous array storage area. These arrays are employed by SLAM II for storing both events with their associated attributes and entities in files with their associated attributes. The term entry is used to refer to both events and entities which are stored in the arrays NSET/QSET. Thus, the dimension of the arrays NSET/QSET determines the maximum number of entries which can be in the system at any one time. The maximum number of entries (MNTRY) that can exist in the system at any one instant in time is limited by the following relationship:

$$MNTRY \leq NNSET/(MATR + 4)$$

where

    NNSET is the dimension of NSET/QSET, and

    MATR is the maximum number of attributes per entry employed in the simulation model.

A discussion of the arrays NSET/QSET and the way in which entries are stored is deferred until Chapter 12. For now the reader should simply note that SLAM II utilizes the arrays NSET/QSET for storing entries, network elements, and plot data points. NSET/QSET must be dimensioned by the user in the main program to provide space for all three uses.

The main program is also used to specify values for the SLAM II variables NNSET, NCRDR, NPRNT, and NTAPE which are in the labeled COMMON block named SCOM1. As discussed above, NNSET is the dimension assigned to the arrays NSET/QSET and must be set accordingly. The variable NCRDR denotes the unit number from which SLAM II input statements are read and would normally be set equal to 5 to denote the input reader unit. Likewise, the variable NPRNT denotes the unit number to which all SLAM II output is to be written and would normally be set to 6 to denote the line printer unit. The variable NTAPE is the number of a temporary scratch file which must be assigned by the user for use by the SLAM II processor for interpreting the free form SLAM II input statements and data. A sample main program is depicted in Figure 11-6.

Two additional user-written subroutines which are commonly employed in discrete event simulation models are subroutines INTLC and OTPUT discussed in Chapter 9. Subroutine INTLC is called by subroutine SLAM before each simulation run and is used to set initial conditions and to schedule initial events. Subroutine OTPUT is called at the end of each simulation and is used for end-of-simulation processing such as printing problem specific results for the simulation.

```
DIMENSION NSET(5000)
COMMON/SCOM1/ATRIB(100),DD(100),DDL(100),DTNOW,II,MFA,MSTOP,NCLNR
1,NCRDR,NPRNT,NNRUN,NNSET,NTAPE,SS(100),SSL(100),TNEXT,TNOW,XX(100)
COMMON QSET(5000)
EQUIVALENCE(NSET(1),QSET(1))
NNSET=5000
NCRDR=5
NPRNT=6
NTAPE=7
CALL SLAM
STOP
END
```

Figure 11-6   Sample main program.

The SLAM II next-event logic for simulating discrete event models is depicted in Figure 11-7. The SLAM II processor begins by reading the SLAM II input statements, if any, and initializing SLAM II variables. A call is then made to subroutine INTLC which specifies additional initial conditions for the simulation. SLAM II has a version of subroutine INTLC which contains only a RETURN statement so that the user need not include subroutine INTLC if no additional initialization is required. The processor then begins execution of the simulation by removing the first event from the event calendar. Events are ordered on the calendar based on low values of event times. The variable I is set equal to the event code and TNOW is advanced to the event time for the next event. Subroutine SLAM then calls the user-written subroutine EVENT(I) which in turn calls the appropriate event routine. Following execution of the user-written event routine, a test is made to determine if the simulation run is complete. A discrete event simulation is ended if any of the following conditions are satisfied:

1. TNOW is greater than or equal to TTFIN, the ending time of the simulation;
2. no events remain on the event calendar for processing; or
3. the SLAM II variable MSTOP has been set in a user-written routine to –1.

If the run is not complete, the new first event is removed from the event calendar and processing continues. Otherwise, a call is made to subroutine OTPUT. (As with INTLC, a version of OTPUT is included in SLAM II.) After the return from OTPUT, the SLAM II Summary Report is printed. A test is then made on the number of runs remaining. If more runs remain, control returns to initialization and the next simulation run is executed. Otherwise, a return is made from the SLAM II processor back to the user-written main program.

The above description provides an overview of the SLAM II framework for simulating discrete event models. To write a discrete event simulation program, the user writes a main program, subroutine EVENT(I) to decipher the event code I

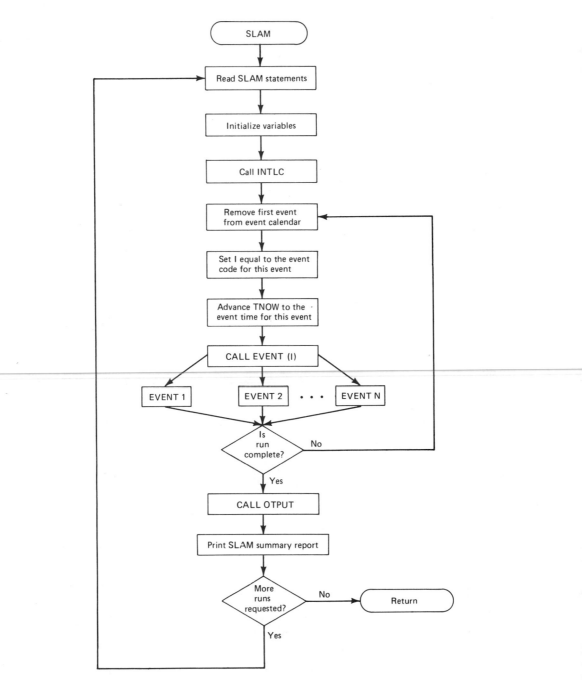

Figure 11-7 SLAM II next event logic for simulating discrete event models.

and to call the appropriate event subroutine, and event routines to specify the changes that occur at event times. Subroutines INTLC and OTPUT are written if special initial conditions or end-of-simulation processing is required. Based on this overview of the SLAM II discrete event framework, we next describe the functional capabilities provided by SLAM II for assisting the analyst in developing the user-written subroutines.

11.4   SCHEDULING EVENTS

The SLAM II processor completely relieves the user of the responsibility for chronologically ordering the events on an event calendar. The user simply schedules events to occur, and SLAM II causes each event to be processed at the appropriate time in the simulation. Events are scheduled by calling subroutine SCHDL(KEVNT, DTIME, A) where KEVNT denotes the event code of the event being scheduled and DTIME denotes the number of time units from the current time, TNOW, that the event is to occur. Attributes associated with an event are specified by passing the buffer array A as the third argument of SCHDL. The buffer array A can be a user-defined vector or the ATRIB vector as was used in network models. The values of the attributes in the buffer array A at the time the event is scheduled are stored on the event calendar with the event. These values are then removed from the event calendar by SLAM II and copied back into the ATRIB buffer at the time the event is executed. Thus, the ATRIB vector at the time of event processing will always contain the values that were in the buffer array A at the time that the event was scheduled.

To illustrate the use of subroutine SCHDL(KEVNT, DTIME, A), consider the scheduling of a customer arrival event with event code 1 to occur at the current time plus a sample from an exponential distribution with mean of 5 and using random stream number 1. In addition, the entity representing the customer has its first attribute value equal to 2. The coding to accomplish this is as follows:

```
A(1)=2
TIMINC=EXPON(5.,1)
CALL SCHDL (1,TIMINC,A)
```

These statements cause subroutine SLAM to call subroutine EVENT(I) with I equal to 1 at time TNOW + EXPON(5.,1) and to set ATRIB(1) equal to 2, prior to the call to EVENT(I).

11.5   FILE MANIPULATIONS

Entities which are grouped together because of a relationship between the entities are maintained in files. In network models, files were used to maintain entities waiting at QUEUE, AWAIT, and PREEMPT nodes. Files are used extensively in discrete event simulation models and they can be shared with network models, that is, file 1 can be associated with a QUEUE node and also be accessed in a discrete event routine. A file provides the mechanism for storing the attributes of an entity in a prescribed ranking with respect to other entities in the file.

In SLAM II, files are distinguished by integer numbers assigned to the files by the user. SLAM II automatically collects statistics on each file and provides the function NNQ(IFILE) which returns the number of entries in file IFILE. For example, NNQ(2) denotes the number of entries in file 2.

Associated with a file is a ranking criterion which specifies the procedure for ordering entities within the file. Thus, each entity in a file has a rank which specifies its position in the file relative to the other members in the file. A rank of one denotes that the entity is the first entry in the file. Possible ranking criterion for files are: first-in, first-out (FIFO); last-in, first-out (LIFO); high-value-first (HVF) based on an attribute value; and low-value-first (LVF) based on an attribute value. The ranking criterion for entries in a file is assumed to be FIFO unless otherwise specified using the PRIORITY statement described in Chapter 8.

SLAM II provides the user with a set of subroutines for performing all file manipulations which are commonly encountered in discrete event simulations. We now describe the use of these subroutines for file manipulations where entries are referenced by their rank in the file. In Chapter 12, we describe the SLAM II filing system in more detail and describe a more flexible approach of referencing entities by their location in the storage array NSET/QSET.

11.5.1   Subroutine FILEM(IFILE,A)

Subroutine FILEM(IFILE,A) files an entry with attributes specified in the buffer array A into file IFILE. The attributes of the entry are stored with the entry and are returned when the entry is removed from the file. As an example of the use of subroutine FILEM(IFILE,A), the following statements cause an entry with its first attribute set equal to the current simulation time to be inserted into file 1.

A(1) = TNOW
CALL FILEM(1,A)

On the LIMITS statement, the user defines the maximum number of entries permitted in all files, MNTRY. If an attempt is made to file more than MNTRY entries in the file storage area, a SLAM II execution error results.

11.5.2   Subroutine RMOVE(NRANK,IFILE,A)

Subroutine RMOVE(NRANK,IFILE,A) removes an entry with rank NRANK from file IFILE and places its attributes into the attribute buffer array A. As an example of the use of subroutine RMOVE, the following statement removes the last entry in file 1 and places its attributes in A.

CALL RMOVE(NNQ(1),1,A)

The dimension of the user-defined array A should be set greater than or equal to the maximum number of attributes per entry, MATR, plus 2. If the user attempts to remove an entry from file IFILE with rank greater than NNQ(IFILE), a SLAM II execution error results.

11.5.3   Subroutine COPY(NRANK,IFILE,A)

Sometimes it is desirable to copy the attributes of an entity belonging to a file without removing it from the file. Subroutine COPY(NRANK,IFILE,A) provides this capability by copying the attributes into the array A of the entry with rank NRANK in file IFILE without removing the entry from the file. For example, the following statement causes the attributes of the first entry in file 4 to be copied into the ATRIB array.

CALL COPY(1,4,ATRIB)

As in the case of subroutine RMOVE, if the user attempts to copy an entry in file IFILE with rank greater than NNQ(IFILE), a SLAM II execution error results.

11.5.4   Function NFIND(NRANK,IFILE,NATR,MCODE,X,TOL)

In some cases the rank of a specific entry of interest may not be known. Function NFIND(NRANK,IFILE,NATR,MCODE,X,TOL) can be used to determine the rank of an entry in file IFILE which contains a value for attribute NATR that bears a relationship designated by the user to a specified value, X. Definitions of the arguments to NFIND for determining the rank of an entry are:

NRANK   The rank of the entry where the search is to begin.

IFILE   The number of the file being searched.

NATR   The attribute number specified to find the entry meeting the conditions specified by MCODE.

MCODE   A code specifying the relationship between X and the attribute value.
The following five options are available.
MCODE =   2: maximum value but greater than X
MCODE =   1: minimum value but greater than X
MCODE =   0: first value within X ± TOL
MCODE = –1: minimum value but less than X
MCODE = –2: maximum value but less than X

X   A specified value used to search for an entry in file IFILE.

TOL   A tolerance value for equality comparisons when MCODE = 0 is specified. (FORTRAN requires that the argument be included for MCODE values of 1, 2, –1, or –2 even though the value is not used.)

If no entry is found in the file which meets the condition, NFIND returns a zero value. Otherwise, NFIND is returned as the rank of the first entry in the file which meets the condition which has a rank of at least NRANK. As an example of the use of function NFIND, the following statement is used to search file 3 beginning with the sixth entry to determine the rank of the entry which has its fourth attribute exactly equal to 10.:

    NRANK = NFIND(6, 3, 4, 0, 10.0, 0.0)

In Chapter 12, we describe procedures by which the user can search for an entry in a file which satisfies two or more conditions simultaneously.

11.5.5   File Manipulations Involving the Event Calendar

The filing system subroutines can also be used for manipulations of entries on the event calendar by specifying the file number as the SLAM II variable NCLNR. For example, the following statement would cause the attributes of the next event to be copied into the ATRIB array.

CALL COPY(1,NCLNR,ATRIB)

As in the case of other files, if the user specifies a rank which is greater than NNQ(NCLNR), a SLAM II execution error results.

When using this feature, the user may want to test and/or change the event code or event time of the event calendar entry. These values are automatically maintained by the SLAM II processor as attributes of the entry with the event code being attribute number MATR + 1 and the event time being attribute number MATR + 2 where MATR is the maximum number of user attributes specified on the LIMITS statement. For example, if the user specified MATR as 5 on the LIMITS statement, then ATRIB(6) and ATRIB(7) would contain the event code and event time, respectively.

As a second example of the use of the filing system subroutines involving the event calendar, consider a simulation model with MATR equal to 5. The following statements would remove the first event from the event calendar, add 10.0 to its event time, and then place it back onto the calendar.

CALL RMOVE(1,NCLNR,ATRIB)
ATRIB(7) = ATRIB(7) + 10.
CALL FILEM(NCLNR, ATRIB)

11.6   STATISTICS COLLECTION

In discrete event simulations, there are two distinct types of statistics which are of interest to the analyst. These are: (1) statistics based on observations; and (2) statistics on time-persistent variables. In this section, we describe the procedures employed in SLAM II discrete event simulations for collecting and estimating values associated with both types of statistics.

11.6.1   Statistics Based on Observations

Statistics based on observations are statistics computed from a finite number of samples. Each sample value is considered as an observation. For example, in the bank teller problem, statistics on the time in the system for each customer are based on observations where each customer processed through the system is considered as one observation. These statistics depend only upon the value of each observation and not upon the time at which each observation is collected.

Each variable for which observation statistics are collected must be defined by the user with a STAT statement following the general format conventions described in Chapter 8. The format for the STAT statement is as follows:

STAT, ICLCT, ID, NCEL/HLOW/HWID;

where ICLCT is an integer code associated with the variable used to distinguish variable types;

ID is an alphanumeric identifier with 16 significant characters which is printed on the SLAM II Summary Report to identify the output associated with the variable; and

NCEL/HLOW/HWID are histogram parameters specifying the number of interior cells, the upper limit for the first cell, and the width of each cell, respectively. If histogram parameters are not specified, then no histogram is prepared.

As an example, the following statement specifies that observation statistics are to be collected on variable type 1, the results are to be displayed as TIME IN SYSTEM, and a histogram is to be generated with 10 interior cells with the upper limit for the first cell equal to 0 and with interior cell widths of 1.

STAT, 1, TIME IN SYSTEM, 10/0/1;

Observations for each variable type defined by the STAT statement are collected within the event subroutines by a call to subroutine COLCT(XVAL,ICLCT) where the variable XVAL contains the value of the observation and ICLCT is the integer code associated with the variable type. For example, the following statement would cause one observation to be collected on the variable TSYS which is coded as COLCT variable type 1.

CALL COLCT(TSYS,1)

Through inputs on the INITIALIZE statement, subroutine COLCT can be used to obtain statistics over simulation runs. The procedure for accomplishing this involves not clearing the variables for which statistics over runs are desired. By calling subroutine COLCT with a value obtained at the end of each run, statistics over runs are obtained. This call to subroutine COLCT could be made from subroutine OTPUT. For example, if six STAT statements are included in the input and we specify that up to the 4th variable type should be cleared (YES/4 on the INIT statement), then variables with codes 4, 5, and 6 would not be cleared at the end of a run. These three variables can then be used to store values collected over multiple runs. Specifically, consider the following input statements:

```
STAT,1,V1;
STAT,2,V2;
STAT,3,V3;
STAT,4,A1;
STAT,5,A2;
STAT,6,A3:
INIT,0,480,YES/4,YES,YES;
```

These statements specify that STAT variables 1, 2, and 3 should be cleared at the end of each run but STAT variables 4, 5, and 6 should not. They also indicate that all SLAM II variables and files should be cleared at the end of a run. The following code in subroutine OTPUT will provide observations on the average value of variables collected as V1, V2, and V3.

```
        SUBROUTINE OTPUT
        DIMENSION A(3)
        DO 10 I=1,3
C****   OBTAIN AVERAGE OF STAT VARIABLE I (SEE CHAP. 12)
        A(I)=CCAVG(I)
C****   SET COLLECT INDEX FOR AVE. OVER RUNS TO I+3
        CALL COLCT(A(I),I+3)
   10   CONTINUE
        RETURN
        END
```

On each summary report printed, statistics based on observations are obtained for the variables A1, A2, and A3 which is the average of the average values for V1, V2, and V3 respectively. Thus, if a summary report is printed for three runs, the average of the three average values obtained for V1 is printed in the row whose descriptor is A1. If ten runs are made, the average of ten average values is obtained.

The above procedure can be used with all SLAM II status variables or with user defined variables. In Chapter 12, the functions for obtaining the average of any SLAM II status variable are described. In the above example, the function CCAVG was used. Similar functions exist for all other SLAM II statistics. As described in Chapter 8, a field on the COLCT node statement allows each COLCT node to have an index code so that the functions described in Chapter 12 can also be used to reference statistics collected at COLCT nodes. STAT statements and COLCT nodes may have the same index code in which case the observations at the COLCT node are combined with the observations obtained through calls to subroutine COLCT.

11.6.2   Statistics on Time-Persistent Variables

Statistics on time-persistent variables refer to statistics maintained on variables which have a value defined over a period of time. For example, the variable BUSY in the bank teller example is used to denote the status of the teller over time. The fraction of time that BUSY equals 1 is the teller utilization. Therefore the utilization of the teller would be a time-persistent statistic.

Statistics on time-persistent variables are obtained in SLAM II discrete event simulations by the use of the TIMST statements. The TIMST input statement causes time persistent statistics to be automatically accumulated by the SLAM II processor. This relieves the user of the burden of calling a subroutine within the event routines to obtain time-persistent statistics. For example, to invoke time-persistent statistics on the variable XX(1), with the output identified as TELLER BUSY TIME, the user simply includes the following statement as part of the SLAM II input.

TIMST, XX(1), TELLER BUSY TIME;

To improve the mnemonics within the coding of the event routines, the FORTRAN EQUIVALENCE statement can be employed. For example, time-persistent statistics could be obtained on the FORTRAN variable BUSY by using the TIMST statement described above in conjunction with the following EQUIVALENCE statement.

EQUIVALENCE(BUSY,XX(1))

This EQUIVALENCE statement must be included in each event subroutine in which the value of BUSY can change.

To obtain a histogram on a time-persistent variable, values for the number of cells, upper limit of the first cell, and a cell width are given in the last fields of the TIMST statement, that is, NCEL/HLOW/HWID. Each cell of the histogram gives the fraction of time that the time-persistent variable was in the range defined by the cell. For the teller utilization, we can obtain the fraction of time idle or busy in histogram form with the statement.

TIMST,XX(1).TELLER BUSY TIME, 2/0/1;

11.7   VARIABLES AND COMMON BLOCK

Variables which are employed in a SLAM II discrete event simulation model can be of two types: SLAM II variables; and user-defined variables. SLAM II variables are those defined by the SLAM II language. Table 11-1 summarizes the important SLAM II variables which are used in discrete event simulation models.

The values of SLAM II variables are transferred between the SLAM II processor and user-written subroutines through the SCOM1 named (labeled) COMMON block. Therefore any user-written subroutine which references a SLAM II variable must contain the SCOM1 COMMON block. The variable list for the SCOM1 COMMON block is shown below.

COMMON/SCOM1/ATRIB(100),DD(100),DDL(100),DTNOW,II,MFA,MSTOP,NCLNR
1,NCRDR,NPRNT,NNRUN,NNSET,NTAPE,SS(100),SSL(100),TNEXT,TNOW,XX(100)

The array sizes selected for the dimensions of COMMON variables are those that have been found to satisfy most SLAM II users without requiring an excessive amount of core storage. Additional information on COMMON blocks and redimensioning for FORTRAN '66 and FORTRAN '77 SLAM II versions is contained in Reference 3.

Table 11-1 Definitions of some important discrete event variables.

Variable	Definition
ATRIB(I)	Buffer for the Ith attribute value of an entry to be inserted or removed from the file storage area
MSTOP	Set by the user to −1 to stop a simulation run before time TTFIN
NCLNR	The file number of the event calendar
NCRDR	The unit number from which SLAM II input statements are read. Normally set to 5 to denote the input device
NNRUN	The number of the current simulation run
NNSET	The dimension of the arrays NSET/QSET
NPRNT	The unit number to which SLAM II output is to be written. Normally set to 6 to denote the lineprinter
NSET/QSET	Equivalenced arrays employed by SLAM II for storing file entries
NTAPE	The unit number of a scratch tape
TNOW	The value of current simulated time
TNEXT	The time of the next scheduled discrete event
XX(I)	The Ith global variable. Time persistent statistics will be collected if XX(I) is specified on the TIMST input statement.

In addition to SLAM II variables, the user may also employ user-defined variables in describing the state of a system. For example, the user variable BUSY was employed in the bank teller problem to denote the status of the teller. The values of user-defined variables are transferred between user-written subroutines by user-defined named COMMON blocks. *The user must not employ unnamed (blank) COMMON for user-defined variables.* As noted earlier in this chapter, all entries in files are stored in the equivalenced arrays NSET and QSET where the array QSET is in unnamed COMMON. The use of additional unnamed COMMON will cause the values stored in the arrays NSET/QSET to be overwritten and will result in an error.

11.8   INPUT STATEMENTS

In discrete event simulation models, input statements are required for specifying a project title, run length, file rankings, monitor events and the like. The input statements employed in discrete event models are identical to those described for network models in Chapter 8. One addition is the STAT input statement described in Section 11.6.1. Thus, a discrete event model must include a GEN statement, LIMITS statement, and FIN statement, plus any additional input statements required by the simulation model.

11.9   OUTPUT REPORTS

The results for each simulation run are summarized by the SLAM II Summary Report which is automatically printed by the SLAM II processor at the end of each simulation run. The SLAM II Summary Report for discrete event models is identical to that produced for network models with the network statistics omitted, and includes statistics on variables based on observations, statistics on time-persistent variables, file statistics, histograms, and plots. Definitions for the statistical quantities provided by the summary report are provided in Chapter 8.

The SLAM II processor also prints a trace report if the TRACE option has been specified using the MONTR statement. The SLAM II Trace Report for discrete-event models includes the event time, the event code, and the attributes of each event processed during the tracing period.

11.10   BUILDING DISCRETE EVENT MODELS

In the preceding sections, we described the discrete event framework of SLAM II and a set of subprograms which allows the user to perform the commonly encountered functions in discrete event simulation models. To construct a discrete event simulation model of the bank teller problem, the user must do the following:

1. Write the main program to dimension NSET/QSET, specify values for NNSET, NCRDR, NPRNT, and NTAPE, and call SLAM.
2. Write subroutine EVENT(I) to map the user-assigned event codes onto a call to the appropriate event subroutine.
3. Write subroutine INTLC to initialize the model.
4. Write event subroutines to model the logic for the events of the model.
5. Prepare the input statements required by the problem.

11.11   ILLUSTRATION 11-1. SERVING BANK CUSTOMERS

In this section, we illustrate how these SLAM II functional capabilities are used in constructing discrete event simulation models by describing the coding of the SLAM II discrete event model of the bank teller example. In the coding which follows, we assume that customers arrive to the bank with the time between arrivals given by an exponential distribution with a mean of 20 minutes and that the teller service time is uniformly distributed between 10 and 25 minutes. The operation of the bank teller system is to be simulated for a period of 480 minutes.

The event model of the bank teller problem contains two events: the customer arrival event and the end-of-service event. We will code the logic for the customer arrival event in subroutine ARVL and assign it event code number 1. The code for the logic for the end-of-service event will be written in subroutine ENDSV and it will be referenced as event code number 2. Thus, subroutine EVENT(I) is as depicted previously in Figure 11-5.

The entities in the system representing customers have only a single attribute. Assuming that there will not be more that 20 customers in the system at any one time, that is, MNTRY = 20, the minimum dimension of NSET/QSET is given by $20*(1+4)=100$. Thus, the main program depicted in Figure 11-6 assigns more than a sufficient amount of file storage area to simulate this problem. For convenience, the main program and subroutine EVENT are shown again in Figure 11-8.

The initialization logic for the bank teller problem is coded in subroutine INTLC as depicted in Figure 11-9. The user-defined variable BUSY is equivalenced to the SLAM II variable XX(1) which is contained in the SLAM II named COMMON block SCOM1 included in the subroutine. The variable BUSY is set to zero to denote that the teller is initially idle. Next, the customer arrival event which has an event code of 1 is scheduled onto the event calendar to occur after 0.0 time units. No attribute values are associated with this scheduled arrival event.

```
      DIMENSION NSET(5000)
      COMMON/SCOM1/ATRIB(100),DD(100),DDL(100),DTNOW,II,MFA,MSTOP,NCLNR
     1,NCRDR,NPRNT,NNRUN,NNSET,NTAPE,SS(100),SSL(100),TNEXT,TNOW,XX(100)
      COMMON QSET(5000)
      EQUIVALENCE(NSET(1),QSET(1))
      NNSET=5000
      NCRDR=5
      NPRNT=6
      NTAPE=7
      CALL SLAM
      STOP
      END


      SUBROUTINE EVENT(I)
      GO TO (1,2),I
C
C *** DEFINE EVENT CODE 1 AS SUBROUTINE 'ARVL'
C
    1 CALL ARVL
      RETURN
C
C *** DEFINE EVENT CODE 2 AS SUBROUTINE 'ENDSV'
C
    2 CALL ENDSV
      RETURN
      END
```

Figure 11-8 Main program and subroutine EVENT for bank teller problem.

```
      SUBROUTINE INTLC
      COMMON/SCOM1/ATRIB(100),DD(100),DDL(100),DTNOW,II,MFA,MSTOP,NCLNR
     1,NCRDR,NPRNT,NNRUN,NNSET,NTAPE,SS(100),SSL(100),TNEXT,TNOW,XX(100)
      EQUIVALENCE (XX(1),BUSY)
C
C *** INITIALIZE THE SERVER TO IDLE
C
      BUSY=0.
C
C *** SCHEDULE EVENT 1 (ARVL) AT TIME 0.
C
      CALL SCHDL(1,0.,ATRIB)
      RETURN
      END
```

Figure 11-9 Subroutine INTLC for bank teller problem.

However, FORTRAN convention requires that all arguments for a subroutine be included when the subroutine is called. Thus, ATRIB is used as the third argument in the call to SCHDL.

The logic for the customer arrival event is coded in subroutine ARVL and is presented in Figure 11-10. Figure 11-3 shows the logic in flowchart form. The values of the SLAM II discrete event variables are passed to the event routine through COMMON block SCOM1 and the SLAM II variable XX(1) is equivalenced to the user defined variable BUSY. The first function performed by the event is the rescheduling of the next arrival event to occur at the current time plus a sample from an exponential distribution with mean of 20. and using random stream number 1. The first attribute of the arriving customer is then set equal to the arrival time, TNOW. A test is then made on the variable BUSY to determine the current status of the teller. If BUSY is equal to 0.0, then the teller is idle and a branch is made to statement 10 where BUSY is set to 1.0 to indicate that the teller is busy and an end-of-service event is scheduled to occur at time TNOW plus a sample from a uniform distribution between 10. and 25. using random stream number 1. Otherwise, the customer is placed in file 1 to wait for the teller. In either case, the customer is identified by his arrival time which is stored as attribute 1 in the entry placed in file 1 or on the event calendar (file NCLNR).

```
      SUBROUTINE ARVL
      COMMON/SCOM1/ATRIB(100),DD(100),DDL(100),DTNOW,II,MFA,MSTOP,NCLNR
     1,NCRDR,NPRNT,NNRUN,NNSET,NTAPE,SS(100),SSL(100),TNEXT,TNOW,XX(100)
      EQUIVALENCE (XX(1),BUSY)
C
C *** SCHEDULE SUBSEQUENT ARRIVALS
C
      CALL SCHDL(1,EXPON(20.,1)ATRIB)
C
C *** ASSIGN ATTRIBUTE 1 THE MARK TIME
C
      ATRIB(1)=TNOW
      IF(BUSY.EQ.0.) GO TO 10
C
C *** IF THE SERVER IS BUSY PLACE THE NEW ARRIVAL IN
C *** FILE 1 TO WAIT FOR THE SERVER
C
      CALL FILEM(1,ATRIB)
      RETURN
C
C *** IF THE SERVER IS NOT BUSY, MAKE THE SERVER BUSY AND
C *** SCHEDULE THE END OF SERVICE (EVENT 2--ENDSV)
C
   10 BUSY=1.
      CALL SCHDL(2,UNFRM(10.,25.,1),ATRIB)
      RETURN
      END
```

Figure 11-10 Subroutine ARVL for bank teller problem.

The logic for the end-of-service event is coded in subroutine ENDSV and is presented in Figure 11-11. Figure 11-4 shows the logic in flowchart form. The variable TSYS is set equal to the current time, TNOW, minus the first attribute of the current customer being processed. When an event is removed from the event calendar, the ATRIB buffer array is assigned the attribute values that were associated with the event when it was scheduled. Since the value of ATRIB(1) for this customer was set to TNOW in the arrival event, the value of TSYS represents the elapsed time between the arrival and end-of-service event for this customer. A call is then made to subroutine COLCT to collect statistics on the value of TSYS as collect variable number 1. A test is made on the SLAM II function NNQ(1) representing the number of customers waiting for service in file 1. If the number of customers waiting is greater than zero, a transfer is made to statement 10 where the first customer waiting is removed from file 1 and placed onto the event calendar. The end-of-service event, event code 2, is scheduled to occur at time TNOW plus the service time. If no customer is waiting, the status of the teller is changed to idle by setting the variable BUSY to 0.

```
      SUBROUTINE ENDSV
      COMMON/SCOM1/ATRIB(100),DD(100),DDL(100),DTNOW,II,MFA,MSTOP,NCLNR
     1,NCRDR,NPRNT,NNRUN,NNSET,NTAPE,SS(100),SSL(100),TNEXT,TNOW,XX(100)
      EQUIVALENCE (XX(1),BUSY)
C
C *** DETERMINE THE TIME THE ENTITY WAS IN THE SYSTEM BY
C *** SUBTRACTING THE MARK TIME FROM THE CURRENT TIME.
C *** COLLECT STATISTICS ON THE 'TIME IN SYSTEM' USING COLCT
C *** CODE 1, DEFINED ON THE STAT STATEMENT.
C
      TSYS=TNOW-ATRIB(1)
      CALL COLCT(TSYS,1)
C
C *** DETERMINE IF ANY ENTITIES ARE WAITING FOR THE SERVER.
C *** IF NOT, MAKE THE SERVER IDLE.
C
      IF(NNQ(1).GT.0) GO TO 10
      BUSY=0.
      RETURN
C
C *** IF ENTITIES ARE WAITING FOR THE SERVER, REMOVE THE FIRST
C *** FROM FILE 1 AND SCHEDULE THE END OF SERVICE 'ENDSV'.
C
   10 CALL RMOVE(1,1,ATRIB)
      CALL SCHDL(2,UNFRM(10.,25.,1),ATRIB)
      RETURN
      END
```

Figure 11-11 Subroutine ENDSV for bank teller problem.

The input statements for this example are shown in Figure 11-12. The GEN statement specifies the analyst's name, project title, date, and number of runs. The LIMITS statement specifies that the model employs 1 file (not including the event calendar), the maximum number of attributes is 1, and the maximum number of simultaneous entries in the system is 20. The STAT statement specifies that collect variable number 1 is to be displayed with the label TIME IN SYSTEM and that a histogram is to be generated with 10 interior cells, the upper limit of the first cell is to be 0, and the cell width of each interior cell is to be 4. The first TIMST statement causes time-persistent statistics to be automatically maintained on the SLAM II variable XX(1) and the results to be displayed using the label UTILIZATION. The second TIMST statement collects statistics on the number in the bank teller queue, NNQ(1), and requests a histogram on the fraction of time there are 0,1,2, $\cdots$ customers in the queue. The INIT statement specifies that the beginning time of the simulation is time 0 and that the ending time is time 480. The FIN statement denotes the end to all SLAM II input statements. This completes the description of the discrete event model of the bank teller system.

```
GEN,I.E.MAJOR,BANK TELLER,11/10/84,1;
LIMITS,1,1,20;
;
;   COLLECT OBSERVATION STATISTICS USING CODE 1 FROM
;      CALLS TO SUBROUTINE COLCT IN THE USER CODE
;
STAT,1,TIME IN SYSTEM,10/0/4;
;
;   COLLECT TIME PERSISTENT STATISTICS ON VARIABLES XX(1)
;      AND NNQ(1), REQUEST A HISTOGRAM ON NNQ(1)
;
TIMST,XX(1),UTILIZATION;
TIMST,NNQ(1),NO. IN QUEUE,10/0/1;
INIT,0,480;
FIN;
```

Figure 11-12   Data statements for bank teller problem .

The coding of the bank teller problem illustrates the general approach of combining SLAM II functional capabilities to code discrete event simulation models. Larger and more complicated models are developed in the same way and differ only in the number and complexity of the event subroutines. In the next sections, we present two additional discrete event examples which are slightly more complicated and further illustrate discrete event modeling using SLAM II.

11.12   ILLUSTRATION 11-2.   BLOCKING OF SERIAL OPERATIONS

In this illustration, a manufacturing process involving four machines in series is modeled. Henriksen used this example to illustrate passive and active resource modeling perspectives(1). The system to be modeled is similar to the one presented in Example 5-1. Four machines in series perform operations on parts in a required sequence, that is, a part goes through machine 1 first, then machines 2, 3, and 4. Parts waiting for processing are placed in buffers which have a capacity to store 50 parts. When the buffer for machine 2 is full and machine 1 completes the processing of a part, machine 1 is blocked. Similarly, when the buffer for machine M is full, machine M-1 is blocked. It is assumed that unless blocked, machine 1 is always processing a part. The output from machine 4 is routed to finishing operations which are considered outside the boundaries of the system to be modeled. The processing time on each of the machines is uniformly distributed between 10 and 190. An analysis is desired for the production of 5000 parts.

A discrete event approach to this blocking model consists of a single event type which is the end of processing by a machine. Except for the start of the processing of the first part on each machine, a machine starts processing a part either when it completes the processing of a part or when the preceding machine completes the processing of a part. An event approach requires only that the logic be developed to specify the disposition of a machine and a part when the machine completes the processing of a part. The logic involves the determination of the next status of: the machine, the part and the immediately preceding and succeeding machines.

In the discrete event model, the following variables of interest are defined by the modeler.

Variable	Definition
NMACH	Number of machines
STATUS(M)	Status of machine M: 0→Idle or blocked; 1→Busy
BLOCK(M)	Blocking status of machine M: 0→Not blocked; 1→Blocked
QUEUE(M)	Number in the queue of machine M
CAPQ(M)	Capacity of the queue for machine M

To initialize the above variables and to start the simulation process, the SLAM II subroutine INTLC is employed. In subroutine INTLC shown in Figure 11-13, the number of machines is set to 4. For machines 2, 3, and 4, the capacity of each queue is set to 50, the number in each queue is set to 40, the status of each machine is set to busy, and the blocking status is set to not blocked.

```
      SUBROUTINE INTLC
C
      COMMON/SCOM1/ATRIB(100),DD(100),DDL(100),DTNOW,II,MFA,MSTOP,NCLNR
     1,NCRDR,NPRNT,NNRUN,NNSET,NTAPE,SS(100),SSL(100),TNEXT,TNOW,XX(100)
C
      EQUIVALENCE (XX(1),STATUS(1)),(XX(11),BLOCK(1)),(XX(21),QUEUE(1))
      DIMENSION STATUS(10),BLOCK(10),QUEUE(10)
      COMMON/UCOM1/CAPQ(10),NMACH
      NMACH=4
C
C**** INITIALIZE QUEUE CAPACITIES TO 50, AND NUMBER IN QUEUE
C**** TO 40 FOR MACHINES 2 THROUGH 4
C
      DO 100 M=1,NMACH
      CAPQ(M)=50.
      IF(M.GT.1) QUEUE(M)=40.0
      STATUS(M)=1.0
      BLOCK(M)=0.0
C
C**** SCHEDULE COMPLETION OF PROCESSING ON EACH MACHINE
C
      CALL SCHDL(M,UNFRM(10.,190.,1),ATRIB)
100   CONTINUE
      RETURN
      END
```

Figure 11-13 Initialization of Server Blocking Model.

For each machine, an end-of-processing event is put on the calendar by a call to subroutine SCHDL with an event code of M. In this model, the event code will be used as an indicator of which machine is completing processing. Thus, although there is only one event type, multiple event codes will be used with the event code identifying the machine number for which the end of processing is occurring.

Since there is only one event type in this model, it is coded in subroutine EVENT. A flowchart and the code for subroutine EVENT are shown in Figures 11-14 and 11-15, respectively.

The EQUIVALENCE statement

EQUIVALENCE (XX(1),STATUS(1)),(XX(11),BLOCK(1)),(XX(21),QUEUE(1))

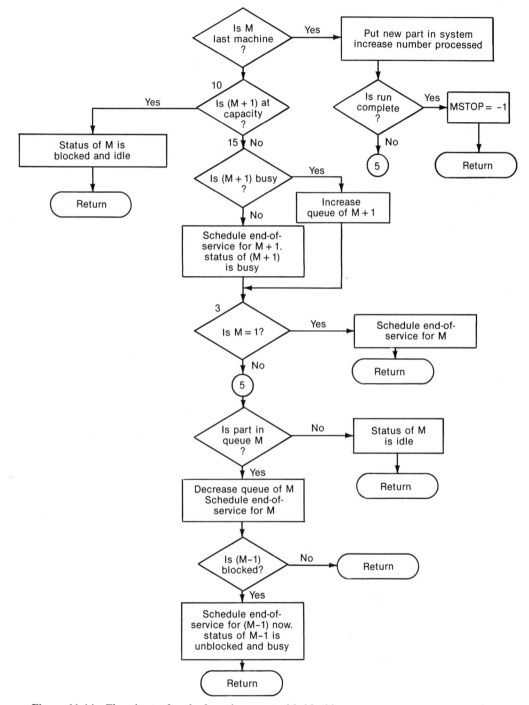

Figure 11-14   Flowchart of end-of-service event with blocking.

is used to obtain statistics on the utilization and blocking percentage for each machine and on the number in the queue before each machine. By equivalencing the first cells of two vectors, the EQUIVALENCE statement makes the vectors equivalent.

The first section of code tests to see if the machine is the last machine in the sequence and, if it is, the number of parts processed is increased by 1. The simulation is stopped when 5000 parts are processed by setting MSTOP to -1.

At statement 10, an evaluation is made as to whether or not the part that just completed processing on machine M can be transferred to the queue of machine $M+1$. If the queue for machine $M+1$ is at capacity, the status of machine M is set to blocked and idle by setting BLOCK(M) to 1 and STATUS(M) to 0. If the part can go to machine $M+1$ then the status of machine $M+1$ is checked at statement 15. If machine $M+1$ is busy or blocked, then the number waiting for machine $M+1$ is increased by 1. Otherwise, an end-of-processing event is scheduled for machine $M+1$ by a call to subroutine SCHDL with an event code of $M+1$. This accomplishes the movement of the part to the next machine.

```
      SUBROUTINE EVENT(M)
C
      COMMON/SCOM1/ATRIB(100),DD(100),DDL(100),DTNOW,II,MFA,MSTOP,NCLNR
     1,NCRDR,NPRNT,NNRUN,NNSET,NTAPE,SS(100),SSL(100),TNEXT,TNOW,XX(100)
C
      EQUIVALENCE (XX(1),STATUS(1)),(XX(11),BLOCK(1)),(XX(21),QUEUE(1))
      DIMENSION STATUS(10),BLOCK(10),QUEUE(10)
      COMMON/UCOM1/CAPQ(10),NMACH
C     STATUS(M)=STATUS OF MACHINE M
C              0 IDLE OR BLOCKED
C              1 BUSY
C     BLOCK(M)=BLOCKING STATUS OF MACHINE M
C              1 BLOCKED
C              0 NOT BLOCKED
C     QUEUE(M)=* IN QUEUE OF M
C     CAPQ(M)=QUEUE CAP.
C     NMACH=NUMBER OF MACHINES
      IF(M.LT.NMACH) GO TO 10
C
C**** LAST MACHINE PROCESSING
C**** UPDATE NUMBER OF PARTS PROCESSED, STOP WHEN 5000
C**** HAVE BEEN PROCESSED.
C
         NUMPROC=NUMPROC+1
      IF(NUMPROC.LT.5000) GO TO 5
        MSTOP=-1
      RETURN
```

Figure 11-15 FORTRAN code for end-of-processing event.

```
C**** CHECK FOLLOWING QUEUE FOR SPACE AVAILABILITY
   10      IF(QUEUE(M+1).LT.CAPQ(M+1)) GO TO 15
           BLOCK(M)=1.0
           STATUS(M)=0.0
           RETURN
C**** ROUTE PART TO NEXT MACHINE
   15      IF(STATUS(M+1).GT.0.0 .OR. BLOCK(M+1).GT.0.0) THEN
               QUEUE(M+1)=QUEUE(M+1)+1.0
           ELSE
               STATUS(M+1)=1.0
               CALL SCHDL(M+1,UNFRM(10.,190.,M+1),ATRIB)
           ENDIF
    3      IF(M.GT.1) GO TO 5
C**** ALWAYS KEEP MACHINE 1 BUSY
           CALL SCHDL(M,UNFRM(10.,190.,M),ATRIB)
           RETURN
C
C**** SET MACHINE TO IDLE IF PARTS ARE NOT WAITING
C
    5      IF(QUEUE(M).GT.0.0) GO TO 7
               STATUS(M)=0.0
           RETURN
C**** SCHEDULE ANOTHER JOB ON MACHINE JUST FINISHING
    7      QUEUE(M)=QUEUE(M)-1.
           CALL SCHDL(M,UNFRM(10.,190.,M),ATRIB)
C**** UNBLOCK IF NECESSARY
           IF (BLOCK(M-1).LT.1.0) RETURN
           BLOCK(M-1)=0.0
           STATUS(M-1)=1.0
C
C**** SCHEDULE UNBLOCKING OF PREVIOUS MACHINE
           CALL SCHDL(M-1,0.0,ATRIB)
           RETURN
           END
```

Figure 11-15 (continued).

The disposition of the machine starts at statement 3. If M is 1, another end-of-processing event is scheduled. For M greater than 1, if there are no parts waiting for machine M, then the status of machine M is changed to idle and no further processing is required. Otherwise, a transfer is made to statement 7 where the number waiting for machine M is decreased by 1. An end-of-processing event for machine M is then scheduled through a call to subroutine SCHDL with the event code set at M.

Since a part has been taken out of the queue for machine M, it is necessary to determine if its preceding machine should be unblocked. If the blocking status of machine M-1 is not blocked, a return is also made. If machine M-1 is blocked, the blocking status is reset to unblocked and the status to busy. An end-of-processing event for machine M-1 is scheduled to occur immediately be a call to subroutine SCHDL with an EVENT code of M-1 and a processing time of 0. By rescheduling an event to unblock a machine, the event routine will automatically take care of the unblocking or prior machines through the evaluation of the logic just described when the event for the end of processing of machine M-1 is considered.

The SLAM II input statements for the discrete event model of the blocking situation are shown in Figure 11-16.

```
GEN,PRITSKER, D.E.BLOCKER, 12/21/85,1;
LIMITS,,,10;
;
;   COLLECT TIME PERSISTENT STATISTICS ON VARIABLES--
;      XX(1), XX(2), XX(3), XX(4), XX(11), XX(12), XX(13),
;      XX(22), XX(23), XX(24)
;
TIMST,XX(1), STATUS1;
TIMST,XX(2), STATUS2;
TIMST,XX(3), STATUS3;
TIMST,XX(4), STATUS4;
TIMST,XX(11), BLOCK1;
TIMST,XX(12), BLOCK2;
TIMST,XX(13), BLOCK3;
TIMST,XX(22), QUEUE2;
TIMST,XX(23), QUEUE3;
TIMST,XX(24), QUEUE4;
FIN;
```

Figure 11-16 Control statement for discrete event blocking.

The LIMITS statement indicates that there are no files and no attributes per entity. The number of concurrent entities in the system is set at 10 to accommodate 10 machines in series. The only entries in the filing structure will be for events and the only events in the system are end-of-processing events for which there can only be one for each machine in the system. The next ten input statements are for collecting statistics on XX variables. A FIN statement then indicates that there are no more SLAM II input statements. A summary report for this illustration is shown in Figure 11-17.

11.13   EXAMPLE 11-1.   A DISCOUNT STORE OPERATION

A discount store has developed a new procedure for serving customers. Customers enter the store and determine the item they wish to purchase by examining display items. After selecting an item, the customer proceeds to a centralized area

S L A M   I I   S U M M A R Y   R E P O R T

SIMULATION PROJECT D.E.BLOCKER   BY PRITSKER

DATE 12/21/1985   RUN NUMBER   1 OF   1

CURRENT TIME   0.5082E+06
STATISTICAL ARRAYS CLEARED AT TIME   0.0000E+00

STATISTICS FOR TIME-PERSISTENT VARIABLES

	MEAN VALUE	STANDARD DEVIATION	MINIMUM VALUE	MAXIMUM VALUE	TIME INTERVAL	CURRENT VALUE
STATUS1	0.9796E+00	0.1414E+00	0.0000E+00	0.1000E+01	0.5082E+06	0.1000E+01
STATUS2	0.9785E+00	0.1451E+00	0.0000E+00	0.1000E+01	0.5082E+06	0.1000E+01
STATUS3	0.9738E+00	0.1596E+00	0.0000E+00	0.1000E+01	0.5082E+06	0.1000E+01
STATUS4	0.1000E+01	0.0000E+00	0.1000E+01	0.1000E+01	0.5082E+06	0.1000E+01
BLOCK1	0.2041E-01	0.1414E+00	0.0000E+00	0.1000E+01	0.5082E+06	0.0000E+00
BLOCK2	0.2150E-01	0.1451E+00	0.0000E+00	0.1000E+01	0.5082E+06	0.0000E+00
BLOCK3	0.2616E-01	0.1596E+00	0.0000E+00	0.1000E+01	0.5082E+06	0.0000E+00
QUEUE2	0.4326E+02	0.5349E+01	0.1600E+02	0.5000E+02	0.5082E+06	0.3900E+02
QUEUE3	0.4062E+02	0.8692E+01	0.1700E+02	0.5000E+02	0.5082E+06	0.3100E+02
QUEUE4	0.4182E+02	0.7117E+01	0.1500E+02	0.5000E+02	0.5082E+06	0.2100E+02

Figure 11-17 Summary report for discrete event blocking model.

where a clerk takes the order and travels to an adjacent warehouse to pick up the item. Clerks will service as many as six customers at a time. The time for the clerk to travel to the warehouse is uniformly distributed between 0.5 and 1.5 minutes. The time to find an item depends on the number of items the clerk must locate in the warehouse. This time is normally distributed with a mean equal to three times the number of items to be picked up. The standard deviation is equal to 0.2 of the mean. Thus, if one item is to be obtained from the warehouse, the time to locate the item is normally distributed with a mean of 3 and a standard deviation of 0.6 minutes. The time to return from the warehouse is uniformly distributed within the range of 0.5 and 1.5 minutes. When the clerk returns from the warehouse, the sale is completed for each customer. The time to complete the sale is uniformly distributed between 1 and 3 minutes. The completion of sales for the customers is performed sequentially in the same order that requests for items were made of the clerk. The time between customer requests for items is assumed to be exponentially distributed with the mean of 2 minutes. Three clerks are available to serve the customers. A schematic diagram of the store operation is shown in Figure 11-18.

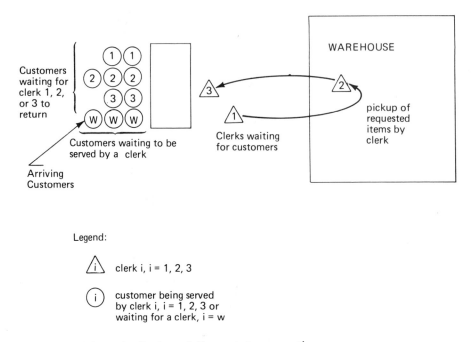

Legend:

⚠ clerk i, i = 1, 2, 3

(i) customer being served
by clerk i, i = 1, 2, 3 or
waiting for a clerk, i = w

Figure 11-18   Schematic diagram of discount store operation.

The objective of this simulation is to determine the utilization of the clerks, the time required to serve a customer from the time at which an item is requested until the completion of a sale, and the number of requests handled by a clerk for each trip to the warehouse. The simulation is to be run for 1000 minutes.

Concepts Illustrated. This example illustrates the use of a user-defined array for filing and removing the attributes of an entity.

SLAM II model. The discrete event model for this example consists of a customer arrival event and an end-of-service event. At first glance, it may also appear that events are required to explicitly model the clerk arrival and departure from the warehouse. However, no specific interactions occur during the time that the clerk is filling a request, and therefore the explicit modeling of these intermediate points by event routines is not necessary. If competition for resources were involved in filling the requests, then a more detailed model of the procedure for locating items would be required. Such a situation would exist if resources were required in the warehouse, say for example, a forklift truck.

There are five files employed in this example. The first four files are employed for storing the attributes of entities representing customers. Customers which arrive to the system when all clerks are busy are stored in file 4. Customers which have placed their orders and are waiting for clerk number 1, 2, or 3 to return reside in files 1, 2, and 3, respectively. File 5 is employed for storing the attributes of entities representing free clerks waiting for customers to arrive. The ATRIB buffer array is utilized for filing and removing customers from the first four files and the user-defined array CLERK is employed for filing and removing clerks from file 5. There is one attribute associated with the customer entities which is used to record the arrival time of the customer to the system. There is also one attribute associated with the clerk entities and it is used to denote the clerk number as 1, 2, or 3. Since a user-defined attribute buffer array must be dimensioned to at least 2 greater than MATR, the variable CLERK is dimensioned to 3.

The customer arrival event is assigned event code 1 and is coded in subroutine ARVL. The end-of-service event is assigned event code 2 and is coded in subroutine ENDSV. Subroutine EVENT(I) for this example is shown in Figure 11-19. It maps the event code onto the appropriate subroutine call.

```
SUBROUTINE EVENT(I)
GO TO (1,2),I
1 CALL ARVL
RETURN
2 CALL ENDSV
RETURN
END
```

Figure 11-19   Sample subroutine EVENT(I).

First, consider the logic associated with the customer arrival event. A flowchart for this event is depicted in Figure 11-20, and the coding is shown in Figure 11-21. Before processing the current arrival, the next arrival is scheduled to occur after an exponential time delay with a mean of 2 minutes by a call to subroutine SCHDL. The arrival time of the current customer is then recorded as ATRIB(1) of the entity. The disposition of the current customer request is determined based on the status of the clerks. A test is made on the number of free clerks waiting in file 5. If file 5 is empty, the customer must wait until a clerk becomes free. File 4 is used to store customers waiting for clerks. If the customer must wait, no further functions can be performed at the arrival event time and a return is made.

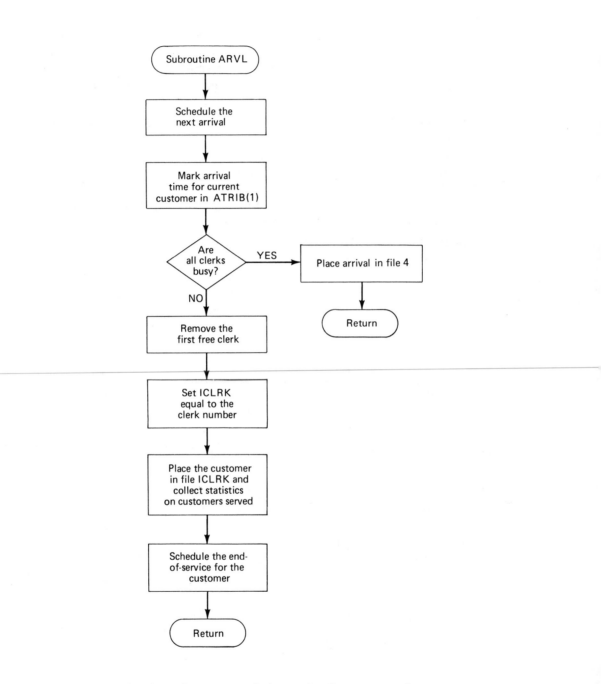

Figure 11-20 Flowchart of customer arrival event for discount example.

If a clerk is available, the order processing can be initiated. The first free clerk is removed from file 5 which causes the attributes of the clerk to be loaded into the array CLERK. The user variable ICLRK is then set to the first attribute of the clerk which corresponds to the clerk number. The attributes of the current customer which are stored in the array ATRIB are then inserted into file ICLRK to wait for the return and end-of-service processing by clerk number ICLRK. Statisitics are collected on the number of customers being served per clerk, which in this case is one. The statistics code for this collect variable is set as 2. The total time spent in traveling to the warehouse, in filling customer requests, returning from the warehouse, and completing the order is then computed as the variable DT. The end-of-service event is then scheduled to occur DT minutes later with the attributes of the clerk associated with the event. A return to SLAM is then made.

```
      SUBROUTINE ARVL
      COMMON/SCOM1/ ATRIB(100),DD(100),DDL(100),DTNOW,II,MFA,MSTOP,NCLNR
     1,NCRDR,NPRNT,NNRUN,NNSET,NTAPE,SS(100),SSL(100),TNEXT,TNOW,XX(100)
      DIMENSION CLERK(3)
C*****SCHEDULE NEXT ARRIVAL
      CALL SCHDL(1,EXPON(2.,1),ATRIB)
C*****MARK ARRIVAL TIME FOR CURRENT CUSTOMER
      ATRIB(1)=TNOW
C*****IF ALL CLERKS ARE BUSY
      IF(NNQ(5).GT.0) GO TO 10
C*****THEN PLACE ARRIVAL IN WAIT FILE AND RETURN
      CALL FILEM(4,ATRIB)
      RETURN
C*****ELSE REMOVE THE FIRST FREE CLERK
   10 CALL RMOVE(1,5,CLERK)
      ICLRK=CLERK(1)
C*****PLACE THE CUSTOMER IN FILE ICLRK AND COLLECT STATS ON NUMBER SERVED
      CALL FILEM(ICLRK,ATRIB)
      CALL COLCT(1.,2)
C*****AND SCHEDULE THE END OF SERVICE FOR THE CUSTOMER AND RETURN
      DT=UNFRM(.5,1.5,1)+RNORM(3.,.6,1)+UNFRM(.5,1.5,1)+UNFRM(1.,3.,1)
      CALL SCHDL(2,DT,CLERK)
      RETURN
      END
```

Figure 11-21 SLAM II code for customer arrivals to discount store.

Next, consider the logic for the end-of-service event. A flowchart for this event is given in Figure 11-22 and the SLAM II coding is shown in Figure 11-23. The attributes associated with the event are passed by SLAM to the event routine in the ATRIB buffer array. Since both clerk and customer attributes will be used in this subroutine, the ATRIB buffer is copied into the user-defined array CLERK. The user variable ICLRK is then set to CLERK(1) and denotes the number of the clerk

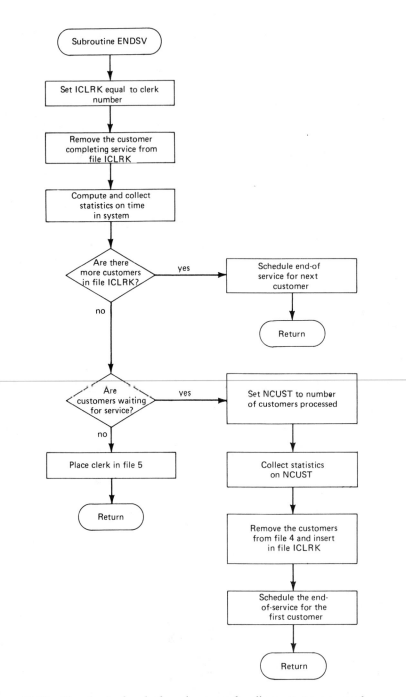

Figure 11-22 Flowchart of end-of-service event for discount store example.

processing the end-of-service for the customer. The customer completing service is then removed from file ICLRK and its attributes are loaded by SLAM into the ATRIB buffer array. The time in the system for the customer is computed and statistics are collected on this value as collect variable number 1. A test is then made on the SLAM II function NNQ(ICLRK) to determine if more customers remain. If NNQ(ICLRK) is greater than zero, the end-of-service event for the next customer is scheduled followed by a return. Otherwise, a test is made on NNQ(4) to determine if there are customers waiting to place their orders with a clerk. If customers are waiting, then the user variable NCUST is set to the number of cus-

```
      SUBROUTINE ENDSV
      COMMON/SCOM1/ ATRIB(100),DD(100),DDL(100),DTNOW,II,MFA,MSTOP,NCLNR
     1,NCRDR,NPRNT,NNRUN,NNSET,NTAPE,SS(100),SSL(100),TNEXT,TNOW,XX(100)
      DIMENSION CLERK(3)
      CLERK(1)=ATRIB(1)
C*****REMOVE THE CUSTOMER COMPLETING SERVICE FROM THE WAIT FILE
      ICLRK=CLERK(1)
      CALL RMOVE(1,ICLRK,ATRIB)
C*****COLLECT STATISTICS ON TIME IN SYSTEM
      TSYS=TNOW-ATRIB(1)
      CALL COLCT(TSYS,1)
C*****IF MORE CUSTOMERS REMAIN IN THE WAIT FILE
      IF(NNQ(ICLRK).EQ.0) GO TO 10
C*****THEN SCHEDULE THE END OF SERVICE FOR THE NEXT CUSTOMER AND RETURN
      CALL SCHDL(2,UNFRM(1.,3.,1),CLERK)
      RETURN
C*****ELSE
   10 CONTINUE
C*****IF CUSTOMERS ARE WAITING FOR SERVICE
      IF(NNQ(4).EQ.0) GO TO 30
C*****THEN SERVICE UP TO SIX CUSTOMERS
      NCUST=NNQ(4)
      IF(NCUST.GT.6) NCUST=6
      XCUST=NCUST
      CALL COLCT(XCUST,2)
      DO 20 J=1,NCUST
      CALL RMOVE(1,4,ATRIB)
   20 CALL FILEM(ICLRK,ATRIB)
C*****AND SCHEDULE THE END OF SERVICE FOR THE FIRST CUSTOMER AND RETURN
      XMN=NCUST*3
      STD=.2*XMN
      DT=UNFRM(.5,1.5,1)+RNORM(XMN,STD,1)+UNFRM(.5,1.5,1)+UNFRM(1.,3.,1)
      CALL SCHDL(2,DT,CLERK)
      RETURN
C*****ELSE PLACE THE CLERK IN THE FILE OF FREE CLERKS AND RETURN
   30 CALL FILEM(5,CLERK)
      RETURN
      END
```

Figure 11-23 SLAM II code for customer end-of-service event at discount store.

tomers to be processed and statistics are collected on this quantity as collect variable number 2. These customers are then removed from file 4 and inserted into file ICLRK to await end-of-service processing. The total time spent in traveling to the warehouse, in filling customer requests, returning from the warehouse, and completing the order for the first customer is then computed as a variable DT. Note that the time to fill requests is obtained as a sample from a normal distribution with a mean XMN equal to the number of customrs, NCUST, times 3. The standard deviation, STD, is two-tenths of the mean, that is, STD = 0.2*XMN. The end-of-service event for the first customer to be processed is then scheduled to occur DT minutes later, followed by a return. In the case where there are no customers waiting to place their orders, the attributes of the clerk are placed in file 5 which is the file used for storing free clerks.

Subroutine INTLC is used to put the three clerks into file 5 as it is assumed that all clerks are idle at the beginning of the simulation run. The first customer arrival event is scheduled to occur at time 0.0 by calling SCHDL with an event code of 1. The listing of subroutine INTLC is given in Figure 11-24. The input statements for this example are shown in Figure 11-25.

```
      SUBROUTINE INTLC
      COMMON/SCOM1/ ATRIB(100),DD(100),DDL(100),DTNOW,II,MFA,MSTOP,NCLNR
     1,NCRDR,NPRNT,NNRUN,NNSET,NTAPE,SS(100),SSL(100),TNEXT,TNOW,XX(100)
      DIMENSION CLERK(3)
C*****ESTABLISH THREE CLERKS IN FILE 5
      DO 10 J=1,3
      CLERK(1)=J
   10 CALL FILEM(5,CLERK)
C*****SCHEDULE ARRIVAL EVENT AT TNOW AND RETURN
      CALL SCHDL(1,0.,ATRIB)
      RETURN
      END
```

Figure 11-24 Subroutine INTLC for discount store example.

```
    1   GEN,PRITSKER,DISCOUNT STORE,1/25/84,1;
    2   LIMITS,5,1,50;
    3   STAT,1,TIME IN SYSTEM;
    4   STAT,2,CUST PER SERVICE;
    5   INIT,0,1000;
    6   FIN;
```

Figure 11-25 Input statements for discount store example.

Summary of Results. The SLAM II Summary Report for this example is given in Figure 11-26. The report reveals that during the 1000 minutes of simulated operation, 517 customers were processed through the system with customers spending an average of 24.49 minutes in the system. The average number of customers per trip by each clerk was 3.040 with a total of 175 trips completed by the clerks. These results indicate that management should investigate the hiring of additional clerks or change the policy regarding the minimum or maximum number of customers a clerk should serve at one time.

Although the average utilization of the clerks is not directly provided by the report, this statistic can be estimated using the average length of file 5. Since idle clerks reside in file 5, the average number of busy clerks is the number of clerks minus the average number in file 5, that is, 2.9274.

```
                    S L A M   I I   S U M M A R Y   R E P O R T


              SIMULATION PROJECT DISCOUNT STORE        BY PRITSKER

              DATE  1/25/1984                          RUN NUMBER   1 OF    1


              CURRENT TIME   0.1000E+04
              STATISTICAL ARRAYS CLEARED AT TIME  0.0000E+00



                  **STATISTICS FOR VARIABLES BASED ON OBSERVATION**

                    MEAN      STANDARD    COEFF. OF    MINIMUM     MAXIMUM    NUMBER OF
                    VALUE     DEVIATION   VARIATION    VALUE       VALUE      OBSERVATIONS

      TIME IN SYSTEM   0.2449E+02   0.1049E+02   0.4282E+00   0.5227E+01   0.5190E+02      517
      CUST PER SERVICE 0.3040E+01   0.1955E+01   0.6429E+00   0.1000E+01   0.6000E+01      175



                  **FILE STATISTICS**

      FILE   ASSOC NODE    AVERAGE    STANDARD    MAXIMUM   CURRENT   AVERAGE
      NUMBER LABEL/TYPE    LENGTH     DEVIATION   LENGTH    LENGTH    WAITING TIME

        1                  3.7121     1.9273        6         6       20.0657
        2                  3.2362     1.8902        6         3       19.1489
        3                  3.3605     1.9725        6         6       18.8794
        4                  2.6481     2.7833       17         1        5.3713
        5                  0.0726     0.2950        3         0        1.8157
        6      CALENDAR    3.9274     0.2950        4         4        3.7262
```

Figure 11-26   SLAM II summary report for discount store example.

11.14 CHAPTER SUMMARY

This chapter presents the SLAM II discrete event simulation modeling procedures. The procedures for scheduling events, file manipulations, and statistical collection routines are described. A list of important discrete event variables is presented in Table 11-2. Only one additional input statement, STAT, is required. Two illustrations and one example of discrete event modeling using SLAM II are presented. The example involves a discount store where clerks must go through multiple operations before completing service on customers.

The organization of a SLAM II program for discrete event modeling is illustrated in Figure 11-27.

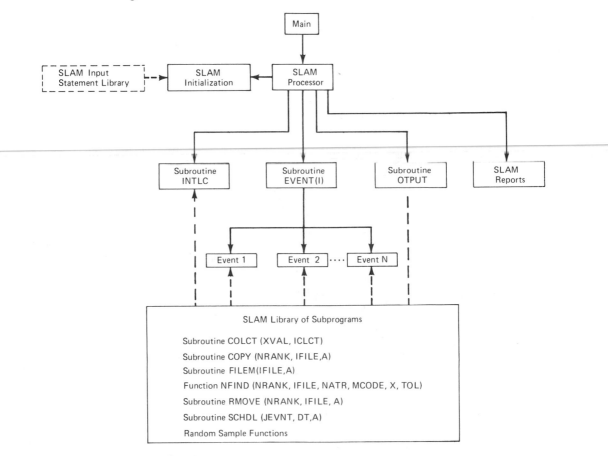

Figure 11-27 SLAM II organization for discrete event modeling.

11.15 EXERCISES

11-1. Simulate a single server system where service is exponentially distributed with a mean service time of one hour and there are two types of arrivals. Arrival type 1 is an item of high priority and waits only for the processing of other high priority items. The time intervals between arrivals of high priority items is exponentially distributed with a mean time between arrivals of four hours. The time between arrivals of low priority items is also exponentially distributed, but has a mean time between arrivals of two hours. Obtain statistics regarding server utilization and information concerning the time in the system and in the queue for high and low priority items. Assume that service on low priority items is not interrupted to process high priority items.

Embellishment: Repeat the simulation with the assumption that high priority items interrupt the servicing of low priority items. Interrupted items are inserted at the beginning of the queue of low priority items. Show that the time remaining for those items interrupted is exponentially distributed with a mean time equal to the mean service time.

11-2. Jobs arrive at a job shop so that the interarrival time is exponentially distributed with a mean time between arrivals of 1.25 hours. The time to process a job is uniformly distributed in the interval 0.75 to 1.25 hours. The time to process the jobs is estimated prior to the time the job is performed. Simulate the processing of jobs in the order of shortest processing time first. Obtain statistics describing the average time jobs are in the shop, the variation of the time the jobs are in the shop, and the utilization of the server. Assume that the *actual* processing times for the jobs are equal to the *estimates* of the processing times. Repeat the simulation with jobs with the longest processing time performed first. Then carry out the simulation with the jobs processed in order of their arrival time. Compare the results.

Embellishment: Repeat the simulation for a situation in which the actual processing time is equal to the estimated processing time plus a sample drawn from a normal distribution with a mean of zero and a standard deviation of 0.2.

11-3. Build a discrete event model for the two work stations in series described in Example 5-1.

Embellishment:

(a) Before subcontracting a unit, check the end-of-service times for the work stations. If the next end-of-service is within 0.1 hours of the current time, the unit is not subcontracted but returns to the work station in 0.2 hour.

(b) Redevelop the work station model under the assumption that the manufacturer only maintains 100 units. The time to failure for each unit is exponentially distributed with a mean of 40 hours. The time to repair a unit if subcontracted is triangularly distributed with a mode of 4, a minimum of 3, and a maximum of 6.

11-4. Develop a discrete event simulation of the PERT network given in Example 7-1. Hint: The simulation of PERT networks involves a single event representing the end of an activity. Perform a simulation of the PERT network for 400 runs.

Embellishment: Estimate the probability that an activity is on the critical path in a run. This probability is referred to as a criticality index.

11-5.   Develop a discrete event model of the inventory situation presented in Example 6-1. Embellishments:

(a) Make the lead time for the receipt of orders lognormally distributed with a mean of 3 and standard deviation of 1.

(b) Include in the model information that each customer may demand more than one radio. The additional number of radios demanded per customer is Poisson distributed with a mean of 2. Thus, the expected number demanded per customer is 3. Presume if there are insufficient radios to meet the total number of radios demanded by a customer, the entire order is backordered with probability 0.2.

(c) Convert the periodic review system to a transaction reporting system where the inventory position is reviewed after every sale or backorder.

(d) Make the mean time between customer demands a function of the number of lost sales and backorders. If the average number of lost sales and backorders per week is less than 1, the mean time between demands is 0.18 week. If the average number of lost sales and backorders is greater than 2, the mean time between demands is 0.22. Otherwise, the mean time between demands is 0.20.

(e) Add the following income and cost structure to the inventory situation. Radios sell for $65 and cost the store $40. An inventory carrying charge of $0.004 per dollar/week is used. The cost of placing an order for radios from the supplier is estimated to be $50. Each review of inventory costs $30. A lost sale results in a loss of goodwill and is estimated to cost $20. The cost for maintaining a backorder is $10. Run a simulation of this inventory situation for 312 weeks.

(f) Evaluate the effect of alternative inventory policies on average profit using the cost and income data provided in embellishment (e).

11-6.   Build a discrete event model for the conveyor situation described in Exercise 9-5.

11-7.   Build a discrete event model of drive-in bank with jockeying, Example 9-1. Embellishment: Enlarge the model by including the departure process of cars into the street next to the bank. Cars departing from the bank tellers can enter the street when a large enough gap exists between cars traveling on the street. The time between such gaps is uniformly distributed between 0.3 and 0.5 minutes. The space following the tellers is only sufficient to allow 3 cars to wait for a gap in the street traffic.

11-8.   Embellishment to Example 11-1. Alter the discount store operation so that clerks wait for 1 minute before traveling to the warehouse to serve a single customer. Determine the effect of this new policy on customer waiting time and clerk utilization. Evaluate alternative time intervals for waiting for additional customers before traveling to the warehouse.

11-9.   Embellishment to Example 11-1. In the simulation of the discount store operation, it has been determined that there is a probability that the clerk will return from the warehouse with an incorrect item. When this occurs the clerk immediately returns to the warehouse to obtain the correct item. The probability of retrieving an incorrect item is 0.15. In addition, there is a probability of 0.1 that the item requested was not in inventory. When this occurs, there is a probability of 0.25 that the customer will request a new item. Evaluate the effect of this new information on clerk utilization and customer waiting time.

11-10. For Exercise 11-4, determine the number of simulation runs that would be required to obtain a 95% probability that: (a) the estimate of the project duration is within 0.1 of a standard deviation of the theoretical project duration: (b) the estimate of the variance of the project duration is within 10% of the theoretical variance of the project duration; and (c) the difference between the observed and theoretical criticality indices is less than or equal to 0.005.

11.16   REFERENCES

1. Henriksen, J. O., "GPSS — Finding the Appropriate World-View," *Proceedings, 1981 Winter Simulation Conference*, 1981, pp. 505-516.
2. Kiviat, P. J., R. Villanueva, and H. Markowitz, *The SIMSCRIPT II Programming Language*, Prentice-Hall, 1969.
3. O'Reilly, J. J., SLAM II® Quick Reference Manual, Pritsker & Associates, 1984.
4. Pritsker, A. A. B., and P. J. Kiviat, *Simulation with GASP II*, Prentice-Hall, 1969.
5. Pritsker, A. A. B., *The GASP IV Simulation Language*, John Wiley, 1974.
6. Schruben, L., "Simulation Modeling with Event Graphs", *Communications of the ACM*, Vol. 26, 1983, pp. 957-963.

CHAPTER 12

Advanced Discrete Event Concepts and Subprograms

12.1 INTRODUCTION

The procedures presented in Chapter 11 provide sufficient information to model complex systems using a discrete event approach. Many support routines are included in SLAM II to perform functions common to a variety of simulation models. In this chapter, we present the subprograms which are of sufficient generality to be included in the standard version of SLAM II. By including these functions in SLAM II, standard names will be employed within simulation models and, thus, program readability will be enhanced. The addition of subprograms, of course, tends to increase the perceived complexity of the language. Herein lies the language developers' design tradeoff. Since SLAM II is FORTRAN based, it is a relatively simple matter to add new functions to the language. We encourage such additions as they will keep the language growing and evolving. We also believe in a natural evolution and expect the survival of only those functions which make significant contributions to simulation modeling.

In this chapter, we present information on: additional file manipulation routines; auxiliary attribute subroutines; report writing subroutines; functions for accessing statistical quantities; and miscellaneous support routines. Illustrations and examples of the uses of these routines are presented throughout this chapter.

12.2 ADDITIONAL FILE MANIPULATION ROUTINES

The primary function of the SLAM II file storage array is to store entries, events and entities, with their associated attributes. Entries are stored in a single, one-dimensional array. By the use of an equivalence statement, EQUIVALENCE (NSET(1),QSET(1)), it is possible to use the same storage array for either real or integer type data. This technique permits entry pointers to be stored as integers, and entry attributes to be stored as reals. To simplify reference to the storage array in this text, the name NSET is used to denote the array when either name applies. Normally the array is not referenced directly by the SLAM II user. However, if the user does reference an attribute value, the QSET variable must be used. If a pointer is referenced, the NSET variable must be used.

NSET is one dimensional and resides in unnamed (blank) COMMON. Because NSET resides in COMMON storage, it is accessible to the SLAM II subroutines without having to be included as an argument in their calling sequence. Since NSET is in unnamed COMMON, the user can dimension NSET in the main program, and all subprograms can be precompiled with NSET dimensioned to one.

12.2.1 Functions for Accessing File and Entry Pointers: FUNCTIONS MMFE, MMLE, NPRED, and NSUCR

Entries in a file can be thought of as items on a list, customers in a queue, days on a calendar, and so forth. The entries stored in a file are considered by SLAM II in a logical order, not a physical order. The first and second entries in a file are not necessarily stored next to one another. Their order in the file is maintained by a pointing system. Each entry potentially has a predecessor and successor entry in the file. Pointers are used to keep track of each entry's predecessor and successor. The pointers establish the location in NSET where the entries are stored. A zero value for a pointer indicates that either no predecessor or no successor exists.

Pointers in SLAM II reference a location in the array NSET where the first word of an entry is stored. The number of words associated with an entry is known; hence all information about an entry can be obtained given the pointer to the starting location of the entry. The pointer to the first entry in file IFILE is obtained by using the function MMFE(IFILE). The function MMLE(IFILE) returns the pointer to the last entry in file IFILE. Thus, the statement NTRY = MMLE(3) defines NTRY as the location in NSET where the first piece of information concerning the last entry in file 3 is stored.

By knowing the pointer to the first entry and by accessing the successor of the first entry, we can obtain the pointer to the second entry. The function NSUCR (NTRY) is used to obtain the pointer to the location of the successor of NTRY. With this function it is possible to proceed in a forward direction from a particular entry in a file until the last entry is encountered. The last entry in a file has no successor and, hence, the function NSUCR would return a zero value when the successor of the last entry is requested. To search through a file from the last entry, MMLE(IFILE), to the first, the function NPRED(NTRY) is provided.

As an example of the use of these functions, consider the problem of locating an entry in file 4 whose second attribute is equal to seven. We desire to locate the entry that is closest to the beginning of the file. The code for accomplishing this search is shown below.

```
C****  SEARCH FILE 4 FROM THE BEGINNING FOR AN ENTRY
       NEXT = MMFE(4)
C****  IF NEXT .EQ. 0, THE SEARCH ENDS
   10  IF (NEXT .EQ. 0) RETURN
       CALL COPY (-NEXT,4,ATRIB)
       IF (ATRIB(2) .EQ. 7.0) GO TO 20
       NEXT = NSUCR(NEXT)
       GO TO 10
C****  AT STATEMENT 20, NEXT IS THE DESIRED ENTRY
   20  CONTINUE
```

In this code, the variable NEXT is used to indicate the next entry to be tested for the desired condition. We start with the first entry in file 4 because the search is to start from the beginning of the file. If NEXT=0 then there are no entries in the file, as function MMFE returns a zero value when the file is empty.

The statement CALL COPY (-NEXT,4,ATRIB) causes the attributes of the entry whose pointer is NEXT to be copied into the ATRIB buffer. In this statement, the negative of the pointer to the location of the entry is used to notify subroutine COPY that a specific location is being passed to it rather than the rank of the entry in the file. As will be discussed later, if the variable used to indicate the rank of an entry has a negative value, the SLAM II subprograms assume that it is a pointer to a location rather than the value of the rank in the file. Subroutine COPY loads the array given as the third argument with the attributes of the entry. In the next statement, the particular attribute number is tested against 7.0 to determine if the prescribed condition is met. If it is, a transfer to statement 20 is made as NEXT has been established as the pointer to the desired entry. If the condition was not met, the successor entry is obtained through the use of function NSUCR. A transfer to statement 10 is made to determine if a successor entry exists.

If it is desired to search file 4 from the end of the file then only minor changes are required in the code. The first entry to be tested is established as the last entry in the file, that is, NEXT=MMLE(4). In addition, it is necessary to obtain the next entry to be tested by using function NPRED, that is, NEXT= NPRED(NEXT).

12.2.2   The Availability File

The space in the file structure that is not used for storing entries can be considered as available locations for storing new entries. This, in essence, is an availability file which is maintained in SLAM II as a linked list. The pointer to the first available space for storing an entry is maintained as the SLAM II variable MFA. The second available space for storing an entry can be accessed by using the function NSUCR, that is, NMFA=NSUCR(MFA) where NMFA is the pointer to the location in NSET where the second entry to be filed will be stored. When subroutine FILEM is called, the entry to be filed is stored in NSET starting at location MFA. The value of MFA is then updated to its successor value. Through knowledge of MFA and its successor, the SLAM II user can ascertain where entries are filed within the file structure. Thus, if two calls to subroutine FILEM are made in succession, the user can obtain the information as to where the second entry associated with the second call to FILEM will be stored. Whenever subroutine RMOVE is called, the value of MFA is updated to be equal to the pointer of the entry that is removed from the file. The successor of this entry is the value of MFA prior to the call to subroutine RMOVE. In this way, the list of available locations for storing entries in the file is linked.

12.2.3   File Rank to Pointer Translation: FUNCTION LOCAT

Since SLAM II allows the referencing of entries in files both by the pointer to the entry and by the rank of the entry, function LOCAT has been included for translating the rank of an entry to the pointer to the location of the entry. The arguments to LOCAT are an entry's rank and file number and, hence, we have: FUNCTION LOCAT(IRANK,IFILE). Thus to obtain the pointer for the entry that is ranked third in file 7, the following statement would be used

NTRY = LOCAT(3,7)

12.2.4   Using SLAM II Routines With Pointer Values

As noted above, whenever a negative value is placed in a subroutine call where SLAM II expects the rank of the entry in a file to be given, SLAM II interprets the value as the negative of the pointer to the location of the entry. This feature pertains to the following subprograms: COPY, NFIND, RMOVE, and ULINK. With this convention either of the following two statements will remove the last entry in file 6 and place the attributes of the entry in the array A:

CALL RMOVE(NNQ(6),6,A)
or
CALL RMOVE(-MMLE(6),6,A)

12.2.5   Direct Accessing of Attributes

If the pointer to an entry in the array NSET/QSET is known, then a specified attribute can be accessed directly. The value of the Ith attribute of the entry with a pointer of NTRY is stored in the QSET array in location NTRY + I. Thus the second attribute of the first entry in file 1 can be directly accessed by using the statements NTRY = MMFE(1) and A2 = QSET(NTRY + 2). In this way an attribute value can be examined or changed without the use of the SLAM II filing subroutines. However, the ranking attribute of an entry should not be changed in this manner.

12.2.6   Entry Removal and Filing Without Attribute Copying: SUBROUTINES ULINK and LINK

An entry can be unlinked from other entries in a file without its attribute values being copied or deleted from NSET. The subroutine that provides this capability is subroutine ULINK(NRANK,IFILE). When subroutine ULINK is called, the value of MFA is updated but the attributes of the entry remain at the same location. The entry is no longer considered as a member of file IFILE. If the next file operation is a call to subroutine LINK(IFILE) then SLAM II presumes that the attributes of the entry to be inserted into file IFILE are already located in QSET with a pointer value of MFA. As an example, consider the removal of the first entry in file 3 and the filing of this entry into file 2. The following statement accomplishes this without removing the attribute values from QSET:

    CALL ULINK(1,3)
    CALL LINK(2)

Since ULINK and LINK do not copy the attributes out of and into NSET/QSET, they execute faster than RMOVE and FILEM.

12.2.7   Two Standard Functions for File Attribute Computation

SLAM II provides functions for obtaining the sum or product of one attribute value for each entry in a file. These functions are described below.

Function SUMQ(NATR,IFILE). Occasionally it is desired to find the sum of all values of an attribute for each entry in a file. For example, if entries in a file represent jobs in a queue waiting for a machine with ATRIB(1) of each entry denoting the processing time for that job, it might be desired to find the total processing time for all jobs waiting in the queue. Function SUMQ(NATR,IFILE) provides this capability by returning SUMQ as the sum of all values of attribute number NATR currently in file IFILE. As an example of the use of function SUMQ, the following statement sets the variable TOTAL equal to the sum of all values of attribute 1 corresponding to the entries currently in file 4.

    TOTAL = SUMQ(1,4)

Function PRODQ(NATR,IFILE). Sometimes it is desired to find the product of all values of an attribute for each entry in a file. For example, if entries in a file represent parallel components in a system with the probability of failure of each entry given by ATRIB(3), then the probability that all components would fail would be the product of all values of ATRIB(3) in the file. Function PRODQ (NATR,IFILE) provides the capability for computing this product by returning PRODQ as the product of all values of attribute NATR currently in file IFILE. For example, the following statement would set the variable PFAIL to the product of all values of ATRIB(3) corresponding to the entries currently in file 2.

PFAIL = PRODQ(3,2)

12.2.8 Summary of File Manipulation Routines

A list of file manipulation functions is given in Table 12-1. It should be noted that a zero value is used as an indicator that no first or last entry exists and that

Table 12-1 List of file manipulation subprograms.

Function	Description†
LOCAT(IRANK,IFILE)	Returns pointer to location of entry with rank IRANK in file IFILE
MMFE(IFILE)	Returns pointer to first entry (rank 1) in file IFILE
MMLE(IFILE)	Returns pointer to last entry (rank NNQ(IFILE)) in file IFILE
NNQ(IFILE)	Returns number of entries in file IFILE
NPRED(NTRY)	Returns pointer to the predecessor entry of the entry whose pointer is NTRY
NSUCR(NTRY)	Returns pointer to the successor entry of the entry whose pointer is NTRY
PRODQ(NATR,IFILE)	Returns the product of the values of attribute NATR for each current entry in file IFILE
SUMQ(NATR,IFILE)	Returns the sum of the values of attribute NATR for each current entry in file IFILE
Subroutine	Description
LINK(IFILE)	Files entry whose attributes are stored in MFA in file IFILE
ULINK(NRANK,IFILE)	Removes entry with rank NRANK from file IFILE without copying its attribute values. If NRANK < 0, it is a pointer

† A pointer value of zero indicates that no entry exists that satisfies the desired function.

an entry has no predecessor or successor. Also note that the following functional equivalents exist:

MMFE(IFILE) = LOCAT(1,IFILE)
and
    MMLE(IFILE) = LOCAT(NNQ(IFILE),IFILE)

For the file processing routines, a positive argument is used to signify a rank value and a negative argument signifies a pointer.

12.3   ILLUSTRATION 12-1. RENEGING FROM A QUEUE

In a standard queueing situation, a customer who joins a queue stays in the queue until service starts. Sometimes a customer decides that it is taking too long to obtain service and leaves the queue before being served. This process of leaving the queue after waiting in it is called reneging.

To model this situation involves embellishing the standard queueing model involving arrival and end-of-service events. For each customer, an event is placed on the event calendar which we will call the renege event. When the renege event occurs, the file of waiting customers could be searched and, if the customer entity is still in the waiting file, the customer would be removed. The disposition of the customer who reneges is then determined. For the customer who starts service before the renege event occurs, no action is necessary at the renege event time.

Note that the process described above employs information about when a customer will renege and uses this information to place a future event on the calendar. This process is different than the thought process used by the customer who upon entering the queue might not even consider leaving it. The modeler, however, is involved in describing the events that can change the status of the system, and it is necessary to project into the future those events that might occur to change system status.

The above approach to modeling a renege event involves the search of the customer queue for each renege event. A more efficient approach is to have an attribute of the renege event point to the entity representing the customer in the waiting file and to have an attribute of the customer entity maintain a reference to the location of the renege event on the event calendar. If this is done, then when the customer entity is removed from the waiting file to start service, the renege event can be removed from the event calendar using the pointer information kept as an attribute of the customer entity. If the renege event occurs before it is deleted from the event calendar, then the customer entity can be removed using the pointer to the location where the customer entity is stored.

The FORTRAN coding for the arrival event described in Chapter 11 including the code for renege processing is shown below. In this code, the time before reneging by a customer is assumed to be lognormally distributed with a mean of 10 and a standard deviation of 2.

```
      SUBROUTINE ARVL
      COMMON/SCOM1/ATRIB(100),DD(100),DDL(100),DTNOW,II,MFA,MSTOP,NCLNR
     1,NCRDR,NPRNT,NNRUN,NNSET,NTAPE,SS(100),SSL(100),TNEXT,TNOW,XX(100)
      EQUIVALENCE(XX(1),BUSY)
C
C**** SCHEDULE SUBSEQUENT ARRIVALS
C
      CALL SCHDL(1,EXPON(20.,1),ATRIB)
C
C**** ASSIGN ATTRIBUTE 1 THE MARK TIME
C
      ATRIB(1)=TNOW
      IF (BUSY.EQ.0.)GO TO 10
C
C**** START OF CODE ADDED FOR RENEGE PROCESSING
C**** MPREN IS POINTER TO RENEGE EVENT
C**** MPCQ IS POINTER TO CUSTOMER IN QUEUE
C
      MPREN=MFA
      MPCQ=NSUCR(MFA)
C
C**** SET ATTRIBUTE 2 OF RENEGE EVENT TO POINTER FOR CUSTOMER
C
      ATRIB(2)=MPCQ
C
C**** SCHEDULE RENEGE EVENT WITH CODE 3
C
      CALL SCHDL(3,RLOGN(10.,2.,1),ATRIB)
C
C**** SET ATTRIBUTE 3 OF CUSTOMER TO POINTER FOR RENEGE EVENT
C
      ATRIB(3)=MPREN
C
C**** END OF CODE ADDED FOR RENEGE PROCESSING
C**** PUT CUSTOMER IN QUEUE (FILE 1)
C
      CALL FILEM(1,ATRIB)
      RETURN
C
C**** IF THE SERVER IS NOT BUSY, MAKE THE SERVER BUSY AND
C**** SCHEDULE THE END OF SERVICE (EVENT 2--ENDSV)
C
   10 BUSY=1.0
      CALL SCHDL(2,UNFRM(10.,25.,1),ATRIB)
      RETURN
      END
```

As discussed in Section 12.2.2, MFA is the pointer to the next available location for storing an entry on either the event calendar or in a file. The renege event is scheduled first, therefore, the renege event is stored in the location pointed to by MFA. After scheduling the renege event, the customer will be put in file 1 and will be stored in the location pointed to by the successor to MFA. Attribute 2 of the renege event is defined as the pointer to the location of the customer in the queue. A renege event is scheduled with the event code 3 by a call to subroutine SCHDL.

Attribute 3 of the customer entity is set equal to the pointer to the renege event, MPREN, and the customer is placed in file 1 by a call to subroutine FILEM. This completes the code that would be placed in subroutine ARVL to establish a customer with an attribute pointing to the renege event and the scheduling of a renege event with an attribute that points to the location of a customer in the queue.

A customer is removed from file 1 when the server has finished processing a customer which is modeled in the end-of-service event. The coding of subroutine ENDSV with the code necessary to delete the renege event from the event calendar when a customer is placed into service is shown below.

```
      SUBROUTINE ENDSV
      COMMON/SCOM1/ATRIB(100),DD(100),DDL(100),DTNOW,II,MFA,MSTOP,NCLNR
     1,NCRDR,NPRNT,NNRUN,NNSET,NTAPE,SS(100),SSL(100),TNEXT,TNOW,XX(100)
      EQUIVALENCE (XX(1),BUSY)
C
C**** DETERMINE THE TIME THE ENTITY WAS IN THE SYSTEM AND
C**** COLLECT STATISTICS ON THAT TIME
C
      TSYS=TNOW-ATRIB(1)
      CALL COLCT(TSYS,1)
C
C**** DETERMINE IF ANY ENTITIES ARE WAITING FOR THE SERVER.
C**** IF NOT, MAKE THE SERVER IDLE.
C
      IF(NNQ(1).GT.0) GO TO 10
      BUSY=0.
      RETURN
C
C**** IF ENTITIES ARE WAITING FOR THE SERVER, REMOVE THE FIRST
C**** FROM FILE 1 AND SCHEDULE AN END OF SERVICE 'ENDSV'.
C
   10 CALL RMOVE(1,1,ATRIB)
      CALL SCHDL(2,UNFRM(10.,25.,1),ATRIB)
C
C**** CODE TO CANCEL RENEGE EVENT
C
      MPREN=ATRIB(3)
      CALL RMOVE(-MPREN,NCLNR,ATRIB)
      RETURN
      END
```

MPREN is set equal to ATRIB(3) which is the pointer to a renege event. Subroutine RMOVE is called with the negative of MPREN with the file identified as the event calendar, NCLNR.

The code for the renege event is written in subroutine RENEGE which is shown below.

```
      SUBROUTINE RENEGE
      COMMON/SCOM1/ATRIB(100),DD(100),DDL(100),DTNOW,II,MFA,MSTOP,NCLNR
     1,NCRDR,NPRNT,NNRUN,NNSET,NTAPE,SS(100),SSL(100),TNEXT,TNOW,XX(100)
      DIMENSION A(5)
C
C**** REMOVE CUSTOMER FROM QUEUE (FILE 1)
C**** ATRIB(2) IS POINTER TO CUSTOMER ENTITY
C
      MPCQ=ATRIB(2)
      CALL RMOVE(-MPCQ,1,A)
C
C**** ADD CODE NECESSARY FOR DISPOSITION
C**** OF RENEGING CUSTOMER WITH ATTRIBUTES IN A( )
C
      RETURN
      END
```

In the renege event, the customer is removed from file 1. If the customer had been put into service prior to this time, then the renege event would have been canceled, that is, subroutine RENEGE would not have been invoked. MPCQ is set equal to ATRIB(2), the pointer to the customer entity in file 1. Before calling subroutine RENEGE, SLAM II has reset ATRIB(2) to be equal to the second attribute of the renege event. Subroutine RMOVE is called to remove this customer entity from file 1 using the negative of the pointer to the customer. For a particular problem, the code necessary to model the disposition of the reneging customer would then be inserted into the subroutine.

This illustration demonstrates how knowledge of the workings of SLAM II's filing system can be used to perform the logic associated with reneging from a queue.

12.4 THE SLAM FILING SYSTEM

The SLAM filing system employs a double link list pointing system for storing entities in files and future events on the event calendar. Each entry consists of a set of real-valued attributes and is linked to other entries in the file by a predecessor and successor pointer. The attribute values and associated pointers are stored as a group in the companion arrays NSET and QSET. By use of the equivalence statement, EQUIVALENCE(NSET(1),QSET(1)), it is possible to use a single contiguous storage area for storing both integer pointers and real-valued attributes.

An entry in SLAM II has the following general form.

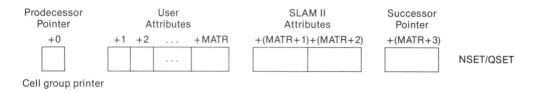

Cell group printer

The quantities stored in the cells of NSET/QSET for a single entry are: a predecessor pointer; user attributes; SLAM II attributes; and a successor pointer. The plus sign before the number above a cell indicates that the index of NSET is computed by adding the value of the cell group pointer to the value after the plus sign. Thus, the pointer to the predecessor of the entry whose cell group pointer is NTRY is obtained from NSET(NTRY). MATR is the value of the number of attributes input by the user on the LIMITS statement. User defined attributes of an entry are stored in QSET at NTRY+1 through NTRY+MATR. Thus, the second attribute of an entity is obtained from QSET(NTRY+2). SLAM II uses two cells for storing internal information about an entry. These are stored following the user-defined attributes. For the event calendar, these two attributes are used to store the event code at cell NTRY+MATR+1 and the event time in cell NTRY+MATR+2. The pointer to the successor of an entry is stored in cell NTRY+MATR+3.

There is no file number associated with a group of cells. The same space is used for storing entries in any file. File membership is determined through the predecessor and successor pointers in conjunction with the pointer to the first entry in a file, MMFE(IFILE), or the pointer to the last entry in the file, MMLE(IFILE).

On initialization, SLAM II organizes NSET/QSET into groups of cells to accommodate the general entry structure described above. Initially all groups of cells are included in the availability file with the availability pointer, MFA, set to 1. The successor pointer for each group of cells is set equal to the cell number following the location of the successor pointer. An illustration of the file structure with emphasis on pointers is given below where Ci is used to indicate a cell group pointer value. If MATR is set to 3 then the values for Ci would be C1=1,C2=7,C3=13, C4=19,C5=25,C6=31 and so on. To simplify the presentation, a series of operations on the file structure is presented without attribute values included. The operations illustrate the changes to the predecessor and successor pointers and first and last entry pointers when calls are made to subroutines FILEM and RMOVE. The illustration is for two files with the priority of file 1 specified as FIFO and the priority of file 2 specified as LIFO.

PRIORITY/1,FIFO/2,LIFO;
<u>OPERATION</u>

1. Initialization   MFA = C1

	1	2		1	2		1	2
MMFE	0	0	MMLE	0	0	NNQ	0	0

Cell Group Pointer	C1	C2	C3	C4	C5	C6
Predecessor Pointer	-1	-1	-1	-1	-1	-1
Successor Pointer	C2	C3	C4	C5	C6	0

2. CALL FILEM(1,A)   MFA = C2

	1	2		1	2		1	2
MMFE	C1	0	MMLE	C1	0	NNQ	1	0

Cell Group Pointer	C1	C2	C3	C4	C5	C6
Predecessor Pointer	0	-1	-1	-1	-1	-1
Successor Pointer	0	C3	C4	C5	C6	0

3. CALL FILEM(1,A)   MFA = C3

	1	2		1	2		1	2
MMFE	C1	0	MMLE	C2	0	NNQ	2	0

Cell Group Pointer	C1	C2	C3	C4	C5	C6
Predecessor Pointer	0	C1	-1	-1	-1	-1
Successor Pointer	C2	0	C4	C5	C6	0

4. CALL FILEM(2,A)   MFA = C4

	1	2		1	2		1	2
MMFE	C1	C3	MMLE	C2	C3	NNQ	2	1

Cell Group Pointer	C1	C2	C3	C4	C5	C6
Predecessor Pointer	0	C1	0	-1	-1	-1
Successor Pointer	C2	0	0	C5	C6	0

5. CALL RMOVE(NNQ(1),1,A)   MFA = C2

	1	2		1	2		1	2
MMFE	C1	C3	MMLE	C1	C3	NNQ	1	1

Cell Group Pointer	C1	C2	C3	C4	C5	C6
Predecessor Pointer	0	0	0	-1	-1	-1
Successor Pointer	0	C4	0	C5	C6	0

6. CALL FILEM(2,A)   MFA = C4

		1	2		1	2		1	2
	MMFE	C1	C2	MMLE	C1	C3	NNQ	1	2

	C1	C2	C3	C4	C5	C6
Cell Group Pointer						
Predecessor Pointer	0	0	C2	-1	-1	-1
Successor Pointer	0	C3	0	C5	C6	0

7. CALL FILEM(2,A)   MFA = C5

		1	2		1	2		1	2
	MMFE	C1	C4	MMLE	C1	C3	NNQ	1	3

	C1	C2	C3	C4	C5	C6
Cell Group Pointer						
Predecessor Pointer	0	C4	C2	0	-1	-1
Successor Pointer	0	C3	0	C2	C6	0

8. CALL FILEM(1,A)   MFA = C6

		1	2		1	2		1	2
	MMFE	C1	C4	MMLE	C5	C3	NNQ	2	3

	C1	C2	C3	C4	C5	C6
Cell Group Pointer						
Predecessor Pointer	0	C4	C2	0	C1	-1
Successor Pointer	C5	C3	0	C2	0	0

9. CALL FILEM(2,A)   MFA = 0

		1	2		1	2		1	2
	MMFE	C1	C6	MMLE	C5	C3	NNQ	2	4

	C1	C2	C3	C4	C5	C6
Cell Group Pointer						
Predecessor Pointer	0	C4	C2	C6	C1	0
Successor Pointer	C5	C3	0	C2	0	C4

10. CALL RMOVE(2,2,A)   MFA = C4

		1	2		1	2		1	2
	MMFE	C1	C6	MMLE	C5	C3	NNQ	2	3

	C1	C2	C3	C4	C5	C6
Cell Group Pointer						
Predecessor Pointer	0	C6	C2	0	C1	0
Successor Pointer	C5	C3	0	0	0	C2

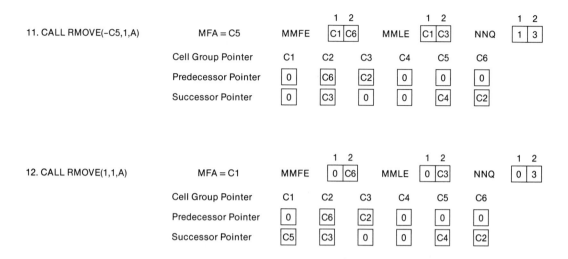

11. CALL RMOVE(−C5,1,A)	MFA = C5	MMFE	1 2		MMLE	1 2		NNQ	1 2
			C1	C6		C1	C3		1 3

Cell Group Pointer	C1	C2	C3	C4	C5	C6
Predecessor Pointer	0	C6	C2	0	0	0
Successor Pointer	0	C3	0	0	C4	C2

12. CALL RMOVE(1,1,A)	MFA = C1	MMFE	1 2		MMLE	1 2		NNQ	1 2
			0	C6		0	C3		0 3

Cell Group Pointer	C1	C2	C3	C4	C5	C6
Predecessor Pointer	0	C6	C2	0	0	0
Successor Pointer	C5	C3	0	0	C4	C2

Operation 1 presumes SLAM II initialization and the illustration shows the various file pointers after initialization is performed. All cell groups are initially in the availability file and the first cell group in the availability file is defined through the MFA pointer which is set equal to C1. In SLAM II initialization, the value of C_i is determined by SLAM II and is dependent on the number of attributes per entry as described previously. In the availability file, all predecessor pointers are initialized to −1 and each successor pointer is set equal to the next cell group pointer. All pointers to the first and last entries in a file are set to 0 and the number in each file is zero.

The second operation, CALL FILEM(1,A), inserts the entity with attributes defined in A into file 1. MFA is updated to the successor of C1 which is C2. The new entry in file 1 is stored in C1 and the pointer to the first and last entry in file 1 is established as C1. The predecessor and successor pointers in C1 are set to 0 as there is only one entry in file 1. The third operation files another entry into file 1. This entry is stored in the cell group pointed to by C2. Since file 1 is a FIFO file, the new entry is the last entry and the pointer to the last entry in file 1 is updated to C2. The predecessor of C2 is the next to last entry in file 1 which is C1. The successor of C1 is then established as C2. MFA is updated to C3.

Operation 4 involves the filing of an entry into file 2. The pointer to the entry is C3. MMFE(2) and MMLE(2) are set equal to C3 and MFA is updated to C4. We now have a file structure which has two files with the first file having two entries and the second file having one entry. The availability file starts at C4 and the successor pointers show that it consists of C4, C5 and C6.

The fifth operation removes the last entry in file 1. The last entry in file 1 was in C2 and this cell group is placed at the top of the availability file. The pointer to the last entry in file 1 is updated to the predecessor of C2 which is C1. The successor of C1 is then set to 0. The sixth operation causes an entry to be inserted into file 2. This new entry is placed in C2 and it becomes the first entry in file 2 because of the LIFO priority rule for file 2. The successor of C2 is the previous first entry in file 2. Operation 7 files another entry in file 2. The updating is similar to that performed for operation 6. Operation 8 inserts an entry into file 1 which is placed at the end of file 1. Operation 9 files an entry into file 2 at C6 which exhausts all the available space for entries in the filing system. At this point, the first availability pointer is set to 0. There are two entries in file 1 and 4 entries in file 2. The first entry in file 1 is in C1 and the successor of C1 is C5. C5 is the end of file 1 as it has no successor. Thus, C5 is the value of the pointer to the last entry in file 1. File 2 starts in C6 and is followed by C4, C2 and C3.

Operation 10 removes the second entry in file 2. The second entry in file 2 is in C4 and C4 becomes the first available cell in the availability file. Since the second entry is removed, the pointers to the first and last entries in file 2 need not be updated. However, both the successor to the first entry, C6, must be changed to the successor of the entry that was in C4. Thus, the successor of C6 becomes C2. The predecessor of C2 must also be updated so that it points to C6 and not to C4. Operation 11 shows the removal of an entry by a pointer value by using the negative of the pointer in the call to subroutine RMOVE. C5 was the last entry in file 1 and it now becomes the first cell group in the availability file. The predecessor of C5, C1, is now made the last entry in file 1. Thus, the successor pointer for C1 is set to 0. Operation 12 removes the first entry in file 1. Since there is only one entry in file 1, this empties the file and the pointers to the first and last entry are set to 0. The file structure now consists of three entries in file 2 and three groups of cells that are available for storing new entries.

This illustration of the operations of the pointers demonstrates how the file processing routines update the cell group pointers and the first and last file pointers. Within SLAM II, a binary search algorithm is used for increasing the efficiency of insertions into the event calendar. This involves a directory and internal tags to perform the binary search. The internal tags are stored in NSET/QSET. From the user's standpoint, the accessing and processing of values as described above is unchanged.

12.5 AUXILIARY ATTRIBUTES

It is sometimes desirable to augment the SLAM II filing system with user-defined arrays that maintain attribute values associated with events and entities that are members of SLAM II files.

Attributes associated with file entries that are not stored in the file array NSET are referred to as auxiliary attributes. Auxiliary attributes are used when one file has many more attributes than the other files or when entities flow through different files and it is convenient not to carry all the attributes along with the entity. For these situations, we provide subroutines to store and retrieve attributes in user-defined arrays and to identify the location where the auxiliary attributes are maintained in the user-defined array. The location is used as an attribute of the entity stored in NSET.

The subroutines for processing auxiliary attributes allow the user to employ different arrays for different sets of auxiliary attributes. In addition, the user can point to the same set of auxiliary attributes from different entries. In the next section, the definition of the subroutines are given. Following the definitions, examples of the use of the subroutines are presented.

12.5.1 Subroutines for Auxiliary Attributes

Four subroutines are available for storing and retrieving auxiliary attributes. These subroutines are:

SETAA: Sets up an array (defined by the user) that will maintain auxiliary attributes

PUTAA: Puts (files) a set of values of auxiliary attributes in an array

GETAA: Gets (removes) a set of auxiliary attributes and establishes a vector of the values of the auxiliary attributes

COPAA: Copies the values of a set of auxiliary attributes without removing the attribute values from the user's auxiliary storage array

Each of the above routines requires a set of arguments, and the definition of these arguments is given below.

AUXF: A one-dimensional array defined by the user to store auxiliary attributes

MFAA: A pointer from which the location of the auxiliary attributes stored in AUXF can be determined. Specifically, AUXF(MFAA+I) will contain the value of the Ith auxiliary attribute

NAUXA: The number of auxiliary attributes associated with each entry in AUXF

NDAUX: The dimension of the auxiliary filing array AUXF
NTRYA: A pointer to an entry to be removed (GET) or copied
VALUE: A vector of the auxiliary attributes to be put into the auxiliary filing array AUXF or to be obtained from the auxiliary file AUXF

12.5.2   Subroutine SETAA(NAUXA,AUXF,MFAA,NDAUX)

This subroutine is used to establish a pointing structure in the array AUXF which has dimension NDAUX and for which each entry has NAUXA elements. MFAA is a pointer to a location in AUXF where the next group of auxiliary attributes will be stored.

12.5.3   Subroutine PUTAA(NAUXA,AUXF,MFAA,VALUE)

This subroutine puts the values stored in VALUE(I),I=1,NAUXA into the filing array AUXF(J) with J=MFAA+1,MFAA+2,...,MFAA+NAUXA. Prior to the call to subroutine PUTAA, the user should establish a pointer to the auxiliary attributes by storing the value of MFAA as an attribute of the entity to be filed in NSET. *PUTAA updates the value of MFAA automatically.*

12.5.4   Subroutine GETAA(NTRYA,NAUXA,AUXF,MFAA,VALUE)

This subroutine gets the values of the auxiliary attributes whose pointer is NTRYA and stores them in VALUE(I)=AUXF(NTRYA+I) for I=1,NAUXA. The value of MFAA is then updated to be equal to NTRYA so that the storage space that was used in AUXF is available for storing a new set of auxiliary attributes.

12.5.5   Subroutine COPAA(NTRYA,NAUXA,AUXF,VALUE)

This subroutine copies the values of auxiliary attributes whose pointer is NTRYA into VALUE(I)=AUXF(NTRYA+I) for I=1,NAUXA. Subroutine COPAA does not change MFAA or the values stored in AUXF.

12.6   ILLUSTRATION 12-2. JOB ROUTING USING AUXILIARY ATTRIBUTES

Consider the situation where a SLAM II file represents a queue for a machine group. If a job cannot gain access to a machine in a machine group, it is stored in the file of the machine group. If the job can be processed by the machine group, it is associated with an event representing an end of service for the machine group. Associated with each job is a routing which sends the job from machine group to machine group. Suppose there is a possibility that the routing consists of up to 10 machine group specifications and there can be as many as 100 jobs being processed simultaneously.

For this setup, it is desirable to store the routing for each job as auxiliary attributes where there can be as many as 10 auxiliary attributes associated with each job as it flows through the system. Since there can be 100 jobs in the system, we will need an auxiliary filing array that is dimensioned to 1,000. For this auxiliary filing array, we will make the following definitions: ROUTS(1000) is the filing array for storing routings of jobs; NROUTA is the number of auxiliary attributes associated with a routing and is set equal to 10; NDROUT is the dimension of the array ROUTS; MROUT is the pointer to the next available set of cells for storing a routing; NTRY is the pointer to the location where the routing for a job is stored.

In subroutine INTLC, the array ROUTS is established to store as many as 100 jobs, each of which has 10 auxiliary attributes, by a call to subroutine SETAA as shown below:

```
NROUTA = 10
NDROUT = 1000
CALL SETAA(NROUTA,ROUTS,MROUT,NDROUT)
```

This call sets up the array ROUTS for use in storing auxiliary attributes and establishes MROUT as equal to zero.

When a job arrives it is necessary to store its routing in the array ROUTS. Assume that the ten machine group numbers have been defined in the vector ROUTE. Further assume that jobs are queued in files corresponding to the auxiliary attributes defining the routing of a job. When the job arrives, it will be sent to the first machine group number as defined by ROUTE(1). Let attribute 2 be the number of machine groups that the job has visited. Further presume that the job must wait at the machine group that is first in its routing. The following statement will accomplish the above.

```
ATRIB(1) = MROUT
ATRIB(2) = 1
MG = ROUTE(1)
CALL FILEM(MG,ATRIB)
CALL PUTAA(NROUTA,ROUTS,MROUT,ROUTE)
```

At a future time, this job is removed from file MG and an end-of-service event is established for it. The values of attributes 1 and 2 described in the above code will be identified wth the end-of-service event. In other words, the job will be removed from the queue and the pointer to the auxiliary attributes will be maintained with the event that is scheduled. When the end-of-service event occurs, SLAM II will remove the event from the event file. One of the functions of the end-of-service event will be to route the job to its next machine group, if there is one. The next machine group in the routing of a job can be obtained by using the following code:

```
NTRY = ATRIB(1)
CALL COPAA(NTRY,NROUTA,ROUTS,ROUTE)
I = ATRIB(2)
NEXTMG = ROUTE(I + 1)
```

In the above code ATRIB(2) represents the number of machine groups that the job has visited and I represents the position in the routing for the next machine group. Of course, the value of NEXTMG could also be obtained by the following code:

```
NTRY = ATRIB(1)
I = ATRIB(2)
IC = NTRY + I + 1
NEXTMG = ROUTS(IC)
```

In this code, the Ith auxiliary attribute is obtained. The pointer to auxiliary attributes points to the location preceding the first attribute, hence, the Ith auxiliary attribute is stored in cell location NTRY + I + 1.

When a job has finished its routing, a call to subroutine GETAA would be made to remove the auxiliary attributes from the system. In this example, the auxiliary attribute subroutines were only used to store attributes in conjunction with one set of user-defined arrays. The subroutines are written in general form to allow the user to employ multiple auxiliary attribute files in a simulation program.

12.7   REPORT WRITING SUBROUTINES

A set of subroutines is contained within SLAM II to enable the user to obtain summary reports or sections of a summary report. These subroutines can be used to obtain summary information of a specific type. The output from a subroutine corresponds to a specific section of the SLAM II summary report (see Chapter 8). The list of subroutines along with the definitions of their arguments is presented in Table 12-2.

Table 12-2   SLAM II report writing subroutines.

Subroutine	Description
SUMRY	Prints the SLAM II Summary Report.
PRNTF (IFILE)	If IFILE>0, prints statistics and the contents of file IFILE. If IFILE=0, prints summary statistics for all files. If IFILE<0, prints summary statistics and contents of all files.
PRNTC (ICLCT)	If ICLCT>0, prints statistics for COLCT variable number ICLCT. If ICLCT≤0, prints statistics for all COLCT variables.
PRNTH (ICLCT)	If ICLCT>0, prints a histogram for COLCT variable number ICLCT. If ICLCT≤0, prints all histograms for COLCT variables.
PRNTP (IPLOT)	If IPLOT>0, prints a plot and/or table for plot/table number IPLOT. If IPLOT≤0, prints all plots/tables.
PRNTT (ISTAT)	If ISTAT>0, prints statistics for time-persistent variable ISTAT. If ISTAT≤0, prints statistics for all time-persistent variables.
PRNTB (ISTAT)	If ISTAT>0, prints a histogram (bar chart) for ISTAT variable number ISTAT. If ISTAT≤0, prints all histograms for time-persistent variables.
PRNTR (IRSC)	If IRSC>0, prints statistics for resource number IRSC. If IRSC≤0, prints statistics for all resources.
PRNTS	Prints the contents of the state storage vectors SS(I) and DD(I).
PRNTA	Prints statistics for activities.
PRNTG (IG)	If IG>0, prints statistics for gate number IG. If IG≤0, prints statistics for all gates.

The terminology used in naming the subroutines is to append a specific letter onto the letters PRNT. Thus, PRNTF is used as the name of a subroutine to print files and PRNTH is used as the name of the subroutine to print histograms. The arguments to these subroutines have been standardized so that an argument value greater than zero requests a specific item to be printed. If the argument is given as a zero value then all items asociated with the subroutine are printed. In the case of subroutine PRNTF(IFILE), the contents of all files can be obtained by specifying a negative value for the argument IFILE. When a negative argument is specified for any of the other subprograms, the output obtained is the same as if a zero value is specified.

Listed below are examples of the use of the report writing routines.

Statement	*Description*
CALL SUMRY	Prints a complete summary report at time of calling (TNOW).
CALL PRNTF(3)	Prints statistics on file 3 and gives a listing of all entries in the file.
CALL PRNTT(0)	Prints statistics on all time-persistent variables.
CALL PRNTA	Prints statistics on all network activities that were given activity numbers.

12.8 FUNCTIONS FOR OBTAINING VALUES ASSOCIATED WITH STATISTICAL ESTIMATES

Functions have been included in SLAM II to allow the user to access information related to statistical estimates during the execution of a simulation model. The names of the functions are all five letters in length with the first two letters being identical and representing one of the six types of variables for which statistics are desired. The last three letters of the function name prescribes the statistic of interest. The six types of variables are: variables based on observation (CC); time-persistent variables (TT); file variables (FF); resource variables (RR); activity variables (AA); and gate variables (GG). The statistical quantities associated with these variables and their three letter codes are: average value (AVG); standard deviation (STD); maximum observed value (MAX); minimum observed value (MIN); number of observations (NUM); time period over which the time-persistent variable was observed (PRD); time of last change to a time-persistent variable (TLC), and percent of time opened (OPN).

The twenty-nine functions included within SLAM II to obtain values associated with the variables and statistical quantities are presented in Table 12-3. These

Table 12-3 SLAM II statistical calculation functions.

Function	Description
Statistics for Variables Based on Observations (COCLT)	
CCAVG(ICLCT)	Average value of variable ICLCT
CCSTD(ICLCT)	Standard deviation of variable ICLCT
CCMAX(ICLCT)	Maximum value of variable ICLCT
CCMIN(ICLCT)	Minimum value of variable ICLCT
CCNUM(ICLCT)	Number of observations of variable ICLCT
Statistics for Time-Persistent Variables (TIMST)	
TTAVG(ISTAT)	Time integrated average of variable ISTAT
TTSTD(ISTAT)	Standard deviation of variable ISTAT
TTMAX(ISTAT)	Maximum value of variable ISTAT
TTMIN(ISTAT)	Minimum value of variable ISTAT
TTPRD(ISTAT)	Time period for statistics on variable ISTAT
TTTLC(ISTAT)	Time at which variable ISTAT was last changed
Queue Statistics	
FFAVG(IFILE)	Average number of entities in file IFILE
FFAWT(IFILE)	Average waiting time in file IFILE
FFSTD(IFILE)	Standard deviation for file IFILE
FFMAX(IFILE)	Maximum number of entities in file IFILE
FFPRD(IFILE)	Time period for statistics on file IFILE
FFTLC(IFILE)	Time at which number in file IFILE last changed
Resource Statistics	
RRAVG(IRSC)	Average utilization of resource IRSC
RRAVA(IRSC)	Average availability of resource IRSC
RRSTD(IRSC)	Standard deviation of utilization of resource IRSC
RRMAX(IRSC)	Maximum utilization of resource IRSC
RRPRD(IRSC)	Time period for statistics on resource IRSC
RRTLC(IRSC)	Time at which resource IRSC utilization was last changed.
Activity Statistics	
AAAVG(IACT)	Average utilization of activity IACT
AAMAX(IACT)	Maximum utilization of activity IACT, or maximum busy time if activity IACT is a single-server service activity
AASTD(IACT)	Standard deviation of the utilization of activity IACT
AATLC(IACT)	Time at which the status of activity IACT last changed
Gate Statistics	
GGOPN(IG)	Percent of time that gate IG was open
GGTLC(IG)	Time at which the status of gate IG last changed

functions can be called from any SLAM II subprogram and the values can be used for decision making within SLAM II models.

The arguments to the SLAM II statistical calculation functions are always a numeric value. File numbers are prescribed for both discrete event and network models and there should be no ambiguity regarding the arguments for the file related functions. For resources, SLAM II or the modeler assigns a resource number to each resource statement in a network statement model. This resource number is used as the argument to functions RRAVG, RRAVA, RRSTD, RRMAX, RRPRD, and RRTLC. For collect variables, numeric values are assigned by the user in discrete event models and by SLAM II or by the user in network models (for each COLCT node). If not specified by the user, SLAM assigns a sequential numeric code for each COLCT statement starting with the first number above the highest user-assigned number prescribed on a STAT statement. Thus, if there are three STAT variables assigned codes 1, 2, and 3 and two COLCT nodes, the average associated with the first COLCT node in the network description would be accessed by the statement.

AVE = CCAVG(4)

For time persistent statistics, both the order in which the TIMST input statements appear in the input and the variable type determine the number to be used in the functions beginning with the letters TT.

SLAM II assigns the statistics index by numbering all TIMST statements referencing XX variables sequentially, then continues the numbering for TIMST statements referencng SS or DD variables.

As an example, consider the following set of statements requesting time-persistent statistics collection:

TIMST,SS(2),HEATING TEMP;
TIMST,XX(4),VALVE STATUS;
TIMST,DD(1),VELOCITY;
TIMST,XX(1),AVAILABILITY;
TIMST,SS(3),PRESSURE;

The average value of each variable would be obtained as follows:

TTAVG(1) = average value of XX(4)
TTAVG(2) = average value of XX(1)
TTAVG(3) = average value of SS(2)
TTAVG(4) = average value of DD(1)
TTAVG(5) = average value of SS(3)

12.9   MISCELLANEOUS INPUT STATEMENTS
AND SUPPORT SUBPROGRAMS

There are many concepts associated with discrete simulation that are supported by SLAM II indirectly through input statement types or by support subprograms that are described in either the network or continuous modeling chapters of this book. In this section, we discuss subroutines UMONT, CLEAR, TRACE, UNTRA, ERROR, UERR, and GTABL, and the functions NNLBL, NNUMB and NNVNT.

12.9.1   Subroutine UMONT(IT)

Whenever a SLAM II trace is on, SLAM II calls subroutine UMONT(IT) to allow the user to prepare a stylized trace and to suppress the standard SLAM II trace. The argument, IT, to subroutine UMONT is used to pass information to the user concerning the type of event or condition that has just been processed. IT is also used by the user to communicate to SLAM II whether or not the standard SLAM II trace is to be suppressed. The numeric codes used for this communication are listed below.

Value of IT passed to SLAM II from user:

  1 if the standard SLAM II trace is to be suppressed.

Value of IT passed to user from SLAM II:

  0 if the current event is a node arrival or a user-scheduled event.
  −2 if an activity keyed to STOPA is beginning.
  −3 if an activity keyed to a node release is beginning.
  −4 if an activity of definite duration is beginning.
  −5 if a conditional branch has just been rejected.

By setting IT to 1, the user suppresses the standard SLAM II trace. IT must be set to 1 every time UMONT is called. If IT is not reset to 1, both user written information and SLAM II generated trace information are generated. The information can be segregated if desired by using different output device numbers. An example of the output from a version of subroutine UMONT written for an analysis of hot metal carriers (submarines) in a steel plant (2) is presented in Figure 12-1. We strongly encourage the use of such tailor-made traces. They assist in program debugging and verification and they support communication about the model. The importance of this communication was discussed in Chapter 1 with regard to implementing results. Additionally, traces support external model validation.

10:57 AM	DEPARTURE FROM DESULFURIZATION WITH 3 SUBS FIRST SUB NO. = 14.
10:57 AM	SUB 9. ARRIVES AT BF AREA
10:57 AM	HML 2. UNLOADS INTO BOF 1.
10:58 AM	BOF 2. ENDS CYCLE
10:59 AM	END UNLOAD OF SUB 23. INTO HML 2.
11:00 AM	SCRAP MELTER COMPLETED A CYCLE
11:01 AM	END CAST ON FURNACE G WITH SUBS 12 5
	HAVING LOADS OF 219. 28
	NO. OF FULL SUBS = 9 NO. OF SUBS IN Q OF BF = 0
11:03 AM	HML 1. UNLOADS INTO BOF 2.
11:03 AM	END UNLOAD OF SUB 17. INTO HML 2.
11:06 AM	END UNLOAD OF SUB 18. INTO HML 2.
11:09 AM	ARRIVAL AT DESULFURIZATION WITH 2 SUBS FIRST SUBS NO. = 2.
11:10 AM	END UNLOAD OF SUB 18. INTO HML 1.
11:11 AM	SUB 11. ARRIVES AT BF AREA
11:13 AM	SUB 15. ARRIVED AT SCRAP MELTER
11:13 AM	START CAST ON FURNACE J WITH SUBS 8 9 5
	HAVING INITIAL LOADS OF 19. 0 28.
11:15 AM	END UNLOAD OF SUB 3. INTO HML 1.
11:16 AM	ARRIVAL AT BOF WITH 3 SUBS FIRST SUB NO. = 14. BOF Q = 1
11:31 AM	END CAST ON FURNACE H WITH SUBS 7 6 16
	HAVING LOADS OF 219. 41. 0
	NO. OF FULL SUBS = 9 NO. OF SUBS IN Q OF BF = 0
11:33 AM	ARRIVAL AT DESULFURIZATION WITH 1 SUBS FIRST SUBS NO. = 2.
11:35 AM	SUB 23. ARRIVES AT BF AREA
11:46 AM	BOF 2. ENDS CYCLE
11:49 AM	DEPARTURE FROM DESULFURIZATION WITH 2 SUBS FIRST SUB NO. = 2.
11:49 AM	SUB 18. ARRIVES AT BF AREA
11:51 AM	HML 2. UNLOADS INTO BOF 2.
11:52 AM	END UNLOAD OF SUB 3. INTO HML 2.
11:56 AM	SUB 17. ARRIVES AT BF AREA
11:59 AM	END UNLOAD OF SUB 21. INTO HML 2.
12:00 PM	START CAST ON FURNACE C WITH SUBS 11 16 23
	HAVING INITIAL LOADS OF 0 0 0
12:00 PM	END UNLOAD OF SUB 14. INTO HML 2.

Figure 12-1 Example of user-trace output produced from UMONT.

Now consider the writing of stylized traces in subroutine UMONT. The input value of IT communicates to the user information concerning a current event and the characteristics of this event. SLAM II functions are provided to obtain additional information concerning the current event. The node label associated with a current event is obtained from function NNLBL. The activity number associated with a current event is available from function NNUMB, and the user-defined event code associated with the current event is obtained from function NNVNT. Each of these functions have a dummy integer argument, IDUM, which is included to have the functions conform with standard FORTRAN programming conventions.

From a tracing standpoint, the SLAM II events are characterized as either being the start of an activity, the arrival of an entity to a node (the end of an activity), or user defined. Table 12-4 characterizes the values returned from each of the three trace support functions for each of these types of events.

Table 12-4   Values Returned From Trace Support Functions

Function†	Event Type		
	Start of Activity	Arrival to Node	User Defined
NNLBL(IDUM)	Label of end node of activity starting	Label of node to which entity is arriving	Blank label
NNUMB(IDUM)	Activity number of activity starting	Activity number of activity just completed	−1
NNVNT(IDUM)	0	Code of EVENT node, otherwise 0	Code of current event

† The function argument IDUM is not used and exists only to conform to ANSI FORTRAN requirements.

Because error and status checking is not performed in SLAM II following a return from subroutine UMONT, it is recommended that no system status changes or file manipulations be performed in subroutine UMONT.

In subroutine UMONT, these functions can be employed so that only selected information is printed on a trace. Thus, by making tests on the node label, the activity number or the event code, only selective lines on a trace would be written. As an example, if it is desired to only print out selected user variables when an entity arrives to a node and the node has the label TRAC, then these two conditions must be met or the standard SLAM II trace will be suppressed by setting IT to 1. The first condition is that the current event is a node arrival, that is, the input value to subroutine UMONT is 0 (IT=0). The second condition is that the node label of the current event must be TRAC. The coding of subroutine UMONT(IT) to accomplish this selective printing is shown below. The code is written so that when the conditions are met the standard trace messages are obtained from SLAM II on one output file and the stylized trace consisting of the current time, the node label, and an array of five values are printed on device number 10.

```
      SUBROUTINE UMONT(IT)
      COMMON/SCOM1/ATRIB(100),DD(100),DDL(100),DTNOW,II,MFA,MSTOP,NCLNR
     1,NCRDR,NPRNT,NNRUN,NNSET,NTAPE,SS(100),SSL(100),TNEXT,TNOW,XX(100)
      COMMON/UCOM1/MYARRAY(5)
      IF(IT.EQ.0) GO TO 100
      IT=1
      RETURN
  100 LABL=NNLBL(IDUM)
      IF (LABL.NE.'TRAC')RETURN
      WRITE(10,200) TNOW,LABL,MYARRAY
      IT=0
      RETURN
  200 FORMAT(5X,'TIME= ',F7.3,'AT NODE',A4,
     1  'ARRAY VALUES ARE',5I10)
      END
```

12.9.2   Subroutines CLEAR, TRACE and UNTRA

Subroutine CLEAR initializes the statistical storage arrays associated with a simulation model. The user invokes this subroutine by the statement

CALL CLEAR

or, by the input statement,

MONTR,CLEAR,TFRST;

When the MONTR input statement is employed, SLAM II calls subroutine CLEAR at the time specified by the variable TFRST. The user calls subroutine CLEAR when statistics clearing is to be based on system status. When subroutine CLEAR is employed in a simulation model, all statistical estimates presented on the SLAM II summary report are based on data collected from the time subroutine CLEAR was called to the time at which the summary report was prepared. Recall that SLAM II summary reports can be prepared periodically through the use of a MONTR input statement which employs the SUMRY option and specifies a time between reports value.

Subroutine TRACE causes the standard SLAM II trace to be turned on at the time that the statement CALL TRACE is used. The use of this subroutine is equivalent to the SLAM II input statement

MONTR,TRACE,TFRST;

where TFRST is the time at which the call to subroutine TRACE is placed.

To stop the trace of events, the user calls subroutine UNTRA. Subroutines TRACE and UNTRA are included in SLAM II to allow the tracing of events based on system status evaluations.

12.9.3   Subroutine ERROR(KODE) and UERR(KODE)

Subroutine ERROR is called when an error is detected in a SLAM II subprogram. It provides useful diagnostic information by listing the error code, TNOW, and current values in ATRIB. A call is made to subroutine UERR(KODE) which permits the user to print out user-specific information when an error is detected. An attempt is then made to print a complete SLAM II Summary Report. Finally, subroutine ERROR has a deliberate FORTRAN error to cause the standard FORTRAN diagnostic and trace-back information to be printed.

12.9.4   Function GTABL(Y,XVALUE,XLOW,XHIGH,XINCR) and Function GGTBL(X,Y,XVALUE,NVAL)

GTABL is a SLAM II table look-up function. The definition of the arguments are given below.

Argument	Definition
Y	The array containing the values of the dependent variable
XVALUE	A value of the independent variable for which the table look-up is desired

XLOW	The lowest value of the independent variable for which Y contains a value for the dependent variable
XHIGH	The highest value of the independent variable for which Y contains a value for the dependent variable
XINCR	The increment of the independent variable between corresponding values for the dependent variable

To use GTABL the user must provide an array and store the appropriate values for the dependent variable in the array. The values must correspond to the equally spaced values of the independent variable between XLOW and XHIGH in increments of XINCR.

GGTBL is a second SLAM II table look-up function. It differs from GTABL in that the independent variable is specified by a set of values corresponding to the values given for the dependent variable. GGTBL is a general version of the network table look-up function GGTBLN in which the independent and dependent vectors are provided as arguments. The definition of the arguments for GGTBL are given below.

Argument	Definition
X	A vector containing the values of the independent variable
Y	A vector containing the values of the dependent variable
XVALUE	A value of the independent variable for which the table look-up is to be performed
NVAL	The number of values in the table

To use GGTBL the user must provide NVAL values in the vectors Y and X which specifies a value of Y corresponding to each value of X, that is, the first value given in Y corresponds to the first value in X. Function GGTBL returns the value of Y corresponding to XVALUE.

The value of the dependent variable returned by GTABL or GGTBL is computed by linear interpolation within the defined range of the independent variable. If the independent variable is outside the range given, the appropriate end point value of the dependent variable is returned. The use of function GTABL is illustrated in Section 9.5 and in Example 13-2.

12.9.5   Subroutines GPLOT(IPLOT) and PPLOT(IPLOT,NREC,XI,XD)

Variables to be tabled and plotted are specified on RECORD and VAR statements. A flexible recording frequency can be obtained by identifying a plot number, IPLOT, with the RECORD statement which specifies that the user desires to record information by calling subroutine GPLOT(IPLOT). The variable IPLOT is referred to as the plot number. Subroutine GPLOT collects values of up to ten dependent variables specified on VAR statements for one independent variable specified on a RECORD statement.

The output for values recorded through calls to subroutine GPLOT is part of the SLAM II Summary Report. The output can also be printed by calling subroutine PRNTP as described in Section 12.7. Presentations of table and plot data are given throughout the text.

To retrieve individual data records stored for plot, IPLOT, SLAM II provides subroutine PPLOT(IPLOT, NREC, XI, XD). Subroutine PPLOT retrieves the NRECth record containing the independent variable XI and the values of the dependent variables in the vector XD recorded with XI. The definitions of the arguments for PPLOT are given below.

Argument	Definition
IPLOT	Plot number as defined on a RECORD statement.
NREC	Number of the next record to be retrieved. Initially, NREC is set to 1. NREC is indexed in subroutine PPLOT and is set to 0 by SLAM II to indicate that there are no more records.
XI	Value of the independent variable for the NRECth record. The variable is specified by the INDVAR field of the RECORD statement for plot IPLOT.
XD	Vector into which SLAM II puts the values of the dependent variables collected with the value of XI. The variables are specified by the DEPVAR field on the VAR statements that follow the RECORD statement.

As an example of the use of subroutine PPLOT, consider the RECORD and VAR statements given below.

RECORD (1),TNOW,TIME,0,P,1.0,0,100;
VAR,XX(1),B,BUSY,0,10;
VAR,NNQ(1),Q,QUEUE LENGTH,0,50;

These statements specify plot number 1 to consist of dependent variables XX(1) and NNQ(1) for which values are to be recorded as a function of TNOW. To retrieve the values that have been recorded during a simulation, the following code could be put in subroutine OTPUT.

```
      SUBROUTINE OTPUT
      DIMENSION Y (2)
      NREC = 1
   10 CALL PPLOT(1,NREC,T,Y)
      IF (NREC.EQ.0) RETURN
C
C**** PROCESS Y(1) AS XX(1) AND Y(2) AS NNQ(1)
C**** AT TIME T
C
         .
         .
         .
      GO TO 10
      END
```

Initially, NREC is set to 1 to indicate that the first record should be processed. A call to subroutine PPLOT is then made with the plot number identified as 1. This specifies that the data to be retrieved was obtained in accordance with the RECORD statement which has the plot number of 1 and the VAR statements following that RECORD statement. In subroutine PPLOT, SLAM II sets NREC to 0 if all records have been processed. Otherwise, NREC is set to the next record number automatically. Upon returning from subroutine PPLOT, NREC is checked by the user and, if equal to 0, a return is made to the calling program. If NREC is not 0, the value of Y(1) corresponds to XX(1) at time T and Y(2) corresponds to NNQ(1) at time T. These values can then be used for plotting, time series analysis or as inputs to a different simulation model. Following the processing of one record, a transfer to statement 10 is made to retrieve the next record.

12.10   Example 12-1.   JOB SHOP SCHEDULING†

This example illustrates how SLAM II can be used for simulating job shops which employ dynamic scheduling practices. The example is based on the article

---

† We wish to thank Ms. Kathy Stecke for suggesting the use of the example.

"Experiments with the SI^X Rule in Job Shop Scheduling" by Eilon, Chowdhury and Serghiou (1). The job shop consists of six machines with each machine performing a different operation. The estimated processing time for each machine is 20 minutes on the average with an exponential distribution (processing times are rounded off to integer values with no value being less than 1). Actual processing time is equal to the estimated processing time plus a random component which is normally distributed with a mean of zero and a standard deviation equal to three-tenths of the estimated processing time (the random component is white noise). Note that the job shop has been balanced so that each machine has the same average processing time.

Jobs arrive to the shop with interarrival times being exponentially distributed with a mean of 25 minutes. The interarrival times are integerized and must be greater than or equal to 1. Each job consists of a set of operations to be performed on the machines in the job shop. The number of operations per job is normally distributed with a mean of 4 and a standard deviation of 1. However, no job can require less than 3 operations nor more than 6 operations. The routing of a job through the machines is determined by random assignment. An illustration of the job shop and the routing of two jobs is given in Figure 12-2.

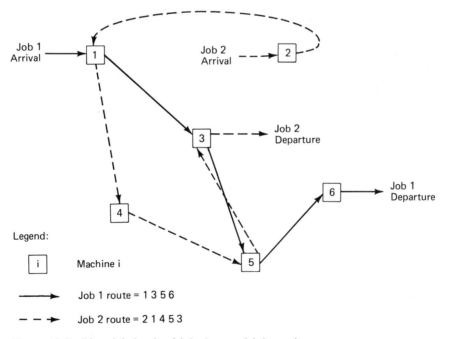

Figure 12-2 Pictorial sketch of job shop and job routings.

As Eilon et al. point out, "If the arrival of jobs, their processing requirements, and the operating facilities are all given, the only control parameter at the disposal of the scheduler is . . . the order in which the job should be processed." The dispatching rule included in this example is the SI^X rule which processes jobs at a machine in the order of the shortest estimated processing time for the job that is in a high priority class. Priority jobs are defined as those jobs whose float is negative. Float is defined for a job as the due date minus the current time minus the estimated time remaining to perform operations on the job minus a safety factor. The SI^X rule divides jobs in front of a given machine into two classes and within each class the jobs are ordered based on shortest estimated processing time. This example will illustrate how SLAM II can be used to evaluate a dispatching rule in a job shop environment.

Concepts Illustrated. This example illustrates the use of advanced file manipulation subprograms for changing the ranking of file members. The accessing of attribute values directly is illustrated. An auxiliary attribute array is used to maintain fixed and dynamic information associated with an entity. Also illustrated is a procedure for sampling from a distribution without replacement. Another procedure illustrated in this example is the use of a derived attribute to obtain complex ranking rules in a SLAM II file.

SLAM II Model. The job shop system is essentially a queueing situation in which jobs arrive, are routed to machines in a prescribed order, and then leave the shop. Basically, there are two events required to model this situation: the arrival of a job to the shop; and end of processing by a machine. When a job arrives, the number of operations to be performed on the job is determined. The order in which the operations are to be performed as well as the estimated time required on each machine is calculated. From this information, a due date is established from which the float associated with the job can be computed. The machine routing, estimated processing times, the number of operations to be performed, the job arrival time, and the job due date will be stored as auxiliary attributes in the job information array XJINF. The job entity will have attribute 1 as a pointer to the location in XJINF of these auxiliary attributes. In the arrival event, the job will be routed to the first machine in its routing. The job will be processed if the machine is idle or will be placed in the queue of the machine if the machine is busy. The variable XX(MACH) will be used to represent the status of machine MACH with a zero indicating an idle machine and a 1 indicating a busy machine. Attribute 2 of the job will be used to rank jobs in the machine queues with a low-value first ranking rule. Ranking will be based on the estimated processing time at the machine if float is negative. If float is positive, then a large constant will be added to the estimated processing time so that a categorization of jobs waiting for the machine can be made. By ranking the jobs in the queue based on low value of attribute 2, the jobs at the front of the queue will be those whose float is negative and they will be ordered by the smallest value of estimated processing time. By adding the large constant to attribute 2 for those jobs whose float is positive, the jobs at the back of the queue will be "non-priority" jobs which will also be ranked by their shortest estimated processing time.

A third attribute will be associated with the job entity which is a modified due date. The modified due date, attribute 3, will be set equal to the due date minus the estimated remaining processing time for the job minus the safety factor. Whenever attribute 3 is greater than the current time, the float is negative and the job classification must be changed. This involves changing the value of attribute 2 for the job. Since decisions regarding which job to process are only made when there is an end-of-service at the machine, it is only necessary to update the classification of the jobs waiting in the queue of the machine at these times. This is accomplished by scanning the file containing the jobs waiting for the machine and updating attribute 2 in accordance with the value of attribute 3 as related to current time as described above. Other details of the SLAM II model for the job shop situation are similar to the queueing example presented in Chapter 11. The detailed description of the variables and subprograms included in this model will now be given.

In this example, six files and the event calendar will be used. For the arrival event, no attributes are required. For the end-of-service event, the following three attributes are used:

1. Pointer to the location in the auxiliary attribute array, XJINF($\cdot$), where job attributes are stored;
2. Machine number for which the end-of-service event is occurring;
3. Modified due date for the job on which service is ending.

The modified due date is computed by subtracting from the due date both a safety factor, SAFET, and the remaining processing time for operations to be performed for the job. By comparing the current time to the modified due date, the priority class for a job can be ascertained.

File I, for I = 1,6, is used as the queue of jobs for machine I. Attributes 1 and 3 of a job entity in these files are the same as the ones described for the event calendar. Attribute 2 will be the estimated processing time for machine I if the job has priority status; otherwise, it will be the estimating processing time for machine I plus 1000.

The definitions of the variables used in this SLAM II model are presented below.

Variable	Definition	Initial Value
AAJOB(I)	Auxiliary attribute buffer array	—
JBPTR	Pointer to location in XJINF where a job's auxiliary attributes are maintained	—
JOBF	Pointer to next set of locations for storing auxiliary attributes	—
NATT	Dimension of AAJOB(·)	15
NLATE	Number of jobs that exceeded their due date	0
NMG	Number of machines in the job shop	6
NUMJOB	Number of jobs that have been processed through job shop	0
PT	Mean processing time for an operation	20.0
SAFET	Safety factor in computation of float	50.0
TBA	Mean time between job arrivals	25.0
XJINF	Auxiliary attribute array	—
XX(MACH)	$\begin{cases}0 \text{ if machine MACH is idle} \\ 1 \text{ if machine MACH is busy}\end{cases}$	0.0

The auxiliary attribute array for job information, XJINF, has NATT = 15 values associated with each job. The values are defined below where the left column specifies the value to be added to the job pointer, JBPTR.

Auxiliary Attribute Number	Definition
I = 1,NMG	Ith machine in routing: if none, 0
I = NMG + 1,2*NMG	Processing time on (I-NMG)th machine in routing
NATT-2	Current operation number for job
NATT-1	Arrival time of job
NATT = 2*NMG + 3	Due date for job

The coding of this example follows the standard SLAM II discrete modeling approach presented in Chapter 11. The main program and subroutine EVENT are in standard form and are shown in Figure 12-3. Subroutine INTLC is used to perform the following: establish constants for the model; initialize to zero the number of jobs processed, NUMJOB, and number of jobs that are late, NLATE; initialize the auxiliary attribute array XJINF and first pointer for extra job information, JOBF; and schedule the first job arrival at time zero. The coding for subroutine INTLC is also shown in Figure 12-3.

```
      PROGRAM MAIN
      DIMENSION NSET(10000)
      COMMON QSET(10000)
      COMMON/SCOM1/ ATRIB(100),DD(100),DDL(100),DTNOW,II,MFA,MSTOP,NCLNR
     1,NCRDR,NPRNT,NNRUN,NNSET,NTAPE,SS(100),SSL(100),TNEXT,TNOW,XX(100)
      EQUIVALENCE (NSET(1),QSET(1))
      NNSET=10000
      NCRDR=5
      NPRNT=6
      NTAPE=7
      CALL SLAM
      STOP
      END

      SUBROUTINE EVENT(I)
      GO TO (1,2),I
C
    1 CALL ARRIV
      RETURN
C
    2 CALL ENDSV
      RETURN
      END

      SUBROUTINE INTLC
      COMMON/SCOM1/ ATRIB(100),DD(100),DDL(100),DTNOW,II,MFA,MSTOP,NCLNR
     1,NCRDR,NPRNT,NNRUN,NNSET,NTAPE,SS(100),SSL(100),TNEXT,TNOW,XX(100)
      COMMON/UCOM1/ XJINF(3000),AAJOB(15),JOBF,NMG,NATT,SAFET,NLATE,
     1NUMJOB
C
C*****XX(MACH) IS BUSY STATUS OF MACHINE MACH
C
      NMG=6
      NATT=2*NMG+3
      SAFET=50.
      DO 10 MACH=1,NMG
   10 XX(MACH) = 0.0
C
C***** SET UP THE AUXILIARY ATTRIBUTE ARRAY AND SCHEDULE
C***** THE FIRST ARRIVAL
C
      CALL SETAA(NATT,XJINF,JOBF,3000)
      CALL SCHDL(1,0.0,ATRIB)
      NUMJOB=0
      NLATE=0
      RETURN
      END
```

Figure 12-3 Main program, subroutine EVENT, and subroutine INTLC for job shop model.

The coding of the arrival event, ARRIV, is shown in Figure 12-4. The first three statements in subroutine ARRIV determine the time until the next job arrival, DT, and schedule the next arrival event to occur. Note that the arrival time is a sample from an exponential distribution which is rounded down to the next lowest integer. The interarrival time is then required to be at least one.

Subroutine DESJOB(AAJOB) is used to establish the routing of the job, the processing time for each machine on the routing, and the due date for the job. These values are returned to subroutine ARRIV through the vector AAJOB. By using a separate subroutine for describing the job, job descriptions can be altered without altering the arrival event routine. The description of subroutine DESJOB will be deferred until after the arrival event has been described.

```
      SUBROUTINE ARRIV
      COMMON/SCOM1/ ATRIB(100),DD(100),DDL(100),DTNOW,II,MFA,MSTOP,NCLNR
     1,NCRDR,NPRNT,NNRUN,NNSET,NTAPE,SS(100),SSL(100),TNEXT,TNOW,XX(100)
      COMMON/UCOM1/ XJINF(3000),AAJOB(15),JOBF,NMG,NATT,SAFET,NLATE,
     1NUMJOB
      DATA TBA/25.0/
C
C*****SCHEDULE NEXT JOB ARRIVAL
C
      IDT=EXPON(TBA,1)
      DT= MAXO(IDT,1)
      CALL SCHDL(1,DT,ATRIB)
C
C*****SET UP ATTRIBUTES AND CHARACTERISTICS OF THIS JOB
C
      CALL DESJOB(AAJOB)
      ATRIB(1) = JOBF
C
C***** STORE ROUTE IN JOB INFORMATION ARRAY 'XJINF'
C
      CALL PUTAA(NATT,XJINF,JOBF,AAJOB)
      ATRIB(3) = DDM(AAJOB)
      ATRIB(2) = AAJOB(NMG+1) +1000.
      IF(ATRIB(3).LE.TNOW) ATRIB(2) = AAJOB(NMG+1)
C
C*****DETERMINE DISPOSITION OF JOB
C
      MACH = AAJOB(1)
      IF(XX(MACH).GT.0.0) GO TO 10
C
C*****SET MACHINE BUSY.  SCHEDULE END OF SERVICE
C
      XX(MACH) =1.0
      ETIME = AAJOB(NMG+1)
      CALL SCHES(MACH,ETIME)
      RETURN
C
C*****PUT JOB IN QUEUE FOR MACHINE
C
   10 CALL FILEM(MACH,ATRIB)
      RETURN
      END
```

Figure 12-4   Subroutine ARRIV for job shop model.

Attribute 1 for the job entity is set equal to JOBF, the pointer to the auxiliary attribute array. The auxiliary attributes as described in the vector AAJOB are then inserted into XJINF by a call to subroutine PUTAA. Attribute 3 of the job is set equal to the modified due date as computed in function DDM. Function DDM will also be described later. Attribute 2 of the job is then set equal to the estimated processing time for the first machine in the routing of the job plus 1000. If the modified due date is less than or equal to TNOW then attribute 2 is reset to the estimated processing time. We are now ready to determine the disposition of the arriving job. The first machine on which the job is to be processed is established as the variable MACH as obtained from AAJOB(1). If machine MACH is busy, a transfer is made to statement 10 where the job is placed in the queue for the machine by a call to subroutine FILEM.

If the machine is idle, it is made busy by setting XX(MACH) equal to 1. The estimated processing time is established as the variable ETIME and subroutine SCHES is called to schedule an end-of-service for machine MACH for a job whose estimated processing time on MACH is ETIME. A return is then made from subroutine ARRIV.

To describe a job in terms of its routing, operation processing times, and due date, a call was made to subroutine DESJOB(A) where A is a vector in which the auxiliary attributes associated with a job are to be defined. Subroutine DESJOB(A) is shown in Figure 12-5. The first statement in subroutine DESJOB establishes the number of operations NOPS to be performed on the job as a sample from a normal distribution whose mean is 4, standard deviation is 1, and stream number is 3. A value of 0.5 is added to the sample from the normal distribution so that the number of operations is rounded to the nearest integer. If NOPS is less than 3, it is set equal to 3. Similarly if it is greater than 6, it is set equal to 6 as the problem statement establishes that the number of operations must be between 3 and 6. This procedure for setting NOPS corresponds to obtaining a sample from a mixed distribution where the probability in the tails of the distribution are lumped at the minimum and maximum points specified. Next, the vector A is initialized to zero as is the variable SUM. SUM is used to obtain a sum of the processing times for all operations for this job.

The next section of the code is used to set the route for the arriving job. First, the vector MA(I) is set to I to indicate that any machine can be the first in routing for the job. A random sampling of the value MA(I) is to be made. The variable TOP is set equal to the number of machines plus 1, which for this illustration will set TOP=7. Next, a DO loop is employed with a running index I set equal to the operation number for the arriving job. BOT is established as the real equivalent of the operation number. INDEX is established as an integer from the operation number to the number of machines. By sampling between the values of BOT and TOP, an integer is selected between I and NMG with each integer being equally likely. Thus, when BOT = 1.0 and TOP = 7.0 each integer from 1 to 6 will have a one-sixth probability of being selected. The machine number selected is stored in MA(INDEX) and this is inserted into the vector A in the Ith position to specify MA(INDEX) as the Ith machine to be visited by the arriving job. The value MA(INDEX) is then redefined to be equal to MA(I). This statement interchanges the machine selected with one that was not selected. By this process, a routing is established which does not include a machine on the routing more than once. This code illustrates the procedure for sampling without replacement from a set of inte-

```
      SUBROUTINE DESJOB(A)
      COMMON/SCOM1/ ATRIB(100),DD(100),DDL(100),DTNOW,II,MFA,MSTOP,NCLNR
     1,NCRDR,NPRNT,NNRUN,NNSET,NTAPE,SS(100),SSL(100),TNEXT,TNOW,XX(100)
      COMMON/UCOM1/ XJINF(3000),AAJOB(15),JOBF,NMG,NATT,SAFET,NLATE,
     1NUMJOB
      DIMENSION A(15),MA(6)
      DATA PT/20.0/
      NOPS =RNORM(4.0,1.0,3) +.5
      IF(NOPS.LT.3)NOPS=3
      IF(NOPS.GT.6) NOPS = 6
      DO 30 I = 1,NATT
   30 A(I) = 0.0
C
C*****SAMPLE TO SET MACHINE ROUTE WITHOUT REPLACEMENT
C
      DO 35 I=1,NMG
   35 MA(I) = I
      TOP = NMG + 1
      SUM = 0.0
      DO 40 I =1,NOPS
      BOT = I
      INDEX= UNFRM(BOT,TOP,4)
      A(I) = MA(INDEX)
      MA(INDEX) = MA(I)
      IETIM = EXPON(PT,4)
      A(I+NMG) = MAX0(IETIM,1)
   40 SUM= SUM + A(I+NMG)
C
C*****SET CURRENT OPERATOR NUMBER TO 1
C
      A(NATT-2) = 1.0
C
C*****SET ARRIVAL TIME OF JOB TO TNOW
C
      A(NATT-1) = TNOW
C
C*****SET DUE DATE TO TWICE THE ESTIMATED PROCESSING TIME
C
      A(NATT) = 2.*SUM+TNOW
      RETURN
      END
```

Figure 12-5 Subroutine DESJOB for describing a job in the job shop model.

gers. Note that INDEX will be a sample from a uniform distribution where the low value is continually increased by 1 since BOT is reset to I for each passage through the DO loop. Thus, when I = 2 there can be five values for INDEX and the five machines still to be selected are stored in MA(2) through MA(6).

An integer value for the processing time is obtained as a sample from an exponential distribution. The estimated processing time is stored in A(I + NMG). Processing times smaller than 1 are not permitted. At statement 40, the end of the DO loop, the sum of the processing times is recorded in the variable SUM. The last three statements of subroutine DESJOB are used to set the current operation number to 1, the arrival time of the job to TNOW, and the due date to be twice the estimated processing time for the job. These values are stored in the last three cells of the A vector. The establishment of the due date as twice the sum of the estimated processing times is taken from the statement of the problem as presented by Eilon.

The function for computing the modified due date DDM is shown in Figure 12-6. The modified due date is the value of the due date minus the estimated processing time minus a safety factor. This involves one statement as shown in Figure 12-6.

```
      FUNCTION DDM(A)
      COMMON/SCOM1/ATRIB(100),DD(100),DDL(100),DTNOW,II,MFA,MSTOP,NCLNR
     1,NCRDR,NPRNT,NNRUN,NNSET,NTAPE,SS(100),SSL(100),TNEXT,TNOW,XX(100)
      COMMON/UCOM1/ XJINF(3000),AAJOB(15),JOBF,NMG,NATT,SAFET,NLATE,
     1NUMJOB
      DIMENSION A(15)
C
C***** SET DUE DATE FOR THE JOB
C
      DDM = A(NATT)-.5*(A(NATT)-TNOW)-SAFET
      RETURN
      END
```

Figure 12-6 Function DDM for computing a modified due date in the job shop model.

The subroutine to schedule an end of service, SCHES, is presented in Figure 12-7. The arguments for this subroutine are the machine number, MG, and the estimated processing time, ET. The actual processing time is equal to the estimated processing time plus a sample from a normal distribution which has a mean of 0 and a standard deviation equal to 0.3 times the estimated processing time. The service time, SERVT, is then set equal to the estimated processing time plus the sample from the normal distribution. A check is then made to insure that the service time is positive (since a negative value greater than the estimated processing time can be obtained from the normal distribution). Attribute 2 is then established as the current machine number MG. The modified due date is then altered to no longer include the estimated operation time of the machine being scheduled. (Alternatively, this updating of the modified due date could be done after the job is processed on this machine.) The end-of-service event is then scheduled by a call to subroutine SCHDL with event code 2, service time SERVT, and attribute buffer ATRIB. Note that ATRIB(1) need not be reset as the pointer to the auxiliary attribute array is established prior to the call to subroutine SCHES. This completes the description of what occurs when a job arrives including the establishment of the job description, the due date, and the scheduling of an end-of-service event if the machine required for the first operation is available for processing the job. Next we consider the second event, an end of service, ENDSV, for a job at a particular machine.

```
      SUBROUTINE SCHES(MG,ET)
      COMMON/SCOM1/ ATRIB(100),DD(100),DDL(100),DTNOW,II,MFA,MSTOP,NCLNR
     1,NCRDR,NPRNT,NNRUN,NNSET,NTAPE,SS(100),SSL(100),TNEXT,TNOW,XX(100)
      SIGMA = 0.3*ET
      SERVT = ET + RNORM(0.0,SIGMA,2)
      IF(SERVT.LT.0.0) SERVT = 0.0
      ATRIB(2) = MG
C
C*****UPDATE THE MODIFIED DUE DATE TO NOT INCLUDE ESTIMATED SERVICE
C*****TIME OF THE CURRENT OPERATION
C
      ATRIB(3) = ATRIB(3) + ET
C
C*****SCHEDULE END OF SERVICE EVENT
C
      CALL SCHDL(2,SERVT,ATRIB)
      RETURN
      END
```

Figure 12-7 Subroutine SCHES for scheduling an end of service for job shop model.

When ENDSV is called, the SLAM II processor has loaded the attribute buffer, ATRIB, with the following attributes:

ATRIB(1) = pointer to the position in the auxiliary attribute array XJINF for information concerning the job on which service was just completed;
ATRIB(2) = machine on which service ended; and
ATRIB(3) = modified due date for job completing service.

In subroutine ENDSV shown in Figure 12-8, JBPTR is set equal to the job pointer (ATRIB(1)) and MACHE is set equal to the machine on which service is ended (ATRIB(2)). Next, the number of operations completed for the job is accessed directly from XJINF(ICOPN) where ICOPN is the thirteenth (NATT−2) auxiliary attribute associated with the job. The next operation number is defined as IOPN which is one more than the current operation number. Since there can be at most 6 operations for a job, an IOPN value of 7 indicates that no further machines in the job shop are to be visited. When this occurs, a transfer to statement 50 is made to delete the job from the job shop.

If IOPN is not 7, the auxiliary attribute array is updated to the new current operation number and the new machine number, NEWM, is accessed. If the new machine number is 0, the job has also completed its tour through the job shop and a transfer is made to statement 50. If the job is to go to another machine, the estimated processing time is obtained from the appropriate cell of XJINF and stored as the variable ET. The code described above obtains the next machine and estimated processing time for the job that just completed service; we are now ready to route the job.

First, a check is made to see if the new machine is busy. If it is, a transfer is made to statement 20 where the value of ATRIB(2) is established. The job is then filed in the queue of the new machine, NEWM. If NEWM was idle, its status is changed to busy by setting XX(NEWM) to 1.0. An end of service for the job on NEWM is then scheduled by calling subroutines SCHES. After determining the disposition of the job that just completed service, a transfer is made to statement 100 where the disposition of the machine that just completed service is determined.

Before going to statement 100, we will describe what happens when a job has been completely processed. The number of jobs completed by the job shop is increased by 1. The attributes of the job are removed from the system by a call to subroutine GETAA which loads the auxiliary attributes into the vector AAJOB and frees up the space that was used for the auxiliary attributes of the job on which service was completed. Statistics are then collected on time in the system by subtracting the job's arrival time (the 14th auxiliary attribute) from TNOW and calling subroutine COLCT with code 1. Next, a check is made to see if the current time is less than or equal to the assigned due date. If not, the number of late jobs is increased by one and statistics are computed on the amount of time the job was late by calling subroutine COLCT with code 2. This completes the description of a job that has completed its routing through the machine shop.

The disposition of machine MACHE is determined starting at statement 100. First, a check is made to see if any jobs are waiting to be processed. If none is waiting, XX(MACHE) is set to 0 to indicate that MACHE is idle, and a return is then made to the SLAM II processor. If a job is waiting, a transfer is made to statement 110. At statement 110, subroutine UPDAT(MACHE) is called to update the priority class of those jobs waiting for machine MACHE. This updating is required since a job's priority could change while it is waiting in the queue and, hence, its position in the queue could change. For example, if the first job in the queue had an estimated processing time of 15.0 and was in the priority class, and

```
      SUBROUTINE ENDSV
      COMMON/SCOM1/ ATRIB(100),DD(100),DDL(100),DTNOW,II,MFA,MSTOP,NCLNR
     1,NCRDR,NPRNT,NNRUN,NNSET,NTAPE,SS(100),SSL(100),TNEXT,TNOW,XX(100)
      COMMON/UCOM1/ XJINF(3000),AAJOB(15),JOBF,NMG,NATT,SAFET,NLATE,
     1NUMJOB
      JBPTR = ATRIB(1)
C
C*****SAVE MACHINE NUMBER ON WHICH SERVICE ENDED
C
      MACHE= ATRIB(2)
C
C*****DETERMINE DISPOSITION OF JOB ON WHICH SERVICE ENDED
C
      ICOPN = JBPTR + NATT - 2
      IOPN = IFIX(XJINF(ICOPN)) + 1
C
C***** IF NEXT OPERATION IS OPERATION NUMBER 7,
C***** THE JOB HAS COMPLETED
C
      IF(IOPN.EQ.7) GO TO 50
      XJINF(ICOPN) = IOPN
      NEWM = XJINF(JBPTR+IOPN)
C
C***** IF NEXT MACHINE IS O, JOB HAS COMPLETED
C
      IF(NEWM.EQ.O) GO TO 50
C
C***** DETERMINE THE ESTIMATED PROCESSING TIME
C
      ET = XJINF(JBPTR + IOPN +NMG)
      IF(XX(NEWM).GT.0.0) GO TO 20
      XX(NEWM) = 1.0
      CALL SCHES(NEWM,ET)
      GO TO 100
C
C***** IF NEW MACHINE IS BUSY, PLACE THE JOB IN QUEUE
C
   20 ATRIB(2) = ET + 1000.
      IF(ATRIB(3).LE.TNOW) ATRIB(2) = ET
      CALL FILEM(NEWM,ATRIB)
      GO TO 100
C
C***** UPDATE COUNTER AND COLLECT STATISTICS FOR COMPLETE JOBS
C
   50 NUMJOB = NUMJOB + 1
      CALL GETAA(JBPTR,NATT,XJINF,JOBF,AAJOB)
      TISYS= TNOW-AAJOB(NATT-1)
      CALL COLCT(TISYS,1)
      TLATE = AAJOB(NATT) - TNOW
      CALL COLCT(TLATE,3)
      IF(TNOW.LE.AAJOB(NATT)) GO TO 100
      NLATE = NLATE +1
      CALL COLCT(TLATE,2)
C
C*****DETERMINE DISPOSITION OF MACHE
C
  100 IF(NNQ(MACHE).GT.0) GO TO 110
      XX(MACHE) = 0.0
      RETURN
C
```

Figure 12-8 Subroutine ENDSV for job shop model.

```
C*****UPDATE PRIORITY OF JOBS WAITING IN QUEUE OF MACHINE
C
   110 CALL UPDAT(MACHE)
       CALL RMOVE(1,MACHE,ATRIB)
       ETIME=ATRIB(2)-1000.
       IF(ETIME.LE.0.0) ETIME = ATRIB(2)
       CALL SCHES(MACHE,ETIME)
       RETURN
       END
```

Figure 12-8   (continued).

the seventh job in the queue of the machine had a processing time of 12 but was
in the non-priority class when the machine last started processing a job, then if
the priority of the job that is seventh in the queue changed, it should be processed
prior to the job that is first in the queue since it has a smaller estimated processing time. Subroutine UPDAT performs the necessary rearrangement of the jobs in
the queue. When the rearrangement is completed, subroutine RMOVE is called to
remove the first job in the queue of MACHE and to load the job's attributes into
ATRIB. The estimated processing time, ETIME, is set equal to ATRIB(2)–1000. If
ETIME is less than or equal to 0 then ETIME is reset to ATRIB(2). Throughout
the code, it is presumed that no estimated processing time is greater than 1000. If
this is not the case, a larger constant must be used to distinguish between jobs of
different priority. Throughout the code, it is also assumed that non-priority jobs
are more frequently encountered than priority jobs and, hence, attribute 2 is more
frequently greater than 1000 than not. An end-of-service event is scheduled for
MACHE and the job by calling subroutine SCHES. This completes the description of the code for the end-of-service subroutine.

```
      SUBROUTINE UPDAT(MACH)
      COMMON/SCOM1/ ATRIB(100),DD(100),DDL(100),DTNOW,II,MFA,MSTOP,NCLNR
     1,NCRDR,NPRNT,NNRUN,NNSET,NTAPE,SS(100),SSL(100),TNEXT,TNOW,XX(100)
      DIMENSION NSET(1)
      COMMON QSET(1)
      EQUIVALENCE (NSET(1),QSET(1))
C
C***** UPDATE THE PRIORITY OF THOSE JOBS IN QUEUE
C
      NTRY = MMLE(MACH)
   11 IF(NTRY.EQ.0) RETURN
      IF(QSET(NTRY+2).LT.1000.) RETURN
      IF(QSET(NTRY+3).GT.TNOW) GO TO 15
      NEXT = NPRED(NTRY)
      QSET(NTRY+2) = QSET(NTRY+2) -1000.
      CALL ULINK(-NTRY,MACH)
      CALL LINK(MACH)
      NTRY = NEXT
      GO TO 11
   15 NTRY = NPRED(NTRY)
      GO TO 11
      END
```

Figure 12-9   Subroutine UPDAT for updating job priority class in the job shop model.

Subroutine UPDAT(MACH) is shown in Figure 12-9 and is used to update the priority class of jobs waiting in the queue for machine MACH. The procedure for updating jobs will be to start at the last job in the file and to proceed toward the first job in the file. As soon as a job is encountered which is in the priority class, that is, its second attribute is less than 1000, we know that no further jobs need be considered for updating. If the second attribute of a job is greater than 1000 then the modified due date is compared to the current time to determine if its priority class should be updated. If it should, the second attribute is reduced by 1000, the job is taken out of its current position in the file, and then refiled in order that its correct position will be determined based on its new value of attribute 2. The removing and refiling of the job is done by calls to subroutine ULINK and LINK. The scanning of the file is accomplished by accessing the pointers to entries. The pointer to the last entry in the file is obtained by using function MMLE. Predecessor entries are obtained by using function NPRED. We will now describe the code explicitly.

The first statement in subroutine UPDAT establishes NTRY as the pointer to the last entry in file MACH. If NTRY is zero, no further entries exist in the file and a return from subroutine UPDAT is made. The second attribute of the entry is stored in QSET(NTRY + 2). Rather than employ subroutine COPY to access the second attribute, we will access it directly. If the second attribute is less than 1000, then no further changes in priority for jobs in the queue need be considered and a return from UPDAT is made. If the second attribute is greater than or equal to 1000, the third attribute is tested to determine if the float is negative. If it is not, a transfer to statement 15 is made where NTRY is set equal to the predecessor of NTRY and the process is started again by a return to statement 11. If the float is negative, the priority class for the job whose pointer is NTRY must be changed. First the pointer NEXT is established as the predecessor of NTRY. It is necessary to save this predecessor pointer since the entry under consideration is going to be unlinked from the other entries and, hence, its predecessor pointer will change. The second attribute is reduced by 1000 which is the procedure used in this example to increase the priority of the job. A call is then made to subroutine ULINK with a negative first argument to indicate that a pointer rather than a rank is being used. The entry is then reinserted into the file by a call to subroutine LINK. ULINK and LINK were employed to illustrate their use. For this example, since the second attribute was changed, the code could have employed subroutine RMOVE, then updated the second attribute and refiled the entry with a call to subroutine FILEM. After linking the entry back into file MACH, the next entry to be considered, NTRY, is set equal to NEXT and a transfer to statement 11 is made to continue the scanning of the file. This completes the description of subroutine UPDAT.

```
      SUBROUTINE OTPUT
      COMMON/SCOM1/ ATRIB(100),DD(100),DDL(100),DTNOW,II,MFA,MSTOP,NCLNR
     1,NCRDR,NPRNT,NNRUN,NNSET,NTAPE,SS(100),SSL(100),TNEXT,TNOW,XX(100)
      COMMON/UCOM1/ XJINF(3000),AAJOB(15),JOBF,NMG,NATT,SAFET,NLATE,
     1NUMJOB
C
C***** PRINT THE NUMBER OF JOBS THAT WERE COMPLETED,
C***** THE NUMBER OF LATE JOBS AND THE FRACTION OF LATE JOBS
C
      AVE= FLOAT(NLATE)/FLOAT(NUMJOB)
      WRITE(NPRNT,10) TNOW,NUMJOB,NLATE,AVE
   10 FORMAT(F10.2,2I10,F10.5)
      RETURN
      END
```

Figure 12-10   Subroutine OTPUT for job shop model.

The code for subroutine OTPUT is shown in Figure 12-10. In subroutine OTPUT, the number of jobs processed and the amount of time required to process the jobs is shown. Also printed are the number of jobs late and the fraction of jobs that were late.

The input statements for the job shop model are shown in Figure 12-11. The GEN statement indicates that only one run is to be made. The LIMITS statement specifies that six files are to be used, that there are three attributes per entry, and at most there will be 200 entries. The 200 entries is our best estimate of the maximum possible number of concurrent jobs. The PRIORITY statement establishes that each file is to be ranked on low-value-first using attribute 2. A PRIORITY statement for the event calendar is also given which specifies that secondary ranking for the event calendar should be based on high-value-first based on the event code. In this way if an arrival event and end-of-service event occur at the same time, the end-of-service event will be processed first since it has the higher event

```
   1  GEN,PRITSKER,ADVANCED JOBSHOP,1/4/1984,1;
   2  LIMITS,6,3,200;
   3  PRIORITY/1,LVF(2)/2,LVF(2)/3,LVF(2)/4,LVF(2)/5,LVF(2)/6,LVF(2);
   4  PRIORITY/NCLNR,HVF(JEVNT);
   5  TIMST,XX(1),MACHINE 1 UTIL;
   6  TIMST,XX(2),MACHINE 2 UTIL;
   7  TIMST,XX(3),MACHINE 3 UTIL;
   8  TIMST,XX(4),MACHINE 4 UTIL;
   9  TIMST,XX(5),MACHINE 5 UTIL;
  10  TIMST,XX(6),MACHINE 6 UTIL;
  11  STAT,1,JOB PROC TIME,20/100./20.0;
  12  STAT,2,JOB TARDINESS,20/-100./5.0;
  13  STAT,3,JOB LATENESS;
  14  INI,0,20000;
  15  FIN;
```

Figure 12-11   Input statements for job shop model.

code. The remaining input statements prescribe the statistics to be collected during the simulation run.

Summary of Results. The SLAM II Summary Report for this example is shown in Figure 12-12 for a simulation length of 20000 minutes. The estimate of the mean job processing time is approximately 147 minutes based on 802 jobs. The standard deviation estimate is approximately 90 minutes and indicates an extremely high variability in the time to process a job through the shop. This can be attributed to

S L A M   I I   S U M M A R Y   R E P O R T

SIMULATION PROJECT ADVANCED JOBSHOP   BY PRITSKER

DATE 1/ 4/1984   RUN NUMBER   1 OF   1

CURRENT TIME   0.2000E+05
STATISTICAL ARRAYS CLEARED AT TIME 0.0000E+00

STATISTICS FOR VARIABLES BASED ON OBSERVATION

	MEAN VALUE	STANDARD DEVIATION	COEFF. OF VARIATION	MINIMUM VALUE	MAXIMUM VALUE	NUMBER OF OBSERVATIONS
JOB PROC TIME	0.1465E+03	0.8994E+02	0.6139E+00	0.1171E+02	0.5892E+03	802
JOB TARDINESS	-0.5003E+02	0.5146E+02	-0.1028E+01	-0.3732E+03	-0.5469E-01	337
JOB LATENESS	0.7524E+01	0.6688E+02	0.8888E+01	-0.3732E+03	0.2112E+03	802

STATISTICS FOR TIME-PERSISTENT VARIABLES

	MEAN VALUE	STANDARD DEVIATION	MINIMUM VALUE	MAXIMUM VALUE	TIME INTERVAL	CURRENT VALUE
MACHINE 1 UTIL	0.5247E+00	0.4994E+00	0.0000E+00	0.1000E+01	0.2000E+05	0.1000E+01
MACHINE 2 UTIL	0.5090E+00	0.4999E+00	0.0000E+00	0.1000E+01	0.2000E+05	0.0000E+00
MACHINE 3 UTIL	0.5486E+00	0.4976E+00	0.0000E+00	0.1000E+01	0.2000E+05	0.1000E+01
MACHINE 4 UTIL	0.4931E+00	0.5000E+00	0.0000E+00	0.1000E+01	0.2000E+05	0.1000E+01
MACHINE 5 UTIL	0.4956E+00	0.5000E+00	0.0000E+00	0.1000E+01	0.2000E+05	0.1000E+01
MACHINE 6 UTIL	0.5143E+00	0.4998E+00	0.0000E+00	0.1000E+01	0.2000E+05	0.1000E+01

FILE STATISTICS

FILE NUMBER	ASSOC NODE LABEL/TYPE	AVERAGE LENGTH	STANDARD DEVIATION	MAXIMUM LENGTH	CURRENT LENGTH	AVERAGE WAITING TIME
1		0.5424	1.0975	7	0	28.9286
2		0.4129	0.8831	7	0	25.2562
3		0.5796	1.1461	9	1	29.7213
4		0.4584	0.9799	7	1	25.0497
5		0.3852	0.8898	7	1	23.9254
6		0.4436	0.8932	6	0	25.9395
7	CALENDAR	4.0852	1.2675	7	6	19.9767

Figure 12-12   SLAM II summary report for job shop model.

the variability in the number of operations required per job (normally distributed with mean of 4) and the variability of each operation time (exponentially distributed with mean of 20). Thus, the job processing time is a random sum of random variables†. The variance of the sample mean is not estimated so we do not have a reliability value for the estimate of the mean processing time. However, the machine utilization statistics indicate a relatively low usage of machines (from 0.49 to 0.55) and the file statistics indicate a small amount of queueing on the average (from 0.39 to 0.58). Thus, the jobs although not independent have less interaction than expected.

Let us investigate other outputs of the model to see if they agree with our expectations. The number of job arrivals in 20000 time units is 810. This includes 802 jobs processed, plus 8 jobs currently at the machines. The mean job interarrival time is given as 25 minutes but is discretized to a rate of 0.0409 jobs/minute. Thus in 20000 minutes, we expect 817 jobs. Since the interarrival times are approximately exponential, the number of arrivals is Poisson distributed and the variance of the number of job arrivals in 20000 minutes is also 817. The standard deviation of number of job arrivals is approximately 28, and the observed number of 810 is within 1 standard deviation of the mean value of 817. From this discussion, the estimates derived from this simulation run for utilization should be lower than average since the number of job arrivals is lower than average. This point is explored further below.

The utilization of a shop is usually estimated as the arrival rate divided by the service rate. The average arrival rate was given above as 0.0409 jobs/minute. The average service rate for the shop is 0.05 operations/minute/machine. Since there are six machines, we have an average capacity of 0.30 operations/minute. To convert the average arrival rate from jobs/minute to operations/minute, we must estimate the expected number of operations/job. The number of operations/job was described as normally distributed with mean of 4 but with the tails of the distribution clipped at 3 and 6. As described in Chapter 18, this is referred to as a mixed distribution. The probability of requiring 3, 4, 5, or 6 operations/job can be obtained from normal probability tables on the next page.

Using this table, the expected number of operations/job is computed as 4.0668. Using the estimates, we compute a hypothesized utilization of 0.554. The average of the six average machine utilization values given in Figure 12-12 is slightly lower (0.514) than the theoretical value computed as anticipated. As explained previ-

---

† The variance of a random sum of independent random variables, σ_J^2, is given by

$$\sigma_J^2 = E[N]\sigma_J^2 + E^2[S]\sigma_N^2$$

where N is the number of operations and S is the service time random variable (see Chapter 2).

Number of Operations/Job	Probability
3	0.3085
4	0.3830
5	0.2417
6	0.0668

ously, a low utilization factor reduces job interference and, hence, statistical dependence. The above analysis does, however, provide guidelines that assist in comprehending the outputs from a simulation.

Returning now to the SLAM II Summary Report, jobs are early by an average value of 7.52. If we desire a zero average lateness then the due-date setting procedure should be adjusted by decreasing the due date by 7.52. This adjustment affects the priority assignments of jobs and we should not expect the average to be precisely zero. From the summary report, it is seen that 337 of the 802 jobs processed were tardy, that is, they were completed after their assigned due day, and the average tardiness was 50.03. Since the shop has a low utilization, the inability to meet due dates is primarily due to the due date setting procedure.

This example illustrates how job shop dispatching rules can be analyzed using SLAM II. Extensions to more complex situations are easily incorporated. The basic structure of the model need not be changed in order to study advanced job shop procedures.

12.11 CHAPTER SUMMARY

In this chapter, advanced file manipulation subprograms are discussed and their use illustrated. Procedures for accessing and using file entry pointers are presented. Four subprograms for associating auxiliary attributes with an entry in the SLAM II filing system are described. SLAM II subprograms for directly obtaining statistical estimates for variables on which data is collected, and for presenting and preparing specialized output reports, are detailed. An example of outputs obtained from a user-developed trace of events by coding subroutine UMONT is provided. Subprograms for clearing statistics, reporting errors, monitoring and tracing events, and obtaining values from tables are described. An example involving a job shop model is presented which illustrates the use of auxiliary attributes and the advanced SLAM II file manipulation subprograms.

12.12   EXERCISES

12-1.   Write the statements required to determine the location of the entry that satisfies the following conditions:

(a) The entry in file 2 that has attribute 3 equal to 10.

(b) The entry with the largest value of attribute 4 greater than 0 in file 1. What subroutine should be used to remove this entry from file 1?

(c) The entry whose third attribute is closest to 10 but does not exceed 10 in file 3.

(d) In file 3, the entry whose third attribute is closest to 10 but is not less than 10.

(e) The entry in file 4 whose second attribute is the largest. Entries with the value of attribute 2 less than 3 cannot be used.

12-2.   Write the statements necessary to obtain the attribute values associated with the second event in the event file. Specify the values of the variables that would be obtained.

12-3.   Specify whether you would use subroutine COLCT or a TIMST input statement to collect statistics on each of the following variables. Give an explanation for your decision.

(a) The age of individuals going to a barber shop.

(b) The time it takes a secretary to type a letter.

(c) The amount of time devoted by employees to breaks.

(d) The amount of water in a reservoir.

(e) The price of a stock during a given day.

(f) The price of a stock over the last 100 days.

(g) The dollars in your bank account.

12-4.   Given that a histogram has 10 cells, the lower limit is specified as 5, and the width of each cell is specified as 1, determine the cell number into which each of the following values would be inserted: 7.2, 9.1, 5.0, 4.1, 22.7, and 3.3.

12-5.   Write a program to compute 400 samples from a normal distribution with a mean of 50, a standard deviation of 10, a minimum value of 0, and a maximum value of 100. In this program, compute the average and standard deviation of the samples, the minimum and maximum value obtained from the samples, and a histogram of the sample values between 30 and 70 with a cell width equal to 4.

12-6.   Develop a function subprogram that determines the total number of entries in all files. Define the function name as NTOTE.

12-7.   Write a subroutine called FIND3 that locates an entry NTRY in file J such that the value of ATRIB(1) is less than X(1), the value of ATRIB(2) is less than X(2), and the value of ATRIB(3) is greater than or equal to X(3). If no entry in file J satisfies the above conditions, return a negative value to the calling routine. Write a statement using FIND3 that locates an entry in file 4 whose first two attributes are less than three and whose third attribute is greater than five.

12-8.   Write a subprogram that will compute the sum of the product of two attributes of a given file.

12-9.   Develop a function subprogram for inclusion in SLAM II that performs a table look-up function for a table whose independent variable is not specified in equal increments, and for which extrapolation outside the table is employed.

12-10. Develop a subroutine for SLAM II called RGRES that will generate samples from any desired regression equation involving terms in x with the exponents $1/2$, 1, 2, and 3, and terms of the form e^x and $\ln(x)$. Assume that the random portion of the regression sample is normally distributed and the coefficients are arguments to the subroutine.

12-11. Write the code to schedule an event into file 1 and to file an entry into file 3. The event and entry have the following attributes:

Attributes of event:

Event time	Current time plus a sample from a uniform distribution between 10. and 20.
Event code	7
Attribute 1	Location of the entry to be placed in file 3

Attributes of entry in file 3:

1	127.
2	TNOW
3	Location of the *event* described above

12-12. (From Schriber(2)) A production shop is comprised of six different groups of machines. Each group consists of a number of identical machines of a given kind as indicated below.

Machine Group Number	Kind of Machines in Group	Number of Machines in Group
1	Casting units	14
2	Lathes	5
3	Planers	4
4	Drill presses	8
5	Shapers	16
6	Polishing machines	4

Three different types of jobs move through the production shop. These job types are designated as Type 1, Type 2, and Type 3. Each job type requires that operations be performed at specified kinds of machines in a specified sequence. All operation times are exponentially distributed. The visitation sequences and average operation times are shown on the next page.

Jobs arrive at the shop with exponential interarrival times with a mean of 9.6 minutes. Twenty-four percent of the jobs in this stream are of Type 1, 44 percent are of Type 2, and the rest are of Type 3. The type of arriving job is independent of the job type of the preceding arrival. Build a SLAM II model which simulates the operation of the production shop for five separate 40-hour weeks to obtain: 1) the distribution of job residence time in the shop, as a function of job-type; 2) the utilization of the machines; and 3) queue statistics for each machine group.

Visitation Sequences and Mean Operation Times for the Three Types of Jobs

Job Type	Total Number of Machines to be Visited	Machine Visitation Sequence	Mean Operation Time (Minutes)
1	4	Casting Unit	125
		Planer	35
		Lathe	20
		Polishing machine	60
2	3	Shaper	105
		Drill press	90
		Lathe	65
3	5	Casting unit	235
		Shaper	250
		Drill press	50
		Planer	30
		Polishing machine	25

Embellishments:

(a) Employ a shortest processing time rule for ordering jobs waiting before each machine group. Compare output values.

(b) Give priority to jobs on the basis of type. Job type 3 is to have the highest priority then Type 2 and then Type 1 jobs.

(c) Change the average job interarrival time to 9 minutes and evaluate system performance.

(d) Develop a cost structure for this problem that would enable you to specify how to spend $100,000 for new machines.

12-13. Generate the job sequences and processing times for the production shop of Exercise 12-12 so that they can be used to test different operating rules and machine configurations.

12-14. Consider a job shop in which there can be as many as 10 machine groups. The number of machines/machine group is an input. The types of jobs processed in the job shop have not been categorized. The number of operation per job, the processing time/job and the routing of a job is established as input. Develop a general SLAM II program to simulate the job shop.

Embellishment: Use this general purpose job shop model to study the production shop described in Exercise 12-12.

12-15. An analysis is desired of an information service system, such as might be found in airports and hotels, which provides a questioner with information about weather or flight arrivals or room reservations. The system consists of six stations that allow requests for information by using a code posted at the station. The six stations are far apart so that each customer has only one choice of a booth and will not jockey for the shortest queue. If a station is being used, a customer joins the queue and waits to place a request. After a request has been placed (dialed), the customer waits until his answer is displayed. The answer may consist of lighted panels, pictures, or words. When the presentation of the answer is made, the customer reads it.

A request for information from a station terminates at a point on a six-point scanner. The scanner rotates and recognizes a request made at a station associated with a scanning point. The request is transferred through the scanner to a buffer unit capable of holding three messages. When the scanner recognizes a request but the buffer is full, the scanner stops and waits until space is available in the buffer.

The buffer unit works in conjunction with a computer to answer the messages. The computer can work on three messages simultaneously. When the information requested is found by the computer, it is placed in the buffer and transferred from the buffer to the originating station. Transfer of the answer takes place directly between the station and the buffer and does not require the use of the scanner.

The system parameters are: customers arrive to each station at a mean rate of 5/6 customers per minute with an exponential distribution of time between arrivals; customer's dialing time and reading time are uniformly distributed in the range (0.3, 05) and 0.6, 0.8) minutes, respectively; scanner rotation time between points is 0.0027 minutes, and scanning time is also 0.0027 minutes. It takes 0.0117 minutes to achieve a transfer from the scanning unit to the buffer, there is a 0.0397 minute delay for transferring from the buffer to the station; and the time required for the computer to locate the answer to a request is uniformly distributed in the range (0.05, 0.10) minutes. Build and analyze a discrete event model of this system to find out if equipment with the capabilities shown by the parameters in the statement of the problem can effectively handle the flow of customers. Two statistics are of primary interest: the distribution of the waiting time a customer encounters between the time of dialing to the time an answer is displayed; and the average size of the queue expected in front of each station. Run the simulation for 480 minutes.

Embellishments:

(a) Determine the critical resource in the system. Redesign the system to make the critical resource less constraining.
(b) Add reneging to the model by having customers leave the system if their waiting time is greater than 5 minutes.

12.13   REFERENCES

1. Eilon, S., I. G. Chowdhury, and S. S. Serghiou, "Experiments with the SI^x Rule in Job Shop Scheduling," *Simulation,* Vol, 24, 1975, pp. 45-48.
2. Schriber, T., *Simulation Using GPSS,* John Wiley, 1974.
3. Wineberger, A. et al., "Use of Simulation to Evaluate Capital Investment Alternatives in the Steel Industry: A Case Study," Bethlehem Steel Corporation. Presented at the Winter Simulation Conference, December 1977.

CHAPTER 13

Continuous Modeling

13.1   INTRODUCTION

Continuous modeling involves the characterization of the behavior of a system by a set of equations. The time-dependent portrayal of the variables described by the equations is one of the desired outputs from such models. The models can

483

consist of sets of algebraic, difference, or differential equations and can contain stochastic components. The status of a system defined in this manner is changing continuously with time. Events may occur, however, and instantaneously affect the status of the system.

The continuous systems modeler has two basic tasks: 1) the development of the equation set and events that describe the time-dependent, stochastic behavior of the system; and 2) the evaluation of the equation set and events to obtain specific values of system behavior for different operating policies. A simulation language for continuous models assists in the first task by defining the format for the equation set. However, it is in task 2 that the simulation language has its greatest impact. The language provides the mechanisms for obtaining the values of the variables described by the equations. It does this by solving the equation set at a single point in time and by recording the values for future reporting. Time is then advanced in a step-wise fashion and the equations are again solved for the values of the variables at this new time, assuming knowledge of the variables at previous times. When events occur, the effect of the event is incorporated into the model and in the evaluation of the variable values. In this manner, the entire time history of the model variables are obtained. The procedures employed by SLAM II in supporting continuous system modeling are described in this chapter.

13.2   SLAM II ORGANIZATION FOR CONTINUOUS MODELING

Models of continuous systems involve the definition of state variables by equations and the definition of state-events based on the values of state variables as presented in Chapter 10. The development of a SLAM II continuous simulation program requires the user to write subroutine STATE for defining state equations, SEVNT input statements to prescribe the conditions that define state-events, and subroutine EVENT(I) for modeling the consequences of the occurrence of state-event I. In addition, the user must write a main program that calls the SLAM II executive. Initial conditions are established by writing subroutine INTLC or through INTLC input statements as described in Chapter 8. Specialized outputs of the system variables can be obtained by writing subroutine OTPUT or through the RECORD and VAR input statements. A block diagram of the SLAM II organization for continuous models is shown in Figure 13-1.

The function of the main program is to call the SLAM II processor which controls the running of the simulation. The SLAM II processor first calls upon the standardized initialization routine to initialize SLAM II variables and to read SLAM II input statements that define the characteristics of the model. Non-SLAM II variables can be initialized in subroutine INTLC which is called after the SLAM II input statements are read. The user can perform additional initialization functions in subroutine INTLC.

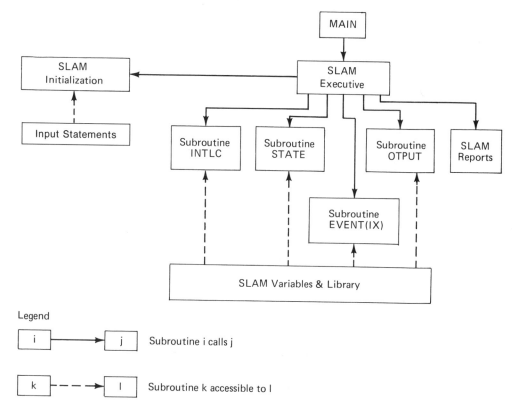

Figure 13-1   SLAM II organization for continuous modeling

After initialization, the executive routine calls subroutine STATE in order to obtain new values for the state variables. The state variables, SS($\cdot$), and their derivatives, DD($\cdot$), were defined in Section 10.2. The function of subroutine STATE is to define the dynamic equations for the state variables. If only difference equations are used in subroutine STATE, the executive routine will call STATE at periodic intervals called steps unless an intervening event or information request is encountered. In subroutine STATE, the state variable values at the end of the step are computed from the user-written equations. If differential equations are included in subroutine STATE, then the executive routine calls subroutine STATE many times within a step. In either case, values for the state variables at the end of a step are computed. The executive routine then tests if conditions defining state-events are included in the model. These conditions are specified in terms of state variables crossing prescribed thresholds. The user specifies such conditions on SEVNT input statements. One input statement is required for each state condition definition. If a state-event is detected, the executive routine calls subroutine EVENT(I) with the state-event code that was specified on the SEVNT input statement. It is the user's responsibility to code the changes to be made to the system when a particular state-event occurs. In SLAM II, the following types of changes are allowed:

1. Changes in the defining equations for the state variables;
2. Discrete changes in the values of the state variables;
3. Changes in the state variables and/or thresholds involved in state condition specifications;
4. Changes to SLAM II or non-SLAM II variables.

To record the performance of the system over time, the executive routine collects system status information throughout a simulation run. The information to be collected is specified by the user on TIMST and RECORD input statements. The frequency with which the information is recorded is user specified during the standardized initialization. The options are to record: at periodic intervals, at the end of each step, at each event time, and at each call to subroutine GPLOT.

The ending of the simulation can be specified either by a state-event condition or by a total simulation time. When the executive routine determines that the simulation is ended, subroutine OTPUT is called. In OTPUT, the user can write a specialized output report and perform any end-of-simulation processing that is required. Following the call to subroutine OTPUT, standardized summary reports are printed. These standardized summary reports include any tables and plots of data requested through RECORD statements, along with any statistical computations of values recorded during the simulation.

As can be seen from the above description and from Figure 13-1, the SLAM II organization has decomposed a problem by specifying the subprograms within which the user must define the state variables and the potential changes to the state variables when state-events occur. Superimposed on these functional elements are information processing statements and routines for detecting state variable crossings (SEVNT), recording state variable values (RECORD), initializing variables (INTLC), and reporting specialized output (OTPUT).

13.3 EXECUTIVE FUNCTION FOR CONTINUOUS SIMULATION MODELS

The executive function in SLAM II establishes the current simulation time, TNOW, and calls an appropriate user-written routine in accordance with the defined SLAM II organizational structure. If state variables are to be updated, the executive function defines the step size, DTNOW, and calls subroutine STATE where the user has written equations that update state variables to time TNOW.

As was the case in discrete models, the executive function is performed by subroutine SLAM which is called by the user from the MAIN program. Subroutine SLAM calls subroutine STATE to update the values of the state variables. Subroutine SLAM then determines if a state-event has occurred, and, if it has, calls subroutine EVENT(JEVNT) to communicate to the user that an event with code JEVNT has occurred. The executive routine also determines if the newly computed values are to be recorded as specified by a RECORD input statement. The next step size is then computed and the above procedure repeated.

13.4 ILLUSTRATION 13-1. CEDAR BOG LAKE

To illustrate the ease with which models of continuous systems can be analyzed using SLAM II, we present a model of Cedar Bog Lake that was developed by Williams (16).

The model includes three species, a solar energy supply (x_s), and the organic matter that forms a sediment on the lake bottom (x_o). These lake variables are modeled in terms of their energy content (calories/centimeter²) and the energy transfers between the various lake variables and losses to the environment (x_e). The three species are plants (x_p), herbivores (x_h), and carnivores (x_c). The differential equations relating these species to the sediment and the solar energy source are shown below.

$$\frac{dx_p}{dt} = x_s - 4.03x_p.$$

$$\frac{dx_h}{dt} = 0.48x_p - 17.87x_h.$$

$$\frac{dx_c}{dt} = 4.85x_h - 4.65x_c.$$

$$\frac{dx_o}{dt} = 2.55x_p + 6.12x_h + 1.95x_c.$$

$$\frac{dx_e}{dt} = 1.00x_p + 6.90x_h + 2.70x_c.$$

The values of the variables at time zero are: $x_p(0) = 0.83$, $x_h(0) = 0.003$, $x_c(0) = 0.0001$, $x_o(0) = 0.0$ and $x_e(0) = 0.0$.

The annual cycle in solar radiation is simulated using the following equation:

$$x_s = 95.9 \ (1 + 0.635 \sin 2\pi t)$$

where t is time in years. These equations represent such processes as the predation of one species by another, plant photosynthesis, and the decaying of dead species. Energy transfers between lake entities and their environment are due to respiration and migration.

We will use SLAM to illustrate the procedure for obtaining the values of the variables x_p, x_h, x_c, x_o, x_e, x_s over time. First, we make an equivalence between the model variables and the SLAM state vector SS(·) as shown below.

$$SS(1) = x_p \rightarrow DD(1) = \frac{dx_p}{dt}$$

$$SS(2) = x_h \rightarrow DD(2) = \frac{dx_h}{dt}$$

$$SS(3) = x_c \rightarrow DD(3) = \frac{dx_c}{dt}$$

$$SS(4) = x_o \rightarrow DD(4) = \frac{dx_o}{dt}$$

$$SS(5) = x_e \rightarrow DD(5) = \frac{dx_e}{dt}$$

and

$$SS(6) = x_s.$$

The entire SLAM II program consists of writing the standard main program (see Figure 11-6), subroutine STATE, and the input statements. The latter are shown in Figure 13-2. In subroutine STATE, the set of differential equations is coded. The translation of the equations from the model to the SLAM II code is direct and normally does not require an excessive amount of work. The input statements for the model involve mainly the definitions of the variables to record on RECORD and VAR input statements, and the CONTINUOUS statement as de-

```
      SUBROUTINE STATE
      COMMON/SCOM1/ ATRIB(100),DD(100),DDL(100),DTNOW,II,MFA,MSTOP,NCLNR
     1,NCRDR,NPRNT,NNRUN,NNSET,NTAPE,SS(100),SSL(100),TNEXT,TNOW,XX(100)
      DATA PI/3.14159/
C
C *** DEFINE THE SOLAR RADIATION AND DIFFERENTIAL EQUATIONS
C
      SS(6)=95.9*(1.+0.635*SIN(2.*PI*TNOW))
      DD(1)=SS(6)-4.03*SS(1)
      DD(2)=0.48*SS(1)-17.87*SS(2)
      DD(3)=4.85*SS(2)-4.65*SS(3)
      DD(4)=2.55*SS(1)+6.12*SS(2)+1.95*SS(3)
      DD(5)=SS(1)+6.9*SS(2)+2.7*SS(3)
      RETURN
      END
 1  GEN,PRITSKER,CEDAR BOG LAKE,3/5/1978,1;
 2  ;
 3  ;   DEFINE 5 DIFFERENTIAL EQUATIONS AND 1 RATE EQUATION
 4  ;   TO BE UPDATED IN A CONTINUOUS MANNER
 5  ;
 6  CONTINUOUS,5,1,.00025,.025,.025;
 7  ;
 8  ;   INITIALIZE THE CONTINUOUS VARIABLES
 9  ;
10  INTLC,SS(1)=.83,SS(2)=.003,SS(3)=.0001,SS(4)=0.0,SS(5)=0.0;
11  ;
12  ;   DEFINE THE INDEPENDENT PLOTTING VARIABLE AS TNOW
13  ;
14  RECORD,TNOW,TIME,0,P,.025;
15  ;
16  ;    DEFINE THE DEPENDENT PLOTTING VARIABLES AS SS(1),
17  ;   SS(2), SS(3), SS(4), SS(5), AND SS(6)
18  ;
19  VAR,SS(1),P,PLANTS;
20  VAR,SS(2),H,HERBIVORES;
21  VAR,SS(3),C,CARNIVORES;
22  VAR,SS(4),O,ORGANIC;
23  VAR,SS(5),E,ENVIRONMENT;
24  VAR,SS(6),S,SOLAR ENERGY;
25  INITIALIZE,0,2.0;
26  FIN;
```

Figure 13-2 SLAM II program of Cedar Bog Lake.

fined in Chapter 10. For this example, the CONTINUOUS statement specifies five differential equations (NNEQD), one state variable equation (NNEQS), a minimum step size (DTMIN) of 0.00025, a maximum step size (DTMAX) of 0.025, and a recording interval (DTSAV) of 0.025. The INTLC statement initializes the SS(·) values as prescribed by the problem statement, and the INITIALIZE statement specifies that the simulation should start at time zero and end at time 2. A segment of the plot requested through the RECORD statement is shown in Figure 13-3. This illustration demonstrates the ease of coding continuous models in SLAM II.

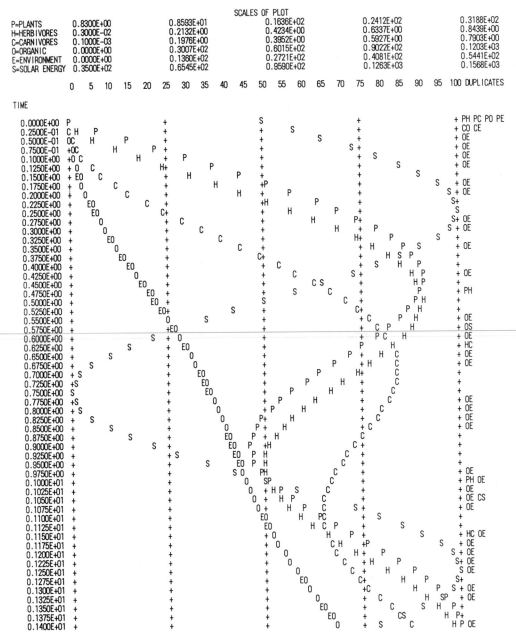

Figure 13-3 Plot of variables for Cedar Bog Lake illustration.

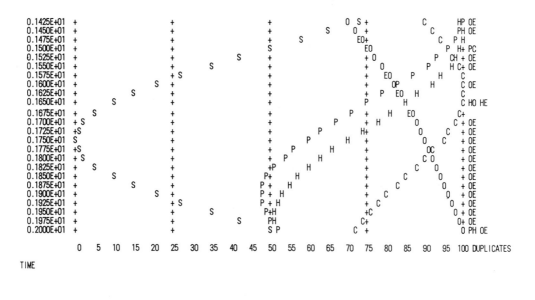

Figure 13-3 (continued).

```
OUTPUT   CONSISTS  OF        83 POINT SETS (     498 POINTS)
STORAGE  ALLOCATED FOR  1426 POINT SETS (  9982 WORDS )
STORAGE  NEEDED    FOR       83 POINT SETS (     581 WORDS )
```

13.5   COLLECTING TIME-PERSISTENT STATISTICS ON STATE VARIABLES

State variables and their derivatives are recomputed at the end of each step; hence, their values can be considered to be changing continuously over time. An average value for the variable can be computed by integrating the state variable and dividing by the time period of the integration. This could be accomplished by the SLAM II user through the definition of a new state variable. For example, if it is desired to obtain the average value of state variable 3, we could define state variable 10 to be the integral of state variable 3. The following statement coded in subroutine STATE would achieve the desired integration:

DD(10) = SS(3)

In subroutine OTPUT, the average of SS(3) could be computed as SS(10)/TNOW.

To avoid the definition of new state variables to compute averages and to obtain second moments about the mean for state variables, SLAM II allows the use of the input statement TIMST to be used with state variables and their derivatives. The format for the TIMST input statement is

TIMST,VAR,ID,NCEL/HLOW/HWID;

When VAR is defined as an SS(·) or DD(·) variable, SLAM II assumes a linear function between the state variable values at the ends of steps and computes an average and standard deviation based on this linear approximation. The statistical estimates are included in the portion of the SLAM II Summary Report that presents information on time-persistent variables.

13.6   THE SEVNT INPUT STATEMENT

To assist the user in specifying state-events, the SLAM II input statement SEVNT is provided. The SEVNT statement is a control statement that is analogous to the DETECT node in network models. SEVNT causes the detection of the crossings of one state variable against a threshold. A tolerance is specified for detecting the crossing. The crossing can be in the positive, the negative, or in both directions. The tolerance and direction of crossing are specified as fields in the SEVNT input statement. The format for the SEVNT input statement is

SEVNT,JEVNT,XVAR,XDIR,VALUE,TOL;

where
JEVNT is a user supplied state-event code;
XVAR is a variable and can be SS(J), DD(J) or other SLAM II status variable;†
XDIR specifies the direction of crossing: X→ either direction; XP→ positive direction; XN→ negative direction;
VALUE is the crossing threshold and can be a constant, SS(I), DD(I) or other SLAM II status variable.
TOL is a numeric value which specifies the tolerance within which the crossing is to be detected.

---

† Other SLAM II status variables are XX(I), NNACT(I), NNCNT(I), NNGAT(I), NNRSC(I), NNQ(I), NRUSE(I) and USERF(N).

Below are three examples of the SEVNT statement:

1. Define state-event 1 to occur when SS(3) crosses 100 in the positive direction with a tolerance of 2.0 (See Figure 10-1(a))
   SEVNT,1,SS(3),XP,100.,2.0;
2. Define state-event 2 to occur when SS(2) crosses SS(1) in the negative direction with a tolerance 0.01. (See Figure 10-1(b))
   SEVNT,2,SS(2),XN,SS(1),0.01;
3. Define state-event 4 to occur when DD(1) crosses 0.0 in either direction with a tolerance of 0.0.
   SEVNT,4,DD(1),X,0.0,0.0;

In this case, state-event 4 detects minimums, maximums, or points of inflection of state variable 1. A tolerance of 0.0 insures that the minimum step size DTMIN will be used to detect the crossing if necessary.

When a crossing within tolerance is determined, the executive function calls subroutine EVENT(JEVNT) to indicate that state-event JEVNT has occurred. If a crossing occurs that is not within tolerance, the step size is automatically reduced and the above process is repeated until a crossing within tolerance is obtained or the minimum step size is used. The user would then code in subroutine EVENT the logic involved and system status changes required when such an event is detected.

In event routines, discrete changes to state variables can be made by changing the value of SS(J). For example, the statement SS(3) = SS(3)+5.0 changes the value of state variable 3 by 5 units. In addition, parameters of equations and system status indicators used in STATE equations can be changed at event times. Thus, the values of coefficients and in effect the entire STATE equation structure could be changed at an event occurrence.

13.7   SIMULTANEOUS STATE-EVENTS

Sometimes the processing logic associated with a state-event depends upon knowledge concerning the possible occurrence of one or more additional state-events at the same instant in time. Because of this, SLAM II provides the modeler with function SSEVT(N) which returns a code defining the status of state-event N at the current time. The state-event number, N, is the value of JEVNT prescribed by the user on SEVNT statements. The status code for state-event N specifies whether the crossing variable crossed the threshold value during the last time advance, and if so, in what direction. In addition, the code denotes whether the minimum step size, DTMIN, permitted the crossing to be isolated within the prescribed tolerance. The codes returned from function SSEVT(N) are listed below.

Code	Definition
+2	crossing in the positive direction exceeding tolerance
+1	crossing in the positive direction within tolerance
0	no crossing
−1	crossing in the negative direction within tolerance
−2	crossing in the negative direction exceeding tolerance

As an example, consider that state-event 4 has occurred and subroutine EVENT has been called. If it is necessary to determine if state-event 6 has also occurred then the value obtained from function SSEVT(6) will provide this information. The value of SSEVT(6) is one of the five codes listed above.

13.8 MODELING USING STATE VARIABLES

Subroutine STATE can be used to model the state variables, employing any combination of DD(·) and SS(·) variables. The use of SS(·) to model difference equations is a straight translation of the difference equations, that is,

$$y_n = A * y_{n-1} + B$$

is equivalent to the SLAM II statement

SS(1) = A*SSL(1) + B

where SS(1) is used to represent the value of y at time n, that is, y_n, and SSL(1) represents the value of y at time n − 1, that is, y_{n-1}. Generalization to allow A and B to be functions of time is also direct. Higher order difference equations can be modeled. The translation of

$$y_n = A * y_{n-1} + B * y_{n-2} + C$$

is made by defining SS(1) to be y_n, then SSL(1) corresponds to y_{n-1}. Letting SS(2) = SSL(1) then SSL(2) corresponds to y_{n-2}. The value of y_n can be obtained as SS(1) using the following two state equations:

SS(1) = A * SSL(1) + B * SSL(2) + C
SS(2) = SSL(1)

State variables can also be used to model differential equations by means of an Euler integration method (1). In this case, the user must provide the integration method to solve the equations for the SS(·) variables. Consider the differential equation

$$\frac{dy}{dt} + A * y = B$$

or

$$y' = B - A * y.$$

Suppose that we know y at the point n−1, and we call this value $y_{n-1}^{\,o}$. Then

$$y_{n-1}' = B - A * y_{n-1}^{\,o} \,.$$

If we assume that y_{n-1}' is a good approximation of y′ between n−1 and n, then

$$y_n = y_{n-1}^{\,o} + h * y_{n-1}'$$

where h is the interval of time between points n−1 and n.

Substituting yields

$$y_n = y_{n-1}^{\,o} + h * (B - A * y_{n-1}^{\,o})$$

and

$$y_n = y_{n-1}^{\,o} * (1 - h * A) + h * B.$$

In terms of SLAM II variables, this becomes

SS(1)=SSL(1)*(1.−DTNOW*A)+DTNOW*B

since DTNOW is the time interval between calculations of state variables. The SLAM II user can employ more advanced Euler type integrators by coding such integrators in subroutine STATE.

If the equations involving the SS(·) variables cannot be written sequentially due to an interdependence of the variables, the user must provide the means for solving them. For example, the Gauss-Seidel procedure (2,11) could be used. SLAM II does not provide such a method since convergence is highly dependent on the set of equations.

13.9 MODELING USING DERIVATIVES OF STATE VARIABLES

SLAM II uses a Runge-Kutta-Fehlberg (RKF) algorithm to integrate the equations of subroutine STATE written in terms of the DD(·) variables. The RKF algorithm is used to obtain a solution to a set of simultaneous first-order ordinary differential equations of the form

$$y_j'(t) = f_j(y_1, y_2, \ldots, y_m, t), \qquad j = 1, 2, \ldots, m$$

where

$$y_j'(t) = \frac{dy_j(t)}{dt}.$$

In SLAM II, these equations would be expressed in the following form:

DD(J)=f$_j$(SS(1),SS(2), . . . , SS(M), TNOW), J=1, 2, . . . , M.

For example, if

$$y_1'(t) = a_1 y_1 + b_1 y_2$$

and

$$y_2'(t) = a_2 y_1 + b_2 y_2,$$

the corresponding SLAM II coding would be

DD(1)=A1*SS(1)+B1*SS(2)

and

DD(2)=A2*SS(1)+B2*SS(2)

Higher order differential equations can be modeled by putting the equations in canonical form. Thus, if an nth order differential equation of the form

$$x^{(n)}(t) = f(x, x', x^{(2)}, \ldots, x^{(n-1)}, t)$$

is to be modeled using SLAM II, N variables SS(J), J = 1,2, . . . ,N, must be defined as follows:

$$SS(1) = x$$
$$SS(2) = x' = DD(1)$$
$$SS(3) = x^{(2)} = DD(2)$$

$$\cdot$$
$$\cdot$$
$$\cdot$$

$$SS(N) = x^{(n-1)} = DD(N-1)$$

By substitution, $DD(N) = f(SS(1), SS(2), \ldots, SS(N), t)$. With these equations, $SS(1)$ is the solution to the nth order differential equation. As an example, consider the second order equation

$$x^{(2)}(t) = Ax'(t) + Bx(t) + C.$$

Let $SS(1) = x(t)$
and $DD(1) = x'(t) = SS(2)$
then $DD(2) = x^{(2)}(t)$

In SLAM II, this would be coded in subroutine STATE as

$$DD(1) = SS(2)$$
$$DD(2) = A * SS(2) + B * SS(1) + C$$

13.10 NUMERICAL INTEGRATION IN SLAM II

The integration method provided by SLAM II is a fourth (fifth) order, variable step size Runge-Kutta routine for integrating systems of first-order ordinary differential equations with initial values. The particular constants and error estimation used are from Fehlberg(4) as further described by Shampine and Allen(12).

The Runge-Kutta-Fehlberg (RKF) method provides the specific capability of numerically integrating a system of first-order ordinary differential equations of the form

$$\frac{dy(t)}{dt} = f(y(t); t)$$

with initial conditions $y(t_o)$. In this section $y(t)$ is a vector, and the above equation represents a system of first-order simultaneous differential equations.

Given the equation for dy(t)/dt, simulation can be used to obtain the time history of y(t). This is accomplished by considering y(t) as a function of the derivatives of y(t) using a Taylor series expansion. A Taylor series expansion allows the writing of a function in terms of the derivatives of the function where the accuracy with which the series represents the function is related to the order of the terms in the series. The error of the series approximation can be estimated by evaluating the first term omitted from the series approximation. In this way be evaluating the derivatives, dy(t)/dt, the values of y(t) can be estimated. This procedure for estimating y(t) is embodied in the RKF algorithm.

The RKF algorithm is a one step procedure involving six function evaluations over a step of size h. Let $t_1 = t_o + h$. The procedure as implemented in SLAM II for computing $y(t_1)$ involves the evaluation of the following equations:

$$a_1 = hf(y(t_o); t_o)$$

$$a_2 = hf(y(t_o) + \frac{1}{4} a_1 ; t_o + 1/4 \, h)$$

$$a_3 = hf(y(t_o) + \frac{3}{32} a_1 + \frac{9}{32} a_2 ; t_o + \frac{3}{8} h)$$

$$a_4 = hf(y(t_o) + \frac{1932}{2197} a_1 - \frac{7200}{2197} a_2 + \frac{7296}{2197} a_3 ; t_o + \frac{12}{13} h)$$

$$a_5 = hf(y(t_o) + \frac{439}{216} a_1 - 8a_2 + \frac{3680}{513} a_3 - \frac{845}{4104} a_4 ; t_o + h)$$

$$a_6 = hf(y(t_o) - \frac{8}{27} a_1 + 2a_2 - \frac{3544}{2565} a_3 + \frac{1859}{4104} a_4 - \frac{11}{40} a_5 ; t_o + \frac{1}{2} h)$$

$$EERR = \frac{1}{360} a_1 - \frac{128}{4275} a_3 - \frac{2197}{75240} a_4 + \frac{1}{50} a_5 + \frac{2}{55} a_6$$

$$y(t_o + h) = y(t_o) + \frac{25}{216} a_1 + \frac{1408}{2565} a_3 + \frac{2197}{4104} a_4 - \frac{1}{5} a_5$$

$$TERR = AAERR + RRERR * |y(t_o + h)|$$

where AAERR and RRERR are the user-specified absolute and relative error values provided on the SLAM II CONTINUOUS statement.

In the above equations, the a_i values can be considered as changes to y(t) at the end of the step based on derivative evaluations at intermediate points within the step. The equation for $y(t_o + h)$ is a weighted average of these changes added to $y(t_o)$. Thus, a_1 is the change that would occur if the derivative at the beginning of the step was used to project the value at the end of the step; a_2 is the change based on the derivative at the quarter step; a_3 is the change based on the derivative at three-eights of the step; a_4 is the change based on the derivative at twelve-thirteenths of the step; a_5 is the change based on the derivative at the end of the step. Note that a_6 is not used in the computation of the value at the end of the step but only used in the calculation of the estimated error, EERR. The allowed error, TERR, is based on the estimated value of the state variable at the end of the step. The computation indicated above involves multiple evaluations of the function f which corresponds to the computation of DD(·) as specified in subroutine STATE. These evaluations are made by SLAM II through calls to subroutine STATE. Prior to each call, the step size, DTNOW, is set as a fraction of h and the values of the state variables, SS(·), are computed. These values are passed to subroutine STATE through the SCOM1 COMMON block. The full step value, h, is available as the SLAM II variable DTFUL.

The RKF algorithm employed in SLAM II estimates the local error by simultaneous computation of fourth and fifth order approximations. This method permits the step size to be changed with little additional computation. A derivation and description of Runge-Kutta methods are contained in most books on numerical analysis (1,12). In the above procedure a_j is a vector and EERR is computed for each state variable defined by a differential equation. Only if the EERR $\leqslant$ TERR for all such state variables will the $y(t_1)$ be accepted.

In the RKF procedure, a variable step size is employed. Let Q be equal to the largest value of $|$EERR$|$ / TERR for any state variable defined by a differential equation. If Q > 1.0, the values are not accepted and h is reduced. If Q $\leq$ 1.0, the values are accepted and, depending on the value of Q, the next step size may be increased. If the step is accepted or rejected, the new step size is related to the value of $h/Q^{1/5}$. The new step size is maintained within the user-specified minimum and maximum allowable step sizes, that is,

$$\text{DTMIN} \leq h_{new} \leq \text{DTMAX}$$

Runge-Kutta integration has three advantages that make it an appropriate method for SLAM II. First, it is widely used and well-documented. For example, most CSSLs provide Runge-Kutta integration. Second, it is easy to change the step size with a Runge-Kutta routine. This is very important in a combined simulation where events are not normally spaced uniformly in time. The third advantage is closely related to the second. Runge-Kutta integration is self-starting, thus there is no loss of efficiency when restarting from an event. This is of critical importance in a combined simulation.

13.11   TIME ADVANCE PROCEDURES

In SLAM II, the amount by which simulated time is advanced depends on the type of simulation (network, continuous, discrete) being performed and the values of specific variables at the current point in simulated time.

In a network or discrete simulation, time is advanced from one event to the next event. In this case, the time interval between events is TNEXT-TNOW where TNEXT is the time of the next event. Time is advanced from TNOW to TNEXT by resetting TNOW equal to TNEXT and assuming that the system status has remained constant between events. Since status at TNOW is always accepted, there is no need to maintain system status values at TTLAS, and TNOW can be used as both the last update point and then as the new event time, that is, TTLAS is neither required nor used.

For a continuous simulation, the variable DTFUL is the value for a full step size. If all the equations for state variables are written in terms of SS(·), the variable representing the time advance increment, DTNOW, will normally be set equal to DTFUL. The time at the beginning of the step is TTLAS, and the time at the end of the proposed step is TNOW, that is, TNOW = TTLAS + DTNOW. DTFUL remains constant at the maximum step size prescribed by the user, DTMAX, unless an event occurs within the step.

The increment in time between the recording of the status of the system is specified through the variable DTSAV. For DTSAV>0, SLAM II will record values as specified on RECORD statements every DTSAV time unit. The variable TTSAV is used to define the next time at which the system status is to be recorded. These time points are called record or save times. If a save time occurs within an interval, that is, TTLAS + DTNOW is greater than TTSAV, the step size is reduced to TTSAV–TTLAS. In this way, the status of the system is updated to TNOW = TTSAV. TTLAS is then updated and TTSAV is reset to TNOW + DTSAV.

The size of a step is also reset (decreased) if SLAM II determines that the step would cause a state-event to be passed. Thus, if the value of a state variable at time TTLAS + DTNOW results in the crossing of a threshold beyond allowable tolerances, the step size is reduced. If the tolerance is still not met, the step size continues to be reduced until the value of DTFUL is set equal to a user-prescribed minimum step size, DTMIN. Note that DTFUL can be less than DTMIN if either TNEXT-TTLAS<DTMIN or TTSAV-TTLAS<DTMIN. Further note that a fixed step size can be specified for a simulation involving only SS(·) variables in which there are no time-events by specifying DTMIN = DTMAX.

The most complex time advance procedure occurs when variables defined by DD(·) equations are included in a SLAM II simulation program. In this case, all the considerations described above pertain; in addition, the full step size, DTFUL, is divided into fractions so that the time advance increment, DTNOW, proceeds as required by the RKF algorithm. The SS(·) variables are evaluated at these intermediate points within the step and used by SLAM II in the integration of the equations for DD(·). These intermediate values for SS(·) and DD(·) allow error estimates to be made and simultaneous differential equations to be evaluated.

The variable DTACC is defined as the next step size to be used, based on allowable error specifications. Initially, DTACC is set equal to DTMAX. Whenever the accuracy of the integration algorithm does not meet the user's prescribed accuracy, as defined by an absolute error value AAERR and a relative error value RRERR, the value of DTACC is recomputed. At the start of each step, DTFUL is set equal to DTACC. The decreasing of DTACC when accuracy is not met is permitted only until DTACC becomes less than DTMIN, at which time DTACC is set equal to DTMIN. If the specified accuracy cannot be achieved using DTMIN, the following conditions specified in the CONTINUOUS statement define the appropriate action:

N →Proceed without printing a warning message;
W →Proceed after printing a warning message;
F →Terminate the simulation after printing a fatal error message.

When all state variables are within the accuracy specifications to a significant extent, DTACC is increased by a factor that is dependent on the ratio of the estimated error to the allowed error. In this way, the step size is increased when good estimates for the state variables are obtained by the integration algorithm. In no case will SLAM II allow DTACC to become greater than DTMAX.

It is obvious that the time advance procedures included within SLAM II involve many variables with many interactions between these variables, SLAM II automatically advances time for the user on the basis of the input values prescribed for DTMIN, DTMAX, DTSAV, and the accuracy requirements (AAERR and RRERR) when DD(·) equations are specified. The calculation of DTFUL, DTNOW, and DTACC, as well as the next discrete event time, TNEXT, and the next save time, TTSAV, are internally computed in SLAM II.

13.11.1   Use of DTNOW

A note of caution is in order regarding the use of DTNOW in subroutine STATE. DTNOW is the value of the time increment through which SLAM II updates the state variables over a full step of size DTFUL. In many cases, DTNOW = DTFUL. In other cases, this is not so. In fact, if two events occur at the same time, DTNOW will assume a value of zero for the second event processed. Thus, DTNOW should not be used in the denominator of any equation in subroutine STATE. However, DTNOW should be used in any equation in subroutine STATE in which the state variable is updated by a rate multiplied by the increment in time for which the update is being performed.

13.11.2 Summary of Time Advance Procedure

A summary of the time advance procedures used in SLAM II is given below. If derivative equations are included (NNEQD > 0), then

DTFUL = min[DTACC; TTSAV−TTLAS; TNEXT−TTLAS; time to next
   state-event],
DTNOW = f(DTFUL),
TNOW = TTLAS + DTNOW,
max DTACC = DTMAX,
min DTACC = DTMIN,
min (full step size to next state-event) = DTMIN.

If derivative equations are not included (NNEQD = 0), but state equations are (NNEQS > 0), then

DTFUL = min[DTMAX; TTSAV−TTLAS; TNEXT−TTLAS; time to next
   state-event],
DTNOW = DTFUL,
TNOW = TTLAS + DTNOW,
min (full step size to next state-event) = DTMIN,
and DTACC is not used.

If NNEQD=0 and NNEQS=0, then DTFUL, DTNOW, DTACC, and TTLAS are not used and TNOW is the time of the current event being processed. In this case only time-events are possible.

13.12   Example 13-1. PILOT EJECTION

The pilot ejection system of an aircraft that is flying level and at a constant velocity is to be simulated. This example is frequently cited in the literature and is referred to as the pilot ejection model. The specific version described here is extracted from Reference 3.

The pilot ejection system, when activated, causes the pilot and his seat to travel along rails at a specified exit velocity V_E at an angle θ_E backward from vertical. After traveling a vertical distance Y_1, the seat becomes disengaged from its mounting rails and at this point the pilot is considered out of the cockpit. When this occurs, a second phase of operation begins during which the pilot's trajectory is influenced by the force of gravity and atmospheric drag. A critical aspect of this phase is whether the pilot will clear the tail of the aircraft. The tail is 60 feet behind and 12 feet above the cockpit. Graphical and mathematical descriptions of the two phases are shown in Figure 13-4 along with a legend for the variables of the model.

The objective of this simulation is to determine the trajectory of a pilot ejected from an aircraft to assess whether he would hit the tail of the aircraft. This information is desired for a fixed ejection velocity and angle for the two aircraft velocities: 900 feet/second and 500 feet/second.

Concepts Illustrated. The purpose of this example is to introduce the use of SLAM II for the preparation of a continuous simulation model. The coding of subroutine STATE is illustrated. The procedures for programming different run termination conditions within a run and for making multiple runs are demonstrated. On each run, two plots are prepared and printed. On one of the plots, altitude versus distance is graphed to illustrate the plotting of a state variable against an independent variable other than time.

SLAM II Model. The pilot ejection model is simulated using continuous variables. The equations describing the state variables and their derivatives are programmed in subroutine STATE. The conditions for state-events are defined on SEVNT input statements. Since this is a continuous model, no time-events are involved. Since multiple runs will be made, the main program will initialize only the non-SLAM II variables that are constant for all runs. Other non-SLAM II variables and the initial conditions for the simulation will be set in subroutine INTLC. A RECORD statement is used to specify the outputs desired to portray the trajectory of the pilot as he leaves the aircraft. A RECORD statement will also be used to obtain the data to plot the pilot's relative position from the aircraft over time, and his speed and direction over time.

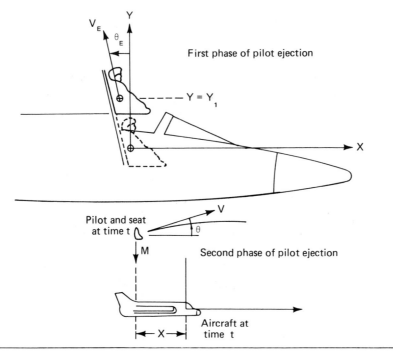

Variable	Definition	Equations
X	Horizontal distance from point of ejection	$\dfrac{dX}{dt} = V\cos\theta - V_A$
Y	Vertical distance from point of ejection	$\dfrac{dY}{dt} = V\sin\theta$
Y_1	Vertical distance above point of ejection where first phase ends	$\dfrac{dV}{dt} = 0, \quad 0 \leq Y < Y_1$
V_A	Velocity of aircraft	$\dfrac{d\theta}{dt} = 0, \quad 0 \leq Y < Y_1$
V	Pilot and seat velocity	
θ	Angle for pilot and seat movement	
V_E	Ejection velocity	
θ_E	Angle of ejection	$D = \dfrac{1}{2}\rho c_d S V^2$
M	Mass of pilot and seat	$\dfrac{dV}{dt} = -\dfrac{D}{M} - g\sin\theta \quad Y \geq Y_1$
ρ, c_d, S	Parameters of the model	$\dfrac{d\theta}{dt} = -\dfrac{g\cos\theta}{V}, \quad Y \geq Y_1$

Figure 13-4 Graphical and mathematical description of a pilot ejection model.

The main program is used only to initialize input/output device numbers, the dimension of the filing array, and the non-SLAM II variables that remain constant over all runs. After this is done, subroutine SLAM II is called to control the running of the simulation. The FORTRAN listing for the main program is shown in Figure 13-5. Problem specific variables are initialized through a READ statement.

```
      DIMENSION NSET(50)
      COMMON QSET(50)
      COMMON/SCOM1/ ATRIB(100),DD(100),DDL(100),DTNOW,II,MFA,MSTOP,NCLNR
     1,NCRDR,NPRNT,NNRUN,NNSET,NTAPE,SS(100),SSL(100),TNEXT,TNOW,XX(100)
      COMMON /UCOM1/ CD,G,RHO,THED,VA,VE,XM,XS,Y1
      EQUIVALENCE (NSET(1),QSET(1))
      NCRDR=5
      NPRNT=6
      NTAPE=7
      NNSET=50
C
C *** READ INPUT DATA
C ***      XM   - MASS OF PILOT AND SEAT
C ***      G    - GRAVITATIONAL CONSTANT
C ***      CD   - COEFFICIENT OF DRAG
C ***      XS   - PROJECTED AREA OF THE PILOT AND SEAT
C ***      Y1   - VERTICAL DISTANCE ABOVE POINT OF EJECTION,
C ***             WHERE FIRST PHASE 1 ENDS
C ***      VE   - EJECTION VELOCITY - DEGREES
C ***      THED - ANGLE OF EJECTION
C ***      RHO  - AIR DENSITY
C
      READ (NCRDR,101) XM,G,CD,XS,Y1,VE,THED,RHO
      CALL SLAM
      STOP
C
  101 FORMAT (7F5.0,E10.4)
C
      END
```

Figure 13-5 Main program for pilot ejection model.

The major programming effort for this example is expended in writing subroutine STATE where the equations describing the pilot ejection model are coded. Table 13-1 presents a listing of the SLAM II variables that are equivalent to the variables presented in Figure 13-4 and discussed in the problem statement. As can be seen from Table 13-1, there is a direct mnemonic relationship between the SLAM II variables and the variables included in the model. The coding of the equations presented in Figure 13-4 using SLAM II variables is shown below.

$$DD(1) = SS(3)*COS(SS(4)) - VA$$
$$DD(2) = SS(3)*SIN(SS(4))$$
$$DD(3) = 0.0 \qquad\qquad SS(2) < Y1$$
$$DD(4) = 0.0 \qquad\qquad SS(2) < Y1$$

Table 13-1 Variables for pilot ejection model.

Problem Statement	SLAM	Initial Value
X	SS(1)	0.0
dX/dt	DD(1)	Computed
Y	SS(2)	0.0
dY/dt	DD(2)	Computed
V	SS(3)	Computed
dV/dt	DD(3)	0.0
θ	SS(4)	Computed
$d\theta/dt$	DD(4)	0.0
C_d	CD	1.0
g	G	32.2 ft/sec^2
ρ	RHO	2.3769x 10^{-3} slug/ft^3
θ_E (rad)	THE	15./57.3
θ_E (deg)	THED	15.0
V_A	VA	Input
V_E	VE	40.0 ft/sec
M	XM	7 slugs
S	XS	10.0 ft^2
Y_1	Y1	4.0 ft

$$DD(3) = -XD/XM - G*SIN(SS(4)) \qquad SS(2) \geq Y1$$
$$DD(4) = -G*COS(SS(4))/SS(3) \qquad SS(2) \geq Y1$$
where
$$XD = 0.5*RHO*CD*XS*SS(3)*SS(3)$$

When SS(2) $\geq$ Y1, the pilot is released from the cockpit and the second set of equations for DD(3) and DD(4) is used. Since the relative position of the pilot during the simulated period of interest will always exceed Y1 after he is released, the test of whether SS(2) is greater than Y1 can serve for specifying when the equations for DD(3) and DD(4) are to be used.

In the coding to follow, the global variable XX(1) will be set to 1.0 to indicate this condition. The listing for subroutine STATE is shown in Figure 13-6. It should be noted that DD(1), DD(2), DD(3), and DD(4) are functions of both SS(3) and SS(4). The RKF integration algorithm of SLAM II simultaneously solves for the desired SS(·) values, and the multiple dependence is taken into account. The order of coding the DD(·) equations will not affect the results obtained in this example.

```
      SUBROUTINE STATE
      COMMON/SCOM1/ ATRIB(100),DD(100),DDL(100),DTNOW,II,MFA,MSTOP,NCLNR
     1,NCRDR,NPRNT,NNRUN,NNSET,NTAPE,SS(100),SSL(100),TNEXT,TNOW,XX(100)
      COMMON /UCOM1/ CD,G,RHO,THED,VA,VE,XM,XS,Y1
C
C *** DEFINE THE DIFFERENTIAL EQUATIONS FOR RATES OF CHANGE OF:
C ***    HORIZONTAL DISTANCE OF PILOT FROM POINT OF EJECTION - DD(1)
C ***    VERTICAL DISTANCE OF PILOT FROM POINT OF EJECTION - DD(2)
C ***    PILOT AND SEAT VELOCITY - DD(3)
C ***    ANGLE OF PILOT AND SEAT MOVEMENT - DD(4)
C
      DD(1)=SS(3)*COS(SS(4))-VA
      DD(2)=SS(3)*SIN(SS(4))
      IF(XX(1).LT.1.) RETURN
  101 XD=.5*RHO*CD*XS*SS(3)*SS(3)
      DD(3)=-XD/XM-G*SIN(SS(4))
      DD(4)=-G*COS(SS(4))/SS(3)
  102 RETURN
C
      END
```

Figure 13-6 Subroutine STATE for pilot ejection model.

In subroutine STATE, DD(3) and DD(4) are not set to zero since this is done in the initialization routines. When SS(2) ≥ Y1, a state-event will be coded to change the value of XX(1) to 1. Initially XX(1) = 0. When XX(1) = 1, the values of DD(3) and DD(4) will be recomputed for each call to subroutine STATE. Three conditions define state-events. A state-event occurs when the pilot achieves a relative height of Y1 with respect to the aircraft. When this occurs, the equations governing the pilot's movement are altered as described above. The other two state-events are for stopping the simulation. When the pilot is 60 feet behind the cockpit, he will be beyond the tail. When the pilot is 30 feet above the airplane, he is well above the 12 foot high tail. In either case, the simulation is halted. In terms of the SLAM II variables, these conditions are SS(1) ≤ -60 and SS(2) ≥ 30. The other condition for stopping involves time exceeding four seconds. This is accomplished through the INITIALIZE input statement by setting TTFIN = 4.

The establishment of the conditions for state-events is made using statement type SEVNT. The input statements for the three conditions are shown below:

    SEVNT,1,SS(1),XN,-60.,0.0;
    SEVNT,1,SS(2),XP,30,0.0;
    SEVNT,2,SS(2),XP,4,0.0;

The first SEVNT statement specifies that state-event 1 occurs when SS(1) crosses the value -60 in the negative direction with zero tolerance. This corresponds to the pilot passing the tail in the X-direction. A tolerance of zero is used to force SLAM II to use the minimum step size when SS(1) exceeds the value of -60. In this way, a precise determination of when the state-event occurs can be obtained. State-event 1 represents one of the conditions by which the simulation run will be terminated. The second statement specifies that state-event 1 will also occur when SS(2) crosses the value 30 in the positive direction with a tolerance of zero. This statement corresponds to the run termination condition that the pilot exceeds a vertical distance of 30 feet above the cockpit.

The third state-event input statement corresponds to the pilot achieving a vertical distance, SS(2), of four feet. The crossing is prescribed to be in a positive direction and again a zero tolerance is specified to force SLAM II to detect the point at which the pilot leaves the aircraft with as much precision as possible. In this manner, the switching from one equation set to another will be accomplished with the precision specified by the input value for DTMIN. This third SEVNT statement prescribes state-event 2 to be the event code associated with the pilot leaving the cockpit.

The effects associated with the occurrence of state-events are modeled in subroutine EVENT(IX). When an event with code one occurs (IX=1), we request a stopping of the simulation by setting MSTOP negative. When IX=2, we desire to change the equations for DD(3) and DD(4). This latter change is prescribed in subroutine STATE to occur when the value of XX(1) is set to 1. The coding for subroutine EVENT is shown in Figure 13-7. Since the code required to model the state-events associated with this problem is extremely short, both state-events are programmed directly in subroutine EVENT. A computed GO TO statement is used to decode state-event codes. When a termination condition is reached, the SLAM II variable MSTOP is set to −1 to indicate to the SLAM II executive that the run should be terminated. When state-event 2 occurs (IX=2), a transfer is made to statement 2 where XX(1) is set equal to 1 to cause the desired change to be made in the equations written in subroutine STATE. This completes the description of subroutine EVENT for the pilot ejection model.

```
      SUBROUTINE EVENT(IX)
      COMMON/SCOM1/ ATRIB(100),DD(100),DDL(100),DTNOW,II,MFA,MSTOP,NCLNR
     1,NCRDR,NPRNT,NNRUN,NNSET,NTAPE,SS(100),SSL(100),TNEXT,TNOW,XX(100)
      GO TO (1,2) ,IX
C
C *** TERMINATE THE SIMULATION
C
    1 MSTOP=-1
      RETURN
C
C *** SET XX(1)=1 FOR SECOND PHASE OF FLIGHT
C
    2 XX(1)=1.
      RETURN
      END
```

Figure 13-7 Subroutine EVENT for pilot ejection model.

```
      SUBROUTINE INTLC
      COMMON/SCOM1/ ATRIB(100),DD(100),DDL(100),DTNOW,II,MFA,MSTOP,NCLNR
     1,NCRDR,NPRNT,NNRUN,NNSET,NTAPE,SS(100),SSL(100),TNEXT,TNOW,XX(100)
      COMMON /UCOM1/ CD,G,RHO,THED,VA,VE,XM,XS,Y1
C
C *** READ INPUT DATA
C ***      VA    - VELOCITY OF AIRCRAFT
C
      READ (NCRDR,101) VA
C
C *** CALCULATE VARIABLES
C ***      THE  - ANGLE OF EJECTION - RADIANS
C ***      VX   - VELOCITY IN THE HORIZONTAL DIRECTION
C ***      VY   - VELOCITY IN THE VERTICLE DIRECTION
C ***      SS(3)- INITIAL VELOCITY OF THE PILOT AND SEAT
C ***      SS(4)- INITIAL ANGLE OF PILOT AND SEAT
C
      THE=THED/57.3
      VX=VA-VE*SIN(THE)
      VY=VE*COS(THE)
      SS(3)=SQRT(VX*VX+VY*VY)
      SS(4)=ATAN(VY/VX)
      XX(1)=0.
      RETURN
C
  101 FORMAT (1F10.0)
C
      END
```

Figure 13-8 Subroutine INTLC for pilot ejection model.

Subroutine INTLC is used to initialize the state variables and non-SLAM II variables that require initialization before each run. For this example, INTLC is written to allow the aircraft velocity to be initialized through data input before each run. In the listing of INTLC given in Figure 13-8, the first statement performs this reading operation. Initial values of the state variables are then established for a run. The initial position of the pilot relative to the aircraft cockpit is at the origin, that is, SS(1) = 0.0 and SS(2) = 0.0. These values are automatically established in the SLAM II initialization process. The pilot's initial velocity vector, caused by pushing the ejection button, is given by the following equations:

$$V_{\text{initial}} = \sqrt{(V_A - V_E \sin \theta_E)^2 + (V_E \cos \theta_E)^2}$$

and

$$\theta_{\text{initial}} = \tan^{-1} \left(\frac{V_E \cos \theta_E}{V_A - V_E \sin \theta_E} \right).$$

These initial values are set equal to SS(3) and SS(4) at the beginning of each simulation run since ejection is initiated at time zero. The initial values of the derivatives of the state variables are obtained when the SLAM II executive calls subroutine STATE at time zero. The last statement in subroutine INTLC sets the value of XX(1) to 0.0 to indicate that the pilot has not left the cockpit at the beginning of each run.

A listing of the input for this example is shown in Figure 13-9. The first input record is for the variables that were initialized through the READ statement in the main program. The second input record describes the general information (GEN statement type) and indicates that two runs are to be made. The INITIALIZE statement specifies that the beginning time for a run (TTBEG) is to be zero and that the ending time (TTFIN) is to be four. Default values for other variables on the INITIALIZE statement are to be used. Information regarding the number of equations, step size, recording interval, and accuracy of the RKF algorithm are provided on the CONTINUOUS input statement. Specifically, the following values are prescribed: NNEQD = 4; NNEQS = 0; DTMIN = 0.0001; DTMAX = 0.01; DTSAV = 0.01; W = *W*arning; AAERR = 0; and RRERR = 0.000005.

With these values for the variables, there are four differential equations and no difference or state equations. The minimum step size is 0.0001 seconds and the maximum step size is 0.01 seconds. The recording or communication interval is also 0.01 seconds. When a state-event cannot be detected within specified tolerances or when the RKF numerical integration algorithm cannot meet accuracy specifications, warning messages are to be printed. The simulation is to be continued even when these conditions are detected. The accuracy requirement for the numerical integration algorithm has a zero value for the absolute error (AAERR) and a five millionth value for the relative error (RRERR).

```
     7. 32.2   1.  10.    4.  40.  15.  .0023769
GEN,PRITSKER,PILOT EJECTION,9/10/1983,1;
INITIALIZE,0,4;
CONTINUOUS,4,,.0001,.01,.01,W,0.,.000005;
RECORD,SS(1),X POS,8,P,-1.0;
VAR,SS(2),Y,Y POS.,0.0,20.;
RECORD,TNOW,TIME,9,B,0.02;
VAR,SS(1),X,X POS.,-70.,30.;
VAR,SS(2),Y,Y POS.,0.0,20.0;
VAR,SS(3),V,SPEED,MIN(100),MAX(100);
VAR,SS(4),T,THETA,MIN,MAX;
SEVNT,1,SS(1),XN,-60.0,0.0;
SEVNT,1,SS(2),XP,30.0,0.0;
SEVNT,2,SS(2),XP,4.0,0.0;
SIMULATE;
        900.
FIN;
        500.
```

Figure 13-9   Data input and statements for pilot ejection model.

The next seven input statements refer to the recording of values for eventual tabling and plotting. The first RECORD statement specifies that the state variable SS(1) is to be the independent variable and it is to be labeled as X POS. The val-

ues are to be stored on peripheral device 8 and only a plot of the dependent variable versus the independent variable is desired. The plotting interval is to be –1,
that is, successive lines on the plot are to be –1 units apart. A negative interval is
used as the X-position of the pilot with respect to the cockpit will be decreasing
during the entire simulation. Default values are prescribed for the last three fields
of the RECORD statement since the statement is terminated prior to the specification of these fields. With default values, the plot will be started at TTSRT =
TTBEG = 0 and completed at TTEND = TTFIN = 4. KKEVT is defaulted to
YES so that variable values at events will be recorded.

There is only one VAR input statement following the RECORD statement which
implies that only one dependent variable is to be recorded for each value of the independent variable. The dependent variable is SS(2) which is to have a plot symbol of Y and a label of Y POS. The scale for the dependent variable is specified
to have a low ordinate value of 0.0 (the left-hand axis) and a high ordinate value
of 20.0 (the right-hand axis).

A second RECORD statement prescribes that a plot and table is desired with
the independent variable being current time, TNOW. The label for TNOW is to be
TIME. The values associated with this RECORD statement are to be stored on
peripheral device 9. A plot interval of .02 is specified. The dependent variables for
the plot are prescribed on the next four statements to be SS(1) through SS(4). The
plot symbols are labels are easily identified from the input statements. A full range
of options is given to define the scale limits for these four dependent variables.
The left- and right-hand scales for the X and Y variables are prescribed to be
(–70.,30.) and (0.0,20.0), respectively. For SS(3), the speed of the pilot, the low
ordinate value is to be taken as the minimum value observed from the simulation
rounded down to the nearest multiple of 100. The high ordinate value is to be
taken as the maximum value from the simulation rounded up to the next higher
multiple of 100. For SS(4), THETA, the minimum and maximum observed values
from the simulation run are to be used as the low and high ordinate values,
respectively.

The next three input statements describe the state-event conditions; these were
described in detail previously. The SIMULATE statement indicates that the execution of the first run is to be initiated. Following the reading of the SIMULATE
statement, a call to subroutine INTLC is made. The next input record is read by
the READ statement in subroutine INTLC, and it specifies the aircraft velocity,
VA, to be 900 feet per second. The next input statement is FIN which specifies
that this is the last SLAM II statement of the program. After the first run is completed, SLAM II attempts to read additional input statements to alter the conditions for the next simulation run. The FIN statement indicates that no changes are

to be made in the SLAM II data values. The input line with the value 500. on it specifies the aircraft velocity for the second run.

The SLAM II echo report for the input presented in Fig. 13-9 is presented in Fig. 13-10. Before examining SLAM II outputs for this example, we will describe the sequence in which the subprograms are invoked. Subroutine SLAM II is called by the main program and reads in the input statements, initializes SLAM II variables, and calls subroutines INTLC and STATE. Subroutine SLAM II then controls the simulation by advancing time in steps. During each step advance, the RKF algorithm integrates the DD($\cdot$) equations coded in subroutine STATE to evaluate SS($\cdot$) values at select time points within the step and at the end of the step.

When accuracy is acceptable, SLAM II tests if any state-events occurred because of the updating of the state variables. The state-events are those that were defined on SEVNT input statements. If a state-event was passed, the step size is reduced and the above process is repeated with a new but smaller step size. If a state-event ends the step, subroutine EVENT is called. If the state-event is one that ends the simulation, MSTOP is set to –1 in subroutine EVENT and subroutine SLAM II ends the run by calling OTPUT and SUMRY. Subroutine SLAM II also checks TNOW against TTFIN and if TNOW $\geq$ TTFIN, the simulation run is ended.

```
                    S L A M   I I   E C H O   R E P O R T
_____

          SIMULATION PROJECT PILOT EJECTION          BY PRITSKER

               DATE  9/10/1983                       RUN NUMBER   1 OF   2

                         SLAM II VERSION FEB 86


     GENERAL OPTIONS

          PRINT INPUT STATEMENTS (ILIST):          YES
          PRINT ECHO REPORT (IECHO):               YES
          EXECUTE SIMULATIONS (IXQT):              YES
          WARN OF DESTROYED ENTITIES:              YES
          PRINT INTERMEDIATE RESULTS HEADING (IPIRH):  YES
          PRINT SUMMARY REPORT (ISMRY):            YES


     CONTINUOUS VARIABLES

          NUMBER OF DD EQUATIONS (NNEQD):                4
          NUMBER OF SS EQUATIONS (NNEQS):                0
          MINIMUM STEP SIZE (DTMIN):             0.1000E-03
          MAXIMUM STEP SIZE (DTMAX):             0.1000E-01
          TIME BETWEEN SAVE POINTS (DTSAV):      0.1000E-01
          ACCURACY ERROR SPECIFICATION (LLERR):    WARNING
          ABSOLUTE ERROR LIMIT (AAERR):          0.0000E+00
          RELATIVE ERROR LIMIT (RRERR):          0.5000E-05
```

Figure 13-10 SLAM II echo for pilot ejection model.

STATE EVENTS

NUMBER	MODE/JEVNT		CROSSING VARIABLE	DIRECTION OF CROSSING	THRESHOLD VALUE	TOLERANCE OF CROSSING
1	EVENT	1	SS(1)	NEGATIVE	-0.6000E+02	0.0000E+00
2	EVENT	1	SS(2)	POSITIVE	0.3000E+02	0.0000E+00
3	EVENT	2	SS(2)	POSITIVE	0.4000E+01	0.0000E+00

RECORDING OF PLOTS/TABLES

PLOT/TABLE NUMBER   1

    INDEPENDENT VARIABLE: SS(1)
    IDENTIFIER: X POS
    DATA STORAGE UNIT: TAPE/DISC 8
    DATA OUTPUT FORMAT: PLOT
    TIME BETWEEN PLOT POINTS (DTPLT): -0.1000E+01
    STARTING TIME OF PLOT (TTSRT):   0.0000E+00
    ENDING TIME OF PLOT (TTEND):   0.4000E+01
    DATA POINTS AT EVENTS (KKEVT):   YES

DEPENDENT VARIABLES

VARIABLE	SYMBOL	IDENTIFIER	LOW ORDINATE VALUE		HIGH ORDINATE VALUE	
SS(2)	Y	Y POS.	VALUE EQUALS	0.0000E+00	VALUE EQUALS	0.2000E+02

PLOT/TABLE NUMBER   2

    INDEPENDENT VARIABLE: TNOW
    IDENTIFIER: TIME
    DATA STORAGE UNIT: TAPE/DISC 9
    DATA OUTPUT FORMAT: PLOT AND TABLE
    TIME BETWEEN PLOT POINTS (DTPLT): 0.2000E-01
    STARTING TIME OF PLOT (TTSRT):   0.0000E+00
    ENDING TIME OF PLOT (TTEND):   0.4000E+01
    DATA POINTS AT EVENTS (KKEVT):   YES

DEPENDENT VARIABLES

VARIABLE	SYMBOL	IDENTIFIER	LOW ORDINATE VALUE		HIGH ORDINATE VALUE	
SS(1)	X	X POS.	VALUE EQUALS	-0.7000E+02	VALUE EQUALS	0.3000E+02
SS(2)	Y	Y POS.	VALUE EQUALS	0.0000E+00	VALUE EQUALS	0.2000E+02
SS(3)	V	SPEED	MIN TO NEAREST	0.1000E+03	MAX TO NEAREST	0.1000E+03
SS(4)	T	THETA	MIN TO NEAREST	0.0000E+00	MAX TO NEAREST	0.0000E+00

RANDOM NUMBER STREAMS

STREAM NUMBER	SEED VALUE	REINITIALIZATION OF STREAM
1	428956419	NO
2	1954324947	NO
3	1145661099	NO
4	1835732737	NO
5	794161987	NO
6	1329531353	NO
7	200496737	NO
8	633816299	NO
9	1410143363	NO
10	1282538739	NO

Figure 13-10   (continued).

```
INITIALIZATION OPTIONS

       BEGINNING TIME OF SIMULATION (TTBEG):    0.0000E+00
       ENDING TIME OF SIMULATION (TTFIN):       0.4000E+01
       STATISTICAL ARRAYS CLEARED (JJCLR):      YES
       VARIABLES INITIALIZED (JJVAR):           YES
       FILES INITIALIZED (JJFIL):               YES


NSET/QSET STORAGE ALLOCATION

       DIMENSION OF NSET/QSET (NNSET):          50
       WORDS ALLOCATED TO FILING SYSTEM:         0
       WORDS ALLOCATED TO VARIABLES:            13
       WORDS AVAILABLE FOR PLOTS/TABLES:        37


INPUT ERRORS DETECTED:   0


EXECUTION WILL BE ATTEMPTED
```

Figure 13-10   (continued).

Summary of Results. Intermediate results for the first simulation, in which the aircraft velocity was 900 feet/second, are shown in Figure 13-11. Because a zero tolerance was specified on the SEVNT input statements for detecting crossings, warning messages that tolerances could not be met are expected. The messages could have been suppressed by specifying an N in the field for warning messages on the CONTINUOUS input statement. It was decided to retain the diagnostics to indicate the time of occurrence of the specified state conditions. The messages indicate that the pilot left the cockpit at time 0.1035 and that the simulation ended at time 0.4339. Since DTMIN was set at 0.0001, these results indicate that the pilot left the cockpit in the time interval from 0.1034 to 0.1035, and that the conditions for ending the simulation occurred in the interval from 0.4338 to 0.4339.

```
                                      **INTERMEDIATE RESULTS**


  SPECIFIED  TOLERANCE  EXCEEDED  FOR  SS(  2)  AT  TIME     .1035+00

  SPECIFIED  TOLERANCE  EXCEEDED  FOR  SS(  1)  AT  TIME     .4339+00
```

Figure 13-11   Intermediate results showing diagnostic messages for pilot ejection model.

Figure 13-12 is the plot of the pilot's position relative to the aircraft cockpit which is obtained in accordance with the first RECORD input statement. Note that the independent variable on the first plot is not time but distance (X-position) and that it is monotonically nonincreasing. With SLAM II, the independent variable must be monotonically nonincreasing if DTPLT < 0 and monotonically non-

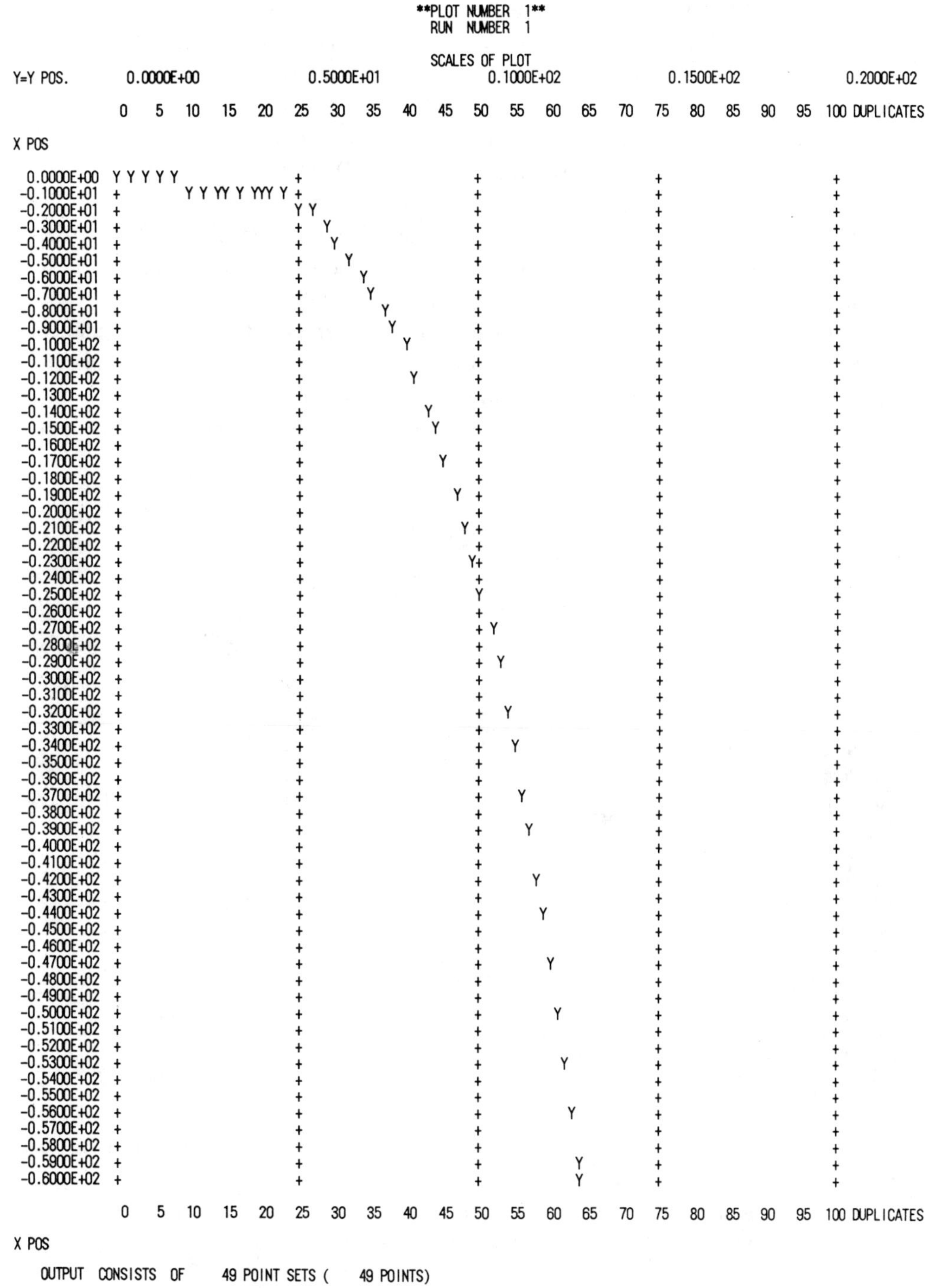

```
                              **PLOT NUMBER  1**
                                RUN  NUMBER  1

                                 SCALES OF PLOT
Y=Y POS.      0.0000E+00          0.5000E+01          0.1000E+02          0.1500E+02          0.2000E+02

              0   5   10  15  20  25  30  35  40  45  50  55  60  65  70  75  80  85  90  95  100 DUPLICATES

X POS

   0.0000E+00  Y Y Y Y Y          +                   +                   +                   +
  -0.1000E+01  +       Y Y YY Y YYY Y +                +                   +                   +
  -0.2000E+01  +               Y Y    +                +                   +                   +
  -0.3000E+01  +                 + Y   +               +                   +                   +
  -0.4000E+01  +                 +  Y  +               +                   +                   +
  -0.5000E+01  +                 +    Y +              +                   +                   +
  -0.6000E+01  +                 +     Y+              +                   +                   +
  -0.7000E+01  +                 +      Y +            +                   +                   +
  -0.8000E+01  +                 +       Y+            +                   +                   +
  -0.9000E+01  +                 +        Y +          +                   +                   +
  -0.1000E+02  +                 +         Y+          +                   +                   +
  -0.1100E+02  +                 +          +          +                   +                   +
  -0.1200E+02  +                 +          Y +        +                   +                   +
  -0.1300E+02  +                 +          + +        +                   +                   +
  -0.1400E+02  +                 +          Y +        +                   +                   +
  -0.1500E+02  +                 +           Y+        +                   +                   +
  -0.1600E+02  +                 +           + +       +                   +                   +
  -0.1700E+02  +                 +            Y+       +                   +                   +
  -0.1800E+02  +                 +            + +      +                   +                   +
  -0.1900E+02  +                 +             Y +     +                   +                   +
  -0.2000E+02  +                 +             + Y     +                   +                   +
  -0.2100E+02  +                 +             + Y     +                   +                   +
  -0.2200E+02  +                 +             +  +    +                   +                   +
  -0.2300E+02  +                 +             Y+      +                   +                   +
  -0.2400E+02  +                 +             + +     +                   +                   +
  -0.2500E+02  +                 +             Y       +                   +                   +
  -0.2600E+02  +                 +             + +     +                   +                   +
  -0.2700E+02  +                 +             + Y     +                   +                   +
  -0.2800E+02  +                 +             + +     +                   +                   +
  -0.2900E+02  +                 +             +  Y    +                   +                   +
  -0.3000E+02  +                 +             +       +                   +                   +
  -0.3100E+02  +                 +             +       +                   +                   +
  -0.3200E+02  +                 +             +   Y   +                   +                   +
  -0.3300E+02  +                 +             +       +                   +                   +
  -0.3400E+02  +                 +             +    Y  +                   +                   +
  -0.3500E+02  +                 +             +       +                   +                   +
  -0.3600E+02  +                 +             +       +                   +                   +
  -0.3700E+02  +                 +             +     Y +                   +                   +
  -0.3800E+02  +                 +             +       +                   +                   +
  -0.3900E+02  +                 +             +      Y+                   +                   +
  -0.4000E+02  +                 +             +       +                   +                   +
  -0.4100E+02  +                 +             +       +                   +                   +
  -0.4200E+02  +                 +             +       Y                   +                   +
  -0.4300E+02  +                 +             +       +                   +                   +
  -0.4400E+02  +                 +             +       + Y                 +                   +
  -0.4500E+02  +                 +             +       +                   +                   +
  -0.4600E+02  +                 +             +       +                   +                   +
  -0.4700E+02  +                 +             +       +  Y                +                   +
  -0.4800E+02  +                 +             +       +                   +                   +
  -0.4900E+02  +                 +             +       +                   +                   +
  -0.5000E+02  +                 +             +       +    Y              +                   +
  -0.5100E+02  +                 +             +       +                   +                   +
  -0.5200E+02  +                 +             +       +                   +                   +
  -0.5300E+02  +                 +             +       +      Y            +                   +
  -0.5400E+02  +                 +             +       +                   +                   +
  -0.5500E+02  +                 +             +       +                   +                   +
  -0.5600E+02  +                 +             +       +        Y          +                   +
  -0.5700E+02  +                 +             +       +                   +                   +
  -0.5800E+02  +                 +             +       +                   +                   +
  -0.5900E+02  +                 +             +       +          Y        +                   +
  -0.6000E+02  +                 +             +       +          Y        +                   +

              0   5   10  15  20  25  30  35  40  45  50  55  60  65  70  75  80  85  90  95  100 DUPLICATES
```

X POS

OUTPUT CONSISTS OF   49 POINT SETS (49 POINTS)

Figure 13-12 Plot of pilot's position relative to aircraft for aircraft velocity of 900 feet/
second.

decreasing if DTPLT > 0. From the first plot, it is seen that the pilot clears the cockpit (Y = 4.0) at about X = -1.0. At X = -30.0, Y is approximately 10.6, and at X = -60.0, Y is approximately 12.8. With the tail being 12 feet high, there are design problems with the ejection system when the aircraft velocity is 900 feet/second. If desired, the precise values for X and Y could have been obtained by requesting a table in addition to the plot in the RECORD statement. Since precise values are printed out for plot-table 2, this was not done.

The table from the second RECORD statement is labeled **TABLE NUMBER 2** and is presented in Figure 13-13. In this table, the state variables are given as a function of time. The state-event representing the pilot leaving the cockpit is seen at time 0.1035. At this time, two sets of values are shown in the table since SLAM II records the values at the end of the step before the event and immediately after the event occurs. This second set of values is recorded in case the event causes a discrete change in the variables. At the end of the table, minimum and maximum values are printed.

```
                                    **PLOT NUMBER  2**
                                      RUN  NUMBER  1

                                       SCALES OF PLOT
X=X POS.    -0.7000E+02         -0.4500E+02          -0.2000E+02         0.5000E+01         0.3000E+02
Y=Y POS.     0.0000E+00          0.5000E+01           0.1000E+02         0.1500E+02         0.2000E+02
V=SPEED      0.5000E+03          0.6000E+03           0.7000E+03         0.8000E+03         0.9000E+03
T=THETA      0.2848E-01          0.3221E-01           0.3594E-01         0.3967E-01         0.4340E-01


            0    5   10   15   20   25   30   35   40   45   50   55   60   65   70   75   80   85   90   95  100 DUPLICATES

TIME

0.0000E+00  Y                      +                   +                   X    +               V  T
0.2000E-01  + Y  Y                 +                   +                   X    +               V  T
0.4000E-01  +      Y Y             +                   +                   X    +               V  T
0.6000E-01  +         Y Y          +                   +                  X     +               V  T
0.8000E-01  +            YY        +                   +                  X     +               V  T
0.1000E+00  +             Y YYY    +                   +                  X     +           V  V T VT
0.1200E+00  +                Y +   +                   +                  X     +         V   T    +
0.1400E+00  +                  Y Y +                   +                 X      +      V V V T T    +
0.1600E+00  +                  +  YY                   +                XX      +   V V  T T        +
0.1800E+00  +                  +     Y Y               +                XX    V V T T               +
0.2000E+00  +                  +         Y Y           +               XX   V  V  T  T              +
0.2200E+00  +                  +           Y Y         +              XX  V  V T  T                 +
0.2400E+00  +                  +              Y Y      +             X X V V T T                    +
0.2600E+00  +                  +                YY  +  X X VT  T      +                            + XV
0.2800E+00  +                  +              YYX+XTV T                +                            + XV
0.3000E+00  +                  +             X  XTYY                   +                          + YV XV XT
0.3200E+00  +                  +          T XTX V  + YY                +                            + XV
0.3400E+00  +              +  T    T  TVXV V    +   YYY               +                            + TX
0.3600E+00  +         T +T X  X  W          +        YY               +                             +
0.3800E+00  +       T  T  X +X  V V         +         YY              +                             +
0.4000E+00  +     T T    X X    +V V        +          YY             +                             +
0.4200E+00  +T  T    X  X     V+V           +           YY            +                             +
0.4400E+00  T        X       V +            +            Y            +                             +


            0    5   10   15   20   25   30   35   40   45   50   55   60   65   70   75   80   85   90   95  100 DUPLICATES
```

Figure 13-14 Plot of state variables over time for pilot ejection model with an aircraft velocity of 900 feet/second.

```
                                      **TABLE NUMBER   2**
                                        RUN   NUMBER    1


       TIME        X POS.       Y POS.       SPEED        THETA

    0.0000E+00   0.0000E+00   0.0000E+00   0.8905E+03   0.4340E-01
    0.0000E+00   0.0000E+00   0.0000E+00   0.8905E+03   0.4340E-01
    0.1000E-01  -0.1035E+00   0.3864E+00   0.8905E+03   0.4340E-01
    0.2000E-01  -0.2070E+00   0.7727E+00   0.8905E+03   0.4340E-01
    0.3000E-01  -0.3106E+00   0.1159E+01   0.8905E+03   0.4340E-01
    0.4000E-01  -0.4141E+00   0.1545E+01   0.8905E+03   0.4340E-01
    0.5000E-01  -0.5176E+00   0.1932E+01   0.8905E+03   0.4340E-01
    0.6000E-01  -0.6211E+00   0.2318E+01   0.8905E+03   0.4340E-01
    0.7000E-01  -0.7246E+00   0.2705E+01   0.8905E+03   0.4340E-01
    0.8000E-01  -0.8282E+00   0.3091E+01   0.8905E+03   0.4340E-01
    0.9000E-01  -0.9317E+00   0.3477E+01   0.8905E+03   0.4340E-01
    0.1000E+00  -0.1035E+01   0.3864E+01   0.8905E+03   0.4340E-01
    0.1035E+00  -0.1072E+01   0.4000E+01   0.8905E+03   0.4340E-01
    0.1035E+00  -0.1072E+01   0.4000E+01   0.8905E+03   0.4340E-01
    0.1100E+00  -0.1167E+01   0.4248E+01   0.8818E+03   0.4317E-01
    0.1200E+00  -0.1422E+01   0.4624E+01   0.8688E+03   0.4280E-01
    0.1300E+00  -0.1805E+01   0.4992E+01   0.8562E+03   0.4243E-01
    0.1400E+00  -0.2312E+01   0.5351E+01   0.8439E+03   0.4205E-01
    0.1500E+00  -0.2941E+01   0.5701E+01   0.8320E+03   0.4166E-01
    0.1600E+00  -0.3686E+01   0.6044E+01   0.8204E+03   0.4128E-01
    0.1700E+00  -0.4546E+01   0.6378E+01   0.8091E+03   0.4088E-01
    0.1800E+00  -0.5517E+01   0.6705E+01   0.7981E+03   0.4048E-01
    0.1900E+00  -0.6596E+01   0.7024E+01   0.7874E+03   0.4007E-01
    0.2000E+00  -0.7780E+01   0.7336E+01   0.7770E+03   0.3966E-01
    0.2100E+00  -0.9067E+01   0.7641E+01   0.7669E+03   0.3925E-01
    0.2200E+00  -0.1045E+02   0.7938E+01   0.7570E+03   0.3882E-01
    0.2300E+00  -0.1194E+02   0.8228E+01   0.7474E+03   0.3840E-01
    0.2400E+00  -0.1352E+02   0.8512E+01   0.7380E+03   0.3796E-01
    0.2500E+00  -0.1519E+02   0.8789E+01   0.7289E+03   0.3752E-01
    0.2600E+00  -0.1695E+02   0.9059E+01   0.7200E+03   0.3708E-01
    0.2700E+00  -0.1880E+02   0.9322E+01   0.7113E+03   0.3663E-01
    0.2800E+00  -0.2073E+02   0.9580E+01   0.7028E+03   0.3617E-01
    0.2900E+00  -0.2275E+02   0.9831E+01   0.6945E+03   0.3571E-01
    0.3000E+00  -0.2485E+02   0.1008E+02   0.6864E+03   0.3525E-01
    0.3100E+00  -0.2703E+02   0.1031E+02   0.6784E+03   0.3478E-01
    0.3200E+00  -0.2929E+02   0.1055E+02   0.6707E+03   0.3430E-01
    0.3300E+00  -0.3162E+02   0.1077E+02   0.6631E+03   0.3382E-01
    0.3400E+00  -0.3403E+02   0.1100E+02   0.6557E+03   0.3333E-01
    0.3500E+00  -0.3652E+02   0.1121E+02   0.6485E+03   0.3284E-01
    0.3600E+00  -0.3907E+02   0.1142E+02   0.6414E+03   0.3234E-01
    0.3700E+00  -0.4169E+02   0.1163E+02   0.6345E+03   0.3183E-01
    0.3800E+00  -0.4438E+02   0.1183E+02   0.6278E+03   0.3132E-01
    0.3900E+00  -0.4714E+02   0.1202E+02   0.6211E+03   0.3081E-01
    0.4000E+00  -0.4997E+02   0.1221E+02   0.6146E+03   0.3029E-01
    0.4100E+00  -0.5286E+02   0.1239E+02   0.6083E+03   0.2976E-01
    0.4200E+00  -0.5581E+02   0.1257E+02   0.6020E+03   0.2923E-01
    0.4300E+00  -0.5882E+02   0.1274E+02   0.5959E+03   0.2869E-01
    0.4339E+00  -0.6002E+02   0.1281E+02   0.5936E+03   0.2848E-01
    0.4339E+00  -0.6002E+02   0.1281E+02   0.5936E+03   0.2848E-01

     MINIMUM -0.6002E+02   0.0000E+00   0.5936E+03   0.2848E-01
     MAXIMUM  0.0000E+00   0.1281E+02   0.8905E+03   0.4340E-01
```

Figure 13-13   State variables versus time for pilot ejection model with an average velocity of 900 feet/second.

The plot corresponding to Table 2 is presented in Figure 13-14. At the top of the plot, the symbols representing the variables are defined with the scale for each variable. The scales used correspond to the input conditions specified. The plots illustrate how the state variables change with time.

The form of the output of the second run is similar to that for the first run. Figure 13-15 presents the plot of the state variables over time when the aircraft is traveling at 500 feet/second. In this case, the pilot is above 12 feet after 0.36 seconds and he is approximately 15.5 feet behind the cockpit.

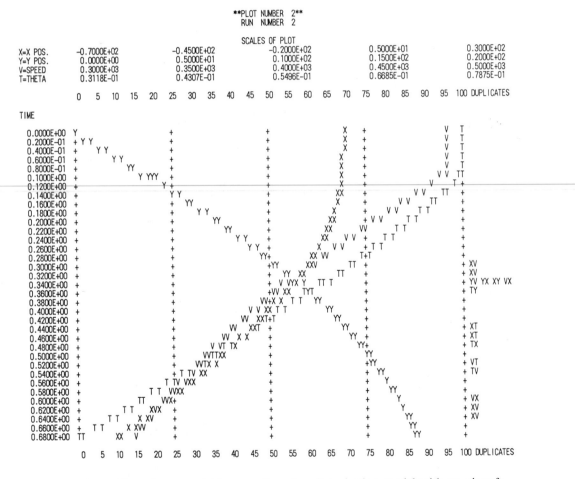

Figure 13-15 Plot of state variables over time for pilot ejection model with an aircraft velocity of 500 feet/second.

13.13   EXAMPLE 13-2. WORLD DYNAMICS

The world model analyzed in this example was defined by Forrester and has been described by him in detail (7). An extensive analysis of an extended version† of this model has been reported by Meadows (8).

For readers interested in the model development and use, these books are highly recommended. World Dynamics is an aggregate model of world interactions to illustrate the behavior of a set of defined variables depending on whether population growth is eventually suppressed by a shortage of natural resources, pollution, over-crowding, or an insufficient food supply. The model portrays the world interactions by five interrelated state variables: population P; natural resources NR; capital-investment CI; capital investment-in-agriculture fraction CIAF; and pollution POL. Table 13-2 defines the variables that are used in this example. The equations for the model are presented in the discussion of subroutine STATE.

Concepts Illustrated. The primary objective of this example is to illustrate the coding of the World Dynamics model in SLAM II. Systems Dynamics (6) models are easily built using SLAM II and extensions to include discrete events are straightforward (9,14). Difference equations are used to define the state variables. The use of EQUIVALENCE statements for improving the readability of subroutine STATE is also illustrated.

SLAM II Model. The major SLAM II coding required for Systems Dynamics problems is for subroutine STATE. State variables are written in terms of rate components and "auxiliary" values that may be required for the computation of the rate components. Using the definitions of Table 13-2, the statements for the rate components as defined by Forrester are:

```
BR = P*BRN*BRFM*BRMM*BRCM*BRPM
DR = P*DRN*DRMM*DRPM*DRFM*DRCM
NRUR = P*NRMM
CIG = P*CIM*CIGN
CID = CI*CIDN
POLG = P*POLN*POLCM
POLA = POL/POLAT
```

---

† These extensions are easily coded in SLAM II. For explanation purposes, the basic model was selected for presentation.

where all variables on the right-hand side are previous values of state variables or values already obtained in subroutine STATE.

The above equations illustrate that Forrester hypothesized multiplicative relationships in the computation of rate components. For example, birth rate, which is a component rate to be used in the computation of population, is equal to the product of the current population, birth rate normal, birth-rate-from-food multiplier, birth-rate-from-material multiplier, birth-rate-from-crowding multiplier, and birth-rate-from-pollution multiplier. Each of the multipliers are in turn computed from other variables. For example, BRFM, birth-rate-from-food multiplier, is obtained from a table function with the independent variable being the food ratio, FR. FR in turn is equal to a product of terms divided by food normal. Each of these relations is included in subroutine STATE.

Table 13-2   Definition of variables for world dynamics model.

Variable†	Definition
BR	Birth rate
BRCM	Birth-rate-from-crowding multiplier
BRFM	Birth-rate-from-food multiplier
BRMM	Birth-rate-from-material multiplier
BRPM	Birth-rate-from-pollution multiplier
CFIFR	Capital fraction indicated by food ratio
CI	Capital investment
CIAF	Capital-investment-in-agriculture fraction
CID	Capital-investment discard
CIG	Capital-investment generation
CIM	Capital-investment multiplier
CIQR	Capital-investment-from-quality ratio
CIR	Capital-investment ratio
CIRA	Capital-investment ratio in agriculture
CR	Crowding ratio
DR	Death rate
DRCM	Death-rate-from-crowding multiplier
DRFM	Death-rate-from-food-multiplier
DRMM	Death-rate-from-material multiplier
DRPM	Death-rate-from-pollution multiplier
ECIR	Effective-capital-investment ratio
F	Food
FC	Food coefficient
FCM	Food-from-crowding multiplier
FPCI	Food potential from capital investment
FPM	Food-from-pollution multiplier

With the above rate components, new values for the state variables can be computed using the following statements:

P = P+DTNOW*(BR−DR)
NR = NR+DTNOW*(−NRUR)
CI = CI+DTNOW*(CIG−CID)
POL = POL+DTNOW*(POLG−POLA)
CIAF = CIAF+DTNOW*(CFIFR*CIQR−CIAF)/CIAFT
QL = QLS*QLM*QLC*QLF*QLP

The equation for population P indicates that the population projected to time TNOW is equal to the population at TTLAS, plus the time interval times the rate of change of population. The time interval is DTNOW = TNOW − TTLAS and the rate of change of population is equal to the birth rate, BR, minus the death

Table 13-2   (Continued).

Variable†	Definition
FR	Food ratio
MSL	Material standard of living
NR	Natural Resources
NREM	Natural-resource-extraction multiplier
NRFR	Natural-resources-fraction remaining
NRI	Natural resources, initial
NRMM	Natural-resources-from-material multiplier
NRUR	Natural-resources-usage rate
P	Population
PD	Population density
POL	Pollution
POLA	Pollution absorption
POLCM	Pollution-from-capital multiplier
POLG	Pollution generation
POLR	Pollution ratio
POLS	Pollution standard
QL	Quality of life
QLC	Quality of life from crowding
QLF	Quality of life from food
QLM	Quality of life from material
QLP	Quality of life from pollution
QLS	Quality of life standard

† The letters, I, N, and T when added to the variable name respectively denote initial, normal, and table, for example, BRN is birth rate normal.

rate, DR. The above equations are coded directly in subroutine STATE and together with the statements for evaluating the "auxiliaries" comprise the World Dynamics model. The listing of subroutine STATE is given in Figure 13-16. In this
example, the state variables are equivalenced to the SS(·) variables.

The listing of the main program for this example is given in Figure 13-17. In the
main program, the input and printer numbers are set and the table functions are
read. Subroutine SLAM II is then called to control the simulation.

Subroutine INTLC initializes the non-SLAM II variables that specify the
starting conditions for the run. The listing for subroutine INTLC is given in
Figure 13-18.

A listing of input statements for this example is shown in Figure 13-19.

Summary of Results. The plotted output for the world model is shown in Figure
13-20. By properly selecting the limits for the plots, five variables can be plotted
primarily in the 0-50% range and the other five variables in the 50-100% range.
Each plot then consists of a top half and a bottom half and the possible confu-
sion from having ten variables on a plot is decreased.

In running this simulation, no external excitation of the system is introduced.
For stable systems, the dynamic behavior exhibited is due to the transients of the
model. When the model is simulated for a longer period of time, all levels reach
their steady state values.

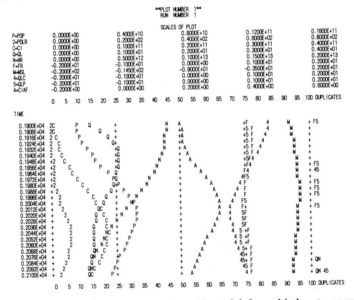

Figure 13-20   Plot of dynamic behavior of world model for a birth rate normal value of
0.04.

```
      SUBROUTINE STATE
      REAL MSL,NREM,NRFR,NRMM,NR,NRI,NRUR,NREMT,NRMMT,LA
      COMMON/SCOM1/ ATRIB(100),DD(100),DDL(100),DTNOW,II,MFA,MSTOP,NCLNR
     1,NCRDR,NPRNT,NNRUN,NNSET,NTAPE,SS(100),SSL(100),TNEXT,TNOW,XX(100)
      COMMON /UCOM1/ BR,BRCM,BRFM,BRMM,BRN,BRPM,CFIFR,CIAFI,CIAFN,CIAFT,
     1CID,CIDN,CIG,CIGN,CII,CIM,CIQR,CIR,CIRA,CR,DR,DRCM,DRFM,DRMM,DRN,D
     2RPM,ECIR,ECIRN,FC,FCM,FN,FPCI,FPM,LA,NREM,NRFR,NRI,NRMM,NRUR,
     3PDN,PI,POLA,POLAT,POLCM,POLG,POLI,POLN,POLS,QLF,QLM,QLS,XNRUN
      COMMON /UCOM2/ BRCMT(6),BRFMT(5),BRMMT(6),BRPMT(7),CFIFRT(5),CIMT
     1(6),CIQRT(5),DRCMT(6),DRFMT(9),DRMMT(11),DRPMT(7),FCMT(6),FPCIT(7)
     2,FPMT(7),NREMT(5),NRMMT(11),POLATT(7),POLCMT(6),QLCT(11),QLFT(5),Q
     3LMT(6),QLPT(7)
      EQUIVALENCE (P,SS(1)),(NR,SS(2)),(CI,SS(3)),(POL,SS(4)),(CIAF,SS(5
     1)),(POLR,XX(1)),(QL,XX(2)),(FR,XX(3)),(MSL,XX(4)),(QLC,XX(5)),
     2(QLP,XX(6))
C
C*****AUXILIARIES.
C
      NRFR=NR/NRI
      CR=P/(LA*PDN)
      CIR=CI/P
      NREM=GTABL(NREMT,NRFR,0.,1.,.25)
      ECIR=CIR*(1.-CIAF)*NREM/(1.-CIAFN)
      MSL=ECIR/ECIRN
      BRMM=GTABL(BRMMT,MSL,0.,5.,1.)
      DRMM=GTABL(DRMMT,MSL,0.,5.,.5)
      DRCM=GTABL(DRCMT,CR,0.,5.,1.)
      BRCM=GTABL(BRCMT,CR,0.,5.,1.)
      CIRA=CIR*CIAF/CIAFN
      FPCI=GTABL(FPCIT,CIRA,0.,6.,1.)
      FCM=GTABL(FCMT,CR,0.,5.,1.)
      POLR=POL/POLS
      FPM=GTABL(FPMT,POLR,0.,60.,10.)
      FR=FPCI*FCM*FPM*FC/FN
      CIM=GTABL(CIMT,MSL,0.,5.,1.)
      POLCM=GTABL(POLCMT,CIR,0.,5.,1.)
      POLAT=GTABL(POLATT,POLR,0.,60.,10.)
      CFIFR=GTABL(CFIFRT,FR,0.,2.,.5)
      QLM=GTABL(QLMT,MSL,0.,5.,1.)
      QLC=GTABL(QLCT,CR,0.,5.,.5)
      QLF=GTABL(QLFT,FR,0.,4.,1.)
      QLP=GTABL(QLPT,POLR,0.,60.,10.)
      NRMM=GTABL(NRMMT,MSL,0.,10.,1.)
      CIQR=GTABL(CIQRT,QLM/QLF,0.,2.,.5)
      DRPM=GTABL(DRPMT,POLR,0.,60.,10.)
      DRFM=GTABL(DRFMT,FR,0.,2.,.25)
      BRFM=GTABL(BRFMT,FR,0.,4.,1.)
      BRPM=GTABL(BRPMT,POLR,0.,60.,10.)
C
C*****RATE COMPONENTS.
C
      BR=P*BRN*BRFM*BRMM*BRCM*BRPM
      DR=P*DRN*DRMM*DRPM*DRFM*DRCM
      NRUR=P*XNRUN*NRMM
      CIG=P*CIM*CIGN
      CID=CI*CIDN
      POLG=P*POLN*POLCM
      POLA=POL/POLAT
C
C*****LEVELS.
C
      QL=QLS*QLM*QLC*QLF*QLP
      P=P+DTNOW*(BR-DR)
      NR=NR-DTNOW*NRUR
      CI=CI+DTNOW*(CIG-CID)
      POL=POL+DTNOW*(POLG-POLA)
      CIAF=CIAF+DTNOW*(CFIFR*CIQR-CIAF)/CIAFT
      RETURN
C
      END
```

Figure 13-16   Listing of subroutine STATE for world model.

```
      REAL MSL,NREM,NRFR,NRMM,NR,NRI,NRUR,NREMT,NRMMT,LA
      DIMENSION NSET(500)
      COMMON QSET(500)
      COMMON/SCOM1/ ATRIB(100),DD(100),DDL(100),DTNOW,II,MFA,MSTOP,NCLNR
     1,NCRDR,NPRNT,NNSET,NTAPE,SS(100),SSL(100),TNEXT,TNOW,XX(100)
      COMMON /UCOM2/ BRCMT(6),BRFMT(5),BRMMT(6),BRPMT(7),CFIFRT(5),CIMT
     1(6),CIQRT(5),DRCMT(6),DRFMT(9),DRMMT(11),DRPMT(7),FCMT(6),FPCIT(7)
     2,FPMT(7),NREMT(5),NRMMT(11),POLATT(7),POLCMT(6),QLCT(11),QLFT(5),Q
     3LMT(6),QLPT(7)
      EQUIVALENCE(NSET(1),QSET(1))
      NCRDR=5
      NPRNT=6
      NTAPE=7
      NNSET=500
C
C *** READ INPUT DATA
C ***      BRCMT  - BIRTH-RATE-FROM-CROWDING MULTIPLIER
C ***      BRFMT  - BIRTH-RATE-FROM-FOOD MULTIPLIER
C ***      BRMMT  - BIRTH-RATE-FROM-MATERIAL MULTIPLIER
C ***      BRPMT  - BIRTH-RATE-FROM POLLUTION MULTIPLIER
C ***      CFIFRT - CAPITAL FRACTION INDICATED BY FOOD RATIO
C ***      CIMT   - CAPITAL-INVESTMENT MULTIPLIER
C ***      CIQRT  - CAPITAL-INVESTMENT-FROM-QUALITY RATIO
C ***      DRCMT  - DEATH-RATE-FROM-CROWDING MULTIPLIER
C ***      DRFMT  - DEATH-RATE-FROM-FOOD MULTIPLIER
C ***      DRMMT  - DEATH-RATE-FROM-MATERIAL MULTIPLIER
C ***      DRPMT  - DEATH-RATE-FROM-POLLUTION MULTIPLIER
C ***      FCMT   - FOOD-FROM-CROWDING MULTIPLIER
C ***      FPCIT  - FOOD POTENTIAL FROM CAPITAL INVESTMENT
C ***      NREMT  - NATURAL-RESOURCE-EXTRACTION MULTIPLIER

C ***      NRMMT  - NATURAL-RESOURCES-FROM-MATERIAL MULTIPLIER
C ***      POLATT - POLLUTION ABSORPTION MULTIPLIER
C ***      POLCMT - POLLUTION-FROM-CAPITAL MULTIPLIER
C ***      QLCT   - QUALITY OF LIFE FROM CROWDING MULTIPLIER
C ***      QLFT   - QUALITY OF LIFE FROM FOOD MULTIPLIER
C ***      QLMT   - QUALITY OF LIFE FROM MATERIAL MULTIPLIER
C ***      QLPT   - QUALITY OF LIFE FROM POLLUTION MULTIPLIER
C
      READ (NCRDR,101) BRCMT
      READ (NCRDR,101) BRFMT
      READ (NCRDR,101) BRMMT
      READ (NCRDR,101) BRPMT
      READ (NCRDR,101) CFIFRT
      READ (NCRDR,101) CIMT
      READ (NCRDR,101) CIQRT
      READ (NCRDR,101) DRCMT
      READ (NCRDR,101) DRFMT
      READ (NCRDR,101) DRMMT
      READ (NCRDR,101) DRPMT
      READ (NCRDR,101) FCMT
      READ (NCRDR,101) FPCIT
      READ (NCRDR,101) FPMT
      READ (NCRDR,101) NREMT
      READ (NCRDR,101) NRMMT
      READ (NCRDR,101) POLATT
      READ (NCRDR,101) POLCMT
      READ (NCRDR,101) QLCT
      READ (NCRDR,101) QLFT
      READ (NCRDR,101) QLMT
      READ (NCRDR,101) QLPT
      CALL SLAM
      STOP
C
  101 FORMAT (7F10.4)
C
      END
```

Figure 13-17 Listing of main program for world model.

```
      SUBROUTINE INTLC
      REAL MSL,NREM,NRFR,NRMM,NR,NRI,NRUR,NREMT,NRMMT,LA
      COMMON/SCOM1/ ATRIB(100),DD(100),DDL(100),DTNOW,II,MFA,MSTOP,NCLNR
     1,NCRDR,NPRNT,NNRUN,NNSET,NTAPE,SS(100),SSL(100),TNEXT,TNOW,XX(100)
      COMMON /UCOM1/ BR,BRCM,BRFM,BRMM,BRN,BRPM,CFIFR,CIAFI,CIAFN,CIAFT,
     1CID,CIDN,CIG,CIGN,CII,CIM,CIQR,CIR,CIRA,CR,DR,DRCM,DRFM,DRMM,DRN,D
     2RPM,ECIR,ECIRN,FC,FCM,FN,FPCI,FPM,LA,NREM,NRFR,NRI,NRMM,NRUR,
     3PDN,PI,POLA,POLAT,POLCM,POLG,POLI,POLN,POLS,QLF,QLM,QLS,XNRUN
      COMMON /UCOM2/ BRCMT(6),BRFMT(5),BRMMT(6),BRPMT(7),CFIFRT(5),CIMT
     1(6),CIQRT(5),DRCMT(6),DRFMT(9),DRMMT(11),DRPMT(7),FCMT(6),FPCIT(7)
     2,FPMT(7),NREMT(5),NRMMT(11),POLATT(7),POLCMT(6),QLCT(11),QLFT(5),Q
     3LMT(6),QLPT(7)
      EQUIVALENCE (P,SS(1)),(NR,SS(2)),(CI,SS(3)),(POL,SS(4)),(CIAF,SS(5
     1)),(POLR,XX(1)),(QL,XX(2)),(FR,XX(3)),(MSL,XX(4)),(QLC,XX(5)),
     2(QLP,XX(6))
C
C*****INITIAL CONDITIONS.
C
      PI=1.65E9
      READ (NCRDR,101) BRN
      ECIRN=1.
      NRI=900.E9
      XNRUN=1.
      DRN=.028
      LA=135.E6
      PDN=26.5
      FC=1.
      FN=1.
      CIAFN=.3
      CII=.4E9
      CIGN=.05
      CIDN=.025
      POLS=3.6E9
      POLI=.2E9
      POLN=1.
      CIAFI=.2
      CIAFT=15.
      QLS=1.
      P=PI
      NR=NRI
      CI=CII
      POL=POLI
      CIAF=CIAFI
      RETURN
C
  101 FORMAT (F5.0)
C
      END
```

Figure 13-18 Listing of subroutine INTLC for world model.

13.14 CHAPTER SUMMARY

This chapter presents the continuous simulation procedures of SLAM II. The definition of state variables is given and the methods for writing differential and difference equations are prescribed. A detailed description of SLAM II's time advance procedure is presented. The use of SLAM II to model a pilot ejecting from an aircraft is presented as Example 13-1. As an example of the use of SLAM II for building Systems Dynamics models, the coding for Forrester's World Dynamics model is presented as Example 13-2.

```
1.05     1.       .9       .7       .6       .55
0.       1.       1.6      1.9      2.
1.2      1.       .85      .75      .7       .7
1.02     .9       .7       .4       .25      .15      .1
1.       .6       .3       .15      .1
.1       1.       1.8      2.4      2.8      3.
.7       .8       1.       1.5      2.
.9       1.       1.2      1.5      1.9      3.
30.      3.       2.       1.4      1.       .7       .6
.5       .5
3.       1.8      1.       .8       .7       .6       .53
.5       .5       .5       .5
.92      1.3      2.       3.2      4.8      6.8      9.2
2.4      1.       .6       .4       .3       .2
.5       1.       1.4      1.7      1.9      2.05     2.2
1.02     .9       .65      .35      .2       .1       .05
0.       .15      .5       .85      1.
0.       1.       1.8      2.4      2.9      3.3      3.6
3.8      3.9      3.95     4.
.6       2.5      5.       8.       11.5     15.5     20.
.05      1.       3.       5.4      7.4      8.
2.       1.3      1.       .75      .55      .45      .38
.3       .25      .22      .2
0.       1.       1.8      2.4      2.7
.2       1.       1.7      2.3      2.7      2.9
1.04     .85      .6       .3       .15      .05      .02
GEN,PRITSKER,WORLD MODEL,1/17/1984,1;
CONT,0,5,.2,.2,8.,W,.00001,.00001;
RECORD,TNOW,TIME,0,P,8.;
VAR,SS(1),P,POP,0.,16E9;
VAR,XX(1),2,POLR,0.0,80.0;
VAR,SS(3),C,CI,0.0,4.0E10;
VAR,XX(2),Q,QL,0.0,4.0;
VAR,SS(2),N,NR,0.0,2.0E12;
VAR,XX(3),F,FR,-2.0,2.0;
VAR,XX(4),M,MSL,-20,2.0;
VAR,XX(5),4,QLC,-2.0,2.0;
VAR,XX(6),5,QLP,-2.0,2.0;
VAR,SS(5),A,CIAF,-.20,0.6;
INITIALIZE,1900.,2100.;
FIN;
   .04
```

Figure 13-19 Listing of input statements of world model.

13.15 EXERCISES

13-1.   Prepare the input statements and experiment with the model of Cedar Bog Lake. Embellishments:

(a) Superimpose a normally distributed random variation with a mean of 0, and a standard deviation of 9 on the solar energy supplied to the lake ecosystem that occurs every 0.025 years.

(b) Determine the effects of a step increase of 20% in the solar energy input to the lake ecosystem.

(c) Suppose that 20% of the energy losses to the lake sediment are considered as fertilizer and reflected in the rate of change in plant energy. Determine the effect of this change in lake ecosystem structure on the behavior of the model.

(d) Superimpose the following control policies on the natural environment: stock the lake with carnivores by 0.3 cal/cm² every tenth of a year; replenish the lake with 0.2 cal/cm² of herbivores when the population of carnivores increases above 0.6 cal/cm²; and spray the lake to reduce the plant population by 70% every 0.5 year. Simulate Cedar Bog Lake under the above control strategies.

13-2.   Use SLAM II to compute a value of y for the following equation

$$y = \int_{3}^{6} 0.0012193 \; x(x-3)^7 dx$$

Integrate to obtain y and compare with the value obtained from SLAM II.

13-3.   Modify the pilot ejection model of Example 13-1 to allow for a two-pilot aircraft. The second pilot ejects from the aircraft 1 sec after the first pilot. The second pilot is located 7 feet behind the first pilot. Simulate the system for an aircraft speed of 500 feet/second to determine if the pilots maintain a separation of 5 feet and if the pilots clear the aircraft. Assume all parameters used in Example 13-1 hold for the two pilot situation.

13-4.   A mass is suspended by a spring and a dashpot. The mass is subjected to a vertical force, f(t), and the vertical movement of the mass is described by

$$m\ddot{y} + k_2 \dot{y} + k_1 y = f(t)$$

where
   y is the spring displacement,
   $\dot{y}$ is the derivative of y with respect to time,
   $\ddot{y}$ is the second derivative of y with respect to time,
   m is the mass,
   k_1 is the spring constant, and
   k_2 is the dashpot constant

Run a 60 second simulation to obtain a plot of the vertical movement of the mass when m = 1, k_1 = 1.0, and k_2 = 0.3, f(t) = 1.0 for all t.
Embellishment:
(a) Rerun the simulation when f(t) = sin 2πt.
(b) Rerun the simulation when f(t) is sampled every 5 seconds from a uniform distribution whose range is 0.5 to 1.5.

13-5.   A bank has a drive-in teller whose service time is exponentially distributed with a mean time of 0.5 minutes. Customer interarrival time is also exponentially distributed with a mean of 0.4 minutes. Only 1 car can wait for the drive-in teller, that is, customers that arrive when 1 car is waiting balk from the system and do not return. Letting $p_n(t)$ = the probability that n customers are in the system (queue plus service) at time t, the following equations can be derived

$$\frac{dp_0(t)}{dt} = -2.5^*p_0(t) + 2.0^*p_1(t)$$

$$\frac{dp_1(t)}{dt} = 2.5^*p_0(t) - 4.5^*p_1(t) + 2.0^*p_2(t)$$

and

$$\frac{dp_2(t)}{dt} = 2.5^*p_1(t) - 2.0^*p_2(t).$$

Assume that at $t=0$ there are no customers in the system and, hence, the teller is idle. Develop a SLAM II model to obtain the expected number in the system, $E[N(t)]$, for $t=0,2,4,6, \ldots , 100$ where

$$E[N(t)] = \sum_{i=0}^{2} i*p_i(t).$$

Prepare a table and plot for $p_i(t)$ and $E_t N(t)_1$ using $DTSAV = DTPLT = 0.05$.

13-6.   Develop a simulation model to portray a single server finite queueing system with exponential interarrival and service times. For a maximum of 20 in the system, plot the transient values for selected probabilities, and the expected value and variance of the number in the system. Also plot the average value of the expected number in the system observed at 5 minute intervals. Let the arrival rate be 4 units/minute and the service rate be 5 units/minute (10).

13-7.   Use SLAM II to analyze Forrester's model of industrial dynamics (5).

13-8.   For the world model presented as Example 13-2, incorporate the following events (9):

Event	*Effect*	*Event Code*
Food shortage, occurs every 20 years and lasts 4 years	Decreases food ratio to 70% of its "normal" value over a 4-year period and population by 10% immediately	1
Discovery of new resources (or equivalent technological development), occurs in 1975	Increases natural resources immediately by 50%	2
Worldwide epidemic, occurs in 1980	Decreases population by 15% and capital investment by 20%	3
Legislative controls go into effect against pollution, occurs 5 years after pollution threshold is reached	Set indicator so that controls are in effect which will decrease POLCM by 75%	4
End of food shortage, occurs 4 years after event 1	Reset indicator to restore food ratio to normal value	5
Initiate zero population growth drive, occurs when population threshold is reached	Decrease birth rate by 15% when popultaion exceeds $2.5*10^9$ (TOL = $1.5*10^8$)	6
Initiate legislation to correct pollution, occurs when pollution threshold is reached	Schedule event 4 to occur in 5 years when pollution exceeds $5*10^8$ (TOL = $5*10^7$	7

Event	Effect	Event Code
Begin conservation measures to protect supply of natural resources, occurs when natural resources threshold is reached	Set conservation indicator that will cause NRMM to be decreased by 10% when natural resources decrease below $7*10^{12}$ (TOL = $1.05*10^{12}$)	8

13-9.   The equation for the current i in a series electrical system is shown below.

$$L\frac{d^2i}{dt^2} + R\frac{di}{dt} + \frac{i}{c} = \omega E \sin \omega t.$$

Develop the equations for simulating the current i in SLAM II form. When the current exceeds the value A, a fuse is blown. Develop the statements that would detect the time at which the fuse would be blown. Assume that the tolerance on the current is $0.1*A$.

13-10.   The equations modeling the height of a dropped ball y_1 are

$$\frac{dy_1}{dt} = y_2$$

$$\frac{dy_2}{dt} = -g$$

where $g = 32.2$. The initial height of the ball is 4 feet and its initial speed is 0. When the ball hits the ground, it reverses direction and has only 0.80 of its speed. Model this bouncing ball to observe the first eight times it hits the ground. Compare with the calculated bounce times (15) of (0.4984, 1.2960, 1.9340, 2.4444, 2.8527, 3.1794, 3.4407, 3.6498). Develop a plot which shows the position of the ball over time.

13-11.   Model a swinging pendulum to determine its position, $y_1(t)$, and the times it reaches the maximum angular displacement. The equation for the pendulum is:

$$\frac{d^2y_1}{dt^2} = 2k^2y_1^3 - (1+k^2)y_1$$

where $k = 0.5, y_1(0) = 0$ and $dy_1(0)/dt = 1$.

13.16   REFERENCES

1. Carnahan, B., H. A. Luther, and J. O. Wilkes, *Applied Numerical Methods,* John Wiley, 1969.
2. Conte, S. D. and C. de Boor, *Elementary Numerical Analysis,* McGraw-Hill, 1972.
3. *CONTROL DATA MIMIC—A Digital Simulation Language Reference Manual,* Publication No. 44610400, Revision D, Control Data Corporation, Minneapolis, Minn., 1970.
4. Fehlberg, E., "Low-Order Classical Runge-Kutta Formulas with Step-Size Control and Their Application to Some Heat Transfer Problems," NASA Report TR R-315, Huntsville, Alabama, April 15, 1969.
5. Forrester, J. W., *Industrial Dynamics,* John Wiley, 1961.
6. Forrester, J. W., *Principles of Systems,* Wright-Allen Press, 1971.
7. Forrester, J. W., *World Dynamics,* Wright-Allen Press, 1972.
8. Meadows, D. H., D. L. Meadows, J. Randers, and W. W. Behrens, III, *The Limits to Growth,* Potomac Associates, 1972.
9. Pritsker, A. A. B. and R. E. Young, *Simulation with GASP_PL/I,* John Wiley, 1975.
10. Pritsker, A. A. B., "Three Simulation Approaches to Queueing Studies Using GASP IV," *Computers & Industrial Engineering,* Vol. 1, 1976, pp. 57-65.
11. Ralston, A., *A First Course in Numerical Analysis,* McGraw-Hill, 1965.
12. Shampine, L. F. and R. C. Allen, Jr., *Numerical Computing: An Introduction,* W. B. Saunders, 1973.
13. Shampine, L. F. et al., "Solving Non-Stiff Ordinary Differential Equations—The State of the Art," *SIAM Review,* Vol. 18, 1976, pp. 376-411.
14. Talavage, J. J. and M. Triplett, "GASP IV Urban Model of Cadmium Flow," *Simulation,* Vol. 23, 1974, pp. 101-108.
15. Thompson, S., "Rootfinding and Interpolation with Runge-Kutta-Sarafyan Methods," *Transactions of SCS,* Vol. 2, 1985, pp. 207-218.
16. Williams, R. B., "Computer Simulation of Energy Flow in Cedar Bog Lake, Minnesota Based on the Classical Studies of Lindeman," in *Systems Analysis and Simulation in Ecology,* B. C. Patten, Ed., Academic Press, 1971.

CHAPTER 14

Combined Modeling

14.1   INTRODUCTION

Systems are often classified as either discrete change, continuous change, or combined discrete continuous change, according to the mechanism by which the state space description of the system changes with time. In discrete change systems, the state of the system changes discretely at event times. As previously described, discrete change systems can be modeled with SLAM II using a network, discrete event or combined network-discrete event orientation. In contrast, continuous change systems are characterized by variables defined through state and derivative equations and the variables are assumed to change continuously with time. Continuous systems are modeled with SLAM II by describing the dynamics of the system as a set of differential or difference equations. In combined discrete-continuous change systems, the state of the system may change discretely, continuously, or continuously with discrete jumps superimposed. Changes to status

531

variables and to the model configuration are made at event times. For combined modeling, an event is defined as a point in time beyond which the status of the system cannot be projected with certainty(6). Thus, status variables can change without an event occurring as long as the change is made in a prescribed manner. Also, an event could occur and no change be made. Events that occur at a specified point in time are defined as *time-events*. They are commonly thought of in terms of next event simulation. Events that occur when the system reaches a particular state are called *state-events*. Unlike time-events, they are not scheduled in the future but occur when state variables meet prescribed conditions. In SLAM II state-events can initiate time-events, and time-events can initiate state-events.

In this chapter, examples of combined modeling are presented. The features of SLAM II that facilitate the use of alternative modeling viewpoints are described. No new modeling elements are presented. The SLAM II interfaces for combined modeling have been introduced in previous chapters and they are reviewed in the next section. The processing logic to integrate discrete and continuous models is described.

14.2   SLAM II INTERFACES

This section summarizes information on the interfaces between network, discrete event, and continuous models. The variables SS, DD, XX, II, ATRIB, and TNOW provide for the transfer of values of status variables throughout a SLAM II model. In discrete event and continuous submodels, the SCOM1 labeled COMMON block is used to transfer these values. The other variables in SCOM1 also provide for the transfer of information between SLAM II submodels. The two dimensional global variable ARRAY generalizes the concept of the XX vector. Subroutines for accessing and changing the values in ARRAY are provided by SLAM II. The functions NNACT, NNCNT, NNQ, NNRSC, NRUSE, and NNGAT provide the method for accessing the status of network elements in discrete event and continuous submodels. The function USERF provides a means for writing FORTRAN code to establish a value to be used in a network model. USERF can replace any SLAM II status variable in a network model. The subroutines FREE, ALTER, OPEN and CLOSX are used to change the status of network elements from user-written FORTRAN subprograms. The subprograms ALLOC, SEIZE, NQS, and NSS provide a method for incorporating advanced decision logic when modeling resource allocations and server and queue selection procedures.

The DETECT node and SEVNT statement provide mechanisms to monitor state variables and to detect state-events. They provide the interface between changes in status variables in any modeling viewpoint and the invoking of subroutine EVENT. The EVENT node is the direct interface for this capability between a network model and discrete event subprograms. Inserting entities into a network from a discrete event is accomplished through the use of subroutine ENTER which places an entity in a network at an ENTER node. Within a discrete event, entities can be manipulated through the use of the file processing routines. This includes entries on the event calendar. Entities in activities can be made to complete an activity from a discrete event by invoking subroutine STOPA.

The above describes the constructs available in SLAM II to interface models developed from alternative viewpoints. Clearly, SLAM II has been designed to foster and facilitate the combined modeling of systems.

14.3   USING ALTERNATIVE MODELING VIEWPOINTS

The question of which modeling viewpoint to use on a particular problem is academic. Experience has shown that the education of an individual is the prime determinant of the viewpoint taken. It is one of the clearest examples of the impact of the educational process on problem-solving capabilities. Electrical engineers, mechanical engineers, chemical engineers and physicists tend to be continuous modelers. Computer scientists tend to be process or network oriented as do management and business people. Operations researchers and industrial engineers are basically discrete simulation modelers with an emphasis more toward the discrete event modeling viewpoint. An underlying hypothesis that establishes the need for SLAM II is that there is not one best viewpoint when modeling a system.

When using SLAM II, the primary question is not which modeling viewpoint to take, but how to start to solve the problem. As demonstrated throughout this book, models developed in SLAM II are easily embellished. The interfaces between modeling viewpoints provide the flexibility to include constructs from each viewpoint into an existing model. Experience has shown that models, just like systems, evolve and change to meet new specifications and conditions. As a model grows, the need for new modeling viewpoints and concepts becomes greater. The flexibility provided by SLAM II to perform combined modeling is a long term advantage of strategic importance.

14.4   SLAM II PROCESSING LOGIC FOR COMBINED MODELS

The processing logic employed by SLAM II for simulating combined models is depicted in Figure 14-1. The processor begins by interpreting the SLAM II input statements. This is followed by an initialization phase including a call to the user-written subroutine INTLC for establishing user-defined initial conditions for the simulation. During the initialization phase, the processor schedules an entity arrival event for each CREATE node to occur at the time of the first release of the CREATE node. These events are filed on the event calendar in the order in which the CREATE nodes appear in the statement model. In addition, entities initially in QUEUE nodes are created and end-of-activity events are scheduled for the service activities following the QUEUE nodes. All attribute values of these created entities are zero. Events placed on the event calendar through calls to subroutine SCHDL in subroutine INTLC are then put on the event calendar.

After processing subroutine INTLC or after any increase in TNOW, a test is made for one of the following end-of-simulation conditions:

1. TNOW is greater than or equal to the ending time specified for the simulation.
2. The value of MSTOP has been set to a negative value.
3. A TERMINATE node has reached its specified terminate count in the network portion of the model.
4. There are no events on the event calendar and there are no continuous variables in the model.

If the simulation run is ended, statistics are calculated, a call is made to subroutine OTPUT, and the SLAM II Summary Report is printed. A test is then made to determine if additional runs are to be executed. If there are, the next run is initiated. If all simulation runs have been completed, SLAM II returns control to the user-written main program.

If none of the end-of-simulation conditions is satisfied, the SLAM II processor determines if there is a time event at the current time. If there is, JEVNT is set to the event code of the current event. A negative value for JEVNT specifies that the event is associated with the network model. JEVNT is positive for user defined events. A small diamond is shown in Figure 14-1 to represent logical tests on internal SLAM II variables. For user events, subroutine EVENT is called directly. In subroutine EVENT, the modeler performs the transfer necessary to process the

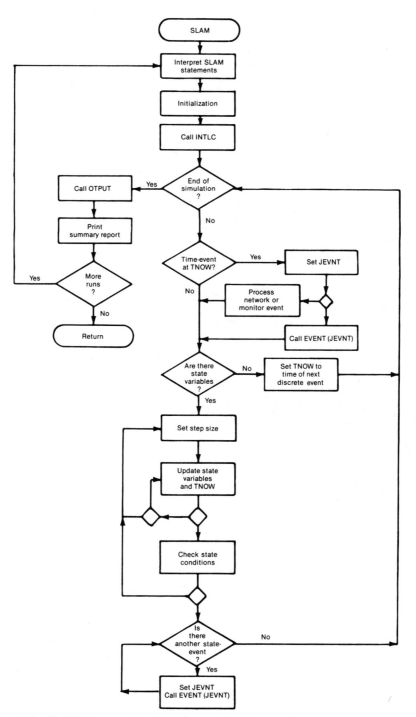

Figure 14-1   SLAM II processing logic for combined networks.

event associated with JEVNT. For network and monitor events, the SLAM II processor performs the necessary operations. All network events are associated with entity arrivals to a node. The functions necessary to process an entity arriving to a node are precoded and the updates to network status variables are made automatically in accordance with the parameters specified for the node. As part of this nodal processing, events are placed on the calendar in accordance with the node type. For example, an entity arrival to a FREE node may reallocate the freed resources to other entities, place the other entities in activities, and schedule an end-of-activity event for the arriving entity to the FREE node.

Following the processing of a time event, continuous variables are updated and state-event conditions checked since threshold crossings can be caused by discrete changes made in the event subroutines. If no continuous variables are included in the model, TNOW is updated to the time of the next discrete event and the end-of-simulation conditions are checked. If continuous variables are included in the model, a step size is established and the values of the SS variables are updated by calls to subroutine STATE. If differential equations are involved, then multiple calls to subroutine STATE are made within the step. If accuracy is not met, the step size is reduced. After the updating of the continuous variables, a check is made on state-event conditions. If there is not a state-event to be processed, a test for the end-of-simulation is made. If a variable exceeds a threshold beyond tolerance then the step size is reduced and the updating process is repeated. If there is a state-event to be processed, the event code JEVNT is set, and a call is made to subroutine EVENT. Following the return from subroutine EVENT, a test for another state-event is processed as multiple state-events can occur at the same time.

The processing logic presented in Figure 14-1 gives an overview of how combined simulation is performed. It is important to note that the values of continuous variables are always updated before a time-event is processed. In this way, decisions in a time-event are based on the current values of continuous variables. Following the processing of a time-event, the state variables are updated in accordance with any changes made to parameters or other state variables in the model. For display convenience, the flowchart does not include the details with regard to the setting of the step size, accuracy checking, out-of-tolerance state-event detection, and the recording of values at save times.

14.5   EXAMPLE 14-1.   SOAKING PIT FURNACE† (1,4,7)

Steel ingots arrive at a soaking pit furnace in a steel plant with an interarrival time that is exponentially distributed with mean of 2.25 hours. The soaking pit furnace heats an ingot so that it can be economically rolled in the next stage of the process. The temperature change of an ingot in the soaking pit furnace is described by the following differential equation

$$\frac{dh_i}{dt} = (H-h_i)*C_i$$

where h_i is the temperature of the ith ingot in the soaking pit; C_i is the heating time coefficient of an ingot and is equal to $X+.1$ where X is normally distributed with mean of .05 and standard deviation of 0.01; and H is the furnace temperature which is heated toward 2600°F with a heating rate constant of 0.2, that is,

$$\frac{dH}{dt} = (2600.-H)*0.2$$

The ingots interact with one another in that adding a "cold" ingot to the furnace reduces the temperature of the furnace and thus changes the heating time for all ingots in the furnace. The temperature reduction is equal to the difference between furnace and ingot temperatures, divided by the number of ingots in the furnace. There are 10 soaking pits in the furnace. When a new ingot arrives and the furnace is full, it is stored in an ingot storage bank. It is assumed that the initial temperature of an arriving ingot is uniformly distributed in the interval from 400 to 500°F. All ingots put in the ingot storage bank are assumed to have a temperature of 400°F upon insertion into the soaking pit. The operating policy of the company is to continue heating the ingots in the furnace until one or more ingots reach 2200°F. At such a time all ingots with a temperature greater than 2000°F are removed. The initial conditions are that there are six ingots in the furnace with initial temperatures of 550, 600, 650, . . ., 800°F. Initially, the temperature of the furnace is 1650°F, and the next ingot is due to arrive at time 0.

The objective is to simulate the above system for 500 hours to obtain estimates of the following quantities:

---

† This example describes a hypothetical situation which illustrates combined modeling concepts using SLAM II. See Exercise 14-1 for a representative description of the operation of soaking pits.

1. Heating time of the ingots;
2. Final temperature distribution of the ingots;
3. Waiting time of the ingots in the ingot storage bank; and
4. Utilization of the soaking pit furnace.

Concepts Illustrated. This example illustrates the stopping of an ACTIVITY in a network model due to the occurrence of a state-event through the use of a STOPA activity duration specification and a call to subroutine STOPA. A RESOURCE is used to model the soaking pits.

SLAM II Model. The furnace problem is described using a combined network-discrete event-continuous model. A network is employed to model the ingot arrival process, the ingot storage bank, the soaking pit activity, and the exit from the system. A time-event initiated through an arrival to an EVENT node is used to insert ingots into the furnace. A state-event ends the soaking activity and causes ingots to be removed from the furnace. Continuous state variables model the temperature of the ingots in the furnace and the temperature of the furnace. The status of the pits in the furnace is maintained in the SLAM II vector XX. If pit I is occupied by an ingot, XX(I) is set to 1; otherwise XX(I) is equal to 0 which indicates that the pit is empty. The continuous portion of the model employs twelve state variables. The first ten state variables are used to model the temperature of the ingot in each of the ten soaking pits. If soaking pit I is occupied, SS(I) equals the temperature of the ingot in pit I, otherwise SS(I) is set equal to 0. The temperature of the furnace is modeled as state variable SS(11). State variable SS(12) is set equal to the maximum temperature of all ingots in the furnace. The equations for the temperatures of each ingot and the furnace are coded in subroutine STATE in terms of their derivatives as shown in Figure 14-2. The value of SS(12), representing the maximum ingot temperature, is initially set to 0. In a DO loop, each ingot temperature is compared to the value of SS(12), and SS(12) is reset to the temperature of the ingot as appropriate. Also each value of DD(I) is calculated as the rate of change of the temperature for the ingot in the Ith pit. Note that if the Ith pit is empty, then XX(I) equals 0., and therefore DD(I) is set to 0. The variable C(I) is the normally distributed heating coefficient for the ingot in pit I which is set in the discrete event portion of the model and is passed to subroutine STATE through user labeled COMMON. It is assumed that the ingot has material properties that affect its heating rate. Following the exit from the DO loop, DD(11) is set equal to the rate of change of the temperature in the furnace.

The network model for this example is depicted in Figure 14-3. The ingot arrivals are created at the CREATE node with the arrival time marked as ATRIB(1).

```
      SUBROUTINE STATE
      COMMON/SCOM1/ ATRIB(100),DD(100),DDL(100),DTNOW,II,MFA,MSTOP,NCLNR
     1,NCRDR,NPRNT,NNRUN,NNSET,NTAPE,SS(100),SSL(100),TNEXT,TNOW,XX(100)
      COMMON/USER/ C(10)
      SS(12)=0.
C
C *** MONITOR THE TEMPERATURE IN EACH PIT AND STORE THE
C *** HIGHEST PIT TEMPERATURE IN SS(12).
C
      DO 100 I=1,10
      IF(SS(I).GT.SS(12)) SS(12)=SS(I)
  100 DD(I)=(SS(11)-SS(I))*XX(I)*C(I)
      DD(11)=(2600.-SS(11))*.2
      RETURN
      END
```

Figure 14-2 Subroutine STATE for furnace example.

The time between arrivals of the ingot entities is exponentially distributed with a mean of 2.25 hours. The ingots continue to the ASSIGN node where ATRIB(2) is set equal to a sample from a uniform distribution between 400 and 500 which represents the initial temperature of the ingot. The ingots then wait, if necessary, in file 1 for one unit of resource PIT. Once the ingot has seized a PIT, it proceeds to the EVENT node labeled LOAD where an event with code 1 is invoked. This event determines the pit number of the available pit, sets the status of the pit to occupied, sets the initial temperature and heating coefficient for the ingot, and sets ATRIB(3) of the entity equal to the pit number. If the ingot has waited in the storage bank, statistics are collected on the waiting time and ATRIB(1) is re-marked as TNOW. In addition, the temperature of the furnace is reduced appropriately as the result of adding an ingot to the furnace. The logic for this event is coded in subroutine LOAD and is described following the discussion of the network model.

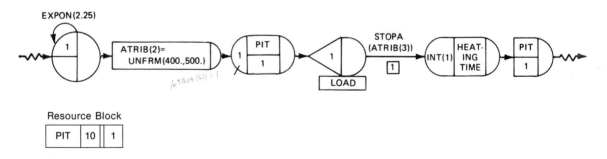

Figure 14-3 Network model for furnace example.

Following processing at the EVENT node, the ingot undertakes ACTIVITY 1 representing the soaking activity. The duration of this ACTIVITY is indefinite and is terminated from within a discrete event when stopping conditions are satisfied.

This is modeled in the event EXITP with a call to subroutine STOPA(NTC) where NTC is equal to the PIT number which was assigned to ATRIB(3) in the LOAD event. The invocation of subroutine STOPA(I) causes the ingot in pit I to complete ACTIVITY 1. ACTIVITY 1 is completed for each ingot I with SS(I) greater than or equal to 2000° in event EXITP. This event is initiated as a state-event whenever the highest temperature of an ingot, SS(12), exceeds 2200°F within a tolerance of 50°F as prescribed on the SEVNT statement depicted below.

SEVNT,2,SS(12),XP,2200.,50;

We defer the description of the coding of subroutine EXITP for the processing logic associated with event code 2 until we complete the discussion of the network model.

Following the completion of the soaking activity, the ingots proceed to a COLCT node where interval statistics are collected on the heating time for the ingot using the time reference contained in ATRIB(1). An ingot entity is routed to the FREE node where a PIT is released. The entity is then terminated from the system.

The discrete events for this example are coded in subroutines LOAD and EXITP. Subroutine EVENT(I) maps the event code, I, onto the appropriate subroutine call and is shown in Figure 14-4. Recall that the LOAD event is initiated from within the network model by ingots arriving to the EVENT node with code 1, and the EXITP event is initiated as a state-event from within the continuous model by the maximum ingot temperature reaching 2200°F.

```
      SUBROUTINE EVENT(I)
        GO TO (1,2),I
    1 CALL LOAD
      RETURN
    2 CALL EXITP
      RETURN
      END
```

Figure 14-4   Subroutine EVENT(I) for furnace example.

The coding for subroutine LOAD is shown in Figure 14-5. The pit number, JJ, in which the ingot is to be inserted, is set equal to the first J such that XX(J) equals 0. The variables associated with pit JJ are then initialized and ATRIB(3) is set equal to the pit number. If the ingot waited in the storage bank, statistics are collected on the wait time using COLCT variable number 2, the temperature of the ingot is set to 400°F, and ATRIB(1) is re-marked at TNOW. The ingot is placed in the pit which causes the temperature of the furnace to be reduced by the difference in temperature between the ingot and the furnace divided by the number of ingots in the furnace (including the new one).

```
      SUBROUTINE LOAD
      COMMON/SCOM1/ ATRIB(100),DD(100),DDL(100),DTNOW,II,MFA,MSTOP,NCLNR
     1,NCRDR,NPRNT,NNRUN,NNSET,NTAPE,SS(100),SSL(100),TNEXT,TNOW,XX(100)
      COMMON/USER/ C(10)
C*****DETERMINE PIT NUMBER OF AVAILABLE PIT
      DO 10 J=1,10
      JJ=J
      IF(XX(J).EQ.0.) GO TO 20
   10 CONTINUE
      CALL ERROR(1)
C*****TURN PIT ON,SET INITIAL TEMP. AND HEATING COEFF., SET ATRIB(3)=PIT NUMBEF
   20 XX(JJ)=1.
      SS(JJ)=ATRIB(2)
      C(JJ)=RNORM(.05,.01,1)+.1
      ATRIB(3)=JJ
C*****IF COLD INGOT - COLLECT STATISTICS, RESET TEMP., AND RE-MARK ATRIB(1)
      WAIT=TNOW-ATRIB(1)
      IF(WAIT.LE.0.) GO TO 30
      CALL COLCT(WAIT,2)
      SS(JJ)=400.
      ATRIB(1)=TNOW
   30 XNUM=NNACT(1)+1
C*****REDUCE FURNACE TEMP. BY TEMP. DIFFERENCE/ NUMBER OF INGOTS
      REDUCT=(SS(11)-SS(JJ))/XNUM
      SS(11)=SS(11)-REDUCT
      RETURN
      END
```

Figure 14-5 Subroutine LOAD for furnace example.

The coding for subroutine EXITP is shown in Figure 14-6. When EXITP is called, one of the ingots has a temperature greater than or equal to 2200°F. The temperature of each pit, J, is tested against the desired temperature of 2000°F. If the ingot in pit J has reached 2000°, the variable XX(J) is set to zero to denote the pit has been emptied, statistics are collected on the temperature of the ingot exiting the pit as COLCT variable number 2, and the temperature of the pit is set equal to 0. A call is then made to subroutine STOPA(J) which causes the entity representing the ingot in pit J to complete ACTIVITY 1 in the network model.

The initial conditions for the simulation are established in subroutine INTLC depicted in Figure 14-7. The temperature of the furnace is initialized to 1650°F and six ingots with initial temperatures of 550, 600, 650, . . ., 800°F are inserted into file 1 of the AWAIT node through multiple calls of the statement CALL FILEM(1,ATRIB).

```
      SUBROUTINE EXITP
      COMMON/SCOM1/ ATRIB(100),DD(100),DDL(100),DTNOW,II,MFA,MSTOP,NCLNR
     1,NCRDR,NPRNT,NNRUN,NNSET,NTAPE,SS(100),SSL(100),TNEXT,TNOW,XX(100)
C
C *** REMOVE INGOTS FROM THE PITS WHEN ONE PIT REACHES 2200 DEGREES.
C *** REMOVE ALL INGOTS THAT ARE 2000 DEGREES OR HIGHER.
C
      DO 10 J=1,10
      IF(SS(J).LT.2000.) GO TO 10
      XX(J)=0.
C
C *** COLLECT STATISTICS ON UNLOADED INGOT TEMPERATURE.
C
      CALL COLCT(SS(J),1)
      SS(J)=0.
      CALL STOPA(J)
   10 CONTINUE
      RETURN
      END
```

Figure 14-6 Subroutine EXITP for furnace example.

```
      SUBROUTINE  INTLC
      COMMON/SCOM1/ ATRIB(100),DD(100),DDL(100),DTNOW,II,MFA,MSTOP,NCLNR
     1,NCRDR,NPRNT,NNRUN,NNSET,NTAPE,SS(100),SSL(100),TNEXT,TNOW,XX(100)
      COMMON/USER/  C(10)
      TEMP=550.
      SS(11)=1650.
      DO 10 I=1,10
   10 C(I)=0.
C
C *** LOAD THE SYSTEM INITIALLY WITH 6 INGOTS
C
      DO 100 I=1,6
      ATRIB(1)=TNOW
      ATRIB(2)=TEMP
      CALL FILEM(1,ATRIB)
  100 TEMP=TEMP+50.
      RETURN
      END
```

Figure 14-7 Subroutine INTLC for furnace example.

The input statements for this example are shown in Figure 14-8. Note that the STAT statement for COLCT variable number 1 specifies that a histogram be generated with 20 interior cells, with the upper limit of the first cell equal to 2000°F, and a cell width of 10°F. In addition, time-persistent statistics are specified for state variable SS(11).

```
 1  GEN,OREILLY,INGOT PROBLEM,7/1/83,1;
 2  LIMITS,1,4,40;
 3  STAT,1,HEATING TEMP,20/2000/20;
 4  STAT,2,WAITING TIME;
 5  CONT,11,1,.1,10,10;
 6  TIMST,SS(11),FURNACE TEMP.;
 7  SEVNT,2,SS(12),XP,2200,50.;
 8  NETWORK;
 9      ;
10          RESOURCE/PIT(10),1;
11          CREATE,EXPON(2.25),,1;              CREATE INGOT ARRIVALS
12          ASSIGN,ATRIB(2)=UNFRM(400.,500.);   ASSIGN INGOT TEMPERATURE
13          AWAIT,PIT;                          AWAIT A PIT
14  LOAD    EVENT,1;                            LOAD INGOT INTO PIT
15          ACT/1,STOPA(ATRIB(3));              SOAKING ACTIVITY
16          COLCT,INT(1),HEATING TIME;          COLLECT STATISTICS
17          FREE,PIT;                           FREE THE PIT
18          TERM;                               EXIT THE SYSTEM
19          ENDNETWORK;
20  INIT,0,500;
21  FIN:
```

Figure 14-8 Input statements for furnace example.

Summary of Results. The SLAM II Summary Report for this example is included as Figure 14-9. The report reveals that during the 500 hours of simulated time, 242 ingots are processed and the average heating time is 19.43 hours. The heating time for ingots varied between 11.26 and 30.24 hours. The variation in the heating time is due to the coupling of the equations of the system through the furnace temperature and to the random component of the equation that is associated with the ingot, that is, the normal sample used to define the heating coefficient, $C(I)$. The average waiting time for ingots in the cold ingot bank is 14.58 hours. Note that the waiting time statistics are collected only for the 186 ingots that waited. The determination of whether to include the zero waiting time of the ingots that went directly into the furnace is a modeling question that must be decided by the analyst. If waiting times with and without zero values are desired, then two calls to subroutine COLCT can be incorporated into the program.

From both the activity and resource statistics sections of the summary report, it is seen that the average number of pits in use was 9.45. At the end of the simulation, all ten pits were being utilized. These results show that the system is overloaded and that new procedures or additional resources are required. From the file statistics output, the average number of ingots in the storage bank is 5.42 and the maximum number 18. Thus, a storage area for at least 18 ingots is required. The state and derivative variables portray the final conditions of the simulation. At 500 hours, the furnace temperature was 1500°F and increasing at 259°F/hour. The corresponding value for each ingot in a pit is shown in the summary report. A histo-

```
                          S L A M   I I   S U M M A R Y   R E P O R T

              SIMULATION PROJECT INGOT PROBLEM          BY OREILLY

              DATE  7/ 1/1983                           RUN NUMBER   1 OF   1


              CURRENT TIME   0.5000E+03
              STATISTICAL ARRAYS CLEARED AT TIME  0.0000E+00
```

STATISTICS FOR VARIABLES BASED ON OBSERVATION

	MEAN VALUE	STANDARD DEVIATION	COEFF. OF VARIATION	MINIMUM VALUE	MAXIMUM VALUE	NUMBER OF OBSERVATIONS
HEATING TEMP	0.2159E+04	0.6708E+02	0.3107E-01	0.2002E+04	0.2246E+04	242
WAITING TIME	0.1458E+02	0.9184E+01	0.6300E+00	0.1000E+00	0.3532E+02	186
HEATING TIME	0.1943E+02	0.3456E+01	0.1779E+00	0.1126E+02	0.3024E+02	242

STATISTICS FOR TIME-PERSISTENT VARIABLES

	MEAN VALUE	STANDARD DEVIATION	MINIMUM VALUE	MAXIMUM VALUE	TIME INTERVAL	CURRENT VALUE
FURNACE TEMP.	0.2145E+04	0.2494E+03	0.6491E+03	0.2508E+04	0.5000E+03	0.1500E+04

FILE STATISTICS

FILE NUMBER	ASSOC NODE LABEL/TYPE	AVERAGE LENGTH	STANDARD DEVIATION	MAXIMUM LENGTH	CURRENT LENGTH	AVERAGE WAITING TIME
1	AWAIT	5.4234	5.0924	18	0	10.8036
2	CALENDAR	1.0000	0.0000	11	1	0.4039

REGULAR ACTIVITY STATISTICS

ACTIVITY INDEX/LABEL	AVERAGE UTILIZATION	STANDARD DEVIATION	MAXIMUM UTIL	CURRENT UTIL	ENTITY COUNT
1 SOAKING ACTI	9.4537	1.2487	10	9	242

RESOURCE STATISTICS

RESOURCE NUMBER	RESOURCE LABEL	CURRENT CAPACITY	AVERAGE UTILIZATION	STANDARD DEVIATION	MAXIMUM UTILIZATION	CURRENT UTILIZATION
1	PIT	10	9.4537	1.2487	10	9

RESOURCE NUMBER	RESOURCE LABEL	CURRENT AVAILABLE	AVERAGE AVAILABLE	MINIMUM AVAILABLE	MAXIMUM AVAILABLE
1	PIT	1	0.5463	0	10

STATE AND DERIVATIVE VARIABLES

(I)	SS(I)	DD(I)
1	0.1774E+04	0.1049E+03
2	0.4000E+03	0.4033E+02
3	0.4000E+03	0.4153E+02
4	0.4000E+03	0.3916E+02
5	0.4000E+03	0.4103E+02
6	0.4000E+03	0.4168E+02
7	0.4000E+03	0.4074E+02
8	0.1716E+04	0.1049E+03
9	0.1745E+04	0.1050E+03
10	0.0000E+00	0.4062E+02
11	0.1500E+04	0.2591E+02
12	0.2229E+04	0.0000E+00

Figure 14-9 SLAM II summary report for furnace example.

gram of the temperatures of the ingots removed from the pit is shown in Figure 14-10. The histogram reveals that 80 percent of the ingot temperatures were greater than 2100°F. This clustering is due to the stopping decision that requires at least one ingot to have a temperature of 2200° before any ingot can be removed from the furnace. Thirty three percent of the ingots reached the 2200°F limit.

```
                            **HISTOGRAM NUMBER  1**

                                  HEATING TEMP

  OBSV      RELA      CUML       UPPER
  FREQ      FREQ      FREQ    CELL LIMIT    0      20      40      60      80      100
                                           +   +   +   +   +   +   +   +   +   +   +
     0     0.000     0.000    0.2000E+04    +                                        +
    13     0.054     0.054    0.2020E+04    +***                                     +
    12     0.050     0.103    0.2040E+04    +**  C                                   +
    11     0.045     0.149    0.2060E+04    +**     C                                +
     7     0.029     0.178    0.2080E+04    +*        C                              +
     4     0.017     0.194    0.2100E+04    +*         C                             +
    11     0.045     0.240    0.2120E+04    +**          C                           +
    11     0.045     0.285    0.2140E+04    +**            C                         +
    10     0.041     0.326    0.2160E+04    +**              C                       +
    34     0.140     0.467    0.2180E+04    +*******            C                    +
    49     0.202     0.669    0.2200E+04    +**********               C              +
    47     0.194     0.864    0.2220E+04    +**********                     C        +
    26     0.107     0.971    0.2240E+04    +*****                               C+
     7     0.029     1.000    0.2260E+04    +*                                       C
     0     0.000     1.000    0.2280E+04    +                                        C
     0     0.000     1.000    0.2300E+04    +                                        C
     0     0.000     1.000    0.2320E+04    +                                        C
     0     0.000     1.000    0.2340E+04    +                                        C
     0     0.000     1.000    0.2360E+04    +                                        C
     0     0.000     1.000    0.2380E+04    +                                        C
     0     0.000     1.000    0.2400E+04    +                                        C
     0     0.000     1.000       INF        +                                        C
   ---                                      +   +   +   +   +   +   +   +   +   +   +
   242                                      0      20      40      60      80      100
```

STATISTICS FOR VARIABLES BASED ON OBSERVATION

	MEAN VALUE	STANDARD DEVIATION	COEFF. OF VARIATION	MINIMUM VALUE	MAXIMUM VALUE	NUMBER OF OBSERVATIONS
HEATING TEMP	0.2159E+04	0.6708E+02	0.3107E-01	0.2002E+04	0.2246E+04	242

Figure 14-10 Histogram of ingot temperatures.

14.6   EXAMPLE 14-2.   A CONTINUOUS FLOW CONVEYOR

A sparkplug packing line modeled by Henriksen(5) consists of a conveyor belt which transports sparkplugs at the rate of 999 per minute over a conveyor belt that feeds five machines. Each machine can process 333 plugs per minute. The flow from the conveyor to the machines is automatic. The time to transport a plug between two adjacent machines is 9 seconds. If a plug is not packed by one of the five machines, it flows into a barrel for future packing. Each packing machine has stoppages due to jams. The jamming of a machine occurs after it has packed between 200 and 600 sparkplugs. The time to repair a jammed machine is uniformly distributed between 6 and 24 seconds. The initial conditions for the system are that all machines are idle. It takes 18 seconds to build up to a rate of 999 sparkplugs per minute or 16.65 plugs per second before the first machine. The processing of plugs by the first machine starts at this time.

Concepts Illustrated. State variables are used to model the amount of production since the last jam for each packing machine. A state variable maintains the number of sparkplugs that fall into the barrel at the end of the conveyor belt. DETECT nodes are used to monitor the number of sparkplugs packed by each machine and to detect when the machine jams. Events contain the logic for stopping a machine, starting a machine, and changing the flow of sparkplugs on the conveyor belt at points along the conveyor belt.

SLAM II Model. The SLAM II model consists of SS(M) variables to model the number of plugs packed since the last unjamming of machine M and the number of plugs falling into the barrel, SS(6). A DETECT node is used to determine when the number of plugs packed by a machine exceeds a threshold value. For this state event, an entity is inserted into the network and routed to an EVENT node to stop the packing of plugs by the machine. One DETECT node is employed for each machine. The discrete event portion of the model involves three events. A stop machine event called STOPM which sets a status variable XX(M+6) to indicate that machine M is in a jammed state and not packing plugs. The flow of plugs to the machine is then diverted and scheduled to arrive at the next machine after a 9 second transport time on the conveyor. A start machine event, START, is scheduled to occur at the failure time plus the time to unjam the machine. The START event changes the status variables for the machine and, if there is a flow of parts to the machine, schedules a decrease in the flow of parts along the conveyor. A third event models the flow changes on the conveyor to a point in front of each machine. Detailed logic is included in this routine to specify if the flow change causes the machine to go to an idle status; a packing status; or that a flow change should be made to a downstream point on the conveyor. The variables and events of the model are listed below.

Variable	Definition
M	Machine Number, M = 1,5
XX(M)	Status of Machine M: 0→NOT WORKING; 1→WORKING.
XX(M + 6)	Jam Status of Machine M: 0→JAMMED; 1→NOT JAMMED.
SS(M)	Number of sparkplugs packed since last jam of Machine M.
SS(6)	Number of sparkplugs in barrel.
XX(6)	Production flow rate into barrel.
SS(M + 6)	Number of sparkplugs that define the threshold for the jamming event for machine M.
FLOW(M)	Flow rate on conveyor before Machine M.
ATRIB(1)	Machine number.
ATRIB(2)	Change in flow rate.
PFLOWR	Flow rate on conveyor, 5.55 plugs/sec.

Event Code	Subroutine Name	Definition
1	START	Start machine because it is unjammed.
2	STOPM	Stop machine because of jam.
3	FLOWCHNG	Change of rate of flow of sparkplugs on the conveyor.

Figure 14-11 gives the FORTRAN code for subroutine STATE. For each machine, the number of plugs packed is prescribed by the following difference equation

$$SS(M) = SSL(M) + DTNOW*XX(M)*PFLOWR$$

where XX(M) is the status of machine M and PFLOWR is the plug flow rate. The production into the barrel is also defined as a difference equation with the rate given by XX(6). Figure 14-12 presents the initial values and conditions for the simulation. Subroutine EVENT is shown in Figure 14-13.

```
      SUBROUTINE STATE
      COMMON/SCOM1/ATRIB(100),DD(100),DDL(100),DTNOW,II,MFA,MSTOP,NCLNR
     1,NCRDR,NPRNT,NNRUN,NNSET,NTAPE,SS(100),SSL(100),TNEXT,TNOW,XX(100)
      COMMON/UCOM1/FLOW(10),PFLOWR,NFLOW(10)
      DO 10 M=1,5
      SS(M)=SSL(M)+DTNOW*XX(M)*PFLOWR
10    CONTINUE
      SS(6)=SSL(6)+DTNOW*XX(6)
      RETURN
      END
```

Figure 14-11 Equations describing sparkplug rate of flow, subroutine STATE.

```
      SUBROUTINE INTLC
      COMMON/SCOM1/ATRIB(100),DD(100),DDL(100),DTNOW,II,MFA,MSTOP,NCLNR
     1,NCRDR,NPRNT,NNRUN,NNSET,NTAPE,SS(100),SSL(100),TNEXT,TNOW,XX(100)
      COMMON/UCOM1/FLOW(10),PFLOWR,NFLOW(10)
      PFLOWR=5.55
      DO 100 M=1,5
      SS(M)=0
      SS(M+6)=UNFRM(XX(20),XX(21),2)
      XX(M)=0.0
      FLOW(M)=0.0
      NFLOW(M)=0
      XX(M+6)=1.0
100   CONTINUE
      ATRIB(1)=1.
      ATRIB(2)=PFLOWR
      CALL SCHDL(3,18.,ATRIB)
      CALL SCHDL(3,18.,ATRIB)
      CALL SCHDL(3,18.,ATRIB)
      XX(6)=0.0
      RETURN
      END
```

Figure 14-12   Initial conditions subroutine for flow conveyor model.

```
      SUBROUTINE EVENT(I)
      COMMON/SCOM1/ATRIB(100),DD(100),DDL(100),DTNOW,II,MFA,MSTOP,NCLNR
     1,NCRDR,NPRNT,NNRUN,NNSET,NTAPE,SS(100),SSL(100),TNEXT,TNOW,XX(100)
      GO TO (1,2,3),I
1     CALL START
      RETURN
2     CALL STOPM
      RETURN
3     CALL FLOWCHNG
      RETURN
      END
```

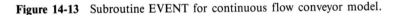

Figure 14-13   Subroutine EVENT for continuous flow conveyor model.

Figure 14-14 gives the FORTRAN code for subroutine START. M is set to the machine number which is passed to subroutine START in ATRIB(1). Machine status is then set to not jammed, that is, $XX(M+6)=1.0$. A test is then made to determine if plugs are flowing to the machine and, if not, the packing status of the machine is not changed. If plugs are flowing to the machine, the status of the machine is set to packing by setting $XX(M)=1.0$. This change causes fewer plugs to be routed on the conveyor and a flow change to the next machine is scheduled to occur in 9 seconds by calling SCHDL with an event code of 3.

The coding for the stop machine event, STOPM, is shown in Figure 14-15. This event is called when a DETECT node determines that the number of parts produced exceeds the jam number defining the threshold. Machine status is set to idle and to jammed. The number of plugs packed since the last jamming is reset to 0 by setting SS(M) to 0. A potential start for the machine is scheduled through a call to subroutine SCHDL with event code 1 to occur in a uniformly distributed time between 6 and 24. When a machine is stopped, a flow change to the next machine is scheduled by a call to subroutine SCHDL with an event code of 3 to indicate a flow change is required.

```
SUBROUTINE START
COMMON/SCOM1/ATRIB(100),DD(100),DDL(100),DTNOW,II,MFA,MSTOP,NCLNR
1,NCRDR,NPRNT,NNRUN,NNSET,NTAPE,SS(100),SSL(100),TNEXT,TNOW,XX(100)
COMMON/UCOM1/FLOW(10),PFLOWR,NFLOW(10)
MACH=ATRIB(1)
XX(MACH+6)=1.0
IF(FLOW(MACH).LT.PFLOWR)RETURN
XX(MACH)=1.0
ATRIB(1)=ATRIB(1)+1
ATRIB(2)=-PFLOWR
CALL SCHDL(3,9.,ATRIB)
RETURN
END
```

Figure 14-14 Subroutine START for restarting a machine.

```
SUBROUTINE STOPM
COMMON/SCOM1/ATRIB(100),DD(100),DDL(100),DTNOW,II,MFA,MSTOP,NCLNR
1,NCRDR,NPRNT,NNRUN,NNSET,NTAPE,SS(100),SSL(100),TNEXT,TNOW,XX(100)
COMMON/UCOM1/FLOW(10),PFLOWR,NFLOW(10)
MACH=ATRIB(1)
XX(MACH)=0.0
XX(MACH+6)=0.0
SS(MACH)=0
CALL SCHDL(1,UNFRM(6.,24.,1),ATRIB)
IF(FLOW(MACH).LT.PFLOWR)RETURN
ATRIB(1)=ATRIB(1)+1.
ATRIB(2)=PFLOWR
CALL SCHDL(3,9.,ATRIB)
RETURN
END
```

Figure 14-15 Subroutine STOPM for stopping a machine due to jamming.

Figure 14-16 presents a flowchart of the flow change event, FLOWCHNG. The local variables M and CHNG are established as the machine number and the amount of change for which the flow change event is occurring. If M is greater than 5, no further machines are on the conveyor and XX(6), the rate into the barrel, is updated by the value of CHNG. Otherwise, the flow change is made on the conveyor before machine M. If machine M is jammed, the flow change is passed on down the conveyor to the next machine. If the flow change is positive and machine M is working, or if the flow change is negative but there is still flow to ma-

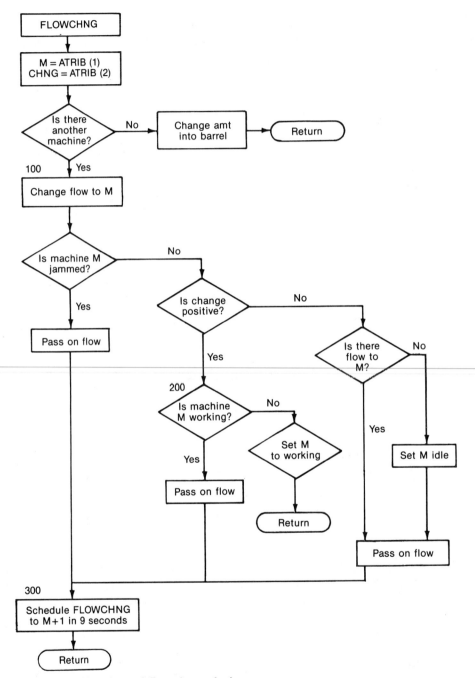

Figure 14-16 Flowchart of flow change logic.

chine M, the flow change is passed on to the next machine. If the flow change is positive and the machine is not working, then the flow change allows machine M to start working. If the flow change is negative and there is no flow to the machine, the flow change is also passed on to the next machine. This is necessary because the flow before a machine could be 0 but positive changes down the conveyor could have been scheduled from earlier flow changes. To further clarify this point, it is possible for all five machines to be working concurrently even though the flow rate on the conveyor can support a maximum of three machines working. This occurs, for example, when machines 1 and 2 are jammed and are then repaired in a short period of time so that machines 4 and 5 begin processing using the overflow during the times that machines 1 and 2 were jammed and are packing when machines 1 and 2 come back on line.

The FORTRAN coding for subroutine FLOWCHNG is shown in Figure 14-17. The code in Figure 14-17 follows the logic presented in the flowchart for subroutine FLOWCHNG.

The network input statements and control statements for this model are shown in Figure 14-18. Network statements are used to monitor and detect when the number of plugs packed at a machine exceeds a value specifying the jamming

```
      SUBROUTINE FLOWCHNG
      COMMON/SCOM1/ATRIB(100),DD(100),DDL(100),DTNOW,II,MFA,MSTOP,NCLNR
     1,NCRDR,NPRNT,NNRUN,NNSET,NTAPE,SS(100),SSL(100),TNEXT,TNOW,XX(100)
      COMMON/UCOM1/FLOW(10),PFLOWR,NFLOW(10)
      M=ATRIB(1)
      CHNG=ATRIB(2)
C*****IS THERE ANOTHER MACHINE?
      IF(M.LT.6)GO TO 100
C*****CHANGE AMOUNT INTO BARREL
      XX(6)=XX(6)+CHNG
      IF(XX(6).LT.0.0) XX(6)=0.0
      RETURN
C*****CHANGE FLOW TO M
100   FLOW(M)=FLOW(M)+CHNG
      IF(FLOW(M).LT.0.0) FLOW(M)=0.0
C*****IS MACHINE M JAMMED?
      IF (XX(M+6).LE.0.0)GO TO 300
C*****IS CHANGE POSITIVE?
      IF (CHNG.GT.0.0) GO TO 200
C*****IS THERE FLOW TO M?
      IF (FLOW(M).LE.0.001) XX(M)=0.0
      GO TO 300
C*****IS M WORKING?
200   IF (XX(M).GT.0.0) GO TO 300
      XX(M)=1.0
      RETURN
C*****SCHEDULE FLOWCHNG TO M+1
300   ATRIB(1)=M+1
      CALL SCHDL(3,9.,ATRIB)
      RETURN
      END
```

Figure 14-17 Implementation of flow change logic, subroutine FLOWCHNG.

threshold. Disjoint subnetworks for each machine are included. For machine 1, a DETECT node is used to detect SS(1) crossing SS(7) in the positive direction. SS(7) is a sample from a uniform distribution in the range XX(20) to XX(21). An INTLC statement is used to set XX(20) to 200 and XX(21) to 600. When the detection of this state-event occurs, an entity is sent to an ASSIGN node which identifies machine 1 as being jammed by setting ATRIB(1)=1. The next threshold value, SS(7), is then set and a transfer to the EVENT node with event code 2 is made. In event 2, machine 1 is stopped and SS(1) is reset to 0 since no plugs have been packed since the last jamming. Also, in event 2 the status of machine 1 is changed to idle by setting XX(1) to 0. This causes the rate of plugs packed by machine 1 to become 0 in the equation for SS(1) which is included in subroutine STATE.

```
 1   GEN,PRITSKER,CONT_CONV_FAIL,12/2/85,1,,N,Y,N;
 2   LIMITS,,2,100;
 3   TIMST,XX(6),RATE IN BAR;
 4   TIMST,XX(1),MTIME FOR 1;
 5   TIMST,XX(2),MTIME FOR 2;
 6   TIMST,XX(3),MTIME FOR 3;
 7   TIMST,XX(4),MTIME FOR 4;
 8   TIMST,XX(5),MTIME FOR 5;
 9   RECORD,TNOW,TIME,0,B,10,2500,3000;
10   VAR,SS(1),1,S1,0,4000;
11   VAR,SS(2),2,S2,-1000,3000;
12   VAR,SS(3),3,S3,-2000,2000;
13   VAR,SS(4),4,S4,-3000,1000;
14   VAR,SS(5),5,S5,-3000,1000;
15   VAR,SS(6),6,BARREL,0,1000;
16   INTLC,XX(20)=200,XX(21)=600;
17   NETWORK;
18        DETECT,SS(1),XP,SS(7),1;
19        ASSIGN,ATRIB(1)=1,SS(7)=UNFRM(XX(20),XX(21),2);
20            ACT,,,EVNT;
21        DETECT,SS(2),XP,SS(8),1;
22        ASSIGN,ATRIB(1)=2,SS(8)=UNFRM(XX(20),XX(21),2);
23            ACT,,,EVNT;
24        DETECT,SS(3),XP,SS(9),1;
25        ASSIGN,ATRIB(1)=3,SS(9)=UNFRM(XX(20),XX(21),2);
26            ACT,,,EVNT;
27        DETECT,SS(4),XP,SS(10),1;
28        ASSIGN,ATRIB(1)=4,SS(10)=UNFRM(XX(20),XX(21),2);
29            ACT,,,EVNT;
30        DETECT,SS(5),XP,SS(11),1;
31        ASSIGN,ATRIB(1)=5,SS(11)=UNFRM(XX(20),XX(21),2);
32   EVNT EVENT,2;
33        TERM;
34        ENDNETWORK;
35   INIT,0,6000;
36   CONTINUOUS,0,11,.1;
37   FIN;
```

Figure 14-18 Control and network statements for continuous flow conveyor model.

The control statements for this model show that there are no files used and that there are 2 attributes per entity. Statistics are collected on the utilization of the machines as specified on the TIMST statements. The average rate of plugs into the barrel is also obtained in this fashion. A plot and table of the number of plugs packed since the last jamming and the number of plugs going into the barrel is obtained through specifications on RECORD and VAR statements. Note that the scale for the dependent variables has been set so that the number of plugs packed since the last jamming appears in different sections of the plot. The INIT statement sets the run length at 6000 seconds. The CONTINUOUS statement indicates that there are no differential equations, 11 state variables, and that the minimum step size is to be one tenth of a second. This model of the conveyor system illustrates the use of network, discrete event and continuous modeling viewpoints.

Summary of Results. The SLAM II summary report for this example is shown in Figure 14-19. The flow of the sparkplugs through the packing machines and into the barrels is shown in Figure 14-20. The time-persistent statistics provide the machine busy time and the rate of sparkplugs into the barrel. The rate of plugs into the barrel is about one plug every 8 seconds. This value is obtained from the reciprocal of the estimate of the mean rate into the barrel. The plug rate into the barrel can also be obtained by dividing the value of SS(6) by the time period for the simulation, that is, 740 by 6000 seconds. This small rate into the barrel is due to having sufficient machines available to pack plugs. The utilization time for machine 5 is a low 11 percent. The utilization for machine 4 is also low, less than 40 percent, which indicates that a design change should be considered.

If we remove machine 5, the average rate of plugs into the barrel will increase by the product of its packing rate and its utilization. Removal of machine 5 does not change the utilization of the other machines. The removal of machine 4 will change the utilization of machine 5. To some extent, the operation of the sparkplug line employs the conveyor as a buffer area for the machines. By removing machine 4, the buffer area on the conveyor for machine 5 can contain up to 18 seconds supply of plugs for processing by machine 5. This will increase the utilization of machine 5 but will also increase the rate of flow into the barrel.

The utilization of machines 1, 2 and 3 are all about 80 percent. Intuitively, the utilization of these machines should be statistically the same as the input rate to the machines after the initial warmup period is 5.55 plugs per second. The variation in utilization values is due to different settings of plug thresholds and the random generation of machine unjamming times. An expected value analysis of the machine utilization time shows that machine utilization should be approximately 82 percent. This value is arrived at by dividing the expected jamming threshold,

400 plugs, by the plug packing rate of 5.55. This yields an expected time until jamming of approximately 72 seconds. The expected time to unjam a machine is 15 seconds which results in a working plus idle cycle for a machine of 87 seconds. Since the input flow on the conveyor is always available to machines 1, 2 and 3, the utilization of these machines should be approximately 72 divided by 87 or 82.8 percent. The values presented in Figure 14-19 are not significantly different from this value.

```
                 S L A M   I I   S U M M A R Y   R E P O R T

           SIMULATION PROJECT CONT_CONV_FAIL        BY PRITSKER

           DATE 12/ 2/1985                          RUN NUMBER   1 OF    1


           CURRENT TIME   0.6000E+04
           STATISTICAL ARRAYS CLEARED AT TIME  0.0000E+00


                 **STATISTICS FOR TIME-PERSISTENT VARIABLES**

                      MEAN        STANDARD    MINIMUM     MAXIMUM     TIME        CURRENT
                      VALUE       DEVIATION   VALUE       VALUE       INTERVAL    VALUE

      RATE IN BAR     0.1233E+00  0.8247E+00  0.0000E+00  0.1110E+02  0.6000E+04  0.0000E+00
      MTIME FOR 1     0.8246E+00  0.3803E+00  0.0000E+00  0.1000E+01  0.6000E+04  0.1000E+01
      MTIME FOR 2     0.8042E+00  0.3968E+00  0.0000E+00  0.1000E+01  0.6000E+04  0.1000E+01
      MTIME FOR 3     0.8267E+00  0.3785E+00  0.0000E+00  0.1000E+01  0.6000E+04  0.0000E+00
      MTIME FOR 4     0.3961E+00  0.4891E+00  0.0000E+00  0.1000E+01  0.6000E+04  0.0000E+00
      MTIME FOR 5     0.1115E+00  0.3147E+00  0.0000E+00  0.1000E+01  0.6000E+04  0.0000E+00


                 **STATE AND DERIVATIVE VARIABLES**

                 (I)      SS(I)        DD(I)
                  1     0.2583E+03   0.0000E+00
                  2     0.3737E+03   0.0000E+00
                  3     0.0000E+00   0.0000E+00
                  4     0.2089E+03   0.0000E+00
                  5     0.2123E+03   0.0000E+00
                  6     0.7396E+03   0.0000E+00
                  7     0.5693E+03   0.0000E+00
                  8     0.4056E+03   0.0000E+00
                  9     0.5438E+03   0.0000E+00
                 10     0.4526E+03   0.0000E+00
                 11     0.4939E+03   0.0000E+00
```

Figure 14-19 Summary report for continuous flow conveyor model

```
                              **PLOT NUMBER  1**
                              RUN  NUMBER  1

                                SCALES OF PLOT
1=S1        0.0000E+00         0.1000E+04        0.2000E+04        0.3000E+04        0.4000E+04
2=S2       -0.1000E+04         0.0000E+00        0.1000E+04        0.2000E+04        0.3000E+04
3=S3       -0.2000E+04        -0.1000E+04        0.0000E+00        0.1000E+04        0.2000E+04
4=S4       -0.3000E+04        -0.2000E+04       -0.1000E+04        0.0000E+00        0.1000E+04
5=S5       -0.3000E+04        -0.2000E+04       -0.1000E+04        0.0000E+00        0.1000E+04
6=BARREL    0.0000E+00         0.2500E+03        0.5000E+03        0.7500E+03        0.1000E+04

          0   5   10  15  20  25  30  35  40  45  50  55  60  65  70  75  80  85  90  95  100 DUPLICATES

TIME

0.2500E+04  +      1            2       6            +         3         5      4            +
0.2510E+04  1         11        2       6            +        33         5     44            +
0.2520E+04  1                  22       6            +        33         5     44            +
0.2530E+04  11                + 2       6            +          33       5      4            + 54
0.2540E+04  +11               + 22      6           3           3        4                  + 45
0.2550E+04  + 111             +  22     6          33                    444                + 45
0.2560E+04  +    11           +    22 6            + 3                    +54                + 45
0.2570E+04  +     11          2     2 6            + 33                   + 4                + 45
0.2580E+04  +        1        2       6            +  3                   + 4                + 45
0.2590E+04  +       11        2       6            +   33                 + 44               + 45
0.2600E+04  +         1       +2      6            +     3                + 5 4              +
0.2610E+04  1          1      + 2     6            +     3                + 5 4              +
0.2620E+04  1               + 222     6            +        33            + 5 4              +
0.2630E+04  11               2    22  6            +        33            + 5 4              +
0.2640E+04  +11              2       6            3           3           + 5 44             +
0.2650E+04  + 111           22       6            3                       + 5    44          +
0.2660E+04  +    11         +22      6            3                       4 5     4          +
0.2670E+04  +     11        + 22   666            333                     4 55               +
0.2680E+04  1        1      + 22     66 66        + 33                    4  55              +
0.2690E+04  1               2    22       6       + 33                    4   5             +
0.2700E+04  +11             2            6        3  33                   4   5             +
0.2710E+04  + 111           222          6        3                       444  5            +
0.2720E+04  +    11         + 22         6        3                       + 44 55           +
0.2730E+04  +     1         + 2        666        +3                      +  44 55          +
0.2740E+04  +     1         +  2         6        + 3                     +   4 5           +
0.2750E+04  +      1        +   2        6        +  3                    +   4 5           +
0.2760E+04  +      1        +    2       6        +  3                    +   4 5           +
0.2770E+04  +       1       +     2      6        +   3                   +   4 5           +
0.2780E+04  1         1     +     22     6        +    3                  +   4 5           +
0.2790E+04  1               2      2     6        +    3                  +   4 5           +
0.2800E+04  +11             2            6        +      33               +   4 5           +
0.2810E+04  + 11            22           6        +       333             +  44 5           +
0.2820E+04  +    1          +22          6        3         33            +   455           +
0.2830E+04  +     11        + 22         6        3                       +    45           + 54
0.2840E+04  +     11        +  22        6        33                      +    44           + 45
0.2850E+04  1        1      +   22       6        + 3                     +    54           +
0.2860E+04  1               +    2       6        +  3                    +    54           +
0.2870E+04  +1              +     3      6        +   3                   +    54           +
0.2880E+04  + 11            +      2     6        +    3                  +    54           +
0.2890E+04  +    1          +       2    6        +     3                 +    5 4          +
0.2900E+04  +     1         +      22    6        +     3                 +    5 4          +
0.2910E+04  +     11        +       22   6        +       33              +    5 4          +
0.2920E+04  +      1        2          26         +       33              +    5 4          +
0.2930E+04  +     11        2            6        3          3            +    5 4          +
0.2940E+04  1         1     2            6        3                       +    5  4         +
0.2950E+04  1               +2           6        3                       +    55  4        +
0.2960E+04  +11             + 222        6        +33                     +     555 4       +
0.2970E+04  + 111           +  22        6        + 33                    +      5 44       +
0.2980E+04  +    11         +   22       6        + 33                    +      5   4      +
0.2990E+04  +     1         2     2      6        +  3                    +      5   4      +
0.3000E+04  +      1        2            6        +  3                    +      5   4      +

          0   5   10  15  20  25  30  35  40  45  50  55  60  65  70  75  80  85  90  95  100 DUPLICATES

TIME

        OUTPUT  CONSISTS  OF      364 POINT SETS (   2184 POINTS)
        STORAGE ALLOCATED FOR    1300 POINT SETS (   9100 WORDS )
        STORAGE  NEEDED  FOR      364 POINT SETS (   2548 WORDS )
```

Figure 14-20   Plot of state variables for continuous flow conveyor model

14.7   CHAPTER SUMMARY

This chapter summarizes the modeling approaches of SLAM II and demonstrates how SLAM II supports alternative world views. Two examples of combined modeling are presented. The first example involves the heating of ingots in a soaking pit furnace, and the second example describes the modeling of a conveyor system using continuous variables.

14.8   EXERCISES

14-1.   A drag of ingots arrives at the soaking pit on the average every 1.75 hours. The distribution of times between arrivals is exponential. A drag consists of ten ingots. All ten ingots in the drag have the same temperature which is lognormally distributed between 300°F and 600°F with a mean of 400°F and standard deviation of 50°F. Upon arrival, the ingots may be charged directly into an available pit or they may be placed in a waiting-to-be-charged queue. The temperature of the arriving ingots decreases with the square root of the time they must wait, that is, temperature = old temperature−SQRT(waiting time) * 157°. When an ingot reaches a temperature of 150°F, it is assumed cold. Charging practice dictates that cold ingots may not be charged into any pit at a temperature of greater than 750°F since this causes surface cracks and, hence, poor quality. Hot ingots (over 150°F) may be charged into any temperature pit. It is assumed that the pit temperature after charging is equal to a weighted average of the pit and ingot temperatures (new pit temperature = 0.7* old pit temperature + 0.3* temperature of ingots in drag), and from then on varies according to the equation $dT/dt = 2600-T$. When the pit temperature reaches 2200°F, the ingots are ready to soak and the pit temperature remains at this temperature for 2 hours. After the soak time, the ingots are removed and the pit is available to be reloaded. If nothing can be loaded, the pit temperature changes according to the equation $dT/dt = 600-T$. There are four independent pits and each pit has a capacity of five ingots. (Note: Ingots can be processed in units of 5 which saves on computer storage and processing.) The initial conditions for the simulation are:
1. Pit temperatures are 1150°F, 800°F, 1500°F, and 1000°F.
2. The pits whose temperatures are 1150°F and 800°F are loaded with 5 ingots. The other pits are not charging.
3. The first drag is scheduled to arrive at time zero. Simulate the soaking pit system for 400 hours to obtain estimates of the following quantities: heating time for ingots; waiting time for ingots; the number of ingots waiting (total and by temperature class); and the number of loaded pits.

Embellishments:

(a) Clear statistics at time 100 and obtain summary reports at times 200, 300 and 400 hours.

(b) Give priority to processing the cold ingots at times 200 and 300, that is, if cold ingots are waiting assign pit(s) to process the cold ingots.

(c) Evaluate the policy of making one of the pits a processor of cold ingots.

(d) Simulate the system if the sum of the temperatures of the pits is restricted to 7000°F due to energy considerations.

14-2. Model and simulate a gas station attendant providing service to arriving customers. The time between arrivals of customers is exponentially distributed with a mean value of 4 minutes. The service station employs one attendant whose service rate is dependent on the number of customers waiting for his service. The relationship between service rate and number in queue is given below

$$S = 1.2\exp(0.173N) \qquad\qquad N < 5$$
$$S = 1.2(0.25\exp(-0.305(N-5))+0.5) \quad N \geqslant 5$$

where N is the number of customers in the queue.

The nominal amount of service time required by each customer is an exponentially distributed random variable with a mean of 3.5 minutes. Analyze this queueing situation for 1000 minutes to obtain statistics on the following quantities: the time a customer spends in the queue, the time a customer spends at the gas station, the fraction of time the attendant is busy, and the number of customers waiting in the queue of the server. A plot is to be made that illustrates the status of the server, the number in the queue, and the remaining service time for a customer (7,8).

14-3. For the continuous flow conveyor example, assume that the time until the next jam on a machine is uniformly distributed between 200/5.55 and 600/5.55 seconds. Discuss how this information could be used in modeling the sparkplug situation.

14-4. Build a model of the sparkplug packing line in which the sparkplugs arrive in trays every 20 seconds and each tray contains 333 sparkplugs. Arrivals are to the first machine with the first arrival scheduled to occur at 18 seconds. When a machine becomes jammed, the sparkplugs remaining in the tray are placed back on the conveyor and routed to the next machine. Perform a simulation of this situation for 6000 seconds and compare the results with those obtained in Example 14-2.

14-5. Redesign the sparkplug packaging line presented in Example 14-2 to eliminate one of the machines. The machines may be placed in any location along the conveyor belt. The input rate of sparkplugs to the first packing machine changes every 1000 seconds and is a sample from the following probability mass function: (0.25,888); (0.5,999); and (0.25,1110).

14-6. For the drive-in bank example with jockeying, Example 9-1, the system description is enlarged to include a limited waiting area for cars departing the bank into the street. Space for five cars exists between the tellers and the street. A bank customer will pull out into the street traffic if a gap of 0.4 time units is perceived in the street traffic until the next car arrives. However, as the customer waits, shorter gaps become acceptable. For wait times of .25, .50, .75, and 1.0 time units, the gap time that is acceptable is 0.4, 0.3, 0.2, and 0.1 time units. If a gap is not acceptable, the customer waits for an acceptable gap. The time between arrivals of cars on the street is exponentially distributed with a mean of 0.3 time units. The departure point from the bank is located within six car lengths of a traffic light. Thus, if six cars are waiting for the light to turn green, the bank exit is blocked. The traffic light has a red/green cycle that stays green for 0.65 time units and then red for 0.5 time units. Assume cars will only go through the light when the light is green. The time to pass through the light is 0.125 time units when a car is waiting for the light to turn green. Time to pass through the light decreases by 80 percent for each successive car waiting in the line. Develop a model that embellishes the bank teller system with jockeying to include the features described above. Consider the time in the system for the bank customer to be from the time of arrival to the bank until the customer passes through the street traffic light. Determine the number of times a customer's anxiety level results in an acceptance of a gap that is less than 0.25 time units.

14-7. A flexible manufacturing system performs machining operations on castings. The FMS consists of ten horizontal milling machines which can perform any of three operations. Castings arrive to the FMS every 22 minutes. For each casting it is necessary to perform operations 10, 20, and 30 and the processing time for these operations are 120, 40, and 56 minutes respectively. The ten milling machines can be dedicated to the performing of one or more operations. A flexible milling machine can do all three operations. One cart is available to move the castings to the milling machines and the central inspection station. The time for the cart to move from the inspection station to station 1 is 0.3 minutes. This is also the time for the cart to move from machine 10 to the inspection station. Movement from mill 5 to mill 6 is around a bend and this takes 0.6 minutes. Movement between all other stations takes 0.2 minutes. The cart always travels from the inspection station to station 1 and then back around to the inspection station. After each operation, the casting is inspected and the inspection time is 10 minutes. There are two inspectors available to inspect the castings. Castings arrive to the inspection station and are immediately assigned a destination upon arrival. Develop a model of this system to determine the throughput of castings for two 40 hour shifts.

14.9   REFERENCES

1. Ashour, S. and S.G. Bindingnavle, "An Optimal Design of a Soaking Pit Rolling Mill System," *Simulation,* Vol. 18, 1972, pp. 207-214.
2. Cellier, F.E., and A.A.B. Pritsker, "Teaching Continuous Simulation Using GASP", *Simulation,* Vol. 34, 1980, pp. 137-139.
3. Duket, S.D. and C.R. Standridge, "Applications of Simulation: Combined Models", *Modeling,* Issue No. 19, December 1983, pp. 20-29.
4. Golden, D.G. and J.D. Schoeffler, "GSL-A Combined Continuous and Discrete Simulation Language," *Simulation,* Vol. 20, 1973, pp. 1-8.
5. Henriksen, J.O., "GPSS - Finding the Appropriate World-View," *Proceedings, 1981 Winter Simulation Conference,* 1981, pp. 505-516.
6. Pritsker, A.A.B., *The GASP IV Simulation Language,* John Wiley, 1974.
7. Pritsker, A.A.B., and R.E. Young, *Simulation with GASP-PL/I,* John Wiley, 1975.
8. Pritsker, A.A.B., "Three Simulation Approaches to Queueing Studies Using GASP IV", *Computers and Industrial Engineering,* 1976, Vol. 1, pp. 57-65.

CHAPTER 15

Simulation Languages

15.1   INTRODUCTION

The widespread use of simulation as an analysis tool has led to the development of a number of languages specifically designed for simulation. These languages provide specific concepts and statements for representing the state of a system at a point in time and moving the system from state to state. In this chapter, we describe other simulation languages and discuss their relationship to the modeling framework of SLAM II. The intent is not to provide manuals on how to write programs in these languages, but to illustrate the similarities and differences between the various simulation languages.

Discussions will center on GPSS and SIMSCRIPT as these languages along with SLAM II are the most widely used languages available. For continuous simu-

559

lation languages, the discussion will be kept to a minimum as the general orientation of the various continuous languages is similar.

15.2 GPSS

GPSS (General Purpose Simulation System) is a process-oriented simulation language for modeling discrete systems. GPSS exists in a number of dialects such as GPSS/360, GPSS V(15) and GPSS/H(16). GPSS/H is more a statement language and does not rely on the block orientation as much as the earlier versions. We limit the discussion to fundamental constructs. Schriber's book, *Simulation with GPSS (38),* is an excellent text for learning GPSS.

The principal appeal of a GPSS is its seemingly modeling simplicity. A GPSS model is constructed by combining a set of standard blocks into a block diagram which defines the logical structure of the system. Entities are represented in GPSS as transactions which move sequentially from block to block as the simulation proceeds. Learning to write a GPSS program consists of learning the functional operation of GPSS blocks and how to logically combine the blocks to represent a system of interest.

The GPSS processor interprets and executes the block diagram description of a system. In many implementations, the language is limited in computing power and lacks a capability for floating point or real arithmetic. For such implementations, the GPSS simulation clock is integer valued. This means that changes in the state of the system can occur only at integer points in time. For example, if we model a single-channel queueing system and we select our unit of time as minutes, then the time between arrivals and the service time for the customers must be an integer number of minutes. Another consequence of an integer valued clock is that simultaneous events frequently occur, and therefore the tie-breaking mechanism takes on added significance. In GPSS, this problem is addressed by assigning each transaction a special attribute called a priority which can be any integer between 0 and 127 with higher values assuming greater priority.

There are over forty basic blocks in GPSS, and in some implementations over 100 statement types. Each block is pictorially represented by a stylized figure which is intended to be suggestive of the operation of the block. Fifteen of the more commonly used blocks are shown in Table 15-1. For each block, there is a corresponding SLAM II network element which performs a similar function. The correspondence is not exact, however, and in several cases elements of GPSS and SLAM II which have the same names perform different functions.

A GPSS block diagram must be translated by the modeler into an equivalent statement form for interpretation and execution by the GPSS processor. Each block specification consists of three categories of information. The first is a symbolic location name and is used as a block label similar to the node labels in SLAM II. The second category of information is the operation and is a verb suggestive of the function of the block. The third category of information is the block's *operands* which provide specific information concerning the operation of the block. The number and interpretation of the operands depends upon the specific block type and are denoted in general by the letters A, B, C, D, E, F, and G. With this brief introduction, we will now describe the operation of the blocks shown in Table 15-1.

Transactions are created in GPSS by use of the GENERATE block with operands A, B, C, D, E, F, and G. The A operand specifies the mean time between creations. If the B operand is specified as a number, then the time between creations is uniformly distributed in the range from A−B to A+B. For example, if the A and B operands are specified as 10 and 2, respectively, then the time between the creations is uniformly distributed over the integers 8, 9, 10, 11, and 12. We defer until later the discussion of the meaning of a nonnumeric B operand. The C operand specifies the time of the first transaction creation and is referred to as the offset interval. The D operand prescribes a limit on the number of transactions which can enter the model through a given GENERATE block. Each transaction created at the GENERATE block has a priority specified by operand E and has F associated parameters where G specifies the type of parameters. The CREATE node in SLAM II is analogous to the GENERATE block.

Transactions are destroyed in GPSS by a TERMINATE block. In addition to destroying each arriving transaction, the TERMINATE block also reduces the termination counter by the value specified as its A operand. The termination counter is an integer value which is initially specified by the A operand of the GPSS control card START. As transactions arrive to TERMINATE blocks, the counter is decremented accordingly. As soon as the counter is reduced to zero or less, the simulation stops. Note that although there may be many TERMINATE blocks in a model, there is only one termination counter. This differs from the TERMINATE node of SLAM II.

Time advance is provided for in GPSS by the ADVANCE block. When a transaction enters the ADVANCE block, its progress is delayed by the time specified by the A and B operands. The A and B operands of the ADVANCE block are defined in the same manner as the A and B operands of the GENERATE blocks. The A operand prescribes the mean delay time and the B operand, when specified as a number, defines the half-width of a uniform distribution.

Table 15-1 Basic GPSS blocks and their SLAM II equivalents.

Block Symbol	Functional Description of Block	SLAM II Network Element Providing Analogous Function
GENERATE A,B,C,D,E,F,G	Creates transactions as prescribed by the operands A, B, C, D, E, F, and G	CREATE node
TERMINATE A	Destroys the arriving transaction and reduces the termination counter by A	TERMINATE node
ADVANCE A,B	Advances simulated time as prescribed by operands A and B	ACTIVITY
SEIZE A	Causes transaction to await and capture facility A	AWAIT node
RELEASE A	Frees facility A	FREE node
ENTER B A	Causes transaction to await and capture B units of storage A	AWAIT node
LEAVE A B	Frees B units of storage A	FREE node

Table 15-1 (continued).

Block Symbol	Functional Description of Block	SLAM II Network Element Providing Analogous Function
QUEUE A B	Increments the number in queue A by B units	†
DEPART A B	Decrements the number in queue A by B units	†
A, B, C ASSIGN	Assigns the value specified as B to parameter number A of the transaction	ASSIGN node
MARK A	Assigns the current clock time to parameter number A of the transaction	ASSIGN node
SAVEVALUE A, B, C	Assigns the value specified as B to savevalue location A	ASSIGN node
TRANSFER A (C) (B)	Causes a transfer to location C with probability A, and location B with probability 1-A	ACTIVITY
A X B TEST (C)	Causes a transfer to location C if A is not related to B according to operator X	ACTIVITY
TABULATE B A	Records an observation for the variable prescribed in table A	COLCT node

† Queue statistics are automatically recorded in SLAM II.

To generate samples from distributions other than the uniform distribution, a user-written table function must be included in the model. A function is defined by a function header card and one or more function follower cards. The function header card specifies the name of the function, the random number stream, and the number of points specified for the distribution. The function follower cards follow immediately after the function header card and contain the entries $X_1,Y_1/X_2,Y_2/$ . . . $/X_n,Y_n$ where X_i and Y_i are the ith cumulative probability and associated function value, respectively.

A sample from a table function is obtained by specifying the name or number of the function as the A operand and defaulting the B operand of the GENERATE or ADVANCE block. The entry FN1 would be used to reference function number 1 and the entry FN$XPDIS would be used to reference a function named XPDIS. To simplify sampling from an exponential distribution, a special provision is made whereby if the B operand is specified as a function, then the sample is taken as the product of A and B. Thus, if the modeler includes a table function for the exponential distribution† with unit mean and prescribes it as the B operand, then a sample from a general exponential distribution can be obtained by specifying the mean of the desired exponential distribution as the A operand. In SLAM II, time advance is modeled using an activity.

In GPSS, resources are modeled as either facilities or storages. Facilities have a capacity of one and need not be defined in a declarative statement. Storages may have a capacity greater than one. The capacity of a storage is defined using a STORAGE definition card. There are two distinctions made between facilities and storages in addition to capacity. Facilities may be preempted using a PREEMPT block whereas storages cannot. Secondly, a facility can only be released by the transaction which seized it. In contrast, any transaction can be used to release one or more units of a storage.

Facilities are seized and released by transactions passing through SEIZE and RELEASE blocks, respectively. The A operand specifies the name of the facility being seized or released. When a transaction arrives to a SEIZE block, it attempts to enter the block and capture the facility. If the facility is busy, the transaction's progress is delayed until the facility becomes free as the result of a transaction passing through a RELEASE block.

The ENTER and LEAVE blocks perform analogous functions for capturing and freeing storages. When a transaction arrives at an ENTER block, it is delayed until B units of storage A are available. The transaction then captures

---

† An approximation of the exponential distribution using a table function is required because there is no logarithm function in GPSS.

the B units of storage A and continues. Similarly, when a transaction arrives to the LEAVE block, it frees B units of storage A. The correspondence in SLAM II for the SEIZE and ENTER blocks is the AWAIT node. The SLAM II FREE node is similar to the RELEASE and LEAVE blocks of GPSS.

Statistics on waiting time and queue length for transactions waiting for facilities or storages can be obtained by using the QUEUE and DEPART blocks. An entity arriving to a QUEUE block causes the number of units in queue A to be increased by B. Likewise, an entity arriving to a DEPART block causes the number of units in queue A to be decreased by B. In both cases, the default value for B is 1. Note that the QUEUE block in GPSS and QUEUE node in SLAM perform distinctly different functions. No waiting of transactions occurs at a QUEUE block; its function is simply to record the number of units entering the queue.

During a simulation, the GPSS processor automatically maintains certain variables which describe the status of the system. These variables are collectively referred to as standard numerical attributes (SNA). A partial listing of SNAs is shown in Table 15-2. If a symbolic name is employed for an SNA, it must be preceded by the symbol $. For example, the GPSS variable F$SRVR denotes the status of the facility which has the name SRVR.

Table 15-2 Selected GPSS standard numerical attributes.

C1	The current value of clock time
Fn	The current status of facility number n.
M1	The transit time of a transaction.
Nn	The total number of transactions that have entered block n.
Pn	Parameter number n of a transaction.
Qn	The length of queue n.
Rn	The space remaining in storage n.
Sn	The current occupancy of storage n.
Wn	The number of transactions currently at block n.
Xn	The value of savevalue location n.

Transactions in GPSS can be assigned attributes which are referred to as parameters. The GPSS variable P_j is used to reference parameter number j of a transaction. Hence, the GPSS variable P_j is equivalent to the SLAM II variable ATRIB(j). The ASSIGN block is used to assign values to parameters of transactions. The ASSIGN block causes the value specified as the B operand to be assigned to the parameter number specified in the A operand. The B operand can be an integer or any SNA. If a + or − sign is appended to the A operand, then the value specified by the B operand is added to or subtracted from the current value of parameter number A. For example, the statement

ASSIGN   P1+,1

would increment parameter number 1 by 1. A separate block called the MARK block is provided for assigning the value of the current time to a parameter. The A operand of the MARK block specifies the parameter number to which the value of current time is to be assigned.

In some cases, it is desired to record values as global variables rather than as attributes of an entity. In SLAM II, this is done by assigning values to XX or ARRAY. In GPSS, this is accomplished by employing the SAVEVALUE block which assigns the value specified by the B operand to savevalue location number A. If a + or − sign is appended to the A operand, then the value specified by the B operand is added to or subtracted from the existing contents of savevalue location A, respectively.

For arithmetic operations beyond the decrementing and incrementing discussed above, the modeler must employ the VARIABLE definition card. This card type can be thought of as defining a user-specified standard numerical attribute. The variable definition card is specified by entering the variable name in the location field, the word VARIABLE in the operation field,† and the arithmetic expression which defines the variable in the operands field. The arithmetic operators available in GPSS are addition (+), subtraction (−), multiplication (*), integer division (/), and modulus division (@).

GPSS provides a series of blocks which are used for directing a nonsequential flow of transactions through the block diagram. The TRANSFER block provides a probabilistic branching capability by causing the arriving transaction to be routed to statement location C with probability A and to statement location B with probability 1-A. If A is defaulted, the TRANSFER block causes deterministic branching to statement location B. Transaction routing can also be based on the system status by use of the TEST block. In addition to the normal operands, this block employs an auxiliary operator, X. The TEST block accepts a transaction if operand A is related to operand B according to relationship X; otherwise the transaction is routed to the location specified by the C operand. If the C operand is not specified, the block operates in the refusal mode and

---

† The use of the word FVARIABLE causes the expression to be calculated and only the result is integerized.

the transaction is held until the condition is satisfied. The permissible codes for the auxiliary operator X are:

G   Greater than
GE   Greater than or Equal
E   Equal
NE   Not Equal
LE   Less than or Equal
L   Less than

These GPSS conditional transfer capabilities are a subset of the conditional branching capabilities available in SLAM.

Observation type statistics are recorded in a GPSS model at TABULATE blocks. The purpose of this block is analogous to the function of the COLCT node in SLAM II. The A operand of the TABULATE block specifies the name of a TABLE card whose A operand defines the variable for which observation statistics are recorded at each transaction arrival to the TABULATE block. Values corresponding to the SLAM II variables HLOW, HWID, and NCEL are specified as the B, C, and D operands of the TABLE card, respectively. The variable for which statistics are to be recorded can be any of the SNA's. The special variable MP_j is provided for collecting interval statistics using the time marked in parameter number j. Hence, by using a MARK block and TABULATE block in conjunction with a TABLE card, the transit time between any two points can be recorded.

To illustrate the modeling features of GPSS, we will present a GPSS model of the following single-server queueing system. Customers arrive with a mean inter-arrival time of 20 minutes, exponentially distributed. The service time for each customer is uniformly distributed between 10 and 25 minutes. Statistical estimates are desired of the mean queue length, utilization of the server, and time in system.

The GPSS model for this illustration is given in Figure 15-1. The GPSS coding is presented in Figure 15-2. The first segment of the program models the flow of customers through the system. The order in which the blocks appear corresponds to the sequence of steps through which each customer progresses. We will describe the model with the coding lines given in parentheses. The transactions representing customers are created at the GENERATE block (line 14) with the time between transactions specified as the product of 20 and function XPDIS. The function XPDIS is entered (lines 4-8) as an approximation to the exponential distribution with unit mean. A transaction proceeds to the MARK block (line 15) where the current clock time is recorded as parameter number 1.

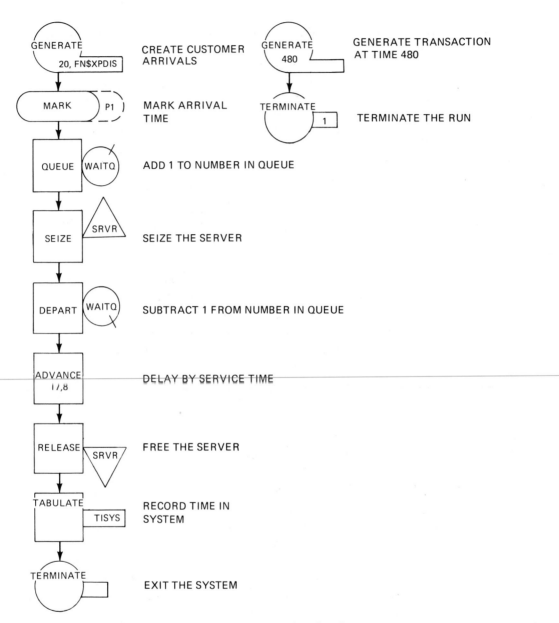

CREATE CUSTOMER ARRIVALS

GENERATE TRANSACTION AT TIME 480

MARK ARRIVAL TIME

TERMINATE THE RUN

ADD 1 TO NUMBER IN QUEUE

SEIZE THE SERVER

SUBTRACT 1 FROM NUMBER IN QUEUE

DELAY BY SERVICE TIME

FREE THE SERVER

RECORD TIME IN SYSTEM

EXIT THE SYSTEM

Figure 15-1 GPSS model of a single-server queueing situation.

The transaction then enters the QUEUE block (line 16) where the number in queue WAITQ is incremented by 1. When the facility SRVR becomes free, it is

```
1       *
2               SIMULATE
3       *
4       XPDIS FUNCTION    RN1,C24
5       0.0,0.0/0.1,0.104/0.2,0.222/0.3,0.355/0.4,0.509/0.5,0.69
6       0.6,0.915/0.7,1.2/0.75,1.38/0.8,1.6/0.84,1.83/0.88,2.12
7       0.9,2.3/0.92,2.52/0.94,2.81/0.95,2.99/0.96,3.2/0.97,3.5
8       0.98,4.0/0.99,4.6/0.995,5.3/0.998,6.2/0.999,7/0.9997,8
9       *
10      TISYS TABLE       MP1,0,5,20
11      *
12      *       MODEL SEGMENT
13      *
14              GENERATE    2,0,FN$XPDIS        CREATE CUSTOMER ARRIVALS
15              MARK        P1                  MARK ARRIVAL TIME
16              QUEUE       WAITQ               ENTER THE WAITING LINE
17              SEIZE       SRVR                SEIZE THE SERVER
18              DEPART      WAITQ               EXIT THE WAITING LINE
19              ADVANCE     17,8                DELAY BY SERVICE TIME
20              RELEASE     SRVR                FREE THE SERVER
21              TABULATE    TISYS               RECORD TIME IN SYSTEM
22              TERMINATE                       EXIT THE SYSTEM
23      *
24      *       TIMING SEGMENT
25      *
26              GENERATE    480                 CREATE TRANSACTION AT TIME 480
27              TERMINATE   1                   TERMINATE THE RUN
28      *
29      *       CONTROL CARDS
30      *
31              START       1                   START THE RUN
32              END
```

Figure 15-2 GPSS coding of single channel queueing problem.

seized at the SEIZE block (line 17) by the first waiting transaction which then continues to the DEPART block where the number in queue WAITQ is decremented by 1. The service activity is modeled by the ADVANCE block (line 19) which delays the transaction by a sample from a uniform distribution between 9 and 25.† At the completion of service, the transaction is routed to the RELEASE block where the facility SRVR is freed. The transaction then continues to the TABULATE block (line 21) which causes the transaction's transit time to be recorded as prescribed in table TISYS. The transaction is then destroyed at the TERMINATE block (line 22).

The second segment of the model is used to control the duration of the simulation. The GENERATE block (line 26) creates a transaction at time 480. This transaction continues to the TERMINATE block (line 27) where it is destroyed and reduces the termination counter by 1. The termination counter was initially

---

† The (A−B,A+B) range format of GPSS does not allow an even number of points. Therefore, the range from 10 to 25 specified in the problem statement cannot be obtained.

set to 1 on the START card (line 31); hence, the transaction arrival causes the simulation to stop at time 480. The timing segment performs the same function as entering a value for TTFIN on the INITIALIZE statement in SLAM II.

Although this example employs only a small subset of the GPSS blocks, it does illustrate the general modeling framework of GPSS. The simplicity of the block orientation of GPSS has made it one of the more popular discrete simulation languages. The commonly cited disadvantages of the language are the lack of computational capability, the difficult procedures for sampling from non-uniform distributions, and long computer execution times. These disadvantages have been alleviated by GPSS/H(16).

15.3   OTHER PROCESS-ORIENTED LANGUAGES

In recent years, many process-oriented simulation languages that employ a statement approach have been developed. The best known of these languages is SIMULA which is a superset of ALGOL. SIMULA has many statements that make it attractive for performing discrete event simulation including advanced list processing capabilities. In addition, it has the capability to be extended and has been augmented to allow the combined modeling concepts of GASP IV and network techniques of GERT to be used within the SIMULA framework (44). Little use of SIMULA has been made in the United States although it has received considerable attention in Europe. A graphic interface for SIMULA called DEMOS has been developed (4). For detailed discussions of SIMULA, we refer the reader to Hills (17), Birtwistle (3), and Franta (14).

Other process-oriented languages that have been developed are: Q-GERT (30), SOL (20); ASPOL (1); INS (33,34); and BOSS (35). In addition, process-oriented features have been added to SIMSCRIPT (36) and GASP (45).

15.4   SIMSCRIPT II

SIMSCRIPT II is a computer language developed by Kiviat, Villanueva, and Markowitz (19). A history of SIMSCRIPT developments is given by Markowitz

(23). The discrete simulation modeling framework of SIMSCRIPT† is primarily event-oriented. The modeling framework of SIMSCRIPT and the event portion of SLAM II are conceptually similar. In SIMSCRIPT, the state of the system is defined by entities, their associated attributes, and by logical groupings of entities referred to as sets. Thus, the notion of a set in SIMSCRIPT is analogous to a file in SLAM II. The dynamic structure of the system is described by defining the changes that occur at event times.

In SIMSCRIPT, two types of entities are considered. An entity which remains throughout a simulation is referred to as a permanent entity. In contrast, the term temporary entity is used to refer to entities which are created and destroyed during execution of the simulation. In the latter case, computer storage space is automatically allocated and freed during execution of the simulation as individual entities are created and destroyed. In a queueing system, each server would be modeled as a permanent entity, and the customers which arrive and depart the system would be modeled as temporary entities. The storage for permanent entities can be considered to correspond to a user-defined array in SLAM II and the storage for temporary entities as the SLAM II filing array storage.

In SIMSCRIPT, the attributes of entities are separately named, not numbered, thereby enhancing model description. For example, we could define a temporary entity named CUSTOMER which has an attribute named MARK.TIME. Sets are also named as opposed to numbered, further enhancing the model description. For example, a set containing customers waiting for service could be named QUEUE.

A SIMSCRIPT simulation model consists of a preamble, a main program, and event subprograms. The preamble is not part of the executable program and is used to define the elements of a model. The preamble also includes declarative statements for defining all variable types and arrays. SLAM II COMMON, EQUIVALENCE and DIMENSION statements correspond to the SIMSCRIPT preamble.

The main program is used for initializing variables, scheduling the initial occurrence of events, and starting the simulation. The event subprograms are used for defining the logic associated with processing each event in the model. The calls to the event subprograms are scheduled by the user but executed by the SIMSCRIPT control program. These SIMSCRIPT subprograms are similar to their SLAM II counterparts.

One of the primary functions of the preamble is to define the static structure of the model by prescribing the names of permanent and temporary entities, their

---

† We will hereafter denote SIMSCRIPT II as simply SIMSCRIPT.

associated attributes, and set relationships. These are declared in one or more EVERY statements by using attribute name clauses, set ownership clauses, and membership clauses. These clauses have the following form where capitalized words indicate required terms and lower case words indicate that user-employed variables are to be used.

> EVERY entity name HAS attribute name list
> EVERY entity name OWNS set name list
> EVERY entity name BELONGS TO set name list

Several clauses can be combined within a single EVERY statement by separating each clause by a comma. If desired, a clause can be preceded by the word MAY or CAN. Also, the items in an attribute or set name list must be separated by commas and one of the words A, AN, THE, or SOME. The following is an example of the EVERY statement:

> EVERY CUSTOMER HAS A MARK.TIME AND MAY BELONG TO THE QUEUE

This statement specifies that each entity in the entity class named CUSTOMER has an attribute named MARK.TIME and that these entities may at one time or another during the simulation be a member of the set named QUEUE. Each set employed in a model must be declared as being owned by either an entity or the system. An example of the latter is the following:

> THE SYSTEM OWNS A QUEUE AND HAS A STATUS

This preamble statement defines the existence of a system owned set named QUEUE and a system attribute named STATUS.

A collection of EVERY statements can be used to define either temporary or permanent entities. Temporary entities are defined by preceding the collection of EVERY statements with the statement

> TEMPORARY ENTITIES

Similarly, permanent entities are defined by preceding the collection of EVERY statements with the statement

> PERMANENT ENTITIES

EVERY statements are also used to define event names and event attributes by preceding the statements with the statement

> EVENT NOTICES

For example, an event named DEPARTURE with an attribute named SERVER could be defined by the following statements

    EVENT NOTICES
    EVERY DEPARTURE HAS A SERVER

The preamble is also used to define global variables through the use of the DEFINE statement which has the following form:

    DEFINE variable name AS A variable type.

Examples of the DEFINE statement are shown below:

    DEFINE X AS A REAL VARIABLE
    DEFINE Y AS A 2-DIMENSIONAL ARRAY

Another variation of the DEFINE statement is

    DEFINE word TO MEAN string of words

In this case whenever the specified word is seen by the SIMSCRIPT compiler, the string of words is automatically substituted for it before the statement is compiled. For example, if the statement

    DEFINE BUSY TO MEAN 1

is included in the preamble, then each occurrence of the word BUSY in the program is compiled as the numeral 1. Thus, the DEFINE statement can be used in a fashion similar to the SLAM II EQUIVALENCE statement.

    Another function of the preamble is to define variables for which statistics are to be collected. This is accomplished using the TALLY and ACCUMULATE statements. The TALLY statement causes observation statistics to be automatically collected on a specified variable. For example, the following statement would cause the variable AVE.TIME to be calculated as the mean of the variable TIME.IN.-SYSTEM:

    TALLY AVG.TIME AS THE MEAN OF TIME.IN.SYSTEM

SIMSCRIPT would automatically collect an observation at each assignment of a value to the variable TIME.IN.SYSTEM during the execution of the program. In contrast, SLAM II requires the user to explicitly include a call to COLCT for this purpose. Time-persistent statistics are obtained in a similar fashion by use of the ACCUMULATE statement. For example, the following statement would compute the variable AVG.UTIL as the time-persistent average of the variable STATUS

ACCUMULATE AVG.UTIL AS THE MEAN OF STATUS

Hence, the use of the ACCUMULATE statement in the SIMSCRIPT preamble is analogous to the use of the TIMST control statement in SLAM II. In addition to the MEAN, other statistical quantities such as the VARIANCE, MINIMUM, MAXIMUM, can also be recorded using the TALLY and ACCUMULATE statements.

Once the description of the static structure is completed by writing the preamble, the next step in writing a SIMSCRIPT program is to code the main program and event subprograms. These are coded by using the general purpose programming statements of SIMSCRIPT in conjunction with special statements for creating and destroying entities, manipulating entities between sets, obtaining random samples, and scheduling events. These statements are similar in function to the SLAM II subprograms for event modeling.

Temporary entities are created within the execution portion of a SIMSCRIPT program by the statement (braces are used to indicate options)

$$\text{CREATE} \begin{Bmatrix} \text{A} \\ \text{AN} \end{Bmatrix} \text{ entity name.}$$

For example, we can create a CUSTOMER entity as follows

CREATE A CUSTOMER

When this statement is executed, SIMSCRIPT searches for a block of space large enough to store the entity record that includes the attributes of the entity. Note, however, that the existence of the entity class named CUSTOMER must be defined in the preamble portion of the program. Temporary entities are destroyed in a similar fashion. For example, the statement

DESTROY THE CUSTOMER

causes the record corresponding to the entity CUSTOMER to be deleted and its storage space to be released. Attributes of entities are assigned and referenced by including the name of the entity in parentheses following the name of the attribute. For example, the following statement would assign the current simulated time in days, denoted by the SIMSCRIPT variable TIME.V, to the attribute MARK.TIME of the entity CUSTOMER:

LET MARK.TIME(CUSTOMER) = TIME.V

SIMSCRIPT provides a number of statements for filing, locating, and removing temporary entities from sets. An entity can be filed into a set by using the statement

FILE entity name IN set name

For example, we could file the entity CUSTOMER in the set named QUEUE using the statement

FILE THE CUSTOMER IN THE QUEUE

Entities can be removed in a similar fashion using the statement

REMOVE $\left\{ \begin{matrix} \text{FIRST} \\ \text{LAST} \end{matrix} \right\}$ entity name FROM set name

For example, we can remove the first CUSTOMER from the set QUEUE using the statement

REMOVE THE FIRST CUSTOMER FROM THE QUEUE

These set manipulation statements are analogous to subroutine calls to FILEM and RMOVE.

Event scheduling in SIMSCRIPT is accomplished using a statement of the form

SCHEDULE $\left\{ \begin{matrix} \text{A} \\ \text{AN} \end{matrix} \right\}$ event name IN variable $\left\{ \begin{matrix} \text{MINUTES} \\ \text{HOURS} \\ \text{DAYS} \end{matrix} \right\}$

where variable can be a constant, a variable, or any of the SIMSCRIPT provided random sampling functions. The following is an example.

SCHEDULE AN ARRIVAL IN EXPONENTIAL.F(20.,1) MINUTES

When attributes are associated with an event, the names of the variables specifying the attribute values are included in the SCHEDULE statement as follows:

SCHEDULE $\left\{ \begin{matrix} \text{A} \\ \text{AN} \end{matrix} \right\}$ event name GIVEN variables IN variable $\left\{ \begin{matrix} \text{MINUTES} \\ \text{HOURS} \\ \text{DAYS} \end{matrix} \right\}$

When passing the attributes of an entity to an event, it is necessary only to pass the name of the entity, which serves as a pointer to the list of attributes associated with the entity. For example, the attributes of the entity CUSTOMER can be passed to the event DEPARTURE as follows:

SCHEDULE A DEPARTURE GIVEN CUSTOMER IN
UNIFORM.F(10.,25.,1) MINUTES

Within the DEPARTURE event, reference to the attribute MARK.TIME of the entity is made using MARK.TIME(CUSTOMER) where CUSTOMER is a local integer variable containing the pointer to the location of the attributes of the entity CUSTOMER. There is an obvious correspondence between the SCHEDULE statement in SIMSCRIPT and subroutine SCHDL in the event-oriented portion of SLAM II.

As an illustration of the simulation features of SIMSCRIPT, we will present a model of the single channel queueing system which was used in Chapter 11 to illustrate the event modeling orientation of SLAM II and for which a GPSS model was built. Customers arrive to the system, possibly wait, undergo service, and then exit the system. The interarrival time is exponentially distributed with mean 20. and the service time is uniformly distributed between 10. and 25. Statistics are desired on the mean queue length, utilization of the server, and time in the system.

The SIMSCRIPT coding for this example is shown in Figure 15-3. The model consists of a PREAMBLE, MAIN program, ARRIVAL event, DEPARTURE event, and STOP.SIMULATION event.

```
1      PREAMBLE
2          THE SYSTEM OWNS A QUEUE AND HAS A STATUS
3          TEMPORARY ENTITIES
4              EVERY CUSTOMER HAS A MARK.TIME AND MAY BELONG TO THE QUEUE
5          EVENT NOTICES INCLUDE ARRIVAL AND STOP.SIMULATION
6              EVERY DEPARTURE HAS A SERVER
7          DEFINE BUSY TO MEAN 1
8          DEFINE IDLE TO MEAN 0
9          DEFINE TIME.IN.SYSTEM AS A REAL VARIABLE
10         TALLY NO.CUSTOMERS AS THE NUMBER,AVG.TIME AS THE MEAN, AND
11         VAR.TIME AS THE VARIANCE OF TIME.IN.SYSTEM
12         ACCUMULATE AVG.UTIL AS THE MEAN, AND VAR.UTIL AS THE VARIANCE OF
13         STATUS
14         ACCUMULATE AVE.QUEUE.LENGTH AS THE MEAN, AND VAR.QUEUE.LENGTH AS
15         THE VARIANCE OF N.QUEUE
16     END
```

```
1      MAIN
2          LET STATUS=IDLE
3          SCHEDULE AN ARRIVAL NOW
4          SCHEDULE A STOP.SIMULATION IN 8 HOURS
5          START SIMULATION
6      END
```

Figure 15-3 SIMSCRIPT II coding for single-channel queueing problem.

The PREAMBLE defines a system owned set named QUEUE which represents the waiting area for the server. The status of the server is maintained as the system attribute STATUS where 1 denotes busy and 0 denotes idle. Customers are modeled as temporary entities named CUSTOMER which have an attribute named

```
1      EVENT ARRIVAL
2         SCHEDULE AN ARRIVAL IN EXPONENTIAL.F(20.,1) MINUTES
3         CREATE A CUSTOMER
4         LET MARK.TIME(CUSTOMER)=TIME.V
5         IF STATUS=BUSY,
6            FILE THE CUSTOMER IN THE QUEUE
7            RETURN
8         ELSE
9            LET STATUS=BUSY
10           SCHEDULE A DEPARTURE GIVEN CUSTOMER IN UNIFORM.F(10.,25.,1)
11           MINUTES
12           RETURN
13     END
```

```
1      EVENT DEPARTURE GIVEN CUSTOMER
2         DEFINE CUSTOMER AS AN INTEGER VARIABLE
3         LET TIME.IN.SYSTEM=1440.*(TIME.V-MARK.TIME(CUSTOMER))
4         DESTROY THE CUSTOMER
5         IF THE QUEUE IS EMPTY
6            LET STATUS=IDLE
7            RETURN
8         ELSE
9            REMOVE THE FIRST CUSTOMER FROM THE QUEUE
10           SCHEDULE A DEPARTURE GIVEN CUSTOMER IN UNIFORM.F(10.,25.,1)
11           MINUTES
12           RETURN
13     END
```

```
1      EVENT STOP.SIMULATION
2         START NEW PAGE
3         SKIP 5 LINES
4         PRINT 1 LINE THUS
            SINGLE CHANNEL QUEUE EXAMPLE
6         SKIP 4 LINES
7         PRINT 3 LINES WITH NO.CUSTOMERS, AVG.TIME, AND VAR.TIME THUS
NUMBER OF CUSTOMERS =         ********
AVERAGE TIME IN SYSTEM =      ****.***
VARIANCE OF TIME IN SYSTEM = ****.***
11        SKIP 4 LINES
12        PRINT 2 LINES WITH AVG.UTIL AND VAR.UTIL THUS
AVERAGE SERVER UTILIZATION = ****.***
VARIANCE OF UTILIZATION =    ****.***
15        SKIP 4 LINES
16        PRINT 2 LINES WITH AVE.QUEUE.LENGTH AND VAR.QUEUE.LENGTH THUS
AVERAGE QUEUE LENGTH =        ****.***
VARIANCE OF QUEUE LENGTH =    ****.***
19        STOP
20     END
```

Figure 15-3 (continued).

MARK.TIME. This attribute is used to record the arrival time of the customer to the system.

The MAIN program initializes the system attribute STATUS to IDLE. The first ARRIVAL event is scheduled to occur at time 0, and the END.OF.SIMULA-

TION event is scheduled to occur in 8 hours. The statement START SIMULA-TION then causes control to transfer to the SIMSCRIPT timing routine.

The ARRIVAL event program defines the logic associated with processing the arrival of a customer to the system. The first action is to schedule the next arrival to occur at the current time plus the interarrival time (line 2). The entity representing the current arriving customer is then created (line 3), and its MARK.TIME attribute is set (line 4) to the current simulated time, TIME.V. The disposition of the arriving customer is based on the status of the server. If the server is busy, the CUSTOMER is filed in the set named QUEUE (line 6). Otherwise, the status of the server is set to busy (line 9) and the DEPARTURE event is scheduled (line 10-11). Note that the pointer to the CUSTOMER entity is passed as the attribute of the DEPARTURE event.

The DEPARTURE event defines the logic associated with the completion of service and departure of the CUSTOMER from the system. The event declaration statement (line 1) specifies that the attribute of the event is assigned to the local variable CUSTOMER. Since this attribute corresponds to a pointer to the entity CUSTOMER, it is necessary to define the local variable CUSTOMER as an integer variable (line 2). The variable TIME.IN.SYSTEM is set (line 3) to 1440 times the difference between the current simulated time, TIME.V, and the MARK.TIME attribute of the CUSTOMER. The conversion factor 1440 is used to convert the time from days to minutes. Note that the TALLY statement of the preamble (line 10-11) causes statistics on the number of observations, mean, and variance to be automatically recorded each time a value is assigned to TIME.IN.SYSTEM. The current CUSTOMER is then destroyed (line 4), thus freeing the storage space allocated to the entity. A test is then made on the state of the QUEUE. If the QUEUE is empty, the status of the server is set to idle (line 6). Otherwise, the first CUSTOMER is removed from the QUEUE (line 9) and the DEPARTURE event for this customer is scheduled (line 10-11).

The STOP.SIMULATION event displays the results for the simulation and returns control back to the main routine following the STOP statement (line 19). As the reader can observe from the coding, SIMSCRIPT provides flexible statement types for modeling and displaying output values but requires mastering a difficult syntax (5).

15.5   CONTINUOUS SIMULATION LANGUAGES

There are a number of simulation languages which have been specifically developed for continuous system simulation. In this section, we will describe a class of equation-oriented languages referred to as CSSL's which are useful for modeling systems described by differential equations. We will also discuss the DYNAMO language which was developed specifically for use with System Dynamics models.

15.5.1 Continuous System Simulation Languages

The family of languages which have been reasonably standardized by the Society for Computer Simulation's Continuous System Simulation Language Committee (40) are referred to as CSSL's. These languages provide a convenient equation-oriented format which allows the user to enter first order differential equations in a mathematical-like form. In contrast to SLAM II continuous models, the user of CSSL's need not know FORTRAN.

Simulation models which are programmed using a CSSL employ a FORTRAN-like symbol convention. However, meaningful symbolic names can be assigned to state and derivative variables without the use of the EQUIVALENCE statement as employed in SLAM II. In addition, special operators are provided for intergating derivatives to obtain state values as well as other useful functions. The equations in the model are automatically sorted by the CSSL for execution in the proper procedural order. The languages normally provide the modeler with a choice of an integration algorithm.

Examples of CSSL languages are CSMP III (8), CSSL III (9), CSSL IV, ASCL (26), and DARE-P (21). As a result of the standardization provided by the CSSL Committee, these languages are similar with differences mainly associated with input and output formatting and special library functions. The CSMP III language is available for IBM 360 and 370 series computers and the CSSL III language is available for the CDC 6000 and 7000 series computers. A special continuous language, MIMIC, was also designed specifically for CDC computers (10). Both CSSL IV and ASCL represent improved versions of CSSL III. DARE-P is a relatively new CSSL derived from DARE III-B. An advantage of the DARE-P system is that the processor is written entirely in ANSI FORTRAN IV and is therefore machine independent. Examples of the use of these languages would be similar to the SLAM II continuous examples and, hence, we will not illustrate their coding practices.

15.5.2   DYNAMO

Systems Dynamics, as developed by Forrester, is a problem solving approach to complex problems which emphasizes the structural aspects of models of systems (13). State variables, called *levels*, are defined in difference equation form and may be nonlinear. Nonlinearities are also included in the model through the use of table functions, delays, and clipping operations. The DYNAMO programming lan-

Table 15-3   Comparison of DYNAMO and SLAM II statement forms.

DYNAMO Statement	SLAM II Statement
1. PER = 52	1. PER = 52.
2. AID = 6	2. AID = 6.
3. ALF = (1000) (RR1)	3. AIF = 1000.$\star$ RRI
4. TIS.K = IAR.K + IAD.K + IAF.K	4. TIS = IAR + IAD + IAF
5. SNE.K = (SIH)SIN((2PI) (TIME.K)/PER)	5. SNE = SIH $\star$ SIN(6.28318 $\star$ TNOW/PER)
6. UOR.K = UOR.J + (DT) (RRR.JK – SSR.JK)	6a. SS(1) = SSL(1) + DTNOW $\star$ (RR – SSR)
	6b. SS(1) = SS(1) + DTNOW $\star$ (RRR – SSR) if NNEQD = 0
	6c. DD(1) = RRR – SSR
7. RSR.K = RSR.J + (DT) (1/DRR) (RRR.JK – RSR.J)	7a. DD(3) = (1./DRR) $\star$ (RRR – RSR)
	7b. SS(3) = SSL(3) + DTNOW $\star$ (RRR – RSR)/DDR
8. STP.K = STEP(STH,STT)	8. IF(TNOW.GT.STT) STP = STH
9. SSR.KL = CLIP (STR.K,NIR.K, NIR.K,STR.K)	9. SSR = STR IF (NIR.LT.STR)SSR = NIR
10. Y.K = SMOOTH(Q.JK,SMTM)	10a. CALL GDLAY (10,10,Q,SMTM)
	10b. SS(10) = SSL(10) + DTNOW $\star$(Q – SSL(10))/ SMTM and Y = SS(10)
11. Y.K = DLINF1(X.K,TRX)	11a. CALL GDLAY (11,11,X,TRX)
	11b. DD(11) = (X – S(11))/TRX
	11c. SS(11) = SSL(11) + DTNOW $\star$(X – SSL(11))/ TRX and Y = SS(11)
12. PSR.KL = DELAY3 (PDR.JK,DCR)	12. CALL GDLAY (18,20,PDR,DCR) and PSR = SS(20)
13. Y.K = DLINF3(X.K,TRX)	13. CALL GDLAY (15,17,X,TRX) and Y = SS(17)

guage (32) was developed to provide a language for analyzing Systems Dynamics models. DYNAMO uses a fixed step size, Euler-type integration algorithm to evaluate the level variables over time.

The translation of DYNAMO statements into SLAM II or FORTRAN is a straightforward process (6). A comparison of statement forms is shown in Table 15-3. In DYNAMO, variables that are a function of time are indicated by a period following the variable name and subscripts denoting time following the period. Single letters denote points in time and double letters denote that the value holds for an interval. Three points in time are used, which are represented by the letters J, K, and L. The intervals between these points are represented by JK and KL. The length of the interval is fixed and defined by the variable DT. Since Systems Dynamics models can be simulated in SLAM II and the CSSL languages, we recommend using these general languages for this purpose.

15.6 SPECIAL PURPOSE SIMULATION LANGUAGES

As the use of modeling and simulation has increased, the need for languages oriented to specific problem types and industries has increased. Many special purpose simulation languages have been developed to meet these needs. Such languages usually are not as well documented nor supported as the general purpose languages. This is to be expected since the range of applications is smaller and the variability in the user population is less. As greater use is made of a specialized simulation language, more flexibility is added to the language and a tendency toward "general" special purpose languages results.

The advantage of a special purpose simulation language is the direct modeling viewpoint included in the language for studying its special class of systems. When the system fits the modeling language, it is easy to build the model and analyze the system. Typically, fewer concepts are required when modeling with a special purpose language and the concepts provide a greater level of detail due to the limited scope of the language. Examples of special purpose simulation languages are: MAP/1 (25), GALS, and MAST (24) for modeling manufacturing systems; BETHSIM (11,39) for modeling and decision support of steel operations; IDSS 2.0 for modeling and decision support of aerospace manufacturing; SAINT (7,47,48), HOS (18) and MOPADS (29) for modeling human operator tasks; SNAP (28) for modeling safeguards at nuclear reactors; NETWORK II.5 for modeling computer networks; CROPS for modeling agricultural systems, and FACTOR for on-line scheduling of jobs and for controlling shop floor operations.

Specialized simulation languages are typically based on general purpose simulation languages and, in turn, provide feedback with regard to enhancement possibilities of the general purpose languages. As will be seen in Chapter 16, concepts employed in BETHSIM and MAP/1 have been used to design and develop add-in capabilities for SLAM II. These add-ins are part of the Material Handling(MH) Extension to SLAM II. The MH Extension contains new nodes, activities and support subprograms for the modeling of material handling systems which include cranes and automated guided vehicles as components. A description of MAP/1 is given in the next section as an illustration of a special purpose simulation language.

15.6.1   MAP/1

MAP/1 is an example of a special purpose simulation language which is used to model manufacturing systems.(25) MAP/1 includes modeling concepts to represent the major components of a manufacturing system such as PARTS, STATIONS, TRANSPORTERS, CONVEYORS, OPERATORS and FIXTURES. PARTS are the entities that are processed in a manufacturing system and may be in different states of manufacture such as raw materials, work-in-process, subassemblies or finished products. PARTS may be modeled to arrive according to a specific production schedule and move through the system according to a process plan or ROUTE. A ROUTE can be deterministic, probabilistic or conditional based on the system status or PART characteristics. Each operation in the ROUTE of a PART is performed at a STATION. The operation time at a station may consist of a setup time and a processing time. The STATION may be modeled to perform an operation on a part or a group of parts, to inspect parts, to assemble one part from many, to produce many parts from one, or to load or unload a FIXTURE.

Material handling equipment moves parts between stations. Conventional material handling devices such as carts, forklifts or trucks are represented as TRANSPORTERS in MAP/1. An explicit representation of CONVEYOR systems is also included to represent straight line segments, conveyor loops, and merge and diverge points. Special resources are defined in MAP/1 to model PERSONNEL, OPERATORS and FIXTURES. MAP/1 also provides a specific capability to model SHIFTS and equipment BREAKDOWN. A summary of the MAP/1 statement types is given in Figure 15-4.

Statement Type   **Statement Parameters**

Statement Type						
PART	PART TYPE NAME	PRIORITY	INTERARRIVAL TIME	FIRST ARRIVAL TIME	ARRIVING LOT SIZE	EXPECTED FLOW TIME
SCHEDULE	PART TYPE NAME	EARLIEST START TIME	NUMBER OF PARTS			
ROUTE	STATION NAME/OPERATION #	SETUP TIME	OPERATION TIME	UNLOAD FIXTURE INFORMATION	LOAD FIXTURE INFORMATION	NEXT STATION IDENTIFICATION
STATION	STATION NAME	SIZE	PREPROCESS INVENTORY STORAGE	POSTPROCESS INVENTORY STORAGE	MATERIAL HANDLING EQUIPMENT	TRANSPORT LOT SIZE
	EXCESS RULE	SHIFT ID	OPERATION MODE	OPERATION MODE PARAMETERS		
RANKING	STATION NAME	RANKING RULE	MINIMIZE SETUP FLAG			
TRANSPORTER	TRANSPORTER NAME	NUMBER OF TRANSPORTERS	MOVE TIME	RESPONSE TIME	SHIFT ID	VELOCITY
DISTANCE	FROM STATION NAME	TO STATION NAME	DISTANCE	TO STATION NAME	DISTANCE	
CONVEYOR	CONVEYOR NAME	VELOCITY	EXCESS RULE	CONVEYOR TYPE	SHIFT ID	PATH LIST
MERGE	JUNCTURE NAME	OUTGOING CONVEYOR NAME				
DIVERGE	JUNCTURE NAME	# OF OUTGOING CONVEYORS	CONVEYOR (STATION LIST)			
PERSONNEL	PERSONNEL CLASS NAME	NUMBER OF PEOPLE	SHIFT ID	SELECTION PRIORITY LIST		
OPERATOR	EQUIPMENT TYPE	EQUIPMENT NAME	CHOICE RULE	PERSONNEL CLASS LIST		
FIXTURE	FIXTURE NAME	NUMBER OF FIXTURES	SHIFT ID	ALLOCATION RULE	LOAD STATION LIST	
BREAKDOWN	EQUIPMENT TYPE	EQUIPMENT SPECIFICATION	FAILURE TIME	REPAIR TIME		
SHIFT	SHIFT ID	WORKING SCHEDULE				

Figure 15-4 MAP/1 input statement summary.

An example of a MAP/1 model of the flexible manufacturing system (27) described in Section 4.4 for which an abbreviated SLAM II model was developed in Illustration 6-3 is given in Figure 15-5. By using the definitions of the statement parameters given in Figure 15-4, the MAP/1 model of Figure 15-5 can be reviewed. The PART that moves through the system is a gear and its interarrival time is a constant 42.67 time units. The time of first arrival of a PART is at 0 and

```
BEGIN,P&A,FMS DESIGN EVALUATION,2,9,84,1,0,6720;
;
; DEFINE A PART AS A GEAR: SPECIFY THE GEAR'S ROUTE
;     MILL MACHINES: MILL1(OP.10),MILL2(OP.20),MILL3(OP.30),MILLU(FLEX)
;
PART,GEAR,,42.67,0,2;
     ROUTE,LATHE,,42.67,,,STORE;
     ROUTE,STORE,,,,,WASH;
     ROUTE,WASH,,4,,LOAD(FIXTUREA,1),MILL1,NPRE(MILL1).NE.5,MILLU;
     ROUTE,MILL1,,133.33,,,WASH/2;
     ROUTE,MILLU,,133.33,,,WASH/2;
     ROUTE,WASH/2,,2,UNLOAD,LOAD(FIXTUREB,1),MILL2,NPRE(MILL2).NE.2,MILLU/2;
     ROUTE,MILL2,,44.44,,,WASH/3;
     ROUTE,MILLU/2,,44.44,,,WASH/3;
     ROUTE,WASH/3,,4,,,MILL3,NPRE(MILL3).NE.2,MILLU/3;
     ROUTE,MILL3,,62.22,,,LASTWASH;
     ROUTE,MILLU/3,,62.22,,,LASTWASH;
     ROUTE,LASTWASH,,4,UNLOAD,,STAGING;
     ROUTE,STAGING,,,,,INSPECT;
     ROUTE,INSPECT;
;
; DEFINE THE STATIONS: LATHE,STORAGE,WASH,MILLS,STAGING,AND INSPECT
;
STATION,LATHE,2,100,9999,C/CONVEYR1,1,SUBCONTRACT,1;
STATION,STORE,5,,9999,ROBOT1,1,,1;
STATION,WASH,10,,9999,CART,1,,1,FIXTURE;
STATION,MILL1,5,5,5,CART,1,,1;
STATION,MILL2,2,2,2,CART,1,,1;
STATION,MILL3,2,2,2,CART,1,,1;
STATION,MILLU,1,1,1,CART,1,,1;
STATION,LASTWASH,5,,9999,ROBOT2,1,,1,FIXTURE;
STATION,STAGING,5,,,,C/CONVEYR2,1,,1;
STATION,INSPECT;
;
; DEFINE THE REGULAR TRANSPORTERS: 2 ROBOTS AND 1 CART
;
TRANSPORTER,CART,1,1,0,1;
TRANSPORTER,ROBOT1,1,0,0,1;
TRANSPORTER,ROBOT2,1,0,0,1;
;
; DEFINE THE TWO CONVEYORS IN THE SYSTEM
;
CONVEYOR,CONVEYR1,1200,ACCUMULATE,OPEN,1,LATHE,45,0,STORE;
CONVEYOR,CONVEYR2,1200,ACCUMULATE,OPEN,1,STAGING,90,0,INSPECT;
;
; DEFINE THE FIXTURES REQUIRED FOR THE MILLING OPERATIONS
;
FIXTURE,FIXTUREA,12,1,,WASH;
FIXTURE,FIXTUREB,12,1,,WASH;
;
; DEFINE THE SHIFT SCHEDULE TO BE ON
;
SHIFT,1,ON,6720;
;
; CLEAR STATISTICS AFTER TWO DAYS (FOUR SHIFTS)
;
CLEAR,1920;
;
END;
```

Figure 15-5 MAP/1 model of a flexible manufacturing system.

the arriving lot size is 2. The ROUTE statement provides information on the way the PART is routed to the STATIONS. The first three stations on the gear's route are the LATHE, STORE, and WASH. At the WASH station the part is loaded onto one unit of FIXTUREA. The next station is to be MILL1 if the number of parts in the preprocess inventory storage for MILL1, NPRE(MILL1), is not equal to 5. The alternative station if the condition is not met is station MILLU which is the flexible station. The remainder of the ROUTE statements are similar to the one just described.

The next portion of the model defines the STATIONS. In addition to the STATION name, the size of the station identifies the number of parts that can be concurrently processed at the STATION. Information is then given on the capacities of the preprocess and postprocess inventory storage, the material handling equipment that serves the station, and the transport lot size for moving parts from the station. Additional information regarding details of the station operation are then provided. For the LATHE station, we see that there are two lathes, a limit of one hundred parts for the preprocess storage, and a large capacity for postprocess inventory. A CONVEYOR moves parts from the LATHE station. The excess rule specifies to subcontract the units when the preprocess storage buffer of the station is full. After the definition of the stations, the MAP/1 model includes statements to define the TRANSPORTERS of which there are three: two for robots, and one for the cart. This is followed by definitions for the two conveyors. CONVEY1 moves parts from the LATHE station to the STORE area and CONVEY2 models the movement of parts from the STAGING station to the INSPECT station. As can be seen from the above description, the MAP/1 model is specialized to manufacturing systems and that it is easy to build a model of manufacturing operations.

The input statements for MAP/1 are nonprocedural in that the sequence in which the statements are listed is not dictated. Since the MAP/1 model deals specifically with manufacturing concepts, output reports can be tailored to provide output statistics in a form that corresponds to reports obtained from an actual system. The MAP/1 output reports provide the type of information discussed in Section 4.2.1 and include a production summary, along with reports on throughput, utilization, station inventory, and part movement times. A forms input processor is available for MAP/1 and TESS can be used to store and process the outputs from MAP/1.

15.7   CHAPTER SUMMARY

In this chapter, we have described the basic features of simulation languages in relationship to the modeling framework of SLAM II. There are many additional programming details which must be mastered in each language, and only a thorough reading of the latest language manuals and much practice can provide such mastery.

The selection of a simulation language is frequently based on knowledge and availability as opposed to a formal comparison of language features. However, if the frequent use of simulation is anticipated, then a comprehensive evaluation of the available languages and anticipated modeling needs is warranted. Banks and Carson provide a table comparing the most commonly used discrete languages (2). Bratley, Fox, and Schrage review the general purpose languages presented in this chapter (5). Table 15-6 is a summary of important factors to consider in comparing simulation languages.

15.8   EXERCISES

15-1. Build a GPSS block model of the inspection and adjustment stations described in Example 5-2. Embellishment: Build a SIMSCRIPT model for Example 5-2.

15-2. Discuss the similarities and differences between the following: The TERMINATE block of GPSS and the TERMINATE node of SLAM; the CREATE statement in SIMSCRIPT and the CREATE node in SLAM II; the SEIZE block of GPSS and the AWAIT node in SLAM II; and the ACCUMULATE statement in SIMSCRIPT and the ACCUMULATE node in SLAM II.

15-3. Discuss the symbol or statement type that causes simulated time to change in SLAM II, GPSS, SIMSCRIPT, a CSSL, and DYNAMO.

15-4. Grade any four languages on each row of Table 15-6 using the scale: E = Excellent; G = Good; F = Fair; P = Poor; and NA = Not Applicable.

15-5. Discuss the organizational structure of simulation languages and how it supports and/or impedes the development and use of simulation models.

15-6. Modify the SIMSCRIPT II queueing example to include two parallel servers. Assume that the service time for each server is uniformly distributed between 15 and 30 minutes and that the selection of free servers is made randomly.

Table 15-6   Features on which to evaluate a simulation language.

TRAINING REQUIRED	EASE OF LEARNING THE LANGUAGE
	TRAINING AVAILABILITY
	EASE OF CONCEPTUALIZING SIMULATION PROBLEMS
CODING CONSIDERATION	EASE OF CODING INCLUDING RANDOM SAMPLING AND NUMERICAL INTEGRATION
	DEGREE TO WHICH CODE IS SELF-DOCUMENTING
PORTABILITY	AVAILABILITY ON OTHER OR NEW COMPUTERS INCLUDING MICROCOMPUTERS AND SUPERCOMPUTERS
FLEXIBILITY	DEGREE TO WHICH LANGUAGE SUPPORTS DIFFERENT MODELING CONCEPTS AND WORLD VIEWS
EXTENSIBILITY	AVAILABILITY OF NEW FEATURES, ADD INS, AND SPECIALIZED MODULES
	COUPLING OF LANGUAGE WITH DATABASE AND GRAPHICS CAPABILITIES
PROCESSING CONSIDERATIONS	BUILT-IN STATISTICS GATHERING CAPABILITIES
	LIST PROCESSING CAPABILITIES
	ABILITY TO ALLOCATE CORE
	EASE OF PRODUCING STANDARD REPORTS
	EASE OF PRODUCING USER-TAILORED REPORTS
DEGUGGING & RELIABILITY	EASE OF DEBUGGING
	RELIABILITY OF COMPILERS, SUPPORT SYSTEMS, AND DOCUMENTATION
RUN-TIME CONSIDERATIONS	COMPILATION SPEED
	EXECUTION SPEED
MAINTENANCE	DEGREE TO WHICH LANGUAGE IS SUPPORTED AND IMPROVED
KNOWLEDGEABLE USERS	USE OF LANGUAGE IN INDUSTRY, GOVERNMENT AND UNIVERSITIES
	AVAILABILITY OF CONSULTING ASSISTANCE
SURVIVABILITY	CAPABILITIES OF CORPORATION AND INDIVIDUALS DISTRIBUTING AND USING THE LANGUAGE

15.9   REFERENCES

1. *A Simulation Process-Oriented Language (ASPOL)*, Publication No. 17314200, Control Data Corporation, Sunnyvale, CA., 1972.
2. Banks, J. and J.S. Carson, II, *Discrete-Event System Simulation*, Prentice-Hall, 1984.
3. Birtwistle, G., O.J. Dahl, B. Myhrhaug, and K. Nygaard, *SIMULA BEGIN*, Auerbach, 1973.
4. Birtwistle, G.M., *DEMOS — Discrete Modelling on SIMULA*, Macmillan, 1979.
5. Bratley, P., B.L. Fox, and L.E. Schrage, *A Guide to Simulation*, Springer-Verlag, 1983.
6. Cellier, F. and A.E. Blitz, "GASP V: A Universal Simulation Package," *Proceedings, IFAC Conference*, 1976.
7. Chubb, G.P., "SAINT, A Digital Simulation Language for the Study of Manned Systems", *Manned Systems Design: Methods, Equipment and Applications*, J. Moraal and K.F. Kraiss, Eds., Plenum Press, 1981.
8. *Continuous System Modeling Program III and Graphic Feature, Program No. 5734-X59 Manual*, IBM Corporation, New York, 1972.
9. *Continuous System Simulation Language, Version 3, User's Guide*, Control Data Corporation, Sunnyvale, CA., 1971.
10. *CONTROL DATA MIMIC - A Digital Simulation Language Reference Manual*, Publication No. 44610400, Revision D, Control Data Corporation, Minneapolis, MN., 1970.
11. DeJohn, F.A., C.W. Sanderson, C.T. Lewis, J.R. Gross, and S.D. Duket, "The Use of Computer Simulation Programs to Determine Equipment Requirements and Material Flow in the Billet Yard," *Proceedings, AIIE Spring Annual Conference*, 1980, pp. 402-408.
12. Delfosse, C.M., *Continuous Simulation and Combined Simulation in SIMSCRIPT II.5*, CACI, Inc., Arlington, VA., 1976.
13. Forrester, J.W., *Principles of Systems*, Wright-Allen Press, 1971.
14. Franta, W.R., *The Process View of Simulation*, North Holland, 1977.
15. Gordon, G., *The Application of GPSS V to Discrete Systems Simulation*, Prentice-Hall, 1975.
16. Henriksen, J.O, and R.C. Crain, *The GPSS/H User's Manual*, (Second Edition), Wolverine Software, Annandale, VA., 1983.
17. Hills, P.R., *An Introduction to Simulation Using SIMULA*, Publication No. S55, Norwegian Computing Center, Oslo, 1973.
18. *Human Operator Simulator (HOS) Study Guide*, Technical Report 1400.22E, Analytics, Willow Grove, PA., March 1982.
19. Kiviat, P.J., R. Villanueva, and H. Markowitz, *The SIMSCRIPT II Programming Language*, Prentice-Hall, 1969.
20. Knuth, D. and J.L. McNeley, "SOL-A Symbolic Language for General Purpose Systems Simulation," *IEEE Transactions on Electronic Computers*, August, 1964, pp. 401-408.
21. Korn, G.A. and J.V. Wait, *Digital Continuous-System Simulation*, Prentice-Hall, 1978.
22. Law, A.M. and W.D. Kelton, *Simulation Modeling and Analysis*, McGraw-Hill, 1982.
23. Markowitz, H., "SIMSCRIPT", *Encyclopedia of Computer Science and Technology*, J. Belzer, A.G. Holzman, and A. Kent, Eds., Marcel Dekker, Inc., 1978.

24. *MAST User Manual*, CMS Research, Oshkosh, WI., 1982.

25. Miner, R.J. and L.J. Rolston, *MAP/1 User's Manual, Version 3.0*, Pritsker & Associates, West Lafayette, IN., 1986.

26. Mitchel, E.E.L. and J.S. Gauthier, "Advanced Continuous Simulation Language (ACSL)," *Simulation*, Vol. 25, 1976, pp. 72-78.

27. Musselman, K.J.,"Computer Simulation: A Design Tool for FMS", *Manufacturing Engineering*, September, 1984, pp. 117-120.

28. Polito, J., *The Safeguards Network Analysis Procedure (SNAP): A User's Manual*, Pritsker & Associates, Report ABQ0013, West Lafayette, IN., March, 1983.

29. Polito, J. and A.A.B. Pritsker, "Computer Simulation and Job Analysis", *Job Analysis Handbook*, S. Gael, Ed., John Wiley, (forthcoming).

30. Pritsker, A.A.B., *Modeling and Analyis Using Q-GERT Networks*, (2nd Edition), Halsted Press and Pritsker & Associates, 1979.

31. Pritsker, A.A.B., *The GASP IV Simulation Language*, John Wiley, 1974.

32. Pugh, A.L., III, *DYNAMO II User's Manual*, M.I.T. Press, 1970.

33. Roberts, S.D. and T.E. Sadlowski, "INS: Integrated Network Simulator," *Proceedings, Winter Simulation Conference*, 1975, pp. 575-586.

34. Roberts, S.D., *Simulation Modeling and Analysis with INSIGHT*, Regenstrief Institute for Health Care, Indianapolis, IN., 1983.

35. Roth, P.F., "The BOSS Simulator — An Introduction," *Proceedings, Fourth Conference on Applications of Simulation*, 1970, pp. 244-250.

36. Russell, E.C., *Building Simulation Models with SIMSCRIPT II.5*, CACI, Inc., Los Angeles, CA, 1983.

37. Sabuda, J., F.H. Grant, III and A.A.B. Pritsker, *The GASP IV/E User's Manual*, Pritsker & Associates, Inc., West Lafayette, IN., 1978.

38. Schriber, T., *Simulation Using GPSS*, John Wiley, 1974.

39. Scribner, Ben, "Operations Research — Out of the Closet and Into the Tool Kit," presented at the IE Manager's Seminar, March, 1981.

40. SCi Software Committee, "The SCi Continuous-System Simulation Language," *Simulation*, Vol. 9, 1967, pp. 281-303.

41. Shannon, R.E., *Systems Simulation: The Art and The Science*, Prentice-Hall, 1975.

42. *SIMPL/1 Operations Guide*, SH 19-5038-0, IBM Corporation, New York, June, 1972.

43. *SIMPL/1 Program Reference Manual*, SH19-5060-0, IBM Corporation, New York, June, 1972.

44. Sol, H.G., "SIMULA(TION) in the Analysis and Design of Information Systems," *Proceedings*, Simulation '77, 1977, pp. 67-71.

45. Washam, W., "GASPPI: GASP IV With Process Interaction Capabilities," unpublished MS Thesis, Purdue University, West Lafayette, IN., May 1976.

46. Wilson, J.W. and A.A.B. Pritsker, "Computer Simulation," *Handbook of Industrial Engineering*, G. Salvendy, Ed., John Wiley, 1982.

47. Wortman, D.B., S.D. Duket, D.J. Seifert, R.L. Hann, and G.P. Chubb, *The SAINT User's Manual*, AMRL-TR-76-22, Aerospace Medical Research Laboratory, Wright-Patterson AFB, OH., June 1978.

48. Wortman, D.B., and J.J. O'Reilly, *New Capabilities in the SAINT Simulation Program*, Pritsker & Associates, West Lafayette, IN., 1982.

CHAPTER 16

Material Handling Extension to SLAM II†

† Steven Duket, Christopher Clapp, Michael Sale, Jean O'Reilly, William Lilegdon and Cathy Stein of Pritsker & Associates provided material for use in this chapter.

16.1 INTRODUCTION

In recent years there has been a realization that a major cause for delay in manufacturing operations is related to the transporting of jobs from one work center to another (1,7). This realization has created a need for improved facility design as well as a requirement for detailed analyses of material handling (MH) equipment. Estimates are that over 85 percent of manufacturing time is spent in a waiting or transporting mode and less than 10 percent in a processing state. Caution should be exercised in basing decisions on such rules of thumb (10). However, new facility layouts are based on group technology concepts where machines are grouped in work cells rather than by machine type to reduce or eliminate the waiting and materials handling delays.

Due to the specialized nature of material handling equipment (4,9) and because the characterization of a facility layout is required to analyze the effectiveness of material handling equipment (7,8), simulation analysis has typically been done with stylized models oriented to a specific situation. In Chapter 15, the discussion on special purpose simulation languages described some of these capabilities. In this chapter, concepts for modeling material handling equipment are presented.

These modeling constructs are an extension or add-in to SLAM II. As will be seen, the analysis of material handling equipment requires constructs for the definition of both the facility and the control logic associated with MH operations. Because of this, many alternative approaches can be taken. The capabilities presented in this chapter represent developments that have proven useful in practice (2,4,5).

16.2   MODELING VIEWS OF THE MH EXTENSION

The modeling orientation of the MH Extension to SLAM II is to provide add-in nodes and capabilities to augment the network worldview. Imbedded within the network constructs are continuous variables to represent the movement of material handling resources, and user-written subprograms for modeling the detailed control features required for directing the operations of MH equipment. Specialized resource definitions are used to represent: the MH equipment; the area for pick-up and delivery of material; and the paths in which the MH equipment can move.

Three types of MH capabilities are presented in this chapter. The first type involves the introduction of a node which combines the capabilities of the BATCH node and the AWAIT node and is called the QBATCH node. At a QBATCH node, a batch is not released until a specified resource is available and a group of entities have arrived that constitute an amount that is to be moved. A QBATCH node can be employed to represent a MH device such as a switch engine that picks up all the entities that are ready to be moved without exceeding the load capacity of the switch engine. The QBATCH node requires no specialized definition of resources nor complex routing rules. It is an add-in node for SLAM II for use in modeling a special type of MH function. It is used in conjunction with UNBATCH and FREE nodes to recover entities from the batch and to make the resource seized at the QBATCH node available.

The second capability models cranelike MH devices where material waits in an area for pickup and dropoff to other locations. An area is a resource which has a physical location and a capacity. An area consists of piles which are used to segregate material in an area. A crane resource should be considered in a general sense as a MH device which picks up, moves and drops off material from one area to another. Crane movements are controlled by the generalized wait and free nodes, GWAIT and GFREE. When an entity requires transportation from one location to another, it waits in a GWAIT node file where it can request a crane, an area, and possibly other user-defined resources. The crane resource is dispatched to the entity's pickup location based on crane control logic built into the MH Extension. The time the crane takes to get to the pickup location is based on the crane and entity locations; crane movement characteristics such as bridge speed, trolley speed, acceleration and deceleration limits; and runway congestion. The pickup and drop-off locations are specified by area names. Piling rules are built into the MH Extension to control the placement of material within an area. The GFREE node is used to release areas, cranes and/or regular resources. It may also specify the beginning of a crane movement.

The MH Extension has built-in control logic associated with crane interference resulting from multiple cranes operating on the same runway. As mentioned above, built-in piling rules provide for the placement of material within individual piles and automatically select from among alternative locations for storing material within an area. The worldview taken in characterizing cranes is to provide built-in capabilities which address facility layout concerns, and the logistics and programmed control logic for moving cranes to areas. These constructs have proven to be invaluable for modelers of metals production, warehousing, and automatic storage and retrieval systems (ASRS).

The third add-in presented in this chapter provides a modeling capability for Automatic Guided Vehicle Systems (AGVS). Performance of an AGVS is highly dependent on the physical layout and the control policies used to direct a fleet of vehicles. Resources in an AGVS include vehicles, vehicle control points, and guidepath or track segments connecting the control points. In an AGVS, the movement of vehicles from control point to control point over segments may be managed or controlled by a central computer, an onboard computer, or a combination of the two. A vehicle may encounter portions of a route which are congested by slower moving traffic, traffic at intersections, loading and unloading vehicles, and interference from other MH devices. Because of the large number of control decisions, the modeling view taken for AGVS is to provide interfaces to user-written subprograms for implementing rules and procedures which can augment standard control logic included in the module. At a control point, a rule for vehicle contention and a rule for routing is necessary. For each type of vehicle, a dispatching rule for selecting from among waiting entities and a rule for idle vehicle disposition are necessary.

To model the logic associated with the movement of vehicles, vehicle wait, VWAIT, and vehicle free, VFREE, nodes are defined. The VWAIT node is used to model entities requesting a vehicle for transportation to its next station or inventory location. A VFREE node is used to model a completed move by a vehicle. The movement of a vehicle is accomplished in a VMOVE activity in accordance with the characteristics of the vehicle fleet and the guidepath system. The guidepath system is defined in terms of control points and segments which are modeled as special resource types. At each arrival to a control point, if needed, the rule for contention and the rule for routing defined for the control point are put into effect. When an entity arrives to a VWAIT node, a rule is required to specify which vehicle should be assigned to transport the entity. When the vehicle arrives to pickup the entity, a rule is required to determine which entity should be released to the vehicle if more than one entity is waiting for that vehicle type.

In summary, the modeling worldviews provided by the MH Extension provide representative approaches to the design and evaluation of MH systems. The details of each approach and examples of the MH Extension are provided in subsequent sections of this chapter.

16.3   QBATCH NODE

The symbol and statement for the QBATCH node are shown below.

QBATCH(IFL),RES,NBATCH/NATRB,THRESH,MAXSIZE,NATRS,
SAVE,RETAIN(NATRR),M;

The QBATCH node is used to accumulate entities before one unit of resource RES is requested. Like the BATCH node, the values contained in attribute NATRS of arriving entities are summed to compute the amount accumulated for the batch. When this sum exceeds THRESH, a batch is ready. If resource RES is available, the batch will be made, one unit of RES will be seized, and the entity will be released from the node. Otherwise, the batch will continue to grow until the resource is available. If the sum of ATRIB(NATRS) exceeds the MAXSIZE specification before the resource becomes available, the entities are still maintained as a single set. However, at the time the resource becomes available, a batched entity is formed consisting of the earliest arriving members such that the addition of the next entity would make the sum of the ATRIB(NATRS) values exceed MAXSIZE.

For a QBATCH node, the attributes of the original entities are always maintained and are pointed to by the value assigned to ATRIB(NATRR) of the batched entity. The RETAIN attribute index, NATRR, must be specified for QBATCH nodes (in contrast to BATCH nodes where the field is optional). The batched entity released from a QBATCH node has the attribute values of the FIRST arriving entity for the batch unless a list of attribute indices is specified. For each index specified, the corresponding attribute will be the sum of the attribute values of each entity in the batch. Statistics concerning the number of entities waiting at the QBATCH node are associated with file number IFL.

A QBATCH node can be used to sort and accumulate entities based on an entity type, that is, accumulate multiple batches. A maximum of NBATCH batches may be maintained simultaneously. Each arriving entity is placed in a batch according to the value in attribute NATRB. Thus, entities with equal values of ATRIB (NATRB) are placed in the same batch. Each batch requires the specified resource to be available before the batch can be released. A maximum of M activities are initiated at each release of the QBATCH node.

As an example of the QBATCH node, consider the following.

QBATCH(3),FORKLIFT,3/6,500,650,2,,ALL(7),1;

Entities arriving to this node are sorted into 3 groups according to ATRIB(6). Each group is accumulated until the sum of the values of ATRIB(2) of member entities is equal to or greater than 500. When the sum reaches this value, a FORKLIFT is seized if available. The FORKLIFT only moves entities with up to 650 units of ATRIB(2) at one time. The entity that emanates from the QBATCH node has the attributes of the first entity in the batch except for ATRIB(7) which is assigned a pointer to the original entities. File Statistics on the number of entities waiting to be batched for all groups are associated with file 3. One activity can be taken at each release of this QBATCH node.

Consider the following QBATCH node.

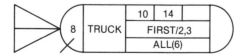

QBATCH(8),TRUCK,,10,14,FIRST/2,3,ALL(6);

Entities arriving to this QBATCH node require the resource named TRUCK to proceed. Because NBATCH, NATRB, and NATRS are defaulted, one group will be maintained and each entity will be counted until the number of arriving entities equals the threshold value of 10. A batch will be released from the node when a TRUCK resource is available and there are at least 10 entities waiting. No more than 14 entities will be released in a single batch. The batched entity will have attribute values of the first entity except that ATRIB(2) and ATRIB(3) will be the sum of the second and third attribute values of all entities in the batch. ATRIB(6) of the batched entity contains a pointer to the original entities.

16.4   ILLUSTRATION 16-1. TRANSFER CAR

This illustration addresses a situation in which steel pipe is routed to an off-line repair area during manufacturing as shown in Figure 16-1. A transfer car is used to transport the pipe to the repair area. The car can handle a maximum of 30 tons of pipe in one trip, but it is not practical for a trip to be made unless there is a load of 20 tons waiting to be moved. The network model of this situation is shown in Figure 16-2, and the statement model is shown in Figure 16-3.

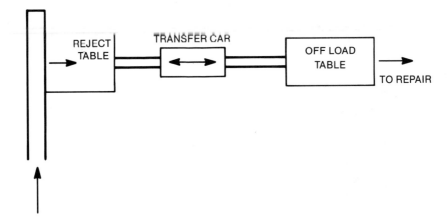

Figure 16-1 Transfer car schematic.

An entity representing an amount of pipe is routed to the repair area according to an exponential distribution with a mean of 2.7 time units. Each such entity represents between 5 tons and 8 tons of material. The weight is defined in ATRIB(1). Entities wait at the QBATCH node until a batch is formed which weighs at least 20 tons. At that time, a TRANCAR is requested. If more than 30 tons of material are waiting when the TRANCAR returns, the entities are removed from the file as they arrived until the sum of the weight of the pipe does not exceed the maximum of 30 tons.

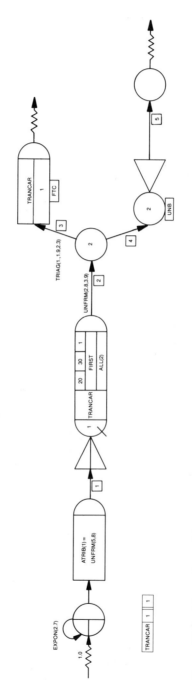

Figure 16-2 Network model of a transfer car operation.

```
 1   GEN,PRITSKER & ASSOC.,QBATCH ILLUSTRATION,4/1/1986,1,Y,Y,Y,N,,72;
 2   ;
 3   ;   TRANSFER CAR EXAMPLE
 4   ;
 5   LIMITS,1,4,50;
 6   INIT,0.,300.;
 7   MONTR,CLEAR,100;
 8   NETWORK;
 9   ;
10   ;   RESOURCE DEFINITIONS
11   ;
12         RESOURCE,TRANCAR(1),1;
13   ;
14         CREATE, EXPON(2.7),1;
15         ASSIGN,ATRIB(1)= UNFRM(5.0, 8.0);      ASSIGN INCOMING WEIGHT
16           ACT/1;
17         QBATCH(1),TRANCAR,,20,30,1,FIRST,ALL(2);
18           ACT/2,UNFRM(2.8,3.9);                TRANSPORT
19         GOON,2;
20           ACT/3,TRIAG(1.0,1.9,2.3),,FTC;       CAR RETURN
21           ACT/4,,,UNB;                         TO UNBATCH
22   ;
23   FTC   FREE, TRANCAR;                         RELEASE CAR
24         TERMINATE;
25   ;
26   UNB   UNBATCH ,2;
27           ACT/5;                               UNBATCHING
28         TERMINATE;
29         ENDNETWORK;
30   FIN;
```

Figure 16-3 Statement model of a transfer car operation.

The entities are put into a batch with the attribute characteristics of the FIRST entity in the batch. ATRIB(2) is set to point to the original entities. Activity 2 models the time for the loaded transfer car to take pipe to the repair area with a duration that is uniformly distributed between 2.8 and 3.9 time units. When the car reaches the repair area, the entity is duplicated at a GOON node. One entity is sent to an UNBATCH node which retrieves the original set of pipe entities. Activity 3 models the return trip of the TRANCAR which takes a triangularly distributed time with a minimum of 1 time unit, a maximum of 2.3 time units and a mode of 1.9 time units. At the completion of this activity the TRANCAR is freed and made available.

16.5   CRANE SYSTEM DEFINITIONS

A crane system consists of a runway and a crane. The crane itself is a MH device that spans across and moves up and down the runway. The cab of the crane moves along the crane rails perpendicular to the runway. Movement along the runway is referred to as movement in the bridge direction and is defined as the x-coordinate direction in the MH Extension to SLAM II. The movement of the cab is referred to as movement in the trolley direction and is defined as the y-coordinate direction. To initiate a pick or a drop, the cab must move to the x and y coordinates of the entity to be transported. Cranes are assumed to have a width of 10 units in the x direction and a maximum of 10 cranes can be placed on a single runway. In the x direction, a crane position is defined by the center point of the crane which requires that five units of x are on either side of the crane.

16.5.1   Crane Movement Characteristics

The time to traverse distances in the bridge direction are based on the acceleration, maximum velocity and deceleration parameters specified for the crane resource in its definition block. Other cranes on the same runway affect movements in the bridge direction. Trolley movements in the y direction are assumed to have instantaneous acceleration and instantaneous deceleration. Bridge and trolley movements are accomplished simultaneously. If the trolley movement time is less than the bridge movement time, the trolley movement time does not affect the time to reach a location. If the trolley movement time is longer than the bridge movement time, the excess trolley movement time is added on to the bridging time.

The raising and lowering of the picking device for pickups and dropoffs are not modeled as part of the crane movement. Vertical move times are modeled using regular activities of a SLAM II network.

Crane movements in the bridge direction are characterized by an acceleration to a maximum velocity, a period of no acceleration, and a period of deceleration. The duration of each of these phases is dependent upon the distance between pickup and dropoff points and the locations of other cranes on a runway. The MH Extension maintains the location in the bridge direction of all cranes and automatically adjusts a crane's velocity to accommodate acceleration and deceleration characteristics specified for the crane resource. If a zero acceleration or deceleration is prescribed, instantaneous startup and stopping for the crane resource is employed.

16.5.2 Crane Movement Priority Rules

Since cranes can share the same runway, situations arise in which some cranes must be moved out of the way or delayed in order that pickups and dropoffs be made by other cranes. A determination of when a crane needs to move or wait to allow another crane to perform its operation is established through crane priorities. These priorities are computed dynamically each time the status of a crane changes. Cranes involved in a pickup or dropoff activity are not allowed to move. For cranes that are moving, the following rules apply:

(1) Cranes moving to drop points have priority over cranes moving to pickup points.
(2) If two cranes are performing the same type of function, the crane which was assigned its task earliest is given priority.
(3) To break ties on 1 and 2 above, the crane closest to its dropoff point is given priority.
(4) Idle cranes are moved out of the way of cranes which have been given assignments.

The above priority rules have been derived from practical experience and detailed observations of crane movements in different environments. The rules tend to keep loaded cranes waiting a minimum amount of time.

The MH Extension automatically keeps statistics on crane interference time, defined as delays caused by other cranes. The interference time is computed as the difference between the time that a crane should have reached its destination and the time that it actually reached its destination. Interference times are recorded for each crane move and are reported independently for move-to-pick and move-to-drop operations.

16.5.3 Area, Pile, and Crane Resources

Definitions of three resource types are used to model crane movements: AREA, PILE, and CRANE. An AREA resource block describes an area in which entities wait to be picked up by a crane or where entities are to be dropped off by a crane. The statement for an area resource block is shown below.

AREA,ARLBL/ARNUM,NPILES,NATRA,IFLs;

An area is identified by a resource label, ARLBL, and an optional area number, ARNUM. NPILES specifies the number of piles contained within the area. Each pile is subsequently described by a PILE resource block which describes the location and size of the area within the crane coordinate structure. The amount of space required by an entity is defined by the NATRA attribute of the entity. A list of files is defined in the AREA resource block which specifies the file numbers at GWAIT nodes where areas can be requested. When units of an AREA resource are freed, the files listed in its resource block are interrogated in the listed order to determine if entities are waiting for the allocation of the AREA resource.

The AREA resource statement

AREA,WAREHOUS,10,2,7,8,9;

specifies an area resource named WAREHOUS which has 10 piles. Attribute 2 of entities requesting space in the WAREHOUS specifies the amount of space desired. Entities requesting space in the WAREHOUS wait in files 7, 8, and 9.

A PILE resource block is used to describe the piles located within a previously defined area. A pile is given a unique identification number, PNUM, and must be identified as belonging to an area by specifying an ARLBL. A label PLBL for the pile is optional. The PILE resource block statement is shown below.

PILE,PNUM/PLBL,ARLBL,CAP,XCOORD,YCOORD;

An entity that requests an area is placed in a pile of the area depending on the available space in a pile and piling logic rules. The amount of space in a pile is defined on the PILE resource block by CAP. The location of the pile is defined by prescribing an x-coordinate value, XCOORD, and a y-coordinate value, YCOORD. Only cranes which can reach these coordinate values are allowed to pickup or dropoff entities to a pile.

Examples of PILE resource block statements are shown below.

PILE,1,WAREHOUS,40,20,10;
PILE,2,WAREHOUS,100,60,80;

These statements identify two piles within the area WAREHOUS. Pile 1 has the capacity of 40 units and the center of the pile is located at position (20,10). Pile 2 is also in the WAREHOUS, has a capacity of 100, and is located at position (60,80).

The CRANE resource block statement is shown below.

CRANE,CRLBL/CRNUM,RUNWAY,XCOORD,YCOORD,MXVEL,ACC, DEC,TVEL,IFL;

A crane is defined by a crane label CRLBL, an optional crane number CRNUM, and a runway number, RUNWAY. Runway numbers must be assigned starting with 1. CRANE resource blocks must be listed in the statement model with consecutively numbered runways and crane numbers.

The initial coordinates for the crane are prescribed by the values XCOORD and YCOORD. As previously discussed, the movement in the bridging direction is accomplished in accordance with a maximum velocity, MXVEL, acceleration, ACC, and deceleration, DEC. Cranes will attempt to move at their maximum velocity until they are required to decelerate in order to reach their destination with a zero velocity. Acceleration and velocity will be restricted by the position of other cranes. Instantaneous acceleration and deceleration is accomplished by specifying zero values for ACC and DEC. Trolley speed, TVEL, is used to determine if the trolley travel time in the y direction is longer than the travel in the bridge direction. Whenever possible, trolley travel time is performed concurrently with bridge travel time. It is assumed that there is instantaneous acceleration and deceleration in the y direction. A user-defined file number, IFL, is required in order that entities waiting for a crane can be maintained in one file.

Examples of CRANE resource statements are shown below.

CRANE,OVERHED1,1,25,10,50,0,0,10,4;
CRANE,OVERHED2,2,60,15,50,20,20,10,5;

The first statement gives the label OVERHED1 to the crane and places it on runway 1. Its initial position is (25,10), and it has a maximum velocity of 50. Acceleration and deceleration in the bridge direction is instantaneous and the trolley velocity is 10. Entities waiting for the crane OVERHED1 wait in file 4. The second crane is given the label OVERHED2 and travels on runway 2. Its initial position is (60,15) and it has the same velocity as OVERHED1 crane. Its acceleration and deceleration are 20, which means that 2.5 time units are required to reach the maximum velocity and 2.5 time units are required to stop. Entities waiting for crane OVERHED2 wait in file 5.

The input order of cranes on the same runway must be such that the x-coordinate values are in increasing order. For example, if there are three CRANE resource statements and, each crane is on runway 1, then the first resource block statement must have an x-coordinate value that is smaller than the x-coordinate value on the second CRANE resource statement which must have an x-coordinate value that is smaller than the value given on the third CRANE resource statement.

16.5.4   Crane Resource Definitions Illustration

As an illustration of the definitions of CRANE, AREA, and PILE resources, we consider the movement and cutting of aluminum coils. A schematic diagram of an aluminum facility for processing coils is shown in Figure 16-4. In the facility, there are three cooling racks, two cranes, two slitting machines, one cut-to-length shear machine, and storage areas for coils and for sheets. Coils from an annealing furnace arrive to the cooling racks. Cranes move the coils one at a time to either the slitter or shear machines. The cranes are also necessary to move the group of coils produced from slitting to coil storage, and the sheets produced from shearing to the sheet storage. Three attributes of the coil entities are listed on the next page.

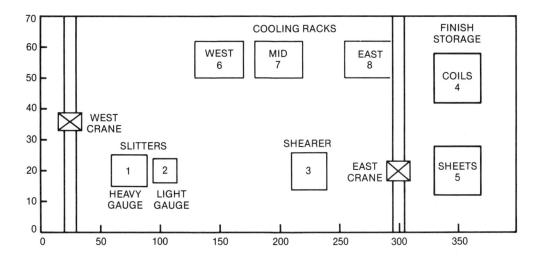

Figure 16-4 Schematic diagram of an aluminum facility for processing coils.

Attribute	Description
1	coil type: 1 → heavy gauge
	2 → light gauge
	3 → sheet
2	coil volume in in^3
3	processing time

The resource block statements for the two crane are:

```
C*CRANE BLOCK  CRLBL RWY   X     Y   BSPD   ACC    DEC  TSPD IFILE
       CRANE,CRNW,   1,   25.0, 35.0, 300.0,   0  ,    0 ,300.0,  1;
       CRANE,CRNE,   1,  300.0, 20.0, 300.0, 2000.0, 2000.0,300.0,  2;
```

The west crane, CRNW, is on runway 1 at location (25, 35). It has a bridge speed of 300 ft/min., and a trolley speed of 300 ft/min. Acceleration and deceleration are not modeled for CRNW. Job entities waiting for CRNW are placed in file 1. The east crane, CRNE, is initially located at (300, 50) and also is on runway 1. Its bridge speed is 300 ft/min., and it has a trolley speed of 300 ft/min. Acceleration and deceleration for CRNE is 2000 ft/min. and 2000 ft/min., respectively. Job entities waiting for CRNE wait in file 2.

Each machine in the facility is modeled as an area in order that a location can be identified for cranes to pick up and deliver entities. Finished goods storage and the cooling racks are also modeled as areas. The five area resource blocks for the coil processing facility are shown below.

```
C*AREA  ARLBL      /ARNUM NPILES  NATRA  IFILE
  AREA, HVYSLITA   /1       , 1  ,        , 4;
  AREA, LGTSLITA   /2       , 1  ,        , 4;
  AREA, SHEARA     /3       , 1  ,        , 4;
  AREA, FINSTOR    /4       , 2  ,   2    , 6;
  AREA, COOLRACK   /5       , 3  ,   2    , 3;
```

The three areas associated with the machines have only one pile and no attribute is specified to identify the amount of area required by requesting entities. The default for this field is 1 unit of the area for each entity. Entities waiting for machine areas are put in file 4. The finished storage area, FINSTOR, has two piles and the space required by an entity is specified by attribute 2 of the entity. Entities waiting for FINSTOR are put in file 6. There are three piles associated with the

COOLRACK area. Attribute 2 of an entity requesting the COOLRACK area establishes the amount of space required by the entity. These entities wait in file 3 for space in the cooling racks.

The pile statements for this illustration are shown below.

C*PILES	PNUM	/ PLBL	ARLBL	CAP	X	Y
PILE ,	1	/ BUFHSLIT	,HVYSLITA ,	1.0,	75.0,	20.0 ;
PILE ,	2	/ BUFLSLIT	,LGTSLITA ,	1.0,	110.0,	20.0 ;
PILE ,	3	/ BUFSHRD	,SHEARA ,	1.0,	225.0,	20.0 ;
PILE ,	4	/ COILSTOR	,FINSTOR ,	1000.0,	350.0,	50.0 ;
PILE ,	5	/ SHETSTOR	,FINSTOR ,	1000.0,	350.0,	20.0 ;
PILE ,	6	/ WESTRACK	,COOLRACK ,	800.0,	150.0,	55.0 ;
PILE ,	7	/ MIDRACK	,COOLRACK ,	600.0,	200.0,	55.0 ;
PILE ,	8	/ EASTRACK	,COOLRACK ,	600.0,	275.0,	55.0 ;

Piles 1, 2, and 3 are for the three machine areas as indicated by the area labels given for these piles. Only one space is available at each machine area. The heavy gauge slitter location is (75, 50); the light gauge slitter location is (110, 50); and the shear machine location is (225, 50). Piles 4 and 5 are in the finish storage area and their names are COILSTOR and SHETSTOR. For the storage areas, space is measured in thousands of cubic inches (tci). There is space for 1,000 tci in each of these piles. They are located at (350, 50) and (350, 20), respectively.

WESTRACK, MIDRACK, and EASTRACK are the three cooling racks and are identified as piles 6, 7, and 8. These piles have a capacity for coils of 800, 600, and 600 tci, respectively. WESTRACK is located at (150, 55); MIDRACK at (200, 55); and EASTRACK at (275, 55).

The definition of the resource block statements contains sufficient information to model crane movements from different locations and to maintain space restrictions included in a model. This type of resource information can be used in conjunction with many SLAM II models. Exercise 16-3 provides additional information on the aluminum coil processing facility and requests that a model be built to analyze the situation.

16.6 GWAIT NODE

The GWAIT node is a generalized AWAIT node where an entity waits for an amount of space in an area, AREA; one crane from a list of crane resources,

CRANE; and/or a set of regular resources, RES. An arriving entity waits in file IFL, and no resource is seized until enough units of all requested resource types are available. When all resources are available and allocated to an entity, the crane begins its move to the pickup location of the entity. The symbol and statement of the GWAIT node are shown below.

GWAIT(IFL),AREA/PLOGIC(NATRM),
CRANE/repeat,RES/UR,repeats,MODE,M;

When an area is requested, the entity specifies the amount of pile space required. The amount of space is defined by attribute NATRA of the entity where NATRA is an input on the AREA resource block. When an area consists of more than one pile, a pile is selected according to a piling logic rule specified by PLOGIC(NATRM) on the GWAIT statement. These rules are discussed later in this section.

If an entity requires a crane, cranes capable of performing the materials handling operation are listed. Only one crane is required for the move and alternative crane labels are separated by slashes. Regular resources requested at the GWAIT node are listed along with the amount of each resource type required, UR. A MODE specification of AND or OR indicates whether all or any one of the listed regular resources can be used to satisfy the request. For the OR specification, the list will be searched starting from the left, and the first resource that has the available number of units will be allocated to the entity.†

A choice may exist between the piles in an area. Pile logic rules can be based on attribute NATRM of the arriving entity if PLOGIC(NATRM) is specified. The input for PLOGIC defines the rule for seizing piles using a three character code. The first character is used to define the order in which the piles are to be polled. *F* indicates a search starting with the first PILE block statement; *L* indicates a search from the last PILE block statement; and *M* indicates a match that is to be made on the value of ATRIB(NATRM). The second character is used to indicate how the incoming entity should relate to the entities currently in the piles. *A* indicates that only space availability should restrict the entity from being put in a pile; *M* indicates a match is to be made between ATRIB(NATRM) of the arriving entity and the value of attribute NATRM for entities currently in the pile; *S* indicates

---

† Several restrictions pertain to the regular resources requested at GWAIT nodes. A maximum of five regular resource types may be allocated at a GWAIT node. At any time, only one resource type may be held by an entity when the MODE is specified as OR.

that ATRIB(NATRM) specifies the pile number, PNUM, into which the entity is to be placed.

The third character specifies whether the entities should wait (*W*) until a pile of the type desired is available or whether the characteristics desired of the pile should be ignored and that any available (*A*) pile can be used if space is not available in the desired pile.

Table 16-1 lists the eight rules and code that can be used to specify PLOGIC. The code FAW in accordance with the above descriptions indicates that a pile should be searched starting with the *F*irst PILE resource block listed under the AREA resource, that any pile with *A*vailable space should be used, and, if no pile has available space, the entities should *W*ait in the file of the GWAIT node. As another example, consider LMA(4) which specifies that the search of piles should

Table 16-1 Rules for seizing piles at GWAIT nodes.

Code	Procedure†
FAW	Search piles within area starting with the first listed pile; seize any available pile; if no available piles, wait.
LAW	Search piles within area starting with the last listed pile; seize any available pile; if no available piles, wait.
FMW(NATRM)	Search piles within area starting with the first listed pile, use first available pile for which entries match incoming entity on ATRIB(NATRM); wait otherwise.
LMW(NATRM)	Search piles within area starting with last listed pile; use available pile for which entries match incoming entity on ATRIB(NATRM); wait otherwise.
FMA(NATRM)	Search piles within area starting with the first listed pile; use first available pile for which entries match incoming entity on ATRIB (NATRM); otherwise use any available pile.
LMA(NATRM)	Search piles within area starting with the last listed pile; use first available pile for which entries match incoming entity on ATRIB(NATRM); otherwise use any available pile.
MSW(NATRM)	Seize pile whose number is the value contained in ATRIB(NATRM); otherwise wait.
MSA(NATRM)	Seize pile whose number is the value contained in ATRIB(NATRM); otherwise select an available pile which is listed closest to the specified pile.

† An empty pile is considered a match if no other matches are found.

start with the last pile statement and that ATRIB(4) of the arriving entity should match the corresponding attribute value of entities in the pile. If space is not available in such a pile, then use any available pile within the area. When looking for matches of attribute values, an empty pile is considered only if no other piles match the designated attribute. As a third example, consider MSA(5) which specifies the desired pile number through the value of ATRIB(5). If that pile is not available, the closest available pile is selected.

When an entity is released from the GWAIT node, it can be routed over at most M activities as specified by the last field of the GWAIT statement.

Suppose an entity requires transport to area AR3 by either crane CRN1, CRN2 or CRN3. To make the move, two units of resource MAN and one unit of resource CHAIN are required. If any of the resources are not available, the entity is to wait in file 6. At most one activity can be taken from this GWAIT node. The GWAIT statement and symbol for this situation are shown below.

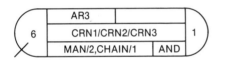

GWAIT(6),AR3,CRN1/CRN2/CRN3, MAN/2,CHAIN/1,AND,1;

In this GWAIT node, the default value for PLOGIC is used which is FAW, that is, search from the first pile and take any pile which has available space. If there are none, wait in file 6. The amount of pile space required for the arriving entity is specified by ATRIB(NATRA) where NATRA is defined in the AREA block definition for AR3.

As a second illustration, consider the following GWAIT statement and symbol.

GWAIT(2),ATRIB(1),,ATRIB(2)/ATRIB(3);

Entities arriving to this GWAIT node require an AREA whose number is given by ATRIB(1). Also, ATRIB(3) units of the resource number ATRIB(2) are requested. If these resources are available, the entity passes through the GWAIT node. Otherwise, the entity waits in file 2. No cranes are required by the entity at this GWAIT node.

The following GWAIT node specifies that entities require area BAY3 and that piles should be searched starting with the first listed pile in the PILE resource block for area BAY3. A match is desired between ATRIB(2) of the arriving entity and the second attribute of entities in a pile of BAY3. If such a pile is not available and if no empty pile exists, the entity should wait in file 11.

GWAIT(11),BAY3/FMW(2);

16.7   GFREE NODE

The GFREE node is used to free an area resource, a crane resource, a number of regular resources, or a combination of these resource types. When an area is specified, the GFREE node will release the amount of area capacity held by the arriving entity as determined from ATRIB(NATRA). The AREA field is used to indicate that the area held by the entity is to be released. Either FRAREA, YES, or the area label can be input for this purpose. NOFRA or NO indicates that an area is not to be freed. If a crane resource is specified, the crane will be released to be dispatched to its next job. Only one crane may be released at a GFREE node. However, more than one crane may be listed if the entity could have been transported by one of several cranes. If regular resources are to be freed, they are listed along with the number of resources of each type to be released. The symbol and statement for the GFREE node are shown below.

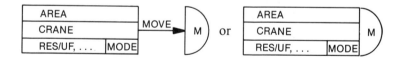

GFREE/MOVE,AREA,CRANE/repeats,RES/UF,repeats,MODE,M;

The MODE field can be AND or OR. If MODE is set to AND, all resources specified will be freed. If the MODE specification is OR, only that regular resource type allocated to the arriving entity is released. Thus, if an OR mode is used on a GFREE node, the entity arriving to that GFREE node must have seized one of the regular resource types listed at a GWAIT node with an OR mode. Because of this, a GWAIT node may specify an OR mode even if the selection list consists of only one resource. The OR mode specification ensures that the specific resource allocated to the entity is released at a GFREE node. The number of units of the resource freed is specified by the value of UF.

The optional MOVE field on the GFREE node indicates that a crane movement is to begin. The entity must have seized a crane resource and its destination area. Upon entering a GFREE node with the MOVE specification, the arriving entity is removed from the network and placed in control of the MH Extension. The MH Extension processes the crane travel to the entity's destination as defined by the area the entity has most recently seized. Upon completion of the crane movement, the entity branches from the GFREE node in accordance with its M value.

As each resource is released, an attempt is made to reallocate the resource. If a regular resource is freed, the files listed in the RESOURCE definition are polled in the order listed. If space within an area is freed, the files listed in the AREA definition are polled in the order listed. A freed crane examines the job list in the file specified in the CRANE definition.

A maximum of M activities are initiated at each release of a GFREE node.

Consider the following GFREE node.

GFREE,AREA1,,RR3/2,RR4/1,OR,1;

This node releases an amount of area resource AREA1 as determined by the value of ATRIB(NATRA) of the arriving entity where NATRA is prescribed on the area resource block. Either two units of resource RR3 or one unit of resource RR4 will be released, depending on which resource was seized at a GWAIT node. One activity will be initiated with each release of the node.

An entity arriving to the GFREE node shown below

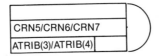

GFREE,,CRN5/CRN6/CRN7,ATRIB(3)/ATRIB(4);

frees either crane CRN5, CRN6, or CRN7 depending on which crane was allocated to the arriving entity. The regular resource whose identifying number is carried in ATRIB(3) is also released. The number of units of the resource to be released is given by the value of ATRIB(4).

The resource block and network statements shown below involve two GWAIT nodes and a GFREE node.

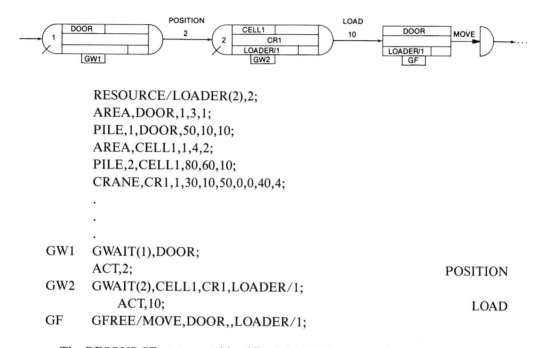

```
RESOURCE/LOADER(2),2;
AREA,DOOR,1,3,1;
PILE,1,DOOR,50,10,10;
AREA,CELL1,1,4,2;
PILE,2,CELL1,80,60,10;
CRANE,CR1,1,30,10,50,0,0,40,4;
    .
    .
    .
```

```
GW1   GWAIT(1),DOOR;
      ACT,2;                                          POSITION
GW2   GWAIT(2),CELL1,CR1,LOADER/1;
         ACT,10;                                      LOAD
GF    GFREE/MOVE,DOOR,,LOADER/1;
```

The RESOURCE statement identifies LOADER as a regular resource which has a capacity of 2 and is allocated to entities waiting in file 2. The AREA statement identifies an area labeled DOOR which has one pile. Entities requesting the DOOR area wait in file 1 and require an amount of space specified by ATRIB(3). The pile associated with DOOR has a space availability of 50 and is located at position (10,10). A second area labeled CELL1 has only one pile. Entities requiring space in CELL1 wait in file 2 and ATRIB(4) defines the space requirement. The pile for CELL1 is identified as pile number 2, it has a capacity of 80, and is located at position (60,10). A CRANE resource CR1 on runway 1 is initially located at position (30,10). The bridge speed for CR1 is 50 and the trolley speed is 40. No acceleration or deceleration is to be modeled for CR1. Entities waiting for crane CR1 are maintained in file 4.

At GWAIT node GW1 arriving entities request the area resource DOOR and wait in file 1 if ATRIB(3) spaces are not available. When sufficient DOOR space is available, the entity proceeds over an activity which takes 2 time units. Upon activity completion, the entity arrives to GWAIT node GW2 where space in area CELL1 is desired along with crane CR1 and one unit of the regular resource LOADER. If all three resources are not available, the entity waits in file 2. The space requirement for CELL1 is defined by attribute 4 of the arriving entity. The entity arriving to node GW2 has the space associated with the area DOOR allocated to it. The space cannot be returned until the crane and loader resources move the entity from DOOR to CELL1. Before a movement to CELL1 is made, space from CELL1 must be allocated to the entity. When the space is allocated to the entity, the entity is removed from file 2 and the loading of the entity onto the crane occurs through an activity which takes 10 time units. The entity then arrives to GFREE node GF which frees the occupied DOOR space and returns ATRIB(3) spaces to pile 1. The entity is then moved to area CELL1 in accordance with the crane characteristics of CR1. At the GFREE node, the LOADER resource is also made available.

To summarize the use of the GWAIT and GFREE nodes in the modeling of a crane movement, an entity must seize an area to obtain an initial location. It then requests both a crane resource and an amount of a desired destination area prior to beginning the crane transport. When all resources requested at the GWAIT node are available, they are seized by the waiting entity. The GWAIT node initiating the crane transport process is followed by an activity representing a load or pick operation. The activity time is to connect the material to the crane and lift it off the ground. This operation may consist of one activity or a sequence of activities and will end at a GFREE node which releases the area resource that the entity held and initiates the crane transport with the MOVE specification.

16.8   MH EXTENSION INPUTS, OUTPUTS AND SUPPORT FUNCTIONS FOR CRANES

The MH Extension is integrated with SLAM II, and there are only a few conceptual changes to be noted. On the LIMITS statement, the number of attributes, MATR, must be specified to be 3 greater than the number of user-defined attributes. Attributes ATRIB(MATR−1) and ATRIB(MATR) are used for storing a resource number allocated when the OR mode is used, and a pile number, respectively. When an entity is duplicated because of more than one activity being taken from a node, the entity released first from the node will carry any crane or space resources that have been allocated to the entity. Entities holding such resources should free resources before being terminated from the network or combined with other entities.

The priority of entities in files associated with QBATCH and GWAIT nodes is the same as in standard SLAM II. Entities are released from GWAIT nodes when all the resources required of an entity are available. The priority rule for a QBATCH file does not affect the entities included in the batch as the sum of ATRIB(NATRS) is always done on a first-in, first-out basis. The file prescribed on the CRANE resource block may also be given a priority. The ranking of jobs in the crane joblist specifies the sequence in which entities will be polled in order to allocate crane resources. This causes all jobs waiting for a crane resource to be compared directly without polling files associated with GWAIT nodes.

A new control statement MHMONTR, is available to produce special purpose traces. The "option" field for the crane trace is CTRACE. The crane trace can be routed to device number IUNIT. The MHMONTR statement is shown below.

MHMONTR/IUNIT,option,TFRST,TSEC;

If IUNIT is specified then a device with number IUNIT needs to be defined by the job control statements. The default value for IUNIT is NPRNT. Subroutine CMONTR(IUNIT) is available to change the device number on which the crane trace is written or to suppress the crane trace output. The crane trace is started whenever subroutine CMONTR is called with IUNIT greater than zero. The crane trace is suppressed by calling CMONTR with IUNIT equal to zero.

Three other support functions are also provided. Function ARAVL(NAREA, NPILE) returns the area available in area number NAREA if NPILE is zero. Function ARAVL returns the area available for pile number NPILE if NPILE is greater than zero.

Function JOBCRN(ICRN) returns the current status of crane number ICRN. If the crane is moving to a pick point, the status is positive; if the crane is moving to a drop point, the status is negative; and if the crane is idle, the status is zero. Thus, the statement ISTAT = JOBCRN(3) sets ISTAT to a positive value if crane 3 is moving to a pick point, to zero if crane 3 is idle, or to a negative number if crane 3 is moving to a drop point.

To obtain the number of a pile in which an entity with specific attributes is stored, the function NPILN(BTRIB) is used. This function checks the various piles to make a match on all the attributes of the entity which are given in BTRIB. When a match is found, the pile number in which the entity resides is returned from function NPILN.

The SLAM II Summary Report has been augmented to provide statistical results for cranes and areas/piles. The statistics for areas/piles include the minimum, maximum and average utilization of the space in the area/piles, and the number of entities that arrived and left each pile. Statistics reported for cranes are the number of pickups, the number of dropoffs, crane utilization for pickups, crane utilization traveling to pickups, interference percentage in traveling to a pickup, utilization of the crane during dropoffs, utilization of the crane moving to dropoff piles, percentage of time there was interference when moving to dropoff piles, and the total utilization of the crane. These statistics are illustrated in the examples to follow.

16.9   ILLUSTRATION 16-2. CRANE TRANSPORT OF STEEL COILS

In this illustration, the movement of steel coils of width 36 and 48 inches is modeled. Coils originate in a bulk storage area and require a gantry crane to move them to a load storage area which supplies coils to two spindles. A schematic diagram of this crane transport operation and the resource definition blocks for areas, cranes, and regular resources is shown in Figure 16-5. The labels given to the two areas are INBAY and LOADSTOR. The INBAY area is a staging area where a gantry crane picks up a coil for delivery to the load storage area. There is no need to identify separate stacks for locations within INBAY and the entire area is modeled as one pile with a capacity of 1600 inches. The center of the pile is located at position (15,10) with the x direction origin at the left bottom of the figure.

The LOADSTOR area has four piles and entities request space in the area based on the value in attribute 1 specifying coil width. Entities wait for the LOADSTOR area in file 2. The four piles associated with area LOADSTOR are given numbers 2, 3, 4 and 5, and have capacities of 40, 38, 55 and 52 respectively. Piles 2 and 3 can only accommodate 36 inch coils whereas piles 4 and 5 can accommodate all coils. The piles are all located at y position 30 with x positions given by 145, 150, 155 and 160. The gantry crane is labeled GANTRY and is on runway 1. Its initial position is (30,30). The maximum velocity of the gantry is 190 which means that it can travel from the INBAY area to pile 5 of the LOADSTOR in 145/190 time units. The trolley speed of 200 causes the trolley time to be well within the bridge movement time. Therefore, trolley movement times are not included in the model. Entities waiting to be picked by the crane are stored in file 4 as defined on the CRANE resource block.

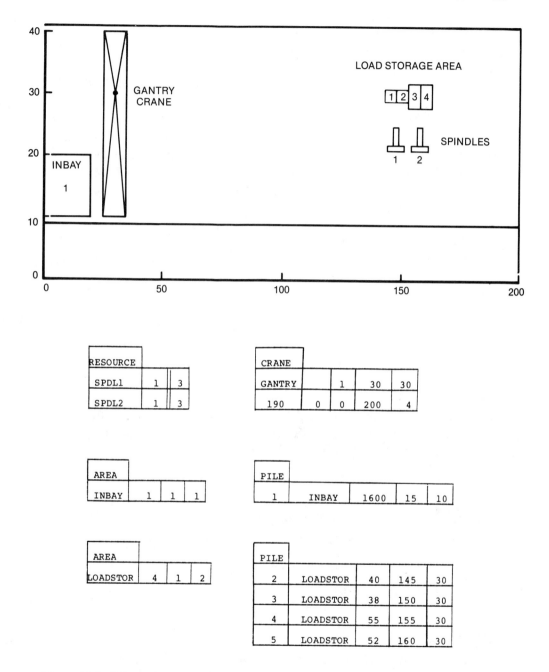

Figure 16-5 Schematic diagram and resource definition blocks of a crane transport operation.

The two spindles are regular resources with a capacity of 1. An entity waiting for a spindle waits in file 3. The network model for this situation is shown in Figure 16-6. The CREATE node routes entities to ASSIGN nodes where attribute 1 is assigned a value of either 36 or 48. It is assumed that coils brought into the INBAY area are equally likely to be of size 36 or size 48. Coils obtain area INBAY at the GWAIT node GW3. Prior to requesting the gantry crane, the coils are positioned as represented by activity 3. At GWAIT node GW4, a pile in area LOADSTOR is requested along with a request for the gantry. The pile logic rule FAW requests the first available pile. Coils wait in file 2 for the LOADSTOR space and gantry crane. When these resources are allocated to the coil, and the crane is positioned overhead, a crane pickup operation is performed as modeled by activity 4. GFREE node FR5 is then used to return the space in area INBAY and to initiate the crane move to the allocated LOADSTOR pile. The move time depends on the pile that was assigned as well as the crane's velocity. When the crane arrives to the LOADSTOR pile, the coil entity is routed from the GFREE node. Activity 5 models the dropoff of the coil. Following the dropoff, the gantry crane is freed at GFREE node FR7. The coil is then routed to GWAIT node GW8 to wait for either spindle 1 (SPDL1) OR spindle 2 (SPDL2) with spindle 1 having preference. When a spindle has been allocated, the coil entity is loaded on the processing unit in 0.17 time units. GFREE node FR9 frees the space in area LOADSTOR. The processing of the coil is uniformly distributed between 2.7 and 3.3 time units for 36 inch coils. For the 48 inch coils the process time is one and one-third times the processing time for 36 inch coils. Following processing, the spindle is freed. The statement model corresponding to the network is shown in Figure 16-7. Figure 16-8 shows a portion of the SLAM II Summary Report for the statistics on area/piles and cranes.

16.10   EXAMPLE 16-1. AUTOMATIC STORAGE AND RETRIEVAL SYSTEM (ASRS)

A local ASRS system is used to serve the storage needs of two machines within a cell. The two machines are a lathe and a mill. The system is shown in Figure 16-9. Pallets of castings arrive to the system every 15 minutes to be moved by stacker cranes into storage. If 10 pallets are waiting for the lathe, arriving pallets are routed to a bypass cell. The storage racks are slotted for pallets with each rack having a capacity to hold up to 5 pallets. Racks 2 and 6 use two slots for machine entry and exit and thus only have 3 slots for storage. When the lathe becomes

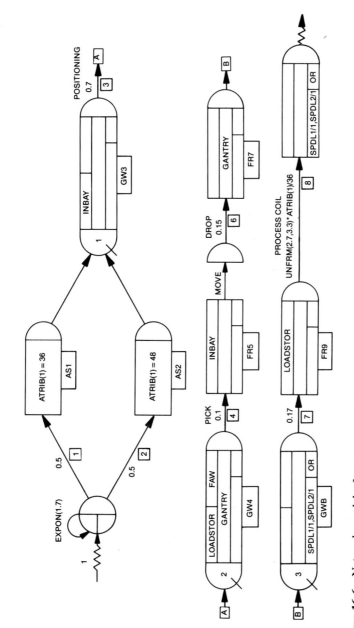

Figure 16-6 Network model of crane transport.

```
1   GEN,PRITSKER & ASSOC.,CRANE ILLUSTRATION,4/1/1986,1,Y,Y,Y,N,,72;
2   ;
3   ;  CRANE EXAMPLE
4   ;
5   LIMITS,4,4,50;
6   INIT,0.,100.;
7   NETWORK;
8   ;
9   ;  MATERIAL HANDLING EXTENSION RESOURCE DEFINITIONS
10  ;
11  ; AREAS
12  ;               AREA        NUMBER      QUANTITY    ASSOCIATED
13  ;               LABEL       OF PILES    ATTRIBUTE   FILES
14  ;               =====       ========    ========    =========
15      AREA,       INBAY,      1,          1,          1;
16      AREA,       LOADSTOR,   4,          1,          2;
17  ;
18  ; PILES
19  ;
20  ;               PILE    AREA    PILE        X-COOR    Y-COOR
21  ;               NUMBER  LABEL   CAPACITY    POSITION  POSITION
22  ;               ======  =====   ======      ========  ========
23      PILE,       1,      INBAY,    1600.,      15,       10;
24      PILE,       2,      LOADSTOR,  40.,      145,       30;
25      PILE,       3,      LOADSTOR,  38.,      150,       30;
26      PILE,       4,      LOADSTOR,  55.,      155,       30;
27      PILE,       5,      LOADSTOR,  52.,      160,       30;
28  ;
29  ; CRANES
30  ;
31  ;       CRANE RUNWAY INITIAL INITIAL  MAXIMUM   NOMINAL NOMINAL TROLLEY JOB F
32  ;       LABEL NUMBER X POS.  Y POS.   BRDG. SPD. ACCEL. DECELL. SPEED   NUMB
33  ;       ===== ====== ======= =======  ========= ======= ======= ======= ====
34      CRA, GANTRY, 1,    30,     30,      190,      0,      0,      200,    4;
35  ;
36  ; SLAM II RESOURCE DEFINITIONS
37  ;
38          RESOURCE, SPDL1(1),3;
39          RESOURCE, SPDL2(1),3;
40  ;
41          CREATE,EXPON(1.7),1;                 ENTITIES ARRIVE
42              ACT/1,,0.5,AS1;                  W36 ARRIVES
43              ACT/2,,0.5,AS2;                  W48 ARRIVES
44  AS1     ASSIGN,ATRIB(1)=36.;
45              ACT,,,GW3;
46  AS2     ASSIGN,ATRIB(1)=48.;
47  ;
48  GW3     GWAIT(1),INBAY;                      SEIZE UNLOADING BAY
49              ACT/3,0.7;                       TO UNLOADING BAY
50  GW4     GWAIT(2),LOADSTOR/FAW,GANTRY;        SEIZE SPACE AND CRANE
51              ACT/4,0.1;                       CRANE PICK
52  FR5     GFREE/MOVE,INBAY;                    RELEASE INBAY SPACE
53  ;                                            AND MOVE TO LOADSTOR
54              ACT/6,0.15;                      CRANE DROP OFF
```

Figure 16-7 Statement model of crane transport.

```
55 FR7    GFREE,,GANTRY;                              RELEASE CRANE AFTER DROP
56 GW8    GWAIT(3),,,SPDL1,SPDL2,OR;                  SEIZE SPINDLE UNIT
57            ACT/7,0.17;                             LOAD COIL
58 FA9    GFREE,LOADSTOR;                             RELEASE STORAGE SPACE
59            ACT/8,UNFRM(2.7,3.3)*ATRIB(1)/36.;      PROCESS
60 FA10   GFREE,,,SPDL1,SPDL2,OR;                     RELEASE SPINDLE UNIT
61 ;
62            TERMINATE;
63            ENDNETWORK;
64 FIN;
;
```

Figure 16-7   (continued).

```
                    S L A M   I I   S U M M A R Y   R E P O R T


        SIMULATION PROJECT CRANE ILLUSTRATION          BY PRITSKER & ASSOC.

        DATE  4/ 1/1986                                RUN NUMBER    1 OF    1


        CURRENT TIME   0.1000E+03
        STATISTICAL ARRAYS CLEARED AT TIME  0.0000E+00
```

```
                              **AREA/PILE STATISTICS**

        AREA AREA   PILE   PILE  CURRENT AVERAGE MINIMUM MAXIMUM NUM. ENTITIES
        NUM. LABEL  NUM. CAPACITY UTIL.   UTIL.   UTIL.   UTIL.  ARRIVE LEAVE

         1 INBAY     1 1600.00 372.00  222.77    0.00  552.00     60     52
         2 LOADSTOR  2   40.00   0.00   17.74    0.00   36.00     14     14
                     3   38.00  36.00   11.18    0.00   36.00      9      8
                     4   55.00  36.00   26.22    0.00   48.00     17     16
                     5   52.00  48.00   19.99    0.00   48.00     12     11
```

```
                              **CRANE STATISTICS**

                     NUM.  NUM.        TO      TO            TO   TO
        CRANE CRANE   OF    OF   PIK  PIK     PIK   DROP   DROP DROP   TOTAL
        NUM.  LABEL  PIKS  DROPS UTIL UTIL  INTERF  UTIL   UTIL INTERF UTIL

         1  GANTRY    52    51   0.05 0.37   0.00   0.08   0.37  0.00  0.87
```

Figure 16-8   SLAM II Summary Report for crane transport of steel coils.

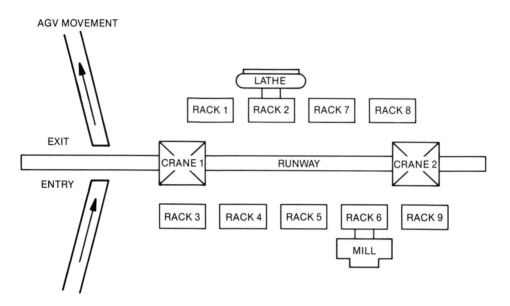

Figure 16-9   Schematic diagram of an ASRS system.

available, a pallet is moved from storage by the stacker crane to the lathe. The lathe automatically indexes the pallet to the workcenter and each casting on the pallet is processed. There is space for two pallets at the lathe.

After all castings on a pallet are processed at the lathe, the pallet is returned to storage to await milling operations. When the milling machine is available, a pallet of lathed castings is moved to the mill and processed. Upon completion of milling, the castings are again returned to storage to await movement to the next process. There is space for only two pallets at the mill. When five pallets of finished castings are accumulated in storage, an Automatic Guided Vehicle (AGV) is requested. When the AGV arrives, which takes between 0 and 20 minutes from the time of request, the pallet of castings is moved to the EXIT, loaded, and transported to the next cell.

Castings at any stage of processing may be stored in any of the racks. However, prelathed castings and finished castings are generally loaded into racks 1-6 and lathed castings when possible are loaded into racks 7-9.

Pallet processing is normally distributed with an average of 10 minutes and a standard deviation of 0.5 minutes at the lathe and 13 minutes with a standard deviation of 0.5 minutes at the mill. After a lathe processes a pallet, there is a 20 percent chance that maintenance is required. On the average, the lathe maintenance time is 20 minutes. Similarly there is a 10 percent chance that the mill requires maintenance after processing a pallet. The average mill maintenance time is

15 minutes. These maintenance times are uniformly distributed with a range of 10 minutes.

Stacker crane pick-and-place operations require 0.5 minutes, indexing movements require 0.25 minutes, and crane speeds are 50 feet per minute.

For this type of system, designers would like to determine the throughput potential of the system and whether or not sufficient storage capacity is available for all pallet groups, especially for finished products. Also, there is concern as to whether two stacker cranes are required to perform all the required moves.

Concepts Illustrated. This example illustrates the use of AREA, PILE, and CRANE resource definitions along with the GWAIT and GFREE nodes. These constructs are integrated with SLAM II network elements including the use of the BATCH and UNBATCH nodes. A priority attribute is defined in order to reduce interference between cranes when performing different material handling moves.

SLAM II Model. The network and statement models representing the ASRS are given in Figure 16-10 and 16-11, respectively. There are 36 nodes in this model with the majority of the nodes used to model material flow using the pick-move-place constructs associated with cranes. Thus, there is a sequence of the following nodes: GWAIT, GFREE, GOON or ASSIGN, and GFREE. For example, after a pallet has been placed in the entry area, it waits for a rack and a crane. When these resources are available, the pallet is picked up, the entry space is freed, the pallet is moved to the designated rack, the pallet is placed in the rack, and the crane is freed. The majority of the nodes in the network are used to depict this sequence of operations. The discussion below concentrates on other aspects of the model and presumes an understanding of the pick-move-place constructs.

There are five area resources in the model: ENTRY, EXIT, LATHA, MILLA, and RACK. The number of piles in each area is one except for RACK which has nine piles corresponding to each individual rack in the ASRS. Each area will reference attribute 2 with regard to the amount of space required by an entity in an area. For this model all entities require one space or one unit of capacity and, therefore, ATRIB(2) for each entity representing a pallet is set equal to 1. The definitions of the files used in this model are given below.

File Number	Definition
1	Pallets waiting for space in entry area
2	Pallets waiting to move from entry to rack
3	Pallets waiting to move from rack to lathe area
4	Queue for lathe
5	Pallets waiting to move from lathe area to rack

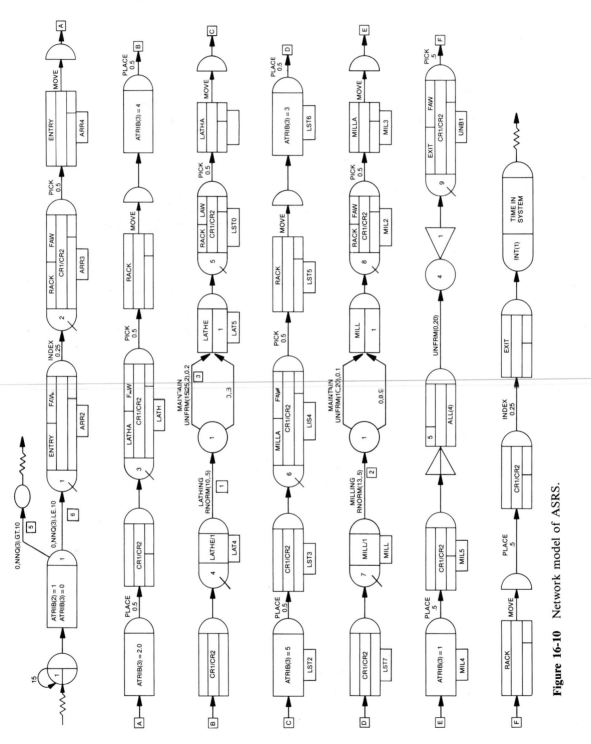

Figure 16-10 Network model of ASRS.

```
 2  GEN, PRITSKER & ASSOC., ASRS EXAMPLE, 3/21/1986,1;
 3  ;
 4  ;  ASRS EXAMPLE
 5  ;
 6  LIMITS,11,6,200;
 7  PRIORITY/10,LVF(3)/11,HVF(3);
 8  INIT,0.,11520.;
 9  MONTR,CLEAR,1440;
10  NETWORK;
11  ;
12  ;  MATERIAL HANDLING EXTENSION RESOURCE DEFINITIONS
13  ;
14  ; AREAS
15  ;                    AREA       NUMBER     QUANTITY    ASSOCIATED
16  ;                    LABEL      OF PILES   ATTRIBUTE     FILES
17  ;                    =====      ========   ========    =========
18       AREA,          ENTRY,        1,         2,          1;
19       AREA,          EXIT,         1,         2,          9;
20       AREA,          LATHA,        1,         2,          3;
21       AREA,          MILLA,        1,         2,          6;
22       AREA,          RACK,         9,         2,          8,5,2;
23  ;
24  ; PILES
25  ;
26  ;                    PILE      AREA     PILE       X-COOR     Y-COOR
27  ;                    NUMBER    LABEL    CAPACITY   POSITION   POSITION
28  ;                    ======    =====    ========   ========   ========
29       PILE,           1,       RACK,       5,         90,        5;
30       PILE,           2,       RACK,       3,        120,        5;
31       PILE,           3,       RACK,       5,         75,        5;
32       PILE,           4,       RACK,       5,        105,        5;
33       PILE,           5,       RACK,       5,        135,        5;
34       PILE,           6,       RACK,       3,        165,        5;
35       PILE,           7,       RACK,       5,        150,        5;
36       PILE,           8,       RACK,       5,        180,        5;
37       PILE,           9,       RACK,       5,        195,        5;
38       PILE,          10,       ENTRY,      1,         45,        5;
39       PILE,          11,       EXIT,       1,         45,        5;
40       PILE,          12,       LATHA,      2,        120,        5;
41       PILE,          13,       MILLA,      2,        165,        5;
42  ;
43  ; CRANES
44  ;
45  ;        CRANE RUNWAY INITIAL INITIAL  MAXIMUM   NOMINAL NOMINAL TROLLEY JOB F
46  ;        LABEL NUMBER X POS.  Y POS.   BRDG. SPD. ACCEL. DECELL. SPEED   NUMB
47  ;        ===== ====== ======  ======   ========= ======  ======  ======  =====
48       CRA, CR1,   1,     60,     5,      50.,       0,      0,      10,     10;
49       CRA, CR2,   1,    140,     5,      50.,       0,      0,      10,     11;
50  ;
51  ; DEFINE REGULAR SLAM II RESOURCES
52  ;
53       RESOURCE,LATHE(1),4;
54       RESOURCE,MILL(1),7;
```

Figure 16-11 Statement model of ASRS.

```
55  ;
56  ; NETWORK STATEMENTS
57  ;
58        CREATE,15.,,0,1;
59  ;
60  ;  PART ARRIVAL PROCESSING
61  ;
62        ASSIGN,ATRIB(2) = 1,
63              ATRIB(3) = 0;             SET PALLET AREA = 1
64          ACT/6,,NNQ(3).LE.10,ARR2;   ENTERING
65          ACT/5,,NNQ(3).GT.10;        BYPASS CELL
66        TERM;
67  ;
68  ;  MOVE PALLETS TO ENTRY AREA
69  ;
70  ARR2 GWAIT(1),ENTRY/FAW;
71        ACT,0.25;                     INDEX
72  ARR3 GWAIT(2),RACK/FAW,CR1/CR2;     GET RACK & CRANE
73        ACT,0.5;                      PICK
74  ARR4 GFREE/MOVE,ENTRY;              FREE ENTRY
75        ASSIGN,ATRIB(3)=2.0;          PALLET: R -> L
76        ACT,0.5;                      PLACE
77        GFREE,,CR1/CR2;               FREE CRANE
78  ;
79  ;  MOVE PALLETS FROM RACK TO LATHE
80  ;
81  LATH GWAIT(3),LATHA/FAW,CR1/CR2;    GET LATHE AREA & CRANE
82        ACT,0.5;
83        GFREE/MOVE,RACK;              FREE RACK
84        ASSIGN,ATRIB(3) = 4;          PALLET: L -> R
85        ACT,0.5;                      PLACE
86        GFREE,,CR1/CR2;               FREE CRANE
87  ;
88  ;  PROCESS PARTS AT LATHE
89  ;
90  LAT4 AWAIT(4),LATHE/1;              REQUEST LATHE
91        ACT/1, RNORM(10.,.5,1);       LATHING
92        GOON,1;
93        ACT/3,UNFRM(15.,25.,2),.2;    MAINTAIN LATHE
94        ACT,,,.8;
95  LAT5 FREE,LATHE/1;                  FREE LATHE
96  ;
97  ;  MOVE PALLETS TO RACK
98  ;
99  LST0 GWAIT(5),RACK/LAW,CR1/CR2;     GET RACK AND CRANE
100       ACT,0.5;                      PICK
101 LST1 GFREE/MOVE,LATHA;             FREE LATHE AREA
102 LST2 ASSIGN, ATRIB(3) = 5;         PALLET : R -> M
103       ACT,0.5;
104 LST3 GFREE,,CR1/CR2;               FREE CRANE
105 ;
106 ;  MOVE PALLETS FROM RACK
107 ;
108 LST4 GWAIT(6),MILLA/FAW,CR1/CR2;   GET MILL AREA AND CRANE
```

Figure 16-11   (continued).

```
109          ACT,0.5;                      PICK
110  LST5  GFREE/MOVE,RACK;              FREE RACK
111  LST6  ASSIGN, ATRIB(3) = 3;         PALLET: M -> R
112          ACT,0.5;                      PLACE
113  LST7  GFREE,,CR1/CR2;               FREE CRANE
114  ;
115  ;   PROCESS PARTS AT MILL
116  ;
117  MILL  AWAIT(7),MILL/1;              REQUEST MILL MACHINE
118          ACT/2,RNORM(13.,.5,1);        MILLING
119        GOON,1;
120          ACT/4,UNFRM(10.,20.,2),.1;  MAINTAIN MILL
121          ACT,,.9;
122        FREE,MILL/1;                  FREE MILL
123  ;
124  ;   MOVE PALLETS FROM MILL TO RACK
125  ;
126  MIL2  GWAIT(8),RACK/FAW,CR1/CR2;    GET RACK AND CRANE
127          ACT,0.5;                      PICK
128  MIL3  GFREE/MOVE,MILLA;             FREE MILL AREA
129  MIL4  ASSIGN,ATRIB(3) = 1;          PALLET: R -> EXIT
130          ACT,0.5;                      PLACE
131  MIL5  GFREE,,CR1/CR2;               FREE CRANE
132  ;
133  ;   GROUP PALLETS AND MOVE VIA AGV
134  ;
135  BTCH  BATCH,,5,,,ALL(4);            BATCH 5 PALLETS
136          ACT,UNFRM(0.,20.,3);          AGV TIME TO ARRIVE
137        UNBATCH,4;                    REESTABLISH PALLETS
138  ;
139  ;   MOVE PALLETS TO EXIT AREA
140  ;
141  UNB1  GWAIT(9),EXIT/FAW,CR1/CR2;    GET EXIT AREA & CRANE
142          ACT,0.5;                      PICK
143        GFREE/MOVE,RACK;              FREE RACK
144  ;      GOON,1;
145          ACT,0.5;                      PLACE
146        GFREE,,CR1/CR2;               FREE CRANE
147          ACT,0.25;                     INDEX
148        GFREE,EXIT;                   FREE EXIT AREA
149        TERM;
150        ENDNETWORK;
151  FIN;
```

Figure 16-11 (continued).

Code	Description
6	Pallets waiting to move from rack to mill area
7	Queue for mill
8	Pallets waiting to move from mill area to rack
9	Pallets waiting to move from rack to exit

There are thirteen piles in this model. The x and y coordinates for each pile are specified in the PILE resource block statements. There are, at most, five pallets in a rack. The pile capacity for each rack is the number of pallets that can be included in the pile which is 3 or 5. It is assumed that the time to move in the y direction is subsumed by the time to move in the x direction, and, hence, it was not necessary to define a different y position for each pallet in the rack. The pile numbers for the rack area have been numbered to facilitate the logic used in assigning pallets to piles. The racks closest to the entry are numbered the lowest (1-6). The FAW piling logic is used to load from the entry area into the rack area and from the mill area into the rack area. This causes the low numbered piles to be used for pallets waiting for processing. The racks farthest from the entry are given the highest numbers and the LAW piling logic rule is used to store pallets containing castings that have completed the lathing operation.

Two regular resources are defined in this model which are the lathe and the mill. These resources are used to restrict the number of pallets that can be processed by the machines and to restrict processing during maintenance operations. The lathe area and mill area represent the entry space to the lathe and the mill, respectively.

The crane resource statements indicate that there are two cranes both on runway 1 with initial positions (60,5) and (140,5). The maximum velocity in the x direction for each crane is 50 feet per minute. Acceleration and deceleration are not modeled in this example. A trolley velocity of 10 is indicated. However, there is no movement in the y direction as all y locations have been given as 5 feet as previously discussed. Jobs waiting for crane 1 wait in file 10, and jobs waiting for crane 2 wait in file 11. Jobs in these files are ranked based on attribute 3 where attribute 3 is defined to prioritize which crane performs a specific type of movement. The following values are used for attribute 3 to define the priority of the next crane move for a pallet.

ATRIB(3)	Next move for entity
0	Entry to rack
1	Rack to exit
2	Rack to lathe area
3	Mill area to rack
4	Lathe area to rack
5	Rack to mill area

File 10 will be ranked on low value first based on attribute 3, LVF(3), and file 11 will be ranked on high values of attribute 3, HVF(3). This priority system for jobs waiting for cranes causes crane 1 to perform moves closer to the entry and exit areas and crane 2 to perform moves at the opposite end.

As mentioned above, attribute 2 is set to the pallet size in terms of the number of units of the capacity that a pallet takes in an area. For this model, it is assumed that the pallet size is 1, and that the area capacity is defined in terms of number of pallets. In the SLAM II model, ATRIB(2) of each entity is set to 1. Attribute 4 is used as the pointer to where the entities in a batch are maintained when accumulating pallets for transporting from the ASRS using an AGV. It is used at both the BATCH and UNBATCH nodes in this example. Below is a summary of the definition of the the attributes used in this model.

Attribute Number	Definition
1	Time of arrival of pallet
2	Space required for pallet
3	Next move descriptor
4	Pointer to entities in a batch.

The above definitions and descriptions should assist in reading the network and statement models. Recall that pile numbers and crane numbers are internally maintained by the MH Extension which records each entities position and location. The SLAM II model only requires that the logic and processing delays encountered by the entities flowing through the system be described.

Summary of Results. The base model was executed for 8 days with statistics cleared after day 1. The resulting output for this run is given in Figure 16-12. System throughput is 94 pallets per day. This throughput of 660 pallets in 7 days is achieved with the following utilizations which includes maintenance times.

Resource	Utilization
Lathe	0.92
Mill	0.95
Crane 1	0.79
Crane 2	0.69

Upon detailed investigation of the use of cranes, it was found that during periods when the machines were in cycle and during periods when the AGV loading was in progress, cranes did not provide adequate service to the machines. For this reason, crane priorities were adjusted. For the move to the lathe and the move from the mill, ATRIB(3) was set to -2 and -1, respectively. The intent was to have crane 1 serve the machines prior to servicing the ENTRY and EXIT areas. While

S L A M   I I   S U M M A R Y   R E P O R T

SIMULATION PROJECT ASRS EXAMPLE   BY PRITSKER & ASSOC.

DATE 3/21/1986   RUN NUMBER   1 OF   1

CURRENT TIME   0.1152E+05
STATISTICAL ARRAYS CLEARED AT TIME 0.1440E+04

FILE STATISTICS

FILE NUMBER	LABEL/TYPE	AVERAGE LENGTH	STANDARD DEVIATION	MAXIMUM LENGTH	CURRENT LENGTH	AVERAGE WAITING TIME
1	ARR2 GWAIT	0.0000	0.0000	1	0	0.0000
2	ARR3 GWAIT	0.0627	0.2425	1	0	0.9524
3	LATH GWAIT	3.0351	3.2221	11	0	46.0054
4	LAT4 AWAIT	0.3793	0.4852	1	1	5.7408
5	LST0 GWAIT	0.0729	0.2600	1	0	1.1053
6	LST4 GWAIT	3.3200	2.3697	9	6	50.1736
7	MILL AWAIT	0.4010	0.4901	1	0	6.1052
8	MIL2 GWAIT	0.0709	0.2566	1	0	1.0791
9	UNB1 GWAIT	0.8231	1.3917	5	4	12.4758
10		2.6445	3.4691	16	6	3.0707
11		2.6445	3.4691	16	6	3.0707
12	CALENDAR	4.4826	0.8138	11	6	0.7242

REGULAR ACTIVITY STATISTICS

ACTIVITY INDEX/LABEL	AVERAGE UTILIZATION	STANDARD DEVIATION	MAXIMUM UTIL	CURRENT UTIL	ENTITY COUNT
1 LATHING	0.6607	0.4735	1	0	666
2 MILLING	0.8530	0.3541	1	1	662
3 MAINTAIN LAT	0.2573	0.4371	1	1	130
4 MAINTAIN MIL	0.0989	0.2986	1	0	68
5 BYPASS CELL	0.0000	0.0000	1	0	8
6 ENTERING	0.0000	0.0000	1	0	665

RESOURCE STATISTICS

RESOURCE NUMBER	RESOURCE LABEL	CURRENT CAPACITY	AVERAGE UTILIZATION	STANDARD DEVIATION	MAXIMUM UTILIZATION	CURRENT UTILIZATION
1	LATHE	1	0.9180	0.2744	1	1
2	MILL	1	0.9520	0.2138	1	1

RESOURCE NUMBER	RESOURCE LABEL	CURRENT AVAILABLE	AVERAGE AVAILABLE	MINIMUM AVAILABLE	MAXIMUM AVAILABLE
1	LATHE	0	0.0820	0	1
2	MILL	0	0.0480	0	1

Figure 16-12 Summary report for ASRS.

AREA/PILE STATISTICS

AREA NUMBER	AREA LABEL	PILE NUMBER	PILE CAPACITY	CURRENT UTILIZATION	AVERAGE UTILIZATION	MINIMUM UTILIZATION	MAXIMUM UTILIZATION	NUMBER OF ARRIVING	ENTITIES LEAVING
1	ENTRY	10	1.00	1.00	0.24	0.00	1.00	665	664
2	EXIT	11	1.00	1.00	0.25	0.00	1.00	661	660
3	LATHA	12	2.00	2.00	1.72	0.00	2.00	665	665
4	MILLA	13	2.00	1.00	1.80	0.00	2.00	661	662
5	RACK	1	5.00	4.00	4.24	0.00	5.00	871	872
		2	3.00	1.00	1.62	0.00	3.00	290	289
		3	5.00	0.00	1.16	0.00	5.00	142	142
		4	5.00	0.00	0.28	0.00	4.00	23	23
		5	5.00	0.00	0.00	0.00	0.00	0	0
		6	3.00	0.00	0.00	0.00	0.00	0	0
		7	5.00	0.00	0.00	0.00	0.00	0	0
		8	5.00	2.00	0.71	0.00	4.00	75	73
		9	5.00	4.00	3.03	0.00	5.00	590	588

CRANE STATISTICS

CRANE NUMBER	CRANE LABEL	NUMBER OF PICKS	DROPS	PICK UTIL	TO PICK UTIL	TO PICK INTERF	DROP UTIL	TO DROP UTIL	TO DROP INTERF	TOTAL UTIL
1	CR1	2187	2186	0.11	0.23	0.12	0.11	0.22	0.01	0.79
2	CR2	1791	1790	0.09	0.22	0.11	0.09	0.17	0.00	0.69

Figure 16-12   (continued).

this strategy might cause some delays for the AGV, it should improve lathe and mill performance.

With these changes, the model was rerun and the results were not as expected. Throughput went down to 74 pallets per day. Upon closer investigation, it was discovered that the new scheduling logic caused an increase in crane interference to about 25 percent due to the second crane's third priority task being to service the EXIT. This increase in interference made the cranes a critical resource. Thus, storage eventually was saturated and the system could not produce outputs.

To eliminate this possibility, the second crane was eliminated from ENTRY and EXIT tasks, the first crane was eliminated from milling machine tasks, and lathe tasks were given a high priority for crane 1 and a lower priority for crane 2. The model was rerun with this alternative schedule. Throughput was increased to 96 pallets per day.

Model outputs also indicate that the capacity of rack storage is larger than needed with little or no usage of racks 4-8. Unless additional machines are planned for this cell, the designers should reduce rack size or the number of racks planned. The model outputs also indicate that crane interference is around 12 percent. Experience shows that this is a reasonable level of crane interference for a system of this type.

16.11 AUTOMATED GUIDED VEHICLE SYSTEM (AGVS) DEFINITIONS

The use of automated guided vehicles in manufacturing facilities is expanding at a rapid rate (5). An AGV provides a means to tightly govern material control policies and can be a cost effective alternative to labor intensive or floor space consuming methods of material handling.

The implementation of an AGVS requires special consideration by system designers and system managers. The impact of an AGVS on the entire production system is critical, especially when the production system has been designed to maintain little in-process inventory. The expense of the vehicles, the guidepath, and the computer control system prohibits excessive over-design.

There have been many applications of automated guided vehicles. The automotive industry has been a prime user, with automobiles assembled directly on an AGV which carries the car from one station to another.

Physically, an AGVS is comprised of vehicles and guidepaths. Guidepaths may be installed as wire under the surface of the floor through which radio signals are sent, tape on the surface of the floor, or other visual or chemical means by which a vehicle's path can be defined. Vehicles travel along these paths to reach destinations at rates depending on the acceleration, maximum speed, and deceleration of the vehicle unit. A vehicle may encounter portions of the guidepath which are congested by slower moving AGV traffic, traffic at intersections, vehicles which are stopped for loading or unloading, and interference from other MH devices.

Vehicles are controlled by either a central computer, an on-board computer, or a combination of the two. The control system instructs a vehicle to travel to a particular position and also describes the route the vehicle is to take to get to that position. The control system also determines system logic rules such as how to dispatch a vehicle for a job when more than one vehicle is available, which vehicle has priority when there is contention for an intersection, and other decisions related to traffic control.

The performance of an AGVS is highly dependent on the physical layout and the control policies. As the AGVS is an integral part of a larger system, the AGVS performance is directly related to the performance of the total system. For example, the number of vehicles selected to operate on the guidepath can cause blocking at a loading station in one of two ways. First, too few vehicles may not be able to remove material from the loading station in a timely manner. On the other hand, adding too many vehicles may cause high levels of congestion which may prevent the vehicles from efficiently removing material from the loading station. Control policies can have a similar effect on total system throughput, utilization of equipment and personnel, and the level of in-process inventory.

16.11.1   AGV Movement Characteristics

The MH Extension provides AGVS modeling capabilities at a systems level. The following is a list of characteristics and assumptions included in the AGVS extension.

1. All vehicles from a vehicle fleet have homogeneous characteristics such as maximum empty speed, maximum loaded speed, acceleration, deceleration, and length. More than one vehicle fleet can operate on the same guidepath.
2. Material to be transported requires a vehicle from exactly one vehicle fleet.
3. A vehicle unit may be in exactly one of the following states:
   - idle stopped,
   - traveling idle,
   - traveling empty to a load point,
   - loading,
   - traveling full to an unload point, and
   - unloading
4. A vehicle occupies a physical location on the guidepath which may cause interference with other vehicles.
5. Acceleration and deceleration rates of a vehicle do not vary with system or vehicle status.
6. Breakdowns and battery charging are modeled as a type of job request for the AGV.
7. Communication with a controller for routing and assignment is restricted to specified positions called control points on the guidepath. Segments connect the control points.
8. Vehicles may travel on a path in either direction although traffic can be restricted to traveling in one direction. Internal default control policies only support deadlock prevention when unidirectional track is used. While internal default control policies attempt to prevent bidirectional deadlock, additional user logic may be required in system specific situations.
9. A vehicle will attempt to take the shortest path to its destination.
10. In situations where one or more vehicles are involved in head-on contention, the higher priority vehicle will be allowed to take the preferred path in bi-directional contention situations. Lower priority vehicles will select an alternative path. These rules result in the following precedence scheme when two vehicles are involved in head-on contention for a segment.
    - A vehicle with only one possible exit will cause the direction of lower priority vehicles to reverse.

- A vehicle traveling full to unload will cause the direction of a lower priority vehicle to reverse.
- A vehicle traveling empty to load will cause the direction of idle vehicles to reverse.

The modeler controls vehicle movement patterns by specifying the logic rules to be used for path selection, precedence at intersections, idle vehicle control, and vehicle and load selection. The standard set of logic rules may be supplemented by the modeler with user-written FORTRAN subprograms.

16.11.2   AGVS Resources: VCPOINT, VSGMENT, and VFLEET

Definitions of three resource types are required for modeling an AGVS in the MH Extension: control points, segments, and vehicle fleets. Vehicle control points are locations on the guidepath network at which an AGV begins traveling on a segment or where it stops to load, to unload, or to wait for instructions from a controlling computer. Control points are also called communication points or broadcast points.

All intersections and load or unload points are inherently control points. Segments begin and end at control points. Vehicles are part of a fleet and are used to pickup and dropoff loads at control points. The movement of a vehicle is in accordance with its velocity and acceleration characteristics. Its routing is determined by specifying rules for selecting segments to take to reach a destination control point. In the MH Extension, the three special resource types VCPOINT, VSGMENT and VFLEET, are used to define control points, segments and vehicle fleets, respectively.

16.11.3   Control Point Resource: VCPOINT

Control points are described by the VCPOINT resource. The resource block statement is shown below.

VCPOINT,CPNUM/CPLBL,RCNTN,RROUT,CHARGE;

The control point is given a reference number, CPNUM, for use in other portions of the model, in FORTRAN code, and on output reports. An optional field, CPLBL, may be specified to provide a label for output reports.

There are two logic rule types for the control point resource; RCNTN, the rule for contention, and RROUT, the rule for routing decisions. Contention for a control point resource occurs when more than one vehicle is waiting to enter the control point. A list of the waiting vehicles is kept in their order-of-arrival to the control point.

The contention rule specifies which vehicle is to be given the right-of-way when the control point becomes available. The options for this rule are FIFO, CLOSEST, PRIORITY, and URCNTN(NR) with FIFO being the default rule.

The CLOSEST rule selects the vehicle with the shortest path distance to its ultimate destination. The PRIORITY rule gives precedence to the vehicle with the highest load status where the load status is based on the bidirectional contention priority described in Section 16.11.1, item 10.

The final option, URCNTN(NR), is available for the user to write selection rule NR for contention. If URCNTN(NR) is entered as the contention rule, subroutine URCNTN is called. The user may write subroutine URCNTN to scan the list of waiting vehicles and select the vehicle which will next capture the control point resource.

The second type of logic rule for control points is RROUT, the rule for vehicle routing. This rule is used when a decision is to be made regarding which path segment a vehicle is to traverse from the control point. There are two options available for RROUT. The default case is shortest path as specified by SHORT. With the SHORT specification, the vehicle is routed on the path segment which starts the shortest feasible path to its destination control point. A feasible path is a series of segments which lead to the destination control point, none of which have directional restrictions that prohibit a vehicle from traveling on it. Dynamic considerations, such as a vehicle currently stopped for loading along a path, are not taken into account with SHORT. When a vehicle is encountered which is traveling in the opposite direction on a selected segment, or stopped at an intermediate control point, a re-evaluation is made of the path selected.

The specification URROUT(NR) for RROUT causes the segment whose identification number is returned from the user-written subroutine URROUT to be taken. The vehicle is stopped if a value of zero is returned. If the vehicle is stopped, a procedure is necessary for determining when the routing decision should be reassessed.

The CHARGE field is used to indicate that a control point has in-line charging capabilities. A YES input for CHARGE causes a vehicle which is stationary at the control point to have its charging flag set to ON. The vehicle accumulates charging time until it begins to exit the control point. The charging flag is then set OFF when the vehicle departs the control point. The user can then use these flag values to calculate and determine vehicle charge levels.

16.11.4   Segment Resource: VSGMENT

The guidepath track between control points is defined as a resource of type VSGMENT. The AGV must travel on a segment to move from one control point to another. One VSGMENT resource is required for each segment connecting two control points. The statement for the VSGMENT resource block is shown below:

VSGMENT,SGNUM/SGLBL,CPNUM1,CPNUM2,DIST,DIR,CAP;

The segment is given an identifying number SGNUM for use in reports and user written FORTRAN subprograms. An optional label SGLBL is used for output reports. The end control points of the segment are specified as CPNUM1 and CPNUM2. The length of the segment is defined as DIST, that is, the distance from one control point to another. The minimum value for DIST is twice the check zone radius plus the vehicle length. The check zone radius is used to control vehicle spacing on a segment and is described in the next section.

The DIR field specifies the directional characteristics of the segment. The options for this field are UNIDIRECTIONAL or BIDIRECTIONAL. If UNIDIRECTIONAL is specified, vehicles may only travel from CPNUM1 to CPNUM2.

The final field, CAP, is used to specify the maximum number of vehicles which may travel on the segment concurrently. A vehicle is on a segment until it has entered the following control point. If the CAP field is defaulted, the AGV extension calculates the maximum number of vehicles allowed on the segment simultaneously based on the length of the segment, the length of the vehicle, the check zone radius required at intersections, and the distance required between each vehicle.

A spur is a special type of bidirectional segment which is used by vehicles to travel from a main guidepath to a load, unload, or idle stopping area. A spur has only one entrance/exit control point and a capacity of one vehicle unit. The MH Extension automatically identifies such segments as spurs. An illustration of a spur segment is given below.

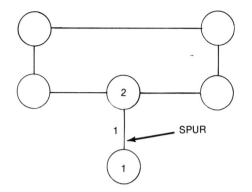

A vehicle positioned at control point 1 in the illustration is off the main guide-path and considered to be on segment 1 and also on control point 1. By defining segment 1 to have a capacity of one, other vehicles will not be allowed to enter segment 1 until the vehicle on the spur has re-entered the main guidepath. This prevents other vehicles from entering the spur, only to reverse direction when the vehicle at control point 1 attempts to exit.

16.11.5   Vehicle Fleet Resource: VFLEET

The VFLEET resource is used to model material handling devices such as driv-erless tractors, wire-guided pallet trucks, and unit-load transporters and platform carriers. Vehicles like forktrucks, which are not automatically guided, can also be modeled with the VFLEET resource.

Each VFLEET resource block defines a fleet of AGV material handling units. The definition includes a vehicle fleet's physical attributes, next job selection rule, rule for idle disposition logic, and the initial location of the individual vehicle units. The statement for the VFLEET resource block is shown below.

VFLEET,VFLBL/VFNUM,NVEH,ESPD,LSPD,ACC,DEC,LEN,DBUF,CHKZ, IFL/RJREQ,RIDL,ICPNUM(NOV,SGNUM)/ . . . repeats . . , REPIND;

The label of the vehicle fleet, VFLBL, is used to reference the vehicle fleet on the network and on output reports. The optional field VFNUM specifies an internal vehicle fleet identification number. This option is provided to allow users access to information about the vehicle fleet in FORTRAN subprograms. If this field is defaulted, an internal number is assigned which is one greater than the highest previously defined vehicle fleet number.

The number of vehicles in the vehicle fleet is NVEH. Each vehicle can travel at a maximum speed ESPD when it is empty and a maximum speed LSPD when it is carrying a load. The acceleration of the vehicles is ACC. The deceleration is DEC.

The fields LEN and DBUF are used to maintain spacing between the vehicles and to control interference around intersections. The length of the vehicle is specified as LEN. DBUF defines the distance buffer which must be maintained between any two vehicles. DBUF should be specified as the on-center distance. On-center refers to the distance between the centers of the vehicles rather than the distance between the front of one vehicle and the back of another. The default for DBUF is LEN. In the default case, vehicles may actually come in contact with one another.

A check zone is the area around an intersection in which only one vehicle may be present simultaneously. The parameter CHKZ is used to define the radius of the check zone area and is the size of the control point resource. The default for CHKZ is one half of LEN. The default case is the minimum acceptable value to insure that a vehicle may be positioned fully on the control point. A vehicle that is loading, unloading, or idle at a control point is typically not considered to be on a segment. In cases where more than one vehicle fleet operates on a guidepath, both DBUF and CHKZ will be altered to the maximum of these parameters for all vehicle fleets. This is necessary to prevent collisions of different sized vehicles.

Job requests waiting to be performed by the vehicle fleet are listed in file IFL. Job requests are not loads but are signal entities. When a vehicle has completed a previous job, it will interrogate the job file to determine if there are outstanding requests. If the rule for job request, RJREQ, is PRIORITY, the vehicle will always take the highest priority request. The ranking of the file is established with the SLAM II PRIORITY statement to allow FIFO, LIFO, HVF(NATR), and LVF(NATR) ranking of job requests. CLOSEST is another option for RJREQ and causes the job that currently is the shortest path distance from the vehicle to be selected. Alternatively, RJREQ may be specified to select a particular request from the file using the user-written subroutine URJREQ.

When an AGV completes delivery of material to a dropoff location, the vehicle is released to perform other tasks. The vehicle will become idle if there are no transport jobs waiting to be performed. The parameter RIDL is used to specify the logic rule to be applied when positioning idle vehicles. Generally, AGV system control logic provides two options:

1. The vehicle will STOP at its current or a specified position on the guidepath network and wait until requested, or

2. The vehicle will CRUISE on the guidepath network in a predefined pattern until requested.

Additional parameters are required if the vehicle must travel to a position other than its current one. The complete RIDL option list is shown in Table 16-2. The default specification is STOP.

Table 16-2 Rules for Idle Vehicle Disposition.

Specification	Description
STOP	Vehicle stays at location where it was made idle and waits for request.
STOP(CPNUM)	Vehicle travels to control point defined by CPNUM, stops, and waits for the next request.
STOP(CPNUM list)	Vehicle travels to the control point listed that has the shortest path from its present position, stops, and waits for the next request.
CRUISE(CPNUM list)	Vehicle travels to the control points listed starting with the closest one and continues in the sequence listed until a request is made.
URIDL(NR)	Vehicle travels to control point whose identification number is specified by subroutine URIDL. NR is a user defined index for the rule.

The initial locations of vehicles are specified on the resource block by the parameters, ICPNUM(NOV,SGNUM) where NOV is the number of vehicles initially located at segment number SGNUM. There are three ways for specifying the initial locations.

First, the vehicles of the fleet can be distributed on the guidepath network at different control points by repeating the control point number field, ICPNUM, and not specifying the optional fields NOV and SGNUM. A vehicle will begin the simulation at each ICPNUM listed. For all vehicles to be positioned initially, there must be a control point for each of the NVEH vehicles defined. For example, if there are n vehicles and each vehicle begins at a different control point, the initial location specification would be as shown below.

VFLEET, . . . , ICPNUM1/ICPNUM2/ . . . / ICPNUMn;

A second alternative for specifying initial locations (in which two or more vehicles can begin the run in the same general location) involves using the optional fields NOV and SGNUM. SGNUM identifies the segment number on which vehicles are positioned while waiting for control point ICPNUM. Therefore, one vehicle begins the run on control point, ICPNUM, and NOV-1 vehicles begin the run on segment SGNUM waiting to get to ICPNUM. Groups of vehicles can be distributed by repeating the parameters as shown below.

VFLEET, . . . ,ICPNUM1(NOV1,SGNUM1)/ICPNUM2(NOV2,SGNUM2)/ . . . ;

The third specification is used if there are more vehicles in the fleet than location specifications. In this case, a vehicle unit remains off the track until a job request is issued which cannot be filled by the vehicles on the track. The entering vehicle will appear at the control point at which the job request is issued.

The final parameter for the VFLEET statement is REPIND (Report Individual). If YES is specified for REPIND, utilization statistics for the individual vehicles are desired as well as the standard statistics for the entire vehicle fleet. The default for REPIND is NO.

16.11.6   AGV Network Elements

Three new network elements have been added to model an AGVS. The VWAIT node allocates a vehicle to an entity. It has a location in the AGVS defined by a control point number. An entity requiring a vehicle waits in a file at a VWAIT node in a manner similar to an entity waiting for a resource at an AWAIT node. When an entity arrives to a VWAIT node, a request for a vehicle is issued and an idle vehicle is dispatched to the control point of the VWAIT node. When a vehicle arrives to the control point of the VWAIT node, it is allocated to an entity waiting at that VWAIT node.

To move an entity which has been allocated a vehicle, the entity is routed to a VMOVE activity. The VMOVE activity assigns a control point destination to the entity and models the traveling of an entity from one control point to another. The duration of the move is calculated automatically by SLAM II, taking into account: 1) the characteristics of the vehicle allocated to the entity; 2) the control point location of the VWAIT node where the entity starts its move; 3) the control point destination assigned at the VMOVE activity; and 4) the congestion and status of control points and segments over which the entity/vehicle travels.

A VFREE node is used to free a vehicle that has been allocated to an entity. After being freed, the vehicle is dispatched from the control point destination of its most recent VMOVE activity. The disposition of the vehicle is determined in accordance with the current list of job requests. If there are no job requests for the vehicle then a rule for idle vehicle movement is invoked. The entity arriving to the VFREE node is routed from it without the vehicle resource. In the following sections, the VWAIT and VFREE nodes and the VMOVE activity are described.

16.12   VWAIT NODE

The VWAIT node is used to model a request for transport by a vehicle from a control point. An entity arriving to a VWAIT node is typically a unit load completing a process and requesting transportation to a station or inventory area. However, entities requesting vehicles could be maintenance or charging tasks. The load entity remains in the VWAIT node file until a vehicle has traveled to the control point specified on the VWAIT node.

When the entity arrives to the VWAIT node, a signal is placed in the vehicle fleet's request file to notify the AGV controller that there is a demand for the vehicle fleet's transportation services. This is comparable to an operator entering a request for vehicle service or an automatic signal being sent to the AGV controller, for example, from a photoelectric device. A load request is satisfied by dispatching an idle vehicle to the load's control point location. The travel time to the control point is based on the location from which the AGV is dispatched, its vehicle dynamics, and the congestion it encounters. A load entity is released from the VWAIT node when the AGV arrives at the control point from which the request was made.

The VWAIT node symbol and statement are shown below.

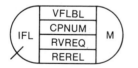   VWAIT(IFL),VFLBL,CPNUM,RVREQ,REREL,M;

Entities wait in file IFL until an idle vehicle from vehicle fleet VFLBL moves to control point CPNUM.

CPNUM may be specified as a positive integer or as a SLAM II variable. When more than one vehicle is idle at the time an entity arrives to the VWAIT node, a rule for vehicle request, RVREQ, is used to select from among idle vehicles. The default for RVREQ is FIFO which causes the vehicle that has been idle the longest to be dispatched. Another option for RVREQ is CLOSEST. In this case, the vehicle with the shortest path distance to the entity requesting pickup will respond to the request. A user rule, URVREQ(NR), can be specified which causes user-written subroutine URVREQ to be called to provide a vehicle number to satisfy the request.

When a vehicle arrives to the pickup control point, CPNUM, an entity is removed from the file, IFL, according to release rule, REREL. The default for REREL is MATCH which specifies that the entity issuing the request for this particular vehicle is to be removed from the file even if the entity is not first in the file. The MATCH specification for REREL is useful when a SLAM II variable is used to specify the control point on the VWAIT node, thus allowing entities in the same file to be at different control points. REREL can also be specified as TOP in which case the entity ranked first in the file is removed. A user-written rule can be used for the selection by specifying UREREL(NR). To release an entity, a vehicle must be located at the same control point at which the entity is waiting.

At each node release, a maximum of M activities are taken from the VWAIT node. If M is greater than 1, care must be exercized by the user. When an entity is assigned a vehicle, one of its internal SLAM II attributes is used to identify the vehicle. If the entity is routed over more than one activity emanating from the same node, then only one of the entities carries the vehicle information with it. This entity is the one released over the first activity in the list of parallel activities The MH Extension only allows an entity to be assigned one vehicle at a time.

When a vehicle is assigned to an entity, the vehicle is considered in a loading state. The state of the vehicle changes to traveling when the entity passes through a VMOVE activity and to unloading when it is released from a VFREE node. When traveling, the vehicle is further classified as being empty or full. Statistics on the utilization of vehicle fleets are provided on the SLAM II Summary Reports according to its state of operation, that is, empty or full and picking up, traveling to or dropping off loads.

Since an entity in a VWAIT node has issued a request for a vehicle, removal of an entity using a FORTRAN subprogram from a VWAIT node can be problematic and should not be done.

Consider the following VWAIT node and statement.

   VWAIT(11),AGV1,32,CLOSEST,TOP,1;

An entity arriving to this node will wait in file 11 and issue a request for a vehicle from fleet AGV1. The entity in the VWAIT node is at control point 32. A request has been made to dispatch the vehicle CLOSEST to control point 32 to move the entity. When the vehicle arrives at control point 32, the entity that has the top priority in file 11 should be picked up. When the vehicle arrives to pick up an entity, one activity emanating from the VWAIT node should be taken.

A second illustration of the VWAIT node is shown below.

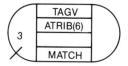   VWAIT(3),TAGV,ATRIB(6),,MATCH;

In this case an entity requests a vehicle from fleet TAGV and waits in file 3 for pickup. The arriving entity is at the control point defined by its ATRIB(6) value; this allows entities in file 3 to be waiting at different control points. The default vehicle request rule, FIFO, is used to dispatch the vehicle which has been idle the longest. When the vehicle arrives at the control point defined by ATRIB(6), it is to pick up the entity that issued the request. This is specified by MATCH for the REREL parameter.

A third illustration of the VWAIT node is shown below.

For this node, entities requesting a vehicle from fleet AGVX wait in file 26 which has a control point defined by the global SLAM II variable XX(1). User rule 2 for vehicle request selection and user rule 3 for entity release are to be employed. Procedures for writing these subroutines for making selections will be described later. When a vehicle picks up an entity that is in file 26, the entity will branch from the VWAIT node and up to two activities following the VWAIT node will be taken. The vehicle assigned to the entity will only travel with the entity over the first branch taken from the VWAIT node.

The following points summarize performance subtleties and assumptions regarding the VWAIT node.

- An entity released from the VWAIT node has ATRIB(MATR-2) set at a value that identifies the individual vehicle which is transporting it. When more than one activity is released following a VWAIT, only the first release carries this identifying vehicle number.
- An entity may hold only one vehicle at a time.

- Entities in VWAIT nodes represent a demand for a vehicle.
- An entity in a VWAIT node should not be removed by the user because SLAM II is maintaining a separate request for a vehicle.
- The rule for vehicle request is used to allocate a vehicle when an entity arrives to a VWAIT node and a vehicle in the desired fleet is available. The rule for job request is used to allocate a vehicle when it is freed at a VFREE node.
- At the time an entity is released from the VWAIT node, the vehicle associated with that entity is considered to be in the loading state. The vehicle remains in that state until the entity passes through a VMOVE activity or VFREE node.

16.13   VFREE NODE

When an AGV completes its transport and unloading operations, it is released to be dispatched to another job request. The VFREE node causes a vehicle from fleet VFLBL to be released. The symbol and statement for the VFREE node are shown below.

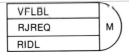   VFREE,VFLBL,RJREQ,RIDL,M;

The vehicle held by an entity is freed and reassigned in accordance with the rule for job requests, RJREQ, specified at the VFREE node. The options for RJREQ are: PRIORITY, CLOSEST, and URJREQ(NR). These options are the same as the ones on the VFLEET statement. The default for RJREQ at the VFREE node is the specification for RJREQ on the VFLEET resource block.

If no requests are outstanding, the vehicle is placed in the idle state, and the rule for idle vehicle disposition, RIDL, is invoked. If RIDL is not specified, the rule given on the VFLEET resource block for VFLBL is used. Entities arriving to the VFREE node must have seized a vehicle at a VWAIT node. Unloading and other activities which require the vehicle should be performed before the entity is routed to the VFREE node.

A maximum of M activities are taken at each node release. An example of the VFREE node is shown below.

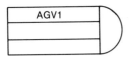   VFREE,AGV1;

A vehicle from vehicle fleet AGV1 is made available for dispatching when an entity arrives to this VFREE node. The entity must have been assigned a vehicle from fleet AGV1. The control point destination assigned to the vehicle is the starting control point for the next movement of the vehicle when it is freed. The control point destination could be a control point location associated with a VWAIT node in which case entities may be waiting for the vehicle at the location. To have this vehicle satisfy a job request at the dropoff control point, the job request rule should be specified as CLOSEST. In such a case, the vehicle is dispatched and arrives to pickup the job in zero time. The vehicle and the entity picked up would then branch from the VWAIT node.

The entity that arrived to the VFREE node branches in accordance with the activities following the VFREE node.

16.14 VMOVE ACTIVITY

To move an entity that has been loaded on a vehicle, it is necessary to specify a destination control point. The movement of the entity is then controlled by the AGVS characteristics as specified through the definition of the VFLEET resource, the guidepath network, and the movement rules. Due to the complexity of such movements, the routing and travel time are automatically calculated by the MH Extension within the VMOVE activity. Since the typical mode of operation involves a loading activity before the move activity and an unloading activity after the move activity, a grouping of symbols to accomplish the vehicle move is used. That is, a GOON node would normally be required before and after the VMOVE activity. To avoid these extra nodes, a node-activity-node symbol is used for the VMOVE activity which is a split GOON node with a branch connecting the two halves of the GOON node as shown below.

VMOVE,CPNUM,M;

The parameter CPNUM is a destination control point number and may be specified as a constant or a SLAM II variable. When the vehicle/entity combination reaches CPNUM, a maximum of M activities following the VMOVE activity may be taken. As explained above, the use of a node-activity-node representation for a vehicle move allows activities such as loading before the VMOVE statement and activities such as unloading following the VMOVE statement.

An illustration of the VMOVE activity is shown below.

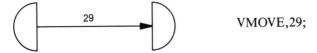

VMOVE,29;

An entity arriving to the VMOVE activity has a vehicle resource assigned to it and is ready to begin its movement toward a control point. The above symbol specifies that the movement is to be toward control point 29. Internal to the MH Extension, the move is made in accordance with the assumptions and rules built into the AGVS system modeled.

It is assumed that when an entity is released from the VMOVE activity, the vehicle associated with the entity is in an unloading state. The vehicle remains in the unloading state until another VMOVE activity or VFREE node is encountered by the entity to which the vehicle has been assigned.

An illustration of a network segment is shown below including a VWAIT node, a VMOVE activity, and SLAM II activities.

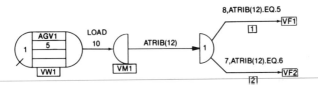

```
VW1   VWAIT(1),AGV1,5;
          ACT,10;
VM1   VMOVE,ATRIB(12),1;
          ACT/1,8,ATRIB(12).EQ.5, VF1;
          ACT/2,7,ATRIB(12).EQ.6, VF2;
```

In this illustration, at VWAIT node VW1 an entity requests and acquires a vehicle from vehicle fleet type AGV1 at control point 5 after waiting in file 1. The movement of the empty AGV to control point 5 is not displayed in the network. A loading activity is then processed which has a 10 time unit duration. The VMOVE activity is used to represent the travel time of both the vehicle and entity to a control point specified by ATRIB(12). When the move is completed, branching occurs. Activity 1 represents the unloading at control point 5 which takes 8 time units; and activity 2 represents the unloading at control point 6 which takes 7 time units.

16.15 MH EXTENSION CONTROLS AND OUTPUTS FOR AGVS

The AGVS portion of the MH Extension includes a new control statement, VCONTROL. The VCONTROL statement provides information on AGVS step size calculations and for output reporting as shown below.

VCONTROL,DTMIN,DTMAX,TRIPREP,CPREP,SGREP;

The definitions of DTMIN and DTMAX are the standard SLAM II definitions and are used to control the continuous features included in the AGVS module. The minimum value of DTMIN and DTMAX are used if both a VCONTROL and a SLAM II CONTINUOUS statement are used. TRIPREP, CPREP, and SGREP are YES/NO variables indicating whether trip counts, control point utilization, and segment utilization reports are to be printed for a particular run. Specifying YES for TRIPREP generates a count of vehicle trips where a trip is defined as a movement from a control point to pickup a load or from a pickup location to a dropoff location. The default value for each of these fields is YES.

The option ATRACE (LEVEL) has been added to the MHMONTR statement to trace selected events that pertain to the AGVS with a LEVEL of detail of either DETAIL or NODETAIL. An example is shown below

MHMONTR,ATRACE(DETAIL),100,200;

where the field DETAIL is optional. The ATRACE option causes a textual trace of detailed vehicle movement and assignments to be printed from time 100 to time 200. This includes a listing of all intersections encountered by a vehicle on the guidepath. A specification of ATRACE(NODETAIL) causes only vehicle assignments and beginning of trip messages to be printed. Multiple MHMONTR statements may be used.

In the MH Extension, ATRIB(MATR-2) is used to store the identification of the vehicle unit allocated to an entity. Recall that two attributes were used for crane information. These must be taken into account when defining the maximum number of attributes per entity, MATR, of the LIMITS statment. The AGVS module uses state equations to model the movement of vehicles. The MH Extension assigns state variable indices which are greater than NNEQD+NNEQS specified by the user on the CONTINUOUS input statement. In addition, state conditions are automatically defined in the MH Extension to detect cranes and vehicles reaching their destination. If a large minimum step size is set on VCONTROL or

CONTINUOUS input statements, the defined tolerance of a crossing may be exceeded. By default, a warning message will then be printed by SLAM II. To suppress warning messages of this type, a NO specification should be made for the field on the CONTINUOUS statement regarding the printing of out-of-tolerance messages.

The following sections have been added to the SLAM II Summary Report: (1) vehicle utilization; (2) control point utilization; (3) segment utilization; and (4) origin and destination trip counts. Illustrations of these outputs are included in Example 16-2.

16.16   EXAMPLE 16-2. MACHINING CELLS WITH AN AGVS

A schematic diagram of a machining cell layout including an AGV guidepath network is given in Figure 16-13. Raw cast parts arrive to a load/unload fixture station from raw material storage. Each part needs to be loaded individually onto a fixture at the fixture station before it can be machined at any one of six machining cells. A request for transport by an AGV for the part is not made until a machining cell is available.

When the fixtured part arrives to the appropriate cell, it is unloaded from the AGV and processing by the machine begins. After delivery, the AGV is released to perform other jobs. If no jobs are outstanding, the vehicle returns to an AGV staging area where it waits for assignment.

Upon completion of the machining process, an AGV is requested to return the part to the fixture station for unloading. Machined parts are given preference over incoming parts at the fixture station. After unloading, the part exits the system.

The vehicles always travel in a clockwise fashion except on the spur segment in front of the fixture station. On the spur, an AGV can travel in both directions. The transfer of a part to and from a machining cell is performed with the AGV on the main guidepath.

There are two vehicles in the system and a maximum of five loaded parts may wait for transport after fixturing. It is desired to analyze the operation of this system for 89,400 seconds. The time between part arrivals is exponentially distributed with a mean of 436 seconds. It takes 220 seconds to load a part on a fixture and 180 seconds to unload it. The time to put a fixtured part on or to take it off the AGV is 45 seconds. Machine processing time for a part is triangularly distributed with the following parameters: a minimum value of 960, a modal value of 1860, and a maximum value of 2720 seconds.

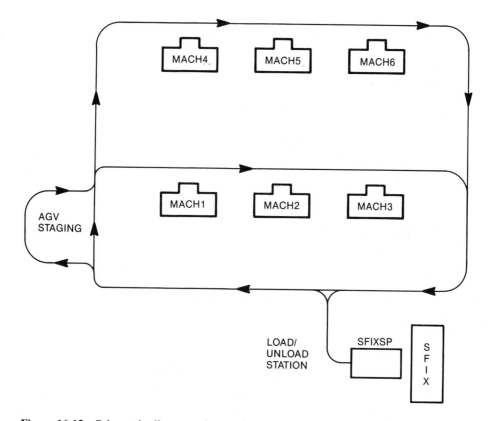
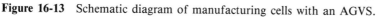

Figure 16-13 Schematic diagram of manufacturing cells with an AGVS.

The objective of the model is to evaluate current rules for using AGVs to transport parts to the manufacturing cells. Statistical information on the time to process a part through the system, resource utilization, and AGV trip and guidepath use are desired.

Concepts Illustrated. In this example, definitions of VCPOINT, VSGMENT, and VFLEET resources for a guidepath are made. The use of the VWAIT node to request an AGV and a VMOVE activity to transport a load by an AGV are demonstrated. The integration of the AGVS nodes and activities with standard SLAM II nodes, activities and resources is described.

SLAM II Model. Figure 16-14 presents a facility model of the manufacturing system including locations, numbers, and labels. All model resources are identified on the diagram by a unique shape, label or number. Vehicle control points on the AGV guidepath are marked as circles with the identification number of the control point in the circle. For convenience, the points are numbered bottom to top; left to right. These control points define locations on the track where vehicles load and unload parts as well as where they stop to wait for an assignment.

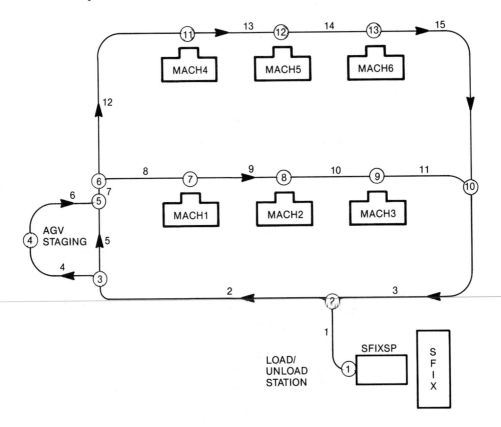

Figure 16-14 Schematic diagram with resource labels and numbers.

Vehicle path segments are identified with a number on the line between control points. Arrows show the direction of movement allowed on a particular segment. The AGV staging loop is modeled as segment 4, control point 4 and segment 6. Control point 4 is the staging area where vehicles are sent when idle. The two vehicle units are initially idle at control point 4. The SLAM II network model for this example is shown in Figure 16-15. The network consists of 19 nodes which model the flow of loads from the fixture station to the manufacturing cells over the AGVS. An ARRAY row is defined as the control point number corresponding to a location of a machine, that is, ARRAY(1,J)=(7,8,9,11,12,13) for J=1,6. For example, the control point corresponding to machine cell 1 is 7 obtained from ARRAY(1,1). The SLAM II EQUIVALENCE statement

EQUIVALENCE/6,NMACH/
          ATRIB(2),MACH/
          ARRAY(1,MACH),CP;

is used to define the number of machine cells, NMACH, to be 6; Attribute 2 of each entity is equivalenced to MACH, the machine cell that is allocated to the entity. The control point to which an AGV is routed is named CP and obtained from ARRAY(1,MACH).

Loads are created at CREATE node CR01 with the time between creations being exponentially distributed with a mean of 436 seconds. The time of arrival of a load is marked in attribute 1. The load is routed to an AWAIT node where it waits for one unit of SFIX, the station for fixturing. Both new parts arrivals and processed parts wait in file 1 for the fixture station. File 1 is ranked low-value-first based on the mark attribute which causes processed parts to have priority as they have earlier mark times. The processed parts waiting to be removed from a fixture wait in file 1 at AWAIT node AW14. After a new part acquires the fixture station, activity 1 is performed which loads the part on the fixture and this takes 220 seconds. The loaded part then awaits a space at the fixture station whose resource name is SFIXSP. If one of the five spaces is available, the fixture station is freed to perform the next fixture operation on an entity waiting in file 1.

The subnetwork from node AS0 to AW05 determines if a machine is available to process the part. The rule employed in this example is to assign machines in ascending order, that is, preference is given to a machine resource with a lower resource number. The entity is first routed to an ASSIGN node AS1 where a proposed machine number, MACH, is set to 0. Next, MACH is indexed by 1 and the SLAM II integer global variable II is set to the machine number. II is used as an argument to other SLAM II variables. The entity then branches from ASSIGN node AS1. If the MACH is greater than the number of machines, the entity is routed to QUEUE node Q6 which has a following service activity with an indeterminant duration that is dependent on the release of a machine at node FR11. This causes a waiting part to receive a machine assignment whenever a machine is made available.

If the proposed machine is available, the entity is routed to AWAIT node AW05. If the proposed machine is not available, the entity returns to ASSIGN node AS1 where the proposed machine number is indexed. When a machine is assigned, MACH is the second attribute of the load entity because MACH is equivalenced to ATRIB(2).

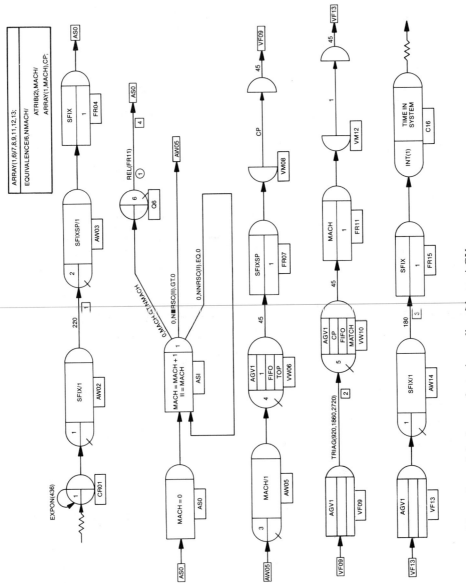

Figure 16-15 Network model of manufacturing cells with an AGV.

The subnetwork from node AW05 to node VF09 models the movement of the part entity using an AGV from the fixture station to the machine. At node AW05, one unit of the machine is seized. The entity then proceeds to VWAIT node VW06 where an AGV is requested at control point 1 which is at the fixture station. The rule for ordering the request for vehicles is first-in, first-out and the rule for releasing entities when a vehicle arrives to the VWAIT node is to take the entity that is at the top of the request list. The part entity waits in file 4 until an AGV arrives then the entity is released to the vehicle. A 45 second delay is incurred to load the entity on the vehicle after which a space at the fixture station is freed at FREE node FR07. Movement to the control point in front of the machine that was assigned to the entity is then accomplished by the VMOVE activity, VM08, where the parameter CP is obtained as the value of ARRAY(1,MACH) defined in the EQUIVALENCE statement. When the entity arrives to this control point, a 45 second delay to remove the fixture and load from the AGV occurs, and the entity is routed to VFREE node VF09.

The AGV1 resource is freed at node VF09 and machine processing of the load is performed at activity 2. Processing can begin immediately as entities are not routed to machines unless they are available. Following processing, the entity waits at VWAIT node VW10 for an AGV to arrive at control point CP. A manufacturing cell only processes one load at a time so there is at most one entity for each machine waiting in file 5 for an AGV. The rule for entity release at node VW10 is specified as MATCH which causes an AGV to pickup the entity that is at the control point to which the AGV was sent. The entity then incurs a 45 second delay for loading onto the AGV after which the machine is freed at FREE node FR11. If there is an entity in activity 4, the release of node FR11 allows it to proceed.

The AGV with the entity is moved to control point 1 by VMOVE activity VM12. This takes the entity back to the fixture station where the part and fixture are taken off the AGV which takes 45 seconds. In traveling to control point 1, the AGV must gain access to segment 1 which is a spur. The internal logic of the MH Extension automatically causes the vehicle to wait to gain access to control point 2 if the other AGV is on segment 1 or at control point 1. Following the removal of the entity and fixture from the AGV, one unit of AGV1 is freed at VFREE node VF13. At this point, movement of the AGV is controlled either by the rule for job requests or the rule for moving idle vehicles as specified on the VFLEET resource statement. In this example, RJREQ is specified as CLOSEST which causes the AGV to pick-up an entity at the fixture station if one is waiting at control point 1. Otherwise, the AGV will go to the closest machine to pickup a load. If no entities are waiting for an AGV, the rule for idle AGVs is invoked. For this example the rule is STOP(4) which causes the idle vehicle to proceed to control point 4 and to stop and wait for a request.

The entity that was delivered to control point 1 awaits the fixture station at node AW14. When the fixture station is available, the entity is removed from the fixture which takes 180 seconds, and the fixture station is freed at FREE node FR15. The load entity then leaves the manufacturing cell system and the time in the system is collected at COLCT node C16.

The statement model for this example is shown in Figure 16-16. As can be seen, a large portion of the statement model is devoted to resource definitions which indicates the degree of dependence that a materials handling model has on facility layout and vehicle characteristics. The small number of nodes required to model this situation demonstrates the versatility of the AGV constructs included in the MH Extension in modeling such systems. A MHMONTR control statement is used in order to monitor the trips taken by the AGV. A VCONTROL statement is used to set minimum and maximum step sizes for the movement of AGVs along the guidepath. The INIT statement specifies that the run length should be 89,400 seconds.

```
 1   GEN,PRITSKER & ASSOC.,AGV EXAMPLE,3/11/84,1;
 2   ;
 3   ;   AGV EXAMPLE MODEL
 4   ;
 5   ;   (ALL TIMES ARE IN SECONDS, DISTANCES IN FEET.)
 6   ;
 7   LIMITS,7,5,100;
 8   INIT,0.0,89400.0;
 9   ;
10   PRIORITY/1,LVF(1);
11   VCONT,0.00001,2.0;
12   ;
13   ;   EQUIVALENCE AND ARRAY DECLARATIONS
14   ;
15   ARRAY(1,6)/7,8,9,11,12,13;
16   EQUIVALENCE /6, NMACH/
17              ATRIB(2),  MACH/
18              ARRAY(1,MACH), CP;
19   ;
20   NETWORK;
21   ;
22   ;   DEFINE MACHINE RESOURCES
23   ;
24       RESOURCE/1,MACH1(1),3;
25       RESOURCE/2,MACH2(1),3;
26       RESOURCE/3,MACH3(1),3;
27       RESOURCE/4,MACH4(1),3;
28       RESOURCE/5,MACH5(1),3;
29       RESOURCE/6,MACH6(1),3;
30       RESOURCE/7,SFIX(1),1;
31       RESOURCE/8,SFIXSP(5),2;
```

Figure 16-16   SLAM II statement model of manufacturing cells with an AGVS.

```
32  ;
33  ;  DEFINE THE VEHICLE CONTROL POINTS
34  ;
35  ;             NUMBER    LABEL
36  ;
37       VCPOINT,    1    /FIXTURE;
38       VCPOINT,    2/;
39       VCPOINT,    3/;
40       VCPOINT,    4    /STAGE;
41       VCPOINT,    5/;
42       VCPOINT,    6/;
43       VCPOINT,    7    /MACH1;
44       VCPOINT,    8    /MACH2;
45       VCPOINT,    9    /MACH3;
46       VCPOINT,    10/;
47       VCPOINT,    11   /MACH4;
48       VCPOINT,    12   /MACH5;
49       VCPOINT,    13   /MACH6;
50  ;
51  ;  DEFINE THE VEHICLE GUIDEPATH SEGMENTS
52  ;
53  ;                            BEGINNING        ENDING
54  ;             NUMBER     CONTROL POINT    CONTROL POINT   LENGTH   TYPE
55       VSGMENT,    1,          1,              2,           20,     BI;
56       VSGMENT,    2,          2,              3,           82,     UNI;
57       VSGMENT,    3,          10,             2,           115,    UNI;
58       VSGMENT,    4,          3,              4,           27,     UNI;
59       VSGMENT,    5,          3,              5,           16,     UNI;
60       VSGMENT,    6,          4,              5,           27,     UNI;
61       VSGMENT,    7,          5,              6,           13,     UNI;
62       VSGMENT,    8,          6,              7,           35,     UNI;
63       VSGMENT,    9,          7,              8,           35,     UNI;
64       VSGMENT,    10,         8,              9,           35,     UNI;
65       VSGMENT,    11,         9,              10,          35,     UNI;
66       VSGMENT,    12,         6,              11,          75,     UNI;
67       VSGMENT,    13,         11,             12,          35,     UNI;
68       VSGMENT,    14,         12,             13,          35,     UNI;
69       VSGMENT,    15,         13,             10,          75,     UNI;
70  ;
71  ;  DEFINE AGV
72  ;
73       VFLEET,AGV1,2,4.5,4.0,,,4.0,4.5,,7/CLOSEST,STOP(4),4(2,4);
74  ;
75  ;  CREATE THE PART ENTITIES
76  ;
77  CR01  CREATE,EXPON(436.),,,1;
78  AW02  AWAIT(1),SFIX/1;                 WAIT FOR FIXTURING STATION
79  ;
80       ACT/1,220.;                       LOAD PART    INTO FIXTURE
81  ;
82  AW03  AWAIT(2),SFIXSP/1;               WAIT FOR OUTPUT BUFFER SPACE
83  FR04  FREE,SFIX/1;                     RELEASE THE FIXTURING STATION
84  ;
85  ;  MACHINE SELECTION LOGIC
```

Figure 16-16 (continued).

```
86  ;
87  AS0    ASSIGN,MACH = 0;              INITIALIZE MACH
88  AS1    ASSIGN,MACH = MACH+1,
89           II = MACH,1;               INCREMENT MACHINE NUMBER
90         ACT,,MACH.GT.NMACH,Q6;       NO MACHINES WAIT IN QUEUE #6
91         ACT,,NNRSC(II).GT.0,AW05;    MACHINE OPEN
92         ACT,,NNRSC(II).EQ.0,AS1;     MACHINE BUSY TRY NEXT MACHINE
93  ;
94  ;  HOLD PARTS FOR NEXT AVAILABLE MACHINE
95  ;
96  Q6     QUEUE(6);
97         ACT(1)/4,REL(FR11),,AS0;  HOLD PART    FOR MACHINE RELEASE
98  ;
99  ;  SEIZE MACHINE AND TRANSPORT PART
100 ;
101 AW05   AWAIT(3),MACH/1;             SEIZE MACHINE NUMBER MACH
102 VW06   VWAIT(4),AGV1,1,FIFO,TOP;    REQUEST A VEHICLE AT CONTROL POINT 1
103        ACT,45.;                     LOAD PART    ONTO AGV
104 ;
105 FR07   FREE,SFIXSP/1;               RELEASE FIXTURE OUTPUT BUFFER SPACE
106 ;
107 VM08   VMOVE,CP;                    MOVE PART TO MACHINE CONTROL POINT
108        ACT,45.;                     UNLOAD PART FROM AGV
109 VF09   VFREE,AGV1;                  RELEASE AGV
110 ;
111        ACT/2,TRIAG(920.,1860.,2720.);   PROCESS PART
112 ;
113 VW10   VWAIT(5),AGV1,CP,FIFO,MATCH;REQUEST A VEHICLE FOR TRANSPORT BACK
114 ;                                   TO FIXTURING STATION
115        ACT,45.;                     LOAD COMPLETED PART ONTO AGV
116 FR11   FREE,MACH/1,                 RELEASE THE MACHINE
117 ;
118 VM12   VMOVE,1;                     MOVE PART TO FIXTURE STATION CONTROL POINT
119 ;
120        ACT,45.;                     UNLOAD PART AT FIXTURE STATION
121 ;
122 VF13   VFREE,AGV1;                  RELEASE AGV
123 ;
124 AW14   AWAIT(1),SFIX/1;             WAIT FOR FIXTURE STATION
125        ACT/3,180.;                  UNLD FIXTURE FROM AGV
126 FR15   FREE,SFIX/1;                 RELEASE FIXTURE STATION
127 ;
128        COLCT,INT(1),TIME IN SYSTEM; COLLECT TIME SYSTEM STATISTICS
129 ;
130        ENDNETWORK;
131 ;
132 FIN;
```

Figure 16-16   (continued).

Summary of Results. Figure 16-17 presents the SLAM II Summary Report. It includes statistics for VWAIT nodes, files 4 and 5, and for job requests waiting for resource AGV1 in file 7. The statistics for the VWAIT nodes indicate that entities do not wait very long for pickup by an AGV at the fixture station or at a machine cell since the average waiting time is 81 seconds and 55 seconds, respectively. The maximum number of loads waiting for a vehicle to arrive at either VWAIT node is 2. From the statistics for file 7, the average number of job requests for vehicle fleet AGV1 is only .07 vehicles. A job request only remains in file 7 until a vehicle is assigned to pickup the entity associated with the job request at a VWAIT node. This number is expected to be smaller than the time the job entity is in the file of the VWAIT node since the latter also includes the time until an AGV arrives at the control point of the VWAIT node. The maximum length indicates that there were three job requests outstanding at one time during the simulation run. From the file statistics for the fixture station, it is a potential bottleneck with an average of two entities waiting for the fixture station for an average waiting time of 493 seconds. This result is consistent with the model data since there is only one fixture station and, for each job, there is 400 seconds of fixture station time required (220 to load part on fixture and 180 seconds to unload the part from the fixture). File statistics show that sufficient spaces are available at the fixture station as there is no waiting time in file 2 for a fixture station space. Since AWAIT node AW05 is only used to seize a machine resource, there is no waiting in file 3. The waiting for a machine is done in QUEUE node Q6 and ACTIVITY 4.

Figure 16-18 presents the vehicle utilization report. Information is given on the travel time of a vehicle fleet to perform a function and statistics on vehicle idleness due to congestion or lack of job requests. In Figure 16-18, the first set of headings relate to the use of the vehicle fleet in satisfying job requests. First, the resource capacity or number of units in the vehicle fleet is given as 2. Statistics on the average number of vehicles used to service jobs is then provided. This includes traveling empty to the load, loading, traveling full to unload control points, and entity unloading at control points. For this output, the average number of vehicles traveling to VWAIT nodes VW06 and VW10 is 0.20. Multiplying this number by the time period over which statistics were computed, 89,400 seconds, yields a result that there are 17,880 vehicle-seconds spent traveling to pickup job entities. The average number of vehicles picking up entities is shown as 0.191 or approximately 17,055 vehicle-seconds are spent in pickup. The second section of the vehicle performance report indicates that there were 379 loads and since each pickup requires 45 seconds there are 17,055 vehicle-seconds involved in pickup. The loading time for an AGV is the time from the AGV arrival to the control point of a VWAIT node to the time it begins its VMOVE activity. The time traveling to unload is the time of the VMOVE activity. The average number of vehicles traveling to unload is the average number of vehicles in the VMOVE activity and is 0.235. The average number of vehicles unloading entities is 0.191 which is the same as

S L A M I I S U M M A R Y R E P O R T

SIMULATION PROJECT AGV EXAMPLE   BY PRITSKER & ASSOC.

DATE 3/11/1984   RUN NUMBER   1 OF   1

CURRENT TIME 0.8940E+05
STATISTICAL ARRAYS CLEARED AT TIME 0.0000E+00

STATISTICS FOR VARIABLES BASED ON OBSERVATION

	MEAN VALUE	STANDARD DEVIATION	COEFF. OF VARIATION	MINIMUM VALUE	MAXIMUM VALUE	NUMBER OF OBSERVATIONS
TIME IN SYSTEM	0.3744E+04	0.1112E+04	0.2969E+00	0.2078E+04	0.7039E+04	188

FILE STATISTICS

FILE NUMBER	LABEL/TYPE	AVERAGE LENGTH	STANDARD DEVIATION	MAXIMUM LENGTH	CURRENT LENGTH	AVERAGE WAITING TIME
1	AW02 AWAIT	2.0936	2.6604	12	0	492.5559
2	AW03 AWAIT	0.0000	0.0000	1	0	0.0000
3	AW05 AWAIT	0.0000	0.0000	1	0	0.0000
4	VW06 VWAIT	0.1731	0.3805	2	0	81.0199
5	VW10 VWAIT	0.1148	0.3355	2	0	54.5701
6	Q6 QUEUE	0.0140	0.1174	1	0	24.4870
7	VEHICLE	0.0674	0.3000	3	0	15.8946
8	CALENDAR	6.1733	1.4694	9	5	112.5628

Figure 16-17 SLAM II Summary Report for manufacturing cells with an AGVS.

REGULAR ACTIVITY STATISTICS

ACTIVITY INDEX/LABEL	AVERAGE UTILIZATION	STANDARD DEVIATION	MAXIMUM UTIL	CURRENT UTIL	ENTITY COUNT
1 LOAD PART	0.4711	0.4992	1	1	191
2 PROCESS PART	3.9422	1.3192	6	3	188
3 UNLD FIXTURE	0.3785	0.4850	1	0	188

SERVICE ACTIVITY STATISTICS

ACTIVITY INDEX	START NODE OR ACTIVITY LABEL	SERVER CAPACITY	AVERAGE UTILIZATION	STANDARD DEVIATION	CURRENT UTILIZATION	AVERAGE BLOCKAGE	MAXIMUM IDLE TIME/SERVERS	MAXIMUM BUSY TIME/SERVERS	ENTITY COUNT
4	HOLD PART	1	0.1126	0.3161	0	0.0000	12852.3535	1446.3086	51

RESOURCE STATISTICS

RESOURCE NUMBER	RESOURCE LABEL	CURRENT CAPACITY	AVERAGE UTILIZATION	STANDARD DEVIATION	MAXIMUM UTILIZATION	CURRENT UTILIZATION
1	MACH1	1	0.8641	0.3427	1	1
2	MACH2	1	0.8135	0.3895	1	1
3	MACH3	1	0.8148	0.3884	1	0
4	MACH4	1	0.7955	0.4033	1	0
5	MACH5	1	0.7290	0.4445	1	0
6	MACH6	1	0.6159	0.4864	1	1
7	SFIX	1	0.8496	0.3575	1	1
8	SFIXSP	5	0.3958	0.5734	3	0

RESOURCE NUMBER	RESOURCE LABEL	CURRENT AVAILABLE	AVERAGE AVAILABLE	MINIMUM AVAILABLE	MAXIMUM AVAILABLE
1	MACH1	0	0.1359	0	1
2	MACH2	0	0.1865	0	1
3	MACH3	1	0.1852	0	1
4	MACH4	1	0.2045	0	1
5	MACH5	1	0.2710	0	1
6	MACH6	0	0.3841	0	1
7	SFIX	0	0.1504	0	1
8	SFIXSP	5	4.6042	2	5

Figure 16-17 (continued).

the average number loading entities. This will always be the case if the pickup and dropoff times are the same and the number of loads and unloads are equal. Adding the average number of vehicles spent in each of these states yields a total productive fraction of 0.817. This indicates that, on the average, approximately eight tenths of one vehicle is used to perform the operations required in moving parts through the AGVS network.

VEHICLE UTILIZATION REPORT

——————— AVERAGE NUMBER OF VEHICLES ———————

VEHICLE FLEET LABEL	NUMBER AVAILABLE	TRAVELING TO LOAD (EMPTY)	LOADING	TRAVELING TO UNLOAD (FULL)	UNLOADING	TOTAL PRODUCTIVE
AGV1	2	0.201	0.191	0.235	0.191	0.817

VEHICLE PERFORMANCE REPORT

——————— AVERAGE NUMBER OF VEHICLES ———————

VEHICLE FLEET LABEL	NUMBER OF LOADS	NUMBER OF UNLOADS	TRAVELING EMPTY BLOCKED	TRAVELING FULL BLOCKED	TRAVELING IDLE	STOPPED IDLE	TOTAL NON-PRODUCTIVE
AGV1	379	379	0.020	0.007	0.466	0.690	1.183

Figure 16-18   Vehicle Utilization Report for manufacturing cell with an AGVS.

The second section of Figure 16-18 shows that the average number of vehicles traveling to the load and becoming blocked is 0.02 and traveling to unload and becoming blocked is less than 0.01. This represents the interference time on the AGVS network due to loading and unloading being performed on the guidepath as well as contention for the control points. The number of vehicles traveling idle to the staging control point is 0.47, and the number of vehicles waiting at the staging control point (stopped idle) is 0.69.

Figure 16-19 provides segment and control point statistics. For each segment resource, the number of vehicles entering the segment, the average utilization of the segment, the maximum number of vehicles on the segment, and the current number on the segment are provided. The average utilization of a segment is defined as the number of vehicle-time units spent on the segment divided by the time period for the simulation run. The time on the segment for a vehicle includes the time it takes for the vehicle to move through its beginning control point if the vehicle does not stop at the control point. If the vehicle stops at a control point then the time on the segment starts when the vehicle exits the control point. For statistical purposes, the time on the segment ends as soon as the vehicle enters the control point at the end of the segment.

Statistics for control points are shown in Figure 16-19. These statistics include the number of vehicles entering the control point, average utilization of the control point, maximum number of vehicles waiting to enter the control point, and the current number of entities at the control point. A vehicle is at a control point as soon as it gains access to the control point and remains there until the entire vehicle is off the control point. In other words, the time the vehicle spends on the control point is equal to the time the tail of the vehicle leaves the control point minus the time that the nose of the vehicle enters the control point. Only one vehicle

SEGMENT STATISTICS

SEGMENT NUMBER	SEGMENT LABEL	CONTROL END POINTS	NUMBER OF ENTRIES	AVERAGE UTILIZATION	MAXIMUM UTIL.	CURRENT UTIL.
1		1 / 2	365	0.226	1	0
2		2 / 3	528	0.113	2	0
3		10 / 2	528	0.184	2	0
4		3 / 4	285	0.317	2	1
5		3 / 5	244	0.011	1	0
6		4 / 5	284	0.016	2	0
7		5 / 6	528	0.018	2	0
8		6 / 7	354	0.033	2	0
9		7 / 8	354	0.032	2	0
10		8 / 9	354	0.032	2	0
11		9 / 10	354	0.031	2	0
12		6 / 11	174	0.035	2	0
13		11 / 12	174	0.016	2	0
14		12 / 13	174	0.015	2	0
15		13 / 10	174	0.034	1	0

Figure 16-19 Segment and control point statistics for manufacturing cell with an AGVS.

CONTROL POINT STATISTICS

CONTROL POINT NUMBER	CONTROL POINT LABEL	NUMBER OF ENTRIES	AVERAGE UTILIZATION	MAXIMUM NUMBER WAITING	CURRENT UTILIZATION
1	FIXTURE	365	0.195	0	0
2		893	0.019	1	0
3		528	0.011	0	0
4	STAGE	285	0.696	1	1
5		528	0.011	1	0
6		528	0.011	0	0
7	MACH1	354	0.043	1	0
8	MACH2	354	0.041	1	0
9	MACH3	354	0.041	1	0
10		528	0.011	1	0
11	MACH4	174	0.037	1	0
12	MACH5	174	0.034	1	0
13	MACH6	174	0.028	0	0

Figure 16-19   (continued).

can be in the control point at a time. The average utilization of a control point is the sum of all the times that vehicles are on the control point divided by the time period for the simulation.

In this example, control point 4 is used as a staging area for idle vehicles. The average number of vehicles at control point 4 is 0.696. From the vehicle performance report, the number of stopped, idle vehicles is 0.690. Since vehicles are only stopped and idle at control point 4, the difference between the two values is due to the inclusion of the time the vehicle is moving onto the control point in the control point statistics.

Figure 16-20 illustrates the vehicle trip report matrix. The trip report matrix only counts origins and destinations and not trips to intermediate control points. From the trip report matrix, it is seen that 234 AGVs originated at the fixture station. Note that in this model, destination control points are 1, the fixture station, and the control points before the machine cells: 7, 8, 9, 11, 12, and 13. The primary origin control points are these same control points and control point 4 which is the staging area. Control points 2, 3, and 10 can also be origination points since a vehicle passing through these control points can be signaled to pickup a job entity. From the trip report matrix, the number of times that this occurred is small. In this model output, the most frequent trip is from the staging control point which occurred in 275 of the 758 origin-destination moves.

*** VEHICLE TRIP REPORT MATRIX ***

TABLE 1

	TO CP –	1	2	3	4	5	6	7	8	9	10	11	12	13	TOTAL
FROM CP															
FIXTURE	1.	14	0	0	0	0	0	41	38	38	0	39	33	31	234
	2.	4	0	0	0	0	0	0	0	2	0	3	0	0	9
	3.	7	0	0	0	0	0	3	1	2	0	0	1	1	15
STAGE	4.	137	0	0	0	0	0	26	26	21	0	23	26	16	275
	5.	0	0	0	0	0	0	0	0	0	0	0	0	0	0
	6.	0	0	0	0	0	0	0	0	0	0	0	0	0	0
MACH1	7.	38	0	0	0	0	0	0	0	1	0	1	0	0	40
MACH2	8.	36	0	0	0	0	0	1	0	1	0	0	0	0	38
MACH3	9.	38	0	0	0	0	0	0	0	0	0	0	0	0	38
	10.	4	0	0	0	0	0	0	0	0	0	0	0	0	4
MACH4	11.	35	0	0	0	0	0	0	1	1	0	0	0	1	38
MACH5	12.	33	0	0	0	0	0	0	0	0	0	0	0	0	33
MACH6	13.	33	0	0	0	0	0	0	1	0	0	0	0	0	34
	TOTAL.	379	0	0	0	0	0	71	67	66	0	66	60	49	758

Figure 16-20 Vehicle trip report matrix for manufacturing cell with an AGVS.

16.17   AGVS SUPPORT SUBPROGRAMS

A set of support subprograms for AGVS modeling is included in the MH Extension. These subprograms provide access to AGVS data which is useful in the writing of user rules. The support subprograms for AGVS modeling can be grouped into three categories as shown in Table 16-3. The first category retrieves the number of a vehicle unit based on information about its current status or location. Each of the names of these subprograms starts with the letters NVU and ends with letters identifying the input information. For example, NVUSG retrieves the number of a vehicle unit that is on a segment. Each vehicle unit of a vehicle

Table 16-3   Support subprograms for AGVS modeling.

Subprogram Name	Description
Number of a Vehicle Unit	
NVUVF	Number of a vehicle unit given a vehicle fleet number and a relative vehicle unit number.
NVUATR	Number of a vehicle unit given attributes of its load.
NVUIDL	Number of a vehicle unit given its idle ranking.
NVUSG	Number of a vehicle unit given a segment number.
NVUWCP	Number of a vehicle unit given a control point where it is waiting.
Status Information for a Vehicle Unit	
ALTSPD	Alter the speed of a vehicle unit.
SIGVU	Signal vehicle unit that it may proceed.
GETASG	Get current assignment of a vehicle unit.
GETCRG	Get changing indicator and charging start time.
GETLOC	Get current location of vehicle unit.
GETREQ	Get information on requesting entity for vehicle unit.
GETATR	Get attribute values of load entity on vehicle unit.
PUTATR	Put new attribute values on load entity on vehicle unit.
Number of Vehicles in a Particular State	
NNVIDL	Number of vehicles idle
NNVSG	Number of vehicles on a segment
NNVWCP	Number of vehicles waiting for a control point.

fleet is given an internal unique number or index. If a vehicle fleet consists of ten vehicle units then there will be ten numbers assigned. The unique vehicle number can be obtained given a vehicle fleet number and a relative vehicle unit number using function NVUVF. A relative vehicle unit number is a reference such as the third vehicle of a vehicle fleet.

The second category of subprograms retrieves status information for a vehicle unit. The first three letters of the names of these subprograms indicate an action followed by letters describing the information desired. For example, GETLOC is used to get the location of a vehicle unit.

The last category of subprograms is used to obtain the number of vehicles in a particular state. For each of these subprograms, the first three letters of the name are NNV indicating that the number of vehicles is to be obtained. Following NNV are letters defining the particular state: IDL for idle; SG for segment; and WCP for waiting for a control point.

Each support subprogram has arguments to provide the information required to perform the subprogram function. Table 16-4 lists the definitions of the variables used as arguments for the support subprograms. Table 16-5 lists the support subprograms including their arguments. Each support subprogram is described below.

Function NVUVF(NVFNUM,NVUNITR) returns the number of a vehicle unit as NVUVF given a vehicle fleet number NVFNUM, and a relative vehicle unit number NVUNITR.

Subroutine NVUATR(BTRIB, NVUNIT,NVFNUM,NVUNITR) returns the vehicle unit number NVUNIT which is carrying an entity whose attributes are given by BTRIB. It also returns the vehicle fleet number NVFNUM and relative vehicle unit number NVUNITR corresponding to NVUNIT.

Subroutine NVUIDL(NRANKV,NVFNUM, NVUNIT) returns the vehicle unit number NVUNIT given the rank of an idle vehicle, NRANKV, and a vehicle fleet number NVFNUM. Vehicles are given priority based on the length of time they are idle so that the vehicle with rank 1 has been idle the longest. If there are no idle vehicles, NVUNIT is returned as 0.

Subroutine NVUSG(NRANKV,NSGNUM, NVUNIT) returns the number of a vehicle unit NVUNIT that has rank NRANKV on segment number NSGNUM. The rank of a vehicle on a segment is based on its closeness to the endpoint of the segment. The vehicle closest to the endpoint has a rank of 1.

Subroutine NVUWCP(NRANKV,NCPNUM, NVUNIT) returns the number of the vehicle unit NVUNIT that has rank NRANKV in the queue of vehicles waiting at control point number NCPNUM. The vehicle that is ranked first in the queue of the control point is the one that has been in the queue the longest.

Subroutine ALTSPD(NVUNIT,RATE,SPEED, IERR) alters the rate and speed of vehicle unit number NVUNIT. The change in speed only applies to the current

Table 16-4 Definitions of variables used as arguments for support programs.

Variable	Definition
BTRIB	Alternate attribute buffer.
CPOS	Current position of a vehicle on a segment.
ICHRG	Charging status indicator:
	0→not charging
	1→charging
IERR	Error indicator.
IFLVW	File number associated with VWAIT node.
IPLOADE	Pointer to a load entity.
ISTATVU	Status of a vehicle unit.
ITRIPF	Start of trip flag for a vehicle unit.
NCPCUR	Number of current control point.
NCPDES	Number of destination control point.
NCPNEX	Number of the next intermediate control point.
NCPNUM	Number of a control point.
NRANKE	Rank of entity in a VWAIT file.
NRANKR	Rank of entity in a vehicle request file.
NRANKV	Rank of vehicle
	For idle vehicles: 1→longest idle
	On segments: 1→closest to end point
	In queue of CP: 1→longest time in queue
NSGNUM	Number of a segment.
NVFNUM	Identification number of vehicle fleet.
NVUNIT	Identification number of vehicle unit.
NVUNITR	Relative identification number of vehicle unit.
RATE	Velocity of a vehicle unit.
SPEED	Maximum speed of a vehicle unit.
STCHRG	Most recent start time for charging a vehicle unit.

travel of the vehicle unit and, hence, has no effect on idle or stopped vehicles. The variable RATE is used to change the current average rate of travel and to indicate in which direction the vehicle is to travel. Travel on unidirectional segments is always positive. For bidirectional segments, travel on a segment from a lower control point number to a higher control point number is in the positive direction and from a higher control point number to a lower control point number is in the negative direction. A maximum speed for the vehicle unit is specified by the variable SPEED. If acceleration and deceleration are not specified for a vehicle fleet, then SPEED has no affect on the vehicle unit's travel. An error flag IERR is also returned. IERR is set to 1 if the vehicle unit number is not correct. Otherwise, IERR is returned as 0.

Table 16-5 Subprograms with arguments listed.

Function/Subroutine	Short Description†
Number of a Vehicle Unit	
Function NVUVF(NVFNUM,NVUNITR)	NVU given Vehicle Fleet
Subroutine NVUATR(BTRIB,NVUNIT, NVFNUM,NVUNITR)	NVU given Attributes
Subroutine NVUIDL(NRANKV,NVFNUM, NVUNIT)	NVU given idle rank
Subroutine NVUSG(NRANKV,NSGNUM, NVUNIT)	NVU given SG number
Subroutine NVUWCP(NRANKV,NCPNUM, NVUNIT)	NVU given WCP number
Status Information for a Vehicle Unit	
Subroutine ALTSPD(NVUNIT,RATE,SPEED,IERR)	Alter Speed
Subroutine SIGVU(NVUNIT,ITRIPF,IERR)	Signal Vehicle
Subroutine GETASG(NVUNIT,NCPDES, IPLOADE,ISTATVU)	Get Assignment
Subroutine GETLOC(NVUNIT,NSGNUM,CPOS, NCPNEX,NCPCUR)	Get Location
Subroutine GETCRG(NVUNIT,ICHRG,STCHRG)	Get Charging
Subroutine GETREQ(NRANKR,NVFNUM, NCPNUM,NRANKE,IFLVW,IERR)	Get Requestor
Subroutine GETATR(NVUNIT,BTRIB,IERR)	Get Attributes
Subroutine PUTATR(NVUNIT,BTRIB,IERR)	Put Attributes
The Number of Vehicles in a Particular State	
Function NNVIDL(NVFNUM)	Number of Vehicles Idle
Function NNVSG(NSGNUM)	Number of Vehicles on SG
Function NNVWCP(NCPNUM)	Number of Vehicles WCP

† NVU is an abbreviation for "number of a vehicle unit", SG is an abbreviation for "segment", and WCP is an abbreviation for "waiting for a control point".

Subroutine SIGVU(NVUNIT, ITRIPF,IERR) signals vehicle unit NVUNIT that it may attempt to proceed after it has been stopped at a control point in a user-written subprogram, typically, subroutine URROUT. Subroutine SIGVU returns the trip flag ITRIPF which is set to 0 if the vehicle could not proceed on the next segment. If the vehicle unit did start to traverse the next segment, then ITRIPF is set to 1. An error flag IERR is also returned. IERR is set to 1 if the vehicle unit number is not correct or if the vehicle unit is not at a control point. Otherwise, IERR is returned as 0.

Subroutine GETASG(NVUNIT, NCPDES,IPLOADE,ISTATVU) is used to get the assignment of vehicle number NVUNIT. The assignment is defined in terms of the vehicle's destination control point number NCPDES; a pointer to the current load entity IPLOADE; and an indicator of the status of the vehicle unit ISTATVU. IPLOADE will be set to 0 if the vehicle unit is not currently loaded. The indicator codes for the status of the vehicle unit are: 0, not yet introduced to guidepath; 1, stopped and idle; 2, cruising idle; 3, traveling to pickup; 4, traveling to dropoff; 5, picking up entities; 6, dropping off entities.

Subroutine GETLOC(NVUNIT, NSGNUM,CPOS,NCPNEX, NCPCUR) identifies the current location of vehicle unit NVUNIT. The current location can be a segment number or a control point number. If the vehicle unit is on a segment then the segment number is returned as NSGNUM and the next control point encountered is returned as NCPNEX. The distance traveled on the segment is also returned as CPOS. If the vehicle unit is not on a segment then NSGNUM is returned as 0. If the vehicle is at a control point then NCPCUR is returned as the current control point number. If the vehicle is not at a control point then NCPCUR is returned as 0.

Subroutine GETCRG(NVUNIT, ICHRG,STCHRG) gets information about the current charging status of vehicle number NVUNIT. If the vehicle is charging, then the charging indicator ICHRG is set to 1. Otherwise, it is returned as 0. If ICHRG is 1, the variable STCHRG is set to the time the vehicle started charging at the control point.

Subroutine GETREQ(NRANKR,NVFNUM, NCPNUM,NRANKE, IFLVW, IERR) gets the characteristics of an entity making a request for a vehicle from vehicle fleet number NVFNUM. NRANKR is the rank of the request in the file defined for requests on the vehicle resource block. Subroutine GETREQ returns the control point number NCPNUM at which the entity issued the request, the rank of the requesting entity NRANKE in file IFLVW of its current VWAIT node. An error flag IERR is also returned. The flag is set to 0 if no error is detected. IERR is set to 1 if NRANKR is too large or NVFNUM is invalid. If an entity request has been removed from the VWAIT node without a vehicle being assigned to it then NRANKE is returned as 0, and the number of the control point and the file number of the VWAIT node are set to the values that were appropriate when the request for the vehicle was made by the entity.

Subroutine GETATR(NVUNIT, BTRIB,IERR) gets the attributes of the entity currently being transported by vehicle unit number NVUNIT. If NVUNIT is not transporting an entity, error flag IERR is set to 1. The attributes of the entity are returned in the vector BTRIB. The user must dimension the variable corresponding to BTRIB to at least MATR + 3.

Subroutine PUTATR(NVUNIT,BTRIB, IERR) puts the values of the vector BTRIB into the attributes of the entity being carried by vehicle unit number NVUNIT. If NVUNIT is not carrying an entity, the error flag IERR is set to 1.

Three functions are used to obtain the number of vehicles in a particular state. Function NNVIDL(NVFNUM) returns the number of idle vehicles in vehicle fleet number NVFNUM. Function NNVSG(NSGNUM) returns the number of vehicles on segment number NSGNUM. Function NNVWCP(NCPNUM) returns the number of vehicles waiting to gain access to control point number NCPNUM.

16.18   USER WRITTEN SUBPROGRAMS FOR AGV RULE SPECIFICATIONS

In an AGVS, there are many situations that require detailed control procedures for selecting among load entities and vehicles. The MH Extension provides interfaces to allow user-written subprograms that specify such rules. These rules have been discussed earlier in the description of the resource blocks and the VWAIT and VFREE nodes. Table 16-6 lists the user-written subroutines for rule specification and defines variables not included in Table 16-4 that are used as arguments.

16.18.1   Subroutine UREREL(NR,NVFNUM,NVUNIT, IFILE,IPREQE, NRANKE)

Subroutine UREREL is called with rule number NR if the user-specifies UREREL(NR) at a VWAIT node. UREREL is called when a vehicle unit arrives to the VWAIT node. The vehicle fleet number NVFNUM, the vehicle unit number NVUNIT, and the file number IFILE of the VWAIT node are passed as arguments to subroutine UREREL. Also provided is a pointer to the entity that is or was in file IFILE which requested the vehicle. In subroutine UREREL, the entities in file IFILE are polled and a specific entity is selected for release to the vehicle. This is accomplished using standard SLAM II subprograms for accessing attributes of entities in a file. When an entity is selected, its rank or the negative of its pointer is to be returned as NRANKE.

If it is desired to release the highest priority entity in the file, then NRANKE is set equal to 1. The user must check that the vehicle unit is from a vehicle fleet which can service the entity released. Setting NRANKE to 1 corresponds to the specification TOP for REREL on the VWAIT statement.

Table 16-6 User-written subroutines for rule specification.

Subroutine Name	Arguments†	Rule Specification
UREREL	NR,NVFNUM,NVUNIT, IFILE,IPREQE,NRANKE	Rule for selecting entity to load on a vehicle at a VWAIT node.
URVREQ	NR,NVFNUM,NCPNUM, BTRIB,NVUNIT	Rule for selecting an idle vehicle to satisfy an entity request at a VWAIT node.
URJREQ	NR,NVFNUM,NVUNIT, IFLREQ,NRANKR	Rule for selecting job entity requests to which a vehicle unit is to respond.
URIDL	NR,NVFNUM,NVUNIT, NCPCUR,NCPDES	Rule for selecting a control point destination for an idle vehicle unit.
URCNTN	NR,NCPNUM,IORDER	Rule for resolving vehicle contention at a control point.
URROUT	NR,NVFNUM,NVUNIT, NCPCUR,NCPDES,NSGNUM	Rule for selecting a segment when routing a vehicle from a control point.

† Definitions for arguments not defined in Table 6-4: NR, number of rule; IPREQE, pointer to request entity in request file; IFLREQ, file number of request entities; and IORDER, list of vehicle numbers in order of priority of release. IORDER(1) is vehicle number with highest priority.

If it is desired to release the entity that requested the vehicle, then NRANKE is set to minus IPREQE. If a combination of rules is used then it is necessary to check that the entity whose pointer is IPREQE has not been released to another vehicle.

16.18.2 Subroutine URVREQ(NR,NVFNUM,NCPNUM,BTRIB, NVUNIT)

Subroutine URVREQ is called with rule number NR when an entity arrives to a VWAIT node and its vehicle request rule is specified as URVREQ(NR). Subroutine URVREQ will only be called if more than one vehicle unit of the desired vehicle fleet is idle. In subroutine URVREQ a selection from the idle vehicle units is to be made. In addition to the rule number NR, the vehicle fleet number, the control point number defining the location of the entity, and the attributes of the entity are passed to subroutine URVREQ. The number of the vehicle unit that is selected to fill the request is returned in the variable NVUNIT.

In selecting an idle vehicle, function NNVIDL is used to get the number of idle vehicles in fleet NVNUM. The vehicle that has been idle the longest is obtained by the following statement:

CALL NVUIDL(1,NVFNUM,NVUNIT)

The location of NVUNIT is obtained by a call to subroutine GETLOC. An assessment can then be made as to whether it is desirable to use vehicle NVUNIT to satisfy the new request. This process is continued until an idle vehicle is selected to satisfy the request or it is decided not to satisfy the request. When NVUNIT is returned as 0, the request is left unsatisfied. This is allowed even if idle vehicles are available. The attributes of the requesting entity can be changed by changing the attribute buffer BTRIB.

16.18.3   Subroutine URJREQ(NR,NVFNUM,NVUNIT,IFLREQ, NRANKR)

Subroutine URJREQ is called with rule number NR when an entity arrives to a VFREE node and frees a vehicle unit from fleet number NVFNUM and rule URJREQ(NR) is specified. In addition to the rule number NR, the vehicle fleet number NVFNUM, the vehicle unit number NVUNIT, and the file number of requests for units from the vehicle fleet IFLREQ are inputs to subroutine URJREQ. The rank of the selected entity in the request file NRANKR is to be returned. A negative value for NRANKR indicates that the negative of the pointer to the entity request is returned. The attributes of the entity in the request file are obtained using standard SLAM II subprograms. Information on the current control point for the entity and its destination control point could be part of the user definition of the attributes of the entity. The current location of the vehicle unit freed can be obtained using subroutine GETLOC. If NRANKR is returned as 0, no request is selected and the vehicle is made idle. The rule for the disposition of idle vehicles is then invoked.

16.18.4   Subroutine URIDL(NR,NVFNUM,NVUNIT,NCPCUR, NCPDES)

Subroutine URIDL is called with rule NR when a vehicle unit is not assigned to a requesting entity after it is freed at a VFREE node and URIDL(NR) is specified as the rule for the disposition of idle vehicles. The vehicle fleet number NVFNUM, the vehicle unit number NVUNIT, and the current control point number NCPCUR are passed to subroutine URIDL when it is called. The control point number of the desired destination for the idle vehicle is returned from the subroutine. If it is desired to leave the vehicle at its current location, then NCPDES should be set to 0.

The standard specifications for the rule for idle vehicle disposition are shown in Table 16-2. A more advanced rule could involve the current status of jobs being processed as represented by the number of entities in processing activities. Standard SLAM II subprograms such as NNACT can be used to access the number of entities in an activity. This information can be used to determine a control point associated with a VWAIT node that follows an activity. The idle vehicle could then be routed to that control point.

16.18.5   Subroutine URCNTN(NR,NCPNUM, IORDER)

Subroutine URCNTN is called with rule number NR when a control point is freed by a vehicle and more than one vehicle is waiting for the control point and the rule for contention specified on the VCPOINT resource block is URCNTN(NR). In addition to NR, control point number NCPNUM is passed to this subroutine. A list of vehicle unit numbers is to be returned in the vector IORDER to be evaluated by SLAM II to determine which, if any, of the vehicles can be given access to the control point. The list of vehicle unit numbers are placed in IORDER with the highest priority vehicle in IORDER(1). A zero is used to indicate the end of the list. Vehicle units included in the list must be waiting for control point number NCPNUM. The vector IORDER must be dimensioned to at least 1 in this subroutine. The vehicle unit numbers waiting for the control point can be obtained from subroutine NVUWCP. The attributes of the load entity on the vehicle can be obtained using subroutine GETATR. With this information, a prioritized list of vehicle unit numbers can be developed.

16.18.6   Subroutine URROUT(NR,NVFNUM,NVUNIT, NCPCUR,NCPDES, NSGNUM)

Subroutine URROUT is called with rule number NR when a vehicle is at a control point and a segment needs to be selected and the rule for routing specified on the VCPOINT resource block is URROUT(NR). Subroutine URROUT is passed the vehicle fleet number NVFNUM, the vehicle unit number NVUNIT, the current control point number NCPCUR, and the control point destination number for the vehicle NCPDES. A selected segment number NSGNUM is to be returned from subroutine URROUT. If NSGNUM is specified as 0, the vehicle will stop at the current control point until a signal is sent to vehicle NVUNIT through a call to subroutine SIGVU. In determining which segment should be selected, the functions which obtain the number of vehicles on a segment NNVSG and the number of vehicles waiting at control points NNVWCP can be used. Information on the current assignment of the vehicle unit can be obtained through a call to subroutine GETASG. This information is useful in determining the route for a vehicle unit.

16.19   CHAPTER SUMMARY

This chapter has presented the MH Extension to SLAM II which includes three approaches to incorporating a specific functionality into SLAM II through the use of add-ins. The QBATCH node demonstrates that a single node with defined capabilities can augment the standard SLAM II symbols. Crane modeling required specialized resource definitions and generalizations of the AWAIT and FREE nodes. The operation of cranes includes advanced decision logic. System specific decision rules are modeled through changes of parameter values associated with the CRANE, AREA, and PILE resource definitions and the new GWAIT and GFREE nodes. The AGVS part of the MH Extension adds new resource types, new nodes and new capabilities for user-specified rules and procedures. Resource types to model control points, guidepath segments, and vehicles are defined. The new network symbols are the VWAIT and VFREE nodes and the VMOVE activity. Numerous support subprograms are included in the MH Extension to support the development of vehicle contention, vehicle routing and load priority specifications. This chapter demonstrates that SLAM II has the inherent flexibility to accommodate the addition of new capabilities for modeling specialized systems.

16.20   EXERCISES

16-1.   Develop a model of the transfer car operation of Illustration 16-1 using the standard SLAM II capabilities.

16-2.   A logging operation involves cutting a tree which takes 25 minutes on the average. This time is normally distributed with a standard deviation of 10 minutes. The weight of a tree varies uniformly from 1,000 lbs. to 1,500 lbs. based on the density of Southern pine. The logging company has five trucks with a capacity of 30 tons each. The truck will not move the trees to the sawmill until at least 15 tons of trees are loaded. The trip to the sawmill requires 2¹/₂ hours and the return trip requires 2 hours. The time to unload at the sawmill is 20 minutes. Develop a SLAM II model of this situation to determine if the logging operation has sufficient capability to deliver all the trees cut in a one month period.

Embellishments:

(a) Model the operation including an unloading unit for trees at the sawmill which breaks down every 40 hours plus or minus 10 hours and it takes 2 hours to repair.

(b) Include in the model the logging of oak trees in addition to pine trees. The oak trees are also cut and segregated at the cutting site. The oak trees have weights which vary between 1,500 lbs. to 2,000 lbs. uniformly distributed.

(c) Vary the minimum requirement of trees that are loaded on a truck from 15 to 20 to 25 to determine the effect on throughput of this decision variable.

16-3.   For the diagram of the aluminum coil processing facility presented in Section 16.5.4, develop a model to estimate machine and crane utilizations and the time required to process aluminum coils and sheets. Assume that all coils from a slitting machine that come from the same input coil are transported together by a crane. Similarly, assume that all sheets that are cut-to-length by the shearing machine are moved together. Coils arrive to the cooling racks every 20 minutes and have a size between 90 and 110 thousand cubic inches. The processing time on the heavy gauge slitter is triangularly distributed in a range of 38 and 42 minutes with a modal value of 40 minutes. The processing time on the light gauge slitter is triangularly distributed between 24 and 28 minutes with a modal value of 26 minutes. The processing time on the shearing machine is triangularly distributed between 22 and 28 minutes with a modal value of 25 minutes. Forty percent, on a random basis, of the incoming coils are routed to the shearing machine with the remainder having an equal likelihood of going to either the heavy gauge or light gauge slitter. Estimates of the utilization of the various areas in the model are to be obtained. Assume either crane can make any move, but it is preferred to use the east crane for the movement of coils at the shearing machine and the finished storage area. It is required to separate the finished products into separate piles of coil and sheet storage in the finishing area. Both an operator and a helper are required to do the processing at each machine. For the base model, assume that there is an operator and a helper at each of the machines, and they are always available to do the processing at the machine. Run the model for 2400 minutes. Obtain a summary report at the end of each 480 minutes.

Embellishment: Determine the impact of having only 2 operators and 2 helpers which service the three machines. The travel speed for the operators and helpers is 200 ft/min.

16-4. For Example 16-1, embellish the model to include stacker crane movements and movement times into each rack. The time to pickup a pallet is 10 seconds and the time to dropoff a pallet is 15 seconds.

16-5. Use the MH Extension to model and analyze the problem given in Exercise 14-7.
Embellishments:
(a) Add a second cart to the system.
(b) Allow bidirection movement of the carts.
(c) Develop rules for the carts so that they process parts in the order of higher dollar value first.

16-6. Integrate Examples 16-1 and 16-2 by requiring the output from the ASRS system to be the input to the machining cell. Develop the integrated model that requires an AGV to pickup the pallet from the ASRS and transport it to the fixture station. Assume a third AGV is added to the system to perform only this function. The distance between the AGV loading of a pallet to the fixture station is 250. Assume no limit on the number of pallets that are waiting for an AGV. The AGV used for movements from the ASRS to the machining cell has the same speed characteristics as the AGVs used in the machining cell.
Embellishment:
(a) Develop a model that could be used to evaluate the use of any of the AGVs for transporting pallets to the fixture station.
(b) Revise the model to allow the manufacturing cell to perform the milling operation if the ASRS becomes congested.

16-7. For Example 16-2, change the model so that idle vehicles cruise on the guidepath looking for a pickup. Include in the model a requirement that AGVs be charged for five minutes after two hours of operation.
Embellishment: For Example 16-2, specify a set of rules for establishing the priority of AGV moves that might improve system performance. Test your set of rules.

16-8. Revise the model of Example 16-2 to pass the machine number directly to an AGV when a machine completes the processing of a part. The revised model should eliminate the need for searching for a machine number.

16-9. Remodel Example 16-2 by writing subroutine ALLOC for use at AWAIT node AW05 that performs the machine allocation logic contained in the network from node AS0 to AW05.
Embellishment: Develop a machine allocation rule in subroutine ALLOC which employs information about the guidepath structure and the current position of idle vehicles.

16-10. Develop a specification for a conveyor module to include in the MH Extension. The specification should include the design of resource statements and new node statements. Specify the user-written rules that should be included in the module. Describe a set of subroutines to support the writing of user rules. Prepare your specification in the form of the tables given in Appendix E.

16-11. Develop a set of specifications for a SLAM II Extension to model food processing plants including reactors, ion exchangers, filters, and control computers.

16.21   REFERENCES

1. Apple, J. M., *Material Handling Systems Design*, John Wiley, New York, 1972.
2. DeJohn, F. A., C. W. Sanderson, C. T. Lewis, J. R. Gross, and S. D. Duket, "The Use of Computer Simulation Programs to Determine Equipment Requirements and Material Flow in the Billet Yard," *Proceedings, AIIE Spring Annual Conference*, 1980, pp. 402-408.
3. Francis, R. L., and J. A. White, *Facility Layout and Location: An Analytical Approach*, Prentice-Hall, Englewood Cliffs, N.J. 1974.
4. Miner, R. J., and L. J. Rolston, *MAP/1 User's Manual*, Pritsker & Associates, Inc., West Lafayette, IN, January 1986.
5. Müller, T., *Automated Guided Vehicles*, Springer-Verlag, 1983.
6. Sims, E. R. Jr., "Materials Handling Systems," *Handbook of Industrial Engineering*, G. Salvendy, ed., Chapter 10.3, John Wiley, 1982.
7. Tompkins, J. A., *Facilities Design*, North Carolina State University, Raleigh, 1975.
8. Tompkins, J. A., and J. A. White, *Facilities Planning*, John Wiley, New York, 1984.
9. White, J. A., and H. D. Kinney, "Storage and Warehousing," Chapter 10.4, *Handbook of Industrial Engineering*, G. Salvendy, ed., John Wiley, 1982.
10. White, J. A., "The 85% Rule of Thumb," *Modern Materials Handling,* July 1980.

CHAPTER 17

Simulation Support Systems

17.1 INTRODUCTION

Throughout this book, a problem solving approach based on modeling and simulation is presented. The fundamental assumption is that additional information is generated through the building of a model that explains or characterizes the operation of a system. Simulation languages such as SLAM II provide both a world view for building such models and a computer program for exercising and analyzing them. However, as described in Chapter 1, these modeling and analysis tasks are only two phases of a problem-solving project.

To support a total simulation project, additional tools and techniques are required. A simulation project involves data management, model management, scenario generation, output analysis, and the presentation of results and recommendations. To support these project functions, database tools and graphics capabilities are required. In this chapter, we present the issues relating to the need for a simulation support system and an overview of TESS, a package which provides such support. Before launching into the details of simulation support, a discussion of who requires support is appropriate.

17.2   WHOM WOULD A SIMULATION SUPPORT SYSTEM SUPPORT?

A major problem in designing a simulation support system is identifying the characteristics of its users. There are many individuals involved in the design, development, and use of a simulation model. In a recent study (10), five potentially overlapping groups were identified as requiring support: the language installers; the language maintainers; the model builders; the model users; and the decision makers. For the type of simulation support system discussed in this chapter, only the model builders, model users, and decision makers are addressed. Pritsker and Sigal (3) characterize these decision makers with respect to how they might use a decision support system. The categorization developed was: the hands-off user; the requestor; the hands-on user; and the renaissance decision maker.

The *hands-off* user reads the reports that are generated but is not in direct contact with the support system either through requests or knowledge of its underpinnings. The *requester* employs an intermediary for modeling and simulation. He or she frames the questions, interprets the results, and uses the answers to make decisions. The requester is not concerned with how the answers are obtained but realizes that they provide additional information that is helpful in making decisions. The *hands-on* user views the support system as an extension of himself or herself and employs direct on-line access to the system using predetermined interfaces and models. The *renaissance* decision maker functions as part of a team, is comfortable talking in terms of database systems and modeling, and is adept at making decisions. The renaissance decision maker knows how to set requirements for information, can prescribe the types of reports he or she wants, and can ask questions concerning the details of a model.

A simulation support system should be oriented to the hands-on user and the renaissance decision maker.

17.3   SUPPORTING THE MODELING TASK

The modeling approach recommended in this book is an iterative one. A model is defined, developed, refined, updated, modified, and extended. One of the secrets to being a good modeler is the ability to re-model. A simulation support system must provide the environment to allow for this approach. To support re-modeling, models should be easily recalled and cataloged. The model building process should be interactive and graphical. Current models should be available as the basis for future models. A documentation trail regarding model development should be maintained. All this suggests that models be stored in a database which provides a referencing, copying, and query capability.

17.4   SUPPORTING THE DATA MANAGEMENT TASKS

The data management activities associated with a simulation project involve the storage, retrieval, and organization of system data, experimental control specifications, and simulation generated outputs (11).

The system data can be used directly in the model or to estimate model parameters. It is frequently used for comparison and for validation purposes. Data from a simulation run is generated in accordance with the control statements which specify the experimental scenario under which the run is made. It is desirable to store the experimental specifications in the database and to make them accessible for use with different models. The simulation outputs should reference both the experimental specifications and the model from which it is generated. Both raw data and summary data are typically stored in the database.

Procedures are required to access, edit, concatenate, and display data stored in the database. This frequently occurs when it is desirable to use operational data as inputs to a cost model. It is also useful to be able to access the simulation outputs for presentation in spreadsheets or as inputs to other models.

Once in the database, the form of the simulation data should not differ from the form of the actual system data. Thus, any analysis performed on system data should be appropriate for model data and vice versa. A major difference is that the simulated data can be reproduced and, if the model is a stochastic one, additional simulated historical records can be obtained by exercising the model repeatedly.

17.5   SUPPORTING ANALYSIS AND REPORTING TASKS

Analysis of data involves the sorting, selection, and computation of values. These are standard functions associated with database management. For a simulation project, data is typically related to time. Analysis of data sets at a particular time for one or more runs is common place. Alternatively, data sets within a single run are sometimes grouped into subintervals or batches in order to study the behavior during the subinterval. Subintervals of special interest are an initial time period and an ending time period. In the former case the system is starting from an atypical state, and it may be of interest to determine how long it takes to transition to a specified or a steady state. The time period just before the ending of a simulation run may also require special processing. In some projects, it is necessary to estimate the dry-up or close-down time for system operations. Procedures for interrogating the data obtained from a simulation run are needed to explore such situations within runs and over multiple runs.

Support should also be provided to transform automatically the outputs of the simulation runs into inputs for statistical analysis procedures such as regression analysis, analysis of variance (ANOVA), autoregressive time series programs, and curve fitting programs.

The support system should provide the capability to output the data in both graphical and tabular form. Graphical forms include histograms, bar charts, pie charts, and plots. Examples of some of these graphical forms are shown on the jacket of this book. A format for a graph can involve hundreds of decisions including such items as heading, heading placement, size of letters, font of letters, color of letters, type of plot line, color of plot line, indication of measurement point, and so on. A key to providing good graphical and tabular outputs is to segregate the format of the graph or table from the data being displayed on the graph or table.

The support system should be capable of working with different output devices such as color graphics terminals, printers, and XY plotters.

In addition to providing a mechanism for storing data in the database, procedures need to be included to easily define the type of data to be put into the database. Since a simulation support system works directly with a simulation language, the data definitions for the database can, in the main, be predetermined. Thus, the data definitions for the simulation model, the controls, and the output formats can be preestablished. However, for system data, a convenient mechanism is necessary to support the data definition process.

17.6 SUPPORTING THE VISUALIZATION TASK USING ANIMATION

Trace data describing system operation includes data that portrays the state changes that occur over time. A dynamic characterization of the state changes can be presented visually as a form of animation in which icons move, change color,and are added or deleted from the picture. In this sense, an animation is a dynamic presentation of the changes in state of the system over time. To obtain an animation, it is necessary to diagram a facility which represents the system and the icons which represent the elements of the system. Such a facility diagram is an abstraction of the real system, hence it is a model of the system. It is a model that is built for the purpose of displaying a visualization of a system and fits the model category previously referred to as explanatory.

A simulation support system should provide the tools to convert either system data or simulation output data into a form which portrays the state changes on a facility diagram. Although this can be done as part of the simulation language, it tends to complicate the modeling process and is not a recommended approach to the visualization task. Any support task that complicates the modeling process should be avoided. Another approach is to use the facility diagram as the modeling language. Although intriguing, experience has shown that facility diagrams tend to lack the detailed description necessary to characterize system operations. A third approach is to use the constructs of the simulation language as the basis for the facility diagram (2,4,5). This approach requires that the viewer of the animation have some familiarity with the simulation language.

An approach which eliminates these disadvantages involves the use of a rule-based system (9). The rules translate the changes occurring in status variables to actions on a facility diagram. By changing the rules, different animations can be obtained which present different features or aspects of a simulation run. The visualization process can be run in a post simulation mode or a concurrent simulation mode. The latter provides for changes during the running of a simulation and can lead to new operating policy alternatives. The post simulation mode is better suited for analyzing the impact of changes on system performance.

17.7   TESS: THE EXTENDED SIMULATION SYSTEM

TESS is a computer package that is specifically designed to support the problem solving process employed on simulation projects.(8,9) TESS contains the following components:

** TESS language
   * Scenario definitions
   * Presentations
   * Visualizations/Animations
   * Statistical analysis
   * Invocation of other TESS components
** TESS database management system
   * TESS library
   * TESS utilities
** Graphics generators
   * Network builder
   * Facility builder
   * Icon builder
** Forms processors
   * Control statements
   * Formats
   * Rules
   * Data and summary occurrences
** Interfaces to
   * Simulation languages
   * Operating systems
   * Terminals

In the following sections, the components of TESS are described to indicate how one package has been developed to meet the simulation problem-solving specifications given in the previous sections of this chapter.

17.7.1   TESS Overview

This overview presents concepts and capabilities without delving into the details of the TESS program. TESS is a large system which has been developed as a generic tool for supporting simulation projects involving different simulation languages. It currently supports SLAM II, MAP/1, and GPSS/H. Detailed discussions of TESS and its uses are presented elsewhere (8,9).

The TESS language consists of commands which either specifies a transfer to a TESS component, develops a new TESS data relation, or causes the execution of a TESS component. There are three basic functions embodied in the TESS language statements: BUILD, GRAPH, and REPORT. The BUILD function is used to create data elements which are stored as relations in the TESS database. The data elements of TESS are: NETWORK, FACILITY, DEFINITION, CONTROL, SCENARIO, DATA, SUMMARY, FORMAT, MACRO, ICON, and RULE.

The GRAPH function displays data element occurrences on graphics devices. These include drawings of networks and facility diagrams; presentations of data and summary information; and animations. The REPORT function displays data occurrences on an alphanumeric terminal or a line printer. These include reports of occurrences of definitions, controls, data, summaries, formats, and rules. From a problem solving orientation, the BUILD function presents the greatest challenge. The GRAPH and REPORT functions are extremely important for achieving the implementation of the results of simulation projects. However, the GRAPH and REPORT functions are easier to grasp and are usually better specified. In this overview, the discussion centers on the BUILD function.

The TESS statements BUILD NETWORK, BUILD FACILITY, and BUILD ICON transfer the user into a graphical model building mode. The network builder module is for SLAM II networks and allows the user to select interactively a SLAM II network symbol and to locate it with respect to other symbols. The network builder requests parameter information for the symbol and populates the TESS database with sufficient information for use by the SLAM II processor.

The graphic builder for facilities provides the capability for drawing rectangles, triangles, stacks, tanks, circles, lines, paths, text, and sectors. Icons are used to represent entities and can be built to any desired level of detail using the graphic builder for icons. Icons can then be included in facility diagrams. Paths define the routes taken by entities and their movement through the facility. Stacks are used to represent queues. Tanks are used for representing continuous variable levels. Circles, lines, rectangles, sectors, and triangles are used to represent standard facility items whose color can be changed to represent status changes. Text is used to annotate the facility diagram.

DEFINITIONS, CONTROLS, DATA, FORMATS, MACROS, and RULES are built using a forms processor. Each TESS command to build one of these data elements transfers control to a forms module containing tables that are used to provide the information displayed on the form. For example, a form for each

SLAM II control statement is included which presents every field. Thus, there is a form in TESS corresponding to each table in Appendix C.

By its design, TESS has a great deal of flexibility. The commands of the TESS language provide access to its components as well as for direct specification of data elements. TESS has a hierarchical structure which facilitates both the description of what is to be done on a simulation project as well as the performing of the project (11,12). This structure is described in the following sections.

17.7.2   TESS Support for the Modeling Task

TESS provides a graphical interface which allows SLAM II networks to be built interactively. The TESS database is used to store the network after it is built. A translator converts the graphical network into the inputs necessary for the SLAM II processor. A SLAM II model stored in the database can be copied and used as a starting point for other models. Each model is given a name which allows for easy reference and reuse. Editing procedures are available to add and delete network elements and for changing and moving symbols and parameters. Zoom and scroll features are included to obtain different views of the network.

The TESS language command to initiate the building of a SLAM II network starts with the words BUILD NETWORK. The simplest form of the statement is

BUILD NETWORK NAMED(FMS);

This command causes TESS to transfer control to the interactive SLAM II network builder. If a network model exists with the name FMS, it is displayed to the modeler. If FMS is a new model, a blank screen is presented. The options available when building a network are shown below.

ADD	MOVE
CHANGE	PROMPT
DELETE	QUIT
EXIT	SCROLL
INCLUDE	VIEW
LABEL	ZOOM
LOAD	

In building models, the most frequently used option is ADD. After selecting the ADD option, a menu of the symbols that can be added to a SLAM II network is displayed. This menu is shown below.

SYMBOL MENU

ACCUMULATE	CLOSE	FREE	OPEN
ACTIVITY	COLCT	FSELECT	PILE
ALTER	CONNECTOR	GATE	PREEMPT
AREA	CRANE	GFREE	QBATCH
ASSIGN	CREATE	GOON	QUEUE
AWAIT	DETECT	GOTO	RESOURCE
BATCH	ENTER	GWAIT	SELECT
BSELECT	EVENT	MATCH	TERM
			TSQUIG
			TEXT
			UNBATCH

When a symbol is selected to be added, the user is prompted for each field of the input statement for the symbol. Alternatively, a type ahead feature allows the model builder to type the SLAM II statement in a single line.

Following the inputting of information for a symbol, TESS requests that a location for the symbol be identified on the screen. When this is done, the symbol is drawn on the screen with the parameters that have been specified for each field. The SLAM II network is built by continuing to add symbols and placing them on the network model. Changes, editing, and labeling are accomplished by selecting options other than ADD.

In this way, TESS supports the interactive building of SLAM II networks, their use as a starting point for building new networks, and the re-modeling through its advanced editing features.

17.7.3   TESS Support for Developing Scenarios

A scenario is the TESS name for a specification of a simulation run or a set of simulation runs. A scenario consists of a model, CONTROL statements, and possibly system data. The CONTROL statements contain the variable names which are used to reference data values obtained from running the simulation. The system data also contains variable names and can be viewed as data that has been generated from an experiment involving the real system. That is, the data comes from a scenario run on the actual system.

The referencing of each of the components for a scenario with the name FMSS1 is accomplished by using the TESS command:

BUILD SCENARIO NAMED(FMSS1) USING NETWORK(FMS)
WITH CONTROL(FMSC) AND DATA(SYSTEM.D1);

The building of network FMS was described in Section 17.7.2. The building of SLAM II control statements using TESS is accomplished through the command

BUILD CONTROL NAMED(FMSC);

This command causes TESS to display a form which has a menu of all SLAM II control statements. When a particular control statement is selected, a form is presented with each of the fields of the control statement listed and the default value for that field displayed as the current value. The form for the GENERAL statement is shown below.

GEN STATEMENT

```
.A  NAME    :
.B  PROJECT :
.C  MONTH   :    1    .D DAY : 1   .E YEAR:   2001
.F  NUMBER OF RUNS      :      1
.G  INPUT LISTING        :      Y
.H  ECHO REPORT          :      Y
.I  EXECUTION            :      Y
.J  WARN OF DEST. ENT.   :      Y
.K  INTERMED.RESULTS     :      Y
.L  SUMMARY REPORT       :      Y
.M  SUMMARY REPORT ON :      1
.N  OUTPUT COLUMNS       :      132
```

Each field that requires a change is then edited to complete the control statement. The set of control statements is continually displayed on the screen so that all SLAM II control statements can be seen at one time.

To obtain additional outputs in TESS from SLAM II runs, new control statements have been developed. These control statements specify additional SLAM II variables that can be stored in the TESS database.

* Summary statistics collected by SLAM II at COLCT nodes or STAT statements are specified to be put in the TESS database by the control statement DSCOLCT. Summary statistics from TIMST control statements are specified to be put in the TESS database by the DSTIMST control statement. Individual observations of either type of variable can be specified to be put in the database through the use of DOCOLCT and DOTIMST control statements. Forms for inputting the fields relating to these new control statements are included in TESS.

Two additional control statements have been developed to obtain trace data from a SLAM II run. This trace data is typically used to drive animations although it could be used for time series analysis (5). The DOEVENT statement specifies event times to be monitored and that data is to be collected and inserted in the database at the time of the event occurrence. Event time specifications are: start of an activity (ACT/S); completion of an activity (ACT/C); the instant before the calling of an event subroutine (EVENT/S); and immediately upon return from an event subroutine (EVENT/C). For the activity-related event times, a range of activity numbers can be prescribed for monitoring. For the event-related times, a range of event codes can be prescribed. In either case, attributes of entities or events can be listed on a DOATRIB control statement. This statement is used to specify a list of SLAM II status variables to be inserted into the database at each occurrence of the event type listed on the DOEVENT control statement. The use of TESS with the MH Extension to SLAM II, GPSS/H, and MAP/1 allows other event specifications to be made.

An additional control statement has been added to obtain animations of the simulation concurrent with the running of the simulation model. The statement name is ANIMATE and it has fields to specify the FACILITY diagram on which the animation is to be displayed, the RULE which translates the event information to actions on the FACILITY diagram, the starting and ending time for the animation, and a time between pauses in the animation to allow for the insertion of new information. The insertion of the new information is accomplished through a pause processor which is further described in Section 17.7.6.

The data definitions for the control statements and the outputs from a simulation are predefined. This is not the case for data obtained from the system under study. For such data, the form of the data and its characteristics are not known to

TESS. Therefore, the data definition for such data must be transmitted to TESS. This is accomplished through a BUILD DEFINITION command. An example of such a command is shown below.

BUILD DEFINITION NAMED(FMSDATA)
DVARIABLE(...) TYPE(DATA);

In the DVARIABLE specification, each variable in the data is given a name, a descriptor, a data type, an integer or character length, and a character field. Multiple variables can be associated with one definition.

The BUILD DEFINITION command provides a definition for a relation in the database and serves as the information source for a form which is presented when the user desires to provide values for the defined variables. The TESS command

BUILD DATA NAMED(FMSS1.FMSDATA)

causes the form to be presented with the names of the variables to be input for scenario FMSS1. As described previously, data of this type can be analyzed and reported using standard TESS commands. This allows data obtained from real world scenarios to be presented with data obtained from simulation runs.

17.7.4   TESS Support for Presentations

In developing a presentation, it is necessary to specify the variables to be presented, the form of the presentation, and the format for the presentation. The form of the presentation specifies the graph type such as a plot, chart, or histogram. The format specifies the color, orientation, and layout to be used within the graph type. Presentations can be obtained for data and summaries. The presentations are either graphical or tabular corresponding to the functions GRAPH and REPORT, respectively. This results in the four TESS language commands: GRAPH DATA; GRAPH SUMMARY; REPORT DATA; and REPORT SUMMARY. Specification of the contents, form and format for a presentation is made through the keywords included in these four commands.

The specification of the variables to be presented involves many decisions. To access the variables in the database, scenario names, data names, and variable names must be given. A wide range of possibilities exists within TESS for specifying the variables through the use of the following keywords: SCENARIO, VARIABLE, REPLICATE, and NAME. Data occurrences can be joined and can be matched according to a time or common variable. A selection of data values can be made using WHERE clauses. The selection of values using the WHERE clause can be made for ranges of values of variables, batch numbers, or replicate (run) numbers.

The form or type of presentation is specified following the TYPE keyword. Possible types of graphical presentation in TESS are: plots, charts, and histograms. For DATA, plots are the only graphical presentation form. Types of plots are discrete, continuous, spike, or scatter. For summaries, the graphical presentation options are a range chart, pie chart, bar chart, and histogram. For many graphical presentations, the form is identified by TESS based on how the data is stored in the database.

The format for a presentation is identified using the FORMAT and DRAW keywords. The DRAW keyword specifies whether the presentation is to contain a grid and the name of the independent variable if other than time. The FORMAT specification is built by the TESS user in a forms mode. The BUILD FORMAT command transfers the user to a forms module which presents the options for defining presentation formats. The list of forms for specifying a format for a discrete plot is shown below.

DISCRETE PLOT

.1 TEXT AND DECISIONS
.2 TITLE PARAMETERS
.3 X AXIS PARAMETERS
.4 Y AXIS PARAMETERS
.5 LABEL AND GRID PARAMETERS
.6 LINE COLORS AND STYLES
.7 FILL COLORS AND STYLES
.8 MARKER COLORS AND STYLES
.9 LEGEND PARAMETERS

For each form, there is a list of characteristics to be filled out. The form for defining text and decisions for a discrete plot is shown on the next page.

.1 TEXT AND DECISIONS

.A	TITLE	=	
.B	X AXIS LABEL	=	
.C	Y AXIS LABEL	=	
.D	PLOT ORIENTATION	=	HORIZONTAL
.E	GRID	=	NO
.F	LINE	=	YES
.G	FILL	=	NO
.H	MARKER	=	NO
.I	PLOT ORDER	=	MAX AREA FIRST
.J	LEGEND TEXT	=	

```
 1 VARIABLE NAME
 2 VARIABLE NAME
 3 VARIABLE NAME
 4 VARIABLE NAME
 5 VARIABLE NAME
 6 VARIABLE NAME
 7 VARIABLE NAME
 8 VARIABLE NAME
 9 VARIABLE NAME
10 VARIABLE NAME
```

In presenting a tabular report, variables or computed measures based on the variable values can be presented. Desired computed values are specified following a MEASURES keyword. The list of measures included in TESS are: MEAN, STANDARD DEVIATION, MINIMUM, MAXIMUM, LOWER BOUND, UPPER BOUND, COUNT, INTERVAL, IDBATCH, REPLICATE, BATCH NAME, SCENARIO NAME. The format for the REPORT function is established through a report specification format which allows the heading, number of rows, and number of columns to be specified for both the body of a page and the footnote of the page. The variables and measures would then be presented in addition to the page, column, and row headings described in the format.

In TESS, the specification of the variables to be presented, the type of presentation, and the format for the presentation are kept separate. By using different values following the keywords in a TESS command, a wide variety of presentations can be made in an effective manner.

17.7.5 TESS Support for Analysis

TESS provides the capability to combine data or summary values through the TESS command BUILD SUMMARY. TESS keywords to identify how the values are to be grouped are: REPLICATE, BATCH, REREPLICATE, REBATCH, and PARTITION. In addition the keywords: IGNORE, WEIGHT, and CELLS provide for the omitting of values, a weighting of values for averaging purposes, and a counting of the number of values within specified ranges. In addition, WHERE clauses can be used within the BUILD SUMMARY command to limit the conditions under which the computations are made. These capabilities permit analyses to be performed in accordance with the statistical procedures described in Chapter 19.

Confidence intervals for data values obtained from batches and replications can be automatically specified for computation and presentation. A thesis has been written (1) describing how the TESS library and database can be used for performing statistical analysis of simulation output in accordance with the procedures outlined in Chapter 19.

17.7.6 TESS Support for Animations

The TESS command

GRAPH FACILITY NAMED(FMSFAC) WITH
DATA(FMSS1.MACH) USING RULE(FMSRULES);

causes TESS to display an animation of the data identified by FMSS1.MACH on a facility diagram named FMSFAC in accordance with the rules, FMSRULES. The facility diagram is built using the graphic facility builder of TESS which is invoked by the TESS command BUILD FACILITY. Each rule for the animation is a set of statements. The statements are developed on a form which is obtained by using the TESS command BUILD RULE. The scenario name which follows the DATA keyword identifies the controls and model (or system) associated with the data to be visualized. As part of the control statements, the trace data collection has been specified. If an ANIMATE statement is included in the control statements, an animation is performed concurrently with the simulation run. TESS's pause processor is used in concurrent animations for interactively setting decision variables. This section describes the graphical facility builder, the forms processor for building RULES, and the capabilities of the pause processor.

A FACILITY is a descriptive model of a system constructed to look like the system. Included in the FACILITY are standard elements and specially built elements called ICONS. The ICONs are developed by the modeler using the TESS graphical ICON builder. To enter the ICON builder, the TESS user employs the BUILD ICON command. ICONs built within the ICON builder are automatically stored in the database and can be recalled for repeated use within the facility. Typical uses for ICONs are as the symbols that flow through the facility diagram. ICONs can also be used as stationary objects to provide a likeness for the component being simulated.

The options available when building a facility are shown below.

ADD	DELETE	EXIT	BMOVE	QUIT	SCROLL
COPY	BDELETE	HELP	NAME	REFRESH	VIEW
BCOPY	DISPLAY	MODIFY	INCLUDE	ROTATE	ZOOM
COUNTER	EDIT	MOVE	PROMPT	BROTATE	

The most frequently used command is the ADD command which is used to place a predefined symbol or user-defined ICON on the facility. The predefined symbols are: circle, line, path, rectangle, sector, stack, and triangle. Each time a symbol is placed on the screen, characteristics of the symbol such as its name, color, edge color, counter and name visibility, and number of sectors are requested. Text for annotating the symbol can also be added.

A path is used to model the flow of an entity. The time that it takes for an entity to move from one location to another dictates the speed at which its ICON or symbol moves along the path. A RULE provides the basis for specifying this flow information. A stack is used to represent a queue of entities. Rectangles can be identified as tanks in which case the rectangle fills up in the same manner that a tank would fill with a liquid. Counters are used to show the number associated with any SLAM II status variable on a facility symbol.

The RULES for translating events in a simulation run to actions on a facility diagram are established by using the TESS command BUILD RULE. A set of rules can include logical IF and ENDIF statements. Rules are available for changing the color of a symbol; increasing a counter to a value or by a value; controlling a flow along a path; putting a symbol in a stack or taking a symbol from a stack; moving a symbol from one position to another over a path; and changing a level in a rectangle (a tank statement). In addition, COMMENT, MONITOR, VIEW and

ZOOM statements can be included in the rules. A PAUSE statement invokes TESS' pause processor. The PAUSE statement allows changes to be made in the model as the animation of the model is displayed on the facility diagram.

The ZOOM and VIEW statements allow a detailed examination of portions of the facility diagram. The symbols and icons are automatically scaled when using these features. There can be many facility diagrams and rules built for a given model. This allows animations for different portions of a model to be viewed from different visual perspectives on several facility diagrams.

TESS has a time scale that relates viewing time to simulated time. For example, 100 time units of simulated time could be viewed in 50, 100 or 1,000 seconds of viewing time. A time line is portrayed on the facility diagram which indicates whether or not the viewing time is being maintained in accordance with this specification. Animations in a post-processing mode provide the capability to examine statistics or variable values over the run and then to relate them to the visualization during different periods within the run.

In summary, TESS supports the visualization of simulation outputs in a structured, comprehensive manner allowing different views and animations for the same simulation run in both a concurrent and a postrun mode.

17.7.7   TESS Library

The TESS library is a collection of FORTRAN subprograms for the storage, retrieval, deletion, and replacement of rows of data and summary occurrences in the TESS database. The TESS database is a relational database which is organized into tables of rows and columns. The library of subprograms can be used within a simulation program to retrieve simulation inputs from the TESS database and to store simulation outputs into the TESS database. This allows the TESS library to be used within SLAM II discrete event code for accessing inputs to describe arrivals or demands and for directly inserting data into the TESS database. Since the TESS library can be used within any FORTRAN program, an analyst has access to all data and summaries. Thus, project-specific programs can be developed to meet any required purpose. Although the TESS commands satisfy almost all of the requirements for querying and using the database, it is comforting to know that the database can be accessed directly if necessary.

The TESS database management system provides subprograms to open and close the TESS database, change, define, delete, and replace rows of relations, and to get, put and set rows of relations. For a detailed specification of the TESS library and an explanation of TESS database management system, the user is referred to the TESS User's Manual (9).

17.8   CHAPTER SUMMARY

For management to accept the outputs of a simulation project, the models developed on the project must augment the managers' understanding of the problem situation. A simulation support system provides a mechanism through which a decision maker can gain knowledge of the definition of an alternative and how its selection may affect performance in a changing environment. The specification for a simulation support system is outlined, and a description of how the TESS package provides simulation support is presented.

17.9   EXERCISES

17-1.   Develop an explanatory model that defines the functions of a simulation support system and how they relate to one another. For each function define its inputs, outputs, resources, and controls.

17-2.   Define each of the data elements discussed in this chapter and build an information model that characterizes the relations between the data elements.

17-3.   Discuss the hypothesis that it is easier to support the analysis and reporting tasks than the modeling tasks in a simulation project.

17-4.   Develop a set of concepts that could display the movement of entities on a SLAM II network.

17-5.   Prepare a detailed specification for a simulation support system.

17-6.   Decision support systems categorize decisions as structural, semistructural and unstructured. Decisions are also described in terms of an organizational level as being part of: strategic planning, management control, and operational control. Discuss how simulation can support decisions in these categories and assess the need for having a simulation support system according to the category of the decision.

17.10 REFERENCES

1. Madhavpeddi, K.V., Procedures for Automatic Simulation Output Analysis, unpublished MS Thesis, The University of Iowa, 1980.

2. Polito, J., *The Safeguards Network Analysis Procedure (SNAP): A User's Manual*, Pritsker & Associates, Report ABQ0013, West Lafayette, IN., March, 1983.

3. Pritsker, A.A.B. and C.E. Sigal, *Management Decision Making: A Network Simulation Approach*, Prentice-Hall, 1983.

4. Sabuda, J., A Study of Q-GERT Modeling and Analysis Using Interactive Computer Graphics, unpublished MS Thesis, Purdue University, December 1977.

5. Sale, M., M. Grant, R. Meyer, E. Casey, and D. Yancey, *IDSS Prototype (2.0) User's Reference Manual*, Pritsker & Associates, Inc., West Lafayette, IN, November 1983.

6. Standridge, C.R., "Performing Simulation Projects with The Extended Simulation System (TESS)", *Simulation*, December 1985, pp. 283-291.

7. Standridge, C.R., and A.A.B. Pritsker, "Using Data Base Capabilities in Simulation," in *Progress in Modelling and Simulation*, ed. by F. Cellier, Academic Press, 1982.

8. Standridge, C.R., L.J. Rolston, D.K. Vaughan, J.R. Hoffman, D.K. LaVal, M.L. Sale, and J.W. Myers, *TESS with MAP/1 User's Manual, Version 2.0*, Pritsker & Associates, Inc., West Lafayette, IN, 1986.

9. Standridge, C.R., D.K. Vaughan, J. Hoffman, D.K. LaVal, J. O'Reilly, M. Sale, and J.W. Myers, *TESS with SLAM II User's Manual, Version 2.0*, Pritsker & Associates, Inc., West Lafayette, IN, 1986.

10. Yancey, D.P., C.N. Busch, and J.P. Whitford, *IDSS Build 1, Final Technical Report*, Volume 1, Pritsker & Associates, Inc. and WPAFB Materials Laboratory, 1984.

11. Zeigler, B.P., *Theory of Modelling and Simulation*, John Wiley, 1976.

12. Zeigler, B.P., *Multifacetted Modelling and Discrete Event Simulation*, Academic Press, 1984.

CHAPTER 18

Random Sampling from Distributions

18.1   INTRODUCTION

Throughout this text, we have dealt with models which contain variables that are characterized by distributions. In Chapter 2, we discussed data acquisition procedures. In Chapter 17, methods for storing data in a database for analysis or for direct use in models is described. In Chapter 16, the special purpose simulation language, FACTOR, is mentioned which relies on the automatic collection of data for the scheduling and control of systems using simulation. The increased use of data acquisition hardware on the manufacturing shop floor establishes new horizons for modelers and establishes a need for further research on data characterization.

A model that is developed for analysis by simulation can employ input data in different forms. The data need not be characterized by a distribution function. The data can be used directly which results in a trace-driven simulation. The data can be grouped, and samples can be obtained from the grouped data. Alternatively, the data can be fit by a theoretical distribution and random samples can be obtained by sampling from the distribution. This last topic is the central concern of this chapter.

First, a discussion is given of the most commonly used theoretical distributions including graphs of their density functions and a rationale for their use in models. The question of when to use a distribution to represent data is then posed. The remainder of the chapter discusses procedures for generating random numbers and random samples.

18.2   DISTRIBUTIONS

Distributions have underlying characteristics which make them appropriate to represent a random process or activity. We refer to a distribution that comes up repeatedly as a standard distribution. We describe these standard distributions in this section. For a more formal discussion and graphical description of distributions, the books "Statistical Methods in Engineering" by Hahn and Shapiro (15) and "Statistical Distributions" by Hastings and Peacock (16) are recommended.

Throughout the discussion, the following variable definitions will be used:

$$X = \text{the random variable}$$
$$f(x) = \text{the density function of } X$$
$$p(x) = \text{probability mass function of } X.$$
$$a = \text{minimum;}$$
$$b = \text{maximum;}$$
$$m = \text{mode;}$$
$$\mu = \text{mean} = E[X]$$
$$\sigma^2 = \text{variance} = E[(X-\mu)^2]$$
$$\sigma = \text{standard deviation}$$
$$\alpha = \text{a parameter of the density function}$$
$$\beta = \text{a parameter of the density function}$$

For the density functions with parameters not prescribed by the mean and standard deviation, we provide equations to relate the mean and standard deviation to the parameters.

18.2.1   Uniform Distribution

The uniform density function specifies that every value between a minimum and a maximum value is equally likely. The use of the uniform distribution often implies a complete lack of knowledge concerning the random variable other than that it is between a minimum value and a maximum value. Thus, the probability of a value being in a specified interval is proportional to the length of the interval. Another name for the uniform distribution is the rectangular distribution. Figure 18-1 gives the density function for the uniform distribution and its graph.

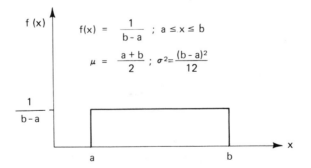

Figure 18-1   Uniform density function and illustration.

18.2.2   Triangular Distribution

The triangular distribution contains more information about a random variable than the uniform distribution. For this distribution, three values are specified: a minimum, a mode, and a maximum. The density function consists of two linear parts: one part increases from the minimum value to the modal value; and the other part decreases from the modal value to the maximum value. The average associated with a triangular density is the sum of the minimum, mode and maximum values divided by 3. The triangular distribution is used when a most likely value can be ascertained along with minimum and maximum values, and a piecewise linear density function seems appropriate. Figure 18-2 gives the density function for the triangular distribution and its graph.

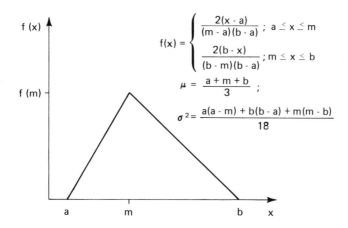

$$f(x) = \begin{cases} \dfrac{2(x-a)}{(m-a)(b-a)} \;;\; a \le x \le m \\[2mm] \dfrac{2(b-x)}{(b-m)(b-a)} \;;\; m \le x \le b \end{cases}$$

$$\mu = \frac{a+m+b}{3} \;;$$

$$\sigma^2 = \frac{a(a-m) + b(b-a) + m(m-b)}{18}$$

Figure 18-2   Triangular density function and illustration.

18.2.3   Exponential Distribution

If the probability that one and only one outcome will occur during a small time interval Δt is proportional to Δt and if the occurrence of the outcome is independent of the occurrence of other outcomes then the time between occurrences of outcomes is exponentially distributed. Another way of saying the above is that an activity characterized by an exponential distribution has the same probability of being completed in any subsequent period of time Δt. Thus, if the activity has been ongoing for t time units, the probability that it will end in the next Δt time units is the same as if it had just been started. This lack of conditioning of remaining time on past time expended is called the Markov or forgetfulness property. There is direct association between the assumption of an exponential activity duration and Markovian assumptions. The use of an exponential distribution assumes a large variability as the variance is the square of the mean. The exponential distribution has one of the largest variances of the standard distribution types. The exponential distribution is easy to manipulate mathematically and is assumed for many studies because of this property.

Figure 18-3 gives the density function for the exponential distribution and its graph.

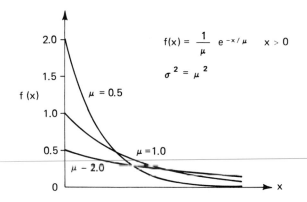

Figure 18-3   Exponential density function and illustrations.

18.2.4   Poisson Distribution

The Poisson distribution is a discrete distribution and usually pertains to the number of outcomes occurring in a specified time period. If the duration of time between outcomes is exponentially distributed and they occur one at a time, the number that occur in a fixed time interval can be shown to be Poisson distributed. Thus, if the interarrival distribution is exponential, the number of arrivals will be Poisson distributed. The Poisson distribution is frequently used as a limiting case approximation to the binomial distribution where the binomial distribution is used to represent a series of independent Bernoulli trials (an outcome of a trial is go-no go, success-failure, yes-no). For a large value of the mean, the Poisson distribution approaches the normal distribution.

Figure 18-4 gives the Poisson probability mass function and illustrates its form.

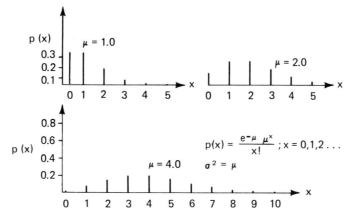

Figure 18-4 Poisson probability mass function and illustrations.

18.2.5   Normal Distribution

The normal or Gaussian distribution is the most prominent distribution in probability and statistics. Justification for the use of the normal distribution comes from the Central Limit Theorem which, as discussed in Chapter 2, specifies that under very broad conditions the distribution of the average or sum of I independent observations from any distribution approaches a normal distribution as I becomes large. Thus, when dealing with phenomena that are related to sums of random variables, approximation by a normal distribution should be considered.

Because of the Central Limit Theorem, it is easy to see why the normal distribution has received a great amount of attention and use in applications of probability and statistics. There is another reason for the heavy use of the normal distribution. The normal distribution also has the advantage of being mathematically tractable and consequently many techniques of statistical inference such as regression analysis and analysis of variance have been derived under the assumption of an underlying normal density function.

As discussed above, for a large mean value, the normal distribution is a good approximation to the Poisson distribution.

Figure 18-5 gives the density function for the normal distribution and illustrates the distribution for selected values of the mean and standard deviation.

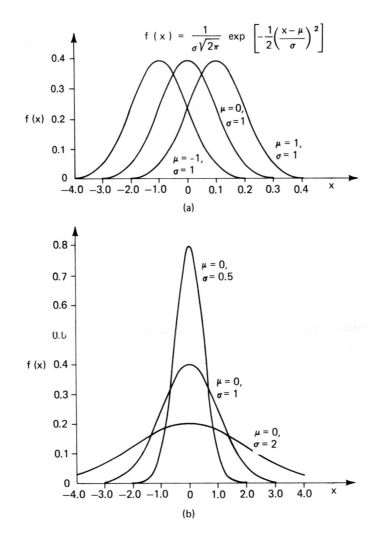

Figure 18-5   Normal density function and illustrations

18.2.6   Lognormal Distribution

The lognormal distribution is the distribution of a random variable whose natural logarithm follows the normal distribution (15). The lognormal distribution is appropriate for a multiplicative type process in the same manner that the normal distribution is applicable for additive type processes. By use of the Central Limit Theorem, it can be shown that the distribution of the product of independent positive random variables approaches a lognormal distribution.

If a set of data is transformed by taking the logarithm of each data point, and if the transformed data points are normally distributed, then the original data is said to be lognormally distributed. The lognormal distribution has been used as an appropriate model in a wide variety of situations from biology to economics. It is an appropriate model for processes where the value of an observed variable is a random proportion of the previous observed value. Examples of such processes include the distribution of personal incomes, inheritances and bank deposits, and the distribution of particle sizes.

Figure 18-6 gives the density function for the lognormal distribution and illustrates the distribution for selected values of the mean and variance.

18.2.7   Erlang Distribution

The Erlang distribution is derived as the sum of independent and identically distributed exponential random variables. It is a special case of the gamma distribution. The density function, illustrations and remarks concerning the gamma distribution therefore apply to the Erlang distribution. The Erlang distribution is used extensively in queueing theory when an activity or service time is considered to occur in phases with each phase being exponentially distributed.

18.2.8   Gamma Distribution

The gamma distribution is a generalization of the Erlang distribution where conceptually the number of sums of exponential variables need not be integer valued. Gamma distributed times can take on values between zero and infinity. By

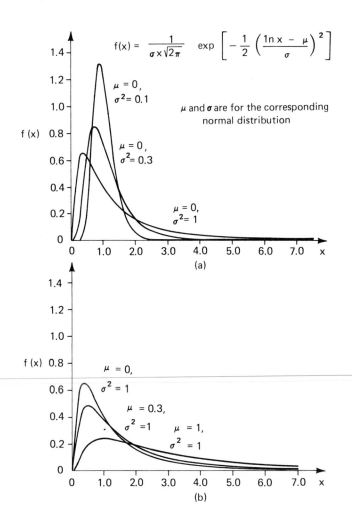

Figure 18-6 Lognormal density function and illustrations.

different parameter settings, the gamma distribution can be made to take on a variety of shapes and, hence, can represent many different physical processes.

The chi-square distribution is also a special case of the gamma distribution where a chi-square random variable is the sum of squared normal random variables. Thus, special cases of the gamma are the chi-square distribution, the Erlang distribution and, hence, the exponential distribution.

Figure 18-7 gives the density function for the gamma distribution and illustrates the density function for selected values of its parameters.

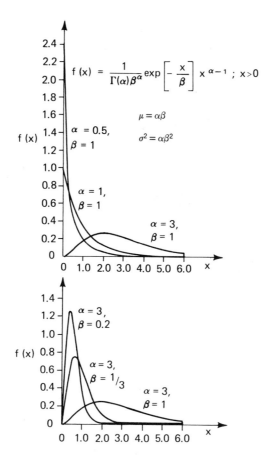

Figure 18-7 Gamma density function and illustrations

18.2.9 Beta Distribution

The beta distribution is defined over a finite range and can take on a wide variety of shapes for different values of its parameters. It can be bell shaped, symmetric or asymmetric, or it can be U-shaped within the finite range. For U-shaped beta functions, the value of the density function goes to infinity as the ends of the range are approached. A simple variant of the beta distribution is referred to as the Pareto distribution which has been used to characterize income distributions.

Figure 18-8 gives the density function for the beta distribution and illustrates the density function for selected values of its parameters.

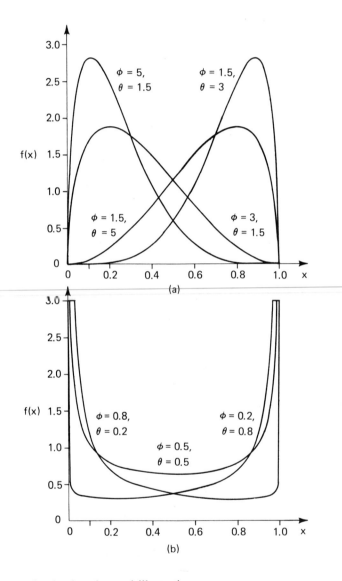

Figure 18-8   Beta density function and illustrations.

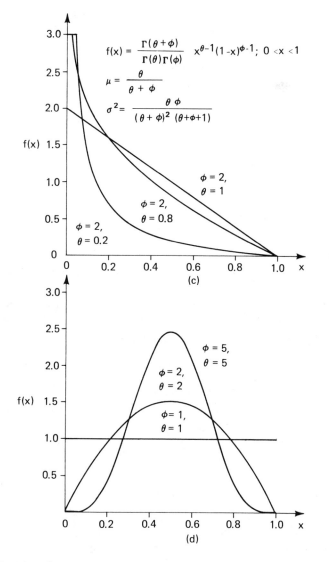

$$f(x) = \frac{\Gamma(\theta+\phi)}{\Gamma(\theta)\Gamma(\phi)} \, x^{\theta-1}(1-x)^{\phi-1}; \quad 0 < x < 1$$

$$\mu = \frac{\theta}{\theta+\phi}$$

$$\sigma^2 = \frac{\theta\phi}{(\theta+\phi)^2\,(\theta+\phi+1)}$$

$\phi = 2,$
$\theta = 1$

$\phi = 2,$
$\theta = 0.8$

$\phi = 2,$
$\theta = 0.2$

(c)

$\phi = 5,$
$\theta = 5$

$\phi = 2,$
$\theta = 2$

$\phi = 1,$
$\theta = 1$

(d)

Figure 18-8   (continued)

Since the beta distribution is defined over a finite interval, it has been used to describe situations which have a finite range. Examples of this are density functions related to percentages and probability estimates. Frequently, the beta distribution is used as the priori distribution of the parameter of a binomial process by Bayesian statisticians. Another use of the beta distribution is as the descriptive density function associated with an activity duration in PERT. Subjective estimates of an activity duration based on optimistic (a), pessimistic (b), and most likely (m) values are combined to estimate the mean and variance of the beta distribution as $(a+4m+b)/6$ and $(b-a)^2/36$, respectively.

Due to the wide variety of shapes obtainable for the beta distribution, it has been used to fit many different types of data.†

18.3   USING DISTRIBUTIONS TO REPRESENT DATA

When the underlying nature of a variable is not understood, then it is necessary to pose the question as to whether a distribution should be used to represent the data. If this question is answered in the affirmative then a procedure for determining the distribution form is required. If it is decided not to use a theoretical distribution then a sampling procedure from the empirical data or a quasi-empirical distribution is required.

The fundamental question associated with using a distribution to represent the data is whether a model of the data, that is, the distribution, provides more information than the direct use of the data. The resolution of this question requires an investigation of the validity of the model used to fit the data and statistical tests of the fit of the model. The advantage of building a model is that it extends the range of the data and allows for values beyond those observed. It also tends to smooth the frequency of occurrence of variable values. Further, it may provide insight into ways of using the variable or improving the system characteristic which the variable represents. The pitfalls of using a theoretical distribution as a model of a variable are well-described by Fox (14) and are included in the discussion below.

If an empirical distribution is used to characterize the data, then sampling from the empirical distribution will in the long run approximate the data. The sampling from a theoretical distribution that models the data allows alternate samples to be obtained which extends the information content of the data. The validity question relates to whether or not the theoretical distribution fits the underlying distribution from which the data come. It also relates to the purpose for modeling and whether the system under study is to be exercised by data values other than those that have been observed.

---

† In this regard, the warning given by Feller (11) with regard to the law of logistic growth should be mentioned: ". . . the only trouble with the theory is that not only the logistic distribution but also the normal, the Cauchy and other distributions can be fitted to the same material with the same or better goodness-of-fit . . . Most contradictory theoretical models can be supported by the same observational material."

Statistical tests of the fit of a theoretical distribution to the data leave much to be desired. Goodness-of-fit tests such as the Chi-square and Kolmogorov-Smirnov have a low probability of rejecting a fit. In addition, distribution fitting procedures typically involve looking at the data and then sequentially attempting to fit theoretical distributions to the data. Bratley, Fox and Schrage (9) point out "statistics associated with tandem goodness-of-fit tests have unknown distributions". They further state that "estimating the tails of distributions accurately from a limited amount of data is impossible". Based on these pitfalls, they recommend fitting a quasi-empirical distribution which is piecewise linear with exponential tails.

If the above pitfalls are not bothersome and, in practical situations they normally are not, then a theoretical distribution can be fit with the aid of a computer package (20,23,25). Typically, a graphical overlay of the theoretical distribution is displayed on a histogram of the data. A visual inspection is then made to assess the fit. At this point, any confidence ascribed to a statistical goodness-of-fit test as previously pointed out has little credence.

18.4   PROCEDURES FOR GENERATING RANDOM SAMPLES

The state-of-the-art for generating random samples on digital computers is quite advanced. A survey of random variate generation techniques by Schmeiser (32) includes over 300 references. The textbooks on simulation usually include one or more chapters devoted to random sampling procedures and algorithms (5,9,12,19). The material in this section follows the excellent survey by Schmeiser.

There are four fundamental approaches for random sample generation:

1. Inverse transformation;
2. Composition;
3. Acceptance/Rejection; and
4. Special Properties.

Each of the approaches use as inputs independent samples from a uniform distribution distributed over the interval (0,1). These uniform samples are called random numbers and they provide the basic source of randomness in simulation experiments. Definitions and procedures for generation random numbers are presented in Section 18.5.

18.4.1   The Inverse Transformation Method

The simplest and most fundamental technique which forms the basis for generating samples is the *inverse transformation method* (1,27). This method uses the information that for the distribution function F(X), the random variable R = F(X) is uniformly distributed on the unit interval [0,1]. Thus, to generate a random sample of X, we generate a random number r and solve the equation

$$r = F(x)$$

for the corresponding value of x = F^{-1}(r). The proof for the validity of the method is straightforward (9,27) and is based on the following reasoning. Let R = F(X) have distribution function G. Then, for 0 ⩽ r ⩽ 1, we have

$$G(r) = \mathbf{P}[F(X) \leqslant r] = \mathbf{P}[X \leqslant F^{-1}(r)] = F(F^{-1}(r)) = r.$$

Thus, R is uniformly distributed on [0,1].

To illustrate the method for a continuous distribution, consider the generating of a sample from the exponential distribution. The distribution function for the exponential is F(x) = 1 − e$^{-\lambda x}$ where 1/λ is the mean of the exponential. Setting F(x) equal to r and then solving for x yields

$$x = - (1/\lambda) \, ln(1-r)$$

Hence, if r is uniformly distributed in the range 0 to 1, then x given by the above equation is exponentially distributed with a mean value of 1/λ.

The method is also applicable to discrete distributions. For example, consider the following probability mass function

$$p(0) = .25$$
$$p(1) = .50$$
$$p(2) = .25$$

The cumulative distribution function, F(X), is depicted in Figure 18-9. To obtain a sample from the above distribution, a random number is generated in the range 0 to 1, and the graph is entered at this ordinate value. The resulting random sample is then obtained by tracing across the graph to the cumulative curve, and then down to the x-axis. For example, the random number 0.81 yields a random sample

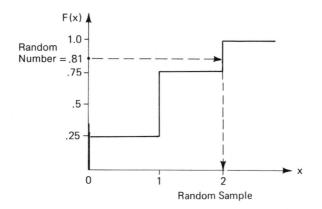

Figure 18-9 Illustration of the inverse transform method for obtaining a sample from a given distribution.

of 2 as shown. The intuitive justification for this procedure is that 25 percent of the random numbers are in the interval (0,0.25), 50 percent in (0.25,0.75) and 25 percent in (0.75,1.00) which is the desired distribution function. An arbitrary but consistent decision should be made at the break points.

The difficulty with the inverse transformation method lies in finding the inverse transformation $F^{-1}(r)$. In some cases, the method leads to a simple explicit transformation as was illustrated for the exponential distribution. However, there are continuous distributions that do not have closed-form inverse functions.

18.4.2 Acceptance/Rejection Method

The acceptance/rejection approach to generating random samples involves generating samples from a distribution and then rejecting (discarding) some of the samples in such a way that the remaining samples have the desired distribution. The procedure involves the generation of two random numbers. The first random number is used to calculate a proposed random sample, x. This sample value is then inserted into the equation for the density function f(x). The second random number is used to obtain a random value t(x) from a function that majorizes or covers the density function. If t(x) is larger than f(x) then x is rejected as a sample from f(x). If the value of t(x) is smaller, x is accepted. In this way, samples are accepted based on the height of the density function which is the property desired from the random sampling procedure. The acceptance/rejection method for generating random samples is similar to the use of the Monte Carlo method to evaluate integrals (1,12).

Acceptance/rejection methods have been developed for both discrete and continuous distributions. The efficiency of the method depends on the amount of rejection required. Research has concentrated on majorizing functions which reduce the value of the test function t(x) used for the acceptance/rejection decision. The acceptance/rejection method has been used for obtaining random samples from most of the standard distributions including normal, gamma and beta (2,3,32).

18.4.3   The Composition Method

The composition method assumes that the density function can be written as a weighted sum of component distribution functions with the sum of the weights totaling one. The inverse transformation method can be used to select one of the subdistributions. A sample is then drawn from that subdistribution and it becomes the sample for the distribution. By repeated sampling, each of the subdistributions are selected in accordance with the weights and, hence, the samples are generated in accordance with the total distribution.

As described above, the composition method may employ the inverse transformation method to select a subdistribution and then any sampling procedure to obtain a random sample from the subdistribution. The acceptance/rejection method is frequently used where majorizing functions are defined for portions of the distribution function. Thus, it should be clear that the methods for generating random samples are not necessarily used independently.

18.4.4   Methods Employing Special Properties

In some cases, a distribution will have a special property or be related to another distribution from which it is easier to obtain a random sample. Examples of this are:

- A lognormal sample computed from e^y where y is a sample from the normal distribution;
- An Erlang sample as the sum of exponential samples;

- A beta sample as the ratio of gamma samples;
- A binomial sample as a sum of Bernoulli trials;
- A negative binomial sample as a sum of geometric samples;
- A normal sample as a sum of uniform samples;
- A Cauchy sample as the ratio of two independent standard normal samples;
- A chi-square sample, if the degrees of freedom are even, can be obtained from an Erlang sample. If the degrees of freedom are odd, add the square of a standardized normal to a chi-square sample which has one less degree of freedom;
- An F-distribution sample as the ratio of two chi-square samples; and
- A t-distribution sample as the ratio of a standardized normal sample to a chi-square sample.

In other situations, relationships are derived in order that random samples may be obtained. One such case relates normal random variables to uniform random variables (8).

Another special situation involves sampling from a distribution with specified limits. If a sample is desired from a truncated distribution between the specified limits A and B then generate a sample and, if it is within the prescribed limits, accept the sample. If it is outside the limits, reject the sample and repeat the procedure. Alternatively, the random numbers can be restricted to values between $F_x(A)$ and $F_x(B)$ and the inverse transformation equation used. A mixed truncated distribution assumes that samples between A and B are desired, and a sample that exceeds either limit is set equal to the limiting value. To obtain samples from a mixed truncated distribution, accept a sample if it is within the limits specified. If the value is larger than B, set the sample to B. If the value is less than A, set the sample to A. Note that the moments of the truncated distribution may be different than those of the underlying distribution.

Lastly, parameterized distributions have been developed to represent systems of distributions, for example, Pearson (24) and Johnson (18). Schmeiser and Deutsch developed a system of distributions for which the inverse transformation can be used to obtain samples (31).

18.4.5 Implementing Random Sampling Procedures

The algorithms for obtaining samples from a distribution are many. Extensive research is being performed in this area. New algorithms are continually being proposed and tested which provide improved computer efficiency. The methods and equations used in SLAM II to obtain random samples are presented in Table 18-1. In the table, R denotes a random number.

The inverse transformation method is used to obtain a sample from the uniform distribution. The inverse transformation method is also used to obtain a sample from the triangular distribution. For the triangular distribution, two equations are involved depending on whether the density function is increasing or decreasing. To obtain samples from the normal distribution, a transformation of variables developed by Box and Muller (8) is used in conjunction with a sample rejection.(2) The procedure generates pairs of normal samples.

A sample from an exponential distribution is obtained using the inverse transformation. In taking the inverse, the value (1–R) is required. Since (1–R) is equivalent to a random number, the value R is used directly in the formula for the exponential sample.

A sample from a Poisson distribution is generated based on the relationship between the exponential distribution and the Poisson distribution. The mean number of events occurring in time T for a Poisson distribution with a mean rate of arrivals U is UT. By setting T equal to one, we can generate exponential samples with mean time 1/U until the sum of the values of the exponential samples exceeds 1. Then one less than the number of samples generated is an observation from the Poisson distribution. The procedure presented in Table 18-1 to obtain a sample from the Poisson distribution has been derived from the following inequality

$$t_1 + t_2 + \ldots + t_N < 1 < t_1 + t_2 + \ldots + t_{N+1}$$

where t_i is a sample from an exponential distribution with mean time 1/U.

Samples from the Erlang distribution are obtained by summing K exponential samples with each sample having a mean of U. The mean of the Erlang distribution would then be KU. The equation to obtain a sample from the Erlang distribution employs the inverse transformation for obtaining samples from the exponential distribution and then summing up the values. The equation shown in Table 18-1 makes use of the information that the sum of logarithms is equal to the logarithm of the product.

A sample from a lognormal distribution is obtained by using the relationship that the natural logarithm of a lognormal random variable is normally distributed. The equations that relate the mean and variance of the normal distribution to the mean and variance of the lognormal distribution (27) are shown in Table 18-1.

Samples from a gamma distribution are obtained using a combined approach. The second parameter of the gamma distribution is a shape parameter. When

Table 18-1 SLAM II random sampling procedures.

Uniform Distribution

> Function Name: UNFRM(A, B, I)
> Method: Inverse transformation
> Equation: UNFRM = A + (B − A) * R

Triangular Distribution

> Function Name: TRIAG(A, D, B, I)
> Method: Inverse transformation (27)
> Equations:

$$\text{For } 0 \leq R \leq \frac{D - A}{B - A}; \text{TRIAG} = A + \text{SQRT}((D - A)*(B - A)*R)$$

$$\text{For } \frac{D - A}{B - A} < R \leq 1; \text{TRIAG} = B - \text{SQRT}((B - D)*(B - A)*(1. - R))$$

Normal Distribution

> Function Name: RNORM(U,S,I)
> Method: Transformation of variables used in conjunction with sample rejection (2,8)
> Procedure: Normal samples are generated in pairs.
> > Let A = 2.*R_1 − 1. and B = 2.*R_2 − 1.
> > Let W = A*A + B*B .
> > If W > 1.0, repeat procedure
> > If W ≤ 1.0 then
> > > RNORM_1= (A*SQRT(−2.*ALOG(W)/W))*S + U
> > > RNORM_2= (B*SQRT(−2.*ALOG(W)/W))*S + U

Exponential Distribution

> Function Name: EXPON(U,I)
> Method: Inverse transformation
> Equation: EXPON = −U*ALOG(R)

Poisson Distribution

> Function Name: NPSSN(U,I)
> Method: The number of exponential samples in a unit interval (27)

Table 18-1 (continued)

Procedure: Set the sample value, NPSSN, equal to the first value of N such that

$$\prod_{n=1}^{N} R_n \geq e^{-U} > \prod_{n=1}^{N+1} R_n$$

where R_n is the nth pseudorandom number.

Erlang Distribution

Function Name: ERLNG(U,K,I)
Method: Sum K exponential samples each having a mean equal to U

Equation: $ERLNG = -U*ALOG(\prod_{i=1}^{K} R_i)$

Lognormal Distribution

Function Name: RLOGN(U,S,I)
Method: Use a sample, N, from a normal distribution in the equation $L = e^N$.
   It can be shown that if N is normally distributed L is lognormally distrib-
   uted (4).
Equation: $RLOGN = EXP(RNORM(\mu_N, \sigma_N, I)$

where $\sigma_N^2 = \ln(S^2/U^2+1.)$

and $\mu_N = \ln U - \frac{1}{2}\sigma_N^2$

Gamma Distribution

Function Name: GAMA(β,α,I)
Method: The method for obtaining a sample from a gamma distribution is a
   function of the parameter α. More efficient methods are employed as α is
   increased. When α is an integer, Function ERLNG should be employed.
       For $0 < \alpha < 1$, the method of Jöhnk is employed (6)
       For $1 \leq \alpha < 5$, the method of Fishman (13) as modified by Tadi-
          kamalla is employed (33, 34)
       For $\alpha \geq 5$, a weighted selection of Erlang samples is employed.

Procedure: For $0 < \alpha < 1$:

   Let $X = R_1**(1./\alpha)$ and $Y = R_2**(1./(1. - \alpha))$

Table 18-1   (continued)

If $X + Y \leq 1$, compute $W = X/(X+Y)$. Otherwise recompute X and Y.

Let $GAMA = W*(-ALOG(R_3))*\beta$

For $1 \leq \alpha < 5$:

Let $a = [\alpha]$ and $b = a - [\alpha]$

Compute $X = (\alpha/a)*(-ALOG(\overset{a}{\underset{i=1}{\amalg}} R_i))$

If $R_{a+1} > (X/\alpha)^b \exp(-b*(X/\alpha-1.))$, recompute X.

Otherwise $GAMA = X*\beta$

For $\alpha \geq 5$:

If $R_1 \geq a - [\alpha]$, $GAMA = ERLNG(\beta,[\alpha],I)$

If $R_1 < a - [\alpha]$, $GAMA = ERLNG(\beta,[\alpha] + 1,I)$

Beta Distribution

Function Name: $BETA(\alpha,\beta,I)$

Method: Transformation of variables where the beta sample is the ratio of two gamma samples (12)

Equation: $BETA = G1/(G1 + G2)$

where $G1 = GAMA(1,\alpha,I)$ and $G2 = GAMA(1,\beta,I)$

and if μ_g and σ_g are given

then $\alpha = \mu(\mu-\mu^2)/\sigma^2-\mu$,

and $\beta = \alpha(1-\mu)/\mu$

where $\mu = (\mu_g-\min)/(\max-\min)$

and $\sigma^2 = \sigma_g^2/(\max-\min)^2$.

Weibull Distribution

Function Name: $WEIBL(\beta,\alpha,I)$

Method: Inverse transformation (16)

Equation: $WEIBL = (-\beta*ALOG(R))**(1./\alpha)$

where β is a scale parameter and α is a shape parameter, as defined by the distribution function $F(W) = 1-\exp(-1/\beta*W^\alpha)$.

Probability Mass Function

Function Name: $DPROB(CPROB, VALUE, NVAL, I)$

Method: Inverse transformation

Procedure: Set $DPROB$ equal to $VALUE(N)$ when

$CPROB(N-1) < R \leq CPROB(N)$ where $CPROB(0) = 0.0$ and $N = 1$ to $NVAL$

the shape parameter is small, an acceptance/rejection technique proposed by Jöhnk (6) is employed. For intermediate values of the shape parameter, a method proposed by Fishman (13) and modified by Tadikamalla (33,34) is used. For shape parameter values greater than 5, a weighted selection of Erlang samples is employed (6). If the shape parameter is an integer, then Function Erlang should be used for efficiency purposes.

A sample from a beta distribution is obtained as the ratio of two gamma samples. Samples from the beta distribution are generated in the interval (0,1). To transform the sample into the interval (A,B), multiply the sample by (B–A) and add A.

Samples from a Weibull distribution are obtained using the inverse transformation method. Note that there are alternative definitions for the Weibull distribution function. The form of the Weibull distribution function used in SLAM II is shown in Table 18-1.

SLAM II obtains samples from a probability mass function by comparing a random number against the cumulative distribution function. When the random number falls in the appropriate range, the corresponding sample value is returned from function DPROB.

18.5   GENERATING PSEUDORANDOM NUMBERS

As discussed in the last section, generating random samples from a distribution on a digital computer normally requires one or more uniform random samples between 0 and 1, and then to transform the uniform sample (or samples) into a new sample from the desired distribution. Independent samples that are uniformly distributed in the interval 0 to 1 are called random numbers.

There are at least three methods for obtaining random numbers for digital simulation. The first method is to read a table of random numbers (29) into the computer and then treat the random numbers as data for the simulation problem. The major shortcomings of this method are related to the relative slowness of computers in reading data from an external device and the need to store large tables. A second method is to employ a physical device such as a vacuum tube which generates random noise. A major objection to this method is that the simulation results are not reproducible, thereby complicating model verification and controlled experimentation with model parameters. The third and preferred method is to employ a recursive equation which generates the (i+1)st random number from previous random numbers. Since the sequence of numbers is produced deterministically by an equation, they are not truly random, and therefore are referred to as "pseudorandom" numbers. It is common practice to shorten the phrase and refer to such numbers as simply random numbers, with the understanding that they are actually pseudorandom.†

The properties desirable in a pseudorandom number generator are:

1. The numbers should be uniformly distributed in the interval (0,1);
2. The numbers should be independent and, hence, no correlation should exist in a sequence of random numbers;
3. Many numbers should be generated before the same number is obtained. This is referred to as the period or cycle length of the generator;
4. A random number sequence should be reproducible. This implies that different starting values or seeds should be permitted to allow different sequences or streams to be generated;
5. The generator should be fast, as many numbers may be required in a simulation; and
6. A low storage requirement is preferred.

The technique which best satisfies these properties is referred to as the congruential method.

A linear congruential method employs the following recursive equation:

$$z_{i+1} = (az_i + b)(\text{mod } c) \qquad i = 0, 1, 2, \ldots$$
$$r_{i+1} = z_{i+1}/c$$

where z_0 is the seed value and r_i is the ith pseudorandom number. This equation denotes that the unnormalized random number, z_{i+1}, is equal to the remainder of $(az_i + b)$ divided by c where z_i is the previous unnormalized random number, z_0 is an initial value or seed, and a, b, and c are constants. In Exercise 18-2, an equation for computing the $(i+1)$st unnormalized random number is given. The assignment of values to the constants a, b, and c has been the subject of intensive research. Fishman (12) presents an excellent review of how to set the

---

† Because they are generated on digital computers by deterministic methods, there is a great deal of controversy over the definition of pseudorandom numbers. From our perspective and approach to simulation, Lehmer's definition is appealing, " . . . a vague notion embodying the idea of a sequence in which every term is unpredictable to the uninitiated and whose digits pass a certain number of tests . . . depending somewhat on the uses to which the sequence is to be put."(21)

constants and the procedures for testing random number generators. In Table 18-2, we summarize the suggested rules for setting a, b, and c for congruential random number generators. Although the rules given in Table 18-2 provide general guidelines for selecting the constants a, b, and c, the overall best values are computer dependent. *We, therefore, recommend that the modeler employ a random number generator that has been specifically designed for the computer on which the simulation model is to be run.* In Chapter 6 of Bratley, Fox and Schrage, the implementation and testing of random number generators is described in detail. Schrage's portable FORTRAN random number generator (9) is also presented which is included in the basic SLAM II package. The SLAM II random number function is DRAND(ISTRM). The function XRN(ISTRM) returns the value of the last random number generated.

In simulation modeling, it is frequently desirable to employ several random number streams within the same model. For example, separate random number streams could be employed in a queueing system to model the arrival and service process. In this manner, the same sequence of arrival times can be generated without regard to the order in which service is performed. Thus, different service procedures could be evaluated for the same sequence of arrivals. Random number generators provide for parallel streams by allowing the modeler to provide a different seed value for each stream to be employed.

18.6   TESTS FOR RANDOMNESS

The statistical validity of the results of a simulation model are dependent upon the degree of randomness of the random number generator employed. Because of this, many statistical procedures have been developed for testing random number generators. However, as noted by Hull and Dobell (17) ". . . no finite class of tests can guarantee the general suitability of a finite sequence of numbers. Given a set of tests, there will always exist a sequence of numbers which passes these tests but which is completely unacceptable for some particular application." This reservation does not present a serious problem as a simulation analyst only desires the properties of randomness that were previously described.

Both analytical and empirical tests have been used to investigate the randomness properties of random number generators. These include the Frequency Test, Serial Test, Gap Test, Sum-of-Digits Test, Runs Test, as well as many others. Empirical results from the use of the tests are contained in Lewis (21) and Fishman (12).

Table 18-2 Linear congruential random number generators.†

Mixed Congruential Generators

A full period of 2^B before recycling will be obtained on a computer that has B bits/word for the generator

$$z_{i+1} = (az_i + b)(\bmod c)$$

when

$c = 2^B$;

b is relatively prime to c; that is, the greatest common factor of b and c is 1;

and

$a \equiv 1(\bmod 4)$ or $a = 1 + 4k$ where k is an integer.

Multiplicative Congruential Generators

A maximal period of 2^{B-2} before recycling will be obtained on a computer that has B bits/word for the generator

$$z_{i+1} = az_i(\bmod c)$$

when

$c = 2^B$;

$a = \pm 3 + 8k$ or $a = 1 + 4k$ for k integer;

and

z_0 is odd.

Fishman refers to these generators as maximal period multiplicative generators.

For multiplicative congruential generators a period of c–1 can be obtained by setting c equal to the largest prime in 2^B and making the coefficient a, a primitive root of c. In some instances 2^B-1 is the largest prime in 2^B. For a to be a primitive root of c, the following equation must be satisfied

$$a^{c-1} = 1 + ck$$

where k is an integer and for any integer $q < c-1$, $(a^q-1)/c$ is nonintegral. These generators are referred to as prime modulus multiplicative congruential generators.

---

† The material contained in this table is based on Fishman (12).

Fishman and Law and Kelton (19) describe the spectral (10) and latticed (7,22) procedures for measuring the performance of congruential random number generators with regard to their departure from the desired randomness properties.

18.7   CHAPTER SUMMARY

This chapter has described standard distributions which are important in modeling random processes. The topics of fitting distributions to data and obtaining samples from distributions is introduced. Extensive research on these topics has been and continues to be performed. Procedures are available to obtain samples from empirical, quasi-empirical, and theoretical distributions.

18.8   EXERCISES

18-1. Use the multiplicative congruential method to generate a sequence of ten random numbers with $c=256$, $a=13$, $b=0$ and $z_0=51$.

18-2. Given that $z_{i+1}=(az_i+b)(\text{mod } c)$, show that z_{i+1} is only a function of z_0, a, b and c, that is, $z_{i+1}=(a^{i+1}z_0+b(a^{i+1}-1)/(a-1))(\text{mod } c)$. Compute z_9 for the values given in Exercise 18-1 using this formula.

18-3. Use the inverse transform method to transform the uniform random numbers from Exercise 18-1 into samples from the continuous distribution whose probability density function is:

$$f(x) = \begin{cases} \dfrac{3x^2}{8}, & \text{if } 0 \leqslant x \leqslant 2 \\ 0, & \text{otherwise} \end{cases}$$

18-4. Use the inverse transformation method to transform the random samples from Exercise 18-1 into samples from the discrete distribution defined by the following probability mass function:

$P(0) = 1/5$; $P(1) = 1/5$; $P(2) = 2/5$; and $P(3) = 1/5$.

18-5. Write a SLAM II function for obtaining a sample from a truncated exponential distribution.

18-6. Show that the maximum of two random numbers is triangularly distributed.

18-7. Develop a procedure for modeling batch arrivals where the time between arrivals is exponential and the number of entities per arrival is Poisson.

18-8. Test to see if the time between arrivals is exponential if 100 arrivals are generated in 1000 minutes by randomly selecting an interarrival time from the current time to the 1000 minute ending time. The current time is the sum of the interarrival times generated.

18-9. Research the question of generating samples from a bivariate normal distribution. Embellishment: Expand your research to the consideration of procedures for obtaining samples from other multivariate distributions.

18-10. Develop a function for use in SLAM II that generates samples from an ARMA process.

18-11. Test Schmeiser's (30) approximate method for generating standarized random normal deviates which uses

RNORMS = (R**0.135−(1.−R)**0.135)/0.1975.

18.9 REFERENCES

1. Abramowitz, M. and I.A. Stegun, Eds., *Handbook of Mathematical Functions*, Applied Mathematics Series 55, Washington, D.C., National Bureau of Standards, 1964.

2. Ahrens,J.H. and U. Dieter, "Computer Methods for Sampling from the Exponential and Normal Distributions," *Comm. ACM*, Vol. 15, 1972, pp. 873-882.

3. Ahrens, J.H. and U. Dieter, "Computer Methods for Sampling from Gamma, Beta, Poisson and Binomial Distributions," *Computing*, Vol. 12, 1974, pp. 223-246.

4. Aitchison, J. and J.A.C. Brown, *The Lognormal Distribution*, Cambridge Press, 1957.

5. Banks, J. and J.S. Carson,II, *Discrete-Event System Simulation*, Prentice-Hall, 1984.

6. Berman, M.B., *Generating Random Variates from Gamma Distributions with Non-Integer Shape Parameters,* The RAND Corporation, R-641-PR, November 1970.

7. Beyer, W.A., R.B. Roof and D. Williamson, "The Lattice Structure of Multiplicative Congruential Pseudo-Random Vectors," *Math. Comp.*, Vol. 25, 1971, pp. 345-363.

8. Box, G.E.P. and M.A. Muller, "A Note on the Generation of Random Normal Deviates," *Annals of Math. Stat.*, Vol. 29, 1958, pp. 610-611.

9. Bratley, P., B.L. Fox, and L.E. Schrage, *A Guide to Simulation*, Springer-Verlag, 1983.

10. Conveyou, R.R. and R.D. MacPherson, "Fourier Analysis of Uniform Random Number Generators," *J.ACM*, Vol. 14, 1967, pp. 100-119.

11. Feller, W., *An Introduction to Probability Theory and Its Applications*, John Wiley, 1950.

12. Fishman, G.S., *Principles of Discrete Event Simulation*, John Wiley, 1978.

13. Fishman, G.S., "Sampling from the Gamma Distribution on a Computer," *Comm. ACM*, Vol. 19, 1976, pp. 407-409.

14. Fox, B.L., "Fitting 'Standard' Distributions to Data is Necessarily Good: Dogma or Myth?", *Proceedings, Winter Simulation Conference*, 1981, pp. 305-307.

15. Hahn, G.J. and S.S. Shapiro, *Statistical Methods in Engineering*, John Wiley, 1967.
16. Hastings, N.A.J. and J.B. Peacock, *Statistical Distributions*, Butterworth, 1975.
17. Hull, T.E. and A.R. Dobell, "Random Number Generators," *SIAM Review*, Vol. 4, 1962, pp. 230-254.
18. Johnson, N.L., "Systems of Frequency Curves Generated by Methods of Translation,", *Biometrika*, Vol. 36, 1949, pp. 149-176.
19. Law, A.M. and W.D. Kelton, *Simulation Modeling and Analysis*, McGraw-Hill, 1982.
20. Law, A.M. and S.D. Vincent, *UniFit User's Guide*, Systems Modeling and Analysis Company, Tucson, AZ, 1983.
21. Lewis, T.G., *Distribution Sampling for Computer Simulation*, D.C. Heath and Co., 1975.
22. Marsaglia, G., "The Structure of Linear Congruential Sequences," in *Applications of Number Theory to Numerical Analysis*, S.K. Zaremba, ed., Academic Press, 1972.
23. Musselman, K.J., W.R. Penick and M.E. Grant, *AID: Fitting Distributions to Observations*, Pritsker & Associates, Inc., 1981.
24. Pearson, K., "Contributions to the Mathematical Theory of Evolution , II. Skew Variations in Homogeneous Material," *Philosophical Transactions of the Royal Society of London, Series A*, Vol. 186, 1895, pp. 343-414.
25. Phillips, D.T., "Applied Goodness of Fit Testing," *AIIE Monograph Series*, AIIE-OR-72-1, 1972.
26. Phillips, D.T., "Generation of Random Gamma Variates from the Two-Parameter Gamma," *AIIE Transactions*, Vol. 3, 1971, pp. 191-198.
27. Pritsker, A.A.B., *The GASP IV Simulation Language,* John Wiley, 1974.
28. Ramberg, J.S., P.R. Tadikamalla, E.J. Dudewicz and E.F. Mykytka, "A Probability Distribution and Its Uses in Fitting Data,", *Technometrics*, Vol. 21, 1979, pp 201-214.
29. RAND Corporation, *A Million Random Digits with 1,000,000 Normal Deviates*, Free Press, 1955.
30. Schmeiser, B.W., "Approximations to the Inverse Cumulative Normal Function for Use on Hand Calculators," *App. Stat.*, Vol.28, 1979, pp. 175-176.
31. Schmeiser, B.W. and S.T. Deutsch, "A Versatile Four Parameter Family of Probability Distributions Suitable for Simulation," *AIIE Transactions*, Vol.9, 1977, pp. 176-182.
32. Schmeiser, B.W., "Random Deviate Generation: A Survey", *Proceedings, Winter Simulation Conference*, 1980, pp. 79-90.
33. Tadikamalla, P.R., "Computer Generation of Gamma Random Variables," *Comm. ACM*, Vol. 21, 1978, pp. 419-421.
34. Tadikamalla, P.R., "Computer Generation of Gamma Random Variables, II," *Comm. ACM*, Vol. 21, 1978, pp. 925-927.

CHAPTER 19

Statistical Aspects of Simulation

19.1   STATISTICAL QUESTIONS FACING SIMULATORS

A simulation model portrays the dynamic behavior of a system over time. A model is built to provide results that resemble the outputs from the real system. Thus, the statistical analysis of the outputs from a simulation is similar to the statistical analysis of the data obtained from an actual system. The main difference is that the simulation analyst has more control over the running of the simulation model. Thus he can design experiments to obtain the specific output data necessary to answer the pertinent questions relating to the system under study.

There are two types of questions that relate to the outputs of simulation models:

1. What is the inherent variability associated with the simulation model?
2. What can be inferred about the performance of the real system from the use of the simulation model?

The first question relates to an understanding of the model and verifying that it performs as designed. The sensitivity of the model outputs to changes in input and model parameters is of interest. The precision of the outputs with respect to the inherent probability distributions employed is a basic part of this type of question.

The second question relates to the validity of the model and to its usefulness. The answer to the second type of question usually involves describing the system performance variables and making statistical computations related to the performance variables. Thus, tables and plots are constructed and viewed as if they were possible outputs from the real system. The computations and statistical analyses made are similar to those performed on data obtained from the real system. If decision-making is based on the probability of occurrence of an outcome or on an average value, such quantities are estimated from the simulation. If the variability of a random variable is important, it is estimated in the same manner as is done in the real system. This mode of simulation analysis is the most common one found in current applications. The fact that a single simulation run represents one sample or time series of a stochastic process is no more bothersome than the fact that an historical record represents only a single time series.

Answering the first type of question involves a detailed statistical analysis to obtain information on the precision and sensitivity of the model. Basically we explore the type of output that would be obtained if the simulation was performed again or run for a longer period of time. In doing this, we recognize that the simulation model is a stochastic one and that the random elements of the model will produce outputs that are probabilistic. This type of analysis can be unfamiliar to the industrial manager since the analysis involves advanced statistical terminology. In addition, more precise responses can be obtained by changing experimental conditions, for example, by performing more runs.

Because the second type of question is system and, hence, model specific, there are no general forms of analysis that can be recommended beyond the standard statistical procedures. The first type of question has been explored extensively, and we provide a description of the techniques that we have found to be useful in this chapter. We would have preferred to present these techniques in a handbook fashion, with detailed examples illustrating each procedure. However, the field has not progressed to such a point, and we can only describe the specific types of problems and current approaches to their resolution. Two excellent surveys of statistical analysis of simulation results are given by Welch (81) and Wilson (87). The books by Bratley, Fox, and Schrage (9) and Law and Kelton (56) provide illustrations of the use of statistical techniques to analyze simulation outputs.

19.1.1   Definition of Terms

During a simulation, observations of variables of interest are to be recorded. Each potential observation is a time-based sample so that the observations can be considered to be random variables. To provide a standard set of terms, we make the following definitions regarding such random variables:

Let

I = the number of intervals, iterations, or individual observations. The word batch or interval will be used in a generic sense in the remainder of this chapter to mean any of the above.

T_i = ending time of the ith interval, i=1,2, . . . ,I with T_o defined as the start time of the first interval.

N_i = the number of observations in the ith interval, i = 1,2, . . . ,I.

$X_i(t)$ = value of X at time t in interval i; t ϵ $[T_{i-1},T_i]$.

$X_i(n)$ = value of X for nth observation in interval i; n = 1,2, . . . ,N_i.

Examples of $X_i(t)$ are the amount of inventory on-hand at time t and the number of customers in a system at time t. These variables were previously referred to as time-persistent variables. Examples of $X_i(n)$ are the time in the queue for the nth customer and the inventory on-hand when the nth receipt of an order arrives. These variables were previously referred to in conjunction with observations. Note that I, T_i, and N_i are usually treated as constants but in some instances may be random variables.

A *stochastic process*† is a set of ordered random variables. Thus $\{X_i(t), t \in [0,\infty)\}$ and $\{X_i(n), n=1, 2, \ldots, \infty\}$ are stochastic processes. A *realization* of a stochastic process is the set of sample paths assumed by the stochastic process. A *time series* is a finite realization of a stochastic process. In simulation terms, each run produces a time series for each stochastic process of interest.

The literature pertaining to stochastic processes and time series is extensive (12,68). Here, we only present a brief, informal background to introduce the topic.

A stochastic process is said to be *stationary* if the underlying joint distribution of the random variables in the process remains the same as time progresses, that is, if the random mechanisms producing the process are time invariant. This is referred to as the strictly stationary property (or *strong stationarity*). A special type of stationarity is referred to as *covariance stationarity* which requires all the means, μ_t, and covariances, R_{st}, of the random variables of the process to be finite and covariances separated by h time units to be equal, that is,

$$\mu_t = E[X_i(t)]$$
$$R_{st} = E[(X_i(s) - \mu_s)(X_i(t) - \mu_t)]$$

and

$$R_{st} = R_{rq} \quad \text{if } |t-s| = |q-r| = h$$

A covariance stationary process is also referred to as stationary in the wide sense or as mean square stationarity or as second-order stationarity.

Tests for the stationarity of a sequence are not well developed. The simplest and most frequently used evaluation is to consider the physics or underlying procedures associated with the phenomenon producing the data. If the basic physical factors which generate the phenomenon are time invariant then typically we accept the stationarity of the resulting data. If we believe trends or seasonality factors are involved, then differencing techniques are employed to remove such time-variant behavior.

An *ergodic* process is one from which the properties of the random variables in the process can be estimated from a single time series. A covariance stationary process is ergodic in the mean and autocovariance if the following two conditions hold (39)

$$\lim_{T \to \infty} \frac{1}{T} \sum_{s=-T+1}^{T-1} R_s = 0$$

and

---

† In this text, we do not differentiate between a stochastic process and a stochastic sequence.

$$\lim_{T \to \infty} \frac{1}{T} \sum_{s=-T+1}^{T-1} R_s^2 = 0 \quad .$$

The following important result regarding sequences of sample means is given by Parzen (68). A sequence of sample means, $\{\overline{X}_s, s = 1, 2, \ldots \infty\}$, may be shown to be ergodic if $\text{Var}[\overline{X}_s] \to 0$ as $s \to \infty$.

The significance of this result for simulation analysts is that the sample mean is approximately equal to the process mean if the variance of the sample mean approaches zero as the length of the sample increases.

19.2 IMPORTANCE OF THE VARIANCE OF THE SAMPLE MEAN, VAR[$\overline{X}_I$]

The sample mean is the average of the I random variables X_i as given below.

$$\overline{X}_I = \frac{\sum_{i=1}^{I} X_i}{I}$$

The notation, $\overline{X}_I$, is employed to indicate that the mean is a random variable that is based on the sum of I random variables. Typically, in simulation studies, we are interested in comparing $\overline{X}_I$ values for different alternatives using a test of hypothesis or in setting confidence limits on the value of $\overline{X}_I$ for a single alternative. To accomplish either of these tasks, it is necessary to calculate the variance of the sample mean denoted by $\text{Var}[\overline{X}_I]$. Extensive research has been performed on methods for estimating $\text{Var}[\overline{X}_I]$ from the time series output associated with a simulation. Procedures have also been suggested for obtaining smaller estimates of $\text{Var}[\overline{X}_I]$ which allow more precise statements about $\overline{X}_I$ to be made. These topics are discussed in Sections 19.3 and 19.4 respectively. In this section, we present the background information and formulas that are pertinent to the understanding of the significance of the variance of the sample mean.

19.2.1 Notation

In our exploration of $\overline{X}_I$, we propose the notation that X_i be a random variable associated with interval or batch i. We will use the term batch throughout this chapter where a batch is an undefined quantity that can be a single observation, a set of observations in a subinterval during a run, or an entire run (replication). How a batch is defined is dependent on the procedures employed in the simulation to compute the sample mean which in turn is based on the test of hypothesis to be performed or the confidence interval to be set. Possible definitions for X_i are given below.

A derived observation: $X_i = \begin{cases} 1, \text{success on batch i} \\ 0, \text{failure on batch i} \end{cases}$

A time-averaged value for batch i:

$$X_i = \frac{1}{T_i - T_{i-1}} \int_{t=T_{i-1}}^{T_i} X_i(t) \, dt$$

An observation-averaged value for batch i:

$$X_i = \frac{1}{N_i} \sum_{n=A_i}^{A_{i+1}} X_i(n)$$

where $A_i = \sum_{j=1}^{i-1} N_j$ and $A_1 = 0$

Note that the latter two definitions involve the computation of an average within a batch. To simplify the presentation of the formulas to be derived, we will not take advantage of this information during the presentation of formulas. In Section 19.3 where specific calculation methods are proposed, this subject will be discussed.

19.2.2   Formulas for Var [$\overline{X}_I$]

Starting with the definition of the variance, we can derive the following expressions (all summations are from 1 to I)

$$\text{Var}[\overline{X}_I] = E[(\overline{X}_I - E[\overline{X}_I])^2]$$

$$= E\left[\left(\frac{\sum_i X_i}{I} - E\left[\frac{\sum_i X_i}{I}\right]\right)^2\right]$$

$$= \frac{1}{I^2} E\left[\left(\sum_i X_i - E\left[\sum_i X_i\right]\right)\left(\sum_j X_j - E\left[\sum_j X_j\right]\right)\right]$$

$$= \frac{1}{I^2} E\left[\sum_i (X_i - E[X_i]) \cdot \sum_j (X_j - E[X_j])\right]$$

$$= \frac{1}{I^2} E \left[\sum_i \sum_j (X_i - E[X_i])(X_j - E[X_j]) \right]$$

$$= \frac{1}{I^2} \sum_i \sum_j E[(X_i - E[X_i])(X_j - E[X_j])]$$

$$= \frac{1}{I^2} \sum_i \sum_j Cov[X_i, X_j] \qquad (19\text{-}1)$$

$$= \frac{1}{I^2} \left(\sum_i Var[X_i] + \sum_i^I \sum_{\substack{j \\ j \neq i}}^I Cov[X_i, X_j] \right)$$

If X_i and X_j are independent for all i and j and $Var[X_i] = \sigma^2$ for all i then

$$Var[\overline{X}_I] = \frac{1}{I^2} \sum_i^I \sigma^2 = \frac{1}{I} \sigma^2 \quad . \qquad (19\text{-}2)$$

From this equation, we note that when independence applies, $I*Var[\overline{X}_I] = \sigma^2$, a constant, and that $Var[\overline{X}_I]$ decreases in proportion to $1/I$. We will return to this observation shortly.

Under the assumption of independence and mild regularity conditions on X_i, the central limit theorem specifies that for large I the distribution of $\sqrt{I}(\overline{X}_I - \mu)/\sigma$ converges to a normal distribution with mean 0 and variance 1, that is, $N(0,1)$. If the X_i are also normally distributed then $\overline{X}_I$ is in fact normally distributed and $(\overline{X}_I - \mu)/\sqrt{S_X^2/I}$ has a t-distribution with $I-1$ degrees of freedom where S_X^2 is an estimator of σ^2. From this information, an exact confidence interval for $\overline{X}_I$ can be constructed. Note that by making X_i a batch average as discussed above, an assumption of normality for X_i is reasonable. Fishman (26) provides the following equation for the variance of S_X^2:

$$Var[S_X^2] = \sigma^4 \left(\frac{2}{I-1} + \frac{\gamma_2}{I} \right) \qquad (19\text{-}3)$$

where γ_2 is the excess kurtosis (fourth central moment divided by the square of the second central moment minus three.) Eq. 19-3 indicates the amount of variability that may be expected in the estimate of the underlying process variance.

When $Cov[X_i, X_j]$ cannot be assumed to be zero, but we can assume a covariance stationary process, then $Cov[X_i, X_j] = R_{j-i} = R_h$. Using this notation in Eq. 19-1 and by combining terms, we can obtain

$$\text{Var}[\overline{X}_I] = \frac{1}{I} \sum_{h=1-I}^{I-1} \left(1 - \frac{|h|}{I} \right) R_h \qquad (19\text{-}4)$$

Substituting $\sigma^2 = R_o$ and $R_{-h} = R_h$ into Eq. 19-4 yields

$$\text{Var}[\overline{X}_I] = \frac{1}{I} \left\{ \sigma^2 + 2 \sum_{h=1}^{I-1} \left(1 - \frac{h}{I} \right) R_h \right\} \qquad (19\text{-}5)$$

Procedures for estimating $\text{Var}[\overline{X}_I]$ using this equation are discussed in Section 19.3.5.

If the autocovariance decays exponentially (a common and reasonable assumption is that $R_h = R_o \alpha^{|h|}$ for $0 < \alpha < 1$) then it can be shown that

$$\lim_{I \to \infty} I \, \text{Var}[\overline{X}_I] = \sum_{h=-\infty}^{\infty} R_h = m \qquad (19\text{-}6)$$

From Eq. 19-6, it is seen that as the number of batches increases the $\text{Var}[\overline{X}_I]$ decreases in proportion to $1/I$. Comparing Eq. 19-6 with Eq. 19-2, we observe that the underlying process variability, σ^2, is related to the sum of all covariances. Throughout this text, we refer to this quantity by the symbol m. Since the value of m is not based on the number of batches I, an estimate of m for a process permits the estimation of $\text{Var}[\overline{X}_I]$ for any I.

19.2.3 Interpreting Var $[\overline{X}_I]$

As mentioned above, $\text{Var}[\overline{X}_I] \to m/I$ for large I and under appropriate assumptions. Based on this, we can picture the distribution of $\overline{X}_I$ over time as shown in Figure 19-1. There are several observations to be made from Figure 19-1. First, $\overline{X}_I$ is a random variable and, hence, the values estimated are sample values. When making a simulation replication, one should expect a different value of $\overline{X}_I$ to result with the precision based on $f(\overline{X}_I)$ and, hence, $\text{Var}[\overline{X}_I]$. The length of a run or the number of batches can change $f(\overline{X}_I)$, that is, $f(\overline{X}_I)$ depends on I and the distribution of X_i. For the same run length, different estimators may be based on a different number of batches. Furthermore, the three distributions shown in Figure 19-1 all are shown with the same $E[\overline{X}_I]$. This need not be the case as biased estimators can be and are used. Thus, when comparing estimators for $\text{Var}[\overline{X}_I]$ under different experimental conditions, we need a basis to compare the estimates. The criterion used for comparison in most research efforts are m and the mean square error of the sample mean (34) denoted by $\text{MSE}[\overline{X}_I] = E[(\overline{X}_I - \mu_x)^2]$.

Unfortunately, equations for the computation of m for non-Markov processes have not been derived. For the number in the system in an M/M/1 queueing situation, Fishman (26) computes $m = 6840$ for the case in which the arrival rate, λ,

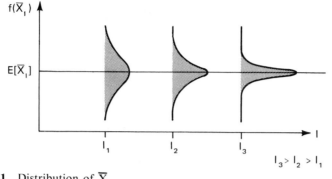

Figure 19-1   Distribution of $\overline{X}_I$.

equals 4.5 and the service rate, μ, equals 5. Also if $\lambda = 2.0$ and $\mu = 2.25$ then $m = 361$. For the M/M/∞ situation (68), $m = 2\lambda/\mu^2$. A procedure has been developed for obtaining values of m for finite state Markov processes† (32,40).

The MSE[$\overline{X}_I$] is a criterion that combines the Var[$\overline{X}_I$] and the bias associated with $\overline{X}_I$. This is seen in the following development:

$$\begin{aligned} \text{MSE}[\overline{X}_I] &= E[(\overline{X}_I - \mu_X)^2] \\ &= E[(\overline{X}_I - E[X_I] + E[\overline{X}_I] - \mu_X])^2] \\ &= E[(\overline{X}_I - E[\overline{X}_I])^2] + E[(E[\overline{X}_I] - \mu_X])^2] \\ &= \text{Var}[\overline{X}_I] + (\text{Bias}[\overline{X}_I])^2 \end{aligned}$$

$$(13\text{-}7)$$

When there is bias‡ associated with $\overline{X}_I$, then the probability that the theoretical mean is covered by (contained in) a confidence interval differs from a prescribed value due to the offset caused by the bias. The "coverage" is defined as the probability that the theoretical mean is covered by an interval centered at $E[\overline{X}_I]$ or $\overline{X}_I$. When estimators are employed which are unbiased, then MSE($\overline{X}_I$) reduces to Var[$\overline{X}_I$] and the coverage is the same as the value associated with the confidence interval. We are now ready to examine proposed procedures for estimating Var[$\overline{X}_I$].

---

† For finite state Markov processes, it is also shown that m is a function of both λ and μ and that $m(a\lambda,a\mu) = (1/a)[m(\lambda,\mu)]$. Glynn proves (32) that this result holds for countable state Markov processes like the M/M/1 queueing situation. Hence, to compute m for the M/M/1 queue with $\lambda/\mu = 0.9$ but with $\lambda = 4$, $\mu = 40/9$ then $a = 8/9$ and $m = 9/8(6840) = 7695$.

‡ A bias could exist if the underlying process was not covariance stationary or if $\overline{X}_I$ was computed using a ratio estimator (see Section 19.3.3).

19.3   PROCEDURES FOR ESTIMATING VAR[$\overline{X}_I$]

As discussed in the previous section, the variance of the sample mean plays a fundamental role in the reliability of simulation output. The estimation of Var[$\overline{X}_I$] from a single simulation run is complicated due to the dependence of the samples that are used in the computation of $\overline{X}_I$. A considerable amount of research has been performed regarding the estimation of Var[$\overline{X}_I$]. This research has resulted in five basic approaches which are listed below:

1. *Replication*—employ separate runs with each run being considered as a batch. From run i, we obtain a value of X_i and the Var[$\overline{X}_I$] is estimated using Eq. 19-2;
2. *Subintervals*—divide a run into equal batches (subintervals) and compute X_i as an average for batch i. Assume each X_i is independent and use Eq. 19-2 for estimating Var[$\overline{X}_I$];
3. *Regenerative cycles*—divide a simulation run into independent cycles by defining states where the model starts anew. Estimate the Var[$\overline{X}_I$] based on observed values in the independent cycles;
4. *Parametric modeling*—fit an equation(s) to the output values or a function of the output values obtained from a run. Derive an estimate of Var[$\overline{X}_I$] from the equations that model the simulation output;
5. *Covariance/spectral estimation*—estimate the autovariance from the sample output and use these in a spectral analysis or directly in Eq. 19-5 to estimate Var[$\overline{X}_I$].

Each of these five procedures will now be presented.

19.3.1   Replications

In this procedure, a value x_i of the random variable X_i is computed on run i. As discussed in Section 19.2.1, the variable X_i could be: the mean number of units in the system on run i; the mean time in the system per customer; or a binomial variable that represents the number of successes in a run. The mean of the X_i values over I runs is used as an estimate of the parameter of interest, that is,

$$\overline{X}_I = \frac{\sum_{i=1}^{I} X_i}{I} \qquad (19\text{-}8)$$

An estimate of $Var[X_i]$, S_X^2, is then obtained using standard procedures as

$$S_X^2 = \frac{1}{I-1} \sum_{i=1}^{I} (X_i - \overline{X}_I)^2 = \frac{1}{I-1} \sum_{i=1}^{I} X_i^2 - \frac{1}{I(I-1)} \left(\sum_{i=1}^{I} X_i \right)^2 \quad (19\text{-}9)$$

Equation 19-9 provides an estimate of the variability associated with a random sample obtained from each run. Since each run is an independent replication, an estimate of the variance of the sample mean can be obtained as shown in Eq. (19-10).

$$S_{\overline{X}}^2 = S_X^2 / I \quad (19\text{-}10)$$

Based on $S_{\overline{X}}^2$ and $\overline{X}_I$ and using the central limit theorem, probability statements about the parameter of interest can be made after observations are taken. Tests of hypotheses can also be made based on these theoretic considerations.

The replication procedure has the desirable property that samples are independent. Another advantage is that it can be used for both terminating and steady-state analysis where a terminating analysis is one that is performed for a specific finite time period.† The disadvantages associated with replications are: 1) each replication contains a startup segment which may not be representative of stationary behavior; and 2) only one sample, X_i, is obtained from each replication which could mean that extensive information about the variable of interest is not being gleaned from the data. This is particularly the case when X_i is computed as a mean value for the run.

19.3.2   Subintervals

The approach to estimating the variance of $\overline{X}_I$ using subintervals is to divide a single simulation run into batches. If each batch has b samples of $X_i(n)$ then a batch sample mean, X_i, is computed from

$$X_i = \frac{\sum_{n=1}^{b} X_i(n)}{b} \quad (19\text{-}11)$$

---

† Law (54) states "we have concluded from talking with simulation practitioners that a significant proportion of real-world simulations are of the terminating type. This is fortunate because it means classical statistical analysis is applicable . . .".

If the subintervals are independent then Eq. 19-8, 19-9, and 19-10 are used to estimate $E[X_I]$, $Var[\overline{X}_I]$, $Var[\overline{X}_I]$, respectively.† The assumption of independence is typically made in simulation analyses even though there exists an autocovariance between the values at the end of one subinterval and those at the beginning of the next subinterval. This variance can cause a positive covariance between batch means. By making the batch size b larger, the covariance between the sample batch means should decrease. Procedures for determining the batch size such that the covariance between adjacent batch means is insignificant have been developed by Mechanic and McKay (59), Law and Carson (55), and Fishman (27,28). Schriber and Andrews (74) have proposed a modification to Fishman's algorithm.

Fishman's proposed procedure involves recomputing the batch values by doubling the batch size b until the null hypothesis that the $X_{i,b}$ for i=1,2, . . . , I_b are iid is accepted, where the subscript b is appended to X_i to indicate the dependence of it on the batch size. He recommends the use of the test statistic

$$C_b = 1 - \left[\sum_{i=1}^{I_b - 1} (X_{i,b} - X_{i+1,b})^2 \middle/ 2 \cdot \sum_{i=1}^{I_b} (X_{i,b} - \overline{X}_{I_b})^2 \right]$$

where I_b = number of batches when the batch size is b.

For large b, C_b is approximately the estimated autocorrelation coefficient between consecutive batches.

If the $X_{i,b}$ are independent and normally distributed then C_b has a mean of zero, a variance of $(I_b-2)/(I_b^2-1)$, and a distribution that is close to normal for b as small as 8. Thus, if these conditions hold, a standard test using normal tables can be applied. If $\{X_{i,b}\}$ has a monotone autocovariance function then a one-sided test is appropriate: otherwise a two-sided test is in order.

Several procedural details are necessary when using the above approach to setting the batch size. The observed values $x_i(n)$ must be recorded in order to compute $x_{i,b}$ for different values of b. Any non-representative values of $x_i(n)$ at the beginning of a run should be truncated before applying the test (see Section 19.5.2). An initial batch size needs to be set. Fishman recommends setting $I_b = 1$ initially. Schmeiser has concluded that an I_b value between 10 and 30 is reasonable for most simulation situations (72). He also states that the initial value of I_b for Fishman's algorithm could be much larger with almost no deterioration in the confidence intervals.

---

† It can be shown that if the subintervals are independent, m associated with X_I, denoted $m(X_I)$, is $\frac{1}{b}*m(X_I(n))$ where $m(X_I(n))$ is the m value associated with individual observations.

The advantages of using subintervals to estimate the variance of the sample mean are that a single run can be used to obtain an estimate and only one transient period is included in the output (or required to be deleted). The disadvantage of the procedure is in establishing the batch size, b, which makes the subintervals independent. Note that for a fixed number of observations, increasing the batch size decreases the number of batches and, hence, could yield larger estimates of $\text{Var}[\overline{X}_I]$. Schmeiser's results (72) alleviate this difficulty by suggesting a number of batches, 10 to 30, that is reasonable. Another disadvantage involves the boundaries of a batch. Care must be taken in computing batch averages when an observation spans more than a single batch, for example, an arrival in batch i that leaves the system during batch $i+1$.

19.3.3   Regenerative Method

The regenerative method (16,29,44) is similar to the subinterval method in that it divides a simulation run into intervals which are referred to as cycles. A cycle starts when a specific state of the system is reached in which future behavior is independent of the past behavior. When a return is made to such a state, the cycle ends and one independent observation of each quantity of interest is obtained. By defining cycles in this manner, independent samples from the model are obtained and the covariance problem encountered when using subintervals is avoided. A different statistical problem arises, however, in that the length of a cycle is not predetermined but is a random variable.

The most commonly used regeneration point in queueing studies is a return to a status where servers are idle and no customers are waiting. If the next customer arrival is processed in a consistent fashion then each time a customer arrives to an empty system is a regeneration point and the start of a regeneration cycle. In inventory models, a possible regeneration point is when the inventory position is equal to a stock control level.

By construction, each cycle of a simulation run will be independent and we can base the estimates of the sample mean on cycle values. Following the development by Crane and Lemoine (17), let

Y_i = the value of interest in the ith cycle, for example, the sum of customer waiting times in the ith cycle†; and

---

† The value of interest could also be a time-integrated variable, that is, the time-integrated number in the system during the ith cycle.

L_i = the length of the ith cycle, for example, the number of customers or the cycle time.

If X_{ik} is the kth sample on the ith cycle and we perform the simulation run until there are I cycles, then the following two equations hold

$$Y_i = \sum_{k=1}^{L_i} X_{ik} \qquad (19\text{-}12)$$

and

$$\sum_{i=1}^{I} L_i = N \qquad (19\text{-}13)$$

where N = total number of samples (a random variable).

The average of all the samples for a simulation run, $\overline{X}_I$, would normally be computed as shown in Eq. 19-14.

$$\overline{X}_I = \frac{\sum_{i=1}^{I} \sum_{k=1}^{L_i} X_{ik}}{N} \qquad (19\text{-}14)$$

By substituting in the variables from Eqs. 19-12 and 19-13 into Eq. 19-14, we illustrate that $\overline{X}_I$ can be considered as the ratio of cycle averages.

$$\overline{X}_I = \frac{\sum_{i=1}^{I} \sum_{k=1}^{L_i} X_{ik}}{N} = \frac{\sum_{i=1}^{I} Y_i}{\sum_{i=1}^{I} L_i} = \frac{\sum_{i=1}^{I} Y_i/I}{\sum_{i=1}^{I} L_i/I} = \overline{Y}_I/\overline{L}_I \qquad (19\text{-}15)$$

Since the number of samples per cycle is a random variable, we cannot specify both the number of cycles and the total number of samples. In such a case, we are using a ratio estimator which can be shown to be biased (26).

An estimate of the variance of $\overline{X}_I$ can be computed using Eq. 19-16

$$S_{\overline{X}}^2 = \frac{S^2}{(\overline{L}_I \sqrt{I})^2} \qquad (19\text{-}16)$$

where

$$S^2 = S_Y^2 - 2 \overline{X}_I S_{YL} + \overline{X}_I^2 S_L^2$$

and

$$S_Y^2 = \frac{1}{I-1} \sum_{i=1}^{I} (Y_i - \overline{Y}_I)^2 \; ,$$

$$S_L^2 = \frac{1}{I-1} \sum_{i=1}^{I} (L_i - \overline{L}_I)^2 ,$$

$$S_{YL} = \frac{1}{I-1} \sum_{i=1}^{I} (Y_i - \overline{Y}_I)(L_i - \overline{L}_I) \; .$$

As noted above $\overline{X}_I$ is a biased estimator. To alleviate the problem of bias with ratio estimators, a Jackknife estimator (36,46) can be used which eliminates the bias term of order $1/I$. The equation for the Jackknife estimator of the sample mean is given in Eq. 19-17

$$\overline{J} = \frac{1}{I} \sum_{i=1}^{I} J_i \tag{19-17}$$

where J_i is referred to as a psuedo-value computed from

$$J_i = I \overline{X}_I - (I-1) \sum_{\substack{j=1 \\ j \neq i}}^{I} Y_j \Big/ \sum_{\substack{j=1 \\ j \neq i}}^{I} L_j \quad .$$

The J_i are considered to be independent and identically distributed so that confidence intervals for $\overline{X}_I$ can be constructed using estimates of $\overline{J}$ and $S_{J_i}/\sqrt{I}$.

The advantages of the regenerative method are that independent and identically distributed random variables for each cycle are obtained. Thus, standard statistical procedures can be used for tests of hypothesis and confidence interval calculations. However, to use the procedure, a regenerative point must be established for which the expected time between returns is finite and for which sufficient cycles are observed to achieve a reasonable confidence interval. As illustrated by the arc sine law presented in Chapter 2, Section 2.13.3, this may not be an easy determination. An additional advantage is that the problems of determining a start-up procedure are avoided as statistical collection can begin when a regeneration point is reached which, if possible, could be the initial conditions specified. Disadvantages of the procedure are the added computations and the bias associated with the estimator for the sample mean.

19.3.4   Parametric Modeling

Parametric modeling involves the building of a model to describe the outputs from a simulation model. Values of the estimates of quantities of interest are then obtained through computations made on the parametric model. The procedure for employing parametric modeling involves the collection of sample values from a simulation and then fitting an equation(s) to the observed data values. This approach is similar to the one used when attempting to describe real world systems by fitting equations to data obtained from the system.

To provide further rationale for using this approach, consider a single server queueing situation in which customers are processed on a first-come, first-serve basis. If the variable of interest is the waiting time of a customer, we could write an equation that describes the waiting time of the $(j+1)$st customer in terms of the waiting time of the jth customer, the interarrival time random variable and the service time random variable. Such an equation can be developed based on the pictorial sketch shown below.

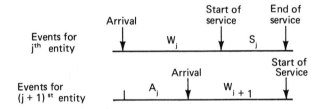

In this sketch, it is assumed that the $(j+1)$st entity arrives before the end of service for the jth entity. By equating the variables from the two lines given in the sketch, we obtain the following equation

$$W_{j+1} = \begin{cases} W_j + S_j - A_j & \text{if } A_j < W_j + S_j \\ 0 & \text{otherwise} \end{cases}$$

where

W_j is the waiting time of the jth entity,
S_j is the service time of the jth entity, and
A_j is the interarrival time between the jth and $(j+1)$st entity.

If we had simulated a queueing system and saved the observed values of W_j, S_j, and A_j and were sufficiently astute to derive a model for W_{j+1} such as the one given above, then we could use it to obtain information concerning the waiting time of customers.†

---

† This model is a convenient one for introducing sampling procedures and for performing research on the waiting time associated with a single server queueing situation. Since it is difficult to embellish, it is not a good model for teaching modeling procedures.

The above model development was presented to provide a rationale for attempting to fit a model to sample data. Past research on parametric modeling of simulation output has primarily been through the use of autoregressive (AR) models. Fishman has done extensive analyses of such models (29). Since the values obtained from the execution of a simulation model can be considered as time series data, we recommend the use of the Box-Jenkins methodology for model identification and estimation (8). An excellent discussion of this methodology is provided by Mabert (58). The Box-Jenkins methodology attempts to derive a parametric model of the sample data using an equation of the following form:

$$Y_t = \sum_{k=1}^{p} \phi_k Y_{t-k} - \sum_{l=1}^{q} \theta_l U_{t-l} + U_t \;; t = 1,2, \ldots ,n \qquad (19\text{-}18)$$

where

$$Y_t = X(t) - E[X(t)]$$

and

U_t is white noise, that is, $E[U_t] = 0$ and

$$E[U_t U_{t-l}] = \begin{cases} \sigma_U^2 & l=0 \\ 0 & \text{otherwise} \end{cases}$$

The model of Eq. 19-18 is referred to as a combined autoregressive and moving average (ARMA) model. If all the terms which have a coefficient θ_l in Eq. 19-18 are deleted, then an autoregressive model of order p is obtained (AR(p)). An autoregressive model expresses Y_t as a linear combination of previous values of the time series plus a white noise component.

If all the terms with a coefficient of ϕ_k are deleted from Eq. 19-18, then a moving average model is obtained. The moving average model expresses Y_t as a linear combination of the past q error terms. This is referred to as a moving average model of order q and is designated as an MA(q) model.

The Box-Jenkins methodology provides procedures for identifying the order of an autoregressive model, p, and the order of a moving average model, q. In addition, procedures have been computerized for obtaining the best estimates of ϕ_k and θ_l. The program for making such computations can be obtained from The Ohio State University Data Center or The University of Wisconsin. The outputs from this program provide values for θ_l and ϕ_k and an estimate of the variance of the white noise, σ_U^2. With these estimates, an estimate of the variance of the sample mean can be computed as shown in Eq. 19-19

$$s_{\bar{X}}^2 = \frac{\widehat{m}}{n} \qquad (19\text{-}19)$$

where

$$\hat{m} = \hat{\sigma}_U^2 \frac{\left(1 - \sum_{l=1}^{q} \hat{\theta}_l\right)^2}{\left(1 - \sum_{k=1}^{p} \hat{\phi}_k\right)^2}$$

In our experience, parametric modeling of the time series obtained from a simulation model has not produced reliable estimates of the variance of the sample mean. This could be related to a non-stationary behavior of the time series and a non-normality of the individual observations. Significantly improved results have been obtained by building parametric models using a time series consisting of batch means. By using batch means, assumptions regarding stationarity and normality are alleviated. Andrews and Schriber (3) have developed a procedure for automatically fitting an ARMA model to estimate m.

Schruben, taking a different approach, computes a standardized time series for the output data by forming cumulative sums of the difference between the average and the observations (75). Statistical output measures are then developed from the standardized time series. This approach has the advantage of transforming the data into a familiar form.

The advantage of using parametric modeling is that an equation describing a variable of interest is obtained. Further analyses using the derived model can provide new insights into the system being simulated. The main disadvantage of parametric modeling is the lack of knowledge of the reliability of the model. Building a parametric model of a simulation model takes the analysis one step further away from the real system and requires much care on the part of the analyst.

19.3.5 Estimating Covariances and the Use of Spectral Analysis

In Section 19.2.2, we showed that

$$\text{Var}[\overline{X}_I] = \frac{1}{I}\left[R_o + 2 \sum_{h=1}^{I-1}\left(1 - \frac{h}{I}\right) R_h\right]$$

Thus, if estimates of the autocovariances, R_h, can be obtained, we could compute

$$s_{\overline{X}}^2 = \frac{1}{I}\left[\hat{R}_o + 2 \sum_{h=1}^{I-1}\left(1 - \frac{h}{I}\right)\hat{R}_h\right]$$

In the literature, the following three alternative equations have been proposed for computing $\hat{R}_h$:

$$A_h = \frac{1}{I-h} \sum_{i=1}^{I-h} (X_i - \overline{X}_I)(X_{i+h} - \overline{X}_I)$$

$$B_h = \frac{1}{I} \sum_{i=1}^{I-h} X_i X_{i+h} - \frac{1}{I-h} \left(\sum_{i=1}^{I-h} X_i \right)\left(\sum_{i=1}^{I-h} X_{i+h} \right)$$

$$C_h = \frac{1}{I} \sum_{i=1}^{I-h} (X_i - \overline{X}_I)(X_{i+h} - \overline{X}_I)$$

When a time series is short and the end points differ significantly from $\overline{X}_I$, B_h has been recommended as the estimator. In simulation studies, the time series is normally long and B_h is not employed.

The estimator A_h has intuitive appeal as it averages $(I-h)$ values. However, A_h has a larger mean square error than C_h. C_h has the disadvantage that it is a biased estimator. The current consensus is that for a time series with large I, C_h should be used to estimate R_h.

The past research on estimating the autocovariances dealt primarily with the sample variables, $X_i(t)$ or $X_i(n)$. These variables are highly correlated and estimates of the autocovariances are highly correlated. Thus, if a large value is obtained for the estimator of R_0, we would expect a high value relatively for the estimator of R_h. In simulation experiments, this phenomenon has been observed by Duket (20). The use of batch values, X_i, as recommended previously will alleviate this correlation in the estimates of the autocovariances.

An alternative procedure is to employ spectral analysis. The *spectrum* is defined as

$$g(\lambda) = \frac{1}{2\pi} \sum_{h=-\infty}^{\infty} R_h e^{-i\lambda h} \quad -\pi \leq \lambda \leq \pi \tag{19-20}$$

and the *spectral density function* as

$$f(\lambda) = g(\lambda)/R_0$$

The inverse transform of $g(\lambda)$ is given by

$$R_h = \int_{-\pi}^{\pi} g(\lambda) e^{i\lambda h} d\lambda$$

and can be used to obtain values of R_h.
For $h=0$, this yields

$$R_0 = \int_{-\pi}^{\pi} g(\lambda) d\lambda$$

Thus, the underlying variance of the process can be thought of as consisting of nonoverlapping contributions at frequency λ. A large value of $g(\lambda)$ indicates variability in the process at frequency λ or periodically at $2\pi/\lambda$.

As discussed in Section 19.2, it can be shown for many systems that

$$\lim_{I \to \infty} I \, \text{Var}[\overline{X}_I] = m = \sum_{h=-\infty}^{\infty} R_h$$

Letting $\lambda = 0$ in Eq. 13-20 yields

$$g(0) = \frac{1}{2\pi} \sum_{h=-\infty}^{\infty} R_h$$

and, hence,

$$m = 2\pi \, g(0)$$

Thus, if $g(0)$ can be estimated, an estimate of m can be obtained.

Extensive research has been performed on obtaining estimates of the spectrum for time series. The main difficulties involve the determination of the number of covariances to include in the computation, and the weighting function (lag window) to apply to the estimated autocovariance obtained from finite observations. Weighting functions have been developed by Bartlett, Tukey, and Parzen; others are referred to as the Rectangular and Variance weighting functions (21).

The advantage of the spectral approach is the extensive research that has been performed on spectral methods. The main disadvantage is that point estimates obtained from the spectrum, that is, at $\lambda = 0$, are known to be unreliable. Employing data grouped into a single batch observation, as previously suggested, should diminish this reservation (41,61,81).

19.4 VARIANCE REDUCTION TECHNIQUES

The variance of the sample mean is a derived measure of the reliability that can be expected if the simulation experiment is repeatedly performed. It has been shown that longer runs should produce smaller estimates of $\text{Var}[\overline{X}_I]$. Thus, in some sense, the value of $\text{Var}[\overline{X}_I]$ is dependent on experimental procedures and calculations. Variance reduction techniques (VRT) are methods that attempt to reduce the estimated values of $\mathrm{Var}[\overline{X}_1]$ through the setting of special experimental conditions or through the use of prior information. An excellent survey of variance reduction techniques has been published by Wilson (88). Nelson and Schmeiser (66) have attempted to characterize the underlying structure of the variance reduction process.

19.4.1   Antithetic Sampling

Eq. 19-1 for $\mathrm{Var}[\overline{X}_1]$ contains terms involving the $\mathrm{Cov}[X_iX_j]$. If $\mathrm{Cov}[X_iX_j]$ can be made negative then the $\mathrm{Var}[\overline{X}_1]$ will be reduced. Since X_i and X_j are functions of pseudorandom numbers, it has been suggested that if $X_i = f(r_1, r_2, \ldots, r_q)$ then letting $X_j = f(1-r_1, 1-r_2, \ldots, 1-r_q)$ will induce a negative covariance between X_i and X_j. Obtaining a negative covariance depends on the function f, which reflects a transformation of random numbers into sample values by the simulation model. Clearly, a general result regarding the use of the antithetic values can not be provided. However, in experiments, a variance reduction has been observed when such antithetic sampling is employed. The generation of the antithetic stream of random numbers: $1-r_1, 1-r_2, \ldots, 1-r_q$, is easily accomplished when using a multiplicative congruence random number generator of the form

$$z_k = az_{k-1} \bmod(c) \quad k=1,2,\ldots$$

and

$$r_k = z_k/c$$

It can be shown (46) that if

$$z'_0 = c - z_0$$

is used as a starting value for a sequence of random numbers then†

$$z'_k = c - z_k$$

and, hence,

$$r'_k = \frac{z'_k}{c} = 1 - r_k.$$

---

† Care is required when doing this with packaged random number generators that may add 1 to a seed value or that store the initial seed value as a real number.

In SLAM II, an antithetic sequence is obtained by specifying a negative initial seed value on the SEEDS control statement, that is, $-z_0$.

The application of antithetic samples within a batch or even within a run is not recommended. The manipulation of batches to produce antithetic samples could cause a distortion of the basic process and seems not to be warranted. Other proposed procedures appear more palatable. For example, perform pairs of independent runs in which antithetic streams are used on the second run of the pair. For a sequence of arrivals, let the kth interarrival time be based on r_k in the first run of a pair and based on $1-r_k$ on the second run of a pair. When doing this, the variance calculation for 2I runs is simplified by combining values across pairs of runs. If X_i' is the antithetic value for X_i, then

$$\text{Var}[\overline{X}_I] = \text{Var}\left(\frac{\sum\limits_{i=1}^{I}(X_i+X_i')}{2I}\right) = \text{Var}\left(\frac{\sum\limits_{i=1}^{I}U_i}{I}\right)$$

where

$$U_i = \frac{X_i+X_i'}{2}.$$

When combined in this fashion, covariance terms between runs need not be computed.

Another suggestion that is in the spirit of antithetic sampling (28) is to switch the streams employed for interarrival times and service times in alternate runs. To see that this induces antithetic behavior, note that long interarrival times reduce potential congestion whereas long service times increase potential congestion. Results of the application of antithetic sampling are summarized by Kleijnen (46). Typically, the results are for small-scale simulation models. Kleijnen defines a measure of variance reduction as a percentage change, that is, if $\text{Var}_R[\overline{X}_I]$ is obtained using a VRT then

$$\text{Percent Variance Reduction} = \frac{\text{Var}[\overline{X}_I] - \text{Var}_R[\overline{X}_I]}{\text{Var}[\overline{X}_I]} * 100$$

Hammersley and Handscomb (39), Tocher (80), and Fishman (26) define variance reduction as a ratio, that is

$$\text{Variance Reduction} = \frac{\text{Var}[\overline{X}_I]}{\text{Var}_R[\overline{X}_I]}$$

Before ending this discussion of antithetic sampling, two important points should be mentioned. First, although the correlation between antithetic random numbers is −1., the correlation between samples based on such numbers may not be −1. If the samples are from a symmetric distribution, the correlation will be −1. If the samples are from an exponential distribution, Fishman (26) shows the correlation between antithetic samples is −0.645. A similar increase in the negative correlation is obtained for other distributions.

The second point involves the simulation model. If the model involves a square or higher order even relation then the introduction of a negative correlation can result in a positive contribution to the variance of the sample mean.

19.4.2   Common Streams

A typical practice when performing a simulation is to employ historical data as the driving force. As an example, the time of arrivals of jobs to a computer could be maintained and used to define the arrival times and job characteristics for a simulation of the computer center. Simulations involving the use of historical data are sometimes referred to as *trace-driven*. Recognizing that the historical arrival pattern is a single time series, it is apparent that its repeated use reduces the variation of the output from a simulation model. By starting different simulation runs with the same random number seed, that is, employing a common stream, a similar variance reduction can be obtained. Care is required when employing trace-driven or common streams in that the complete variability associated with the system being modeled is not incorporated in the model. The analyst should ensure that the single time series employed by such a practice is representative of the stochastic process being modeled.

A more appealing use of common streams is when comparing alternatives. In this situation, the variance of a difference between sample means is of interest, that is

$$\text{Var}[\overline{X}_I^{(1)} - \overline{X}_I^{(2)}] = \text{Var}[\overline{X}_I^{(1)}] + \text{Var}[\overline{X}_I^{(2)}] - 2\text{Cov}[\overline{X}_I^{(1)}, \overline{X}_I^{(2)}]$$

where $\overline{X}_I^{(k)}$ is the sample mean for the alternative k. By using common streams, the $\text{Cov}[\overline{X}_I^{(1)}, \overline{X}_I^{(2)}]$ should be positive and a reduction in the variance of the difference should be obtained. The use of a common stream here only presumes that the time series generated affects both alternatives in a similar manner. Extreme care must be taken if common streams are employed in conjunction with antithetic sampling techniques as a variance increase has been observed under several situations (46).

19.4.3   Prior Information

The Rao-Blackwell theorem presented in Chapter 2 can be interpreted as specifying that a variance reduction can be obtained by estimating the sample mean based on a conditioned random variable. One procedure for implementing this approach is to employ analytic results in the estimation process. We present two illustrations of this procedure.

It is well known that for a wide class of queueing situations, Little's formula (57,79) holds

$$L = \lambda W$$

where

L is the expected number in a system

W is the expected time in the system

λ is the effective arrival rate to the system, that is, the number of arrivals that are eventually served per unit time.

In a run, the average number of entities in a queue, $\overline{N}$, will be equal to the product of the observed arrival rate, λ_0, in the run and the average waiting time of all entities passing through the queue, $\overline{T}$. Notationally, we can write $\overline{N} = \lambda_0 \overline{T}$. This equation can be developed by observing that the time integrated number in the queue is equal to the sum of the waiting times (assuming at the end of the run all entities leave the queue). Based on this information, we could compute

$$Var[\overline{N}] = \lambda^2 \, Var[\overline{T}]$$

assuming λ is a known value. This equation provides the basis for an indirect estimator of the variance of the average number in the system by multiplying the estimate of $Var[T]$ obtained from a simulation by the value of λ^2. The rationale for the variance reduction is the use of the theoretical arrival rate λ in the estimation of $Var[N]$. In the actual simulation, a sampling process is used and a sample arrival rate would be drawn for the entire simulation. If λ is treated as an independent random variable, the relevant equation would be

$$Var[\overline{N}] = Var[\lambda]Var[\overline{T}] + E^2[\lambda]Var[\overline{T}] + Var[\lambda]E^2[\overline{T}]$$

and, hence, we expect a variance reduction by assuming that λ is a constant, that is, $Var[\lambda] = 0$.

The above establishes that a variance reduction should occur but another question still remains. Should we estimate $Var[\overline{N}]$ directly and use it to obtain an indirect estimate of Var[$\overline{T}$] or vice versa? Law (53) considered five equations that relate the first moments of number in system, number in queue, time in system, time in queue, and work content in system. Using estimates obtained from regenerative procedures, he showed analytically, assuming steady-state values are of interest, that for the M/G/1 queueing situation, it is more efficient (less variance of the sample means) to use indirect estimators based on the variance of the time in the system.

A second example of the use of prior information is based on the work of Carter and Ignall (10). This study involved the analysis of an inventory situation in which backorders were allowed. For such studies, the expected number of backorders must be estimated. However, backorders may be infrequent and long simulations may be required to obtain low variance estimates of the average number of backorders. The Carter and Ignall approach was to derive an expression for the expected number of backorders given the inventory position at the beginning of a period prior to demands being met. For period t, they showed that

$$E[B_t|A_t] = E[D_t] - A_t + \sum_{d=o}^{A_t} (A_t - d) P[D_t = d]$$

where

   B_t is the number of backorders in period t

   A_t is the inventory position at beginning of period t

   D_t is the demand in period t.

An estimate of the average backorders in the simulation was obtained by observing the values of A_t and solving the above equation for $E[B_t|A_t]$ for t=1,2, . . . ,T. Average values for the T periods could then be obtained.

The above procedure resulted in variance reduction ratios of 3.89 and 8.79 for two different parameter settings. The procedure is a direct application of the Roa-Blackwell theorem in which the prior information concerning the distribution of the demand, D_t, is used in estimating the average number of backorders.

The use of prior information as a variance reduction technique is appealing because it allows the combining of analytical and experimental procedures. Since direct estimation is always possible, a check on the variance reduction is easily made. Since the reliability of results is being considered, the question of why there should be multiple estimates of the variance of the sample mean should not be bothersome. Remember the question raised in the first section of this chapter was related to the variability expected if the simulation experiment is repeated? This implicitly assumes that the same procedures for statistics collection and analysis are used when computing the variance.

The difficulty associated with the use of prior information involves the derivation of equations upon which to base the computations of the sample values.

19.4.4   Control Variates as a VRT

The concept associated with control variates is the identification of a variable, say Y, that has a positive covariance with the variable of interest, say X. If such a control variable exists and if we can derive the theoretical expectations associated with the control variable, then a variance reduction for the variable of interest can be obtained. To see how this is accomplished, consider the following equation that combines the sample means $\overline{X}_I$ and $\overline{Y}_I$ to form a new random variable, $\overline{Z}_I$:

$$\overline{Z}_I = \overline{X}_I + (E[Y] - \overline{Y}_I)$$

clearly, $E[\overline{Z}_I] = E[\overline{X}_I]$ if an unbiased estimator for $E[Y]$ is used. Looking now at variances, we have

$$\text{Var}[\overline{Z}_I] = \text{Var}[\overline{X}_I] + \text{Var}[\overline{Y}_I] - 2\,\text{Cov}[\overline{X}_I, \overline{Y}_I]$$

From this equation, we see that $\text{Var}[\overline{Z}_I] < \text{Var}[\overline{X}_I]$ if $\text{Var}[\overline{Y}_I] < 2\,\text{Cov}[\overline{X}_I, \overline{Y}_I]$.

Extensive research has been performed on the theoretical aspects of control variates (46,63). Generalizations to multiple control variates with weighting coefficients have been explored, that is,

$$\overline{Z}_I = \overline{X}_I + \sum_{k=1}^{K} w_k (E[Y_k] - \overline{Y}_{Ik})$$

However, little practical application of control variates has been reported. Typical control variates suggested are input variables (assuming the output variables are positively correlated with the input variables) and models derived by applying limiting assumptions to the simulation model.

The control variate procedure is easy to comprehend and should be considered further. However, application experience is necessary in order to properly evaluate its significance (85,86).

19.4.5   Other Variance Reduction Techniques

In a section on variance reduction techniques, we feel called upon to mention stratified sampling and importance sampling procedures. These VRT have been used in Monte Carlo studies (39,80) and in standard sampling experiments (14). An excellent review of the procedures and attempts to apply them is contained in

Kleijnen (46). Based on a review of the literature, it is our opinion that these techniques require further refinement before they can be applied in advanced simulation applications. Therefore, only a brief review of the techniques is presented.

Stratified Sampling Procedure.   Stratified sampling procedures involve the definition of a variable y from which the stratification classes G_k can be defined. The random samples for X_i are then stratified by examining y_i corresponding to the ith observation and classifying X_i to be in the kth strata if $y_i \in G_k$. It is assumed that $p_k = P(y \in G_k)$ is known and the sample mean based on stratification is computed as

$$\overline{X}_{ST} = \sum_k p_k \overline{X}_k$$

where $\overline{X}_k$ is the sample mean for the kth strata. It can be shown that $\overline{X}_{ST}$ is an unbiased estimator of μ_x. It can also be shown that (14,46)

$$\text{Var}[\overline{X}_{ST}] = \sum_k p_k^2 \text{Var}[\overline{X}_k] \leq \text{Var}[\overline{X}]$$

so that a variance reduction may be obtained through stratification. Greater variance reductions are obtained when the absolute differences between the strata means, μ_k, and population mean, μ_x, are large.

Importance Sampling Procedure.   Importance sampling involves a redefinition of the variable of interest by defining a new density function that gives more weight to the values of X that contribute the most to the expected value. For example, assume we are going to estimate the expected value of g(X) where X is a random variable whose density function is f(x). From the "law of the unconscious statistician" (71), we have

$$E[g(X)] = \int_x g(x)f(x)dx$$

Rewriting the above by defining the density h(x), we have

$$E[g(X)] = \int_x \frac{g(x)f(x)}{h(x)} h(x)dx$$

In importance sampling, this result is used to estimate the sample mean by selecting values of x in accordance with h(x) and then computing a sample value equal to [g(x)f(x)]/[h(x)] for the value of x selected. The estimate of the sample average is obtained by summing these values and dividing by the number of values generated.

19.5   START-UP POLICIES

The initial conditions for a simulation model may cause the values obtained from the model to be different from those obtained after a startup period. If the system being modeled has a natural termination time, then such a transient response is anticipated and the values obtained during the startup period, although different, would be representative of the outputs obtained from the real system. However, when steady-state performance is to be estimated, the initial responses can adversely influence the estimators of steady-state performance. This latter problem is the one discussed in this section.

Startup policies are used for setting the initial conditions for the simulation model and specifying a procedure for establishing a truncation point, d, at which sample values should begin to be included in the estimators being computed. Basically, the initial-condition setting attempts to provide a starting point that requires only a limited amount of data truncation to be performed, that is, one that allows a small value of d to be employed. The truncation point specification involves two considerations. The deletion of initial values tends to reduce the bias of the output estimators. Deletion of values may, however, increase the estimate of the $Var[\overline{X}_I]$ as it would be based on fewer observations. This latter assertion assumes that the deleted values are samples that have a variability similar to the variability associated with steady-state samples. This may not be the case.

From the above discussion, it appears that a trade-off is required in evaluating startup policies between bias reduction and variance reduction. Thus, it seems natural to employ the mean square error and the coverage as evaluation measures for startup policies. A procedure has been developed to make such an evaluation and the reader is referred to Wilson and Pritsker (83,84) for a summary of past research and for further details regarding this trade-off analysis.

In this section, we present various proposed initial condition rules and truncation procedures. Before doing so, several observations are in order. The use of startup policies should be considered in conjunction with the estimation procedure. If estimators are to be obtained using regenerative methods then the startup policy decision is an easy one, that is, start the run in the regenerative state so that the first cycle starts immediately and no truncation is required. If the estimation procedure is based on a single time series then the startup policy is only applied once and it is not too inefficient to truncate. However, if replication is used, the startup policy is used repetitively and great care is needed in establishing it.

Another observation involves past research on startup policies. Theoretical results (7,25) are only available for small, well-behaved models. For such models. the variability associated with sample values during startup is not too different from the steady-state variability. Thus, the theoretical research tends to indicate that no truncation should be performed. Practical applications, however, indicate this is not the case and that truncation is a reasonable policy to follow. This is especially true when dealing with job shops or conveyor systems in which many sequential operations must be performed before the system is "loaded". These points bring the discussion back to initial condition setting.

19.5.1   Initial Condition Setting

The ideal initial condition setting would be to sample from the steady-state distributions that underlie the simulation model and set the initial conditions based on the sample values obtained. Repeated use of this procedure would ensure sound statistical estimates of steady-state performance. This is clearly a "catch 22" situation as knowledge of the steady-state distribution would preclude the need for the simulation model. To avoid such a situation, three basic rules have been proposed for setting the initial state of the system:

1. Start the system "empty and idle";
2. Start the system at the steady-state mode; and
3. Start the system at the steady-state mean.

Rule 1 has the advantage of being easy to implement. It has the disadvantage in application studies of not being a good representative state. For small scale models, such as the $M|M|1$ queueing situation, it is a good representative state as "empty and idle" is the modal state.

Rule 2 specifies that the most likely state, that is, the state with the highest probability of occurring, should be the starting condition. Through experimental analysis, it was selected as the best initial condition for the models evaluated (83). The main disadvantage is the inability to determine the modal state for a large model.

Rule 3 recommends that the starting state be the expected or average state. The advantage of this rule is that the average state can be approximated through the making of a pilot study or by analyzing a related but analytically tractable model. Intuitively, starting in the expected state should provide initial samples that have a representative variability. However, there have been no published results which indicate that starting a simulation in the "average" state produces better statistical estimators.

19.5.2   Truncation Procedures

The simplest truncation procedure is to specify a time at which the collection of sample data is to be initiated. Actually, in simulation models, such a rule is implemented by discarding all sample values collected up to the truncation point. This is the case in SLAM II, and truncation is implemented by either including a MONTR statement in the input with the CLEAR option specified or by directly calling subroutine CLEAR at the time truncation is desired.

The question that arises for setting the truncation point is how to specify the time at which truncation should be made. One approach, which is perhaps the most common in applications, is to make a pilot run and select a time based on the pilot run. Although not normally done on a formal basis, the analyst considers such quantities as the number of consecutive times sample values have increased or decreased, differences between successive batch averages and successive cumulative averages, and crossings of averages by the sample values. An approach proposed by Welch (81) is to perform a set of pilot runs and select a time based on the sequence of averages computed over the runs at equally spaced points in time. Graphing the sequence of averages helps to select the truncation time. Welch (81) graphically shows the effect of using 5, 25, and 100 runs in determining a truncation point for an example problem. He also advocates smoothing the data using a moving average procedure. Many authors have attempted to formalize these concepts and to provide a truncation rule that can be used in the simulation model for detecting when the conditions for truncation are met by the time series values for a given run.

Papers that survey and evaluate some of these rules are available (31,83,84). A limited summary of these rules is presented in Table 19-1. Most evaluations have been made on small models and indicate that truncation should not be employed. For the reasons cited earlier, these results may not apply for large-scale models.

Table 19-1   Truncation rules.†

Proposer	Rule
Conway(15)	Set d so that x(d+1) is neither the maximum nor the minimum of the values $\{x(n) : n=d+1, \cdots, N\}$
Fishman(25)	Set d=n when $\{\text{sgn}(x(t) - \overline{x}_n) : t=0,1,2, \cdots n\}$ contains k runs where k is a parameter to be specified. This rule corresponds to setting d=n when the time series $\{x(t) : t=0,1,2, \cdots n\}$ has crossed $\overline{x}_n$ at least (k – 1) times.
Schriber(73)	Set d=n when the batch means for the k most recent batches of size b all fall within an interval of length ε.
Fishman(25)	Set d so that the number of observations deleted is "equivalent" to one independent observation where the number of dependent observations to independent observations is given by m/R_0. Thus set $d=m/R_0 - 1$.
Gordon(35)	Make k replications to compute Var $[\overline{x}_n], n=1,2, \cdots$. Set d= n for which Var $[\overline{x}_n]$ begins to fall off as 1/n.
Gafarian(31)	Set d equal to the smallest n for which x(n) is neither the maximum nor minimum of all preceding observations $\{x(t) : t=0,1,2, \cdots, n\}$.

† The notation used in this table suppresses the batch number subscript and employs lower case letters as the rules depend directly on the sample values observed.

In the application of the proposed truncation rules, four issues should be kept in mind:

1. The expected value of a sample average lags the expected values of the process variable if the system is initially empty and leads it if the system is loaded to capacity (21);
2. Crossings of averages are not as likely as anticipated;
3. Truncation rules are extremely sensitive to parameter settings. Also, parameter setting procedures are not available for many proposed rules (82); and
4. For a long initial startup period, the application of a truncation rule can be time-consuming and, hence, expensive.

These issues account for the lack of use of truncation procedures in applications and the direct use of a truncation time for clearing statistics. A promising technique that alleviates these difficulties is to use cumulative sums as proposed by Schruben (76). Truncation rules have been explored by Schruben based on this work and quality control procedures.

19.6   STOPPING RULES

Determining the length of a simulation run as specified in terms of the number of batches is a complex problem. If we are willing to assume that $\overline{X}_I$ is unbiased and that $Var[\overline{X}_I] = \sigma_{\overline{X}}^2/I$ then the number of batches I required to obtain a $(1-\alpha)$ confidence that the mean μ_X is contained in a prescribed interval can be computed using standard statistical formula. Symbolically, suppose we desire

$$P[\overline{X}_I - g \le \mu_X \le \overline{X}_I + g] \ge 1 - \alpha$$

where g is a prescribed half-length for the confidence interval. Letting $Z = \sqrt{I}(\overline{X}_I - \mu_X)/\sigma_X$, we have

$$P\left[|Z| \le \frac{g\sqrt{I}}{\sigma_X} \right] \ge 1 - \alpha$$

with equality holding for the smallest value of I, say I*. Assuming I* is large enough so that the central limit theorem applies, we have

$$I^* = \left(\frac{\sigma_X}{g} Z_{\alpha/2} \right)^2$$

where

$$Z_{\alpha/2} \text{ is such that } \frac{1}{\sqrt{2\pi}} \int_{Z_{\alpha/2}}^{\infty} e^{-y^2/2}/2dy = \alpha/2$$

This equation for I* requires knowledge of σ_X. A common trick is to specify g in relative terms of σ_X, that is, let $g = v\sigma_X$ for v>0. In this case, I* can be computed without knowledge of σ_X. Values of I* for combinations of v and α are given in Table 19-2.

From Table 19-2, we see that it requires almost 400 batches to obtain a 95 percent confidence interval that μ_X is within $(\overline{X}_I - 0.1\,\sigma_X, \overline{X}_I + 0.1\,\sigma_X)$. Similar analyses can be performed for determining the sample size to have a prescribed confidence interval on the variance or on a probability value.

Throughout this chapter, we have proposed the use of a batch mean as the sample value X_I. Because of this, the assumptions required in the above procedure are

Table 19-2 Values of I* for combinations of v and α.

I*	α		
v	0.02	0.05	0.10
0.01	54093	38416	27060
0.10	541	384	271
0.20	135	96	68
0.50	22	15	11

tenable. If the independence assumption is not appropriate, then we can use $\text{Var}[\overline{X}_I]$ = m/I and replace σ_X in the above equations by m.

Typically, s_X is used in place of σ_X (or $\hat{m}$ for m) in which case $\sqrt{I}(\overline{X}_I - \mu)/s_X$ has a t-distribution. In simulation studies, I is usually large enough to assume that the normal approximation to the t-distribution holds. To set I* before the simulation is started, a value of s_X is required. In some cases, pilot studies are performed to obtain a value for s_X from which I* is estimated. A more general approach is to use a sequential stopping rule.

A sequential stopping rule specifies a condition that when satisifed will yield the desired objective. Starr (78) has shown that if the X_i are iid normal random variables that

$$P\{\overline{X}_{I*} - g \leq \mu \leq \overline{X}_{I*} + g\} \geq \begin{cases} 0.928 \text{ for } 1-\alpha = 0.95 \\ 0.985 \text{ for } 1-\alpha = 0.99 \end{cases} \qquad (19\text{-}21)$$

when I* is set according to

$$I* = \min\{I : I \geq 3 \text{ and odd}; s_X^2 \leq Ig^2/t_{a/2,I-1}^2\} \qquad (19\text{-}22)$$

where $t_{a/2,I-1}$ corresponds to the $1-\alpha/2$ fractile of the student t-distribution with I−1 degrees of freedom, for example, $t_{.025,10} = 2.228$.

The degradation in the confidence interval occurs because the test is being sequentially applied. Fishman (28) proposes the use of Eq. 19-22 without the requirement for I to be odd, since the requirement for I to be odd is due to the intractability of the analysis when I is even. The use of Eq. 19-22 in a simulation experiment requires a table of t-values, a prescription for g, a batch size specification and then the periodic testing of computed values of s_x until its value is below that required in Eq. 19-22. When this occurs, the confidence interval as specified by Eq. 19-21 holds.

When a relative specification is desired, that is, $-v\mu \leq \overline{X}_I - \mu \leq v\mu$ where $v > 0$, Nadas (64) has shown that the stopping rule

$$I^* = \min \{I : s_x^2 \leq [(Iv\overline{x}_I/t_{a/2,I-1})^2 - 1]/(I-1)\} \qquad (19\text{-}23)$$

will result in a limiting confidence interval of

$$\lim_{v \to 0} P[\overline{X}_{I^*}/(1+v) \leq \mu_X \leq \overline{X}_{I^*}/(1-v)] = 1-\alpha$$

Note that for large I and $g = v\overline{x}_I$, the stopping rule given by Eq. 19-23 approximates the rule given by Eq. 19-22.

Since there is a degradation in the coverage associated with the use of the stopping rules specified by Eq. 19-22 and Eq. 19-23, we recommend for important decisions that I* be established not as the minimum number of batches for which the condition on s_x^2 holds, but as the value of I for which the condition holds a second time. Typically, this should only require one additional batch to be obtained but it may require more. Using this rule should help to compensate for both the degradation expected from Eq. 19-21 and the inherent optimism (smaller variance) estimates) associated with the calculation of s_x^2 based on the assumption of iid batch observations.

In addition to determining the sample size to meet desired confidence interval specifications, there are practical issues associated with the stopping of a simulation run. Such questions involve the consideration of what to do about entities in the model at the end of a run. The answers to such questions are problem specific. If such entities are representative of the other entities on which statistics were collected, then the further processing of them should not matter. However, if they are atypical or if some information has been collected on them, then their processing should be considered. For example, in a job shop where a shortest processing time rule was employed, the jobs remaining at the end of a run could be those jobs whose processing times are extremely long. Not processing such jobs would lead to a bias in the statistics on time in system. Care must be taken to avoid such a situation.

A more general procedure for establishing a stopping condition involves the concept of marginal return. It has been proposed that a run should be stopped when the marginal improvement in potential profits based on the run decreases below the marginal costs associated with continuing the run (30,44). Although this is a good general concept, assessing potential profits and calculating marginal costs can be difficult.

19.7   DESIGN OF EXPERIMENTS

A simulation run is an experiment in which an assessment of the performance of a system is estimated for a prescribed set of conditions. In the jargon of design of experiments, the conditions are referred to as factors and treatments where a treatment is a specific level of a factor. The literature in the field of design of experiments is extensive (13,49,65). The purpose of this section is to present the issues relating to the design of experiments, but not to present the details as to how one should design a simulation experiment. The statistical techniques associated with the design of experiments are well documented. Applications of the procedures of analysis of variance (ANOVA); the Shapiro-Wilk test for testing normality assumptions; or the Newman-Keuls test for investigating all pairs of means are not considered to be significantly different in simulation studies from their use in other areas (2).

The major problem involved in simulation experiments is associated with the definition of the inference space associated with the simulation model. Making *a priori* assessments of how widely the results obtained from the simulation model are to be applied, and developing a thorough understanding of the inferences that can be made, are the most neglected aspects of the design of experiments associated with simulation studies. A possible reason for this is the inclusion of factors in the experiment that relate to the multitude of alternatives open to the analyst and the extensive number of experimental controls that must be set when performing the experiment. In previous sections of this chapter we have discussed some of these experimental controls such as: starting conditions; sampling procedures; run length; batch size; and estimation procedures. Documented examples which include all these factors are not available; however, the survey by Kleijnen is extensive and highly recommended (49). Also there is an excellent article which presents the details of an experimental design and analysis of simulations to evaluate scheduling rules (11).

In general, the objectives of simulation experiments are to:
1. obtain knowledge of the effects of controllable factors on experimental outputs;

2. estimate system parameters of interest;
3. make a selection from among a set of alternatives; and
4. determine the treatment levels for all factors which produce an optimum response.

When multiple factors are involved the approach to the first two items listed above is to select one of the many possible experimental designs and to hypothesize a model for the analysis of variance for the experimental design selected. The experimental design specifies the combination of treatment levels along with the number of replications for each combination for which the simulation model must be exercised. Using the data obtained from the experiment, the parameters of the hypothesized model are determined along with the estimation of the error terms. Interaction plots are then drawn to ascertain the joint effects of the various factors. The significance of each factor is then judged based on the derived model, and from this estimates of system parameters of interest can be calculated. This procedure is reminiscent of the parametric modeling approach described earlier in this chapter for a single performance measure of interest. Kleijnen has used regression techniques to develop metamodels of simulation outputs. The metamodels are then used to investigate alternatives characterizing different parameter settings (48,50,51). Kleijnen has applied this approach with success (49).

In the problem of making a choice among alternatives, the statistical procedures or ranking and selection are used. Kleijnen(47) and Dudewicz(19) present state-of-the-art reviews that summarize past research in this area and how it can be used in simulation analysis. Many procedures have been developed for specifying the sample size required in order to select the alternative whose population mean is greater than the next best population mean by a prescribed value with a given probability. The test procedures involve the computation of the sample mean based on the sample size specified and the selection of the largest sample mean observed. Bechhofer developed this approach which is referred to as the indifference zone approach (4,5)

An alternative approach involves grouping the alternatives into statistically equivalent subsets. The procedures involved in making subset selections are given by Gupta who also compares subset selection with the indifference zone approach (37,38).

A final topic relating to the design of experiments is the selection of a best alternative. This problem differs from those previously described in that we are trying to determine the values for the controllable variables which either maximize or minimize an objective function. For example, in the analysis of a periodic review inventory system, we might wish to employ simulation to determine the values for the stock control level, reorder point, and time between reviews which minimize the average monthly cost of the inventory system.

Although the principles of optimization using simulation experiments are essentially the same as for optimization of mathematical expressions, there are some differences which must be considered. Since the response from a simulation typically involves random variables, the objective function or constraint equations written as a function of the simulation response will also be random variables. As a consequence, it is necessary to formulate response constraints as probability statements and to make statistical interpretations of the objective function value.

There have been two basic approaches to optimization using simulation models. The first approach involves a direct evaluation of the independent variables using the simulation model. Farrell (22) divides these techniques into three categories: mathematically naive techniques such as heuristic search, complete enumeration, and random search; methods appropriate to unimodal objective functions such as coordinate search, and pattern search; and methods for multimodal objective functions.

The second approach to optimization using simulation is response surface methodology (60). In this method we fit a surface to experimental observations using a factorial design in the vicinity of an initial search point. We then apply an optimization algorithm such as the gradient method to determine the optimum values of the controllable variables relative to the fitted equation. The optimum values for the fitted surface are then used to define the next search point. Biles has applied this procedure sequentially in a search for optimal decision values (6). A modular FORTRAN program for simulation optimization using first and second order response surfaces has been developed by Smith (77).

The implementation of an optimization procedure within the GASP IV and SLAM simulation languages has been reported by Pegden and Gately (69,70). In SLAM II, the determination of the next experimental condition in a search for the optimum can be performed in either subroutine INTLC or OTPUT.

19.8 CHAPTER SUMMARY

Two distinct aspects of simulation output analysis involve the accuracy and reliability of the sample values obtained. The main emphasis of this chapter is on reliability. The importance of $\text{Var}[\bar{X}_1]$ in simulation studies is established. It is recommended that the reliability of simulation outputs be based on observations of batch or cycle averages rather than on individual sample values. Five methods for estimating the variance of sample means, $Var[\overline{X}_I]$, based on I batches are presented. Variance reduction techniques, startup policies, stopping rules, and the design of simulation experiments are described. Overall, this chapter provides both detailed practical results and suggestions for the important statistical problems facing a simulation analyst.

19.9   EXERCISES

19-1.   Define the following terms: Reliable; Batch; Stochastic Process; Ergodic; Stationary; Steady State; Time Series; Sample Mean; Average; Expectation; Mean Square Error; Kurtosis; Spectrum; Regeneration Point; Parametric Modeling; Spectral Density Function; ARMA Model; White Noise; VRT; Stratified Sampling; Catch-22; and Bias.

19-2.   A drive-in bank has two windows, each manned by a teller and each has a separate drive-in lane. The drive-in lanes are adjacent. From previous observations, it has been determined that the time interval between customer arrivals during rush hours is exponentially distributed with a mean time between arrivals of 0.25 time units. Congestion occurs only during rush hours, and only this period is to be analyzed. The service time is exponentially distributed for each teller with a mean service time of 0.4 time unit. It has also been shown that customers have no preference for a teller if the waiting lines are equal. At all other times, a customer chooses the shortest line. After a customer has entered the system, he may not leave until he is serviced. However, he may change lanes if he is the last customer in his lane and a difference of two customers exists between the two lanes. Because of parking space limitations, only nine cars can wait in each lane. These cars, plus the car of the customer being serviced by each teller, allow a maximum of twenty cars in the system. If the system is full when a customer arrives, he balks and is lost to the system.

The initial conditions are as follows:

1. Both drive-in tellers are busy. The initial service time for each teller is exponentially distributed with mean of 0.4 time unit.
2. The first customer is scheduled to arrive at 0.1 time unit.
3. Two customers are waiting in each queue.

Theoretical Steady-State Results
Steady-state probabilities: (0.1123, 0.1796, 0.1437, 0.1150, 0.0920, 0.0736, 0.0589, 0.0471, 0.0377, 0.0301, 0.0241, 0.0193, 0.0154, 0.0123, 0.0099, 0.0079, 0.0063, 0.0050, 0.0040, 0.0032, 0.0026)

Number in system: Mean = 4.232; Variance = 15.83; and m = 177.4.

Determine the simulation run length required to obtain estimates that are within 10 percent of the theoretical values.

19-3. Simulate the system described in Exercise 19-2 to obtain estimates of the variance of the sample mean and the value of m for each of the following random variables:

1. The number of customers in the system, and
2. The time a customer spends in the system.

Develop estimates for $Var(\bar{X}_I)$ and m using both replication and subinterval sampling procedures. Select a total amount of simulation time on which to base your estimates and give a rationale for your selection. The same amount of simulation time should be used for each procedure.

Embellishments: (a) Perform a spectral analysis of the experimental data. (b) Use regenerative techniques to obtain the requested estimates. (c) Use parametric modeling techniques to build a model of the simulated data. (d) Obtain 95% confidence limits for m using the information that $(k-1)\hat{m}/m$ is Chi-square distributed with $(k-1)$ degrees of freedom.

19-4. Compare the use of the observed data values and batched observations in the embellishments to Exercise 19-3.

19-5. Evaluate the effect of making the initial conditions for the system described in Exercise 19-2 to be the modal state. Evaluate the effect of starting in the expected state.

19-6. Evaluate three truncation rules for the system described in Exercise 19-2.

Embellishment: Develop a truncation rule and evaluate it.

19-7. Perform four experiments each consisting of ten runs of length 200 time units on the bank teller simulation model, Exercise 19-2. In experiment 1, use random number stream 1 for arrival times and stream 2 for service times. In experiment 2, use stream 1 again for arrivals but the antithetic values from stream 2. In experiment 3, employ antithetic values from both streams 1 and 2. In experiment 4, use stream 1 for arrival times and stream 3 for service times.

Define $X_k^{(1)}$ to be the average computed on run k of experiment i where X could be either the average number in the system,$\bar{N}$, or the average time in the system, $\bar{W}$.

(a) Calculate $Var[\bar{Z}]$ where $\bar{Z}_k = [\bar{X}_k^{(1)} + \bar{X}_k^{(j)}]/2$ using pairs of runs from experiments 1 and 2 and experiments 1 and 3.
(b) Calculate $Var[\bar{X}_k^{(1)} - \bar{X}_k^{(j)}]$ for $j = 2$ and 4.
(c) For each experiment, use the prior information that $\bar{N} = \lambda_E \bar{W}$ to obtain a variance reduction where λ_E is the effective arrival rate after balking occurs.

19-8. Apply the sequential stopping rule suggested in Section 19.6 to the bank teller model, Exercise 19-2, to obtain a 95 percent confidence interval that has a half length, g, of 0.25. Embellishment: Apply the sequential stopping rule procedure to obtain a 95 percent confidence that the half length is 10 percent of the true mean value.

19-9.   Perform Exercises 19-2 through 19-8 for Example 5-2, the inspection and adjustment of television sets model.

19-10.   For the single channel queueing situation, evaluate the variance reduction obtained from switching the streams used for generating arrival and service times.

19-11.   For the bank teller simulation model, stratify customers based on their service time and then compute the variance of the average time in the system. Compare these results with those obtained without stratification.

19-12.   Develop and apply an optimization procedure for setting the reorder point, stock control level and time between reviews for the inventory model and cost values given in embellishment (e) of Exercise 11-5.

19-13.   Discuss the issues involved in performing an analysis of variance on experiments that involve the use of common streams and antithetic samples.

19.10   REFERENCES

1. Anderson, T. W., *The Statistical Analysis of Time Series,* John Wiley, 1970.

2. Anderson, V. L. and R. A. McLean, *Design of Experiments: A Realistic Approach,* Marcel Dekker, 1974.

3. Andrews, R. W. and T. J. Schriber, "ARMA Based Confidence Intervals for Simulation Output Analysis," *American Journal of Mathematical and Management Science,* Vol. 4, 1984, pp. 345-374.

4. Bechhofer, R. E., "A Single-Sample Multiple Decision Procedure for Ranking Means of Normal Populations with Known Variances," *Ann. Math. Stat.,* Vol. 25, 1954, pp. 16-39.

5. Bechhofer, R. E., "Selection in Factorial Experiments," *Proceedings, 1977 Winter Simulation Conference,* 1977, pp. 65-70.

6. Biles, W. E., "Integration-Regression Search Procedure for Simulation Experimentation," *Proceedings, 1974 Winter Simulation Conference,* 1974, pp. 491-497.

7. Blomqvist, N., "On the Transient Behavior of the GI/G/1 Waiting-Times," Skandinavisk Aktuarietidskrift, Vol. 53, 1970, pp. 118-129.

8. Box, G. E. P. and G. M. Jenkins, *Time Series Analysis: Forecasting and Control,* Holden-Day, 1970.

9. Bratley, P., B. L. Fox, and L. E. Schrage, *A Guide to Simulation,* Springer-Verlag, 1983.

10. Carter, G. and E. J. Ignall, "A Variance Reduction Technique for Simulation" *Management Science,* Vol. 21, 1975, pp. 607-616.

11. Chang, Y., R. S. Sullivan, J. R. Wilson, and U. Bagchi, "Experimental Investigation of Real Time Scheduling in Flexible Manufacturing Systems," *Annals of OR,* Vol. 3, 1985, pp. 355-377.

12. Cinlar, E., *Introduction to Stochastic Processes,* Prentice-Hall, 1975.

13. Cochran, W. G. and G. M. Cox, *Experimental Designs,* John Wiley, 1957.
14. Cochran, W. G., *Sampling Techniques,* Third Edition, John Wiley, 1977.
15. Conway, R., "Some Tactical Problems in Digital Simulation," *Management Science,* Vol. 10, 1963, pp. 47-61.
16. Crane, M. A. and D. L. Iglehart, "Simulating Stable Stochastic Systems I: General Multiserver Queues," *J. ACM,* Vol. 21, 1974, pp. 103-113.
17. Crane, M. A. and A. Lemoine, *An Introduction to the Regenerative Method for Simulation Analysis,* Technical Report No. 86-23, California Analysis Corporation, Palo Alto, CA., October 1976.
18. Diananda, P. H., "Some Probability Limit Theorems with Statistical Applications," *Proceedings, Cambridge Phil. Soc.,* Vol. 49, 1953, pp. 239-246.
19. Dudewicz, E. J., "New Procedures for Selection Among (Simulated) Alternatives," *Proceedings, 1977 Winter Simulation Conference,* 1977, pp. 58-62.
20. Duket, S., *Simulation Output Analysis,* unpublished MS Thesis, Purdue University, December 1974.
21. Duket, S. and A. A. B. Pritsker, "Examination of Simulation Output Using Spectral Methods," *Mathematics and Computers in Simulation,* Vol. XX, 1978, pp. 53-60.
22. Farrell, W., "Literature Review and Bibliography of Simulation Optimization," *Proceedings, 1977 Winter Simulation Conference,* 1977, pp. 116-124.
23. Feller, W., *An Introduction to Probability Theory and Its Applications,* Vol. 1, 2nd Edition, John Wiley, 1957.
24. Feller, W., *An Introduction to Probability Theory and Its Applications,* Vol. 2, John Wiley, 1972.
25. Fishman, G. S., "A Study of Bias Considerations in Simulation Experiments," *Operations Research,* Vol. 20, 1972, pp. 785-790.
26. Fishman, G. S., *Concepts and Methods in Discrete Event Digital Simulation,* John Wiley, 1973.
27. Fishman, G. S., "Grouping Observations in Digital Simulation," *Management Science,* Vol. 24, 1978, pp. 510-521.
28. Fishman, G. S., *Principles of Discrete Event Simulation,* John Wiley, 1978.
29. Fishman, G. S., "Statistical Analysis for Queueing Simulation," *Management Science,* Vol. 20, 1973, pp. 363-369.
30. Fishman, G. S., "The Allocation of Computer Time in Company Simulation Experiments," *Operations Research,* Vol. 16, 1968, pp. 280-295.
31. Gafarian, A. V., Ancker, C. J., and Morisaku, T. *The Problem of the Initial Transient with Respect to Mean Value in Digital Computer Simulation and the Evaluation of Some Proposed Solutions,* Technical Report No. 77-1, University of Southern California, 1977.
32. Glynn, P. W., "Some Asymptotic Formulas for Markov Chains with Applications to Simulation," *Journal of Statistical Computation and Simulation,* Vol. 19, 1984, pp. 97-112.
33. Glynn, P. W., "On the Role of Generalized Semi-Markov Processes in Simulation Output Analysis," *Proceedings, Winter Simulation Conference,* 1983, pp. 39-42.
34. Goldenberger, A. S., *Econometric Theory,* John Wiley, 1964.
35. Gordon, G., *System Simulation,* Prentice-Hall, 1969.
36. Gray, H. L. and W. R. Schucany, *The Generalized Jackknife Statistic,* Marcel Dekker, 1972.

37. Gupta, S. S. and S. Panchapakesan, "On Multiple Decision (Subset Selection) Procedures, *Journal of Math and Physical Sciences,* Vol. 6, 1972, pp. 1-71.

38. Gupta, S. S. and J. C. Hsu, "Subset Selection Procedures with Special Reference to the Analysis of 2-Way Layout: Application to Motor Vehicle Fatality Data," *Proceedings, 1977 Winter Simulation Conference,* 1977, pp. 80-85.

39. Hammersley, J. M. and D. C. Handscomb, *Monte Carlo Methods,* Methuen, 1964.

40. Hazen, G. and A. A. B. Pritsker, "Formulas for the Variance of the Sample Mean in Finite State Markov Processes," *Journal of Statistical Computation and Simulation,* Vol. 12, 1981, pp. 25-40.

41. Heidelberger, P. and P. D. Welch, "A Spectral Method for Confidence Interval Generation and Run Length Control in Simulations," *Comm. ACM,* Vol. 24, 1981, pp. 233-245.

42. Hoel, P. G., *Elementary Statistics,* Second Edition, John Wiley, 1966.

43. Hogg, R. V. and A. T. Craig, *Introduction to Mathematical Statistics,* Macmillan, 1970.

44. Kabak, I. W., "Stopping Rules for Queueing Simulations," *Operations Research,* Vol. 16, 1968, pp. 431-437.

45. Karlin, S. and H. Tayler, *A First Course in Stochastic Processes,* Academic Press, 1975.

46. Kleijnen, J. P. C., *Statistical Techniques in Simulation: Part I,* Marcel Dekker, 1974.

47. Kleijnen, J. P. C., *Statistical Techniques in Simulation, Part II,* Marcel Dekker, 1975.

48. Kleijnen, J. P. C., A. J. van den Burg, and R. Th. van der Ham, "Generalization of Simulation Results," *European Journal of Operational Research,* Vol. 3, 1979, pp. 50-64.

49. Kleijnen, J. P. C., *Statistical Tools for Simulation Practioners,* Marcel Dekker, 1986.

50. Kleijnen, J. P. C., "Regression Metamodel Summarization of Model Behaviour," *Encyclopedia of Systems and Control,* Pergamon, Press, 1982.

51. Kleijnen, J. P. C., "Regression Metamodels for Generalizing Simulation Results," *IEEE Transactions on Systems, Man, and Cybernetics,* Vol. SMC-9, 1979, pp. 93-96.

52. Law, A. M., *Confidence Intervals for Steady-State Simulations, I: A Survey of Fixed Sample Size Procedures,* Technical Report 78-5, University of Wisconsin, 1978.

53. Law, A. M., "Efficient Estimators for Simulated Queueing Systems," *Management Science,* Vol. 22, 1975, pp. 30-41.

54. Law, A. M., *Statistical Analysis of the Output Data from Terminating Simulations,* Technical Report 78-4, University of Wisconsin, 1978.

55. Law, A. M. and J. S. Carson, III, "A Sequential Procedure for Determining the Length of a Steady State Simulation," *Operations Research,* Vol. 27, 1979, pp. 1011-1025.

56. Law, A. M. and W. D. Kelton, *Simulation Modeling and Analysis,* McGraw-Hill, 1982.

57. Little, J. D. C., "A Proof of the Queueing formula $L = \lambda W$," *Operations Research,* Vol. 9, 1961, pp. 383-387.

58. Mabert, V. A., *An Introduction to Short Term Forecasting Using The Box-Jenkins Methodology,* AIIE Monograph Series, AIIE-PP C-75-1, Atlanta, Georgia, 1975.

59. Mechanic, H. and W. McKay, "Confidence Intervals for Averages of Dependent Data in Simulations II," Technical Report 17-202, IBM Advanced Systems Development Division, August 1966.

60. Meyer, R. H., *Response Surface Methodology,* Allyn & Bacon, 1971.

61. Moeller, T. L. and P. D. Welch, "A Special Based Technique for Generating Confidence Intervals from Simulation Outputs," *Proceedings, 1977 Winter Simulation Conference,* 1977, pp. 176-184.
62. Moran, P. A. P., "Some Theorems on Time Series, I," *Biometrika,* Vol. 34, 1947, pp. 281-291.
63. Moy, W. A., "Practical Variance Reducing Procedures for Monte Carlo Simulations," in *Computer Simulation Experiments with Models of Economic Systems* by T. H. Naylor, John Wiley, 1971.
64. Nadas, A., "An Extension of a Theorem of Chow and Robbins on Sequential Confidence Intervals for the Mean," *Ann. Math. Stat.,* Vol. 40, 1969, pp. 667-671.
65. Naylor, T. H., Editor, *The Design of Computer Simulation Experiments,* Duke University Press, 1969.
66. Nelson, B. L. and B. W. Schmeiser, "Variance Reduction: Basic Transformations," *Proceedings, Winter Simulation Conference,* 1983, pp. 255-258.
67. Page, E. S., "On Monte Carlo Methods in Congestion Problems; II; Simulation of Queueing Systems," *Operations Research,* Vol. 13, 1965, pp. 300-305.
68. Parzen, E., *Stochastic Processes,* Holden-Day, 1962.
69. Pegden, C. D. and M. P. Gately, "Decision Optimization for GASP IV Simulation Models," *Proceedings, 1977 Winter Simulation Conference,* 1977, pp. 127-133.
70. Pegden, C. D., and M. P. Gately, "A Decision-Optimization Module for SLAM," *Simulation,* 1980, pp. 18-25.
71. Ross, S., *A First Course in Probability,* Macmillan, 1976.
72. Schmeiser, B., "Batch Size Effects in the Analysis of Simulation Output," *Operations Research*, Vol. 30, 1982, pp. 556-568.
73. Schriber, T., *Simulation Using GPSS,* John Wiley, 1974.
74. Schriber, T. J. and R. W. Andrews, "Interactive Analysis of Simulation Output by the Method of Batch Means," *Proceedings, Winter Simulation Conference,* 1979, pp. 513-525.
75. Schruben, L. W., "Confidence Interval Estimation Using Standardized Time Series," *Tech Report 518,* School of O.R.I.E., Cornell University, 1982.
76. Schruben, L. W., "Detecting Initialization Bias in Simulation Output," *Operations Research,* Vol. 30, 1982, pp. 569-590.
77. Smith, D. E., *Automated Response Surface Methodology in Digital Computer Simulation (U), Volume I: Program Description and User's Guide (U),* Office of Naval Research, Arlington, VA: September 1975.
78. Starr, N., "The Performance of a Sequential Procedure for the Fixed-Width Interval Estimation of the Mean," *Ann. Math. Stat.,* Vol. 37, 1966, pp. 36-50.
79. Stidham, S., Jr., "L=λW: A Discounted Analog and a New Proof," *Operations Research,* Vol. 20, 1972, pp. 1115-1126.
80. Tocher, K. D., *The Art of Simulation,* Van Nostrand, 1963.
81. Welch, P. D., "The Statistical Analysis of Simulation Results," in *Computer Performance Modeling Handbook,* ed. by S. S. Lavenberg, Academic Press, 1983.
82. Wilson, J. R., *A Procedure for Evaluating Startup Policies in Simulaton Experiments,* unpublished M. S. Thesis, Purdue University, December 1977.
83. Wilson, J. R. and A. A. B. Pritsker, "A Survey of Research on the Simulation Startup Problem," *Simulation,* Vol. 31, 1978, pp. 55-58.

84. Wilson, J. R. and A. A. B. Pritsker, "A Procedure for Evaluating Startup Policies in Simulation Experiments," *Simulation,* Vol. 31, 1978, pp. 79-89.
85. Wilson, J. R. and A. A. B. Pritsker, "Experimental Evaluation of Variance Reduction Techniques for Queueing Simulation Using Generalized Concomitant Variables," *Management Science,* Vol. 30, 1984, pp. 1459-1472.
86. Wilson, J. R. and A. A. B. Pritsker, "Variance Reduction in Queueing Simulation Using Generalized Concomitant Variables," *Journal of Statistical Computation and Simulation,* 1984, pp. 129-153.
87. Wilson, J. R., "Statistical Aspects of Simulation," Proceedings, IFORS, 1984, pp. 825-841.
88. Wilson, J. R., "Variance Reduction Techniques for Digital Simulation," American Journal of Mathematical and Management Sciences, Vol. 4, 1984, pp. 277-312.

APPENDIX A

User Support and
Callable Subprograms of SLAM II

FUNCTION AAAVG(IACT)

Returns average utilization of activity IACT

FUNCTION AAMAX(IACT)

Returns maximum utilization of activity IACT, or maximum busy time if activity IACT is a single-server service activity

FUNCTION AASTD(IACT)

Returns standard deviation of the utilization of activity IACT

FUNCTION AASTD(IACT)

Returns standard deviation of the utilization of activity IACT

FUNCTION AATLC(ACT)

Returns the time at which the status of activity IACT last changed

SUBROUTINE ALTER(NRES,IU)

Changes the capacity of resource number NRES by IU units

FUNCTION BETA(THETA,PHI,IS)

Returns a sample from a beta distribution with parameters THETA and PHI using random number stream IS

FUNCTION CCAVG(ICLCT)

Returns average value of variable ICLCT

FUNCTION CCMAX(ICLCT)

Returns maximum value of variable ICLCT

FUNCTION CCMIN(ICLCT)

Returns minimum value of variable ICLCT

FUNCTION CCNUM(ICLCT)

Returns number of observations of variable ICLCT

FUNCTION CCSTD(ICLCT)

Returns standard deviation of variable ICLCT

SUBROUTINE CLEAR

Reinitializes the statistical storage areas when it is called. COLCT, TIMST, histograms, files, resources, and network statistical variables are cleared

767

SUBROUTINE CLOSX(NGATE)	Closes gate whose number is NGATE
SUBROUTINE COLCT(ZZ,ICLCT)	If ICLCT > 0, records value ZZ as an observation on variable number ICLCT; if ICLCT = 0, computes and reports statistics on all NNCLT variables; if ICLCT < 0, computes and reports statistics on variable − ICLCT
SUBROUTINE COPAA(NTRYA,NAUXA, AUXF,VALUE)	Copies the values of a set of auxiliary attributes without removing the attribute values from the user's auxiliary storage array
SUBROUTINE COPY(NRANK,IFILE,A)	Copies the values of the attributes of an entry into the vector A. If NRANK is positive then the NRANK*th* entry is to be copied. If NRANK is negative, then the entry with pointer − NRANK is to be copied
FUNCTION DPROB(CPROB,VALUE, NVAL,IS)	Returns a sample from a user-defined discrete probability function with cumulative probabilities and associated values specified in arrays CPROB and VALUE, with NVAL values using random stream IS
FUNCTION DRAND(IS)	Returns a pseudo-random number obtained from random number stream IS
SUBROUTINE DDUMP(N1,N2)	Prints the storage array NSET/QSET between locations N1 and N2
SUBROUTINE ENTER(IN,A)	Releases ENTER node whose number is IN with an entity whose attribute values are in the vector A
FUNCTION ERLNG(EMN,XK,IS)	Returns a sample from an Erlang distribution which is the sum of XK exponential samples each with mean EMN using random number stream IS
SUBROUTINE ERROR(KODE)	Prints reports and error messages
FUNCTION EXPON(XMN,IS)	Returns a sample from an exponential distribution with mean XMN using random number stream IS
FUNCTION FFAVG(IFILE)	Returns average number of entities in file IFILE
FUNCTION FFAWT(IFILE)	Returns the average waiting time in file IFILE
SUBROUTINE FFILE(IFILE,A)	Files an entry with attributes stored in A into file IFILE without checking the status of network activities.

FUNCTION FFMAX(IFILE)	Returns maximum number of entities in file IFILE
FUNCTION FFPRD(IFILE)	Returns time period for statistics in file IFILE
FUNCTION FFSTD(IFILE)	Returns standard deviation of the number of entities in file IFILE
FUNCTION FFTLC(IFILE)	Returns time at which number in file IFILE last changed
SUBROUTINE FILEM(IFILE,A)	Files an entry with attributes stored in A into file IFILE
SUBROUTINE FREE(NRES,IU)	Frees IU units of resource number NRES
FUNCTION GAMA(BETA,ALPHA,IS)	Returns a sample from a gamma distribution with parameters BETA and ALPHA using random number stream IS
SUBROUTINE GDLAY(IFS,ILS,XIN,DEL)	Variable order exponential delay for Systems Dynamics problems. IFS and ILS are the first and last state variables used to maintain the delay; the order of the delay is $ILS - IFS + 1$. NNEQD *must be greater than ILS to use GDLAY;* XIN is the input to the delay and DEL is the delay constant
SUBROUTINE GETAA(NTRYA,NAUXA, AUXF,MFAA, VALUE)	Gets (removes) a set of auxiliary attributes and establishes a vector of the values of the auxiliary attributes
FUNCTION GETARY(IR,IC)	Returns the value of row IR, column IC from the global table, ARRAY
FUNCTION GGOPN(IG)	Returns the percent of time that gate IG was open
FUNCTION GGTBL(X,Y,XVALUE,NVAL)	Performs a table look-up to obtain a value of a dependent variable from the vector Y by interpolating the quantity XVALUE between the independent variable values in the vector X. NVAL is the number of values in X and Y.
FUNCTION GGTLC(IG)	Returns the time at which the status of gate IG last changed
SUBROUTINE GPLOT(IPLOT)	If IPLOT>0, IPLOT is the RECORD statement code and GPLOT stores values of the dependent variables specified on VAR input statements (up

	to 10 values per plot) for a value of the independent variable specified on a RECORD input statement; if IPLOT = 0, prints table and/or graph for all plots; if IPLOT < 0, prints table and/or graph for code = −IPLOT
FUNCTION GTABL(Y,XVALUE,XLOW, XHIGH, XINCR	Performs a table look-up to obtain a value of a dependent variable from values stored in the vector Y for a given value of the independent variable XVALUE; XLOW is the value corresponding to Y(I) and XHIGH is the largest value for Y; XINCR is the difference in the independent variable for successive values of the dependent variable
FUNCTION INTRN(ISTRM)	Returns the current unnormalized value for random number stream ISTRM
SUBROUTINE LINK(IFILE)	Files an entry whose attributes are pointed to by MFA in file IFILE
FUNCTION LOCAT(NRANK,IFILE)	Returns the pointer to the location of the entry whose rank is NRANK in file IFILE
FUNCTION MMFE(IFILE)	Returns pointer to first entry (rank 1) in file IFILE
FUNCTION MMLE(IFILE)	Returns pointer to last entry (rank NNQ(IFILE)) in file IFILE
FUNCTION NFIND(NRANK,IFILE,JATT, MCODE,XVAL,TOL)	Locates an entry with rank ≥ NRANK in file IFILE whose JATT attribute is related to the value XVAL according to the specification given by MCODE as shown below:
	MCODE = 2: maximum value but greater than XVAL
	MCODE = 1: minimum value but greater than XVAL
	MCODE = 0: value within XVAL±TOL
	MCODE = −1: minimum value but less than XVAL
	MCODE = −2: maximum value but less than XVAL
FUNCTION NNACT(NACT)	Number of active entities in activity NACT at current time

FUNCTION NNBLK(NACT,IFILE)

Returns the number of entities currently in activity NACT and blocked by the node associated with file IFILE

FUNCTION NNCNT(NACT)

The number of entities that have completed activity NACT

FUNCTION NNGAT(NGATE)

Status of gate number NGATE at current time: 0→open, 1→closed

FUNCTION NNLBL(IDUM)

Returns the node label associated with a current event. IDUM is not used

FUNCTION NNQ(IFILE)

Returns number of entries in file IFILE

FUNCTION NNRSC(NRES)

Current number of resource type NRES available

FUNCTION NNUMB(IDUM)

Returns the activity number associated with a current event. If none, returns a value of -1. IDUM is not used

FUNCTION NNVNT(IDUM)

Returns event code associated with current event. If none, returns a value of 0. IDUM is not used

FUNCTION NPRED(NTRY)

Returns pointer to the predecessor entry of the entry whose pointer is NTRY

FUNCTION NPSSN(XMN,IS)

Returns a sample from a Poisson distribution with mean XMN using random number stream IS

FUNCTION NRUSE(NRES)

Current number of resource type NRES in use

FUNCTION NSUCR(NTRY)

Returns pointer to the successor entry of the entry whose pointer is NTRY

SUBROUTINE OPEN(NGATE)

Sets the status of gate NGATE to OPEN and empties all files containing entities waiting for the gate to be opened

SUBROUTINE PPLOT(IPLOT,NREC,XI,XD)

Obtains record NREC from plot IPLOT. XI is value of independent variable and XD is vector of dependent variables.

SUBROUTINE PRNTA

Prints statistics for all activities

SUBROUTINE PRNTB(ISTAT)

If ISTAT>0, prints a histogram for time-persistent variable ISTAT. If ISTAT≤0, prints all ISTAT histograms

SUBROUTINE PRNTC(ICLCT)	If ICLCT>0, prints statistics for COLCT variable number ICLCT. If ICLCT≤0, prints statistics for all COLCT variables
SUBROUTINE PRNTF(IFILE)	If IFILE>0, prints statistics and the contents of file IFILE. If IFILE=0, prints summary statistics for all files. If IFILE<0, prints summary statistics and contents of all files
FUNCTION PRNTG(IG)	If IG>0, prints statistics for gate number IG. If IG≤0, prints statistics for all gates
SUBROUTINE PRNTH(ICLCT)	If ICLCT>0, prints a histogram for COLCT variable number ICLCT. If ICLCT≤0, prints all histograms
SUBROUTINE PRNTP(IPLOT)	If IPLOT>0, prints a plot and/or table for plot/table number IPLOT. If IPLOT≤0, prints all plots/tables
SUBROUTINE PRNTR(NRES)	If NRES≤0, prints statistics for resource number NRES. If NRES>0, prints statistics for all resources
SUBROUTINE PRNTS	Prints the contents of the state storage vectors SS(I) and DD(I)
SUBROUTINE PRNTT(ISTAT)	If ISTAT>0, prints statistics for time-persistent variable ISTAT. If ISTAT≤0, prints statistics for all time-persistent variables
FUNCTION PRODQ(NATR,IFILE)	Returns the product of the values of attribute NATR for each current entry in file IFILE
SUBROUTINE PUTAA(NAUXA,AUXF, MFAA,VALUE)	Puts a set of values of auxiliary attributes into an array
SUBROUTINE PUTARY(IR,IC,VAL)	Sets the value of row IR, column IC of the global table, ARRAY, to VAL.
FUNCTION RLOGN(XMN,STD,IS)	Returns a sample from a lognormal distribution with mean XMN and standard deviation STD using random number stream IS
SUBROUTINE RMOVE(NRANK, IFILE,A)	Removes an entry defined by the variable NRANK from a file defined by the variable IFILE. If NRANK is positive, it defines the rank of the entry to be removed. If NRANK is negative, it

	points to the negative of the location where the entry to be removed is stored. RMOVE loads the vector A with the attributes of the entry removed. The value of MFA is reset to the pointer of the entry removed
FUNCTION RNORM(XMN,STD,IS)	Returns a sample from a normal distribution with mean XMN and standard deviation STD using random number stream IS
FUNCTION RRAVA(NRES)	Returns the average number of units available for resource NRES.
FUNCTION RRAVG(NRES)	Returns average utilization of resource NRES
FUNCTION RRMAX(NRES)	Returns maximum utilization of resource NRES
FUNCTION RRPRD(NRES)	Returns time period for statistics on resource NRES
FUNCTION RRSTD(NRES)	Returns standard deviation of utilization of resource NRES
FUNCTION RRTLC(NRES)	Returns time at which resource NRES utilization was last changed
SUBROUTINE SCHDL(KEVNT,DTIME,A)	Schedules event type KEVNT to occur at TNOW + DTIME with event attributes as stored in A
SUBROUTINE SEIZE(NRES,IU)	Sets IU units of resource number NRES busy
SUBROUTINE SET	Initializes file pointers and file statistics arrays
SUBROUTINE SETAA(NAUXA,AUXF, MFAA,NDAUX)	Sets up an array (defined by the user) that will maintain auxiliary attributes
SUBROUTINE SETARY(IR,VALUE)	Sets the values of row IR of ARRAY to VALUE
SUBROUTINE SLAM	The SLAM executive
SUBROUTINE STOPA(NTC)	Stops all network activities whose duration is specified by STOPA for which an entity is currently ongoing with an entity code of NTC
FUNCTION SSEVT(JEVNT)	Returns status of state-event JEVNT as follows: $+2$, positive crossing beyond tolerance; $+1$, positive crossing; 0, no crossing; -1, negative crossing; -2, negative crossing beyond tolerance

FUNCTION SUMQ(NATR,IFILE)	Returns the sum of the values of attribute NATR for each current entry in file IFILE
SUBROUTINE SUMRY	Prints the SLAM II Summary Report
SUBROUTINE TRACE	Initiates the standard SLAM II trace
FUNCTION TRIAG(XLO,XMODE, XHI,IS)	Returns a sample from a triangular distribution in the interval XLO to XHI with mode XMODE using random number stream IS
FUNCTION TTAVG(ISTAT)	Returns time-integrated average of variable ISTAT
FUNCTION TTMAX(ISTAT)	Returns maximum value of variable ISTAT
FUNCTION TTMIN(ISTAT)	Returns minimum value of variable ISTAT
FUNCTION TTPRD(ISTAT)	Returns time period for statistics on variable ISTAT
FUNCTION TTSTD(ISTAT)	Returns standard deviation of variable ISTAT
FUNCTION TTTLC(ISTAT)	Returns time at which variable ISTAT was last changed
SUBROUTINE ULINK(NRANK,IFILE)	Removes entry with rank NRANK from file IFILE without copying its attribute values. If NRANK < 0, it is a pointer
FUNCTION UNFRM(ULO,UHI,IS)	Returns a sample from a uniform distribution in the interval ULO to UHI using random number stream IS
FUNCTION WEIBL(BETA,ALPHA,IS)	Returns a sample from a Weibull distribution with parameters BETA and ALPHA using random number stream IS
SUBROUTINE UNTRA	Stops the standard SLAM II trace
FUNCTION XRN(ISTRM)	Returns the last pseudorandom number obtained from stream ISTRM

User-Written Subprograms

SUBPROGRAM†	DESCRIBED IN CHAPTERS
SUBROUTINE ALLOC(N,IFLAG)	9
SUBROUTINE EVENT(JEVNT)	9,11
SUBROUTINE INTLC	8,11
PROGRAM MAIN	11
FUNCTION NQS(N)	9
FUNCTION NSS(N)	9
SUBROUTINE OTPUT	8,11
SUBROUTINE STATE	10,13
SUBROUTINE UERR(KODE)	12
SUBROUTINE UMONT(IT)	12
FUNCTION USERF(N)	5,9

---

† These subprograms need only be written if a SLAM II model requires special processing. SLAM II contains a general or dummy version of each of these subprograms. The argument N in the above listing signifies a user-defined index to differentiate between calls to the subprograms.

APPENDIX B

Network Input Statement Descriptions

This appendix contains a complete description of the network input statements of SLAM II. In Table B1, a summary of SLAM II input statements and defaults is given. In Tables B2-B25, a detailed description of each network element is provided. A list of tables is given below for ready reference.

Terminology Used In Appendix B

In this appendix, variable names are used to simplify the presentation of the description of each SLAM II node symbol. The largest file number is referred to as MFIL. This value is defined by the user on the LIMITS statement. Throughout the tables, reference is made to SLAM status variables which are SS(I), DD(I), XX(I), NNACT(I), NNCNT(I), NNGAT(I), NNRSC(I), NRUSE(I), and NNQ(I). In addition, II, ARRAY(I,J), and USERF(N) may be used in any field with the option identified as a SLAM II variable.

The letter I is used as an index and must always be a positive integer. When ATRIB(I) is used, the value of the Ith attribute specifies the value for the field. The symbolism ATRIB(I) = J,K specifies that the value is to be taken from ATRIB(I) and the allowed range of values is $J, J+1, J+2 \ldots K$. This specification prescribes the value of an entity's I*th* attribute as the value for a network symbol. NATR is used to refer to a specific attribute number.

In the descriptions, RLBL is used to refer to a resource label which is the name of the resource prescribed in the resource block. Similarly, GLBL is the gate label defined in the gate block. NLBL is used to refer to node labels which are used in columns one through five of any SLAM II network statement (only the first four characters are significant). Each resource and gate label is given a numeric equivalent by the SLAM II program. These numeric equivalents are assigned in a sequential order and correspond to the position of the RESOURCE (GATE) block with respect to other RESOURCE (GATE) blocks. Thus, the first resource block is assigned a numeric equivalent of 1 where the first resource block is the one that is closest to the NETWORK statement. In a description which specifies "RLBL or ATRIB(I)", the latter indicates that an attribute of an arriving entity is to specify the resource. If ATRIB(4) is specified and if ATRIB(4) = 2 for an arriving entity then the second resource defined by resource block statements is the one indicated. This same concept applies to GATES. The names RES and GATE indicate that a label or numeric value can be used to specify the value for a resource or gate field.

In the column whose heading is 'default', the values assigned to the statement fields when the user does not prescribe a value for the field is given. In some cases, it is mandatory that a value be prescribed. When this is the case, the entry listed in the default column is "error". This condition will cause the SLAM II program to not execute. In other situations, no default value need be given and the entry in the default column is "none". For some fields, the default value is a large number and this is indicated in the tables through use of an infinity sign (∞).

Table B1 SLAM II network statement types.

Statement Form	Statement Defaults(ND = no default)
Nodes	
ACCUM,FR,SR,SAVE,M;	ACCUM,1,1,LAST,∞;
ALTER,RES/CC,M;	ALTER,ND/ND,∞;
ASSIGN,VAR = value,VAR = value,...,M;	ASSIGN,ND = ND,ND = ND,...,∞;
AWAIT(IFL/QC),RES/UR or	AWAIT(first IFL in RLBL's or GLBL's
GATE,BLOCK or BALK(NLBL),M;	list/∞),ND/1,none,∞;
BATCH,NBATCH/NATRB,THRESH,	BATCH,1/none,ND, entity
NATRS,SAVE,RETAIN,M;	count,LAST,NONE,∞;
CLOSE,GATE,M;	CLOSE,ND,∞;
COLCT(N),TYPE or VARIABLE,	COLCT(ordered),ND,blanks,no
ID,NCEL/HLOW/HWID,M;	histogram/0./1
CREATE,TBC,TF,MA,MC,M;	CREATE,∞,0,no marking,∞,∞;
DETECT,XVAR,XDIR,VALUE,TOL,M;	DETECT,ND,ND,ND,0,∞;
ENTER,NUM,M;	ENTER,ND,∞;
EVENT,JEVNT,M;	EVENT,ND,∞;
FREE,RES/UF,M;	FREE,ND/1,∞;
GOON,M;	GOON,∞;
MATCH,NATR,QLBL/NLBL,...,M;	MATCH,ND,ND/no routing,ND/no
	routing,...∞,
OPEN,GATE,M;	OPEN,ND,∞;
PREEMPT(IFL)/PR,RES,SNLBL,	PREEMPT(first IFL in RLBL's list)/no
NATR,M;	priority,ND,AWAIT node where
	transaction seized resource, none,∞;
QUEUE(IFL),IQ,QC,BLOCK or	QUEUE(ND),0,∞, none,none,none;
BALK(NLBL),SLBLs;	
SELECT,QSR/SAVE,SSR,BLOCK or	SELECT,POR/none,POR,none,ND;
BALK (NLBL),QLBLs;	
TERMINATE,TC;	TERMINATE,∞;
UNBATCH,NATRR,M;	UNBATCH,ND,∞;
Blocks	
GATE/GLBL,OPEN or	GATE/ND,OPEN,ND/repeats;
CLOSE,IFLs/repeats;	
RESOURCE/RLBL(IRC),IFLs/repeats;	RESOURCE/ND(1),ND/repeats;
Regular Activity	
ACTIVITY/A,duration,PROB or	ACTIVITY/no ACT number,0.0,take
COND,NLBL; ID	ACT,ND; blank
Service Activity	
ACTIVITY(N)/A, duration,PROB,	ACTIVITY(1)/no ACT number,0.0,1.0,ND;
NLBL; ID	blank

Table B2 ACCUMULATE node description summary.

Node Type: ACCUMULATE   *Symbol:*

Function: The ACCUMULATE node is used to combine entities. The combining of entities is controlled by the specification of the release mechanism consisting of the number of arrivals required for the first release (FR), the number of arrivals required for subsequent releases (SR), and the attribute holding criterion for entities to be routed (SAVE). A maximum of M emanating activities are initiated.

Input Format: ACCUMULATE,FR,SR,SAVE,M;

Specifications:

Field	Options	Default
FR	positive integer or ATRIB(I)	1
SR	positive integer or ATRIB(I)	1
SAVE	save criterion specified as: FIRST, LAST, LOW(NATR), HIGH(NATR),SUM,MULT	LAST
M	positive integer	∞

Table B3   ALTER node description summary.

Node Type:ALTER	Symbol:	

Function:	The ALTER node changes the capacity of resource RES by CC units. In the case where the capacity is decreased below current utilization, the excess capacity is destroyed as it becomes freed. The capacity can be reduced to a minimum of zero with additional reductions having no effect. At each release, a maximum of M emanating activities are initiated.

Input Format: ALTER,RES/CC,M;

Specifications:

Field	Options	Default
RES	RLBL or ATRIB(I)	error
CC	SLAM II variable, SLAM II random variable, or a constant	error
M	positive integer	∞

Table B4   ASSIGN node description summary.

Node Type: ASSIGN	Symbol:	

Function:	The ASSIGN node is used to assign values to SLAM variables (VAR) at each arrival of an entity to the node. A maximum of M emanating activities are initiated.

Input Format: ASSIGN,VAR = value,VAR = value,...,M;

Specifications:

Field	Options	Default
VAR	ATRIB(I),SS(I),DD(I),XX(I),STOPA,ARRAY, or II, where I is a positive integer or II	error
value	an arithmetic expression containing constants, SLAM II variables, or SLAM II random variables. Up to 10 addition, subtraction, multiplication and division operations may be performed in an expression. Multiplication and division will be performed before addition and subtraction. Parentheses are allowed only to denote subscripts.	error
M	positive integer	∞

Table B5 AWAIT node description summary.

Node Type: AWAIT	Symbol:

IFL	RES/UR	
QC	or GATE	M

Function: The AWAIT node operates in two modes. In the resource mode, the AWAIT node delays an entity in file IFL until UR units of resource RES are available. The entity then seizes the UR units of RES. The user may determine the disposition of the arriving entity by coding subroutine ALLOC. In the gate mode, the AWAIT node releases the entity if the gate status is open and delays the entity in file IFL if the gate status is closed. At each release of the node, a maximum of M activities are initiated.

Input Format: AWAIT(IFL/QC), RES/UR or ALLOC(I) or
GLBL, BLOCK or BALK(NLBL),M;

Specifications:

Field	Options	Default
IFL	integer between 1 and MFIL or ATRIB(I) = J,K	first IFL in RLBL's list
QC	positive integer	∞
RES or GATE	RLBL or ATRIB(I) or ALLOC(I) or GLBL	error
UR	positive SLAM II variable, SLAM II random variable, or constant	1
BLOCK or BALK(NLBL)	BLOCK or BALK to the node labeled NLBL when QC entities already reside in the file. (This entire field may be omitted.)	none
M	positive integer	∞

Table B6 BATH node description summary.

Node Type: BATCH   *Symbol:*

Function: The BATCH node is released when the sum of ATRIB(NATRS) for each batch member is equal to or greater than a threshold THRESH. The default for NATRS is to count each entity arrival for a batch. NBATCH different batches can be accumulated concurrently. ATRIB(NATRB) defines an entity's batch membership. If ATRIB(NATRB) is the negative of the batch number, a batch is released irregardless of the value of THRESH. The batched entity has attributes as defined by the holding criterion (SAVE). Individual entities included in a batch can be RETAINed by specifying ALL(NATRR) where NATRR is the attribute number to retain an internal SLAM II pointer to the entities in the batch. A maximum of M emanating activities are initiated for the batched entity.

Input Format: BATCH,NBATCH/NATRB,THRESH,NATRS,SAVE,RETAIN,M;

Specifications:

Field	Options	Default
NBATCH	positive integer	1
NATRB	positive integer	no sorting
THRESH	constant or ATRIB(I), where I is a positive integer	∞
NATRS	positive integer	count of entities
SAVE	FIRST, LAST, LOW(I), or HIGH(I), where I is a positive integer. May be followed by /i, j, . . . , a list of attributes for which the primary criterion is overridden by summing the specified attributes of all members of the batch.	LAST
RETAIN	NONE or ALL(NATRR), where NATRR is a positive integer	NONE
M	positive integer	∞

Table B7 CLOSE node description summary.

Node Type: CLOSE   *Symbol:*   (GATE⎪M)

Function: The CLOSE node changes the status of GATE to closed and releases a maximum of M emanating activities.

Input Format: CLOSE,GATE,M;

Specifications:

Field	Options	Default
GATE	GLBL or ATRIB(I)	error
M	positive integer	∞

Table B8 COLCT node description summary.

Node Type: COLCT   *Symbol:*   (TYPE ⎪ ID,H ⎪M)

Function: The COLCT node is used to collect statistics that are related to: either the time an entity arrives at the node (TYPE); or on a VARIABLE at the entity's arrival time. An index N may be assigned to the COLCT node to provide a reference number. ID is an identifier for output purposes and H is a histogram specification for the number of cells (NCEL), the upper limit of the first cell, (HLOW) and the cell width (HWID). A maximum of M emanating activities are initiated.

Input Format: COLCT(N),TYPE or VARIABLE,ID,NCEL/HLOW/HWID,M;

Specifications:

Field	Options	Default
N	positive integer	next sequential index
TYPE or VARIABLE	FIRST,ALL,BETWEEN,INTVL(NATR) or SLAM II variable	none
ID	maximum of 16 characters beginning with an alphabetic character	blanks
NCEL	positive integer	no histograms
HLOW	constant	0.0
HWID	positive constant	1.0
M	positive integer	∞

Table B9   CREATE node description summary.

Node Type: CREATE	*Symbol:*

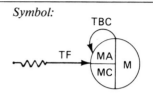

Function: The CREATE node is used to generate entities within the network. The node is released initially at time TF and thereafter according to the specified time between creations TBC up to a maximum of MC releases. At each release, a maximum of M emanating activities are initiated. The time of creation is stored in ATRIB(MA) of the created entity.

Input Format: CREATE,TBC,TF,MA,MC,M;

Specifications:

Field	Options	Default
TBC	constant, SLAM II variable, or SLAM II random variable	∞
TF	constant	0.
MA	positive integer	no marking
MC	positive integer	∞
M	positive integer	∞

Table B10   DETECT node description summary.

Node Type: DETECT   *Symbol:*

| XVAR | XDIR | VALUE | TOL | M |

Function: The DETECT node is used to generate entities whenever the state of system as defined by a crossing variable, XVAR, crosses a prescribed threshold value, VALUE. The node is released whenever a crossing occurs in direction XDIR. The value of TOL specifies the desired interval beyond the VALUE for which a detection of a crossing is desired. In addition, any entity arriving to a DETECT node causes it to be released. A maximum of M emanating activities are initiated at each release.

Input Format: DETECT,XVAR,XDIR,VALUE,TOL,M;

Specifications:

Field	Options	Default
XVAR	SLAM II status variable	error
XDIR	X,XP, or XN	error
VALUE	Constant or SLAM status variable	error
TOL	positive constant	∞
M	positive integer	∞

Table B11   ENTER node description summary.

Node Type: ENTER   *Symbol:*

| NUM | M |

Function: The ENTER node is provided to permit the user to enter an entity into the network from a user-written event routine. The node is released at each entity arrival and at each user call to subroutine ENTER(NUM). A maximum of M emanating activities are initiated at each release.

Input Format: ENTER,NUM,M;

Specifications:

Field	Options	Default
NUM	positive integer	error
M	positive integer	∞

Table B12 EVENT node description summary.

Node Type: EVENT   *Symbol:*

Function: The EVENT node causes subroutine EVENT to be called with event code JEVNT at each entity arrival. This allows the user to model functions for which a standard node is not provided. A maximum of M emanating activities are initiated.

Input Format: EVENT,JEVNT,M;

Specifications:

Field	Options	Default
JEVNT	positive integer	error
M	positive integer	∞

Table B13 FREE node description summary.

Node Type: FREE   *Symbol:*

Function: The FREE node releases UF units of resource RES. The resource is made available to waiting entities according to the order of the wait files specified in the RESOURCE statement. A maximum of M emanating activities are initiated.

Input Format: FREE,RES/UF,M;

Specifications:

Field	Options	Default
RES	RLBL or ATRIB(I)	error
UF	positive SLAM II variable, SLAM II random variable or constant	1
M	positive integer	∞

Table B14   GOON node description summary.

Node Type: GOON	*Symbol:*

Function: The GOON node provides a continuation node where every entering entity passes directly through the node. It is a special case of the ACCUMULATE node with FR and SR set equal to one. A maximum of M emanating activities are initiated.

Input Format: GOON,M;

Specifications:

Field	Options	Default
M	positive integer	∞

Table B15   MATCH node description summary.

Node Type: MATCH	*Symbol:*

Function: The MATCH node is used to delay the movement of entities by keeping them in QUEUE nodes (QLBLs) until entities with the same value of attribute NATR are resident in every QUEUE node preceding the MATCH node. When a match occurs, each entity is routed to a route node NLBL that corresponds to QLBL.

Input Format: MATCH,NATR,QLBL/NLBL,QLBL/NLBL,...;

Specifications:

Field	Options	Default
NATR	a positive integer	error
QLBL	a QUEUE node label	error if less than 2 QLBLs specified
NLBL	a node label for any type of node	destroy the entity

Table B16   OPEN node description summary.

Node Type: OPEN	^{c}	*Symbol:*

Function: The OPEN node changes the status of GATE to open and releases all waiting entities in the AWAIT files specified in the GATE statement. A maximum of M emanating activities are initiated.

Input Format: OPEN,GATE,M;

Specifications:

Field	Options	Default
GATE	GLBL or ATRIB(I)	error
M	positive integer	∞

Table B17 PREEMPT node description summary.

Node Type: PREEMPT   *Symbol:*

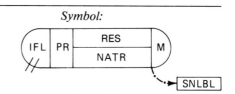

Function: The PREEMPT node is used to preempt a resource specified by RES having a capacity of one. Entities attempting to seize a resource at a PREEMPT node have preemptive priority over entities which seized the resource at an AWAIT node. If a priority (PR) is specified at a PREEMPT node, then preempt entities are preempted by preempt entities with higher priorities. A preempted entity is routed to node SNLBL with the remaining processing time placed in ATRIB(NATR). If a send node, SNLBL, is not specified, the preempted entity resides in the await file of the node where the entity originally seized the resource until it can be reactivated at its preempted location in the network. Entities which are not on regular activities cannot be preempted. A maximum of M of the emanating activities are released.

Input Format: PREEMPT(IFL)/PR,RES,SNLBL,NATR,M;

Specifications:

Field	Options	Default
IFL	integer between 1 and MFIL or $ATRIB(I) = J,K$	first IFL in RLBL's list
PR	HIGH(NATR) or LOW(NATR)	non-priority mode
RES	RLBL or ATRIB(I)	error
SNLBL	the label of a node to send the preempted entity	AWAIT node where entity seized the resource
NATR	positive integer	none
M	positive integer	∞

Table B18   QUEUE node description summary.

Node Type: QUEUE   *Symbol:*

Function: The QUEUE node is used to delay entities in file IFL until a server becomes available. The QUEUE node initially contains IQ entities and has a capacity of QC entities. The specification of blocking causes incoming servers to be blocked whenever the queue is at capacity. The specification of balking causes arriving entities to balk whenever the queue is at capacity.

Input Format: QUEUE(IFL),IQ,QC,BLOCK or BALK(NLBL),SLBLs;

Specifications:

Field	Options	Default
IFL	integer between 1 and MFIL or ATRIB(I) = J,K	*error*
IQ	non-negative integer	0
QC	integer greater than or equal to IQ	∞
BLOCK or BALK(NLBL)	BLOCK or BALK(NLBL)	none
SLBLs	the labels of SELECT or MATCH nodes separated by commas, if any	none

Table B19 SELECT node description summary.

Node Type: SELECT   *Symbol:*

QSR | SSR

Function: The SELECT node is used to select among queues and available servers based upon the queue selection rule (QSR) and the server selection rule (SSR). Blocking or balking may be specified similar to the QUEUE node.

Input Format: SELECT,QSR,SSR,BLOCK or BALK(NLBL),QLBLs;

Specifications:

Field	Options	Default
QSR	CYC,POR,RAN,ASM†,LAV,SWF,LNQ, LRC,SAV,LAF,SNQ,SRC,NQS(I)	POR
SSR	CYC,POR,RAN,LBT,SBT,LIT,SIT, NSS(I)	POR
BLOCK or BALK(NLBL)	BLOCK or BALK(NLBL)	none
QLBLs	labels of QUEUE nodes separated by commas	error

---

† The attribute holding criterion may be specified following a slash as HIGH(NATR),LOW(NATR), SUM or MULT. If the holding criterion is defaulted, the attributes of the entity in the first assoiated QUEUE node are held.

Table B20 TERMINATE node description summary.

Node Type: TERMINATE   *Symbol:*

TC
TC ⟩—⋀⋀⋀→   or   —⋀⋀→

Function: The TERMINATE node is used to destroy entities and/or terminate the simulation. All incoming entities to a TERMINATE node are destroyed. The arrival of the TCth entity causes a simulation run to be terminated.

Input Format: TERMINATE,TC;

Specifications:

Field	Options	Default
TC	positive integer	∞

Table B21   Resource block description summary.

Block Type: RESOURCE   *Symbol:*

RNUM	RLBL(CAP)	IFL1	IFL2

Function: A RESOURCE block defines a resource by its label RLBL a resource number RNUM, and its initial capacity or availability CAP. The file numbers, IFLs, associated with AWAIT and PREEMPT nodes are where entities requesting units of the resource are queued. The IFLs are listed in the order in which it is desired to allocate the units of the resource when they are made available.

Input Format: RESOURCE/RNUM, RLBL(CAP), IFLs;

Specifications:

Field	Options	Default
RNUM	Resource number or omit	sequential index
RLBL	Resource label	error
CAP	positive integer	1
IFLs	integers between 1 and MFIL	error

Table B22   GATE block description summary

Block Type: GATE   *Symbol:*

NUM	GLBL	OPEN or CLOSE	IFL1	IFL2

Function: A GATE block defines a gate by its label GLBL and a gate number NUM. The initial status of the gate is set through an OPEN or CLOSE prescription. The file numbers, IFLs, reference the AWAIT nodes where entities waiting for the gate to open are queued.

Input Format: GATE/NUM,GLBL, OPEN or CLOSE, IFLs;

Specifications:

Field	Options	Default
NUM	Gate number or omit	sequential index
GLBL	Gate label	error
OPEN or CLOSE	'OPEN' or 'CLOSE'	OPEN
IFLs	integers between 1 and MFIL	error

Table B23 REGULAR activity description summary.

Activity Type: REGULAR	*Symbol:*

DUR, PROB or COND
────────────────▶
┌─────┐
│ A │
└─────┘

Function: A REGULAR activity is any activity emanating from a node other than a QUEUE or SELECT node. The REGULAR activity is used to delay entities by a specified duration, perform conditional/probabilistic testing, and to route entities to non-sequential nodes. If the activity is numbered, statistics are provided on the activity utilization, and the number of active entities and the total entity count are maintained as SLAM II variables NNACT(A) and NNCNT(A), respectively. An activity ID, to be printed with activity statistics, may be attached to numbered activities. The ID is entered as a comment following the semicolon which ends the activity definition. A second semicolon within the common field will eliminate following characters from the ID.

Input Format: ACTIVITY/A, duration,PROB or COND,NLBL; ID;

Specifications:

Field	Options	Default
A	positive integer between 1 and 100 or $ATRIB(I) = J,K$	no statistics
duration	constant, SLAM II variable, SLAM II random variable, REL(NLBL) or STOPA(NTC) or an expression.	0
PROB or	probability: constant, SLAM II variable or SLAM II random variable. Must be between 0 and 1.	always take activity
COND	condition: value.OPERATOR.value where value is a constant, SLAM II variable, or SLAM II random variable, and OPERATOR is LT, LE, EQ, GE, GT, or NE. Two or more conditions can be specified that are separated by .AND. or .OR.	
NLBL	label of the node which is at the end of the activity	sequential node
ID	identifying name. The first 12 characters are significant.	blank

Table B24   SERVICE activity description summary.

Activity Type: SERVICE   *Symbol:*

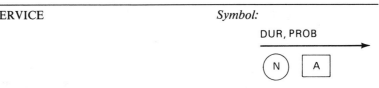

The service activity is any activity emanating from a QUEUE or SELECT node. The service activity is used in conjunction with the QUEUE node to model single channel queues or queues with N identical servers. The service activity is used in conjunction with the SELECT node to model multiple channel queues with non-identical servers. Statistics are collected on all service activities. If the activity is numbered, the server status (number of busy or blocked servers) and the total entity count are maintained as SLAM II variables NNACT(A) and NNCNT(A), respectively. An activity ID, to be printed with activity statistics, may be attached to numbered activities. The ID is entered as a comment following the semicolon which ends the activity definition. A second semicolon within the common field will eliminate following characters from the ID.

Input Format: ACTIVITY(N)/A,duration,PROB,NLBL; ID;

Specifications:

Field	Options	Default
N	positive integer	1
A	positive integer between 1 and 100 or ATRIB(I) = J,K	none
duration	constant, SLAM II variable, SLAM II random variable, REL(NLBL), or STOPA(NTC) or an expression	0.
probability†	constant, SLAM II variable, or SLAM II random variable. Must be between 0 and 1.	1.
NLBL	label of the node at the end of the activity	sequential node
ID	identifying name. The first 12 characters are significant	blank

---

†Used only to represent identical servers emanating from a queue node as a set of probabilistic service activities. When used in this way, each activity must have the same N and A values.

Table B25 UNBATCH node description summary.

Node Type: UNBATCH   *Symbol:*

Function: The UNBATCH node is used to retrieve individual entities that were previously batched at a BATCH node or to split one entity into several identical entities. The NATRR specification is the index of the attribute of the arriving entity which points to the individual entities to be unbatched, or it carries the number of identical entities to be released from the UNBATCH node. All such entities are identical to the entity which arrived at the UNBATCH node. A maximum of M activities are initiated for each unbatched entity.

Input Format: UNBATCH,NATR,M;

Specifications:

Field	Options	Default
NATRR	positive integer	error
M	positive integer	∞

SLAM II Control Statements

This appendix contains a complete description of the SLAM II control statements. In Table C1, a summary of SLAM II control statements is given. In Table C2, a detailed description of each statement is provided including a definition of variables, input value options for the variables, and SLAM II default values assigned.

Table C1 SLAM II control statments.

ARRAY(IROW,NELEMENTS),initial values;
CONTINUOUS,NNEQD,NNEQS,DTMIN,DTMAX,DTSAV,W or F or
    N,AAERR,RRERR;
ENTRY/IFL,ATRIB(1),ATRIB(2),...,ATRIB(MATR)/repeats;
EQUIVALENCE/SLAM II variable,name/repeats;
FIN;
GEN,NAME,PROJECT,MO/DAY/YEAR/,NNRNS,ILIST,IECHO,IXQT/IWARN,
    IPIRH,ISMRY/FSN,IO;
INITIALIZE, TTBEG,TTFIN,JJCLR/NCCLR,JJVAR,JJFIL;
INTLC,VAR = value,repeats;
LIMITS,MFIL,MATR,MNTRY;
MONTR,option,TFRST,TSEC,variables;
NETWORK,SAVE or LOAD,device;
PRIORITY/IFL,ranking/repeats;
RECORD(IPLOT),INDVAR,ID,ITAPE,P or T or B,DTPLT,TTSRT,TTEND,KKEVT;
SEEDS,ISEED(IS)/R,repeats;
SEVNT,JEVNT,XVAR,XDIR,VALUE,TOL;
SIMULATE;
STAT,ICLCT,ID,NCEL/HLOW/HWID;
TIMST,VAR,ID,NCEL/HLOW/HWID;
VAR,DEPVAR,SYMBL,ID,LOORD,HIORD;

Table C2 Definitions and defaults for variables on SLAM II control statements.

Statement Type	Variable	Definition	Options	Default
ARRAY	IROW	Row number	Positive integer	None
	NELEMENTS	Number of elements in this row	Positive integer	None
	initial values	Values for elements in this row	Constants or integer*constant	0.0
CONTINUOUS	NNEQD	Number of differential equations	Positive integer	0
	NNEQS	Number of state equations	Positive integer	0
	DTMIN	Minimum step size	Positive real	.01*DTMAX
	DTMAX	Maximum step size	Positive real	DTSAV
	DTSAV	Time between recording	$0 \rightarrow$ every step; $> 0 \rightarrow$ DTSAV; $< 0 \rightarrow$ at event times	∞
	W or F or N	Warning, Fatal, No warning	W, F, N	W
	AAERR	Allowed absolute error	Positive real	.00001
	RRERR	Allowed relative error	Positive real	.00001
ENTRY	IFL	File number	Positive integer	none
	ATRIB(I),I=1, MATR	Attribute I	Constant	0
EQUIVALENCE	SLAM II variable name	SLAM II variable	SLAM II variable	None
		character string for use in place of variable	12 alphanumeric characters	None
FIN	--	--	--	--

Table C2 (continued).

Statement Type	Variable	Definition	Options	Default
GEN	NAME	Analyst's name	20 alphanumeric characters	Blanks
	PROJECT	Project name	20 alphanumeric characters	Blanks
	MONTH	Month number	Positive integer	1
	DAY	Day number	Positive integer	1
	YEAR	Year number	Positive integer	2001
	NNRNS	Number of runs	Positive integer	1
	ILIST	Request for input listing	Y or N	Y
	IECHO	Request for echo summary report	Y or N	Y
	IXQT	Request for execution	Y or N	Y
	IWARN	Request for destroyed entity warning	Y or N	Y
	IPIRH	Request for intermediate results heading	Y or N	Y
	ISMRY	Request for SLAM Summary Report	Y or N	Y
	FSN	Summary report on: first, first and last, every Nth	F, S, or positive integer	1
	IO	Number of columns in output reports	72 or 132	132

Table C2 (continued).

Statement Type	Variable	Definition	Options	Default
INITIALIZE	TTBEG	Beginning time of run	Constant	0.0
	TTFIN	Finishing time for a run	Constant	∞
	JJCLR	Request to clear statistical arrays	Y or N	Y
	NCCLR	Value up to which JJCLR pertains	Positive integer	all
	JJVAR	Request for SLAM variable initialization	Y or N	Y
	JJFIL	Request for file structure initialization	Y or N	Y
INTLC	VAR	XX(N), SS(N) or or DD(N) where N is an integer	See definition	None
	value	Initial value of VAR	Constant	None
LIMITS	MFIL	Largest file number	Positive integer	100
	MATR	Maximum number of attributes per entity	Positive integer	0
	MNTRY	Maximum number of entries in all files	Positive integer	0
MONTR	option	Monitoring option	SUMRY FILES STATES CLEAR TRACE (nodelist)	None
	TFRST	Time of first monitoring option	Positive real	0.0
	TSEC	Time between monitoring option or end of monitoring	Positive real	∞
	variables	For traces, a list of SLAM variable names. In addition, integers request attributes and negative integers, XX variables	SLAM variables or integers	attributes

Table C2 (continued).

Statement Type	Variable	Definition	Options	Default
NETWORK	option	Save decoded network Load previously decoded network	SAVE LOAD	None
	device	Logical unit number for decoded network	positive integer	error
PRIORITY	IFL	File number	Positive constant $\leqslant$ MFIL + 1	None
	ranking	File ranking	LVF(NATR) HVF(NATR) FIFO LIFO	FIFO
RECORD	IPLOT	Plot number	Positive integer	Next integer
	INDVAR	Independent variable	SLAM II variable	None
	ID	Independent variable identifier	16 alphanumeric characters	Blanks
	ITAPE	RECORD storage medium	$0 \rightarrow$ NSET/QSET > 0 peripheral device number	0
	P or T or B	Plot, table, or both	P, T, B	P
	DTPLT	Interval for independent variable	Constant	5.0
	TTSRT	Time to start recording	Constant	TTBEG
	TTEND	Time to end recording	Constant	TTFIN
	KKEVT	Event record printing	Y or N	'Y'

Table C2 (continued).

Statement Type	Variable	Definition	Options	Default
SEEDS	ISEED	Starting random number seed. If negative, use complement of value	Integer	See data statement in DRAND
	IS	Stream number	Positive integer ≤ 10	Next sequential stream
	R	Reinitialization of seed value for next run	Y or N	N
SEVNT	JEVNT	State event code	Positive integer	Error
	XVAR	Crossing variable	SLAM status variable	Error
	XDIR	Direction of crossing	X→ both directions XP→ positive crossing XN→ negative crossing	Error
	VALUE	Threshold value for crossing	Constant or SLAM status variable	Error
	TOL	Tolerance for crossing	Positive constant	∞
SIMULATE	– –	– –	– –	– –
STAT	ICLCT	Statistics code used in calls to subroutine COLCT	Positive integer	None
	ID	Collect variable identifier	16 alphanumeric characters	Blanks
	NCEL	Number of interior cells for histogram	0→ no histogram; positive integer	0
	HLOW	Upper limit of first cell of histogram	Constant	0.0
	HWID	Width of each cell	Positive constant	1.0

Table C-2 (continued).

Statement Type	Variable	Definition	Options	Default
TIMST	VAR	Time-persistent variable	SLAM II time-persistent variable or USERF	None
	ID	Identifier	16 alphanumeric characters	Blanks
	NCEL	Number of interior cells for histogram	0→no histogram; positive integer	0
	HLOW	Upper limit of first cell of histogram	Constant	0.0
	HWID	Width of each cell	Positive constant	1.0
VAR	DEPVAR	Dependent variable for RECORD statement	SLAM II status variable	None
	SYMBL	Plotting symbol	1 character	Blank
	ID	Identifier	8 alphanumeric characters	Blanks
	LOORD	Low ordinate specification	Constant, MIN or MIN(Real)	MIN
	HIORD	High ordinate specification	Constant, MAX or MAX(Real)	MAX

SLAM II Error Codes and Messages

Error messages are printed by SLAM II when an error condition is detected. Errors detected from input statements are discussed and illustrated in Section 8.7.1. Error messages detected during the execution of a simulation run include an error code, the current simulated time, and, if feasible, the value in error. The error routine then attempts to take the square root of a negative number in order to force an abnormal exit and cause system trace-back information to be printed.

SLAM II execution error codes are numbered beginning with 102. Error codes 1 through 100 are reserved for user calls to subroutine ERROR. Error codes 102 through 999 flag conditions which should have been detected during input and, therefore may have arisen due to variable values being overwritten. The most common cause of overwritten data is the assignment of values to user-defined arrays without checking to see that array indices are within bounds. Another possibility for this type of error is to pass a user-defined array that has too few elements in calls to FILEM, RMOVE, COPY, SCHDL, or ENTER. A user-defined array for attributes must be dimensioned to at least the number of attributes requested on the LIMITS statement plus 2. It is also possible to destroy data by calling RMOVE or ULINK with a pointer to an entity which is actually not in the referenced file.

Error codes greater than 1000 are usually due to errors in setting variable parameters or in arguments to support functions, that is, conditions that could not have been detected during input processing.

This appendix provides a list of all error codes generated by SLAM II along with a message describing the possible cause of the error.

Error Code	Description
102	Number of state equations incorrectly specified
103	Number of derivative equations incorrectly specified
106	Variable defining number of events on the event calendar is negative
211	Number of collect variables incorrectly specified, or NNCLT value destroyed due to an index of an array out of range
212	Number of time-persistent statistics incorrectly specified, or the variable NNSTA destroyed due to an index out of range
213	Incorrect specification for a histogram number, or value of NNHIS destroyed due to an index out of range
214	Incorrect specification for number of files, or MFIL destroyed due to an index out of range
221	Initialization of file structure requested when maximum number of entries in the file was specified as zero
312	Incorrect pointer to next event
411	File pointer structure has an inconsistency which was detected during a search for an entry
412	File pointer to the next entry to be considered during a search was found to be negative
498	The processing of a value assignment or computation for an activity duration involved too many operations
499	Incorrect specification when attempting to make a value assignment or a duration specification
502	Number of observations for a collect variable is negative ($SSOBV(J,3) < 0$).
511	Number of time-persistent variables greater than maximum allowed
512	Current simulated time is less than the time at which a time-persistent variable was last changed
521	Histogram number out of range
531	Independent variable of plot is not monotonic
541	Incorrect file values when attempting to print information on a file
551	Incorrect number of equations for state variables when attempting to print state variable values
568	Incorrect network values when attempting to print information on activities
701	Argument out of range or not specified for collect variable, stream number, etc.
811	Request for zero or negative number of resources
901	Pointer to node description incorrect
999	Incorrect index for computing time-persistent statistics on a state variable
1001	Gate number out of range
1002	Resource number out of range
1004	Collect variable number out of range
1005	Activity number out of range
1006	File number out of range
1007	Rank of entry in file out of range
1008	Attribute number out of range
1009	Enter node number out of range

Error Code	Description
1010	Incorrect specification for MONTR option
1011	Code for stopping activity (STOPA) out of range
1012	TIMST statistics index out of range
1013	State-event code in call to SSEVT out of range
1014	Plot number out of range in call to PRNTP or PPLOT
1015	Number of plot points out of range in call to PPLOT
1016	ARRAY row number out of bounds
1017	ARRAY column number out of bounds
1018	Subscript specified as II, out of range
1020	Entity arrival to MATCH node
1021	Attempt to preempt resource with capacity not equal to 1
1023	Incorrect specification for cumulative probability distribution
1024	Incorrect parameter specification for distribution function
1025	Rank of file member referenced is greater than the number of entries in file
1026	Space allocated to files is insufficient
1027	Arguments to subroutine GDLAY incorrectly specified (IFS > ILS)
1028	Attempt to run simulation backwards in time
1029	Specified integration accuracy cannot be met
1030	Number of activities exceeds allowed limit (Redimension array JJVEC)
1031	Number of activities exceeds allowed limit (Redimension array IIACT)
1032	Number of resources increased beyond prescribed capacity
1033	State-event cannot be detected within the prescribed tolerance value
1034	Incorrect specification for auxiliary attributes
1035	Number of tags for binary search algorithm has been exceeded
1036	Too many batches at a BATCH node
2001	In NQS, a full queue was selected for entity routing
2002	In NQS, an empty queue was selected for obtaining an entity
2003	In NSS, a busy server was selected
2004	An attempt was made to seize more resources than are currently available from subroutine SEIZE

MH Extension Subprograms, Statements, and Error Messages

This appendix contains a complete description of the subprograms, statements, and error messages for the MH Extension. The terminology used in Appendix E is similar to that used in Appendix B and the Description of the Terminology Section of Appendix B should be reviewed. In particular, the option "8 characters maximum" requires the first character to be alphabetic and none of the characters can be a SLAM II delimiter or arithmetic operation. Table E1 is a listing of the MH Extension support subprograms and Table E2 is a listing of subprograms for implementing user rules within the AGVS module. In Chapter 16, Tables 16-3, 4, and 5 provide definitions for the arguments of these subprograms and more complete information on the use of the subprograms. Table E3 is a listing of the MH Extension resource block statements and the default value for each field. Table E4 is a listing of MH Extension node and activity statements and the default value for each field. Tables E5 through E16 provide a detailed description of resource block and network statements. These tables are presented in the alphabetical order as shown on the next page.

Network element	Table number
AREA RESOURCE BLOCK	E5
CRANE RESOURCE BLOCK	E6
GFREE NODE	E7
GWAIT NODE	E8
PILE RESOURCE BLOCK	E9
QBATCH NODE	E10
VCPOINT RESOURCE BLOCK	E11
VFLEET RESOURCE BLOCK	E12
VFREE NODE	E13
VMOVE ACTIVITY	E14
VSGMENT RESOURCE BLOCK	E15
VWAIT NODE	E16

Tables E17 and E18 present information on the VCONTROL and MHMONTR statements. A listing of the MH Extension error messages is given in Table E19.

Table E1 MH Extension Support Subprograms.

Function/Subroutine	Short Description†
Function NVUVF(NVFNUM,NVUNITR)	NVU given Vehicle Fleet
Subroutine NVUATR(BTRIB,NVUNIT, NVFNUM,NVUNITR)	NVU given Attributes
Subroutine NVUIDL(NRANKV,NVFNUM, NVUNIT)	NVU given idle rank
Subroutine NVUSG(NRANKV,NSGNUM, NVUNIT)	NVU given SG number
Subroutine NVUWCP(NRANKV,NCPNUM, NVUNIT)	NVU given WCP number
Subroutine ALTSPD(NVUNIT,RATE,SPEED,IERR)	Alter Speed
Subroutine SIGVU(NVUNIT,ITRIPF,IERR)	Signal Vehicle
Subroutine GETASG(NVUNIT,NCPDES, IPLOADE,ISTATVU)	Get Assignment
Subroutine GETLOC(NVUNIT,NSGNUM,CPOS, NCPNEX,NCPCUR)	Get Location
Subroutine GETCRG(NVUNIT,ICHRG,STCHRG)	Get Charging
Subroutine GETREQ(NRANKR,NVFNUM, NCPNUM,NRANKE,IFLVW,IERR)	Get Requestor
Subroutine GETATR(NVUNIT,BTRIB,IERR)	Get Attributes
Subroutine PUTATR(NVUNIT,BTRIB,IERR)	Put Attributes
Function NNVIDL(NVFNUM)	Number of Vehicles Idle
Function NNVSG(NSGNUM)	Number of Vehicles on SG
Function NNVWCP(NCPNUM)	Number of Vehicles WCP

† NVU is an abbreviation for "number of a vehicle unit", SG is an abbreviation for "segment", and WCP is an abbreviation for "waiting for a control point".

Table E2 User-written subroutines for rule specification.

Subroutine	Arguments	Rule Specification
UREREL	NR,NVFNUM,NVUNIT, IFILE,IPREQE,NRANKE	Rule for selecting entity to load on a vehicle at a VWAIT node.
URVREQ	NR,NVFNUM,NCPNUM, BTRIB,NVUNIT	Rule for selecting an idle vehicle to satisfy an entity request at a VWAIT node.
URJREQ	NR,NVFNUM,NVUNIT, IFLREQ,NRANKR	Rule for selecting job entity requests to which a vehicle unit is to respond.
URIDL	NR,NVFNUM,NVUNIT, NCPCUR,NCPDES	Rule for selecting a control point destination for an idle vehicle unit.
URCNTN	NR,NCPNUM,IORDER	Rule for resolving vehicle contention at a control point.
URROUT	NR,NVFNUM,NVUNIT, NCPCUR,NCPDES,NSGNUM	Rule for selecting a segment when routing a vehicle.

Table E3   MH Extension resource block and control statement types.

```
AREA, ARLBL/ARNUM, NPILES, NATRA, IFLs;
AREA, ND   /NSI    , ND    , 1 unit , ND ;

CRANE, CRLBL/CRNUM, RUNWAY, XCOORD, YCOORD, BSPD, ACC, DEC, TSPD, IFL;
CRANE, ND   /NSI   , ND    , ND    , ND    , 0  , 0  , ∞   , ND;

PILE, PNUM/PLBL, ARLBL, CAP, XCOORD, YCOORD;
PILE, NSI /blank, ND   , ND , ND    , ND   ;

VCPOINT, CPNUM/CPLBL, RCNTN, RROUT, CHARGE;
VCPOINT, ND   /blank , FIFO , SHORT, NO   ;

VFLEET, VFLBL/VFNUM   , NVEH, ESPD                   , LSPD  , ACC, DEC, LEN, DBUF, CHKZ ,
        IFL  /RJREQ   , RIDL, ICPNUM(NOV,SGNUM)/ repeats, REPIND;
VFLEET, ND   /NSI     , 1   , 1.0                    , 1.0   , 0  , 0  , 1.0 , LEN , .5*LEN,
        ND   /PRIORITY, STOP, placement at first request , NO  ;

VSGMENT, SGNUM/SGLBL, CPNUM1, CPNUM2, DIST, DIR, CAP   ;
VSGMENT, NSI   /blank , ND    , ND    , ND  , UNI, f(DIST);

VCONTROL, DTMIN, DTMAX, TRIPREP, CPREP, SGREP;
VCONTROL, ND   , ND   , YES    , YES  , YES  ;
```

ND specifies no default. NSI specifies next sequential index. Infinity (∞) specifies a high value.

Table E4   MH Extension node and activity statements.

```
GFREE/MOVE , AREA       , CRANE/repeats, RES/UF, repeats , MODE, M;
GFREE/No move, no area freed, no crane freed  , no resource freed/1, AND , ∞;

GWAIT (IFL), AREA /PLOGIC(NATRM), CRANE/repeats, RES/UF, repeats, MODE, M;
GWAIT (ND), no area/FAW         , no crane     , no resource/1   , AND , ∞;

QBATCH (IFL), RES, NBATCH/NATRB, THRESH, MAXSIZE, NATRS     , SAVE/i,j..., RETAIN, M;
QBATCH (ND), ND , 1      /none , ND    , ND      , entity count, FIRST     , ND    , ∞;

VFREE, VF, M;
VFREE, ND, ∞;

VMOVE, CP, M;
VMOVE, ND, ∞;

VWAIT (IFL), VF, CP, RVREQ, REREL, M;
VWAIT (ND), ND, ND, FIFO , TOP  , ∞;
```

ND specifies no default. NSI specifies next sequential index. Infinity (∞) specifies a high value.

Table E5   AREA resource block summary.

Input Format: AREA, ARLBL/ARNUM, NPILES, NATRA, IFLs;

Symbol:

AREA			
ARLBL/ARNUM	NPILES	NATRA	IFLs

Specifications:

Field	Description	Options	Default
ARLBL	area label	8 characters maximum	error
ARNUM	area number	positive integer	NSI
NPILES	number of piles contained in area	positive integer	error
NATRA	attribute number of entity that contains area requirement	integer between 1 and (MATR-3)	error
IFLs	file numbers where entities wait for this AREA	integer between 1 and MFIL	error

Table E6   CRANE resource block summary.

Input Format: CRANE, CRLBL/CRNUM, RUNWAY, XCOORD, YCOORD, BSPD, ACC, DEC, TSPD, IFL;

Symbol:

CRANE			
CRLBL/CRNUM	RUNWAY	XCOORD	YCOORD
BSPD	ACC	DEC	TSPD	IFL

Specifications:

Field	Description	Options	Default
CRLBL	crane label	8 characters maximum	error
CRNUM	crane number	positive integer	NSI
RUNWAY	runway on which crane operates	positive integer	error
XCOORD	initial x-location of crane	positive constant	error
YCOORD	initial y-location of crane	positive constant	error
BSPD	maximum bridging speed	positive constant	error
ACC	acceleration	positive constant $0 \rightarrow$ instantaneous starts	0
DEC	deceleration	positive constant $0 \rightarrow$ instantaneous stops	0
TSPD	trolley speed	positive constant	∞
IFL	file number for all waiting jobs	integer between 1 and MFIL	error

Table E7 GFREE node description summary.

<u>Input Format:</u> GFREE/MOVE, FRAREA, CRANE/repeats, RES/UF, repeats, MODE, M;

<u>Symbol:</u>

<u>Specifications:</u>

<u>Field</u>	<u>Description</u>	<u>Options</u>	<u>Default</u>
MOVE	start crane movement to last area seized	MOVE	no move
FRAREA	area resource at current location to free	YES, NO, ARLBL or ATRIB(I)	NO
CRANE	crane to free	CRLBL or ATRIB(I)	no crane freed
RES	regular resource to free	RLBL or ATRIB(I)	no regular resource freed
UF	number of units of RES to free	positive integer or SLAM II variable	1
MODE	regular resource release condition	AND or OR	AND
M	maximum number of activity selections	positive	∞

Table E8   GWAIT node description summary.

Input Format: GWAIT(IFL), AREA/PLOGIC(NATRM), CRANE/repeats, RES/UF, repeats, MODE, M;

Symbol:

Specifications:

Field	Description	Options	Default
IFL	file number for waiting entities	integer between 1 and MFIL	error
AREA	area resource required	ARLBL or ATRIB(I)	no area required
PLOGIC	pile selection rule	alphanumeric pile logic rule (see Table 16-1)	FAW
NATRM	attribute number for matching piles	integer between 1 and (MATR−3)	no matching
CRANE	crane resource required with alternatives separated by slashes	CRLBL or ATRIB(I)	no crane required
RES	regular resource required	RLBL or ATRIB(I)	no regular resources required
UR	number of units of RES required	positive integer or SLAM II variable	1
MODE	regular resource seizure condition	AND or OR	AND
M	maximum number of activity selections	positive integer	∞

Table E9   PILE resource block summary.

Input Format: PILE, PNUM/PLBL, ARLBL, CAP, XCOORD, YCOORD;

Symbol:

PILE				
PNUM/PLBL	ARLBL	CAP	XCOORD	YCOORD

Specifications:

Field	Description	Options	Default
PNUM	pile number	positive integer	error
PLBL	pile label	8 characters maximum	no label
ARLBL	area label in which the pile is located	8 characters maximum	error
CAP	capacity of the pile	positive constant	error
XCOORD	x-location of the pile	positive constant	error
YCOORD	y-location of the pile	positive constant	error

Table E10   QBATCH node description summary.

Input Format: QBATCH(IFL), RES, NBATCH/NATRB, THRESH, MAXSIZE, NATRS, SAVE,
RETAIN(NATRR), M;

Symbol:

Specifications:

Field	Description	Options	Default
IFL	file number	integer between 1 and MFIL	error
RES	regular resource required	RLBL	error
NBATCH	number of batches	positive integer	1
NATRB	attribute number defining batch	integer between 1 and (MATR−3)	none
THRESH	threshold for forming a batch	positive constant	error
MAXSIZE	maximum size of a batch to release	positive constant	THRESH
NATRS	number of attribute to sum t ward threshold	integer between 1 and (MATR−3)	entity count
SAVE	criterion for attributes of batched entity with list of attributes to sum	FIRST/i,j,...	FIRST
RETAIN	attribute number to contain pointer to entities in a batch	RETAIN(NATRR) or ALL(NATRR)	error
M	maximum number of activity selections	positive integer	∞

Table E11   VCPOINT resource block summary.

Input Format: VCPOINT, CPNUM/CPLBL, RCNTN, RROUT, CHARGE;

Symbol:

VCPOINT			
CPNUM/CPLBL	RCNTN	RROUT	CHARGE

Specifications:

Field	Description	Options	Default
CPNUM	control point number	positive integer	error
CPLBL	control point label	8 characters maximum	blank
RCNTN	rule for contention	FIFO, CLOSEST, PRIORITY, URCNTN(NR)	FIFO
RROUT	rule for routing	SHORT or URROUT(NR)	SHORT
CHARGE	in-line charge indicator	YES or NO	NO

Table E12   VFLEET resource block summary.

Input Format: VFLEET, VFLBL/VFNUM, NVEH, ESPD, LSPD, ACC, DEC, LEN, DBUF, CHKZ, IFL/RJREQ, RIDL, ICPNUM(NOV,SGNUM)/repeats, REPIND;

Symbol:

VFLEET								
VFLBL/VFNUM	NVEH	ESPD	LSPD	ACC	DEC	LEN	DBUF	CHKZ
IFL/RJREQ	RIDL			ICPNUM(NOV,SGNUM)				
				repeats			REPIND	

Specifications:

Field	Description	Options	Default
VFLBL	vehicle fleet label	maximum of 8 characters	error
VFNUM	vehicle fleet number	positive integer	NSI
NVEH	number of vehicle units	positive integer	1
ESPD	maximum speed empty	positive constant	1.0
LSPD	maximum speed loaded	positive constant	1.0
ACC	acceleration of the vehicle	nonnegative constant 0→instantaneous starts	0
DEC	deceleration of the vehicle	positive constant 0→instantaneous stops	0
LEN	length of vehicle	positive constant	1.0
DBUF	distance buffer	positive constant	LEN
CHKZ	check zone radius	positive constant greater than or equal to LEN×0.5	LEN×0.5
IFL	file number for all vehicle requests	integer between 1 and MFIL	error
RJREQ	rule for job request selection	PRIORITY, CLOSEST, or URJREQ(NR)	PRIORITY
RIDL	rule for idle vehicle routing	STOP(CPNUM list), CRUISE (CPNUM list), or URIDL(NR)	STOP
ICPNUM	initial control point position	positive integer	vehicles inserted where first requested
NOV	number-on	positive integer	
SGNUM	segment number for overflow	positive integer	
REPIND	report individual vehicle statistics	YES or NO	NO

Table E13   VFREE node description summary.

Input Format: VFREE, VFLBL, RJREQ, RIDL, M;

Symbol:

Specifications:

Field	Description	Options	Default
VFLBL	vehicle fleet of unit to free	8 characters maximum must be vehicle fleet label of unit assigned to entity	error
RJREQ	rule for job request selection	default only (to be PRIORITY, CLOSEST, or URJREQ(NR))	RJREQ on VFLEET statement
RIDL	rule for idle vehicle routine	default only (to be STOP(CPNUM list), CRUISE(CPNUM list), or URIDL(NR))	RIDL on VFLEET statement
M	maximum number of activity selections	positive integer	∞

Table E14   VMOVE activity description summary.

Input Format: VMOVE, CPNUM, M;

Symbol:

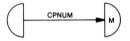

Specifications:

Field	Description	Options	Default
CPNUM	control point destination	positive integer or SLAM II variable	error
M	maximum number of activity selections	positive integer	∞

Table E15   VSGMENT resource block summary.

Input Format: VSGMENT, SGNUM/SGLBL, CPNUM1, CPNUM2, DIST, DIR, CAP;

Symbol:

VSGMENT					
SGNUM/SGLBL	CPNUM1	CPNUM2	DIST	DIR	CAP

Specifications:

Field	Description	Options	Default
SGNUM	segment number	positive integer	error
SGLBL	segment label	8 characters maximum	blank
CPNUM1	identification number of first control point	positive integer	error
CPNUM2	identification number of second control point	positve integer	error
DIST	distance a vehicle travels from one control point to the other	positive constant	error
DIR	direction of vehicle flow	UNIDIRECTIONAL or BIDIRECTIONAL	UNI
CAP	maximum number of vehicles on segment simultaneously	positive integer	calculated based on segment length, vehicle length and distance buffer

Table E16   VWAIT node description summary.

Input Format: VWAIT(IFL), VFLBL, CP, RVREQ, REREL, M;

Symbol:

Specifications:

Field	Description	Options	Default
IFL	file number for waiting entities	integer between 1 and MFIL	error
VFLBL	vehicle fleet required	8 characters maximum	error
CP	control point location of VWAIT node	positive integer or SLAM II variable	error
RVREQ	rule for vehicle request selection	FIFO, CLOSEST, OR URVREQ(NR)	FIFO
REREL	rule for entity release from IFL	TOP, MATCH, UREREL(NR)	MATCH
M	maximum number of activity selections	positive integer	∞

Table E17   VCONTROL statement summary.

Input Format: VCONTROL, DTMIN, DTMAX, TRIPREP, CPREP, SGREP;

Specifications:

Field	Description	Options	Default
DTMIN	minimum step size	positive constant	.01×DTMAX
DTMAX	maximum step size	positive constant	TTFIN−TTBEG
TRIPREP	print trip origination - destination report	YES or NO	YES
CPREP	print control point utilization report	YES or NO	YES
SGREP	print segment utilization report	YES or NO	YES

Table E18   MHMONTR statement summary.

Input Format: MHMONTR/IUNIT, option, TFRST, TSEC;

Specifications:

Field	Description	Options	Default
IUNIT	output device number	positive integer	NPRNT
option	type of trace desired	CTRACE or ATRACE(LEVEL) where LEVEL is DETAIL or NODETAIL	error
TFRST	time to start tracing	positive real	TTBEG
TSEC	time to end tracing	positive real>TFRST	TTFIN

Table E19   MH Extension error codes and descriptions.

Error Code	Description
4000	Crane requested before previous crane freed.
4001	"OR" resource requested before previous one freed.
4002	"AREA" resource requested before previous one freed.
4006	Accuracy error in crane velocity profile.
4007	Negative crane move time calculated.
4008	Invalid pile location being carried.
4010	Arrival to GWAIT node, nothing requested.
4011	Crane requested but no area is held.
4020	Attempt to occupy more than two area resources.
4030	Too many regular resources requested at GWAIT node.
4040	Attempt to MOVE *and* free crane at a GFREE node.
4050	Attempt to free areas not held.
4060	Attempt to maintain more than allowable number of batches at a QBATCH node.
4062	Support routine called for wrong node type.
4065	Attribute-based resource identifier out of range.
4070	Unrecognized node type.
4080	Attribute based crane identifier out of range.
4090	Piling logic cannot find matching pile.
4091	Matching piling logic required but matching attribute value is negative.
4092	Pile number in the matching attribute does not match any pile number in the area.
4100	Area number or pile number passed to function ARAVL not valid.
4110	Batch member not saved, location attributes lost.
4120	Attempt to unbatch while location attributes are carried.
5201	Vehicle and the entity to load have mismatched control point locations.
5202	Illegal entity selection made in UREREL.
5203	Illegal attempt to seize two vehicles.
5204	Invalid segment identifier returned from URROUT.
5205	Illegal attempt to move vehicle from control point.
5213	An invalid job entity was selected for vehicle.
5214	An invalid idle tasking decisions made in URIDL.
5215	Invalid control point destination requested.
5216	AGVS state variables modified in subroutine STATE.
5301	Illegal status change for vehicle made in URVRFQ or NRJREQ.
5302	Illegal status change for vehicle made in UREREL.
5303	Illegal status change for vehicle during an attempt to unload the vehicle at a control point.
5304	Unrecognized condition encountered for idle vehicle.
5305	Unrecognized vehicle status for vehicle unit when the unit ended traversal of control point.
5306	Invalid vehicle speed calculation during travel.
5307	Invalid exit attempt by vehicle from segment.
5308	Illegal status change for vehicle made in URROUT.
5309	Unrecognized vehicle has reached the end of a check zone.
5311	Unidentified vehicle attempt to free a control point.
5312	Illegal status change for vehicle at VFREE node.
5313	Illegal status change for vehicle made in URIDL.
5314	Mismatched segment numbers for colliding vehicles.
5315	Too many vehicles on segment during initialization.
5316	No control point destination selected for vehicle.
5317	Failure to successfully depart a control point.
5318	Entity arrival at VFREE node without a vehicle assignment.
5319	Entity arrival at VFREE node with incorrect vehicle.

Author Index

819

Subject Index